SUSTAINABILITY SCIENCE

Sustainable development is becoming the guiding principle for the 21st century. It is about quality of life: how to develop it and how to sustain it within planetary boundaries. 'Sustainability science' has emerged recently as a new academic discipline and is a growing area of both research and teaching. Sustainability science seeks to:

- advance basic understanding of the dynamics of human-environment systems and forge bridges between the natural and social sciences and between science and policy;
- appreciate the variety of perspectives on sustainable development and the variety of contexts for its design, implementation, and evaluation in particular situations.

Bert J. M. de Vries has taught a course on sustainability science at Utrecht University for many years, in connection to his research at the Netherlands Environmental Assessment Agency (PBL). This textbook is based on that course. The contents have been rigorously class-tested by his students. The book provides a historical introduction into patterns of past (un)sustainable development and into the emergence of the notion of sustainable development. It systematically surveys the key concepts, models and findings of the various scientific disciplines with respect to the major sustainability issues: energy, nature, agro-food systems, renewable and non-renewable resource systems and economic growth. System analysis and modelling are introduced and used as integrating tools. Stories and worldviews are used throughout the text to connect the quantitative and the qualitative and to offer the reader an understanding of relevant trends and events in context. The reader is explicitly invited to engage at a personal level into the interpretation of what sustainable development means and what implications this has for ideas and actions.

Sustainability Science is an ideal textbook for advanced undergraduate- and graduate-level courses in sustainable development, environmental science and policy, ecology, conservation, natural resources and geopolitics.

Bert J. M. de Vries is co-founder of the Institute for Energy and Environment (IVEM) at the University of Groningen in the Netherlands, where he received his Ph.D. on sustainable resource use. Since 1990, he has been a senior scientist at the Netherlands Environmental Assessment Agency (PBL, formerly MNP and RIVM). He has been actively involved in modelling and scenario construction for the Intergovernmental Panel on Climate Change (IPCC). Since 2003, he has also been a Professor of Global Change and Energy at Utrecht University in the Netherlands. His research expertise and publications are in resource and energy analysis, modelling and policy; climate and global change modelling; and complex systems modelling for sustainable development. He has co-edited several books, including *Perspectives on Global Change: The TARGETS Approach* (Cambridge University Press, 1997) and *Mappae Mundi: Humans and Their Habitats in a Long-Term Socio-Ecological Perspective* (2002).

Advance Praise for *Sustainability Science*

"Achieving some sort of sustainability will be THE focus of global societies in the 21st century. To be successful, our leaders will need a perspective of centuries, the full breadth of scientific insights, system thinking skills, great cultural sensitivity and an awareness of spiritual values. All of these are offered in this wonderful, unique text, which will be useful for decades."

– Dennis Meadows, co-author of The Limits to Growth

"This textbook is one of the first truly all-encompassing introductions to sustainability science. It is methodical, clearly written and well-illustrated, truly a pleasure to handle and to read. It sets a standard for the discipline and solidly educates the generation of students that will most directly have to deal with the challenges of creating a sustainable Earth system."

– Sander van der Leeuw, Dean, School of Sustainability, Arizona State University

"In this important new book Bert de Vries has adopted a systems approach to examining all the issues that collectively amount to the determinants of sustainability. It is an excellent, comprehensive and up-to-date text dealing with not only the underlying biophysical science but also human behaviour. His use of interesting examples throughout makes it both instructive and enjoyable to read. I highly recommend it."

– Brian Walker, CSIRO Ecosystem Sciences, Australia

"Bert de Vries' *Sustainability Science* is particularly welcome as it breaks ground in a new field, which so far lacks a proper systematic treatment. No wonder! The challenge is overwhelming: the book covers a series of disciplines and fields – geography, social and economic sciences, physics, chemistry and biology – using a systems description and system dynamics as the main tool. De Vries not only succeeds in this overwhelming task but spices up the text with multiple excursions into history, philosophy, literature, not to forget the key issue of ethics. Justice and how we would like a future world to look is always present. Bert's book is impressive, rich and inspiring."

– Lars Rydén, Centre for Sustainable Development, Uppsala University

Sustainability Science

Bert J. M. de Vries
Utrecht University

CAMBRIDGE
UNIVERSITY PRESS

CAMBRIDGE UNIVERSITY PRESS
Cambridge, New York, Melbourne, Madrid, Cape Town,
Singapore, São Paulo, Delhi, Mexico City

Cambridge University Press
32 Avenue of the Americas, New York, NY 10013-2473, USA

www.cambridge.org
Information on this title: www.cambridge.org/9780521184700

First published 2013

Printed in the United States of America

A catalog record for this publication is available from the British Library.

Library of Congress Cataloging in Publication Data

Vries, Bert de, 1948–
Sustainability science / Bert J. M. de Vries. – 1st ed.
p. cm.
Includes bibliographical references and index.
ISBN 978-1-107-00588-4 (hardback : alk. paper) – ISBN 978-0-521-18470-0
(pbk. : alk. paper) 1. Sustainable development – Research. I. Title.
HC79.E5V735 2012
338.9′27072–dc23 2012024935

ISBN 978-1-107-00588-4 Hardback
ISBN 978-0-521-18470-0 Paperback

Additional resources for this publication at www.cambridge.org/devries

"In bytsje minder as men lêst,
Dat bikomt jin fierwei 't bêst."
(A little bit less than you crave for,
Makes you enjoy so much more.)
Fries spreekwoord

Contents

Preface

This book is the outcome of eight years of teaching the course Sustainability Science for students of the M.Sc. in Sustainable Development at Utrecht University. Its aim is to give a broad overview of what the sciences have to say about sustainable development. To this purpose, it offers a mixture of concepts, theories, models and facts and an invitation to the student to become a critical, independent thinker and to act accordingly.

The book can be used at the B.Sc. level as part of an introductory course or at the M.Sc. level as context for other courses and M.Sc. theses. For most chapters, a high school or college background should be sufficient, but some capacity for abstract thinking is needed. As an introduction into the concept of sustainability and sustainable development issues, it is useful for people in NGOs, government agencies and business who are interested in framing the discourse from multiple perspectives and different scientific disciplines. It can help them to make better decisions for life on a finite planet.

The content is based on three personal convictions. The first one is that humanity faces a transition period in which many ideas, habits and expectations will be challenged and scientists should, therefore, offer an integrated perspective on how developments are connected and may unfold. I follow others in using the term *sustainability science* to summarise this effort. The second conviction is that all scientific disciplines can and should contribute to the content of sustainability science. The conceptualisation of sustainable development as a guiding principle for the 21st century is still fragmented, and this should change. We need a new science, one that uncovers the unity of science and mobilises understanding and offers context. One that uses the novel ways to access information (Internet, Wikipedia, etc.) and to engage in the real world (stories, simulation games, etc.). A third conviction is that sustainable development can best be defined in terms of *quality of life* and that the pluralism in people's values and ideas about what quality of life is and in their circumstances should be acknowledged explicitly. It implies the framing of sustainable development as a global challenge within local diversity, capacity and contingency.

How to realise this ambitious goal is not self-evident. The first part of the challenge is to find observations, concepts, theories and models that are relevant in context. I first introduce system dynamics and influence diagrams and simulation models as an inherently integrating language and toolbox. Next, I summarise history insofar as it is relevant for understanding sustainable development: rise and decline of past civilisations, the transition to the industrial era and the emergence of the idea of sustainable development. Subsequently, I introduce worldviews, as combinations of values and beliefs, in order to communicate and appreciate the diversity in the notions of sustainable development and quality of life. I offer epistemological reflections on the nature of scientific knowledge in order to appreciate the inherent complexity and uncertainty of the issues being dealt with. Some basic natural science, notably on energy, is introduced in Chapter 7.

In the second part of the book, I discuss the subsystems in which the concerns about (un)sustainable development are most pressing: natural ecosystems, agro-food systems, renewable and non-renewable resources and the economy-technology system. Each chapter covers what I consider important concepts, theories and models and their link with the observations in the real world. Each chapter has a natural connection with specific scientific disciplines, but ecology and geography are core disciplines. There are a few stories in each chapter to illustrate and complement the theory and the models and invite the student to engage with real-world experiences and situations. At the end of each chapter, five statements about the subsystem are offered to the reader in order to practice worldview pluralism. A few quotes are given to provoke 'out-of-the-box' thinking.

There are many different angles from which one can look at development, quality of life and sustainability. Moreover, the flood of data, theories and models keeps growing. Therefore, I can and do claim neither completeness nor representativeness. No doubt, the treatment of some topics is biased because I was trained in physics and chemistry and because I know certain persons and books and do not know others. I am aware that the treatment of some topics is incomplete or inadequate. Ethics, psychology, political science and subdisciplines in geography and economics, for instance, remain underexposed. Besides my limited knowledge, it reflects the inherent complexity of these disciplines and their still limited input in the sustainability discourse. Also, engineering details and legal aspects are hardly addressed. As to the narratives, I do not yet offer a balanced mix of stories. In particular, stories about the many social, economic and technological initiatives and innovations for a more sustainable world are missing. I hope that the book is a start for a more comprehensive, web-based 'atlas for a sustainable world', in which the latest scientific insights merge with relevant stories from all over the world and, together, point at the right actions. The website www.sustainabilityscience.eu is a first, simple start towards this goal.

This book could, of course, only be written by building on the thoughts, feelings and intuitions of many other people. Some of them are voices from a distant past or from people I never met, but much of it is easily traced. My promoter Jan Kommandeur offered the inspiration and courage that helped me to follow my own path. With my colleagues at the University of Groningen, I spent many years in the 1970s and 1980s in an atmosphere of great intellectual curiosity and freedom. In those years, I also joined the Balaton Group, founded by Donella and Dennis Meadows. Ever since, the annual meetings have been a source of inspiration, friendship and joy. A considerable part of the ideas in this book have taken shape during exchanges with Balaton Group members. Donella Meadows taught me the virtue of combining clear systems thinking and compassionate engagement. From Dennis

Meadows, I learned the usefulness and fun of communicating insights in the form of games. Aromar Revi nurtured my search for cosmic dimension and wisdom. I refrain from mentioning the many other members of the group – but they know.

In the 1990s, I had the privilege to work in a rather unique group of Global Change Modelling pioneers: the TARGETS and IMAGE teams at RIVM (later MNP and still later PBL – the English name turned out to be more sustainable: Netherlands Environmental Assessment Agency). I thank Klaas van Egmond and Fred Langeweg for the opportunities to follow uncharted paths, and the many colleagues for the instructive and pleasant collaboration. Another source of inspiration and insight were the meetings and discussions as part of writing the book *Mappae Mundi*, with Joop Goudsblom, on the 250th anniversary of the Hollandsche Maatschappij der Wetenschappen. Since 2003, I occupy for one and later two days a week the chair Global Change and Energy at the Copernicus Institute for Sustainable Development and Innovation at Utrecht University. The chair is financed by PBL, and I am most grateful for the permission I got to spend time on the teaching and writing that has led to this book.

As to the actual writing of the book, I received useful and constructive input from many people: Hans Deuss, Tom Fiddaman, Peter Janssen, Eric Lambin, Sander van der Leeuw, Erik Lysen, Evert Nieuwlaar, Martin Patel, Charles Redman, Max Rietkerk, Lars Ryden, Mark Sanders, Isak Stoddard, Yoshi Wada and Bob Wilkinson. I would also like to acknowledge a few persons in particular. The constructive and humorous criticisms of Jodi de Greef on anything I wrote about complexity and sustainability have protected me from an overdose of sincerity. The chapter on worldviews got a major boost from the inspiring discussions with Klaas van Egmond, my colleague at Utrecht University since 2008. The cooperation with Markus Brede since 2006 brought an intellectual depth to parts of the book that would otherwise have been absent. Cristina Apetrei has been of tremendous help with comments and layout, and her practicality and determination kept me afloat in times of despair. My brother Joop gave with great precision the right books at the right moment, my son Tom applied his management skills for me to figure out what the message of the book is and my daughter Marieke assisted with cover design. I owe much inspiration to the silence and darkness, the skies and clouds, and the mountains, trees and humans of the Vallespir. Finalement, Annelize, sans toi j'avais eu ni le courage ni l'espace de persévérer dans un monde où, comme Proust l'a dit, 'les forts, . . . ont seuls cette douceur que le vulgaire prend pour de la faiblesse'.

I suffer from the injustice and violence in the world, the destructiveness of the relentless pursuit of 'stuff', the lack of awareness among large numbers of people and the cynicism and hypocrisy among many members of the powerful and wealthy. But I also discern and enjoy the germs of rejuvenation, the search for meaning, the genuine compassion that can be seen all over the world. I am grateful that I had the opportunity to work on this book. I am grateful to share with my students and others what I have learned up to now about the road forward.

1 Introduction

1.1 Roots

When you open a newspaper, any newspaper, there is a big chance you will encounter the term *sustainable development*. Introduced to a broader public in the 1980s with the publication of the UN's report *Our Common Future*, sustainable development has become common vocabulary. The word 'development' is commonly used to indicate growth, not only in quantity, but primarily in quality. The word 'sustainable' refers to something that can or should last. The idea of sustainable development reflects one of the leading aspirations of humankind in the 21st century, not unlike the idea of socialism in the early 20th century. It has become a modern equivalent of, and complement to, the Declaration of Human Rights, formulated shortly after the devastating Second World War. Civil society organisations have pushed sustainable development forward; respected business and government leaders now hail it as the foremost challenge for the 21st century.

Inevitably, such an aspiration or ideal accommodates a large variety of explanations, objectives and proposals. These are intertwined with personal and collective values and perceptions, which are in turn rooted in millennia of developments shaping human experiences, knowledge, technical skills and social arrangements. Given the human population's continous growth and its use of the planet as a source of resources and a sink of waste, humanity needs an ongoing dialogue that slowly converges to a widely shared vision on the theory and practise of sustainable development.

The word *sustainable* has been known in European languages since the early Middle Ages. It is rooted in the Latin verb *sus-tenere, sub* meaning 'up from below' and *tenere* meaning 'to hold'. In the physical sense, the verb *to sustain* is equivalent to bearing, or carrying the weight of something to keep it from falling by support from below. However, early on, the word had a meaning beyond a simple mechanical act, as is already evident in the words of the Roman philosopher Seneca (3 BCE–65 CE): 'The society of man is like a vault of stones, which would fall if the stones did not rest on another; in this way it is sustained'.

One of the oldest and most common connotations of the verb *to sustain* is to keep a person, a community or the spirit from failing or giving way, to keep it at the proper level or standard. It can be active, as 'to support (life)' and being capable and willing to go on. It can also be passive: 'to undergo' or 'to endure' and is then equivalent to bearable or defensible. Which of the two meanings apply depends on the role, attitude and circumstances of the actor. As he or she may succeed or fail, the verb *to sustain* reflects the human condition: ranging from willpower, duty and pride to fate, pain and suffering. As early as 1290 CE, this connotation was known in the English language.

A closely related connotation stems from the archetypical notion of some force or god, which 'keeps the world running'. In Chinese and Indian cosmology, the forces sustaining the world reflect a dynamic equilibrium between opposites; in Greek cosmology, it is Atlas who kept the earth and the heaven separated. In this sense, the verb *to sustain* gets a transcendent connotation, as in Milton's words:

> *Whatever was created, needs*
> *to be sustained and fed.*

This is even more outspoken in 'sustenance'. The word sustenance has become equivalent to nourishment, food and more generally the means of living or of sustaining life – without any specification of 'that which sustains'. This usage, as in Tennyson's verse:

> *Water is one part,*
> *and that not the least of our sustenance,*

comes close to the ecological meaning, which the word sustainable has acquired in recent times.

An English equivalent of the verb *to sustain* is 'to last', meaning to go on existing or to continue. Interestingly, it used to be associated with performance and duty. The verb 'to endure' is an English equivalent of *to sustain* and is rooted in the Latin verb *durare*. It is common in other European languages. In German, the word *dauerhaft* is the common word for sustainable, with *nachhaltig* as a synonym. The Dutch equivalents are *duurzaam* and *houdbaar*. In French, the word *durable* is most common – and is also used in English as a synonym of lasting or permanent. The words *soutenable* and *viable* are also used in French as synonyms to indicate something that is bearable, can survive or is feasible.

Present-day usage of *sustainable* refers to an act, a process or a situation, which is capable of being upheld, continued, maintained or defended. It has a largely active disposition, in the context of sustainable resource use or management. The word *sustainability* expresses the presence of such a capacity and is a recent coinage. The words rooted in *durare* suggest a more passive connotation than those rooted in *sustenere*.

The word *development* comes from the *des* meaning 'undo' and *veloper* meaning 'to wrap up' in old French and is possibly of Celtic origin. In present-day use, the verb *to develop* means to (help) strengthen and enlarge. In particular, it is a progression from earlier to later stages of a life cycle or a process from simpler to more complex stages of evolution. It is about growing by degrees into a more advanced or mature state. Development is considered to be broader than quantitative growth. It involves maturing, ripening or bringing from latency to or towards fulfilment and fullness. It refers to a dynamic process of (causing to) grow and differentiate along lines natural to its kind, of improving the quality and of (causing to) become more complex or

> **Box 1.1.** *Sustainability science.* 'A new field of sustainability science is emerging that seeks to understand the fundamental character of interactions between nature and society.... [it] needs to move forward along three pathways. First, there should be wide discussion within the scientific community regarding key questions, appropriate methodologies, and institutional needs... Second, science must be connected to the political agenda for sustainable development.... [and] third (and most important), research itself must be focused on the character of nature-society interactions, on our ability to guide those interactions along sustainable trajectories, and on ways of promoting the social learning that will be necessary to navigate the transition to sustainability' (Kates *et al.* Science 292(2001) 641–642).
>
> 'Sustainability science is not yet an autonomous field or discipline, but rather a vibrant arena that is bringing together scholarship and practice, global and local perspectives from north and south, and disciplines across the natural and social sciences, engineering, and medicine' (Clark and Dickson PNAS 100(2003) 8059–8061). In the same article, the link with a new kind of science is emphasised: 'Post-Normal Science has been developed to deal with complex science related issues. In these, typically facts are uncertain, values in dispute, stakes high, and decisions urgent, and science is applied to them in conditions that are anything but "normal"'.

intricate. Development is an evolution from simple to complex in terms of technical and managerial skills and of social-cultural connections and institutions.

The concept of *sustainable development* supposedly combines the ideas of a process or situation that can be continued and one that is growing in complexity and maturing towards 'natural' fulfilment. Introduced in the 1970s, it was applied initially with reference to an ecological or environmental desire, target situation or state. A measure or indicator of sustainability is then the difference between the actual and the desired situation and the timepath towards it. Often, the desired or target situation is related to some reconstructed preindustrial 'natural' situation and serves as a reference. Practitioners often held the view that 'sustainable development' could thus be given an objective interpretation.

In the 1990s, the interference of social scientists and notably economists made it clear that formulating such a desired or target situation for a sustainable development trajectory cannot be legitimised solely on ecological-environmental criteria. First, which indicators for decision making are chosen? Should not economic and social aspects be included as well? Second, if there is agreement on the choice of indicators, the desired or target indicator levels must be the outcome of a societal negotiation process and be open for renegotiation. Even if a command has to be obeyed under penalty of complete and irreversible loss and, therefore, should be considered an unconditional moral obligation – Kant's categorical imperative – it may still have to be renegotiated in the light of new information, a shift in values or new and other pressing needs.[1] Accordingly, economists argue that the quest for

[1] Examples are an unconditional interdiction of slavery and a complete ban on entering Antarctica.

sustainable development can and should be founded on welfare economics and societal cost-benefit analyses. Other social scientists brought their own and different observations, concepts and theories. In any event, it is now widely acknowledged that the interpretation of sustainable development includes the subjective, value-laden elements that are inherent in the words *sustainable* and *development*. The inputs from a variety of scientific disciplines have led to a flux of ideas, concepts and observations, which together lead to a new branch of science in the making: *sustainability science*.

1.2 Sustainability Science

This book has also the word *science* in the title. The word science comes from the Latin noun *scientia* and the associated verb *scire* meaning 'to know'. Originally, it may have come from the Latin verb *scindere* meaning 'to cut, divide'. Thus, an old connotation related the word already to the act of distinguishing, that is, of dividing in the mind. The word science has undergone a gradual evolution in European history. In the 17th century, it became separated from more practical and artistic knowledge – and practises and skills – then became understood as theoretical truth. Gradually, science got the meaning of a 'body of regular or methodical observations or propositions... concerning any subject or speculation'. The overlap with philosophy became smaller, and science evolved into an expanding set of compartments of systematised knowledge about particular objects. When it concerns objects in the physical world and its phenomena, science is natural science – but the word natural is often omitted. The focus is on general truths and laws as well as on particular methods of enquiry. However, the first and common meaning of science still is 'a state of knowing: knowledge as distinguished from ignorance or misunderstanding' (Encyclopedia Britannica).

Connecting sustainable development to science stems from the idea that sustainable development is an aspiration that can and should be realised only on the basis of scientific knowledge. Does it justify the establishment of a new branch of science? Yes, for several reasons. An overriding reason, in our view, is that humanity is confronted at an ever larger scale and rate with the consequences of its own success as a species. The complexities and uncertainties of the human adventure are becoming such that insights from all the sciences are needed to deal with them if we are serious about the aspiration to improve quality of life (development) that endure (sustainable). Understanding and acting upon the causal mechanisms and behavioural responses across several time- and space-scales is a great challenge for an increasingly fragmented science. A transdisciplinary approach is called for, in which the quantitative and the qualitative, the natural and the social and also theory and practise (or science and policy) are reconciled and creatively combined. Such an integrating and synthesising approach deserves the name *sustainability science*.

This new science started with research and education on change processes across all scales in space and time. Known as Global Change Science or Earth System Analysis, the focus is on large clusters of observations and phenomena, and new scientific branches develop along the borders of classical disciplinary sciences. Much of it emanated from the environmental sciences, which branched into many directions such as atmospheric and ocean science, marine biology, human ecology, ecological economics and so on (Boersema and Reijnders 2009). Gradually, it became clear that there were rather general laws underneath the phenomena in the different fields of study. Also, the emphasis shifted from description of states to understanding changes in states. *System theory* and the methods of system analysis and system dynamics became the most coherent expression of this insight (Bossel 1994). The new field increasingly expanded beyond the environmental sciences. The time has come to reinforce the unified approaches and unifying tendencies in science and to liberate the study of real-world processes from the confines of artificial, 19th-century boundaries between the scientific disciplines. Sustainability science is perhaps the most clear and desirable expression of this endeavour.

What is a meaningful definition and working program for sustainability science? During the last few decades, scientists from different backgrounds have formulated attempts at founding statements. The following are a good start (Kates *et al.* 2001; Clark and Dickson 2003):

- sustainability science is about understanding the dynamics of evolving, coupled *social-ecological systems* (SES);
- sustainability science is *transdisciplinary*: solutions to the problems have to acknowledge that the world is/becomes more integrated, more complex and more uncertain;
- the focus of sustainability science is on the *interactions* between the resource system (earth/life sciences), its users and the governance system (social sciences); and
- sustainability science is *problem-driven* to manage complex coupled SES in order to have them deliver what people value.

This book uses the term *sustainability science* to indicate the efforts from all corners of science to construct a framework for understanding and acting in relation to (un)sustainable development. The aim is to provide foundations for and essential concepts and methods in such a sustainability science.

In sustainability science, human individuals and societies are studied in a biosphere context. The key notion here is that of SES, which are defined as integrated human–nature systems with reciprocal feedbacks and interdependences (www.resalliance.org). It appears to be a 'natural' unit of study. We reflect explicitly on the ways and limitations of a transdisciplinary approach and pay attention to models and the process of modelling. We introduce *system dynamics* and, more broadly, *complex systems science* as one set of methods and techniques ('tools') for integration. As part of the transdisplinarity, we consider explicitly the great variety and richness in people's values and beliefs about the world and about sustainable development. The idea of *worldview pluralism* is applied in various places as a way to deepen personal understanding and engagement.

The problem-driven nature of sustainability science is controversial. If one accepts it as a constitutive element, two questions arise: Which problems should be addressed and where does the overlap begin with practical policy and management insights and skills? Because there are very different appraisals of the problems humanity faces in the 21st century, the first question raises the issue of worldview pluralism. Natural scientists formulate the problem largely in terms of human-induced changes at a planetary scale. Some of these processes have become well known, such as the ozone layer depletion ('ozone hole') and anthropogenic climate change ('global warming'). Others are still less known but possibly equally serious (Rockström *et al.* 2009). These 'problems' were already addressed one year after human beings set foot on the Moon, in the report *Man's Impact on the Global Environment* (SCEP 1970): 'the concept of the earth as a "spaceship" has provided many people with an awareness of the finite resources and the complex natural relationships on which man depends for his survival... Some... have warned of both imminent and potential global environmental catastrophe. Theories and speculations of the global effects of pollution have included assertions that the building of CO_2 from fossil fuel combustion might warm up the planet... the possibility of... particles emitted into the air... lower[ing] global temperature... the effects on ocean and terrestrial ecosystems of systematically discharging... heavy metals, oil, and radioactive substances; or nutrients such as phosphorus'. It was repeated and supported with a system dynamics computer model in the report *Limits to Growth* (Meadows *et al.* 1971) to the Club of Rome. Since then, the world has in terms of population and economic activity largely followed the 'business-as-usual' path in that report (Turner 2008). Nevertheless, there are still widely divergent interpretations and appreciations of what happened during the intermediate forty years.

The second question points to the demarcation between science and technique, επίστημη and τεχνη. Sustainability science differs from sustainably managing SES, although the two have a fruitful and creative connection. Knowledge of how the world works and how we think the world works is only one, and not necessarily an essential or sufficient, ingredient of acting upon the world. This book follows a pragmatic course. First, it considers models of systems important in understanding, assessing and communicating about sustainable development. Therefore, Chapters 2, 8 and 10 deal rather extensively with the construction and use of models and with some analytical and modelling tools. Aware of sometimes large, or even impregnable, limitations of (mathematical) modelling, Chapters 7, 9 and 11–13 present stories and case studies in order to bridge the concrete and the abstract, the data and the theory. Finally, Chapters 6, 10 and 12 introduce other ways of engaging people's insights and participation, such as value surveys, participatory modelling and simulation games. In this way, the scientific insights presented provide a context for action – technical, political or other – without explicit discussion of the myriad options for action.

Box 1.2. *Exactness.* Don't expect a Unified Theory of SES or Social-Ecological Systems. It does not exist. Do not expect exactness either... precision is not always what we are looking for:

Joachim went to the famous Museum of Natural History where, he was told, one could see many dinosaurs.

Seeing one of them, the provost standing nearby told him 'This dinosaur is very old, sixty-five million and fifteen years'.

'Very old indeed', replied Joachim, '... but how do you know its age with such precision'?

'Well', the man replied, 'when I came here to do this work, it was sixty-five million years old and that was fifteen years ago.... '

1.3 Sustainable Development Is About Quality of Life

We have spoken about the roots of the words sustainable and development. But what *is* sustainable development? The word sustainable is used in many different contexts, and not always sensibly: sustainable cities, sustainable traffic, sustainable water, sustainable livelihoods, sustainable banking, sustainable technology – and sustainable growth. The French newspaper *Le Monde* celebrated the start of the 'Semaine du développement durable' in 2005 with a special supplement saying: 'Let's forget the words "sustainable development", because they provoke indifference . . . and, worse, they make people smile. Let's take care of its contents, though: producing what people need without destroying their environment. . . . ' They are not the only ones who give this advice. Yet, this book does not follow the advice and sticks to the notion of sustainable development as the core concept. The objective is to deepen its content both in the abstract and the concrete.

For the moment, we use the most widely known definition of sustainable development, given in the UN's World Commission on Environment and Development (WCED) report *Our Common Future* (1987):

> *Sustainable development* is development that meets the needs of the present generation without compromising the ability of future generations to meet their own needs.

Within this definition, the goal of sustainable development expresses a quest for developing and/or sustaining qualities of life. Framing sustainable development in terms of quality of life introduces the subjective and objective dimensions of human well-being and invites a truly transdisciplinary approach. It has an intergenerational and an international dimension: people should act *here and now* in such a way that the conditions for a (decent/high) quality of life *elsewhere and later* are not eroded.

Did we only shift the problem – because: What is quality of life? And which qualities, for whom and for how long? Should these be experienced or imagined qualities of life, for the lucky few or for all, for humans only or for all species, for our children or for the next seven generations or even longer?[2] Another question poses itself: Is it development towards a static situation, a world that can be sustained because there is no change? Such a blueprint approach is seductive, as the many utopias in literature testify (Achterhuis 1998; de Geus 2003). But science tells us that everything is always and everywhere in flux and that development is an evolutionary path of success and failure.

Quality of life is an experience that stretches out over large domains in space and time, in our individual life space (Figure 1.1a,b).[3] In first instance, I as an individual person relate to it in the here and now: material and immaterial well-being. Do I have enough to eat? Do I have shelter? Can I avoid or cure diseases? Can I have sex and experience love? Can I learn or apply skills? Can I communicate and relate? What we experience as quality of life is, through our actions and emotions, our beliefs

[2] Most people will adhere to an anthropocentric view, but some people want to extend it to all life or even the planet as a whole. Such an ecocentric view can in theory be defended – but the defense is, in practise, always by a human being.

[3] Of course, each individual lives in a larger, social-cultural context and the two schemes in Figure 1.1 are flat-world simplifications of a complex reality.

a)

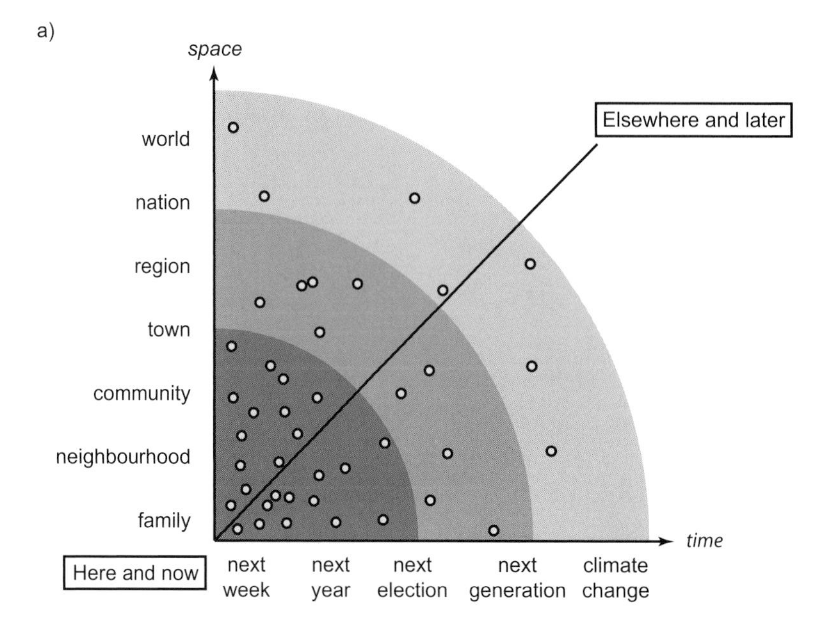

b)

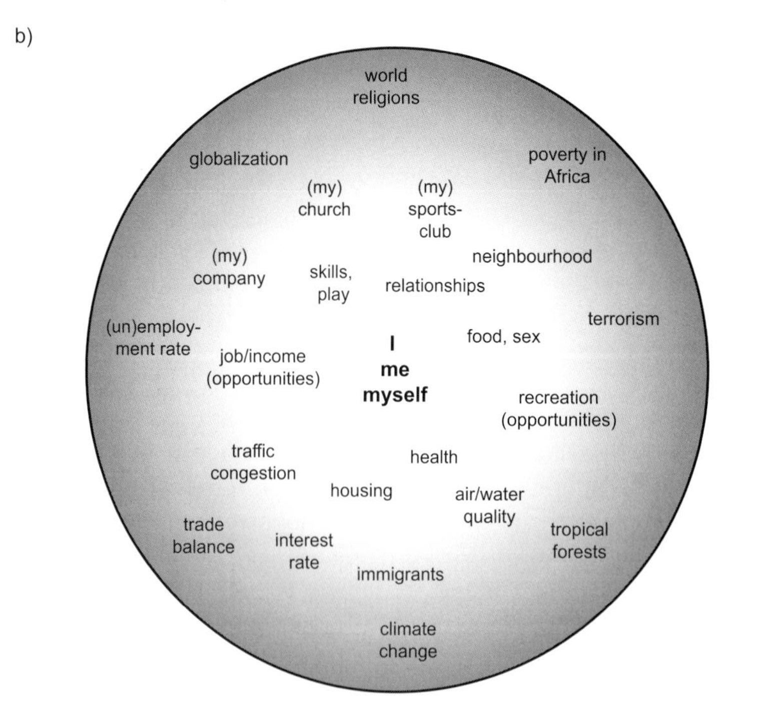

Figure 1.1a,b. Simplified schemes of time and space in the perception of sustainable development issues: (a) the space and time scales in which we experience the world; (b) items in the centre-periphery of our daily consciousness.

and thoughts, also something of others and elsewhere, of past and future (Figure 1.1b). Is there food for the whole family? Can I pay my children's school fees? Will there be riots in town? Will my husband's job disappear next year? Can I still enjoy last week's celebration or forget last year's insult? Will there be a good harvest this year? Clearly, quality of life is not easily condensed in one or a few quantitative measures.

We have seen that development is growth towards a more complex and mature state. Growing up, maturing as a human individual implies a widening of one's perspective on what life is about and on what 'the good life' can or should be. It can coincide with a deepening of experiences, from the sensate and emotional to the mental and spiritual. Thinking and acting can become more inclusive, with caring for the larger scale and the longer term.The motivation for inclusive thinking can be an expression of biological pragmatism and functionalism or of solidarity with wider kin and of tribal 'commandments' and social norms for the benefit of the group. Perhaps, it arises out of spiritual yearning.

Whatever the motives, in the manifestations of inclusive thinking, the aspiration of sustainable development is recognised as more than a scientific quest: It is an ethical and transcendent endeavour. It challenges us as individuals and groups in how we manage our needs and wants *and* how we organise and manage resource use that goes with it. *Ethics* enters the discourse, with questions about fairness, solidarity, justice, egoism and altruism. *Transcendence* appears when we inquire into the meaning and dignity of human life. I can give many references to books by philosophers, artists and mystics who have struggled with these questions and attempted to express it in words (Elgin 1993). It is probably better if you reflect on your own life to see if it makes sense for you. Subsequent chapters offer openings for such reflections.

In a practical sense, we must link sustainable development to the aspirations of human beings for a good quality of life while respecting the plurality of worldviews. Science can contribute here with adequate beliefs about the world and its workings and with methods and tools to make those beliefs effective in action. It can also teach how to complement intuition with rationality in managing uncertainty. It can inspire feelings of respect and joy in the face of the world's rich diversity, beauty and complexity. And, finally, it invites us to reflect and give knowledge and mind their proper place in our lives.

Immanuel Kant (1724–1804) offered a guide in the search for sustainable development when he summarised his philosophy as the search for an answer to three questions: '*What can I know? What ought I to do? What may I hope for?*' In the present context, these questions can be reinterpreted. What can science tell (and what not)? What, then, is an individual's responsibility and duty? And which dreams and destinies can he or she expect? This book is primarily about Kant's first question: *What can I know?*, but it opens a panorama on answers to the other two questions.

1.4 Guidelines for the Reader

Set up as a textbook and an introductory guide, this book is intrinsically transdisciplinary: The student is not expected to become an expert in a particular domain or method but, instead, is invited to learn about and appreciate the different perspectives on (un)sustainable development. Not only are those perspectives following the different scientific disciplines, but they also reflect the different individual worldviews. In this way, the book prepares the student for an independent yet personal position on ideas and actions regarding sustainable development. The most important word in this respect is probably *context*. After studying the book, the student should be able to put into context what is read or seen in journals, newspapers, books or on TV or the Internet about ecological, economic and social aspects of sustainable development. Equally important, the reader should feel better equipped to engage with his or her own values and beliefs in the aspiration for a more sustainable world.

There are four threads in this book. The first is a *historical* and *qualitative* description of past developments and civilisations in order to get an idea of 'where we are and what the problem is'. It is based on reconstructions by archaeologists and historians and comprises Chapters 3 and 4. In most chapters, there are also 'relevant (hi)stories'.[4] These are personal accounts of people living through particular events, which illustrate concepts or models discussed in the chapter. Of course, it is merely a selection of the numerous stories about (un)sustainable development around the world.

The second thread centres around the notion of *worldviews* and is the focus of Chapters 5 and 6. I briefly examine the historical background of the concept of sustainable development. Next, I discuss the objective and subjective aspects of needs and capabilities and the variety in value orientations and beliefs that describe and explain the difference in views about sustainable development and quality of life. After this groundwork, I offer a theory and a framework that explore the centrifugal forces behind unsustainable developments, which are applied throughout the book to categorise and understand the different views people have on important issues such as population, resources, technology and economic growth.

The third thread is the *systems* approach and the methods and techniques that have been developed over the years in order to understand a system's behaviour over time. Chapters 2 on system dynamics and the thematic Chapters 7 and 9–14 use simple simulation models to explore basic mechanisms of change. These models are constructed with software packages such as Stella®, Vensim® and NetLogo® (Mathematical details are set apart in the appendices for the interested student). To appreciate the role of mathematical models in sustainability science, there is a brief orientation on the philosophy of science, in particular on the nature of knowledge and models and on uncertainty and complexity.

A fourth thread summarises the major insights and findings of the *scientific disciplines* that shed light on (un)sustainable development. Each scientific discipline contributes to the search for a sustainable world. It may be argued that ecology and geography are the core of sustainability science, but the natural and engineering sciences and the economic and social sciences may be of equal importance. Largely following the reductionist-empiricist paradigm, later chapters survey observations ('facts'), concepts, methods and theories ('laws') in contemporary science with respect to environment and development. What is the input from classical thermodynamics and mechanics and, broader, energy science (Chapter 7)? What insights do we gain from ecologists and demographers who study nature's evolution and species populations and what is the image of man in the economic and social sciences (Chapters 9 and 10)? What does science tell us about the dynamic processes

[4] We will also use the words narratives or, sometimes, case studies or anecdotes. The website www.sustainabilityscience.eu is collecting stories with relevance for sustainability science issues.

behind resource exploitation and environmental degradation (Chapters 11–14)? Of course, these areas are far too large for a single book or a single mind to comprise. Therefore, the focus is on what are considered the most important concepts and insights, illustrated with data and stories and connected through models and worldviews.

The book's fifteen chapters include the cross-reference of ideas and concepts within chapters, in order to enhance coherence, integration and the benefit of cross-fertilisation. At the end of each chapter is a list of useful books for further reading and of selected websites that offer, as of 2011, relevant data, models, reports and papers. At the end of the book, there is a list of acronyms and a glossary that briefly explains keywords. Most chapters can also be read on their own, particularly those on rather well-defined topics such as agro-food systems and renewable and non-renewable resource use.

One consequence of the integration is that the reader may encounter parts that are already familiar to him or her. There are two options then, apart from getting annoyed or angry and stop reading. The first one is to skip those texts. The second one is to read it again with the objective to refine one's insights through the discovery of new links and vistas and new connections to personal values and beliefs. 'Creativity is the power to connect the seemingly unconnected' (thinkexist.com/quotes/william_plomer/). I invite you to choose the second option.

> There is more to learning than the mind. Learning requires discipline and patience. The Persian Sufi-Master Idries Shah says: *There are some things which you have to do for yourself. These include familiarizing yourself with study-materials given to you. You can only really do this – and thus acquire real qualities – if you suspend the indulgence of desire for immediate satisfactions.* An attitude of empathy, of love is needed as well. As the Persian poet Rumi said:
>
> If you're in love,
You need no proof.
If you're not in love,
What good is a proof?

SUGGESTED READING

Suggestions for a research program from a Global Change perspective.
Clark, W., and N. Dickson. Sustainability science: The emerging research program. *PNAS* 100 (2003): 8059–8061.

A sequel to the outline started in Kates et al. (2001).
Clark, W., ed. Sustainability Science: A room of its own. *Special Issue PNAS* 104 (2007):1737.

A first outline of what sustainability science should address. Since then, quite a few papers have been published on this topic.
Kates, R., et al. Sustainability Science. *Science* 292 (2001): 641–642.

The journal PNAS now has a special section on sustainability (see websites list).
Perrings, C. Future challenges. *PNAS* 104 (39) (2007):15179–15180.

An evaluation of the 1972 Limits to Growth report on the basis of historical data since then.
Turner, G. A comparison of the Limits to Growth with 30 years of reality. *Global Environmental Change* 18 (2008): 397–411.

USEFUL WEBSITES

- www.igbp.net/ is the site of the International Geosphere-Biosphere Program (IGBP) and at www.igbp.net/documents/resources/science-4.pdf one can download the report *Global Change and the Earth System* (Steffen *et al.* 2003).
- www.pnas.org.proxy.library.uu.nl/site/misc/sustainability.shtml is the website of the Sustainability Science section of the Proceedings of the National Academy of Sciences of the United States of America (PNAS), with special issues on a series of topics (land change, health, food, climate change, poverty and others).
- www.awakeningearth.org/ is the site of Duane Elgin, who discusses sustainable development in the perspective of the evolution in human consciousness.

Appendix 1.1 United Nations Decade of Education for Sustainable Development

In December 2002, Resolution 57/254 on the United Nations Decade of Education for Sustainable Development (2005–2014) was adopted by the United Nations General Assembly. UNESCO was designated as lead agency for its promotion (portal.unesco.org/education/en). The rationale for this Decade is stated as:

> Education as the foundation of sustainable development was reaffirmed at the Johannesburg Summit, as was the commitment embodied in Chapter 36 of Agenda 21 of the Rio Summit, 1992. The Plan of Implementation establishes the linkages between the Millennium Development Goals on universal primary education for both boys and girls, but especially girls, and the Dakar Framework for Action on Education for All. The creation of a gender-sensitive education system at all levels and of all types – formal, non-formal and informal – to reach the unserved is emphasized as a crucial component of education for sustainable development. Education is recognized as a tool for addressing important questions such as rural development, health care, community involvement, HIV/AIDS, the environment, and wider ethical/legal issues such as human values and human rights.
>
> There is no universal model of education for sustainable development. While there will be overall agreement on the concept, there will be nuanced differences according to local contexts, priorities and approaches. Each country has to define its own priorities and actions. The goals, emphases and processes must, therefore,

Table 1.1. *Key action themes of the UN decade of education for sustainable development*

Quality education	Cultural diversity
Overcoming poverty	Indigenous knowledge
Gender equality	Media & ICTs
Health promotion	Peace & human security
HIV/AIDS	Governance
Environment	Climate change
Water	Biodiversity
Rural development	Disaster reduction
Sustainable consumption	Sustainable urbanisation
Sustainable tourism	Corporate responsibility
Human rights	Market economy
Intercultural understanding	

> be locally defined to meet the local environmental, social and economic conditions in culturally appropriate ways. Education for sustainable development is equally relevant and critical for both developed and developing countries.

The list of key action themes covers an impressive twenty-three items, revealing how broad the topic of sustainable development is being conceived today (Table 1.1). It not only reflects the understanding that 'everything depends on everything else', but also that many different stakeholders are involved. Full of good intentions and large aspirations, these global initiatives show, on the one hand, the emergence of a global consciousness of the human predicament – and, on the other hand, the risk of creating worldwide bureaucracies that become as much part of the problem as of a solution. Whatever the judgment, they provide the setting for this course.

2 The System Dynamics Perspective

2.1 Introduction

To develop a proper framework for sustainability science, we must learn to think and model across disciplines and in terms of complex systems. In the last couple of centuries, many mathematical methods and techniques have been developed, which have become the hallmark of the natural sciences and their successes. Since the 1960s, the evolution of these methods and their applications became closely linked because of the advent of the computer and its rapid advances in performance. Mathematical modelling has become standard practise in the natural and engineering sciences, and increasingly in the life and social sciences. Scientists in many disciplines now have access to simulation software and models.

The roots of *system theory* go back to the early days of the Enlightenment, in natural science as well as in social science (Richardson 1991). Its present-day form emerged in the 1930s, inspired by insights in biology and ecology. In the second half of the 20th century, system analysis got its specific content from control engineering (cybernetics) and electro-mechanical engineering, which emphasises 'the larger picture' by shifting the focus from events to behaviour to structure. System analysis has been applied in a variety of engineering, management, economic, social and resource contexts.

This chapter gives a brief introduction into the theory and practise of system analysis and of modelling and simulation of systems. These concepts and methods are used throughout the book. System analysis is an excellent method to conceptualise and, to a certain degree, simulate the dynamic behaviour of systems. It is inherently transdisciplinary in its search for general or universal principles governing systems. The classical method in system analysis is the mathematical language of calculus, for example, integral-differential equations. Calculus is an essential tool in understanding behaviour of systems over time. For more complex systems, it has limitations, even for the skilled mathematician. However, rapidly expanding sets of computer simulation tools provide novel and complementary ways to model and examine system behaviour. Before I give a more formal introduction, I illustrate the system's view with the example of the world car system, which is of great importance from a sustainability perspective.

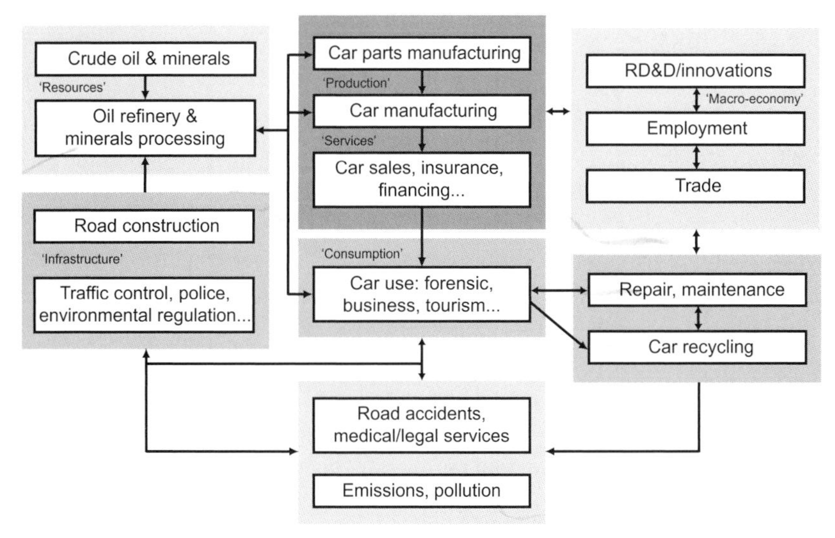

Figure 2.1. The car system and related subsystems in society.

2.2 The World Car System

Many of us use a car for a variety of purposes. In industrialised regions, mobility is an intricate part of life: commuting to work, visiting family and friends as well as allowing for leisure and business trips. Between 1970 and 2002, the number of passenger-kilometres (pkm) travelled in Europe increased from 1,600 to almost 4,000 billion per year – or more than 10,000 kilometres (km) per person per year. Here, we restrict the analysis of the *transport system* to the automobile. This means omitting other transport modes such as trucks, trains, ships and airplanes. Also, the bicycle, motorcycle and bus are not considered, although they are still very important in many low-income regions. How does an integrated systems perspective on *the car* in the context of (un)sustainable development look?

Figure 2.1 shows a diagram of the system illustrating key variables and their relationships. Such a diagram is the first step in a conceptual model. The boxes represent important subsystems and the arrows indicate material, monetary and informational interactions. The graph in Figure 2.2 shows the growth of the world fleet of registered cars since 1960. The total number of cars in use exceeded 700 million in 2010 – more than one for every tenth person on earth. This total is expected to increase to 1.25 billion by 2020. The rate of manufacturing is now around 70 million per year; the rate of scrapping is about 15 million per year. The production and use of this 'vehicle stock' has enormous large-scale and long-term impacts, both directly and indirectly.

The private car's positive aspects for people of all professions and classes are well-known: more mobility and social contacts, better access to schools and medical services, more and cheaper travel and tourism, to mention the most important ones. The private car has thus satisfied a variety of needs and desires – for jobs, status, education, health and so on – and has created new ones. In high and medium income countries, the majority of people are now dependent on the car for their jobs, shopping and entertainment. In the United States, more than 50 percent of

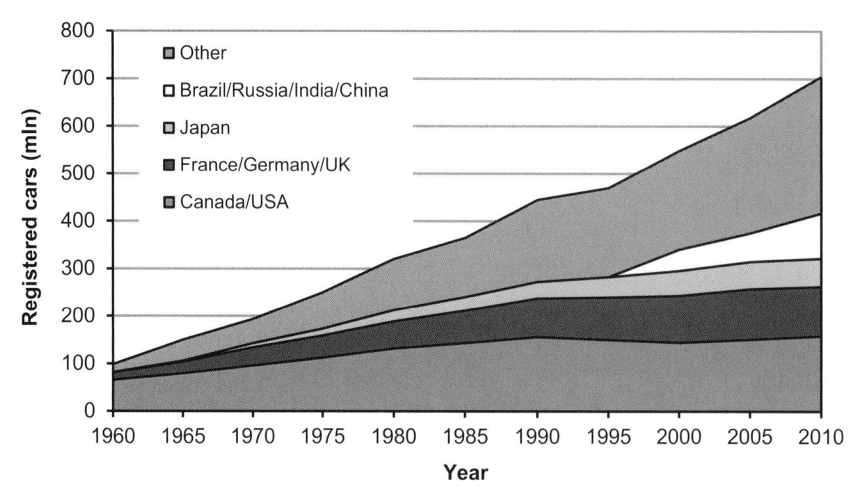

Figure 2.2. Fleet of cars in the world for several regions, 1960–2010. Data are in million of registered cars (source of data: TEPD 2010).

households have two or more cars, and households spend on average almost one-sixth of their income on their cars and its use. Average mileage, measured as the number of km's driven per vehicle per year, of U.S. households has stabilised in the last decades to about 33,000 km per year, with almost one-fourth used for journey to work (TEPD 2010). The average mileage of households in Europe and Japan is lower, as they are in other world regions, but the trends point in the same escalating direction.

The car system also has negative sides that make its present form unsustainable. Initially, these were hardly noticed, because in the early 20th century, cars often replaced horse carriages, solving in some urban areas the huge problem of horse manure in the streets and its odor. It also relieved the less obvious problem of the large amount of land needed to feed the horses. Alternatives such as the electric tramway lost out to the car in the early stage of transportation change, partly because

Box 2.1. *The world motor vehicle industry.* The International Organization of Motor Vehicle Manufacturers' (OICA) website, www.oica.net, gives data on the economic importance of car manufacturing. The motor vehicle industry (cars, vans, trucks and buses) has become a powerful player in the world economic system. It is one of the major employers: More than 8 million people are directly engaged in it while another 40 million are indirectly employed in making vehicles and parts. In 2005, car manufacturing comprised 5 percent of worldwide manufacturing jobs and 3 percent of the world GDP. These numbers do not include the employment and expenditures for road infrastructure.

The automobile industry is also a major innovator, investing almost €85 billion in research, development and production. It is also a major contributor to government revenues around the world, contributing over €430 billion in 26 countries. As the OICA site says, 'If auto manufacturing were a country, it would be the sixth largest economy'.

of active interventions by the automobile industry. The disadvantages of the car system have become more visible since the second half of the 20th century: notably impacts on health from accidents and air pollution and the destruction of landscapes and ecosystems. With the 'oil crises' of 1973 and 1979, the dependence on finite oil resources and the risk of geopolitical conflict became another concern, although they had always been around (Yergin 1990). In the 1990s, the growing evidence that carbon dioxide (CO_2) emissions from burning oil contribute to climate change became another negative side effect.

There are millions of deadly and non-deadly *traffic accidents* every year.[1] In 2005, there were between 50 and 250 fatalities per million inhabitants in the medium- to high-income regions (Figure 2.3a). In emerging economies, the number of cars increases rapidly, and so do traffic casualties.[2] Although there is a clear trend towards declining numbers of fatalities in high-income countries – which sometimes means less people killed but more people wounded and associated medical costs – most low-income regions in the world still report increases in the number of fatalities in absolute and relative terms.

Another health-related side effect is *air pollution*, not only from cars but also from motorcycles and trucks. This has effectively been reduced in the high-income regions, but urban populations in Europe are still exposed to concentrations of nitrogen- and sulphur-oxides and other particulate matter above the recommended World Health Organization (WHO) levels (Table 2.1). Urban pollutant levels in India and China, especially particulate matter concentrations, are significantly above the recommended values (Table 2.1 and Figure 2.3b). They pose already severe health risks and are expected to increase further. One other side effect is *noise pollution*. About 55 million people in the large agglomerations in the EU-27 are exposed to long-term average noise levels above 55 decibel (dB), with 85 percent of them from road traffic noise (EEA 2010). In the Netherlands, nuisance from traffic noise is amongst the three biggest environmental problems perceived by citizens.

The car has a large *impact on landscapes and spatial planning* in the form of land use for roads and parking space (around 1–2 percent) and, indirectly, because of the fuels and materials needed to operate the system. Indeed, the car has largely structured the way in which modern urban areas are built and inhabited. In Europe, the existence of ancient city centres has moderated the rate and extent of adjustment, but elsewhere urban areas are almost completely determined by the car system logic and requirements. This causes additional pressure on land.[3] In many parts of the world, *traffic congestion* is a major problem, with loss of quality of life and of working hours. More roads may diminish the congestion, but at increasing cost and further landscape deterioration.

Another strain on the system's sustainability is *oil use*. In 2000, an estimated 77 EJ was used for transport, nearly half of world crude oil production and the

[1] The numbers in this section are from the OECD Factbook (www.oecd.org; tab, Statistics).

[2] Absolute numbers are misleading because it depends on where a country is in the industrial transition. For instance, the OECD-countries and India had, on average, the same number of road fatalities per million inhabitants (97) – but in India it climbed between 1997 and 2007 with 55 percent, whereas in the OECD-countries, it declined in this period with 27 percent.

[3] It adds to the direct and indirect costs of car use: Parking costs are now in the order of 2–3 €/hr in urban areas in Europe.

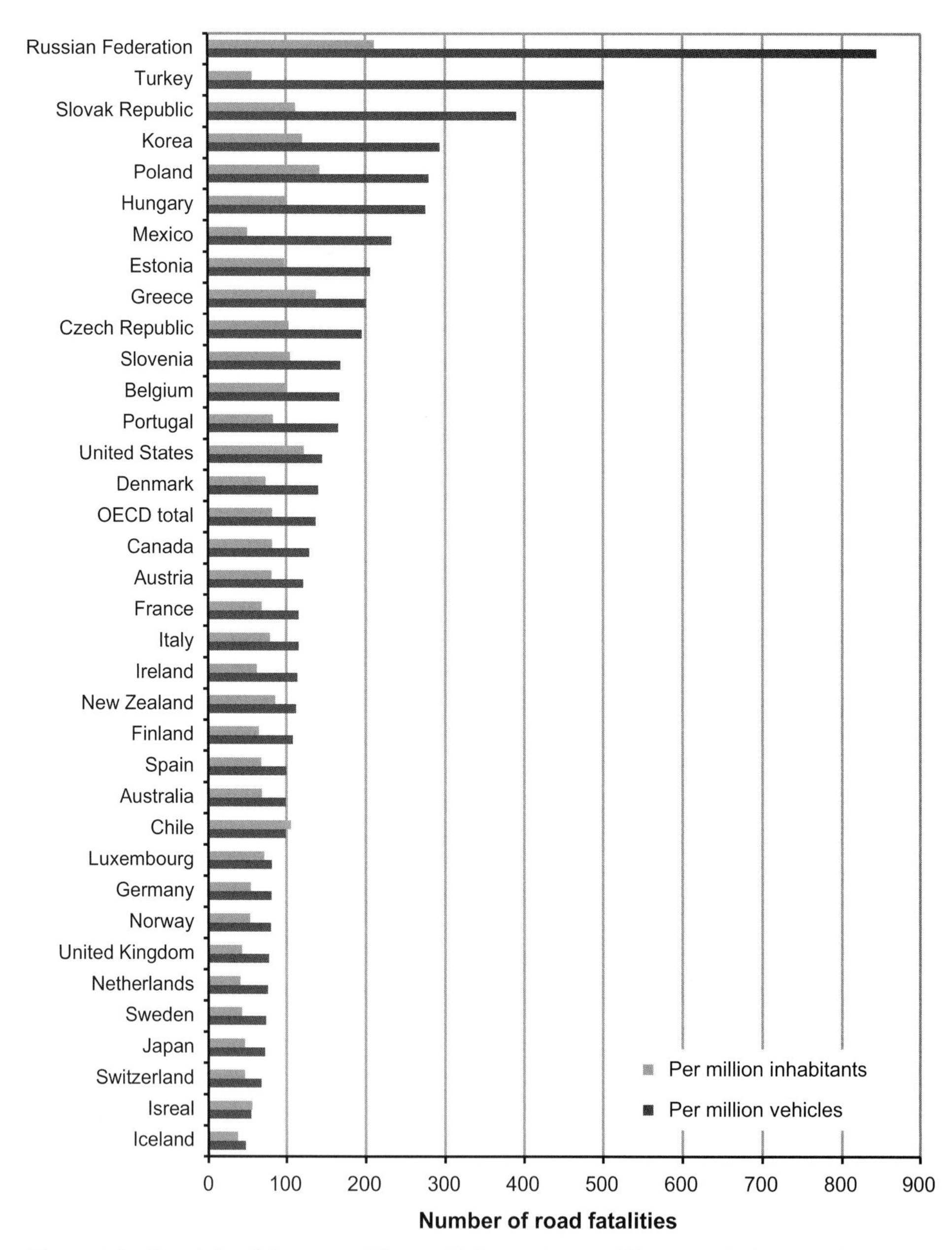

Figure 2.3a. Road fatalities per million vehicles and per million people in 2005 (source of data: OECD Factbook 2007).

equivalent of 2,200 billion litres of gasoline.[4] About three-quarters of this was for road transport. Therefore, fossil fuel use for transport is largely responsible for the fast depletion of oil fields around the world and the geopolitical implications. From 2003 to 2004, the United States spent an estimated \$50 to 100 per ton of oil used (7.5 to 15 \$/bbl) on military expenditures to defend oil supplies from the Middle

[4] EJ is a unit of energy and the equivalent of 10^{18} Joule. See Section 7.2.1 for details on energy units.

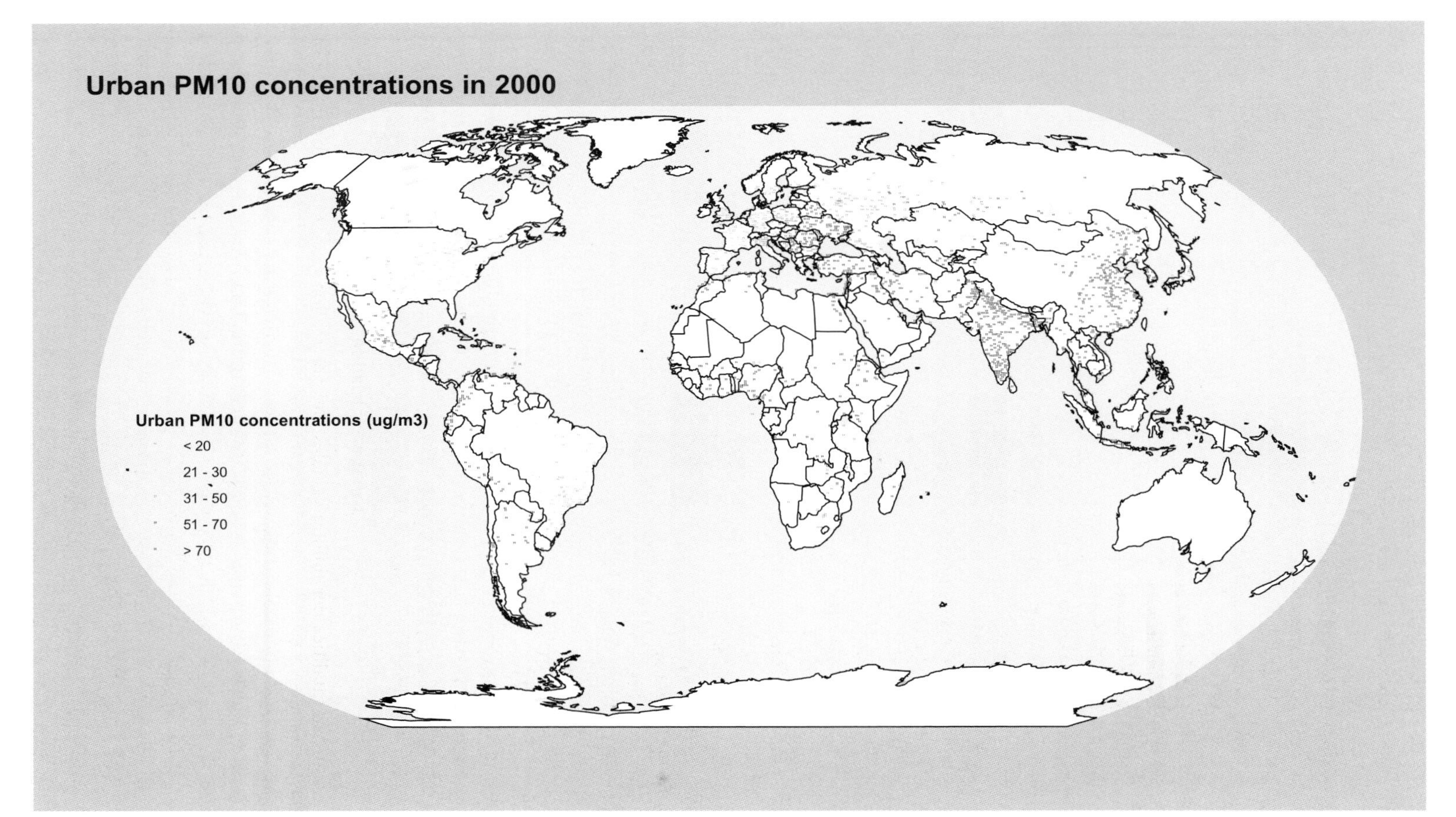

Figure 2.3b. Air quality in terms of Particulate Matter (PM10) concentrations in the cities of the world. GUAM concentration and population data are coupled to city locations (x,y). The values are aggregated to an average value for IMAGE-model $0.5^{\circ} \times 0.5^{\circ}$ grid cells. The cells are converted into square points. The darker the colour, the worse urban air quality (source of data: PBL). (See color plate.)

Table 2.1. *Air quality in large cities in Western Europe and India. WHO guideline values are 40 μg/m³ (annual mean) for NO_2, 20 μg/m³ (24-hr mean) for SO_2 and 20 μg/m³ (annual mean) for PM_{10} (source of data: PBL)*

	City	Population (million)	NO_2 (annual mean)	SO_2 24-hr mean	PM_{10} annual mean
Europe	Paris	9.32	43	8	21
	London	7.65	44	4	25
	Berlin	3.45	22	4	27
	Milano	3.29	55	8	54
	Athens	3.07	32	11	41
	Madrid	3.01	43	11	29
	Rome	2.70	41	2	–
	West Midlands (UK)	2.30	30	3	23
	Greater Manchester	2.28	43	2	23
India	Greater Bombay	12.59	19	7	77
	Calcutta	11.02	53	9	237
	Delhi	8.42	57	9	432
	Madras	5.42	6	5	96
	Hyderabad	4.34	29	6	178
	Bangalore	4.13	61	7	173
	Ahmedabad	3.31	23	15	231
	Pune	2.49	53	31	340
	Kanpur	2.03	20	9	413
	Lucknow	1.67	33	16	391

East.[5] Producing an estimated 14 percent (passenger cars) to 23 percent (all transport) of global CO_2 emissions, the transport system is a major contributor to the enhanced greenhouse effect (OECD/IEA 2009). The car system is also an avid consumer of mineral resources, notably iron, aluminium and plastics. In fact, recycling car parts has become a major industry as car scrap is an increasingly valuable resource.[6]

Many analysts agree that the car system is not sustainable in its present form and that a transition towards a sustainable transport system is needed. Such a *transport transition* should address the previously mentioned adverse consequences. But there are many ways to Rome, as the saying goes. The appeal of the private car and its accessibility with rising incomes will continue to stimulate rapid growth of the car fleet in low-income regions. Cross-country and time-series data suggest a logarithmic relationship between income and car ownership and use, although the curve has a high branch for U.S.-style countries with low population density such as New Zealand and a low branch for European-style countries (Figure 2.4).[7]

[5] One bbl is another unit of energy, namely, the equivalent of one barrel of 159 litre of crude oil. Depending on the type of crude oil, a bbl is equivalent to approximately 6.1 GJ (1 GJ = 10^9 Joule).

[6] A recent analysis, for instance, estimated that automobiles in use in the USA contain 19 million tons of aluminium, which is equivalent to almost one year of world production (Buckingham/USGS 2010).

[7] Characteristically, the countries with exceptionally low car density are the city-states Hong Kong, Macao and Singapore – densely populated places with intense public transport.

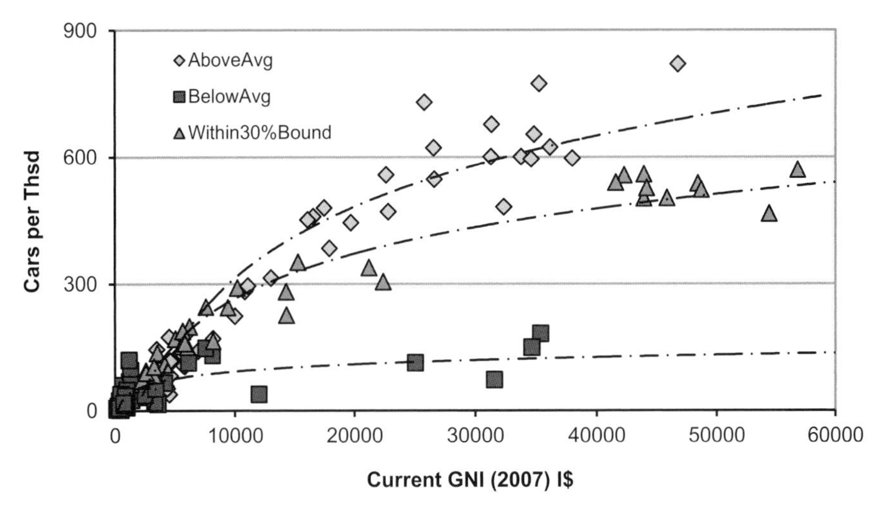

Figure 2.4. Car ownership in number of cars per thousand inhabitants as a function of income (GDP/cap in 2007 I$) for 131 countries in 2007. A logarithmic curve fits fairly well ($R^2 = 0{,}72$). Amongst the high-income countries, the United States, Canada and Australia are more than 30 percent above this curve and countries like Hong Kong and Singapore fall more than 30 percent below this curve. The middle band are mostly countries in northern Europe (source of data: TEPD 2010).

Most efforts to deal with negative impacts focus on car technology. Sustainability is interpreted within an engineering context as more efficient fuel use and lower specific emissions. There is still much room for efficiency improvements and there are good prospects for new engine and new fuel types. Biomass-based fuels supply a growing fraction of demand in countries like Brazil and the United States. The electric/hybrid drive and fuel cell can replace the internal combustion engine, but their introduction requires system changes and behavioural adaptations. There are still many uncertainties and controversies about costs, efficiency and environmental aspects of these novel techniques. Innovative alternatives have to compete with a mature technology – the combustion engine – which is currently produced at astonishingly large scale and low cost.

Any forecast has to consider the system as a whole. Predictions are difficult because it is a complex system with growth-promoting as well as stabilising and counteracting forces. Traffic congestion, space limitations and the cost of infrastructure slow down the growth of car use in most densely populated regions. A penetration rate towards the U.S.-lifestyle number of 800 cars per thousand inhabitants is hard to realise in densely populated parts of Europe, China and India. With rising income, there is also less willingness to accept traffic casualties and urban air and noise pollution. Oil availability and the growing need to reduce emissions can also put a brake on growth. Dependence on finite and unevenly spread oil resources is expected to grow in almost all official scenarios, and so is the oil price (OECD/IEA 2009). As was shown in the 1970s and 1980s, oil price is an important determinant: The hike first led with some delay to increasing fuel efficiency, and the subsequent fall stopped, if not reversed, the trend with the advent of the sports utility vehicles (SUVs). Mounting concerns about climate change and a subsequent imposition of carbon taxes will have great impact. Another uncertain part of the transition concerns the modal split, such as the role of transport modes other than the car (e.g., faster trains and air transport). Without regulation to internalise health expenditures, oil depletion, climate change and so on in the transport cost, the transition will be slow and difficult. These are the ingredients for an integrated system analysis of the car system and its (un)sustainability. Let us now have a closer and more formal look at systems.

2.3 System Dynamics: The Basics

2.3.1 What Is a System?

What is a system? One definition is:

> A *system* is an interconnected set of elements that is coherently organised around some purpose.

Therefore, the three key attributes are: elements, (inter)connections and purpose (Bossel 1994, Meadows 2008). It is important to explicitly define a system boundary. In this definition, a heap of sand or an arbitrary group of pedestrians is not a system. But all around us are examples of (sub)systems:

- a car, a house or an airplane, making up connected parts in order to provide mobility, shelter or movement;
- a forest with its connected elements such as trees and animals, and the purpose to sustain their metabolism;
- a marketplace where buyers and sellers exchange fish, fruits and flowers according to a set of conventions and relationships;
- a football team or a school, with rules being the connections, which permit to play the game or teach the children;
- a country with its people and the customs, contracts, laws and institutions that keep them more or less together around a social-cultural identity.

These systems are organised around a purpose or function, whether it is simple and straightforward or complex and elaborate. A characteristic of these systems is their integrity or their wholeness. They are 'more than the sum of their parts', as the saying goes.

Often, the *elements* are the easiest part to observe and classify. A car has an engine and wheels; a house has windows and doors; a forest has trees and animals; a market has vendors and buyers; a football team has players; and a country has individual people. Element identification depends on scale or resolution and, in turn, on the objective of the analysis. Car engines and tree leaves, for example, are intricate subsystems with their own elements, connections and purpose.

Interconnections or relationships are a more difficult affair. They represent the physical, monetary and informational flows, as well as the laws and rules that govern these flows. In a forest, these are, for example, the connections between the soil and the roots, but at a smaller scale, they are the laws governing diffusion and reaction processes. In a market, buyers and sellers are connected in a process of comparing and bidding from which a price and a transaction result. In football teams, the players are connected via their actions and signals within the constraints of the rules of the game. In the car system, the drivers are connected to the road system via traffic regulatons and other drivers as well as to car manufacturers via car performance and built-in safety devices. Evidently, in more complex systems such as organisms and institutions, the relationships are often hard to observe and classify – and it may be even harder to find the underlying laws and rules.

The third item, *purpose*, or goal, is usually not included in the analysis of 'inanimate' matter in physics and chemistry. Scientific inquiry has increasingly gone to the micro-scale of atoms and molecules with their physical connections and apparent absence of purpose. Most natural scientists will not consider purpose a system property. After all, what is the purpose of a piece of rock, a lake or a volcano? It is only in a more holistic, phenomenological perspective that the notion of purpose makes sense.

The purpose of living organisms is undeniable, but it is still difficult to grasp system property. It manifests itself in all kinds of behaviour and in a variety of settings such as the car driver going home, the wolf trying to catch deer, or human beings protecting themselves against the cold. In a broad sense, the purpose is to survive and to sustain. The individual and, in a larger context, the species tries to maintain a certain structural integrity against the forces of decay and, in order to do so, evolves. The same can be said about human institutions: They sustain and evolve themselves. For inanimate artifacts, it is less ambiguous what is meant with purpose. Tools and appliances are designed with a clear purpose that is usually evident in a given context. For example, think of an oven, bicycle or radio. Or, think of the car system: The purpose of the car is usually to transport people or goods, whereas the purpose of the larger road system is to facilitate individual car movements.[8] Even in technology, however, purpose becomes less obvious and quite diverse with generic machinery such as computers. The human-designed objects – including such things as the Internet – resemble organisms in their complexity. Artificial life may be less distant than we think. From the opposite side, the natural sciences are uncovering the complexity of macromolecules and microorganisms into such depth that the boundaries of what an organism is become ever more obscure.[9]

The purpose of a system is often even less visible than the interconnections between the elements. In fact, it is only meaningful to talk about purpose within a cognitive understanding, which makes such a purpose contextual and anthropocentric. It helps to ask oneself, what would happen if the purpose of a system is changed? For instance, if a car is used to live in, the connections between its elements change and one may end up with something else – a house, possibly? Changing the purpose of a football game or a stock market would create a new system with novel behaviour. Playing with changes in relationships, rules or purposes is precisely what makes fiction novels sometimes so interesting.

[8] Of course, individual car drivers and the designer of the car system consider rules and have purposes that are quite different from these purely functional rules and purposes.

[9] The fading divide made between living ('organic') and non-living ('inorganic') has a long history. It reflects a fault line that for a long time was thought to exist between the 'inanimate' objects of physics and chemistry and the *élan vital* of organisms with their intuitively obvious purpose.

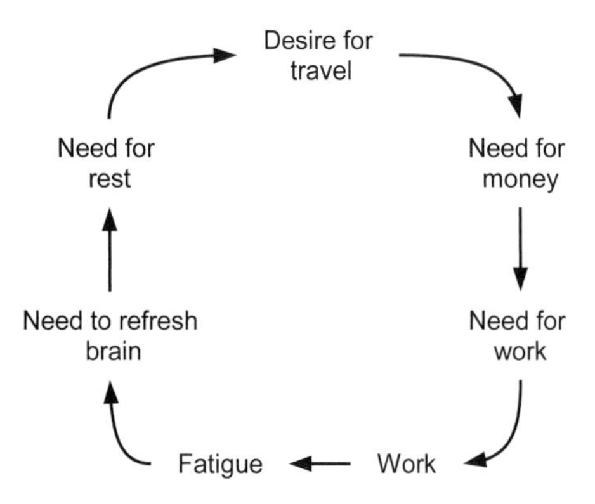

Figure 2.5. An influence or causal loop diagram (CLD), picturing Balzac's vicious circle (Richardson 1991).

2.3.2 Stocks and Flows

Our interest in systems is because we are interested in *change* within the context of sustainable development. There are various ways in which one can look at change in systems. Classical calculus – integral-differential equations – has become the language par excellence to study change and to analyse the dynamics of real-world processes. It has become an indispensable tool in the natural and engineering sciences because it uses the language of mathematics, which implies more consistency and transparency as well as simplifications and limitations.

System dynamics practises an open enquiry into the object system, starting with a conceptual model and a focus on qualitative relationships: A influences B or, even stronger, A causes B.[10] The resulting schemes are influence diagrams or causal loop diagrams (CLDs) that represent what is thought to be the essence of the dynamic structure of the system under consideration. For example, nowadays a French resident travels on average 40 km a day, whereas around 1840, residents averaged only 4 km a day. The French writer Balzac noticed 160 years ago a paradox about travelling: "I'd need rest to refresh my brain, and to get rest, it's necessary to travel, and, to travel, one must have money, and, in order to get money, you have to work, create, etc.: I am in a vicious circle (cercle vicieux), from which it is impossible to escape" (Balzac 1850:32, quoted in Richardson 1991). Such an observation can also be translated into a CLD. Balzac's account then looks like the loop in Figure 2.5. The arrows indicate a presumed cause-effect relationship. This simple observation with diagram has an important message to convey: sometimes we are caught in a loop. If it goes in the wrong direction, as in this case, one speaks of a vicious circle. The opposite is the virtuous circle. The construction of such conceptual models from observations on cause-effect sequences is at the core of system dynamics.

The next step is to formalise the conceptual model into equations and search for solutions. Most real-world systems exhibit nonlinearities, delays and feedbacks, and analytical solutions are hard to come by. The use of mathematical techniques (calculus) often leads to extreme model simplifications and, increasingly, simulation is used to analyse otherwise intractable systems. One method for simulation that is easily accessible is via dedicated software package for system dynamics simulations

[10] We will usually speak of system dynamics as an equivalent of system theory, system analysis and, more generally, the theory of dynamical systems.

(see the list of suggested software packages at the end of this chapter). It allows the modeller to do model experiments and explore the dynamics without advanced mathematical skills and too severe simplifications.

An important distinction in system dynamics is between stocks and flows. *Stocks* (or levels) are referring to the content of reservoirs or compartments in which something is stored. They are the equivalents of integrals in calculus and usually indicated as state variables in system theory.[11] They represent accumulated 'stuff' (things, individuals, matter, etc.) within a given system boundary. For example, water in an underground aquifer, metal in a mineral deposit, phosphates in a lake, carbon in the atmosphere, biomass in a forest, and so on. Or, think about them in relation to the number of cars in the world (Figure 2.1), people in a town, students in a classroom, television sets in a retail shop or money on a bank account. Sometimes, also nonphysical variables are treated as stocks in the sense that they accumulate over time. One can simulate, for instance, happiness or hunger as something that increases or decreases over time and serves as persistent indicator of the state of the system. But such a variable can also be considered a stock characteristic or property, like temperature or density. Changes in a stock are called *flows* (or rates or fluxes). Their equivalent in calculus are derivatives.[12]

In ecological and environmental science, the stocks are those parts of natural ecosystems that provide services for human use, such as fertile soils, animal populations, forest biomass or forested area and aquifers (source function). They can also be air, water and soil compartments for disposal (sink function). In agriculture, relevant stocks are grain stores and live*stock*. In engineering and economic science, the stocks refer to the goods in an inventory or, more broadly, the capital stocks in the form of machinery and other productive facilities. In finance, it refers to a sum of money or a fund, which tends to become the most widely known meaning since the advent of stock markets.[13] Stocks can be lumped together in aggregate stocks, as in 'economic capital', 'financial capital', 'natural capital' or even 'human capital' and 'social capital', although there is a risk that such aggregate variables become meaningless.

The spatial and temporal scale have to be consistent across the definition of a stock and its associated flows. If physical interactions in space are considered, stocks are usually defined as densities (individuals per unit area). The stock of cars, for example, has to be linked to the surface area (ha) where they are parked or drive around if the objective is to make sense of the traffic system. If physical interactions are weak or absent, as in monetary or informational exchanges, variables are often expressed on a per person or per unit cost basis. The temporal aspect, in particular the choice of the relevant time interval, is important because one may only have information on changes in the stock during certain periods. Such a change per period represents the net flow. The chosen time interval in a simulation influences the outcome. Measurements in stock-flow diagrams are sometimes indirect because information about flows is the only way to estimate the size of the stock. An example

[11] See Appendix 2.1 for a brief introduction on integral-differential calculus.

[12] Also, flows can be considered as variables that change over time. In calculus, this is the second derivative.

[13] The word stock in English also refers to a document that confers share of ownership, but with a volatile value – as in the stock market.

is how demographers estimate medieval populations in European cities from baptism and funeral data as proxies of birth and death rate (Braudel 1979).

To do model simulations, stocks identified in an influence or causal loop diagram must be quantified and, therefore, be assigned a unit of measurement. Cars are counted in numbers and roads and railways in km length. Elephants in a natural park can, as a stock, be measured in number of individuals or in biomass-weight, depending on the purpose of the study. Identifying a clear definition and unit of measure can be difficult, notably with aggregate stocks of great heterogeneity such as pollution, machinery or information. Scientists usually look for quantities that can be measured and related to another, more interesting but difficult to measure variables – such a substitute quantity is called a *proxy*. For instance, a proxy of economic capital is the accumulated investments in manufacturing. Conceptualisation, quantification and measurement are related but not the same. It is possible to conceptually enter love and hate in a model, for example, to facilitate thinking about their relationship and dynamic. They can be quantified as normalised variables, that is, fluctuating between 0 (absent) and 1 (full intensity presence). To actually measure them is next to impossible and one has to use proxies.

The level of a stock is determined by an inflow and outflow. A widely used example, which also serves as an analogue, is the bathtub. It can be written mathematically in the form:

$$\frac{\Delta[WaterInBathtub]}{\Delta t} = WaterInflowRate - WaterOutflowRate$$

with WaterInBathtub the stock variable, which should be given an initial value in a simulation. WaterInflowRate and WaterOutflowRate are the associated flows. The time interval Δt is the period over which the change in the state variable due to the inflow and outflow are considered. This type of equation is called a difference equation and, for Δt approaching zero ($\Delta t \rightarrow 0$), a differential equation (see Appendix 2.1). As long as the inflows equal the outflows, the level in the stock does not change and the system is in a state of dynamic equilibrium.[14] It is also called a *stationary* or *steady state*. If the sum of the inflows exceeds the sum of the outflows, the stock level rises and vice versa. It sounds trivial, yet it is often an essential and revealing first step in system analysis.

Stock levels can only be influenced or managed by changing the inflows and outflows – one cannot change their level directly. In the example of the bathtub, this is clear. But there are many other, less evident situations. For instance, the gradual emission of chemical compounds has raised their concentration in coastal sea sediments and this can, at best, slowly disappear through natural mechanisms. As a result, a system with stocks has *inertia*. The stock level cannot change or be changed faster than the maximum difference between inflow and outflow. Stocks are a kind of memory and are the source of delays. Sometimes, depending on the function, this may not pose any problem, as with the slow refilling of a glass of beer for a patient customer. But inertia may be a serious problem in situations when fast change is desired or needed – think of the rate at which people can get out of a cinema in case of fire. Inertia is experienced as an usually unexpected and undesired delay and is

[14] If inflow and outflow are also zero, the system is in a state of static equilibrium.

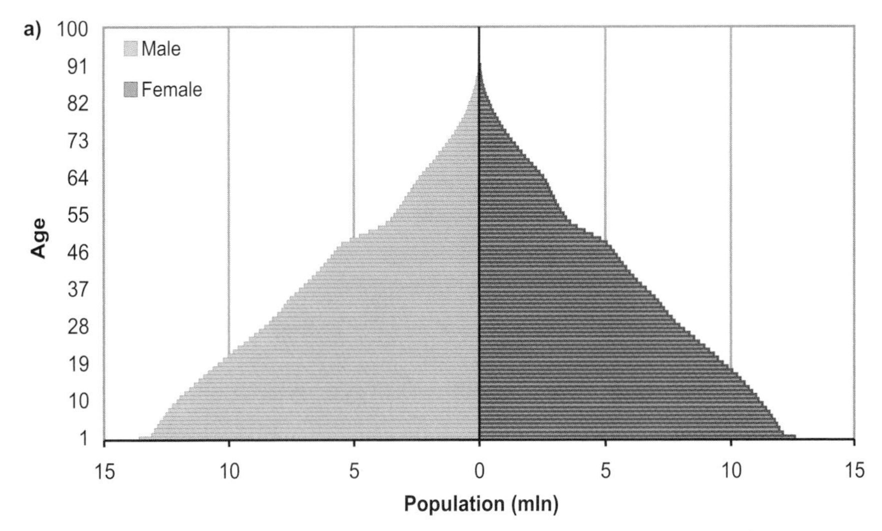

Figure 2.6a. Population pyramid for India in 2000: number of male (left) and female (right) persons in a particular (1-year) age cohort (source of data: PBL).

often not adequately perceived and interpreted. The fact that in 2000 more than half of India's population was comprised of people less than twenty-six years old implies a large and almost inevitable growth of the population in the first two decades of the 21st century – given the slow changes in reproductive behaviour and the probable rise in life expectancy (Figure 2.6a). This is known as the *demographic momentum*. It is in this sense that one can say that a moving oil tanker, extensive infrastructure or social movement has a large momentum – it has much kinetic, economic or social energy that is difficult to stop overnight. Similarly, it is the age distribution as well as other characteristics, such as rent/lease or ownership, of dwellings and cars that are important qualities in an assessment of energy savings potential and implementation rate (Figure 2.6b). In environmental issues, for example, the climate change issue, misunderstanding this can cause great confusion.

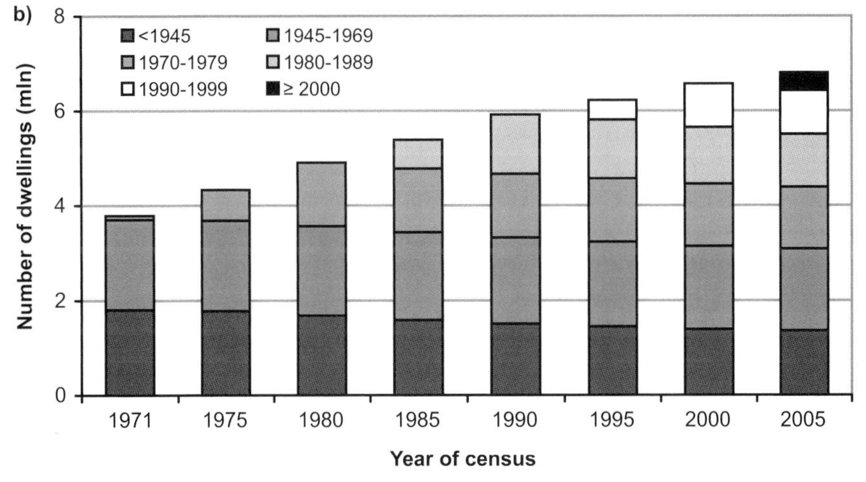

Figure 2.6b. Dwellings in the Netherlands: distribution according to age classes and type (source of data: VROM/ABF 2009).

Box 2.2. *Understanding climate change.* An illustration of the difficulty to appreciate stock-flow delays is provided by Sterman and Booth Sweeney (2002). A widely supported hypothesis amongst scientists is that emission of carbondioxide (CO_2) into the atmosphere leads to warming of the earth surface ('greenhouse effect'). The natural mechanisms to remove CO_2 from the atmosphere are slow and a CO_2-molecule remains on average for about 100 years in the atmosphere. Because the emission rate exceeds the removal rate, the concentration rises. This causes a change in radiative forcing and warming.

Focusing on the climate system, Sterman and Booth Sweeney investigated whether the (American) public understands the physical facts. To this purpose, a group of graduate students were presented descriptions of greenhouse warming drawn from the nontechnical reports of the Intergovernmental Panel on Climate Change (IPCC) and subsequently asked to identify the likely response to various scenarios for CO_2-emissions. The only understanding required for this task is the basic stock-flow dynamics of the climate system and, in particular, the fact that at present the rate of influx of greenhouse gas is twice the removal rate. It was found that the students' performance was poor: 'Many believe temperature responds immediately to changes in CO_2-emissions or concentrations. Still more believe that stabilizing emissions near current rates would stabilize the climate, when in fact emissions would continue to exceed removal, increasing GHG concentrations and radiative forcing'.

2.3.3 Feedback Loops

There are a few elementary processes, which are basic in many systems and are referred to as positive and negative feedback loops. Their essence is that a stock influences, via a series of signals and decisions, the inflow and outflow rates of the stock and hence its own level. The first one is the *positive (or reinforcing or amplifying) feedback loop* (PFL). It is a dynamic growth process in which the rate of change of a quantity is positively proportional to that quantity.[15] Examples of such processes are an autocatalytic chemical reaction (explosion), an unconstrained population growth or the money on a bank account that 'grows' with the rate of interest. Another example is a capital stock producing an output that is partly added to the existing stock – it is considered the engine of economic growth in economic theory. Such exponential growth processes go toward infinity for large t(ime). In other words, infinity (∞) is the attractor, fixed point or equilibrium point for $t \rightarrow \infty$. Of course, this is impossible for a physical system on a finite planet.

The PFL can for a stock variable X in its simplest form be phrased in mathematical terms as a difference equation:

$$X_{t+1} = X_t + a \cdot X_t = (1 + a) \cdot X_t \ (a > 0) \qquad (2.1a)$$

with a discrete time step of one between the events and a fractional (per unit) growth rate a ($a > 0$). Such a discrete formulation is often appropriate, because

[15] There is also the possibility that a system (organism) responds to an anticipated or projected future trajectory, in which case one speaks of a feedforward loop (Rosen 1985). It can make system behaviour significantly more complex.

many processes consist of a series of separate events. The amount of money on a bank account with regular interest being added is a perfect example of continuous exponential growth. Unconstrained population growth and a reproducing capital stock are less perfect examples, because the change events in these processes are usually not regular and can only in the aggregate be described by equation 2.1a.

There are also many processes in which the rate at which a quantity changes is negatively proportional to that quantity. These are called a *negative (or stabilising or balancing) feedback loop* (NFL). An example of such a decline process is radioactive decay. The mathematical formulation is the same as for the PFL but with a negative growth rate a. In its simplest discrete form it is the function $X_{t+1} = (1 + a)X_t$ with the parameter *a* being a fractional (per unit) decline rate $(-1 < a < 0)$. The equilibrium state to which it eventually will go for large t(ime) may differ from zero. If the system takes on the value K for for $t \rightarrow \infty$ in equilbrium, one can write it in mathematical form as:

$$(X - K)_{t+1} = (X - K)_t + a \cdot (X - K)_t = (1 + a) \cdot (X - K)_t \; (-1 < a < 0) \quad (2.1b)$$

The system drives the stock variable towards the state X = K where, once reached, the stock remains constant. K is the attractor. An example is the opening of a sluice after which the high water level will exponentially decline to the low level. Mathematically, the attractor represents a steady-state and as such a system state that can intuitively be associated with a sustainable state. Examples are when the birth rate of a population equals the death rate or the influx of a pollutant is equal to the outflux. In the car system, it can be the state at which the purchase rate equals the discard rate and the number of cars per thousand inhabitants remains constant, also known as market saturation (Figure 2.5). In a management context, K can be interpreted as a goal or target. If the stock is disturbed, the NFL will tend to restore the system state to K.

The classical mathematical approach to describe exponential growth and decay processes is in the form of a differential equation:

$$\frac{dY}{dt} = bY \quad (2.1c)$$

with dt an infinitesimally small timestep $(\Delta t \rightarrow dt \rightarrow 0)$, b the rate of change parameter and Y the state variable.[16] This equation can be solved analytically by taking the integral over time of $(dY/dt)/Y = b$. This yields:

$$Y_t = Y_0 e^{bt} + c \quad (2.1d)$$

with c an integration constant. Exponential decline can be similarly expressed.[17] The advantage of an analytical solution is that the formulas and rules of calculus can be used to explore the behaviour of the system.

[16] In system theory, one usually speaks of the state equation, with Y the state variable. We denote the state variable with Y in a differential equation. It is the same as the state variable X in a discrete equation. See Appendix 2.1 for a brief introduction on the equivalence of continuous and discrete equations. For a negative loop and a non-zero attractor K, Y should be replaced by Y-K. It can be shown that b in equation 2.1d equals ln(1+a) in equation 2.1a.

[17] For exponential decline, $b < 0$ and $c = K$.

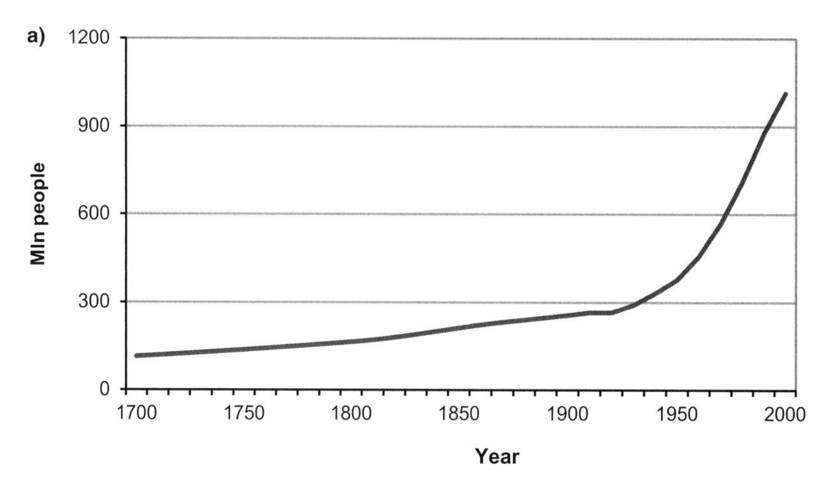

Figure 2.7a. Growth of the number of people in India, 1700–2000 (source of data: PBL).

Exponential growth and decline processes can be characterised by the doubling and halving time, respectively, provided that the growth rate parameter a is constant, which is not usually the case. The *doubling time* (DT) in exponential growth process is the number of timesteps during which the state variable doubles. Doubling happens in T timesteps if:

$$X_T = 2X_0 = (1+a) \cdot X_{T-1} = (1+a)(1+a)X_{T-2} \cdots = (1+a)^T X_0 (\mathrm{a} > 0) \quad (2.1e)$$

Thus, $(1 + \mathrm{a})^{\mathrm{T}} = 2$, from which it is seen that DT = log(2)/log(1 + a) with a > 0. The rule of thumb that the doubling time is roughly seventy divided by the percentage growth rate (DT ~ 70/a%). It is easily checked. For instance, the number of cars in China and India grows with about 10 percent/year, and it will, therefore, double in about seven years.

The exponential decline process can be characterised by the number of timesteps in which the stock halves: the *halving time* (HT). Similar to the doubling time, the halving time is derived from the condition that $X_T = X_0/2$ at time T. Therefore, HT = log(0,5)/log(1 + a) = −log(2)/log(1 + a) with a < 0. The same rule of thumb is valid. For example, if the number of traffic accidents decreases with 2 percent/year, it will be almost halved after thirty-five years.

There are many real-world processes that follow, at least for a while, an exponential growth path. Usually, the growth rate parameter a is not constant. Figure 2.7a,b shows three examples: the growth of the human population in India between 1700 and 2000, the number of Internet users since 1990 and the global wind-power capacity installed since 1996. It should be noted here that there is a difference between the examples. Population growth happens because each woman can give birth to children so change depends directly on the population size. Internet users and wind turbines do not breed young ones. Their growth is driven by more complex phenomena, in which reinforcing feedback loops via, for instance, word-of-mouth infrastructure networks and expected high profits are operating. There are also many examples of exponential decline, such as the decay of the two radioactive elements

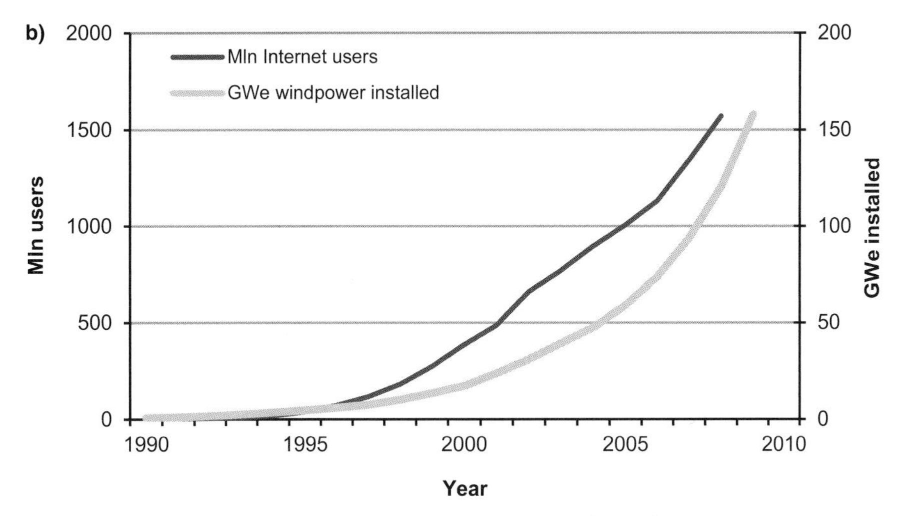

Figure 2.7b. Growth of the number of Internet users (upper) and of windpower capacity installed in the world (lower), 1990–2009 (source of data: World Bank and www.ewea.org).

^{14}C and ^{99}Tc (Figure 2.8). The latter is an ingredient of high level waste from nuclear reactors that is presently generated worldwide at a rate of about 12,000 tons per year.[18]

Often, time-series consist of combinations of growth and decline. The curve in Figure 2.9 depicts the number of casualties in traffic in the Netherlands between 1950 and 2005, with the government target for 2020. Initially, it increased steeply in association with exponential growth in car ownership in the 1950s and 1960s. It was driven by other, interconnected PFLs: population, affluence and technology. Increasingly, the large and rising number of traffic deaths was felt to be unacceptable and political targets were formulated and countermeasures were introduced. This was an NFL with the implicit goal of zero accidents. With the obligation of safety belts in new cars (1971–1975) and helmets for motorbikes (1975), a significant decline set in despite a further growth of the number of vehicles and traffic. In 2009, the number of traffic deaths was down to 720, some 5 percent below the target for 2010. Policy has been very successful – as far as deadly accidents are concerned, as it does not count the wounded – but the rate of decline slows down because a further reduction costs proportionately more policy effort due to other negative feedbacks such as aging and risk perception changes (Adams 1995).

Real-world systems will have both positive and negative feedback loops. The relative dominance of positive and negative feedbacks will then determine whether, in the time period considered, the system exhibits growth or decline or a combination of both. A simple but characteristic dynamic process, in which both a PFL and an NFL operate, is the *logistic growth* process. In system dynamic terms, the outflow rate is a function of Y and approaches the inflow rate for Y approaching K. This

[18] Such waste contains many different elements, most of which are very radioactive and have halving times (half lives) between 5 years and 15 million years. The graph is representative for one of the more important long-lived elements in nuclear fission reactor waste, technetium ^{99}Tc, with a halving time of 211,000 years.

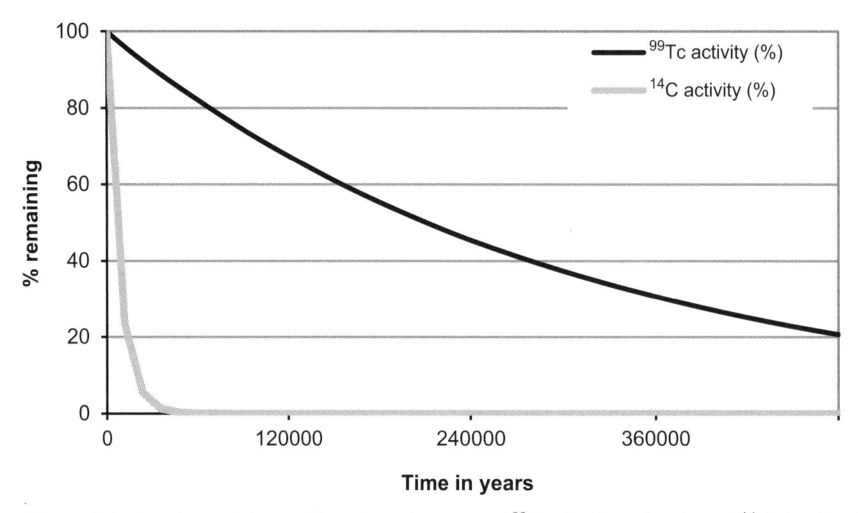

Figure 2.8. Decline of the radioactive element of ^{99}Tc (technetium) and ^{14}C (carbon).

process can be formulated as the differential equation for exponential growth with the growth rate b* approaching zero for Y ≈ K:

$$\frac{dY}{dt} = b^{*}Y = b\left(1 - \frac{Y}{K}\right)Y \tag{2.2a}$$

Effectively, the growth rate itself is part of a negative feedback. It is easily seen that the formula is the one for exponential growth for Y ≪ K. There are two attractors, that is, Y-values for which dY/dt = 0, namely Y = 0 and Y = K. At Y = K, there is a steady-state at which inflow cq. growth rate equals outflow cq. decline rate. The functional solution to equation 2.2a is found by integrating and gives for the state variable Y as a function of time:

$$Y = \frac{K}{1 + e^{-b(t-t_0)}} \tag{2.2b}$$

with t_0 the value at which Y = ½K (Figure 2.10). It is an S-shaped form, reflecting a transition from positive feedback to negative feedback, hence its other name

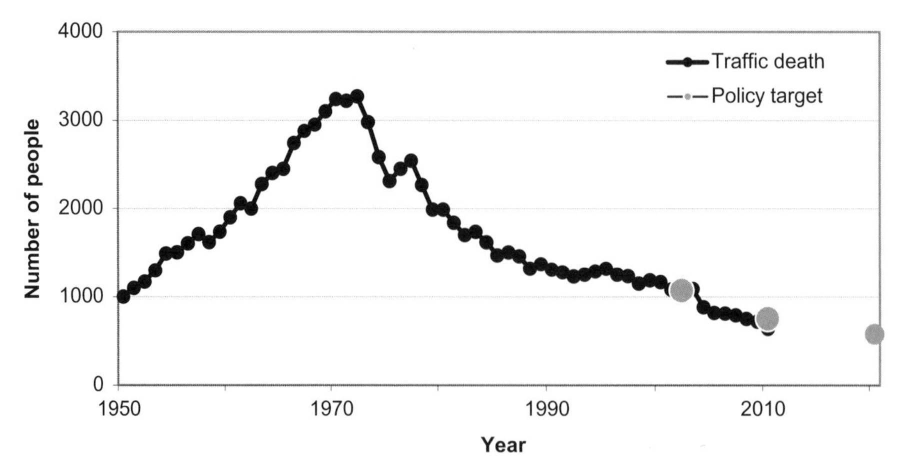

Figure 2.9. Number of casualties in traffic in the Netherlands, 1950–2007 (source of data: CBS).

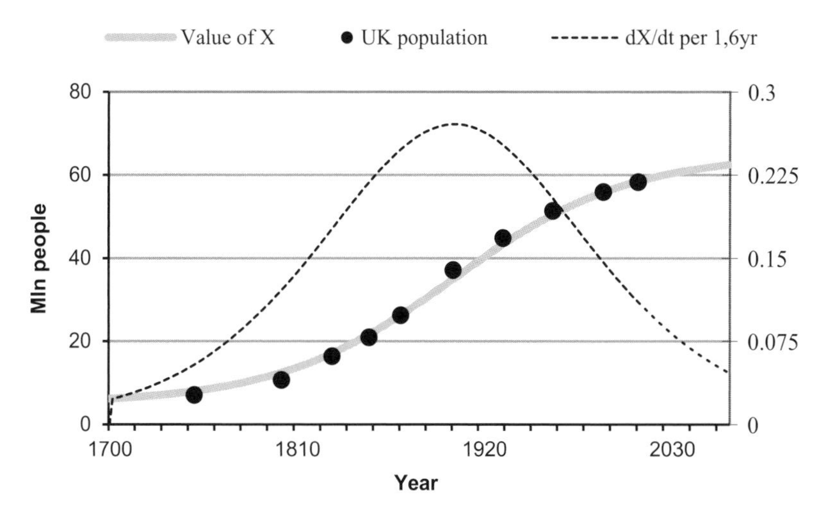

Figure 2.10. Example of a logistic growth process and estimates of the United Kingdom population since 1750. The dotted curve shows the growth rate (source of data: Maddison 2006; Schandl and Krausmann 2007).

sigmoid curve. The value of the attractor K is associated with the carrying capacity because it is the maximum value to which the variable Y ($0 < Y < K$) can grow. The logistic growth equation is used in many branches of science to describe widely different phenomena, such as population or biomass growth and market penetration of products. We later discuss this equation in more detail.

In complex natural and societal processes, more and more interdependent positive and negative feedback processes take place than in the logistic growth model. In environmental systems, there is a mix of natural growth and decay processes and interfering processes induced by humans. In a technical and management/policy context, the negative feedback forces are an essential control mechanism and the equilibrium value is the desired or target value of the state variable. A technical example from everyday life is the thermostat. It is set at a desired value, after which the heating installation will start delivering heat at a rate that is proportional to the difference between the measured and the desired temperature. Provided it functions well, the rate of heat production will decline to zero when $T_{measured}$ approaches $T_{desired}$. Another example is a market with the price of a good or service – presumably the same and known throughout the system – as an information signal. If price goes up, demand tends to go down. Producers shut down the plants with the highest marginal unit cost, which lowers the price and, in turn, stabilises or increases demand again. Often, oscillations will occur and the demand-supply pattern looks like a cobweb – which is why this simple model of the price mechanism is known in economics as the *cobweb model.* Price regulation can be introduced to stabilise prices again and prevent windfall profits or stimulate use.

2.3.4 An Illustrative Simulation Experiment

Two simple model experiments illustrate the phenomena of delay and inertia. A car manufacturing company is selling cars to customers at a constant rate. To be prepared for a change in demand, a desired inventory of ten days of perceived sales is maintained. A larger inventory would be too costly, a smaller one too risky.

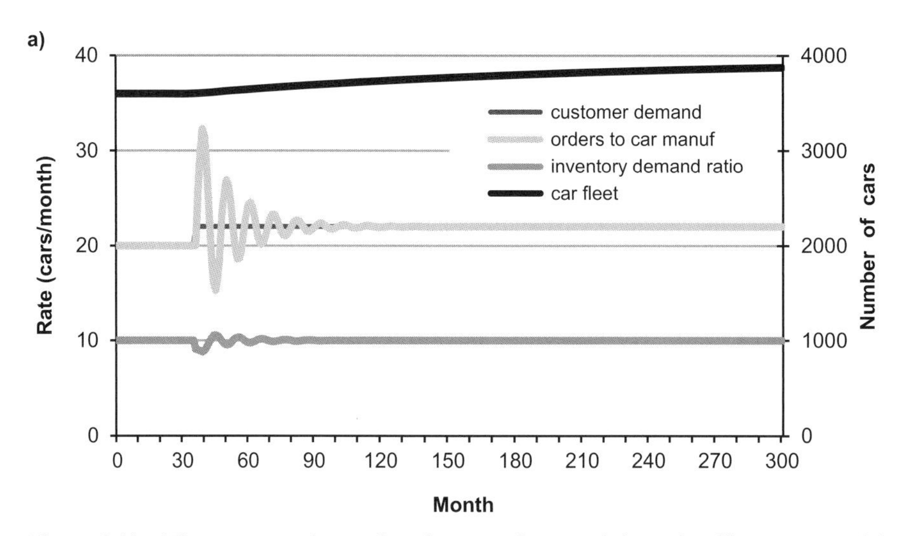

Figure 2.11a. The customer demand, order rate of cars and the ratio of inventory and demand in response to a 10 percent increase in demand in year 25 (timestep $\Delta t = 1$, Runge-Kutta-4).

The output of cars is based on orders for new cars, which is calculated from sales and the difference between desired and actual inventory. If demand suddenly goes up and the manufacturer responds instantly, there will be a sudden jump in cars ordered; however, all customers will still be served because the inventory buffers the change. Real-world people will not and cannot respond instantly and mechanically. They decide on the basis of past experience and forward expectations and, therefore, postpone or adjust their response in line with a rather complex mix of goals (minimise cost and maximise revenues, keep good relations with customers and suppliers, and so on). Therefore, we add to the model:

- *perception delay:* the manufacturer uses some average of past sales instead of yesterday's sales for this decision;
- *response delay:* the inventory level is not restored immediately and completely, instead only a fraction of what would be needed is ordered;
- *manufacturing and delivery delay*: once ordered, it takes time before the new cars are actually produced and arrive at the sales/inventory location.

Some delays are strategic and/or informational whereas other delays are of a more physical-technical nature. How does the system respond with delays to a change in demand?

Let us assume that the car market is initially in equilibrium, with a fleet of 3,600 cars, an average lifetime of 15 years and a monthly delivery of 20 new cars. What happens if in year 3 (month 36) demand suddenly jumps from 20 to 25 new cars/month? It turns out that a perception delay of 2 months diminishes the spike in orders significantly and a response delay of 2 months adds to a further smoothing of the spike. Both can be seen as good management. However, a delivery delay of 2 months causes fluctuations in orders of a factor 2 and reverberate in the system for more than 3 years (Figure 2.11a). Another finding is that it takes more than 25 years (month 300) before the system is again in equilibrium. Experiments with different parameter settings show that the system is easily destabilised.

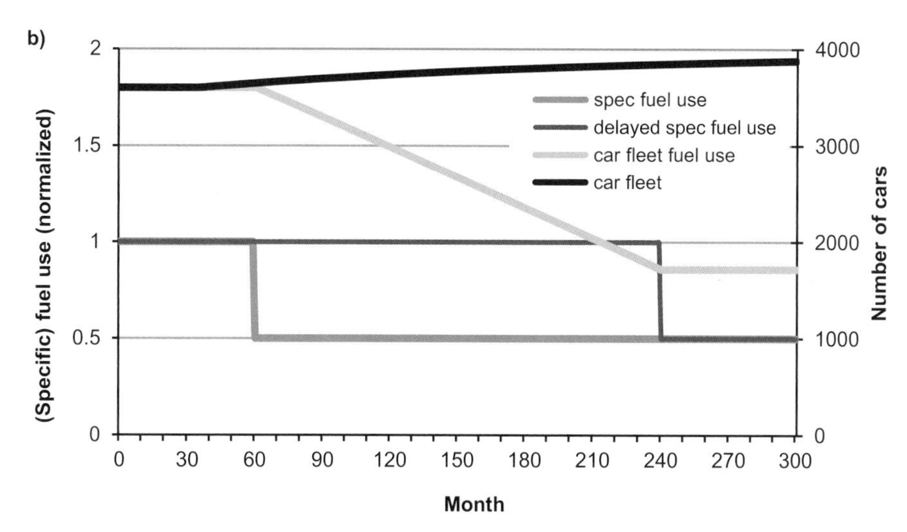

Figure 2.11b. The fuel use of the car fleet in response to a sudden doubling of fuel efficiency in year 72 (timestep $\Delta t = 1$, Runge-Kutta-4).

This simple model run shows the importance of system delays in a changing environment. What is said here for cars is also valid for the emission of pollutants that are gradually broken down or for a chain of beer users, distributors and manufacturers (see the famous Beer Game, beergame.mit.edu/). It is valid for commodity chains in general: apples and pears, chemical and steel plants and products, and for financial products and services as the financial crises show. The core idea is that the world is permanently in a state of dis(turbed) equilibrium.

A second experiment illustrates the *inertia* of a stock. The car market in the model has a complete turnover in fifteen years. Suppose the fuel use of new cars suddenly drops with 50 percent to half its value. Such a sudden change is implausible to say the least, but it is instructive for this model experiment. If the drop takes place in year 5 (month 60), it takes the full 15 years until year 20 (month 240) before the average fuel efficiency has come down to the value of the new car released in year 5 (month 60) (Figure 2.11b). It is recognised in the historical data: The average fuel use km/litre of passenger cars in the United States and Europe doubled between 1978 and 2008, but individual new car performance went up much faster. Again, what is said for cars is true for long-lived substances in air, water and soil, and for other capital goods with long lifetimes, such as dwellings, office buildings and electric power plants with on average lifetimes of more than thirty years. The lesson from a sustainability perspective is: decisions about emissions and investments throw a large shadow into the future and that cleaning up environmental compartments or making a capital stock more resource-efficient will necessarily take time.

2.4 System Dynamics Modelling

2.4.1 The Rules of the Game

Making a dynamic model of a system of interest is as much an art as a science. It starts with genuine interest and careful observation. Although disciplined and

Box 2.3. A *definition of sustainability* in the vein of system dynamics is the one proposed by the ecological economist Daly, in the form of three rules:

- 'Its rate of use of renewable resources do not exceed their rates of regeneration;
- its rates of use of nonrenewable resources do not exceed the rate at which sustainable renewable substitutes are developed; and
- its rates of pollution emission do not exceed the assimilative capacity of the environment' (Meadows *et al.* (1991) 209).

Because of the nonlinearity of the exponential and logistic growth functions, a small change in one variable can easily cause big changes in the same or other variables over time. This key feature of nonlinear reinforcing loops, famous now under the name *tipping point*, was already known long ago as is shown in this old English proverb:

For lack of a nail, the shoe was lost;
For lack of a shoe, the horse was lost;
For lack of a horse, the rider was lost;
For lack of the rider, the battle was lost;
For lack of the battle, the kingdom was lost –
And all for the lack of a horseshoe nail!

formal analysis is always part of it, qualitative stories and accounts are important too. If people talk about (un)sustainable development, they often come up with a story they have heard or read or experienced. Seldomly, such a story 'tells the whole story' and usually it is anecdotal. Yet, such narratives do often reveal essential insights about what (un)sustainability is about. Moreover, a good story tells it in the words of those involved and in the form of concrete, specific events. This makes stories a precious part of sustainability science. This book uses them on various occasions, as experiences to be investigated for their concreteness and lessons. In a system dynamic modelling setting, they are in combination with statistical data the basis for the first steps. These can be summarised with the following guidelines (Bossel 1994; Sterman 2000):

- Develop a clear statement of the problem to be addressed, that is, of the purpose of the model, and choose the time horizon of interest;
- make qualitative and quantitative descriptions of the system's behaviour in the past, and hoped-for or feared behaviour in the future, using a.o. the stories and accounts;
- develop hypotheses about the dynamics driving the system's behaviour and express them as an influence diagram or a causal loop diagram (CLD).

At this stage, one should have a conceptual model that forms the basis for more formal modelling. The next steps are to identify stock variables, that is, the relevant inflows and outflows and those variables that change as a consequence of these flows (or fluxes). The value of a stock variable is referred to as its level and of a flow as its rate of change. The identification of the variables and their interrelations and their representation in a stock-flow CLD is important. The notion of causal loop is not unproblematic, but here we stick to a rather straightforward interpretation: It is when a change in a variable (stock, flow) A causes changes in another variable B, which in turn changes A. Be aware of the notation: A connection between two variables (stocks, flows) is indicated with a + *sign* if a positive (negative) change in variable A causes a positive (negative) change in variable B. If the reverse happens: a positive (negative) change in variable A causes a negative (positive) change in variable B, we use a – *sign*. If a loop has only + signs, it is always a positive (reinforcing) feedback loop. If there are an uneven number of – signs, it is a negative (stabilising) feedback loop.[19] In Figure 2.12, a few simple CLDs are shown, all representing combinations of positive (growth) and negative (decline) feedback loops, with the exception of fossil fuel depletion because its formation rate (inflow) is assumed to be zero. We will meet these elementary CLDs again in the thematic Chapters 10–14.

If the system is complex, one can eliminate some detail and divide the system into subsystems to make analysis manageable. An important next step is to identify the structure of the decisions or relationships that govern the rates of change, which will create feedback loops amongst the stocks. In the process, the conceptual model is adjusted and, hopefully, converging to a more definitive form. It can then be translated into a mathematical simulation model. Variables are given values from available observations and data and one can start experiments with the model in order to determine typical dynamic behavioural modes. This is usually done by simulation experiments and by comparison with other systems thought to be analogues.

Table 2.2 gives a list of stocks and associated qualities and flows relevant in a sustainability context. The examples are from all scientific disciplines. In the process of stock and flow identification, one should consider their *properties*.[20] For instance, the ocean and the atmosphere are simulated in climate models as layers of different temperature. A stock of mineral or fossil fuel resources is divided into size/quality classes according to grade, location and cost. A stock of cars can be classified according to age, purchase price, power and energy efficiency. A stock of people can be distributed over age and gender classes (Figure 2.6a). Whether and how to disaggregate stocks into classes or categories depends on the objective of the analysis. A functional disaggregation is often useful because stocks provide certain goods and services, such as how mineral deposits provide metals, groundwater reserves provide water, forests provide timber, and fertile soils provide food. Similarly, machinery delivers goods and females deliver offspring. The stock properties are usually an important determinant of its productivity, that is, unit of good or service delivered per unit of stock.[21] Stock productivity is usually influenced by the inflow and outflow rates. Indeed, a key question in sustainable resource use is whether resource productivity declines and, if so, how it can be increased or at least maintained and at which effort/cost.

Disaggregation into *(size/quality) classes* should be done up to the level that is still functional for the analysis. Age distribution in a population may be important in considering long-term consumption and savings trends or the health situation.

[19] Another convention is to indicate the positive loop (+) with the s of same and the negative loop (–) with the o of opposite.

[20] Also, flows can be considered as variables that change over time. In calculus, this is the second derivative.

[21] The reverse is called the resource intensity, that is, unit of stock per unit of output.

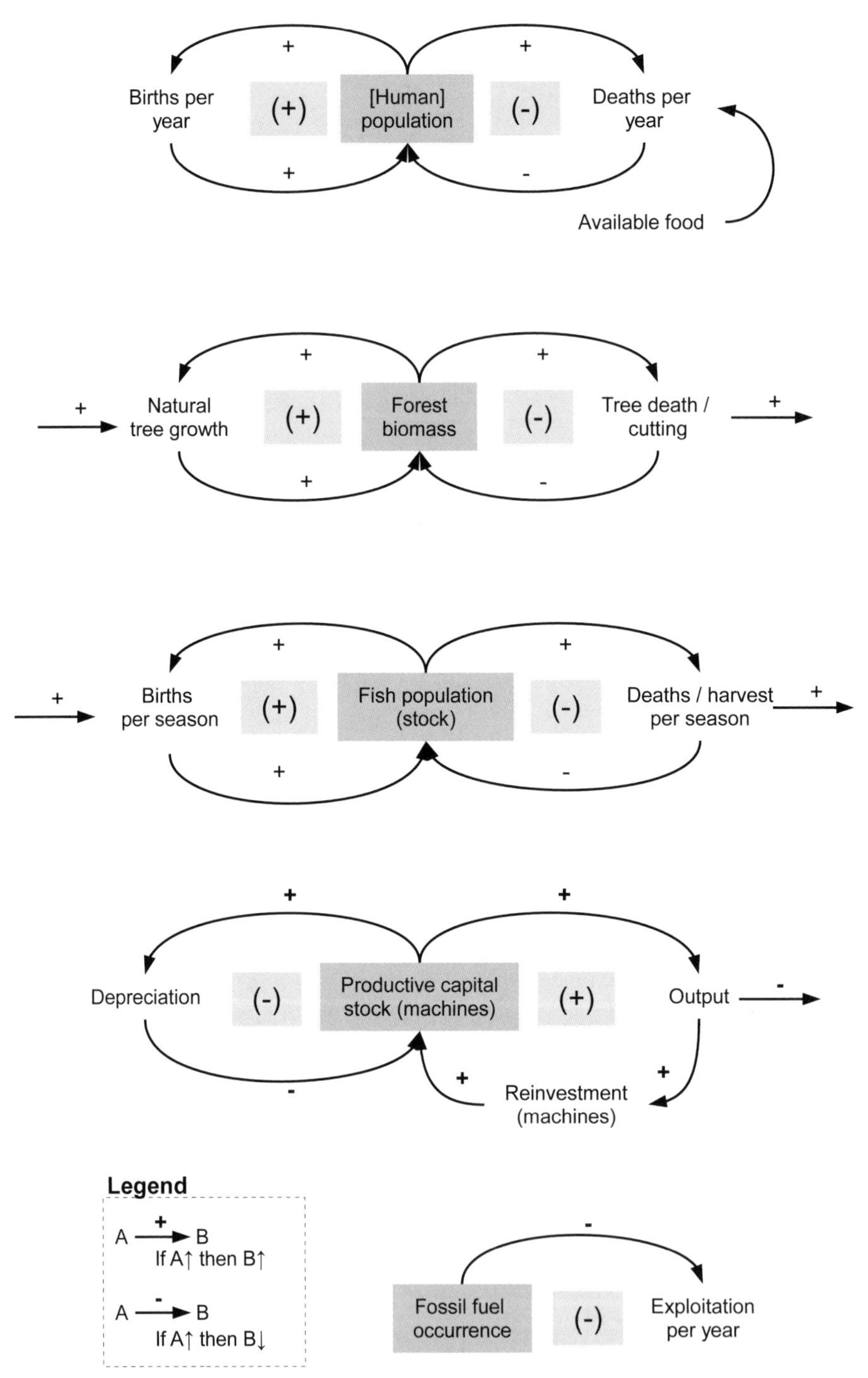

Figure 2.12. Examples of causal loop diagrams of basic dynamic processes: population growth, capital stock growth, forest growth and fossil fuel depletion.

The age distribution (or vintage, in economic jargon) of economic capital influences, amongst other factors, the potential rate of innovation. Distributions across space are often crucial in analysing land change processes. For example, a distinction between rural and urban populations can be important because of different birth

Table 2.2. *Examples of stocks, associated flows and stock qualities*

Stock Units of quantity	Associated flows Units of quantity/time	Stock properties
Mineral deposits and fossil fuel deposits (stocks in tons, flows in tons/yr)	Discovery rate Exploitation/production rate	Mineral ore grade Depth/extension Reservoir characteristics Distance from consumers
Forest: standing biomass (stocks in ton or area, flows in ton/yr or area/yr)	Re/afforestation rate Harvest/exploitation rate	Species characteristics Age cohorts
Agricultural soil (stocks in area or depth, flows in area or depth change per yr)	Sedimentation rate Erosion rate Crop output/harvest rate Agricultural inputs	Organic matter content Yield/productivity
Groundwater occurrences (stocks in ton or m^3, flows in ton/yr or m^3/yr)	Replenishment rate Withdrawal rate	Depth/extension
Population size (animals/livestock) (stocks in number, flows in number/yr)	Birth rate Death rate Hunting/slaughter rate	Age cohorts Metabolism Position in foodweb
Population size (humans) (stocks in number, flows in number/yr)	Birth rate Death rate Migration rate	Age cohorts Health/morbidity Family size Income
Industrial/service capital stock size (factories, offices etc.) (stocks in €, flows in €/yr)	(Expansion/replacement) investment rate Depreciation/demolition/ scrapping rate (Retrofit) investment rate	Age cohorts (vintage) Lifetime Productivity
Household capital stock size (houses, cars etc.) (stocks in €, flows in €/yr)	(Expansion/replacement) investment rate Depreciation/demolition/ scrapping rate (Retrofit) investment rate	Age cohorts (vintage) Lifetime (Average) specific energy use

and death rates and different income (growth) rates. Spatial gradients in resource quality or pollutant accumulation can be important in exploitation or restoration, but it may be irrelevant in other analyses. As a general rule, disaggregation should be done parsimoniously because it makes the model more complex and more difficult to parameterise. It is usually best to start simple and add detail and complexity as understanding advances.

A basic ingredient of models are *relationships* between variables. These are usually derived from experiments, observations and statistical data on the one hand and intuition and 'educated guesses' on the other. Sometimes there is a direct causal relationship that can be put into a formal model. For example, the water level in the bath has a simple causal relationship with the inflow and the average annual fuel use in car travel is causally related to the distance driven, although the relationship is not necessarily known exactly. Often, comparison of datasets suggests a relationship for which no obvious causal mechanism is available. Such a correlation can be considered

Box 2.4. *Causality.* The notion of causality is strongly present in the natural sciences. It implies that, even with uncertainties in the state description, the existence of a (one-directional) causal linkage exists and constrains the possible future states; in other words, successive consistency. The inclusion of human beings with conscious goal-seeking behaviour may suggest that a system develops towards an end – a teleological explanation. However, such goal-seeking behaviour has to be rooted in past experiences. In this sense, it does not contradict the initial-state assumption in system science that the future state of the system has no influence on process dynamics.

A causal chain is a way of representing a hypothesised sequence of causally linked events. The events follow from the behaviour of interrelated and interacting variables. It is important to distinguish between causality and correlation. '... one of the central issues in the quantitative study of two variables: is there a relationship between them?... The statistical term for such a relationship or association is correlation, and the measure of the strength of that relationship is called the correlation coefficient... It must be stressed from the outset that correlation does not imply causation. It is only some appropriate theory that can provide a hypothesis that might lead us to believe that there is a causal relationship between two variables' (Feinstein and Thomas (2002) 71–72).

as a phenomenological 'law' that connects two observable quantities in ways that are not understood at the level of the underlying system dynamics. A well-known example in transport economics is Zahavi's hypothesis, which states that the average travel time budget (TTB) and travel money budget (TMB), that is, the time (hr/day) and money (€/yr/cap) spent on travel by the average person, is remarkably constant across cultures and over time (Schäfer *et al.* 2010). But correlations usually have no direction of causality and no explanatory mechanism. What a modeller hopes to find are regularities, where two variables that are far apart causally have a consistent relationship due to the system's equilibrium tendencies. From that perspective, a regularity like Zahavi's law has to be examined by building a model of individual decision making that yields the law as an outcome. Indeed, proposing hypothetical mechanisms to explain observed regularities and subsequently testing them with (new) data is the core task and challenge of every modeller.

2.4.2 Archetypes

One of the rules mentioned in the preceding list is to identify typical dynamic modes of behaviour. We have already seen exponential growth and decline, and logistic growth. Senge has listed in his book *The Fifth Discipline* (1990) a series of such generic dynamic mechanisms. He refers to them as system dynamics archetypes or templates and applies them in strategic management and in policy formulation and implementation. Constructing archetypes recognizes and exploits the fact that systems that appear on the surface to be quite different may exhibit similar behaviour because they share a common feedback structure. The archetypes can be used as building blocks for a larger, more comprehensive understanding of system behaviour.

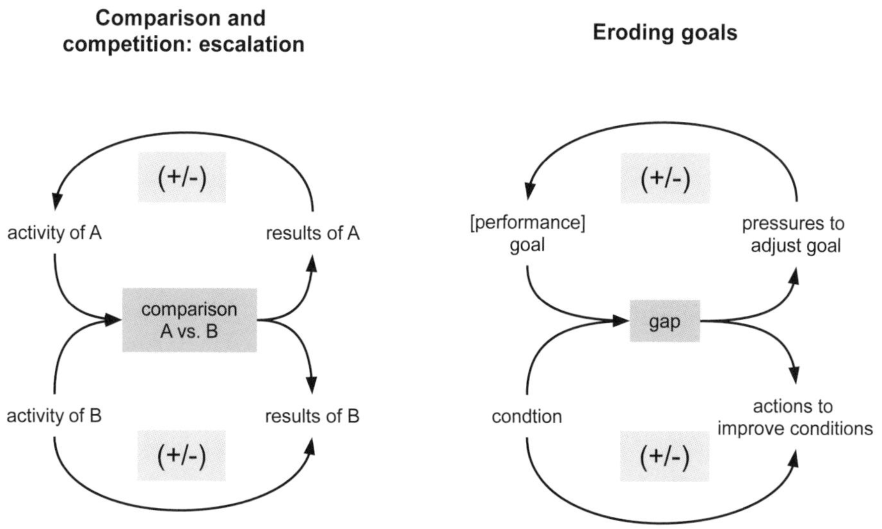

Comparison and competition: escalation

activity of A
(+/-)
results of A
comparison A vs. B
activity of B
(+/-)
results of B

Eroding goals

[performance] goal
(+/-)
pressures to adjust goal
gap
condtion
(+/-)
actions to improve conditions

Figure 2.13. System archetypes: escalation (left) and eroding goals (right) (redrawn from Senge 1990).

Usually, they are constructed and discussed in the form of conceptual models, but they have an analytical equivalent in the form of coupled differential equations. They offer a filter that can clarify complex processes – not more and not less.

A first archetype are the physical and social *Limits to growth*. A PFL causes a (stock) variable to grow exponentially, but at some point counteracting forces start to operate and slow down the positive feedback. Such growth-weakening or stabilising forces start to operate when some condition in the system triggers an until-then hidden mechanism. For instance, food shortages or declining resource quality act as physical limits and rising prices or public dissatisfaction as social limits to growing population. The onset of limits may happen suddenly due to the very nature of exponential growth, and more so if there are nonlinear thresholds in the way the system responds to such growth. The logistic growth phenomenon represents the essence of this archetype. In a context of business or political competition, the *Limits to growth* archetype can manifest itself as a brake on successful operation. A way to avoid it is to anticipate the possible consequences of the rapid growth and respond before the limitations start operating – a recipe for sustainable development.

A second archetype is *Escalation* (Figure 2.13 (left)). The positive feedback consists in this case by setting one's goals or targets for (the growth of) a (stock) variable via comparison and competition with someone else who has the same ambition. The action of a person or organisation or country A is driven by the difference between the own situation and the situation of the other. The other is chosen as the yardstick for comparison or the opponent for competition.[22] Such a mechanism is beneficial, if the desired goal or target benefits also the larger system. It may be the basis for a successful expansion of an economy or penetration of a technology and an important instrument in policies aiming at sustainable development. It may also be harmful,

[22] The word *benchmark* is used to indicate a shared point of reference or standard by which others may be measured or judged.

Box 2.5. *Arms race.* A famous example of the escalation archetype is the arms race model. In 1919, it was first formulated by Richardson to understand one of the causes of the First World War. The first assumption is that the arms expenditures of nation A depend on the perceived expenditures on arms of nation B, and vice versa. This fuels the escalation in arms expenditures. The second assumption is that there is a 'pacifist' tendency in each nation to spend on 'butter, not arms', which counteracts the reinforcing spiral of armaments expenditures. This model has been tested for several historical situations – no evidence of an arms race was found for the Greece-Turkey conflict but the India-Pakistan interaction suggests an arms race (Dunne *et al.* 1999). Extensions of this admittedly oversimplistic model have been proposed. It has heuristic value as a metaphor for corporate business in capitalist economies. For instance, an extensive review of McKinsey & Company of pharmaceutical sales force effectiveness stated (In Vivo, October (2001) 74): 'the leading pharmaceutical companies have driven that [phenomenal] growth by engaging in an increasingly intense commercial "arms race" to shift share to new, more efficacious therapies'. A British journalist called the first decade of the 21st century one of 'a consumption arms race'.

if the original goal or target negatively influences the larger system or gets lost and means become ends. A crucial issue in escalation is to evaluate the measure used for the judgment of each other's actions and the assumptions and delays involved.

Two other archetypes describe the phenomenon of resistance against change, which is a common feature in (social) systems: '. . . resistance is a response by the system, trying to maintain an implicit system goal. Until this goal is recognized, the change effort is doomed to failure . . . Whenever there is "resistance to change", you can count on there being one or more "hidden" balancing processes . . . It almost always arises from threats to traditional norms and ways of doing things. Often these norms are woven into the fabric of established power relationships' (Senge (1990) 88). The first one rests on the idea of *Eroding or drifting goals* (Figure 2.13 (right)). When a person, organisation or country A has a certain goal it wishes to achieve, indicators are formulated to measure the performance of the plans and actions in reaching the goal. If a gap is observed between the desired and the actual situation, one should take corrective action: Improve the actual situation until the goal is met and the gap is closed. Unfortunately, such action takes time and may go against ingrained habits and interests. It is tempting to go for an easier solution: lowering the goal. Possible solutions to such erosion are an understanding of what are the driving forces behind the goal and the allocation of the goal and its monitoring outside the system.

A related archetype is the phenomenon of *Shifting the burden (to the intervenor)*. To manage a system towards a desired goal usually turns out to be difficult because there are so many unanticipated and unintended side effects. Analysis of the problem may point at a 'fundamental' solution, which is rather drastic and uncertain in its consequences. There are usually also lower-risk strategies, which are less drastic and easier to implement. Such symptomatic solutions and their inevitable side effects may draw all the energy away from the real solution. In the process, the real cause

Box 2.6. *Eroding goals and shifting the burden* are occurring in more or less outspoken form in environmental issues during stages of denial and controversy. In 2005, the U.S. government decided to subsidise the production of ethanol and biodiesel from corn in order to reduce the dependence of the U.S. transport sector on imported oil. Most of the alternative fuel is produced from corn and ethanol production increased sevenfold between 2000 and 2009 to almost 40 billion litres. The aggregate worth of the subsidy for the biofuels industry was an estimated $92 billion between 2006 and 2012 (www.globalsubsidies.org). The effectiveness of this policy from an energy security and environmental point of view is doubtful. One interpretation is that the real problem – addiction to cheap oil – is addressed by an ineffective solution, which does benefit special-interest groups. The biofuels industry becomes stronger and the symptomatic solution may weaken attempts to address the real problem through a fundamental solution, like raising the price of oil. It is in this sense that subsidy can create addiction. Another example of eroding goals or shifting the burden is development aid. Although this is surely not the whole story about development aid, it is true that aid may hinder more structural solutions to problems of poverty and corruption (Moyo 2009). Often, the burden is shifted by blaming an external intervenor.

of the problem may move out of sight – the burden has shifted. What is needed is a step back in order to formulate a more comprehensive perspective on what was and is going on.

2.4.3 An Example: Modelling Car and Public Transport Use

In this last section, I return to the car system and construct a conceptual model of a few important loops. The car system has gained a dominant position, not in the least because the costs of pollution, noise, accidents and congestion are not or only partly internalised. Also the opportunity costs of road area and the hidden subsidies on parking and fuel are at best partly accounted for in the costs of car driving. It makes other modes of transport less attractive, not only economically but also regarding other quality of life aspects (health, safety, esthetics). In the longer run, it causes people to move to suburbs, which further increases the need for vehicles, erodes the tax base for mass transit systems and makes urban centres even less attractive. In the low-density suburbs, neither walking and bicycling nor public mass transit transportation systems are attractive options and the car mode becomes a lock-in. I focus in this complex process, which has unfolded in the United States to its extreme, on two subsystems only: the interaction between car ownership, travel time and road construction, and the interference between the car and public transport (Sterman 2000). Figure 2.14 shows CLDs that operate in the transport system. I indicate three stocks explicitly: roads, car fleet and public transport fleet. Each of these causes inertia in the system.

There are four loops in the CLD. The first one is the *road-congestion loop*. It portrays the incentive to construct new road capacity in response to increasing travel time due to congestion and the subsequent public pressure to do something about it. It is in essence a *Limits to Growth* archetype. Given a desired travel time, the

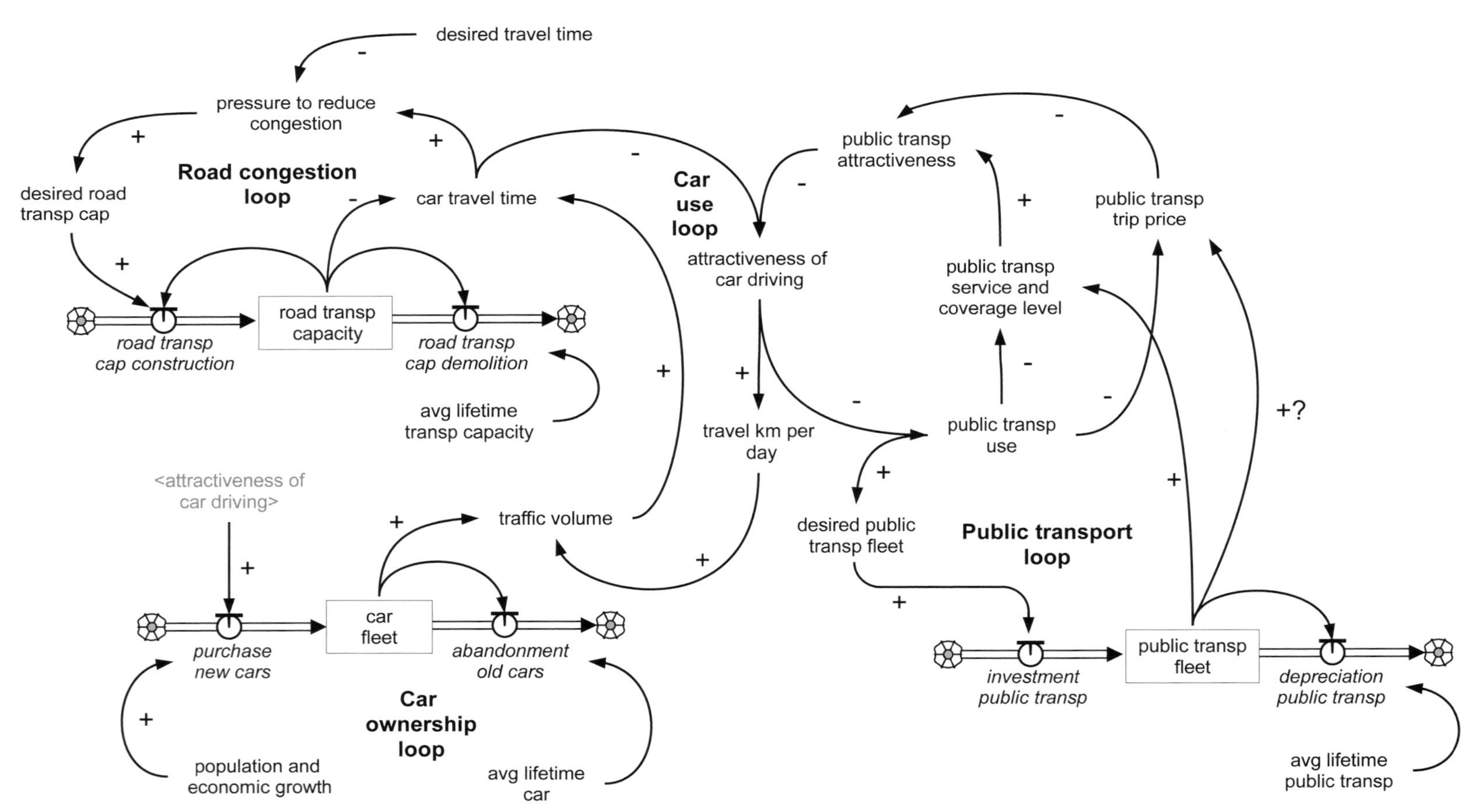

Figure 2.14. A causal loop diagram (CLD) of parts of the transport system. The four loops are indicated in bold (based on Sterman 2000).

system would experience an increase in road capacity over time. Because there is a long delay in road construction and because desired travel time may change due to external factors,[23] one can expect the system to show overshoot and undershoot behaviour (Figure 2.12).

Congestion decreases the attractiveness of car travel, for most people. Once the actual travel time is at or even below the desired travel time, average car speed will go up and the number and length of car trips will probably go up too (*car use loop*). This will increase travel volume. It will also induce people to buy more cars, which, in combination with increasing population and economic activity in the region, is another drive towards larger traffic volume (*car ownership loop*). As a consequence, the system will start to again exhibit more congestion and longer travel times.

Finally, there is a fourth public or *public transport loop*. If the actual travel time by car exceeds the desired travel time, the attractiveness of car driving goes down. People will look around for alternatives and choose the available bus, tram, metro or train or a combination of these and the car (or even walking or bicycling). It increases the use of public transport, which, at least in the short term, leads to lower service levels due to overcrowding, seat capacity problems and the like. This exerts a downward pressure on use (stabilising), but it also tends to increase utilisation rate and, therefore, decrease the cost and, hopefully, the price of a trip (reinforcing). In the longer term, the response is to invest in additional public transport capacity. If this translates into better coverage and service levels, the public transport attractiveness increases relative to car driving.

Clearly, quite different pathways are possible, as the large differences in transport infrastructure between European cities and American cities show. A lock-in for public transport at a low level is sketched by Sterman (2000) and labelled the *Mass Transit Death Spiral*. Because public transport systems have rather high fixed costs (infrastructure, personnel), the operating company is faced with a budget deficit once the number of passengers drops. If ticket prices increase and service and frequency levels reduce in response, the number of passengers drops even further and a downward spiral sets in. Constructing more roads reinforces the negative spiral. The mechanism is particularly hard to overcome if the public transport service degrades because of car congestion. In many urban areas in the world, people are accepting ever longer car *and* bus travel times in a dynamic process that is characteristic of an eroding goals archetype (Figure 2.13).

A very different process can unfold, too. The public transport authority plans for and construct new infrastructure. It takes time, as with road construction. It needs public hearings and construction permits. The construction of buslanes in the city of Utrecht took more than ten years from inception to construction. It also requires upfront investments for which loans and, therefore, the prospect of steady revenues are needed. At this point, failure often sets in due to political and institutional obstacles – for instance, lack of financing due to creditor's desire for high and fast profits. But if the local authorities and the public or private operator succeed in overcoming these obstacles – for example, by attracting more people – a positive loop starts. The utilisation rate goes up, revenues go up and there is room for

[23] One can think here of the availability of audio/video, mobile phones and GPS-navigation, which influence the perception of time spent in the car and the options to control it.

new infrastructure such as bus lanes and greater schedule/route coverage, and other ways to attract passengers such as increased reliability, cheaper tickets, electronic payment, additional comfort and so on. Such a positive loop can be reinforced by popular concern about local air pollution and inner city destruction and by high parking fees as the price of limited space goes up. These dynamic processes create a lock-in effect for public transport at a high level. It can be observed in many European cities.[24] This conceptual scheme, simple as it may be, highlights some of the behavioural responses in the transport system that are overlooked in purely technical analyses. It explains why such systems often show oscillatory behaviour over time and why well-intended policy measures are sometimes ineffective or even counterproductive. It also reveals some of the intricacies of operating a 'public goods' system, an issue of great concern in sustainability science.

2.5 Structure, Space and Time

One last word about concepts. In media accounts of what happens in the world, there is usually a focus on events. For instance, it is observed that in an area in the Mediterranean oak trees die prematurely – an *event*. It is often followed by more systematic observations and measurements. For instance, the tree die-off is traced to declining groundwater levels, which in turn are caused by intensified exploitation for kiwi and citrus in nearby regions. Thus, the *behaviour* (policies, mechanisms) behind the event operates within the wider system of farming and groundwater. But why and how does it operate? A combination of experiment and theory has to investigate the underlying rules and relationships in the system and propose a system *structure*. As it turns out, kiwi and citrus farming is driven by perverse subsidies and the need to compete with other producers, which favours short-term exploitation of groundwater over strategies that are more sustainable in the long-term. System dynamics modelling aims at understanding system *behaviour* and *structure*. Structure will often be influenced by changes outside the system: the system's *environment*.

Let us look at another example. In early 2008, the steep increase in the price of some staple foods on the world market is an *event*, which triggered all kinds of *behaviour*: price controls and rationing to prevent riots, speculation by large international companies, export quota and food aid initiatives. Behind these are structural trends such as chronic underinvestment in agriculture, which may have been reinforced by policies like food subsidies and trade policies. An essential part of systems thinking and systems understanding is to consider the larger system, in which long-term trends and large-scale changes can be investigated as the drivers of behaviour and as the background of events. In short: *inclusive thinking*.

Events, behaviour and structure all have to be considered at various scales in space and time. The choice of *scale* is an important part of the modelling activity and usually directly related to the modeller's objectives. In *space*, every item can be investigated at a micro, smaller scale and at a macro, larger scale – its grain or resolution. Everyday observations are usually in a range of 10^{-3} to 10^{6} meter (m) and can be indicated as phenomenological and coarse-grained. There are also many

[24] The attempts in many European cities to introduce bicycles, with electronic ticketing and road use privileges, is an illustration of the use of system dynamic insights.

phenomena at much smaller than the human scale, such as biological processes in living cells ($<10^{-3}$ m). Assuming a nested and hierarchical organisation of existence, observations and analyses at such a more fundamental and fine-grained level are necessary to understand the human scale processes. In science and technology, it is the domain of bio- and nanotechnology. There is also a level of observations and analysis at much larger than the human scale, for instance, about some global change processes ($>10^6$ m) or stellar nebulae ($>10^{14}$ m). These can also be studied at a fine-grained and at a coarse-grained level, depending on the objectives of the analysis.

As to *time*, every event can be investigated at an earlier, previous time interval and at a later, posterior time interval – the time scale or time horizon. It is important to realise that only the here and now is lived reality. All else is memory, imagination and mental construct. But in commerce and science, time has been abstracted and objectified: It is a line with events on it as markers. We call the time elapsed in-between events a *period*. Everyday experience deals with periods in the range of 1 to 10^8 seconds (sec). As in space, there are coarse-grained and fine-grained observations. The coarse-grained ones make sense for slower processes, for instance, demographic change ($>10^8$ sec) or geological phenomena ($>10^{13}$ sec). The fine-grained descriptions are needed if one wishes to consider the faster processes, such as some chemical reactions ($\sim10^{-3}$ sec) or electromagnetic and atomic particle phenomena (10^{-6} sec).

Table 2.3 summarises some space-time scales for various relevant domains. The categorisation is indicative, because most domains have their specific ranges in space and time, and there can be changes in the categorisation itself, as with technology or politics. With spatial scale, the choice of timescales has to be commensurate with the timescales inherent to the system. If one is interested in the long-term future of the car system, for instance in relation to climate change, one is naturally interested in the macroscale in space and the slow variables in time. At the macroscale, there are clusters of phenomena such as ocean warming/cooling, glacial epochs and tectonic shifts (Turner *et al.* 1990). To understand the corresponding systems, however, one often has to go ‘down’ to the faster processes at the smaller scale, such as the urban road construction or vegetation change. This, in turn, may lead the analysis to yet higher resolution in space and time, but there is a logical end to this for a given objective. For instance, the daily trajectory of an individual car driver or the hourly features of a large storm or a forest fire are hidden in more aggregate descriptions. In the background, lurk questions like the universality of the various laws and the degree to which ‘history matters’ in complex nested dynamic systems.

2.6 Summary Points

This chapter introduced system thinking as a tool to look in an inherently transdisciplinary way at the world. The world car system is used to show the many interrelated aspects of a system. Points to remember from this chapter:

- the essence of system analysis/dynamics is to see the world as ensembles of interrelated and interacting elements;
- a system description contains: system boundary, elements, interconnections and purpose or goal (although on the latter there is difference of opinion);

Table 2.3. *Phenomena at coarse-grained and fine-grained scales and at different scales in space and time*

Phenomenological coarse-grained	Fundamental fine-grained
Temperature, pressure	Atom/molecule dynamics
Phase transition	Minimisation of molecular energy
Tsunami (storm surge), cyclone	Atmospheric particle dynamics
Spiral cloud	Star dynamics
Forest; ecosystem	Trees; plants and animals
Markets	Individual transactions

Fast, local	Intermediate	Slow, global
Earth: day-night and seasonal fluctuations	Erosion and salinisation (wind/water); River formation/change and sedimentation	Earth: orbital forcing, tectonic forcing; mineral ore/fossil fuel formation
Weather: thunderstorm, tsunami (storm surge), cyclone; particle dispersion,	El Niño/Southern and Northern Atlantic Oscillation	Climate: glaciation, global warming; atmospheric circulation patterns
Ecosystems: photosynthesis, metabolism	Habitat change, succesional change, forest dynamics, most ecosystem service dynamics	Life: evolution, species extinction
Population: birth, death	Age distribution	Genetic change
Economy: consumption/ production, market prices and transactions	Capital goods lifetime, labour-capital substitution, business cycles	Economy: system paradigm; capital stocks (buildings, canals, rail/roads, etc.)
Technology: incremental innovations, marketing	R&D induced change, process/product substitution, breakthrough innovations	Major technological waves (Kondratiev)
(Re)sources: production process, market price	Mineral and fossil fuel mining/depletion; long-term environmental sink (greenhousegases, ozone layer) dynamics	Sinks: substitution processes, most environmental sink (pollution) dynamics
Political: events, elections, office time of most CEOs and members of government	Value changes in population, change in governance structure (laws, customs, etc.)	Formation of nations, long-term historical/ cultural change, empires

- key notions in system dynamics are stocks, flows, feedbacks, delays, inertia and causal loop diagrams (CLDs);
- the behaviour of systems over time can be understood by identifying the relevant stock variables and the inflows and outflows for a chosen system boundary and by considering *both* the energy/material aspects *and* the behavioural/social aspects of the system elements and their interactions;

- two elementary modes of system behaviour are exponential growth (positive feedback) and exponential decline/growth towards an attractor (negative feedback);
- real-world systems constitute of interacting positive and negative feedback loops, with the logistic growth as one simple but frequently used model;
- the important steps in the construction of a system dynamics model are: formulate the problem, make a conceptual (stock-flow) model, collect relevant data, implement the model in equations/software and run and test the model;
- some phenomena are apparently quite different but turn out, from a system perspective, to share generic dynamic properties, and can be represented in archetypical models; and
- system dynamic understanding and modelling skills are helpful for the remainder of this book.

> Systems thinking is a discipline for seeing the 'structures' that underlie complex sitiuations, and for discerning high from low leverage change. (Senge (1990) 69)
>
> One of the central insights of systems theory, as central as the observation that systems largely cause their own behaviour, is that systems with similar feedback structures produce similar dynamic behaviours, even if the outward appearance of these systems is completely dissimilar. (Meadows (2008) 51)
>
> Wo aber Gefahr ist, wächst das Rettende auch.
> –Friedrich Hölderlin (1802), from the poem Patmos

> When the people of the Earth
> All know beauty as beauty
> There arises ugliness
> Lao Tze, Tao Te Jing

SUGGESTED READING

An introduction into dynamic models: a gradual build-up from elementary models to more sophisticated ones, with an emphasis on population-environment issues.

Bossel, H. *Modeling and Simulation*. Ltd./Vieweg, 1994.

A practical introduction to system dynamics modelling, with emphasis on natural systems and policy intervention.

Ford, A. *Modeling the Environment*. Washington, DC: Island Press, 2009.

An elementary introduction in system thinking, in clear language and with examples to practice it in a sustainability context.

Meadows, D. *Thinking in Systems*. Vermont: Chelsea Green Publishing, 2008.

A historical investigation into the sources of system thinking in the social sciences.

Richardson, G. P. *Feedback Thought in Social Science and Systems Theory*. Philadelphia: University of Pennsylvania Press, 1991.

Introduction to mathematical modelling, from simple to advanced.

Robinson, J. *Ordinary Differential Equations*. New York: Cambridge University Press, 2004.

An introduction into system thinking applied to management issues and extensive discussion of the nature and role of mental models in decision making.

Senge, P. *The Fifth Discipline – The Art & Practice of The Learning Organization*. New York: Doubleday Currency, 1990.

The most thorough and comprehensive book on how system thinking can and should be applied in a variety of applications, from physics to management, with ample sample models and discussion of the underlying mathematic and simulation features.

Sterman, J. *Business Dynamics – Systems Thinking and Modeling for a Complex World.* Boston: McGraw Hill, 2000.

USEFUL WEBSITES

- www.sutp.org/index.php?option=com_content&task=view&id=426&Itemid=72&lang=en offers a Sourcebook on Sustainable Urban Transport for developing cities in a truly integrated fashion.
- www.iseesystems.com/ is a site about the Systems Thinking software packages Stella™ and iThink.
- thesystemsthinker.com/systemsthinkinglearn.html is a site on the background and content of systems thinking.
- www.systemswiki.org/index.php?title=Category:Model is useful to explore and run models, if the simulation software package is available.
- www.calculusapplets.com/growthdecay.html is a site where you can play interactively with various growth functions: type in the expression for an exponential or logistic function and explore its behaviour.
- www.aw-bc.com/ide/idefiles/navigation/main.html is a site that permits exploration of differential equations interactively, for instance, exponential growth and decay and logistic growth models.
- www.metasd.com/sdbookmarks.html is a site operated by Tom Fiddaman with many environment-energy-economy links and presentations and a system dynamics model library.
- www.systemdynamics.org/ is the site of the system dynamics society with an explanation of the most important concepts.

SOFTWARE PACKAGES

- Stella™ (www.hps-inc.com) is the earliest and simplest-to-use simulation software package for system dynamics modeling.
- Vensim™ (www.vensim.com) is similar to Stella™, has a free demo download, is somewhat less simple to use and has in its professional version interesting model analysis features.
- forio.com/simulate/e.pruyt/ is a freely downloadable system dynamics simulation software package. The site is operated by Erik Pruyt (Delft University) and has a variety of models and simulations.
- www-binf.bio.uu.nl/rdb/grind.html is a freely downloadable software package to analyse ordinary differentiual equations, with application in ecology. The site is operated by Rob de Boer (Utrecht University).
- www.simulistics.com/products/gallery/one.htm is the site of the commecial simulation software package Simile. The site has a model gallery.
- www.wolframalpha.com/ is a site that gives detailed information on any mathematical function.

Appendix 2.1 Integral-Differential Calculus

Exploring sustainable development, we are primarily interested in change. Systems analysis postulates that every system, given a well-defined boundary and initial and boundary conditions, can be described in the form of 1..n state variables at time t:

$$\vec{X} = X_1(t), X_2(t) \cdots X_n(t) \tag{A2.1a}$$

X is a vector in n-dimensional state space. Observations of the system during a period T provide a dataset ***X*** that represents system states at the moments in time $\{t_0, t_1 \ldots T\}$. If the intervals between two observed states are the same and equal to Δt, then the change of the system can be described with a set of discrete difference equations of the form:

$$X_i(t + \Delta t) = F_i(X_1(t), X_2(t) \cdots X_n(t); \Phi; t)\ \mathrm{i} = 1..\mathrm{n} \qquad \text{(A2.1b)}$$

with X the state variables and Φ a set of constitutive and environmental parameters. The F_i express how each state variable depends on itself and the other state variables and on external parameters (Φ) and time (t). If we assume that the interval Δt becomes ever smaller ($\Delta t \rightarrow 0$), the set of difference equations becomes a set of (simultaneous) differential equations:

$$\frac{dX_i}{dt} = F_i(X_1, X_2 \cdots X_n; \Phi; t)\ \mathrm{i} = 1..\mathrm{n} \qquad \text{(A2.1c)}$$

with the d indicating an infinitesimally small change and F_i a set of analytical functions. The representation of all possible states of the system is called *phase space*. Without exogenous or autonomous change and only two state variables, equation A2.1c becomes:

$$dX_1/dt = F(X_1, X_2)$$
$$dX_2/dt = G(X_1, X_2) \qquad \text{(A2.2a)}$$

with F and G functions. dX_i/dt are called (time-)derivatives. The system is in steady-state or stationary state, when $dX_i/dt = 0$ for $i = 1,2$. Such a state is called an *attractor*. It is often depicted qualitatively as a ball in a landscape, in which the lowest point is the attractor because that is where one intuitively expects the ball to move to (Figure A2.1). In an attractor, the rate of change is zero and the tangent is, therefore, a horizontal line. Wherever the rate of change has an extremum, the system is in a state that can flip either way due to a minor perturbation. The landscapes a and b in Figure A2.1 indicate the difference between a stable and an unstable attractor. The landscape c represents the more complex systems one can expect in real-world systems.[25] The attractor is found for $F(X_1, X_2) = 0$ and $G(X_1, X_2) = 0$ in equation A2.2a. Upon perturbation, the system will move in first approximation according to the laws of simple positive/negative feedbacks towards the nearest attractor.

For simplicity, assume that the functions F and G are linear in the state variables. Equation A2.2a becomes a set of linear first-order ordinary differential equations (ODEs):

$$dX_i/dt = \sum_{j=1}^{n} \alpha_{ij} X_j\ \mathrm{i} = 1, 2 \qquad \text{(A2.2b)}$$

and the steady-state occurs for:

$$\frac{dX_1}{dt} = \alpha_{11}X_1 + \alpha_{12}X_2 = 0 \qquad \text{(A2.2c)}$$

$$\frac{dX_2}{dt} = \alpha_{21}X_1 + \alpha_{22}X_2 = 0$$

[25] The situation is often even more complex: the landscape itself changes in response to the movements of the ball – one of the topics in evolutionary modelling.

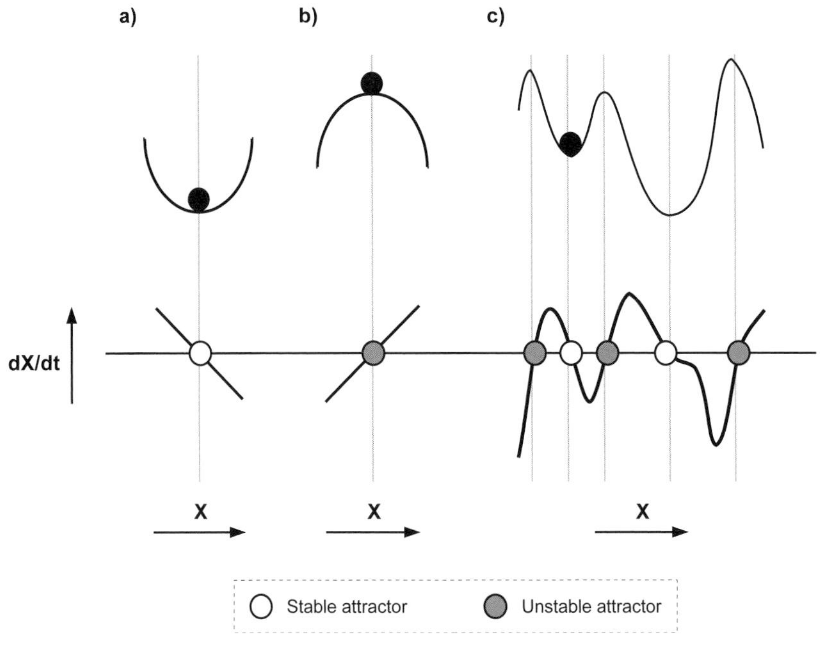

Figure A2.1. Landscapes with attractors. The upper graphs represent the system state as a ball in a (gravitational) forcefield. The lower graphs show the corresponding phase space diagrams of position X and rate of change in position (velocity) dX/dt.

The curves of X_2 as function of X_1 in phase space are called *isoclines*. The point at which the two isoclines intersect indicates the attractor. For the system according to equation A2.2c, the two linear *isoclines* are shown in Figure A2.2. This simple graph can tell much about the system behaviour over time – at least, inasfar as the model is correct. Points on the isocline $dX_i/dt = 0$ are system states for which X_i is constant and $X_{non\text{-}i}$ is *not* constant. Inspection of the different system states show that the phase space is divided in four separate parts, in each of which the system has a certain dynamic tendency indicated with the arrows. Thus, for any (initial) state

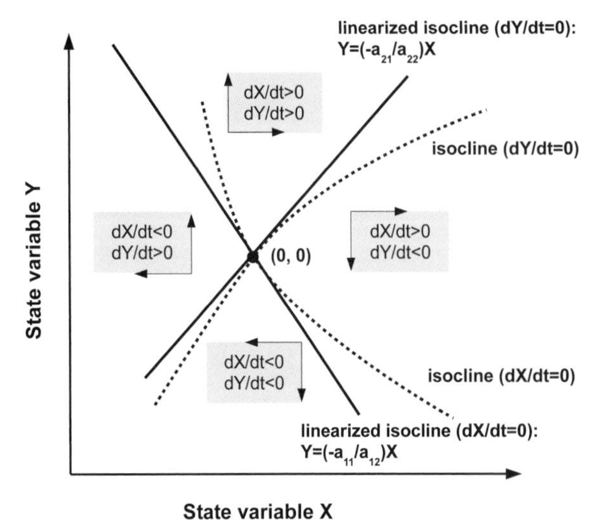

Figure A2.2. Representation of the two isoclines in phase space. The arrows indicate the dynamic tendency of the system. The intersection of the two isoclines is the steady-state. This particular representation is drawn on the assumption that $\alpha_{ij} > 0$ for all i,j.

of the system, we can derive how the state variables will evolve over time. At the intersection of the two isoclines is the attractor of the system $\{X_1^*, X_2^*\}$. It is the state where both X_1 and X_2 do not change – a sustainable state in the sense that the *net* change of the state variables is zero. If no intersection point is found within the permissible domain of X_i-values, there is no finite, meaningful attractor.

An important question is whether a steady-state is a stable or unstable equilibrium. *Stability analysis* is an important part of integral-differential calculus. The coefficients α_{ij} can be written in matrix form and represent the so-called *Jacobian matrix* **A**. In the general description of equation A2.2a, it is defined as:

$$\underline{A} = \begin{bmatrix} \partial F/\partial X_1 & \partial F/\partial X_2 \\ \partial G/\partial X_1 & \partial G/\partial X_2 \end{bmatrix}$$

with ∂ indicating partial derivation. Under the assumption of linearity (equation A2.2b) it becomes:

$$\underline{A} = \begin{bmatrix} a_{11} & a_{12} \\ a_{21} & a_{22} \end{bmatrix}$$

The diagional elements α_{ii} are equivalent to the growth/decline parameters and determine the *internal feedback* processes of X_i. The non-diagonal elements represent *interactions* between the X_i. The nature of an attractor can be seen from the value of the trace and the determinant of the Jacobian matrix. Matrix algebra can be used to solve equation A2.2. It gives the characterictic (or eigen) equation and values. The solutions are sums of exponential functions or products of oscillatory functions with exponentials. If one or both eigenvalues are positive, there is growth over time. If both are negative, there is decline over time. If there are complex eigenvalues, there are oscillations and the real part of the solution determines whether the amplitude increases, remains constant or decreases over time.

Only the simplest phenomena can be described by linear models. Most real-world phenomena are more complex and have to be approximated with nonlinear models. To explore system behaviour, one technique is to expand the state equation around the equilibrium state and retain only the linear terms. Such a *linearised stability analysis* is only valid near the equilibrium state, but that is the most interesting state anyway. The dashed curves in Figure A2.2 indicate the isoclines for a nonlinear system. In that case, the linear curves are the linearised isoclines. Refer to the suggested literature for further study.

3 In Search of Sustainability: Past Civilisations

3.1 Introduction

Early human groups were completely dependent upon their natural environment. They were confronted with changes of catastrophic immediacy such as earthquakes, and slow but no less impactful issues, such as changing courses or drying out of rivers and changes in climate and vegetation. These changes presented threats as well as opportunities and risks as well as challenges. Populations responded with outmigration into new areas with better opportunities or with attempts at increasing control of plants and animals. Indeed, the capacity to adapt to a variety of environmental situations may well be the most remarkable characteristic of the *homo faber sapiens*: 'All mankind shares a unique ability to adapt to circumstances and resolve the problems of survival. It was this talent that carried successive generations of people into many niches of environmental opportunity that the world has to offer – from forest to grassland, desert, seashore and icecap. And in each case, people developed ways of life appropriate to the particular habitats and circumstances they encountered. Farming, fishing, hunting, herding and technology are all expressions of the adaptive talent that has sustained mankind thus far' (Reader (1988) 7–8).

With the broad brush of Big History, one can distinguish several regimes in the existence of *homo* on earth. The first one was the fire regime, in which the control of fire was the determining feature (Goudsblom 1992). The transition from hunting-gathering to agriculture and herding was the second large regime shift. This transition is called the *agrarianisation* process, a better name than agricultural revolution because it was a rather slow process with local characteristics. Humans gradually expanded into nearly all corners of the earth, a phenomenon we call *extensive growth*, and created increasingly larger opportunities for more efficient exploitation of resources, the equivalent of *intensive growth*. It led to stronger interactions between humans and their natural environment, with a more sedentary lifestyle and new techniques and forms of social organisation. In particular, urbanisation did allow activities and exchanges that prompted relationships and innovations that might not otherwise have occurred.

Prior to 15000 (Before Present) BP[1], almost the entire interior of southwest Asia was covered by a desert-like steppe. Around that time, vegetation started to become richer across the Fertile Crescent as a consequence of climatic change. Wild cereals and grasses spread and were followed by oak-dominated woodlands. It offered local foragers a diverse array of wild plant-foods that could provide the necessary ingredients for a healthy human diet at relatively little energy expenditure. The prevailing current view is that agriculture originated at least 12000 BP in the so-called Levantine Corridor, near the Jordan valley lakes, in the form of the domestication of cereals and pulses. Within a millennium, there was also domestication of pigs, goats and sheep. But there were also early agro-pastoral economies outside the Levant, for instance in the semi-arid areas surrounding the Iranian Plateau around 10000 BP. There is increasing evidence of very early, probably independent origins of agriculture in China and in other places in Asia and the Americas.

Recent data from pollen diagrams seem to confirm the view of Sahlins (1972) that the life of early hunter-gatherers may have been more pleasant than that of the early farmers. Why then did the shift from gathering wild plant-foods (foraging) to crop production (horti-/agriculture) and from hunting to protective herding and raising livestock (pastoralism) happen? One theory is that agriarianisation was in part a response to deteriorating conditions for foraging. It initiated a series of changes. People became less mobile, birth-spacing was reduced and women had more children. The resulting higher fertility rates reinforced the formation of settlements and the need for food. Mortality rates probably went up, too, because of higher incidence of diseases due to rising temperature and more intense human-animal contact. The net result was a slow but persistent population growth. The causal loop diagram (CLD) in Figure 3.1 shows elements of the agrarianisation process.

In characteristic economic reasoning, Weisdorf (2005) suggests that hunting and foraging had originally a higher labour productivity (food provision per day) than growing crops, but that it started to decline due to population growth and a subsequent increase in competition for food and depletion of the food resource (game, vegetation). As a consequence, farming became in some places at the margin more productive than hunting-gathering. Changes in organisation and skills did play a role as well. For instance, exogenous improvements in food procurement technology may have triggered the shift to farming and created room for nonfood-related skills and the seed of economic growth.

This chapter explores the rise and fall of human civilisations in order to understand better the mechanisms of (un)sustainable development. The focus is on the role of the natural environment and the interaction with social and economic developments in both growth and decline of states and civilisations. Historians, philosophers and social scientists have thought and written extensively on 'the rise and fall' of civilisations. In the last decades, natural science methods and data collection and processing equipment have enormously expanded the data on sediments, ice cores,

[1] BP: years Before Present. The reference for Present is the year 1950 CE: Christian Era. Some scientists prefer the consistent use of CE and BCE: Before Christian Era. In older texts, one still finds BC: Before Christ, and AD: Anno Domini. Dating is often based on carbon isotope analysis, which is a statistical procedure with a considerable margin of error.

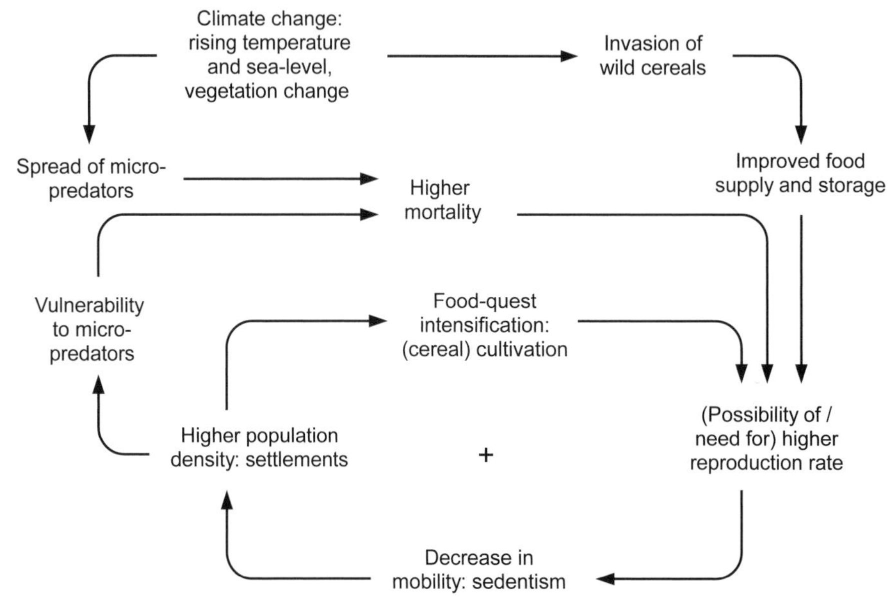

Figure 3.1. A causal loop diagram (CLD) sketching some major dynamic factors in the early agrarianisation process. Intensified food procurement was the determining positive feedback.

tree rings and so on. It allows reconstruction of present and past climate and vegetation characteristics and, in some regions, it has been possible to derive rather detailed reconstructions of people's diets and diseases from the environmental conditions. In combination with archaeological and historical analyses and the use of computer simulation methods, understanding of past human-environment interactions has deepened significantly.

3.2 The Beginnings: Two Environmental Tales

3.2.1 The Hohokam People in Arizona[2]

Early in the 3rd millennium BP, maize farmers settled in the river valleys in places suitable for floodwater farming in central and southern Arizona. The large empty surroundings provided a sustainable flow of wild food and material resources and the localised nature of accessible surface water and farmland stimulated settlement. Known as the Hohokam people, these early farmers depended on the nutrients brought down in flood periods by the two rivers traversing the Phoenix lowland basin. Around 1250 BP, they developed irrigation agriculture, which tied households together into community networks with ritual and belief systems. By 800 BP, larger villages with platform mounds began to appear and canal systems linked communities into larger networks. Hundreds of canals led the water and sediments to the fields, thus regenerating soil fertility and enabling the use of the productive potential of their immediate surroundings. The labour investments in canals and population growth reinforced living in settlements. Social institutions and community networks

[2] This account is largely based on Redman (1999).

evolved in order to maintain the canals and to spread the risks of food shortages. But around 550 BP, the Hohokam culture had disappeared. Why and how?

The Hohokam way of life – no domesticated animals and modest use of wood – tended to spare the vegetation. However, despite a way of life apparently based on principles of sustainability, archaeological evidence suggests that the increase in population led to a scarcity of protein-rich food such as fish and rabbits. The first factor at work was climatic change. Tree ring analyses indicate a rather high climate predictability in the period 1250 to 900 BP without extreme floods or droughts. Communities developed along the feeder canals that brought the water downstream to the distribution channels. Population, organisation and trade all increased. Tree ring evidence suggests that in subsequent centuries flood variability became larger but these communities were resilient enough to handle droughts and floods through storage, repair and trading. However, after 700 BP, climate became apparently more erratic and the frequency of floods and droughts increased. More labour was needed to repair and maintain irrigation structures. Crop losses in the valleys led farmers to overplant in the good fields, reduce fallow periods and start using marginal soils. This further lowered soil fertility and productivity.

Intense irrigation may have been a second cause of trouble. Because the Hohokam remained in the same location, their activities unavoidably caused environmental changes such as increased runoff volume and velocity, subsequent soil erosion and canal silting. Although they maintained soil fertility through various conservation methods and by supplementing local food with goods brought in by exchange systems, these climatic changes put additional pressure on the system. Investing more labour in order to extract the maximum from the land made the system probably more, not less vulnerable to climatic extremes. Socio-political changes towards more centralised control and ceremonial activities may have been a third factor in weakening resilience. Their disappearance, it seems, was due to a dynamic interplay of environmental and social forces working on different time-scales. For several other groups of people in the southwestern United States, such as the Anasazi ('the Ancient Ones'), similar past histories have been reconstructed (Diamond 2006).

3.2.2 Easter Island[3]

In 1722 (Christian Era) CE on Easter Sunday, the Dutch admiral Roggeveen 'discovered' a small island in the Pacific Ocean that he gave the name 'Easter Island'. It has become an archetypical example of how a human population on a finite island with a fragile environment can overexploit its environment and as a consequence collapse. When Roggeveen set foot on the island, it had a society 'with about 3000 people living in squalid reed huts or caves, engaged in almost perpetual warfare and resorting to cannibalism in a desperate attempt to supplement the meagre food supplies on the island' (Ponting 1991). Scattered across the island were more than 600 massive stone statues of, on average, some 6 metres (m) high and weighing more than 50 tons. The inhabitants of the island seemed incapable of carving and moving statues and indicated they had no knowledge of how to do so. It was a mystery how

[3] This account is largely based on Ponting (1991) and Diamond (2006).

they had ever reached such an outlying place: They only had small leaky canoes. It was an even bigger mystery how they could ever have erected such huge statues in large numbers.

Many explanations have been offered, some much more speculative than others. By now, there is enough archaeological and palynological evidence to reconstruct a plausible storyline (Diamond 2006). The first human occupants of Polynesian origin arrived to the island around 900 CE. The land mass of about 180 km^2 has a gentle topography and a mild climate, with fertile soils of volcanic origin. But the island's windiness, little rain by Polynesian standards and relative isolation made it less than an attractive place for humans. Estimates of human population numbers during the heyday range from 6,000 to 30,000 (33 to 165 per km^2); therefore, 15,000 inhabitants might well have been possible. The inhabitants lived on a diet of mainly sweet potatoes and chickens, which was nutritionally adequate and not demanding in terms of labour. The dense and diverse vegetation – amongst them palm trees probably related to the huge Chilean wine palm – provided trunks for canoes, construction material for household goods, pole and thatch houses, firewood for heating and cooking and for cremations, as well as delicious oily nuts.

Structured like other Polynesian islands, there were chiefs and commoners, organised in rival clans. The combination of plenty of free time for ceremonial activities and enduring competition and conflict between the tribal clans led to the construction of the huge stone statues. The statues (*moai*) were hewn out of the rock in quarries in the centre of the volcanic island, and transported over the still visible roads to the coastal areas, where they were erected on top of stone platforms (*ahu*). This technical *tour de force* absorbed not only enormous amounts of peasant labour – obsidian stone axes being the only tools – but also required prodigious quantities of timber rollers as there were no draught animals.

In the end, the increasing numbers and cultural ambitions overexploited the limited resources and collapse followed. Deforestation started upon arrival of the first settlers and by 1700 CE most forest had disappeared. Soils deteriorated. Land birds and open-ocean fish also disappeared from the diet, because of combinations of deforestation, overhunting and predation by rats. Chickens then became more important, and defensive chicken houses were erected. Social and ceremonial life came to a standstill. With no more statues being built, belief systems fell apart and with it the legitimacy of social organisation. Slavery, war and cannibalism followed suit. Social mechanisms probably played an important role. In one interpretation, the chiefs and priests, claiming relationship with the gods and promising prosperity, '... buttressed that ideology by monumental architecture and ceremonies designed to impress the masses, and made possible by food surpluses extracted from the masses... As [their] promises were being proved increasingly hollow, [their] power was overthrown around 1680 by military leaders... and Easter's formerly complexly integrated society collapsed in an epidemic of civil war' (Diamond 2006). Were the people on Easter Island stupid or evil? A comparative analysis of more than eighty Pacific islands suggests that they were, in fact, unfortunate: Easter Island is highly susceptible to erosion and thus amongst the most environmentally fragile islands in the Pacific Ocean.

The Easter Island history may have had quite some precursors elsewhere. One can read a similar sequence of events in the Megalithic memorials and burial mounds

Box 3.1. *Peasants, priests and soldiers.* The rise of social organisation can be interpreted in terms of the risks early agrarian communities were facing. Priests fulfilled a mediating role between ordinary people and the extrahuman world. They also induced the self-restraint required for a farming life of hard work and for the exigencies of food storage and distribution: '... rites conducted by priests helped to strengthen the self-restraint which could keep people from too readily drawing upon their reserves' (Goudsblom *et al.* (1996) 42). Harvest feasts and sacrifices were social institutions to manage the pressures of frugality. Priests were resource-managers *avant-la-lettre* – not a strange idea when one knows about the rules in Christian and Buddhist monasteries.

In many societies, another alliance proved more durable: between peasants and warriors. The priest-led, religious-agrarian regimes were probably first, but came almost everywhere in competition with warrior-led, military-agrarian regimes. Arguably the most crucial force behind the latter was the bonding of warriors and peasants: 'The warriors needed the peasants for food, the peasants needed the warriors for protection. This unplanned – and, in a profound sense fatal – combination formed the context for the great variety of mixtures of military protection and economic exploitation that mark the history of the great majority of advanced agrarian societies... wherever in agrarian societies rural settlements developed into city-states which were subsequently engulfed by larger empires, the priests became subservient to the warriors' (Goudsblom *et al.* (1996) 59). The emergence of a warrior class, that is, of professional killers and pillagers, should be seen as one stage in the monopolisation of violence and cannot be explained solely in terms of their discipline, equipment and organisational skills.

constructed some five to eight thousand years ago on the islands of Malta, Corsica and the Orkneys. The Easter Island tale teaches us several lessons. First, it is possible to give an evocative and persuasive account of past events, even if evidence is scarce and controversial. Some scholars argue that a collapse involving starvation, major warfare and cannibalism is neither supported by the 18th-century journals nor by the scientific evidence (Boersema 2002). An alternative narrative, with apparently strong empirical evidence, is that Polynesian settlers arrived rather late and that Polynesian rats played a major role by preventing forest regeneration. In this reconstruction, there was a rather smooth transition from a 'rich' Moai culture to a 'poor' bird-culture and the real collapse was caused by the raid on the island in 1862 by Peruvian slave traders. A second lesson is that, even in these early primitive communities, both economic (resource) and social-cultural (organisation, rituals) aspects played an important role in rise and decline. Third, it is possible and useful to construct an explicit model of what happened in order to debate the different hypotheses. Following the renewable resource economics approach of the 1950s, Brander and Taylor (1998) have constructed a simple analytical model of an Easter Island-like population that harvests and consumes a renewable resource (food) and some 'other good'. The positive link between net population growth rate and harvest rate causes a 'feast and famine' pattern of population overshoot and endogenous resource degradation and has become a widely used metaphor for unsustainable development.[4]

3.3 Emerging Social Complexity: State Formation

3.3.1 Early Mesopotamia: Urban Centres and Their Elites

From the earliest stages, human groups have shown signs of self-organisation. In many cases, this process gradually intensified into stratification between the rulers and the ruled, differentiation into castes and guilds, and the emergence of towns and trade markets. Tribes and chiefdoms evolved into one dominant authority in larger regions, the *state*. The early state had a 'very strong, usually highly centralised government, with a professional ruling class, largely divorced from the bonds of kinship... highly stratified and extremely diversified internally, with residential patterns often based on occupational specialisation rather than blood or affinal relationship... The state attempts to maintain a monopoly of force... while individual citizens must forgo violence, the state can wage a war; it can also draft soldiers, levy taxes, and exact tribute' (Flannery 1972). Mythical and legendary charters, war and terrorism were amongst the methods in which states enforced legitimacy, displayed power and exerted control. States emerged all over the world, under a variety of environmental conditions, and they are still a vibrant reality in the 21st century. The state represents in its different appearances the most complex organisational form by which human groups resolve the perennial tension between the individual and the group, between competition and cooperation.

The earliest state has probably developed around the urban centres in the plains of southern Mesopotamia. Already in the 6th millennium BP the city of Uruk extended more than 100 hectares (ha). The rural peasant families produced the food such as wheat, barley, vegetables, milk, meat and the flax, wool, hides, dung, reeds, clay for nonfood needs such as clothing and housing. The early towns facilitated all kinds of exchange as part of a 'tributary economy... dependent to a significant degree on the mobilisation of tribute, in the form of goods or the labour used to reproduce them, from producers to a political elite' (Pollock 2001). Hierarchy and status started to replace egalitarian forms of organisation. Building temple platforms and serving as priests, even if on a voluntary basis, became a source of social and political prestige and of material gain. As time progressed into the Uruk Period (6000 BP), the elite divorced itself increasingly from the material forms of production and appropriated ever larger portions of the surplus food and goods. Institutions for political control emerged and long-distance expeditions were set up to procure exotic goods.

There is general agreement that environmental deterioration, mainly in the form of salinisation, played a role in the collapse of the later Ur dynasties. Agricultural productivity declined continuously from the middle of the 5th millennium BP, with a shift from wheat to more salt-tolerant barley. Deforestation as a result of woodcutting for domestic and industrial hearths, with further destruction by goats, was another detrimental force. But social responses were an inherent part of the decline.

[4] A nonlinear model of the underlying dynamics may explain the collapse of ancient societies with the concept of sunk costs (Janssen *et al.* 2003).

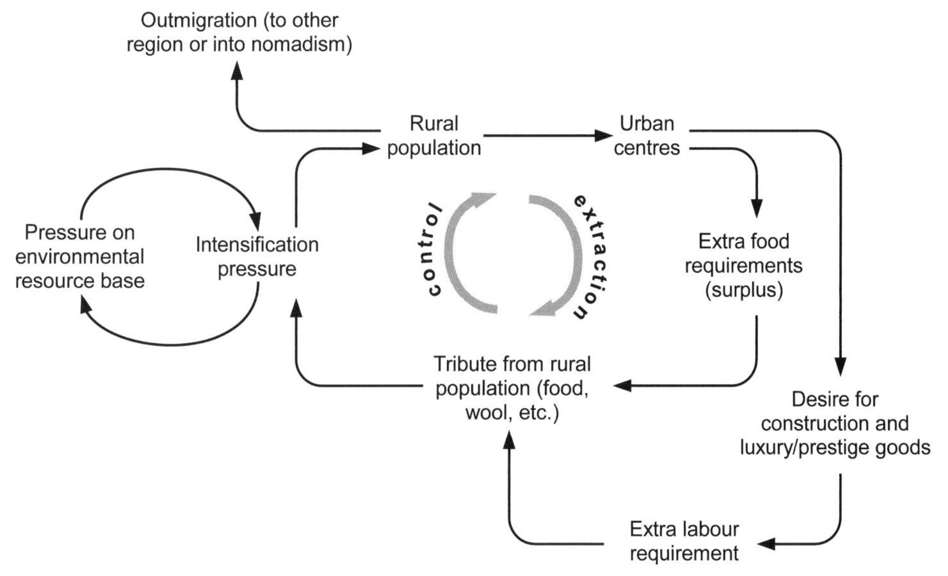

Figure 3.2. Possible factors in the rural-urban dynamics in Ubaid and Early Uruk society in Mesopotamia.

The rise in centralised control with its policy of maximising the surplus was a major force behind the declining agricultural productivity and environmental damage (Redman 1999). When food supply further deteriorated, probably in combination with climatic change, increasing numbers of debt-ridden peasants fled to the cities. To feed the growing urban populations, increasing tribute had to be paid. This intensified the flight to the towns and out of the region, with a further reduction in food supply. A vicious circle of extraction and control evolved in order to sustain the centres of power and prevent social revolt (Figure 3.2). At some point, the elite structures collapsed and gave way to more local production modes that were geared again to local needs and resources.

3.3.2 Egypt: The Nile and Its Rhythms

During the 7th and 6th millennium BP, episodes of severe drought, dwellers of the desert drifted towards the Nile valley in search of water, food and fodder. In subsequent millennia, the basis for Egyptian civilisation was laid. Nile water levels have always mirrored climatic events throughout the region. Long-time historical water level records in combination with archaeological finds show that episodes of high Nile flood discharge are correlated with greater rainfall in East Africa and a warmer climate in Europe, whereas low levels correlate with cooler temperatures and drier conditions in East Africa. The Nile river is Egypt's artery and at the root of its economic prosperity and environmental fragility.

Extreme water levels had large consequences for human populations in Egypt. Very high floods led to the destruction of dikes and houses and prevented the planting of crops. Very low discharge levels reduced the cultivable area as natural irrigation failed. A variety of responses to such risks have emerged: building, maintaining and repairing dikes, irrigation canals and drains. It led to a centralised administrative

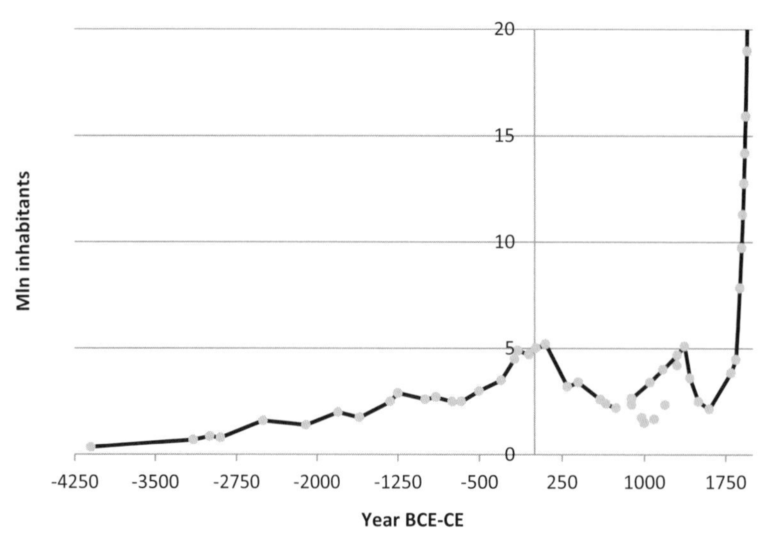

Figure 3.3. Estimated population in Egypt. In 2007, the population amounted to 80 million (source of data: Turner *et al.* 1990).

state. Droughts in the Nile Valley around 5200 BP may have been responsible for the unification of Egypt and the rise of the Old Kingdom. Early documents show administrators shifting from wheat to the more salt-resistant barley to combat increasing salinisation resulting from irrigation. One millennium later, around 4200 BP, low Nile River levels with catastrophic droughts probably caused the collapse of the Old Kingdom. They triggered a series of civil uprisings, widely recorded during this time: 'The Old Kingdom... was a time of tremendous royal power... and it saw a big increase in population... This placed great reliance on maximising the use of the land flooded and fertilised each year by the inundations. Around 4250–4150 BP, the same prolonged dry period that caused such problems for the Ur III kings in Iraq brought a series of consistently low floods and precipitated half a century of famine. This helped pull a declining order. The monarchy was overthrown... ' (Wood 1999).

Social order was regained by the end of the Ninth Dynasty, about 4000 BP, with the administrative centre now at Thebes. The rapid recovery can be interpreted as a sign of the strength of Egyptian civilisation after so many centuries. During the New Kingdom, the population of Egypt as a whole increased again, possibly to as many as 5 million during the Eighteenth and Nineteenth Dynasties (3489–3136 BP). Up to half of the population may have lived in cities such as Memphis and Heliopolis, the former one being perhaps the world's first city with more than 1 million people. Since then, the population has fluctuated between 2.5 to 5 million people, until it started around 1800 to surge to its present value of more than 80 million people (Figure 3.3).

Figure 3.4 shows a CLD with possible mechanisms of social-ecological evolution. As ruling elites were depending on the tax revenues from farmers, a series of extremely high or low Nile water levels were critical for governance. Farmers were confronted with starvation, urban craftsmen and artisans died in massive numbers[5]

[5] It is reported that less than 10 percent of the weavers in Cairo survived the 1200–1201 CE famines (Hassan 2000).

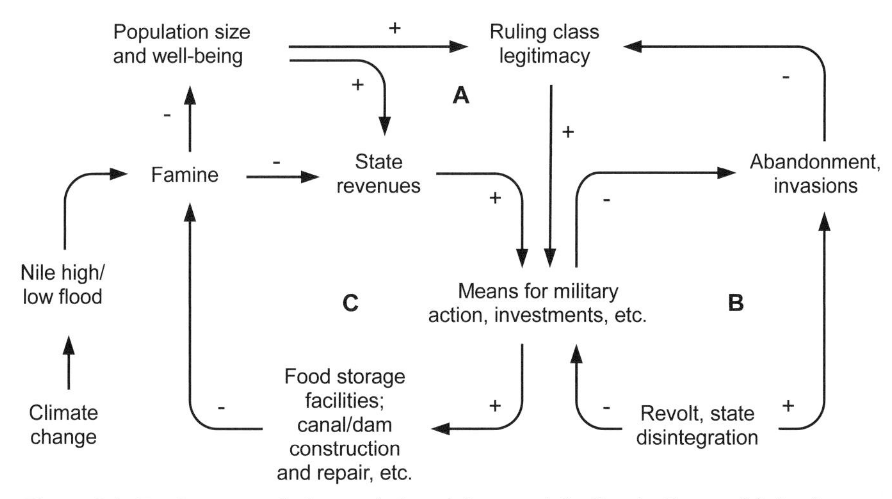

Figure 3.4. Environmental change induced forces of decline in Egypt: (a) famine can cause a positive feedback loop towards disintegration, (b) accelerated by subsequent tendencies towards disintegration. Investment in adaptive capacity and restoring ruling class legitimacy (c) can halt or reverse the decline.

and the king saw his income shrink rapidly. One can imagine a bifurcation in such periods of economic distress. If irrigation channels and dams had been maintained and enough food had been stored for the 'seven bad years', as recounted in the Bible, famines could be avoided, which in turn would make it possible to sustain ruler legitimacy and military support (loop C). If, on the other hand, the infrastructure had been neglected and urban administrators, merchants and other nonfood producers tried to stick to their 'good life' habits, a vicious cycle started (loops A and B). The legitimacy of the ruling class came under threat. Rivals would emerge, denouncing royal excesses such as construction of large monuments. Because of declining revenues, the king could pay neither for military and political support nor for large agricultural projects that would redistribute food and secure against future droughts. The 'power range' of the central government – an estimated 400 km with the available transport capacities – decreased and the state disintegrated in smaller, provincial units. External enemies saw opportunities to invade the country. It is quite probable that such downward spirals happened during the periods of low Nile discharge around 4200, 3100 and 2700 BP. The outcome of such a disintegration process was not necessarily disastrous: the newly emergent, smaller units may actually have been more resilient.

3.3.3 South Asia: The Indus-Sarasvati Civilisation

The first signs of an agrarian civilisation in South Asia emerged in the valleys of the Indus river and the – now largely dried up – Sarasvati river. Ever since British explorers discovered the famous sites of Harappa and Mohenjo-Daro, the civilisation that flourished here some 4,000 years ago has been called the 'Indus civilisation'. Its origin can be dated back to ancient settlements in Baluchistan around at least 7000 BP. In the past decades, remains of this civilisation have been found in a much larger area than originally thought, possibly the size of Western Europe.

Apparently, this civilisation has known three phases. In the Early Phase, 5100–4800 BP, it still wore the traces of the early farming communities and pastoral camps.

Box 3.2. *Priests, no warriors?* There are – and will be – different interpretations of the Harappan settlements' social complexity. It is clear that in the mature stage these settlements were part of a larger political and administrative framework. Some have speculated about priest-kings not unlike those in Sumer and Akkad – but Chakrabarti writes that 'An Egyptian or Mesopotamian kind of kingship need not be envisaged in the Indus context. In later Indian history, the king... was a much more humble figure... he does not strut around in sculptural reliefs towering above ordinary mortals and cutting the heads of his enemies... he functioned within the well-formulated concept of the royal duty of looking after the well-being of his subjects.... It is futile to look for the remains of a royal palace... for the simple reason that there will not be any way of identifying it, going by the later Indian examples. Priesthood is far more sharply visible... The concept of a yogin, one who sits in meditation, is writ large... Remains... unmistakably imply the services of priests – priests of a type that a practising Hindu would engage for performing his household rituals even today' (Chakrabarti (1997) 123). Such a view suggests much more continuity between this mysterious past and the Indian present – and more diversity in the behaviour of early priest-kings – than previously thought. This issue is still an important item in the debate on Indian identity.

The Mature Phase, 4800–3900 BP, is the period of at least a dozen large cities. A variety of crops were grown, including rice and cotton, and there were extensive metal mining and smelting activities. The culture was characterised by sophisticated architecture, a distinctive own writing style and fire worship. A remarkable feature was, apparently, the absence of personality or ruler cults such as those that emerged in contemporary Mesopotamia, Egypt and China. There is evidence of cultural interaction with Rajastan, Central India and Maharashtra and of trade with populations in Mesopotamia through ports such as Lothal.

The Late Phase was around 3900–3400 BP, during which it mysteriously disappeared. Until recently, it was thought that the decline came rather abruptly – the causes being a mixture of climatic change with more frequent droughts and changes in the Saraswati river bed, more or less gradual invasion by northern Aryan nomads and/or internal cultural erosion. However, new data suggest that a combination of adverse changes in rainfall patterns and a massive earthquake caused river courses to change. The important rivers in the region – the Sutlej, the Yamuna and the Sarasvati or Ghaggar-Hakra – have changed course several times in the last four millennia. Whatever the precise mechanisms, there is increasing evidence that, between 4200 and 3800 BP, a global change in climate took place with a particularly fast change around 4200 BP known as the '4.2 K Event' (Costanza *et al.* 2007). This can explain why several early civilisations declined simultaneously in this period. The downfall may have been accelerated by a collapse in trade caused by the synchronicity.

3.3.4 The Aegean and Mesoamerica: The Role of Ecological Diversity

The civilisations of the Euphrates-Tigris, Nile and Indus river basin were centred around a large river. The fertile soils and irrigation potential was a determining factor, it is thought, in the way they developed and were organised into large, centralised 'hydraulic civilisations'. Some civilisations emerged under different circumstances. One example is the Aegean Sea region, with a very different biogeographic situation. It covers mainland Greece in the north and west, the coast of Asia Minor in the northeast and the island of Crete in the south. In between are hundreds of large and small islands. It is an intensely researched and interesting area with regard to state formation.

The oldest civilisation in the region is Crete, an island of about 8,000 km^2 of largely marginal mountain country. It had established itself by the end of the 5th millennium BP. Characteristic features of this culture were the large palaces with their peculiar orientations, the mountain peak sanctuaries and the conspicuous absence of fortified citadels or defensive walls and earthworks. The written script, institutionalised trade- and sacred rituals gave powers of organisational and psychological control to the classes of priests and kings. The palaces were destroyed a few centuries later, probably by earthquakes, and rebuilt. The 'New Palaces' were destroyed violently or abandoned around 3400 BP, probably due to conquest by the less trade- and more war-oriented Mycenaean people in the southern part of Greece.[6] Mycenaean cities collapsed in around 3100 BP, due to invasions and possibly climatic change or natural disasters.

Against the resulting background of petty lords practising small-scale redistribution as part of a strategy of ironing out localised shortages, the later Greek civilisation emerged and dominated the southern Aegean Sea for some centuries. The southern Aegean consists largely of coasts and islands with a large ecological diversity. People experienced interannual harvest fluctuations that were large, unpredictable and asynchronous in these highly diverse environments. In such a situation, *trade* with neighbouring settlements was an important buffer against food shortages. This ecological diversity probably contributed to the emergence of a social ranking with a prominent role for merchants and for symbolic transactions. It is tempting to relate this early landscape-inspired evolution to the well-known features of the later Greek city-states: politically autonomous units resisting unification and an extensive trading network and colonies all along the Mediterranean coasts.[7]

The reconstructions of another civilisation, the Maya in Mesoamerica, also suggest an important role for the biogeographical situation. The Maya were amongst the larger and most advanced civilisations in Meso-America. It lasted from 4000 BP (Early Pre-Classic) to 500 BP (Late Post-Classic). The Yucatán Peninsula was its heartland. It had significant gradients in elevation, soil thickness, rainfall and water accessibility. Three river basins on the peninsula provided transition zones – or ecotones – which created and sustained cultural diversity.

A possible relationship between environment and social organisation is suggested from a comparison between Yucatán and the highland valleys of Oaxaca and Mexico. The mountains in the highland valleys are natural barriers that obstruct

[6] Possibly, a huge volcano eruption on Thira – modern Santorini – did wipe out the Minoan civilisation on Crete in a complex interplay of factors each operating on different time-scales: flood waves, interruption of trade, climatic change and erosion of dominant belief systems and invasions.

[7] An interesting network-based perspective on the dynamics of Aegean civilisation is developed by Knappett *et al.* (2008) (§10.6). The Bronze Age maritime network is used to explore the balance between local resource cultivation and trade.

communication and have limited potential for agriculture. Their agricultural systems tended to be uniform and hierarchical: 'a continual preoccupation with comparatively little horizontal complexity – the vast majority of parts are alike, and they tend to stay that way. The replication of uniformity is usually accomplished by means of a strong vertical differentiation – a powerful hierarchy manages most affairs' (Blanton *et al.* 1993). The Yucatán lowlands, on the other hand, are much vaster and unbounded. There were many interacting socio-political units with permeable boundaries and large effort went into regulating flows of things and people across the boundaries. People in these circumstances tended to employ a wide and diverse spectrum of production techniques. It led to diversity and horizontal complexity, with coordination in matters such as access to resources, allocation of time and labour and exchange of products. Economic institutions developed early and provided a consistent basis for survival. The evolutionary problem was the 'organisation of diversity': how to keep the components in these horizontally complex systems doing their jobs even though they have very different roles and interests? This problem is still with us.

Why did Mayan civilisation decline? Archaeological evidence clearly shows that Mayan farmers had to cope with a capricious and fragile environment. It explains their elaborate calendar system with a superior ability to prognosticate patterns of cyclical environmental changes. It also explains why the intensive farming necessary for an ever-growing population could not be sustained. Climatic change seems to have caused population shifts between the three river basins in the Early and Middle Pre-Classic period. Later, elevated cities were abandoned in the period of great droughts (1050–1250 BP). Only in one river basin are there signs that human activity, notably deforestation with subsequent larger river discharge volumes, played a role in the decline. Yet, the general view is that population outstripped the available resources. Deforestation and erosion degraded the resource base and, at least partly as a consequence, warfare intensified and made the situation worse. Climatic change and short-term egotism of the elite probably were the final blows (Diamond 2006).

3.4 Empires

3.4.1 The Roman Empire[8]

Some state-based civilisations managed to further develop their technical and organisational skills, towards population sizes and areas of control that led to the epitaph 'empire'. Whereas the word *state* refers to political organisation and government, the word *empire* is etymologically rooted in rule and authority. Empires are characterised not only by autocratic rule but also by a territory far beyond the one of the original rulers or nation. The word was used first for Rome after it had been weakened by civil wars and Augustus seized power in 27 BCE. Many books have been written on the rise and fall of the Roman Empire. The focus on the human-environment nexus is rather recent (van der Leeuw 1998). This section presents some narratives and insights about the mechanisms that led to growth and decline of this and other empires. The Roman Empire is of special interest, not only because it is a precursor to the Western or European civilisation, but also because much is

[8] This section is largely based on van der Leeuw and de Vries (2002).

Box 3.3. *Environmental degradation: a record of collapse.* The pioneering Central Petén Historical Ecology Project uncovered the environmental history of a Maya site in lowland Guatemala (Redman (1999) 141–145). In this region, a swidden – or *milpa* – agricultural system with a three- to five-year fallow period can support population densities in the order of 25 people/km^2. How did the Maya manage to sustain in some places and times up to 250 people/km^2? Examining sediment core from lake bottoms, a significant increase in phosphorus and silica deposition during the Maya occupation was found. Accelerated phosphorus deposition points at more phosphorus in the soil due to human activity (waste, food, disintegration of bodies and stoneworks) and to more soil erosion. Similarly, silica deposition increased severalfold during the period, which archaeologists think had the highest population – another indicator of increased erosion rates. Apparently, the Maya appreciated the tropical forest ecosystem well enough to thrive for centuries by managing land clearance, control water flows and transport food to cope with localised shortages.

However, the lake sediment data indicate that 'the high forest that prevailed in much of [the] region was largely removed by the farming and settlement building activities of the Mayas as early as 3000 to 4000 years ago. This resulted in a shift toward more open vegetation . . . with the maximum deforestation between 1000 and 2000 years ago. The basic drain on the land . . . increased to the point where the system was no longer sustainable. . . . By the end of the 10th century AD, most of the large settlements of the Mayan uplands and southern lowlands had been abandoned or at least seriously depopulated' (Redman (1999) 145). Coincident was a relatively dry period, which put crop productivity under additional pressure in an ecosystem already strained by human activities. There are an increasing number of environmental history reconstructions on the local/regional scale, which give empirical and model-based insights on how human activities interfered with long-term and short-term environmental change (Dearing 2006). The reconstructions may help to understand better the present situation.

known about it over a long period and across several geographical scales. Besides, many traits of (pre)modern society were already part of Roman society, such as rapid colonisation of most or all of the known world, an elaborate military and civil organisation to control the Empire, an urban base and major infrastructure.

Let us first have a look at the biogeographical situation. The Mediterranean basin has relatively mild winters, hot summers and precipitation in spring and autumn. Most Mediterranean landscapes were suitable for human habitation and had become entirely dependent on interruptions of the natural cycles by humans by the time the Romans colonised the region. Without human intervention, landscapes would have rapidly changed due to erosion, growth of scrubland (*garrigue*) and recolonisation by forest and become unsuitable for the human activities that had shaped it. The variety of environments in Mediterranean landscapes lend themselves to the small-scale cultivation of different crops, together with herding, rather than to the large-scale farming of homogeneous crops. Already long before the advent of the Romans, the cultivation of cereals in combination with that of olives and vines made it possible for farmers to harvest more than one crop. This natural *resilience through diversity* increased when the original territory of the Roman Republic, which was relatively homogeneous, expanded with the conquest of other provinces. It distributed the risk of food shortage. Moreover, Roman occupation added inventions in the areas of food storage, treatment and trade. It implied a shift from local environmental resilience to national/regional social-ecological resilience, which is an essential characteristic for civilisations to be sustainable.

When the Republic, and later the Empire, expanded, the Romans imposed with a varying degree of success a commercial system of exploitation on the existing, highly varied ways in which the local populations exploited their lands. Exploitation was initially organised in a very dense pattern of farms of variable size from before the advent of the Romans, or in country houses (*villa*) and large, highly efficient farms (*latifundia*). As time went on, the different exploitation systems became more dependent upon each other and had larger impacts on the landscape. In a next stage, the agricultural landscape started to change profoundly under the influence of mass-production for the export of agricultural products. It meant an increase in scale – larger farms, larger installations, larger bovines. Many smaller farms were either abandoned or became part of the larger ones that had grown out of the first Roman establishments in the region, located on the best land and with the most direct access to the markets. The forests on many hilly slopes were cut down and had to be terraced or otherwise protected against erosion. It made these landscapes extremely dependent on human intervention and vulnerable to erosion in its absence.

The commercialisation and rationalisation of agriculture were made possible by effective administration of rural areas (taxes and markets). An important role in this was played by the organisation of the land in a system of square-mile blocks that were laid out and mapped by a professional class of surveyors.[9] Because this 'square' administrative point of view often prevailed over any practical considerations, the resulting organisation of the landscape was regularly at odds with the natural dynamics of an area. For instance, in the Tricastin in southern France the roads and ditches were all laid out in north-south and east-west directions, whereas the natural drainage of the area generally went from north-east to south-west. When the number of soldiers drastically declined, there was insufficient manpower to keep the ditches open. The drainage scheme fell apart, and the land became neglected and ultimately began to erode (van der Leeuw 1998). Such a mismatch between the local reality and the practises of the exploiter can be an important cause of decline, which was also experienced during European colonisation in the 16th to 20th centuries.

In parts of northern Africa, the local situation created another pattern of colonial exploitation. More and more simple Roman farmsteads emerged between the 1st and 3rd centuries CE. It pushed the desert pastoralists farther and farther away from the coast and reduced the grazing land available to them. Eventually, it forced them to begin herding camels instead of sheep and goats. Their impoverishment increased the contrast in wealth. It led to more frequent raids on the rich Roman

[9] The land was divided into square blocks measuring one by one mile. Mile comes from *milia*, thousand – one mile being a thousand steps with a Roman step being two of our steps: left-right. The imposition of order and uniformity as an instrument of control is a characteristic of the hierarchical state (Scott 1998).

Box 3.4. *Views of nature in European Antiquity.*[10] How nature was seen depended in antiquity – and in many places still – on its actual or potential threat to humans. Nature, with its unpredictability and threats, was in no way always a friend. Therefore, it was permitted, and even a duty, to fight it, as can be seen in a variety of cultural motifs. The disappearance of forests was seen as a sign of civilisation, at least in the dominant view.

Clearing *forests* for agriculture, wood, firewood and animal grazing was the main cause of deforestation. Timber demand was another one. Thirgood (1981), quoting Strabo, mentions the mountains in southeastern Spain 'covered with thick woods and gigantic trees', being cut for shipbuilding. North Africa was another important timber-producing region for the Romans, leading to temporary depletion of Moroccan forests. Wars had a devastating effect as it accelerated the felling of trees to be used for the warships and made people flee into the mountains and abandoned their land. *Wild animals* were another aspect of nature. Heracles battled with near-invincible mythical wild animals, becoming a symbol of courage. The Roman Emperors wanted not only slaves but also wild animals to be taken from the conquered lands and to be emblems of total dominance over man and animal. More intense hunting brought some species to extinction. The enormous demand for animal skins, from as far as the Baltic and northern Russia, further increased the pressure.

The *Pax Romana* acquired a specific connotation through the mass killing of wild animals. Thousands of wild animals were slaughtered in the *venationes* during the games or *ludi*. These took place in every garrison town, but Rome had by far the largest. With the inauguration of the Colosseum in 70 AD, some 5,000 animals were 'used' in a few days time. It must have given rise to huge transport problems. Only recently, it has been acknowledged that the grand scale of these killings may have led to the extinction of some of these species. The games may indirectly have contributed to the expansion of agriculture in the Mediterranean by strongly reducing the threat from wild animals – an idea that fits into the view that humans should 'civilise' the world and was, therefore, approvingly supported by scholars until recently.

settlements until, eventually, they were fortified to protect the colonists. Again, it preceded similar developments in European colonies almost two thousand years later.

Between the 4th century BCE to the 1st century CE, the surface of the Roman Empire increased from less than 1,000 to more than 267,000 km^2, and the number of Roman citizens – excluding members of other nations and slaves – from less than 1 million to more than 4 million. After that, the expansion stalled.[11] How did the

[10] This box is based upon a contribution from Jan Boersema, Vrije Universiteit Amsterdam, in de Vries and Goudsblom (2002).

[11] The average population density of Italy – about 20 inhabitants/km^2 – was probably of the same order of magnitude as that of other parts of the Mediterranean world. In contrast, for Gaul at the time of Caesar, the equivalent number was about 9 inhabitants/km^2 – implying about 5.7 million people for all of Gaul, including *Gallia Narbonensis*.

Romans manage to conquer such large areas so quickly and durably? Part of the answer is organisational talent and technology: They were able to establish and control an appropriate infrastructure, which depended on the spread of towns. Urban areas were easier to control and could via their administration and commerce be connected directly to Rome's own. It was essentially an *exploitative, pyramidal pattern of administration*, based on links with the centre of different form and intensity and with a variable degree of Roman impact on the various components of its territory.

There are at least three other elements in how the Romans managed to hold together an Empire the size of the Mediterranean basin. First, the *location* of Rome at a central position in the Mediterranean basin gave it an advantageous position for trade and governance. Second, the *road system* with the major roads and the many (sub)regional roads provided dependable and efficient means to move people around and thus to exchange information and ensure administrative and military control. It also deliberately broke through the unity of the tribal territories, and rightly was the pride of the Empire. The third element was a *trade system*, which greatly exceeded previous commercial activities.

Yet, in the second century of its existence, the Empire started to show signs of crisis. Roman authors used to blame invasions, epidemics, incompetent Emperors or, in a more structural vein, the weak and overstaffed administrative infrastructure and the heavy burdens of maintaining a government and an army over such a wide territory. Lead poisoning, soil exhaustion, famine and possibly climatic change were mentioned, too.[12] Most of these factors were real and did contribute to the difficulties. One plague epidemic lasted from 165 CE to about 180 CE, with the loss of up to a third of the population in certain areas. The 3rd century saw a rapid succession of emperors lasting between a few months and three years. There were always one or more provinces in revolt, independent or out of control. The emperor Diocletian (284–305 CE) restored a semblance of order and his reign initiated half a century of relatively stable government and relative wealth. The administration and the army were reformed into larger, more complex and more powerful institutions; the Empire was divided into an Eastern and a Western part with a separate administration, different languages – Greek and Latin, respectively – and different cultures. The provinces were subdivided into smaller ones. People were conscripted to perform duties in the maintenance of infrastructure and many other activities and taxes were increased. It was not durable: The structure of the Roman Empire collapsed in 395 CE. The periphery increased in independence as the core began to lack the means to impose itself. Cities retracted within their walls, and the flow of traded goods and of people to and from areas farther from Rome began to dry up. Tribes outside the Empire were attracted by its riches and fame and increasingly raided the areas where wealth could be found. The Western Empire was more vulnerable than the culturally and linguistically homogenous Eastern Empire and collapsed in less than a century. The Eastern Empire was able to maintain itself for another millennium.

[12] In areas where these data can be compared – principally in southeastern France – there are no indications of *major* changes in climate between about 400 BCE and about 600 CE. At the time of the Roman expansion, precipitation may have been slightly lower and average temperatures somewhat higher than either before or after that period.

Many books have been written about the causes of the decline. The theories tend to reflect the fashion of the day. There certainly was not one single cause. The decline was a long, protracted sequence of mutually connected causes and consequences. One determinant was the unbalanced financial situation. *Money* was important in sustaining the empire and conquest of new territory provided it: 'In 167 BC the Romans seized the treasury of the King of Macedonia, a feat that allowed them to eliminate taxation of themselves. After the Kingdom of Pergamon was annexed in 130 BC, the state budget doubled, from 100 million to 200 million sesterces. Pompey raised it further to 340 million sesterces after the conquest of Syria in 63 BC. Julius Caesar's conquest of Gaul acquired so much gold that this metal dropped 36% in value... By the last two centuries BC, Rome's victories may have become nearly free of cost, in an economic sense, as conquered nations footed the bill for further expansion' (Tainter 1988). After Augustus became emperor in 27 BCE, there were no longer any rich territories to be conquered in the periphery of the Empire. This amputated the income of the State, a problem well known to modern governments. Most of the emperors between Augustus and Diocletian complained of fiscal shortages and/or had to resort to new kinds of taxes. Faced with crises or ambitious to leave their mark on history, they resorted to the debasement of the currency. Inflation must have demanded a heavy toll.

While revenues declined, ever-larger sums of money went towards the maintenance of the administration and the army, bread and games (*panem et circenses*) for hundreds of thousands of Roman citizens, and the maintenance of the physical infrastructure (roads, wharves, etc.). Written documents and archaeological evidence give the impression that the structure of taxation was to blame, as well as the increasing independence of large *latifundia* and the loss of administrative control over the countryside. There was no money left for crisis management, and the low-margin and highly decentralised economy was not able to generate additional value and provide a flexible source of income to the core.

A second cause of the disintegration was *organisational*. The Empire suffered from insufficient capacity to share and negotiate information between the different members of society. The Romans managed to enhance integration and reduce risks by setting up a road network, control local elite groups and organise a bureaucracy and a monetary system and regular flows of goods to all parts of the Empire. It was directed towards facilitating the flow of goods, energy and information throughout the system, notably from the periphery to the centre and vice versa. In the process, the centre became more and more dependent on interaction with the periphery. However, people in the periphery learned Roman ideals, techniques, goods, attitudes and concepts, and Rome lost its advantage over the periphery in terms of information processing. The political and cultural entities in the periphery came to deal with Rome on a more equal basis and became more difficult to control. Fragmentation was inevitable.

A third and related cause of decline stems from the *accumulation of long-term risks* as a consequence of frequent and intense intervention of people in their environment. Many of these risks resulted from short-term solutions to problems. For instance, deforestation provided new agricultural land in order to produce more food, but in the long term, food production suffered from erosion, which was then solved by increasing food imports. The long-term risks were less easily understood and perceived and were also considered less relevant because they were far away in time and/or space. These changes in 'risk spectrum' differed from region to region, but their temporal coincidence caused regional and local stagnation and ultimately blocked the expansion of the Empire as a whole. The exact domain in which the stagnation – climate, soil exhaustion, social or economic difficulties or invasions – first manifested itself is less relevant than the inability to cope with all the issues simultaneously and effectively. In retrospect, one may wonder why innovation did not emerge to solve these constraints. It is tempting to speculate what would have happened if the Empire had, for example, found new sources of energy and adopted steam engines...

Viewed as a self-organising system, the Roman Empire showed a capacity for conceptualisation and information processing that was in combination with its natural diversity at the essence of its resilience. However, it had to expand for its survival and increasingly suffered from the inherent limitations in correctly handling complex physical and informational networks and observing and interpreting environmental change. The resulting societal crises triggered local overexploitation in the already highly human disturbance–dependent Mediterranean basin. The lesson is that the resilience of social structures is determined by their capacity to innovate in response to the changes in circumstances that they themselves generate. The slow degradation of the Pax Romana of the 1st and 2nd century CE seems highly relevant for the interpretation of the rise and decline of later Empires and of the past Pax Brittanica and the vanishing Pax Americana.

3.4.2 Other Empires: China, India and Russia[13]

The world has seen several other empires. In China, one speaks of an empire since the unification under the king of the State of Ch'in in 221 BCE. The caliphates and dynasties in the Middle East and the southern Mediteranean between the 7th and 15th centuries, and the later Mughal Empire in India (16th–19th centuries) and Ottoman Empire in Turkey (14th–20th centuries) were all rooted in Islamic culture. Although Spain and England had less autocratic imperial rule than these empires, historians do speak of the Spanish Empire (15th–20th centuries) and British Empire (16th–20th centuries). I briefly discuss three empires/periods where the system had come so close to its *carrying capacity* that it could no longer sustain the number and welfare of its people: China, India and Russia.

Chinese history has been influenced greatly by peoples coming in from the large Eurasian steppes. The basin of the Wei River, the largest tributary to the Huang (Yellow) River, in particular has been a corridor in and from which the earliest recorded states have come into being. It had a natural outlet into the large Huang-Huai-Hai plain, covering some 350,000 km^2 and largely built up from 400–500-metres (m)-thick sediment layers from the Huang River (Zuo Dakang 1990). The area has hot summers and cold winters, and the distribution of settlements up to 2000 BP, mostly in the hilly areas between the mountains and the plain, suggest the inhospitable nature of the plain as a human habitat. Yet, it was here in these northern valleys and plains and much earlier than in southern China that state development started.

[13] The accounts of China, India and Russia are based on chapters in de Vries and Goudsblom (2002).

China has a diverse and peculiar geography. From the southwest to the northwest it is surrounded by high mountains, in the north by the cold Mongolian grasslands and mountain ranges, and in the northeast, east and southeast by seas. The natural environment has always pushed its imprints on humans: 'Frequent major and minor river shifts have affected the geomorphology of the plain; it is recorded that during the past 3000 to 4000 years, the Huang river has broken its banks 1593 times' (Zuo Dakang (1990) 473). People practised a mix of hunting and slash-and-burn farming. The king sent out farmer-soldiers to colonise and hunting was a military activity. The word for hunting and agriculture was the same, *t'ien*. Environmental and climatic change may be one reason that *war* was a most prominent feature of life in these parts from the early Xia and Shang period (4200–3100 BP) onwards (Elvin 1993). There was a close connection between sacrifice, war and agriculture with elaborate rituals related to droughts. The king had unique divine powers and divined personally about war and agriculture. The legitimacy of his power depended on his success in manipulating the weather and war spirits.

Only after the unification of the northeastern part of present-day China under the Ch'in Dynasty (221–206 BCE), soon succeeded by the Han Dynasty (206 BCE–220 CE), one can speak of an empire. Famines remained a recurrent phenomenon because of a variety of factors, many of them linked to the endemic warfare, such as field abandonment resulting from corruption and violence and destruction of the farming population and harvests in military campaigns.[14] At the same time, the introduction of iron farming implements and animal husbandry made agriculture more productive and increased the food supply. These higher yields led to higher population growth. During the Ch'in-Han period, the larger part of the estimated 60 million inhabitants of China lived in northern China and most cultivable land was used. Millions of people were forced or sent out to cultivate grasslands and forests. The surging construction of military and civil works – such as the Great Wall – required massive amounts of timber and brick and led to large-scale deforestation and subsequent desertification and increased flooding. The large differences in access to food supplies were another source of tension.

According to Elvin (1993), one of the early scholars on the environmental history of China, the social structure of power was the most important single factor controlling what happened to the natural environment. A most intriguing aspect of the ancient Chinese civilisations was the coexistence of a political philosophy that put conservation of a well-ordered nature at its centre, with a political reality that led to warfare in the ruthless pursuit of personal wealth.[15] Disturbance of the natural order was thought to bring catastrophe and seen as a sign of moral decline – yet, 'in order to mobilize resources and build up populations to man the armies that were engaged in warfare that was rapidly shifting from a ritualized combat to merciless

[14] After the decline of the Roman Empire, the gradual fragmentation of power, administration and land ownership became the start of a similar period in European history, with centuries of almost endemic feudal warring. Yet, there are also some remarkable differences, for instance, in religious practise and technological development.

[15] Of course, the tension between a society's ideology, on the one hand, and the reality of greed and power, on the other hand, are found throughout all civilisations. There is ample historical evidence, for instance, that a similar mismatch between the stated (Christian and chevaleresque) ideals and everyday feudal practises existed in medieval Europe.

destruction . . . statesmen turned to economic development . . . [which became] both the consequence and the cause of intensified warfare' (Elvin (1993) 17–18). In practise, this meant limited access for commoners to non-agricultural land, violation of nature conservation regulations and the creation of an advantageous state monopoly. People were forced to settle down and become a source of taxes and soldiers. Taxes on mountains, marshes, dams and reservoirs were under the control and to the benefit of the Imperial Court. In the official response to the famines and disasters, one recognises the 'powerless wisdom': the rulers were morally obliged to put the people's welfare first and many administrative rules and techniques were introduced, but often only revolt or the fear for it would bring practical relief.

In later times, the Chinese population has known more prosperous periods, but reconstructions of the size of the human population indicate large fluctuations resulting from recurrent clashes between food needs and food supplies and, relatedly, sequences of war and peace, order and chaos (Turner *et al.* 1990). Judgment on the causes of these events is partly a matter of focus and interpretation. Whereas Elvin (1993) summarises Chinese reality as 'fine sentiments, dosed action', Yates (1990) finds it hard to imagine that the Chinese population could have grown to the present size without the technical and moral foundations for a system of state involvement in providing for peoples' basic needs.

Several other historical accounts reinforce the impression that large populations of humans were living close to the environment's carrying capacity and went through periods of severe food shortages and associated violence and hardship (Revi *et al.* 2002). The fertile lands of the *Indian subcontinent* led to rather large population densities long before European colonisation started, with probably more than 100 million people around 1500 CE. Until the 17th to 18th centuries, the population of 100–150 million people lived at a rather low-level equilibrium between food needs and food supplies. Pressure on the land was high because of a variety of sociocultural factors. Variation in rainfall – sometimes as part of larger climatic/monsoon changes – occasionally triggered a cycle of bad harvests, famine and disease. With the decline of the Mughal Empire and the advent of the East India Company and British colonialism, these cycles appear to have been negatively influenced by high land taxes due to the pressures for profit in far away Britain, to lack of reinvestment in land maintenance and to frequent wars, to mention the most important ones. The incidence of famine increased, disease epidemics often followed. Only by the late 19th and early 20th centuries did better nutrition and hygiene and more effective relief measures lead to a decline in mortality. Indian history illustrates that the more recent environmental history can only be understood in a larger geo-political context.

When Czar Peter the Great called himself Emperor of all of Russia in 1721, it was after three centuries of rapid territorial expansion (Revi *et al.* 2002). Although the territory was vast, its population was small: some 15 million people around 1700 CE or an average density of only three to four people/km^2. The peasantry in the upper Volga region with poor forest soils heavily relied on subsistence farming on small plots of land. The severe climatic conditions of Russia required more fodder to be stored than the farmers could provide, which made cattle herding a risky and unprofitable venture. There was no potential for expansion and production for the market. Getting children was also risky, so population growth was modest.

The situation changed when the vast uncultivated territories of the southern steppes became accessible for cattle breeding to Russian peasants. Food shortages were solved by cultivating new land and not by increasing productivity. Population growth went up in line with the expansion of cultivated area. As a result, the population of European Russia had increased by the second half of the 19th century to 49.6 million in 1885 and 67.3 million in 1900, and its birth rate approached its biological maximum. In many regions of European Russia, there was rural overpopulation and scarcity of cultivable land reached dramatic proportions. In combination with large cereal exports to Western Europe, there were serious food shortages. Millions of Russian peasants were forced to leave their villages out of poverty. Unable to improve farming practises and failing the resources to migrate to new regions, they started searching for seasonal work in towns and other regions. By the end of the 1890s, their number had reached 9 million. It is an indicator of rural overpopulation, which was estimated at 23 million in 1901 by a special commission under the government of the Russian Empire. It is hardly surprising that 670 peasant riots were observed in the European part of Russia in the years 1902–1904.

One way to alleviate overpopulation was to resettle Russian peasants from the European part of the country in the southern areas of Siberia and Central Asia. By the end of the 19th century, millions of Russians were already living in these vast and barely populated territories. Many more millions followed, but outmigration remained below 5 percent of the population and did not solve the much larger overpopulation. Thus, one can discern the social-ecological contours of a period, which, in the collective memory, is primarily associated with the Russian Revolution.

3.5 Mechanisms, Theories and Models

3.5.1 Mechanisms

The sociologist Elias distinguishes in his book *The Civilizing Process* (1969/2000), three dangers threatening human groups:

- *the extra-human world*: droughts and floods, wild beasts and pests, earthquakes and volcanic eruptions;
- *inter-human relationships*: hostile neighbours, invading warriors; and
- *intra-human nature*: mismanagement due to negligence, ignorance, lack of self-restraint or discipline.

Initially, the first of these dangers was dominant and the basic options were to adapt or migrate. With the rise in social complexity, the second and third danger became ever more important. Development was a process of continuous attempts at control, in the form of conquest, unification, exploitation and adaptation in the physical as well as the mental realms of life (Figure 3.5). This 'domestication of the wild' was part of the transition to a settled life. It created new dependencies and subsequent needs for further control. For instance, once an irrigation canal has been constructed, it has to be maintained and protected and the food system becomes dependent on experts and on soldiers. It also necessitated more and new forms of cooperation, which often proved to be more advantageous than war and competition between smaller units (Wright 2000). This in turn created social arrangements and hierarchical institutions.

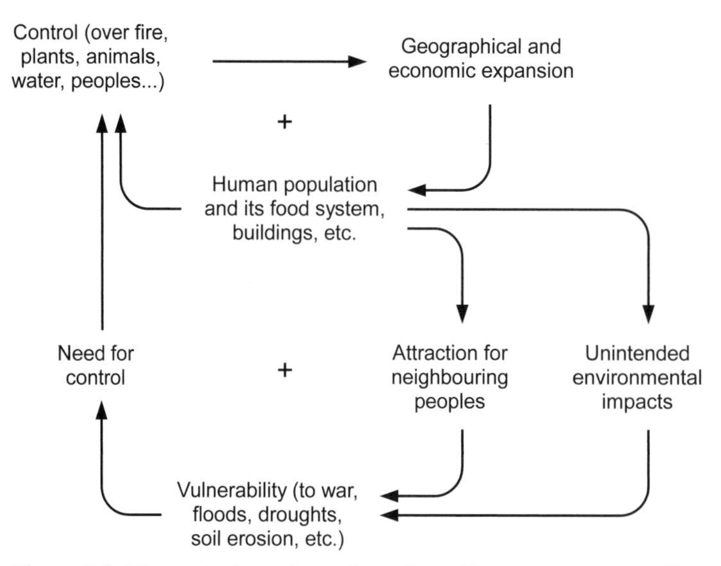

Figure 3.5. The control and dependency loop. Responses to natural hazards have led to control mechanisms, which in turn brought new dependencies and risks.

Environmental historians understandably placed more emphasis on the role of the natural environment as a necessary resource for human life than social scientists tend to do. The latter's neglect is vocally expressed by Reader in his book *Man on Earth*: 'As city states have expanded into nations, kingdoms and empires, civilization has repeatedly become obsessed with the acquisition and wielding of power; an activity that has left millions of people dead in the streets and dominated the pages of history so much that the significance of the substance bases upon which civilizations are founded have all but ignored' (Reader 1988). The environmental histories in previous sections highlight a few points about the human-environment interaction:

- human populations tended to grow towards the maximum number ('carrying capacity') that could be sustained by the regional resource base;
- this occurred at a rate different from the degradation rate of the environment, and could, therefore, usually not be stopped at the moment people became aware of the consequences;
- when the resource base got overexploited as a result, the struggle for the stagnant or declining (food) surplus intensified;
- the resulting tensions could turn outward, in the form of outmigration and conquest, or inward, in the form of oppression and sacrificial rituals or technical and social innovations and changing environmental practises; and
- the capability, or lack of it, of the elite to understand, anticipate and act for the good of the larger system became a more important or even dominant factor in whether a civilisation could sustain itself.

Most of the time, a mix of all these sometimes postponed, sometimes reverted, and sometimes accelerated the processes of decline and collapse. In system dynamics language, developments were driven by positive and negative feedback loops. The positive loop was manifested in growing size and density of human populations and in productivity increases as a function of accumulated experience and its manifestation in technical artefacts. A reverse, negative loop set in when certain thresholds in the use of environmental resources were exceeded and the resource base got degraded as a consequence. The social and cultural dynamics became a more integral part with the rise of economic and social complexity.

For instance, the ornithologist Diamond (2006) lists human-induced environmental change, climatic change, external enemies, external friends causing dependence, and endogenous cultural dynamics as the main determinants of decline. The last one brought him to the sobering reflection that '... we have to wonder why the kings and nobles failed to recognise and solve the[se] seemingly obvious problems undermining their society. Their attention was evidently focused on short-term concerns of enriching themselves, waging wars, erecting monuments, competing with each other, and extracting enough food from the peasants to support all those activities' (Diamond (2006) 177).

Some archaeologists describe past processes of social organisation in system dynamics terms (Flannery 1972). Increasing differentiation and specialisation of subsystems (segregation) and increasing linkages between the subsystems and the highest-order controls (centralisation) are seen as driving forces of the rise of special-purpose institutions like the military, and the over-ruling of lower-order controls by higher-order controls as in centralised irrigation schemes. Such systemic phenomena could develop into socio-cultural 'pathologies', causing stress that might erode a system's resilience and accelerate disintegration. Numerous histories illustrate how such pathologies, in interaction with natural environmental dynamics, caused stagnation and decline. Other archaeologists have emphasised social interactions such as warfare, competitive emulation, cultural absorption, innovation dispersal and the exchange of goods and information as part of the evolution of social-ecological systems (Renfrew and Cherry 1985). For instance, warfare – itself often induced by resource scarcity – was a perennial phenomenon in the early civilisations of the Middle East and Greece with sometimes devastating impacts. Competitive emulation, that is, imitation amongst competing elites, may have been a factor in the decline of some cultures, with the statues on Easter Island but also the famous medieval towers of San Gimignano and present-day skyscrapers in Middle Eastern oil states as examples.

The evolutionary unfolding of social complexity that came with agrarianisation manifested itself in various ways:

- social stratification, as a way to combine technical and organisational skills with social coherence;
- specialisation in certain skills, in order to permit and sustain higher and more diverse production; and
- trading as one of the strategies to reduce risks of famine and war.

An urban-rural divide started to develop, with large variations in population density. A large and stable food surplus, and an adequate means and infrastructure for distributing it, were a precondition for the rise of urban settlements. It required capital goods such as tools, terraces, dams and canals, granaries, roads and transport equipment that were susceptible to deterioration or destruction. The rise of social complexity was thus largely a sequence of cause-and-effect chains in order to control the natural and human environment. With it came new risks and vulnerabilities. The new techniques and practises introduced unknown long-term environmental impacts and started or accelerated detrimental feedback loops. Often, they became visible only by the time the system came under internal or external stress. Learning to cope with long-term and more erratic events is difficult. The human perception and interpretation of what is going on is always simplified and incomplete. As long as change is within the variations experienced before and stored in the collective memory, such simplified understanding may still be adequate. For change outside these variations – many environmental processes operate on centennial or millennial scales – the capacity for adequate response is less. This limited response effectiveness had important consequences for how people in ancient civilisations dealt with environmental change – as it will have in the 21st century.

3.5.2 Theories and Models

Many theories have been proposed to explain the decline and collapse of states and empires. Most spectacular are the Big Catastrophes, such as earthquakes, volcanic eruptions or climatic change. They dismiss the possibility that '. . . the world ends, Not with a bang but a whimper', in the words of T.S. Eliot. Another set of explanations point at population pressure per se as the cause of famine and collapse. The number of people grows exponentially and food availability does not, so people inevitably run against food limits, as the English scholar Malthus (1798) proclaimed. Such (neo-)Malthusian theories invoke resource overexploitation and mismanagement as co-determinants. The Danish economist Boserup introduced in her book *Population and Technological Change* (1966) a feedback that invalidates Malthus' theory: Increasing population pressure forces people to produce more food by putting in more labour, usually at lower labour productivity. The transition to agriculture was thus a response to scarcity and its outcome in the form of improved technical and organisational skills was 'progress born out of necessity' (Wilkinson 1973).[16]

A third set of theories focuses on internal political processes: overstretch in material and cultural terms, erosion or lack of spiritual values. For instance, the British historian Toynbee states in his book *Mankind and Mother Earth*: 'Man is a psychosomatic being, acting within a world that is material and finite . . . But Man's other home, the spiritual world, is also an integral part of total reality; it differs from the biosphere in being both immaterial and infinite; and, in his life in the spiritual world, Man finds that his mission is to seek, not for a material mastery over his non-human environment, but for a spiritual mastery over himself' (Toynbee 1976. It seems prudent to be aware of the degree to which theories big and small are a reflection of dominant scientific fashions, schools, disciplines and individuals.

[16] This is also the view, though from a different angle, of the French agronomist Gourou (1947): '. . . des populations denses sont généralement douées d'une civilisation supérieure: elles ont en effet su résoudre les problèmes économiques, techniques, sociaux et politiques posés par les fortes densités'.

Tainter concluded in his book *The Collapse of Complex Societies* (1988) from the available literature that explanations of the collapse of civilisations can be categorised into three groups:[17]

- resource- and environment-related changes, fully exogenous or partly endogenous, in the sense of human-induced;
- interaction-related changes in the form of conquest or other, less dramatic forms of invasion; and
- internal changes in socio-political, cultural, and religious organisation and worldview, diminishing the adequacy of response to external events.

In his view, the key concepts are energy and complexity (Tainter 1988, 2000). First, socio-political systems require physical energy sources to maintain the investment in infrastructure that keeps the members of the society acting together. Second, the increase of social complexity is a strategy to solve societal problems such as diminishing productivity. Initially, such a response has a high return but in due course, the returns may decline to zero or even become negative. The first cooperative efforts to store food or share irrigation channels were very effective, whereas adding another organisational unit to an already centralised administration may have only costs and no benefits. If social and political complexity increases in an attempt to solve problems and spread risks in a variable and uncertain environment, it may at some point start to be counterproductive and merely serve to reproduce the institutions necessary to deal with the complexity. This is a vicious cycle that leads to an ineffective bureaucracy and, ultimately, collapse.

In this interpretation, the infrastructure costs of integrating Spain, Gaul and other territories into the Roman Empire were relatively low because it was deliberately based on incorporating existing infrastructures of local origin. But around the 3rd century CE, the imperial organisation had reached a point of declining marginal returns. Further expansion and the tapping of new sources of energy became unprofitable. Once expansion had halted, the Empire lacked sufficient reserves of energy to maintain the existing infrastructure when it came under pressure. Disintegration became inevitable, either as a planned devolution or as a more or less gradual breakdown. The strategy of Diocletian and his successors in the 4th century to cut the Empire into two, and later four, parts can in retrospect be considered as one of the failed attempts at avoiding collapse by simplification and devolution. The Eastern part, the later Byzantine Empire, was an example of a successful devolution, based as it was on a planned simplification of society.

This theory raises a number of questions. For instance, there is the problem that (social) complexity is not easily measured and that the provenance and role of innovation is only implicitly addressed. The ecologist-historian Turchin has proposed in his book *Historical Dynamics* (2003) an ethno-kinetic theory of rise and fall of agrarian states that is based on population ecology and social dynamics. His source of inspiration is the 14th-century Arab sociologist *avant la lettre* Ibn Khaldun, who developed a theory of political cycles from studying the history of the contemporary Arab world. He observed that human individuals cannot live outside a group and

[17] This classification matches the categories of extra-, inter- and intra-human change introduced by Elias.

that cooperation and a communal way of life are necessary ingredients of human life. Different groups have different abilities for concerted action, which led Ibn Khaldun to introduce the concept of *asabiya*. It can be translated as 'group feeling' or 'collective solidarity'.[18] In his view, life in the desert tends to increase *asabiya*, whereas life in the civilised urban centres, with their luxuries and intra-elite conflicts, leads to its diminution. Thus, a human group initially has a large collective solidarity or *asabiya* and lives austerely in great ethnic unity. It can manifest itself in an outward expansion and conquest of the urban centres.

Using data and theories from a variety of social sciences, Turchin offers a formal framework for state growth and decline. *Asabiya* is a state variable, which tends to grow towards some fixed value, but at a rate dependent on the territorial size. Its growth rate is high when the territory is still small and there is a large sense of solidarity, for instance, in the form of ethnicity. Territorial area, the other state variable, tends to grow also logistically towards some fixed value. In the model, interaction between individuals is a key element and the centre-periphery gradient in wealth and skills and solidarity increases until the downfall sets in. Such a mathematical model of a complex phenomenon should not be seen as prediction in any sense; its purpose is to yield insights into the relative importance of phenomena. But it incorporates the notion of cultural identity as a driver of change, which is against most expectations back on the scene in early 21st-century world history.

Box 3.5. *Colonists, Inuit and the church.*[19] Danish colonisation of Greenland started in 1721, but a Norse colony in west Greenland had already been an outpost of Europeans throughout the period 985–1500 AD. There had been contacts between North American hunters and European farmers. It seems that the Norse community managed quite well for the first 150 years, with a maritime-terrestrial economy with small animal herds, seasonal hunting of walrus, polar bears and seals and occasional trading with Iceland. However, they were living on a knife edge, with great skills being required to survive the long cold winters. There is ample evidence that the Little Ice Age fluctuations in temperature, sea ice conditions and faunal resources put the communities under stress from 1270 AD onwards. Moreover, the southward migration of the indigenous Inuit peoples – in response to climatic change – and the decline in trading relations with Europe further complicated their subsistence strategies. The western settlements seem to have collapsed rather suddenly around 1350 AD.

In this story, it has been argued, one should not treat the human response to climatic stress as a minor and dependent variable. The Inuit peoples survived these harsh times. The Norse farmers had several options to adapt, for instance, orienting themselves more towards the oceanic resources and de-emphasising pasture and cattle-rearing. They also would have benefited from Inuit practises and technology related to boats, fishing gear and clothing. 'Rather than exploring the possibilities of new technology and searching out alternative resources, Norse

[18] Asabiya produces 'the ability to defend oneself, to offer opposition, to protect oneself, and to press one's claims', as stated by Ibn Khaldun and quoted by Turchin (2003).

[19] This paragraph is largely based on McGovern (1981) in de Vries and Goudsblom (2002). For another account, see Diamond's Collapse (2006).

society in Greenland seems to have resolutely stuck to its established pattern, elaborating its churches rather than its hunting skills' (McGovern (1981) 425). Whence this conservatism and loss of adaptive resilience in the face of rising economic costs and declining returns?

Under the influence of Iceland, the mediaeval church in the Norse communities in Greenland had become more powerful spiritually as well as materially in the 12th century. Between 1125 AD and 1300 AD, spectacular church construction had taken place, small communities building amongst the largest stone structures in the Atlantic Islands. Economic, political, religious and ideological authority appears to have come into the hands of a lay and clerical elite. If a society such as this is confronted with increasing fluctuations in resource abundance and use, it has to invest in additional data collection and improve its interpretation of these data in order to survive. This constitutes an overhead cost, which is often resisted by the population and has to be enforced by military force or by ideology.

The elite-sponsored expensive elaboration of ceremonial architecture and ritual paraphernalia in Norse Greenland may indicate the successful ideological conditioning of the population. Administrators may not only have declined Inuit superior technology but also have sustained erroneous beliefs – for example, that lighting candles had more impact on the spring seal hunt than more and better boats. As stresses mounted, elite groups may have pushed up the necessary overheads in their obsession with conformity and the suppression of dissenters and detached themselves farther and farther from the phenomenal world, adding the pathology of hypercoherence to the pathology of auto-mystification.

3.6 Summary Points

What do these lessons from history tell us about the present and future? Do people inevitably run against food limits, as Malthus believed, or are there always timely feedbacks that prevent collapse, as Boserup argued? It is probably a complex mix. The natural environment has always played an important if not decisive role in the growth and decline of human societies. Yet, environmental determinism is not permitted for all but the simplest human societies. Several civilisations during the agricultural regime were able to sustain themselves more or less continuously for periods up to 1,000 years or 35 generations. In view of the recent duration of the industrial era – say, 300 years – this is impressive. The relatively stable climate during the Holocene and the cyclical agro-ecological patterns and social-religious beliefs that dominated human societies are part of the explanation. Many people in those civilisations will have experienced their societies, if not their individual lives, as sustainable. Such a stability is hard to imagine in the early 21st century, and perhaps we should search the past not only for the causes of unsustainability but also of sustainability. Other points to remember from this chapter:

- development was primarily a dependency-control spiral of outside and inside opportunities and threats. The mechanisms behind growth and decline transitions are a positive feedback loop of expansion, reinforced by population density, followed by negative feedbacks such as overwhelming complexity and environmental constraints;

- the decline/collapse of several past civilisations can be assigned to overexploitation of the resource base in an environmentally fragile environment with (large) natural variability;
- with agrarianisation, the interaction between humans and the natural environment became more intense and socio-cultural processes gained in importance;
- with development, the spectrum of environmental risks widened from nearby and short-term to indirect and long-term – another element in the dependency-control spiral;
- environmental deterioration eroded the economic-financial basis of the ruling elites, creating social tensions and other disturbances such as sectarian violence, invasion by hostile neighbours and dependence on other states;
- short-term thinking of societal elites aggravated the impending crisis in the form of greed and hoarding, ineffective rituals in the form of sacrifices or monuments and resource-demanding oppression or warfare.

The history of past civilisations teaches us some lessons that are the more important now that humanity has no more escape routes into large and unsettled areas and some of the environmental changes become global and irreversible.

> War has always been part of life. The quest for living space and for power and exploitation has taken many forms. This is how Bertold Brecht remembered his brother who left for the Spanish Civil War:
>
> Mein Bruder war ein Flieger,
> Eines Tages bekam er eine Kart,
> Er hat seine Kiste eingepackt
> Und südwärts ging die Fahrt.
> Mein Bruder ist ein Eroberer,
> Unserm Volke fehlt's an Raum,
> Und Grund und Boden zu kriegen, ist
> Bei uns alter Traum.
> Der Raum, den mein Bruder eroberte
> Liegt im Guadarramamassiv,
> Er ist lang einen Meter achtzig,
> Und einen Meter fünfzig tief.
>
> – Bertold Brecht, 1937

> Complex societies, it must be emphasized again, are recent in human history. Collapse then is not a fall to some primordial chaos, but a return to the normal human condition.
>
> – Tainter, The Collapse of Complex Societies, 1988

SUGGESTED READING

Overview of the IHOPE Dahlem conference with accounts of past civilisation developments.

Costanza, R., L. Graumlich, W. Steffen, C. Crumley, J. Dearing, K. Hibbard, R. Leemans, C. Redman, and D. Schimel. Sustainability or collapse: What can we learn from integrating the history of humans and the rest of nature? *Ambio* 36 (2007) 522–527.

Models, maps and myths about the role of the natural environment in ancient civilisations.
de Vries, B., and J. Goudsblom, eds. *Mappae Mundi – Humans and Their Habitats in a Long-Term Socio-ecological Perspective. Myths, Maps, and Models.* Amsterdam: Amsterdam University Press, 2002.

Review of scientific data and theories.
Dearing, J. Climate-human-environment interactions: Resolving our past. *Clim. Past.* 2 (2006) 187–203.

In-depth account of a number of collapse events in human history (Easter Island, Maya, Iceland a.o.).
Diamond, J. *Collapse – How Societies Choose to Fail or Succeed.* New York: Penguin Books, 2006.

An accessible, archaeological investigation into the role of the natural environment in ancient societies.
Redman, C. L. *Human Impact on Ancient Environments.* Tucson: University of Arizona Press, 1999.

USEFUL WEBSITES

- www.pages-igbp.org, about Past Global Change research.
- www.igbp.net is the site of the International Geosphere-Biosphere Program (IGBP) research projects.
- www.ihdp.unu is of the International Human Dimensions Program (IHDP) research projects.
- www.ncdc.noaa.gov/paleo/ is a website with lots of information and links to palaeo-reseach.

A simple STELLA® model, with coconut palm trees and statues as state variables, gives insights into the dynamics and the sensitivity of the outcome for the assumptions. To run the Easter Island model in Stella, visit www.iseesystems.com/community/downloads/EducationDownloads.aspx.

4 The World in the Past 300 Years: The Great Acceleration

4.1 Introduction

Although past civilisations are a source of imagination and insight, present-day concern about (un)sustainability is anchored largely in the exponential growth of population and economic activity in the last few centuries. These growth processes are part of what is known as the *Industrial Revolution*. Industrialisation occurred in many places throughout Western Europe at roughly the same time, although with locally specific features. It is rooted in the commercial and trade capitalism of medieval Europe (Braudel 1979). A series of events and trends since 1700 mutually interacted and boosted manufacturing and trading of goods in a successful mixture of science, technology and capitalism. It reinforced the process of European colonial expansion. The European 'offshoots' in America and Australia underwent similar transformations as Europe. A collectivist form of industrialisation took place in Russia after the revolution. The larger part of the human population, however, still lived a traditional agricultural life at modest levels of population growth and economic output until the middle of the 20th century. Only after 1950, they started to experience similar processes of change.

This chapter explores in some detail the important changes that came with the Industrial Revolution and modernity. Perhaps Churchill was right when he said, 'The further backward you look, the further forward you can see'. In any event, some knowledge and understanding of what happened in the last 300 years is essential for an interpretation of our present situation and an exploration of what a sustainable future might look like. For that reason, I briefly survey demographic and economic trends, aspects of governance and globalisation, and socio-cultural trends. In the last section, two generic concepts are introduced as descriptive tools that can put the changes in a different perspective. The topic is vast, so I refer the reader to the suggested literature for further reading.

4.2 The World in the Last Three Centuries

4.2.1 Accelerating Growth: Population and Economic Activity

The first major change to be mentioned is the number of human individuals. Between the year 1000 and 1820, the *world population* increased from an estimated 268 to

Table 4.1. *Key economic indicators 1913–1998*[a]

	Income (GDP 1990 I$/cap)	Avg life expectancy at birth (yr)	Working hours (hr/yr)	Labour productivity (GDP 1990 I$/hr)	Govt expenditures (% of GDP)
Western Europe	3,473	46[b]	2,553–2,624	3.12[e]	12
1913	4,594	67	1,926–2,316	5.54	29.8
1950	17,921	78	1,428–1,664	28.53	45.9
1998					
USA 1913	5,301	47[b]	2,605	5.12	8
1950	9,561	68	1,867	12.65	21.4
1998	27,331	77	1,610	34.55	30.1
Japan 1913	1,180	44[b]	2,588	1.08	14.2
1950	1,926	61	2,166	208	19.8
1998	20,084	81	1,758	22.54	38.1
Russia 1913	1,488	32			
1950	2,834	65			
1998	3,893	67			
China 1913	552	24			
1950	439	41			
1998	3,117	71			
India 1913	599	24			
1950	619	32			
1998	1,746	60			
Latin America		35			
1913	2,554[c]	51	2,042[d]	2.48[d]	
1950	5,795	69	1,841[d]	7.87[d]	
1998					
Africa 1913	–	24			
1950	852	38			
1998	1,368	52			

[a] *Source:* www.theworldeconomy.org/MaddisonTables/Maddisontable3–9.pdf
[b] 1900
[c] Forty-four Latin American countries (incl. Mexico)
[d] Brazil
[e] Weighed average for twelve Western European countries

1041 million (0.16%/yr). Since 1820 it increased almost sixfold to nearly 7000 million (0.95%/yr) (Table 4.1; Figure 4.1).[1] Perhaps most remarkable was the large emigration out of Europe towards the Americas, Oceania and Africa (Figure 4.2). Because the rapid growth happened in Europe and its offshoots, the geographical *distribution* of the population over the world changed, too. If European Russia and the European offshoots elsewhere are included, the European population increased from an estimated 15 percent of world population in the year 1000 to 23 percent by 1820. Since then, the fraction has declined to 19 percent in 2000. Asian populations made up about two-thirds of world population between 1000 and 1820. The population of China reached an estimated high 38 percent by 1820, after which it declined

[1] The numbers in this paragraph are from Maddison (1995, 2006) and www.ggdc.net/. Most numbers cited in the chapter have rather large uncertainties, especially those before 1900 and for non-European regions.

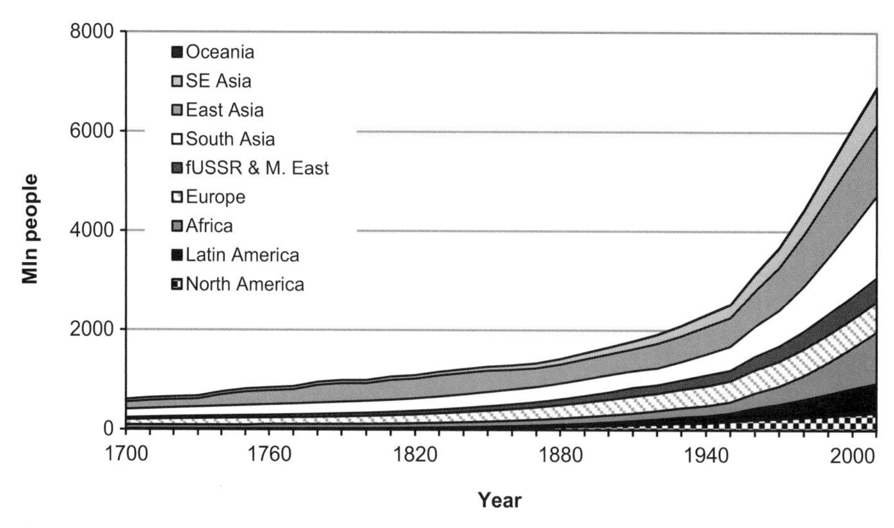

Figure 4.1. Global trends in human population (Source of data: Maddison 2006).

to about 21 percent in 2000. The population in the rest of Asia rose from about 29 percent by 1820 to 37 percent in 2000.

The different population growth rates between Europe, its offshoots and the rest of the world result from the process known as the *demographic transition* (§10.3). The first stage occurred in the 19th century when a spectacular decrease in the mortality rate happened in several European countries. It has led to a phenomenal rise in *life expectancy* at birth, from less than forty years in 1820 to more than seventy-six years by 1990. It was the consequence of ongoing improvements in food quantity and quality and of better sanitation, hygiene, medicine and medical practice. In the second stage, the birth rate started to decline but the delay with the mortality rate decline caused a period of rapid population growth. In the third stage, where most industrialised nations are now, population sizes stabilise or even decline. The transition can be mathematically described with a logistic growth curve (§2.3).

Most countries in the world are in one of these stages of the demographic transition. For instance, estimated average life expectancy was still below thirty-five years in 1900 in Spain and Russia but had risen above sixty-two years in 1950. In many other countries, including Mexico, Brazil, China, India and most African countries, life expectancies were still below fifty years in 1950. Since then, however, almost all went into the transition and as a result experienced a rapid growth in population size. In the fifty years since 1960, many countries in Asia, Latin America and Africa – Philippines, Mexico, Nigeria, to mention a few – experienced a growth of their population that is equivalent to what the United Kingdom experienced in the 150 year period between 1750 and 1900 (Figure 2.10). In other words, three times faster. These high growth rates in an increasingly connected world put an enormous strain on these countries, not only in terms of food and infrastructure but also in terms of employment and governance.

During these centuries, the process of *urbanisation* led to ever more geographical concentration. A bit dependent on how it is defined, the fraction of humans living in cities increased from 3 percent in 1700 to more than 47 percent in 2000.

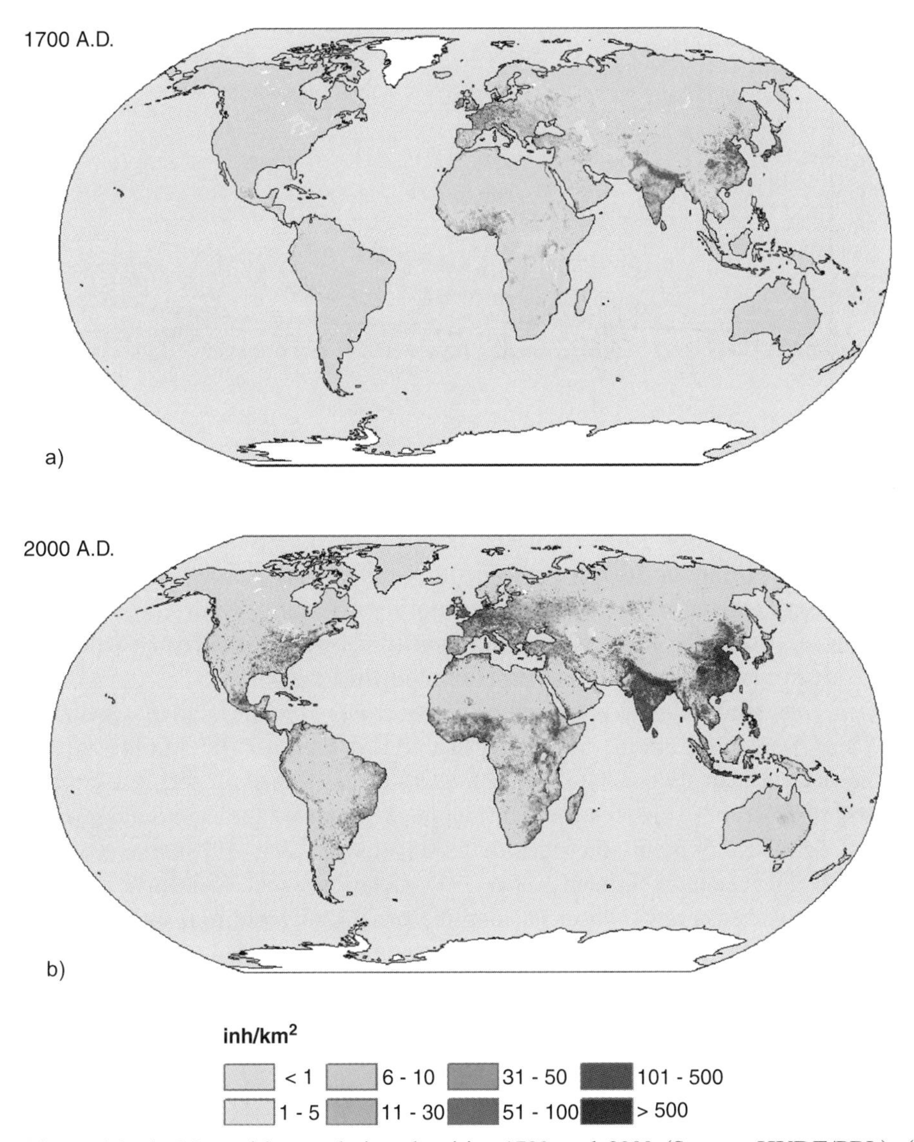

Figure 4.2.a,b. Map with population densities 1700 and 2000 (Source: HYDE/PBL). (See color plate.)

The urbanisation process is still going on, at rates higher than population growth. It leads in the low-income regions of the world to megacities of unprecedented size. Not only do they rapidly become new centres of economic activity and technology, with a burgeoning middle class – but also places where millions of people live in poverty and with inadequate public services such as water, sanitation and electricity. Congestion and air pollution are another increasing problem (Figure 2.3b). Quality of life is deteriorating, and signs of unsustainable developments, such as food riots and political conflict, are most visible in these megacities. The major attractor is employment and income opportunities, and the only stabilising feedback is a

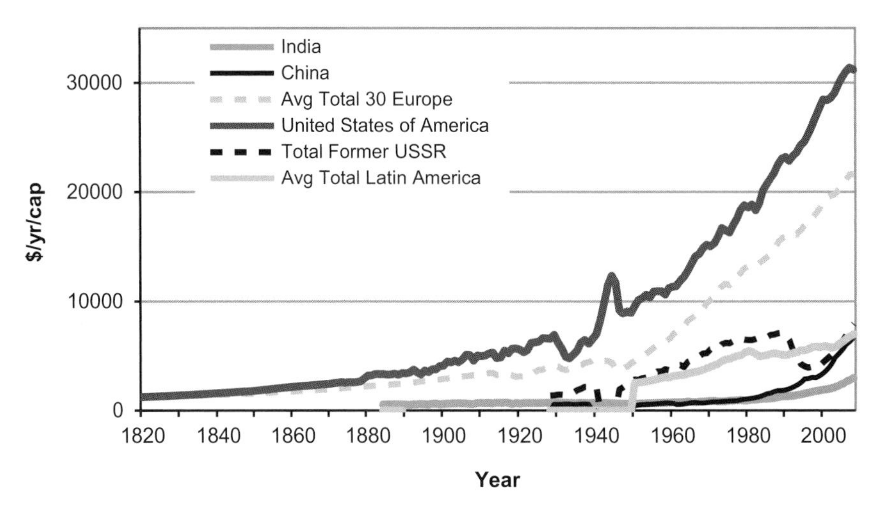

Figure 4.3. Global trends in income in 1990 (International Geary-Khamis $/yr/cap; source of data: www.ggdc.net and www.conference-board.org/economics/database.cfm).

slowdown in immigration from rural areas. Although the stories about London, Paris and New York in the 19th century picture similar dismal circumstances, there are also differences: The 21st-century metropoles have not 1–2 but 12–15 million inhabitants, and they have extensive communications with the outside world and overexploited natural resources in the surrounding areas.

The second and related large change concerns *economic activity*. Average income has risen thirteenfold since 1820 (Table 4.1; Figure 4.3).[2] Maddison distinguishes five growth 'phases': 1820–1870; 1870–1913; 1913–1950; 1950–1973; and 1973–1992. These phases represent identifiable segments of the economic growth process in the world. Notably the periods 1870–1913 and 1950–1973 were times of high growth rates. Long-term analyses reveal oscillations – the Kondratiev cycles, after the Russian economist who first identified them – which more or less coincide with the four growth phases. It appears that science-driven technology is the dominant force behind these growth cycles: They are at least partly explained by the interaction between waves of innovations and investments in capital-intensive industries (Sterman 1986; Tylecote 1992).

Because economic growth since 1820 has been very uneven across the world, *income distribution* did change drastically. Before 1820, differences in average income amongst world regions were less than a factor 2. By 1950, incomes in Western Europa and in particular its offshoots in North America and Australia had increased fivefold to tenfold since 1820. In the same period, average incomes in Asia rose with at most 40 percent and in Africa with an estimated 50 percent. After World War II, the economic take-off of countries such as Japan and Korea led to a sixfold increase in average income between 1950 and 1992. Other Asian countries follow, but African countries lag behind (Figure 4.3).

[2] GDP is defined as the total of monetary transactions in an economy. GDP per inhabitant is known as *income* and is a flow in €/yr/cap. Time-series are usually corrected for inflation. For international comparisons, the US$ can be converted to international dollars I$ in order to account for differences in purchasing power between countries (PPP; see pwt.econ.upenn.edu/php_site/pwt_index.php). GDP/cap is not an adequate measure of well-being, but one of the best if not only ways to picture long-term trends in human activity (§14.4).

Box 4.1. *Mechanisms behind income distribution.* A number of scientists have applied network theory in order to get a better understanding of the causes of income inequality. Boucheaud and Mézard (2000) apply the 'directed polymer' problem in physics to economics. They assume that the world is inhabited by agents with a few simple characteristics: The agents can exchange wealth through trading and the amount of money they earn or spent is proportional to their wealth. This can be formulated in a stochastic dynamic equation for the wealth $W_i(t)$ of agent i at time t. It can be shown that, if assuming that all agents exchange with all others at the same rate, the system ends up in a situation where a small number of agents has a large share of the wealth. This is observed in the real world and known in economic science as the Pareto power-law distribution.

The authors explore the effects of income and capital tax, suggesting that the former may decrease income inequality and the more so if redistributed. The latter can, surprisingly, cause a lower spread of wealth if combined with income tax and without redistribution. If the agents have exchanges according to some social network configuration, again a Pareto distribution is found. It is concluded that wealth tends to be very broadly distributed when exchanges are limited and that favouring exchanges and, less surprisingly, increasing taxes seem to be efficient ways to reduce inequalities.

Of course, many other factors determine income distribution. There is, for instance, a systematic positive relationship between asset holdings and the return on these holdings, which explains why household wealth distribution is much more concentrated than income distribution. The existence of some extremely wealthy households is possibly explained by the disproportionate large number of entrepreneurs in the top income class. Also, the increasing role of unique talents to be exploited on an increasingly global market for information and services is giving rise to large income and wealth differentials.

In preindustrial societies, most people lived a subsistence life in economies dominated by agriculture and commerce, and only a small part of the population belonged to the rich upper class. The large income inequalities *within* these societies declined since the early 19th century, particularly in Europe and its offshoots. But the differences in income growth have created significant disparities in income and wealth *between* countries. Between 1950 and 1992, income disparities did not change much, and world income distribution became even slightly more equal between 1970 and 2000 (Barro and Sala-i-Martin 2004).[3] Since 2000, they appear to increase again. The extremes are drifting apart: Since 1960, the income ratio between the twenty richest and the twenty poorest countries in the world has increased from five to twelve. The acclaimed benefits of globalisation for the poor are apparently undone by the opportunities for the rich to accumulate more wealth. The estimates of how many people live in absolute poverty – which is based on a poverty line established by the World Bank and presently set at $1.25 US/cap/day – indicate a decline as percentage of the world population but a stable or even increasing level in absolute numbers.

[3] If *within* region income differences are included in the analysis, world income inequality increased rapidly and continuously since 1820, with a pause between 1910 and 1929. Thereafter, it decreased (Bourguignon and Morrison 2002).

Trends in population and economic activity are key indicators of the processes of change. However, the rise of modernity was an extremely complex process with many interlinked phenomena, which are not easily disentangled from the whole and for which often no adequate quantititive indicators are available. One such phenomenon is what economists call *structural change*, that is, the gradual shift in sectoral shares in GDP and employment from agriculture to manufacturing and then to services.

The fraction of the population employed in agriculture declined in the last centuries all over Europe and so did agriculture's share in GDP (Figure 4.4). This change, with important consequences for natural resource use, is well documented for the present high-income countries in the world.[4] The surplus of labour from agriculture was absorbed by industrial activities, and the share of industry in employment and GDP rose. Key manufacturing sectors were initially mining, ironworks, textiles, ceramics, machine tools, food processing and base chemicals, and later transport (ships, trains, automobiles), electrical equipment, plastics and electronics. A next stage of development started in the 1960s, in the industrialised regions with a shift in GDP and employment to the services sector. It represents the advent of the service-, the information- or the experience-economy, depending on what is seen as the most crucial component. Whether the low-income countries will follow this trend is as yet unclear.

Another series of trends can be characterised with a single word: *globalisation*. The spreading of people and goods and the dissemination of ideas across the continents started millennia ago. The Roman Empire and later the Asian, Arab and African Empires had important trade routes. From the 17th century onwards, the process accelerated as part of European expansion and colonialism. Apart from the stagnation period of 1910–1945, world trade tonnage has grown exponentially. Seaborne international freight in tons loaded increased more than fourfold between 1880 and 1910, with a further tenfold increase between 1945 and 1980 (Chisholm 1990). As fraction of Gross World Product (GWP) exports made up about one sixth in 2000, a fourfold increase since the early 20th century.

The degree to which countries are connected in the world economy through *trade* (im- and exports) varies considerably. Table 4.2 gives some key indicators for a few countries for the year 2009 – a snapshot view because trade flows can fluctuate significantly.[5] The four richest economic countries/blocks (USA, EU-27, Japan and Korea) account for 35 percent of world merchandise and 43 percent of commercial service trade. China is rapidly gaining a larger role. For all these countries, trade flows as fraction of GDP are above 27 percent. The interconnectedness and specialisation is clearly visible in the numbers. In Saudi Arabia, Nigeria and other fossil fuel or mineral-rich countries, more than 85 percent of merchandise exports are fuel/mining

[4] Different sector classifications and names are used. For instance, agriculture, forestry and mining are also called primary sector, with manufacturing the secondary and services the tertiary sector. There are discussions about the correct of measuring sector shares. For instance, if changes in productivity are considered, the share of the service sector has risen much less (Teives, Henriques and Kander 2010).

[5] In the Netherlands, for instance, there was a surge in exports during the wave of globalisation between 1870 and 1913, with more than 55 percent of GDP exported. It dropped to less than 20 percent in the 1930s but was again above 50 percent by the end of the 20th century.

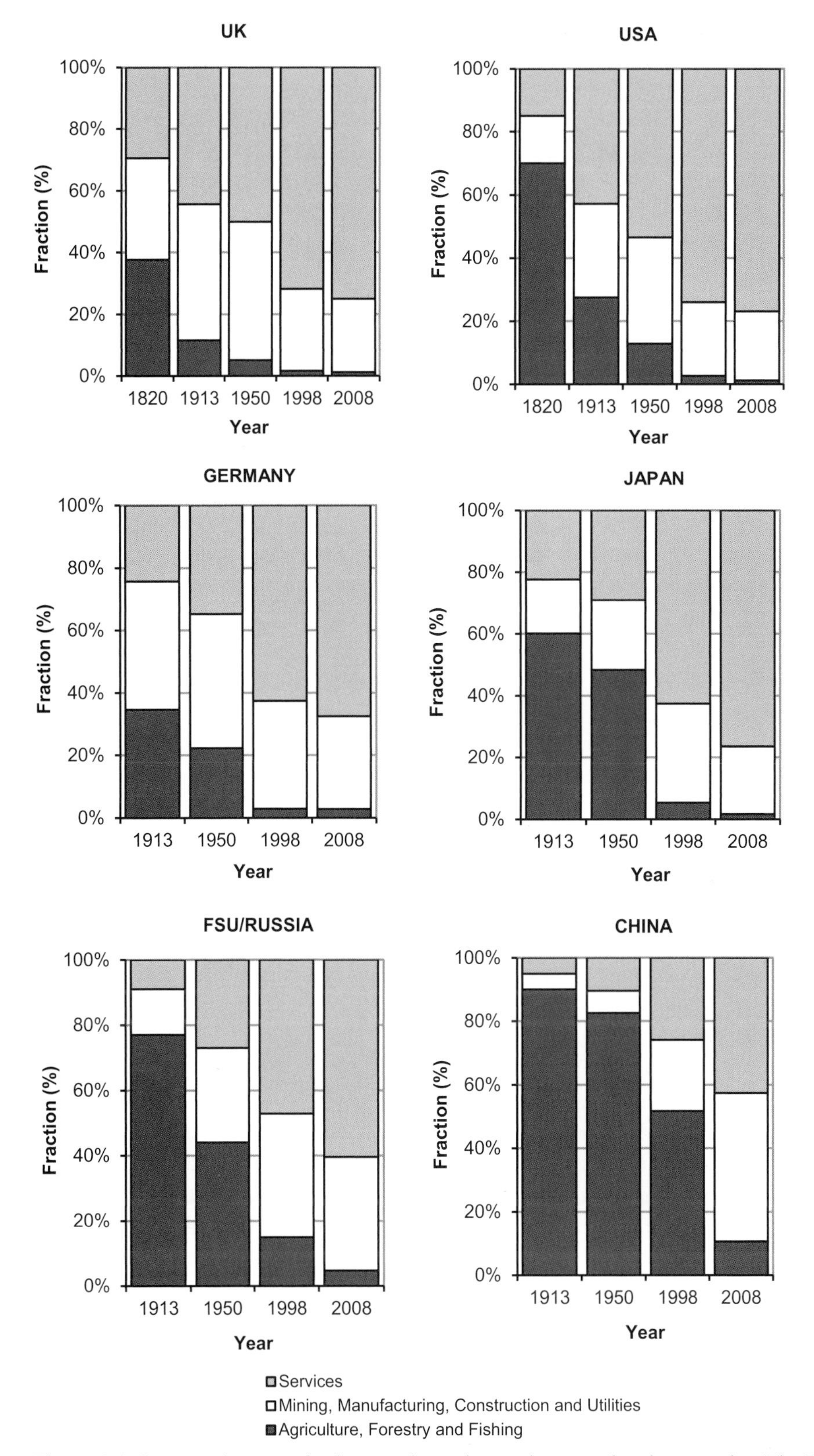

Figure 4.4. Structural economic change: shares in employment for six countries. The United Kindgom and the United States were leading; France (not shown) was almost identical to Germany (Source of data: Maddison 2006, www.ggdc.net and www.conference-board.org/economics/database.cfm).

Table 4.2. *Trade characteristics for some countries/regions in the world in descending order of income, 2009*[a]

2009	USA	Japan	EU-27	Korea, Republic	Saudi Arabia	Russian Federation
Population (mln)	307	128	499	49	25	142
Income (PPP US$/cap, 2009)[b]	**46,437**	**32,433**	**30,543**	**27,195**	**23,421**	**18,938**
Income (current US$/cap, 2009)	46,437	39,714	32,851	17,095	14,535	8,673
Frac Trade pc (2007–2009) in Income	27	30	30	109	110	58
Share Merch Ex/imports in world trade	11	5	17	3	1	2
Frac ServTrade in TotTrade	23	19	27	16	16	17
2009	**Mexico**	**Brazil**	**China**	**India**	**Vietnam**	**Nigeria**
Population (mln)	107	194	1,332	1,155	87	155
Income (PPP US$/cap, 2009)	**14,341**	**10,429**	**6,838**	**3,275**	**2,956**	**2,149**
Income (current US$/cap, 2009)	8,146	8,116	3,744	1,134	1,052	1,092
Frac Trade pc (2007–2009) in Income	67	24	51	45	150	61
Share Merch Ex/imports in world trade	2	1	9	2	1	0
Frac ServTrade in TotTrade	7	20	11	29	9	4

[a] *Source:* WTO.

[b] Purchasing power parity (PPP) values are corrected for differences in purchasing power between countries.

products. More than one third of merchandise exports in Brazil and many African countries is agricultural product. India and EU-27 rank with 29 percent and 27 percent highest in the fraction of commercial service trade in total trade. Commercial service trade in low-income countries is nearly all in travel and transportation, which is an indication of the importance of tourism.

The increase in global trade has many and, according to economists, mostly beneficial consequences. Its net effect in facilitating the penetration of a market-oriented capitalism into the farthest corners of the world is not easily assessed. Trade and growth are not directly or simply connected (Helpman 2004). From an ecological sustainability perspective, the use of fossil fuel in the ocean bulk and container ships and the, sometimes on purpose but often unintentional, dissemination of bacteria, plants and animals across the continents are among the negative side effects (Crosby 1993; Diamond 2006).

The *migration flows* of people across the globe increased significantly in absolute numbers. In the 19th and early 20th century, there was a massive emigration flow out of Europe as a consequence of famines, unemployment and wars. There were also massive migration flows within countries, notably within Russia and China. In the past, migration was caused by rural overpopulation and famines such as the infamous Irish Potato Famine in 1845–1852. In some cases, it was the result of deportations and wars. There is still large-scale migration but different. Emigration flows to North America and Australia and New Zealand were in 1998 twice the size of the flows in 1900, yet they no longer originated from Europe but from Asia and Africa. In Europe, the flow changed sign in this period, from emigrants to immigrants (Maddison 2006). Present-day migrants are mostly economic migrants, that is, people looking for better opportunities to earn a living. It is still for the larger part a consequence of rural overpopulation and natural disasters and wars. In the early 21st century, the flow of refugees is estimated to be several tens of millions of people. Yet, the interregional flow of migrants is still a small fraction of the labour force.

Global *capital flows* have increased enormously with the deregulation trend since the 1980s. Capital, and with it knowledge, is no longer an immobile production factor in the 21st century. Its mobilisation was instrumental in the explosive growth of economic activity. International financial flows have reached unprecedented levels, but only about 10 percent of the transactions are related to the 'real economy' of trade and production, the remainder being essentially speculative dealings (Dicken 2009). In the first decade of the 21st century, global financial system has become one of the most important factors in the rate and direction of economic growth. In its unregulated 'casino capitalism' form, it contributes to unsustainable development because of its relentless search for the highest returns and its inherent instability resulting from speculative and herding behaviour. In the process, the traditional role of the state in creating and controlling money flows is being eroded.

4.2.2 Social-Cultural Changes

Equally important as the numbers about growth are the more difficult to quantify social, cultural and political changes, which were enablers as well as outcomes of the above-sketched change processes. The common view is that the philosophical and societal basis of western civilisation was laid in the 18th-century period of the Enlightenment. It was then that the idea of *modernity* emerged: the gradual replacement of feudal, political and administrative power by commercial interests and market forces; the emergence of democratic forms of governance with separation between church and state; and the empiricist-rational approach to the world expressed in science and technology. The United States Constitution of 1787 and the French Revolution that began in 1789 are two of the hallmarks of these changes.

Cultural and political developments are visible in a society's *institutions*. In the course of evolution, human beings have learned to organise themselves in groups to get the benefits of cooperation (Wright 2000). Past and present institutions, foremost the nation-state but also armies, churches, ministries, corporations, sport clubs and so on, evolved out of the early tribal groups in order to benefit from cooperation and coordination and restrain behaviour that is harmful for the group. They are cornerstones of societal as well as their own stability and continuity, in their role of repositories of learning and customs and of conflict mediators. They have proven to be the most durable structures in the human world: flexible yet resilient and able to sustain themselves for centuries.

An example is the Catholic Church, possibly one of the longest surviving institutions. Monastic orders such as the Cisterciënzers have survived wars and revolutions for more than eight centuries. Nation-states, as political and ethnic/cultural units tied to a particular geographical territory, have succeeded to maintain a certain coherence for many hundreds of years. Some multinational corporations have been sustaining an essentially unchanged identity for many decades.[6] Evidently, such institutions need to have the capability of self-renewal and adaptation, in combination with a degree of inertia and opportunism. A hierarchical structure may be a precondition, provided it does not become too rigid.

The other side of institutional resilience is resistance to change. So how was industrialisation possible against the vested institutional interests? There are no easy answers, but economic historians suggest some clues (Heilbroner and Milberg 2005). Before 1700, economic life in Europe was largely traditional and had no incentives to expand. There were all kinds of rigidities impeding the production and use of goods, services, labour and capital: 'gild regulations, monopolies, an excessive schedule of holy days (ninety-two in France in 1666), sumptuary legislation, monasticism, settlement laws, price controls, and taboos and religious sanctions on economic behaviour or even on the study of science and technology' (Jones (2003) 96). These were large social and religious barriers to industrialisation, often actively defended by elites and institutions with their preference for privileges and command-and-control ('conservatives').

With the advent of merchant capitalism and later industrial capitalism, the real force of the *market mechanism* in combination with science and technology became apparent. The organising entrepreneur, the enabling capitalist and the industrial labourer came to dominate economic life. Economic growth and innovation were the outcome of personal ambitions to grasp opportunities and reap large monetary rewards. There was no central control, coordination was done by the 'invisible hand' of the market and democracy was the political reflection – at least that was the myth. Industrialism generated new forces of resistance, notably socialism in response to the human sufferings of early industrialism and romanticism as opposition against the new rationalism and commercialism. In the process, European society created some of its best social science and art. But the capitalist system and its liberal ideology have turned out to be resilient, as they fulfilled expectations of a better quality of life, adapted to all kinds of change and economically outperformed the 20th-century state-led command-and-control economies.

Partly in conflict and partly in association with these changes, the role of the *(European) nation-state* in life did increase, not decrease. With the rise in social and economic complexity, collective arrangements became more important in order to protect citizens with legal codes of conduct and their enforcement against fires and floods, epidemic diseases, external enemies – and against each other. With the rise of European colonialism since the 16th century, the state served as protector of exploitation and trade in the colonies. In the 19th century, state interventions broadened as a consequence of scientific and technological developments and of enlightened ideas about societal progress. Sanitation and sewage systems, medical care and schooling, clean air and water, social security and income redistribution all became to some extent state responsibilities.

[6] According to the Forbes 100 list of United States companies, most had gone bankrupt or merged between 1917 and 1987. Eighteen companies were still in the elite group: "These eighteen companies... were grand-champion survivors, weathering the storms of the Great Depression, World War II, the inflationary 1970s... " (Beinhocker 2005). Interestingly, almost all of these companies underperformed in terms of stock market value.

With this came the need for larger state revenues. In ancient and medieval times, random confiscations and political arbitrariness by the elite were quite common in order to finance their luxuries and rivalries. This misinterpretation of what is in the interest of the state was aptly expressed by the French King Louis XIV when he tried to secure money for his wars in Parlement: 'L'etat, c'est moi!'. Gradually, state revenues have become organised in administered trade and taxation schemes (Jones 2003). In Europe, government expenditures as fraction of GDP increased in the course of the 20th-century from an estimated 8–18 percent to 40–52 percent (Maddison 2006; Table 4.1). Part of it is for the delivery of public goods and services like infrastructure and health and education. Another part is redistribution from the rich to the poor and unemployed – social security arrangements that became the hallmark of the 20th-century *welfare state*. Elsewhere, the fraction of government expenditures in GDP is smaller: less than 40 percent in the United States and Japan and 20–30 percent in Africa.

By the end of the 20th century, the role of the state became less prominent in the wake of the fall of communism and the rise of neoliberalism. New societal actors such as the large and increasingly multi-/transnational corporations (MNCs/TNCs) and a variety of Non-Governmental Organisations (NGOs) enter the public arena and create, in combination with the Internet, an ever-more interconnected and volatile world. At the same time, the American-European postwar model of economic growth and its modernist worldview are challenged by the emerging economies of China and other countries. The pressure on finite resources increases, while the expectations of hundreds of millions of people keep rising. Governments find it more and more difficult to balance expenditures and revenues, with the resulting problems of economic debt and inflation and social unrest. The deregulated financial system, with the right to create money, owes its existence partly to this tendency. These are the ingredients of economic (un)sustainability.

Perhaps the least tangible and yet most important and enduring change has to do with *values*. Social scientists have attempted to grasp the cultural aspects of change, with an important role for the distinction between individualism and collectivity (Durkheim) and for religion (Weber). More recently, extensive empirical data have been collected in the *World Value Survey*, a large investigation of attitudes, values and beliefs around the world in four waves (1981–1983, 1989–1991, 1995–1997 and 1999–2001). The results have been analysed and presented by Inglehart and Welzel (2005) as a (revised) Theory of Modernisation. The empirical data, it turns out, suggest two crucial value dimensions: traditional versus secular/rational, and survival versus self-expressive.

If the data are confronted with indicators of socioeconomic development, such as income and economic sector shares, it is found that the process of industrialisation coincides with a clear value shift from *traditional to secular/rational*. The shift correlates with the employment share of industry (Figure 4.4). A process of secularisation and rationalisation but also bureaucratisation and centralisation sets in, during which people become less dependent on the caprices of nature and science, and technology takes the place of religion as a source of authority. Social life gets organised in the uniform way of mechanised mass production, with rigid social classes and uniform standards. It appears this is the price people willingly pay for the enlarged control of technology over nature.

With the further rise in income that happened in the advanced economies of Europe and its offshoots in the second half of the 20th century, there is another shift *from survival to self-expression* values. Survival needs are largely met and people's values move towards self-expression. This change coincides fairly well with the employment share of the service sector (Figure 4.4). Existential security concerns are replaced by the desire for personal autonomy, self-fulfillment and other forms of self-expression. Keywords in the postindustrial society are the provisions of the welfare state, the intellectual mobilisation through advances in information and communication technology (ICT), and liberation from the social constraints of family and community. In the context of environmental or, more broadly, sustainability concerns, it is worth mentioning that issues such as environmental protection, recycling, education, gender and income equality are amongst the postindustrial self-expression values. Whereas industrialisation liberated people from the constraints posed by the natural environment with secularisation *of* authority, postindustrialisation led to an emancipation *from* authority (Inglehart and Welzel 2005).

The value survey results indicate distinct clusters in the two-dimensional space. For instance, Protestant Europe ranks high on both dimensions, whereas low-income countries in Africa and South Asia rank low on both. Former communist countries rank high on secularisation but low on self-expression, an obvious legacy of communism. Catholic and Protestant countries show marked differences. While Protestant societies rank high on interpersonal trust, which is positively correlated with local control, Catholic societies tend to put higher value on hierarchical, centralised control and rank lower on the secularisation dimension.

It seems that socioeconomic development tends to propel people all over the world from a survival-oriented traditional worldview towards a more secular and rational worldview with emphasis on expression of the individual self. Yet, the survey results also show convincingly that the cultural traditions of individuals and nations do play an important role, too. Unlike some social scientists have argued, modernisasion is not the same as Westernisation and the United States is not the model for the cultural changes happening around the world. Cultural modernisation results from socioeconomic developments, but it is also influenced by the traditional cultural heritage and may also be reversed if the economy collapses.

4.3 Accelerating Impacts: The Natural Environment

4.3.1 The Source Side

There were not only economic and social but also natural obstacles to industrialisation. However, the natural environment posed less and less resistance to the control over natural that science and technology brought. The growth of population and economic activities influenced the natural environment, slowly but steadily, but it hardly raised widespread concern about the consequences for human populations: 'restraint in the interest of the biosphere was rarely given any consideration at all until the emergence of a more or less worldwide environmental movement in the 1960s' (McNeill (2007) 302).

Box 4.2. *The European Miracle.* Many books have been written about the reasons for Europe's ascendancy in the last 300 years, amongst them *The European Miracle* by Jones (2003). It is a study of an historian on Europe's comparative advantage in becoming the world's first and leading industrialist region. Jones argues that environment, market and state have to be analysed to understand the reason. Europe's industrialisation is of much earlier origin than the rise of Britain and cannot be attributed mostly to the exploitation of colonies. In Jones' view it is a mix in which the relatively small-scale, political decentralisation, markets for land and labour and rather small income differences makes the difference with the steppe imperia (Mongol, Moghul, Ottoman), which were command hierarchies imposed upon customary agriculture. Also, European 'governments' had more concern about public goods such as disaster relief and incentives for private investment. 'A relatively steady environment and above all the limits to arbitrariness set by a competitive political arena do seem to have been the prime conditions of growth and development. Europe escaped the categorical dangers of giant centralised empires as these were revealed in the Asian past. Beyond that, European development was the result of its own indissoluble, historical layering' (Jones (2003) xxxvii).

The economist Maddison (1995) has a different emphasis in the list of factors that have favoured Europe's development: rational scientific investigation and experiment, the ending of constraints on property transactions, more trustworthy and regular financial institutions and taxing, and the system of nation-states and family traditions, which induced more modest population growth than elsewhere.

The most visible changes had to do with *land* (Figure 4.5a,b). Agriculture was and still is the greatest force of land cover and land use change. Next come forest cutting for timber, expansion of urban lands and mining. Between the year 600 CE and the early 13th century, the population in Europe went from 18 million to 76 million while the land area covered by forests and swamps decreased from roughly 80 percent to 50 percent in this period. Similar changes happened in other densely population regions of the world. In 1700, roughly 2–3 percent of the global land surface was cultivated. In 2000, it was an estimated 37 percent.[7] Cropland expanded in the last three centuries from 3–4 million km^2 to 15–18 million km^2, largely at the expense of forested lands. Global forest area declined from an estimated 53 million km^2 in 1700 to 43–44 million km^2 today. Deforestation accelerated in the 20th century with large-scale cutting of tropical forests for cropland and grazing land. In high-income regions, cropland area hardly changes anymore or even declines. Wood production also stabilises since 1990. However, water use, irrigated area, production of crops and meat and the catch of fish continue to grow (Figure 4.6a,b). At present, an

[7] Most of the date cited are based on Lambin and Geist (2006) and Klein Goldewijk (2004). It should be noted that there are considerable uncertainties, due to different definitions and limited data. Besides, not only areas but also quality, e.g., of forests, has to be considered (§ 11.3 and § 12.2).

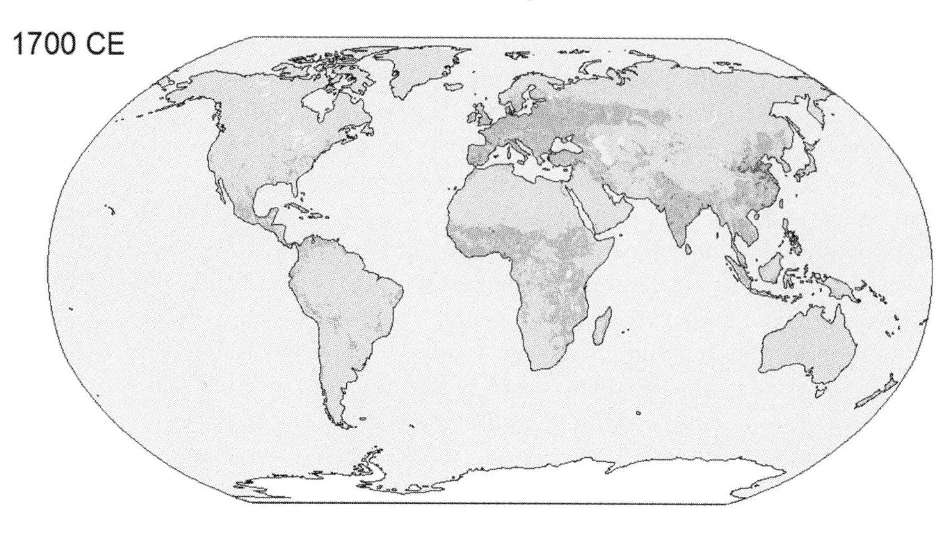

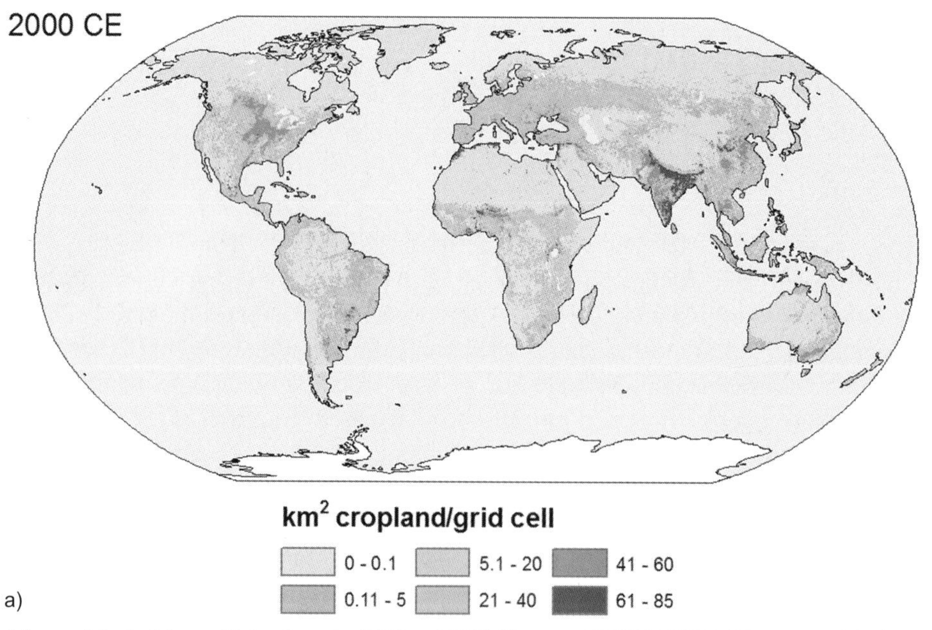

Figure 4.5a,b. Map with land use in 1700 and 2000 (Source: HYDE/PBL). The upper (a) shows the fraction of a gridcell used as cropland; the lower (b) shows the fraction used as pasture.

estimated 25 percent to 39 percent of the net biomass production on earth is under human control (Haberl 1997).

The development perspective on nature is that of a resource, but the resources have come under increasing pressure. One fishery after another is depleted through overexploitation and habitat degradation. World fish catch from open fisheries has been in decline since the 1990s and sustaining supply for an increasing demand is only possible by large-scale expansion of aquaculture. Non-renewable resources

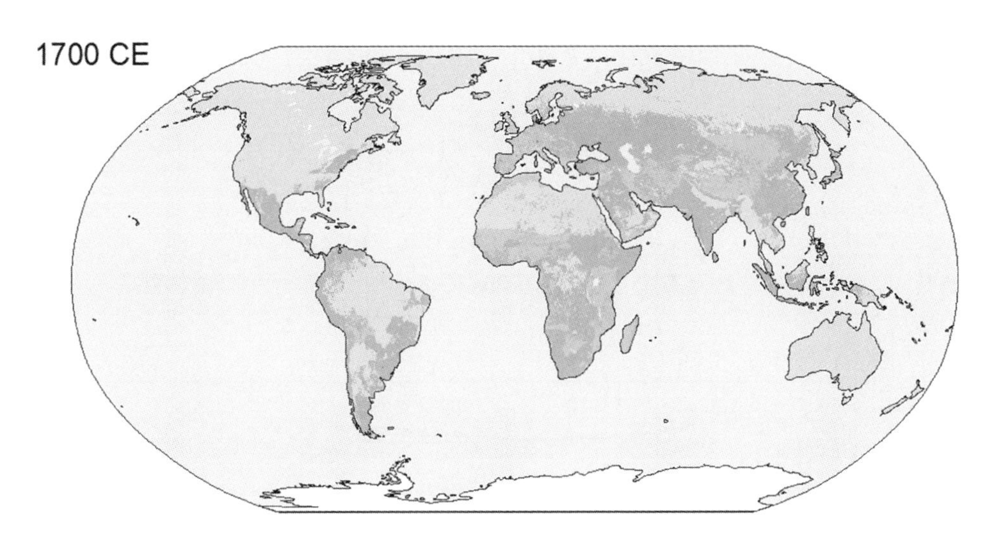

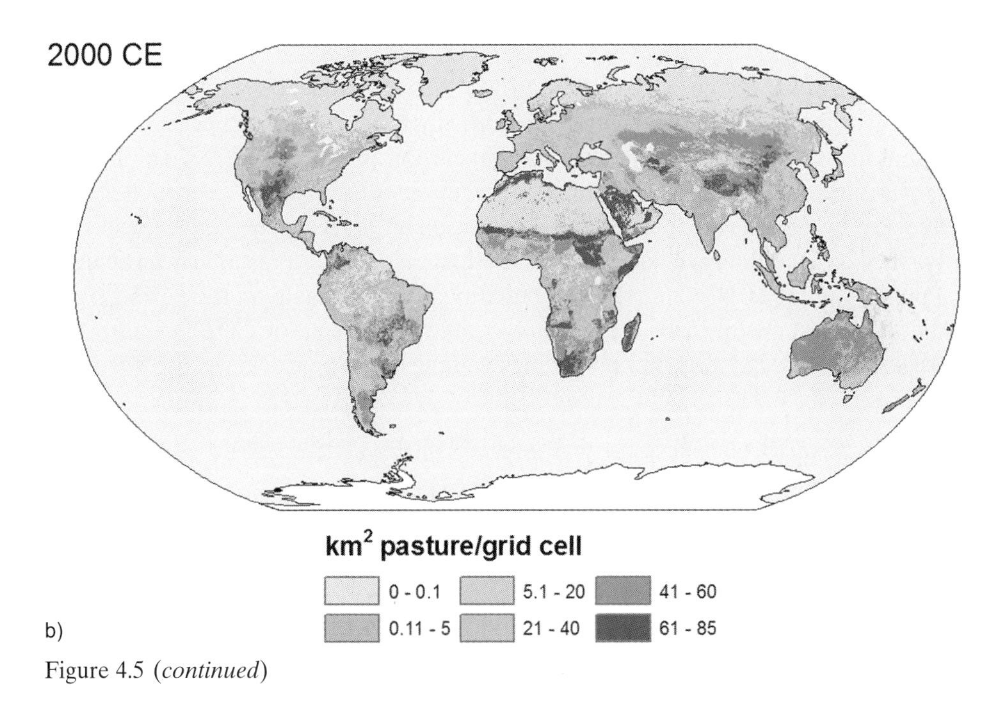

Figure 4.5 (*continued*)

such as high-grade mineral ores and low-cost oil and gas deposits are depleted to the extent that newly discovered ones have lower ore grades and are farther away and/or deeper and more fragmented. Their exploitation causes additional burdens on the environment, as is evident from, for instance, the exploitation of tar sands in Canada as a source of oil. There are also more insidious changes going on. In many parts of the world, soil fertility has declined because of a combination of agricultural and pastoral exploitation and the natural forces of water and wind erosion.

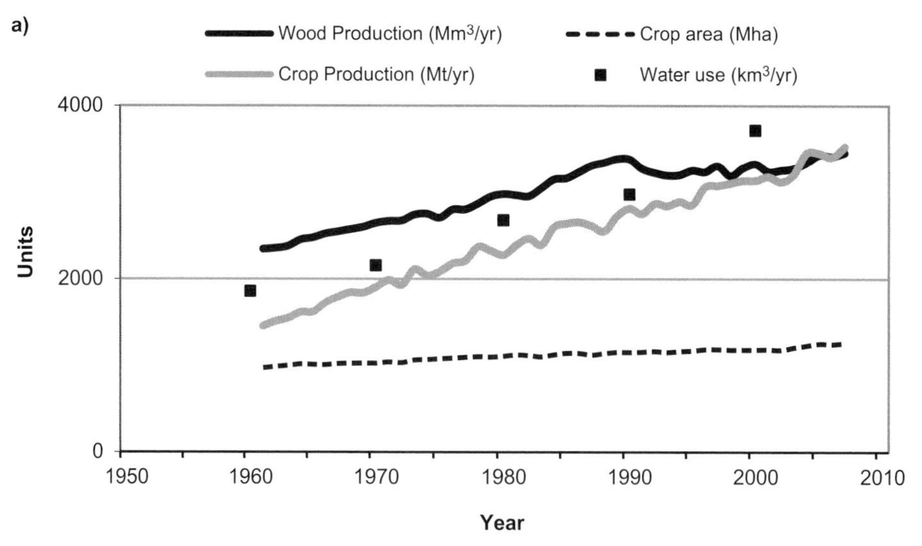

Figure 4.6a. Global trends in wood and crop production, crop area and water use since 1950 (Source of data: FAO/PBL). (See color plate.)

4.3.2 The Sink Side

The changes on the source side show up on the sink side of system Earth in the form of emissions to air, water and soil and subsequent interference with natural element flows across the planet. The natural carbon (C), nitrogen (N), sulphur (S) and phosphorus (P) flows are to an even larger extent, and at an ever higher rate, influenced by human activities (Figure 4.7). Exponential growth in the combustion of fossil fuels and the change in land use and production of cement resulted in an average 9.3 billion tons of carbon emitted into the atmosphere annually in the period 2006–2009. This is responsible for a continuous rise of the atmospheric CO_2-concentration from an estimated 284 parts per milion on a volume basis (ppmv) in 1832 to 389 ppmv

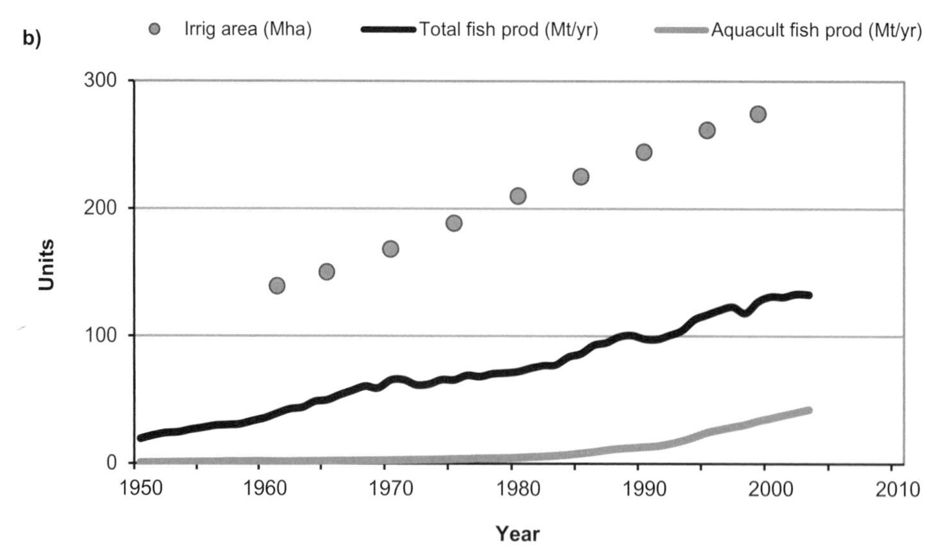

Figure 4.6b. Global trends in total fish production, aquaculture fish production and irrigated area (Source of data: FAO/PBL). (See color plate.)

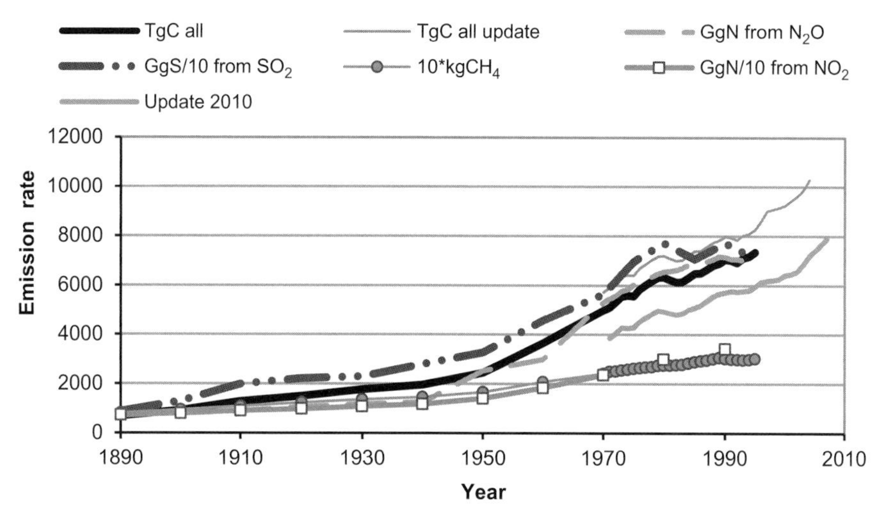

Figure 4.7. Global trends in the emissions of the greenhousegases CO_2, N_2O and CH_4 and the acidifying gases SO_2 and NO_2 since 1890 (Source of data: PBL).

in 2010. In combination with the rise in other greenhousegases such as methane, this is thought to pose serious risks of disruptive climatic changes in the course of the 21st century (IPCC 2007). The combustion of fossil fuels is also an important source of SO_2- and NO_x-emissions into the atmosphere and a cause of acidification of lakes and soils. These emissions have been reduced drastically in the last decades, which illustrates the effectiveness of technological change in combination with regulatory policies. The emission profile of ozone-depleting CFCs, which caused depletion of the stratospheric ozone layer, is another illustration of how effective policy cannot completely eliminate, but can at least significantly mitigate harmful impacts (Haas *et al.* 1993; Miles *et al.* 2002).

The food system is another important link in this part of the story. Until the 20th century, the provision of nitrogen had been done by natural fertilisation such as planting clover. In 1908, Haber en Bosch invented an industrial process by which nitrogen, which constitutes 78.1 percent of ambient air, reacts with hydrogen to form ammonia (NH_3). This can be converted into nitrates and nitrites, which became the feedstock for the large-scale production of industrial fertiliser. In this way, food output increased and now sustains a larger world population than the British scholar Malthus had ever contemplated to be possible. However, one of the consequences is the eutrophication and algae growth in many surface waters all over the globe (Bouwman *et al.* 2009). Another element in the food system on the source and sink side is phosphorus (P). It is an essential ingredient for plant growth and naturally occurs in the form of phosphates. Phosphates are industrially produced and more than two-thirds are used in agriculture in fertilisers. Its widespread use is contributing to eutrophication.[8] Finally, biodiversity has been declining in most places – a trend that is difficult to measure and has complex causes and consequences.

In a system dynamics frame with stocks and flows, the preceding trends are unsustainable if following the definition outlined in Section 2.3. The rate of exploitation

[8] It is also a source side issue: Phosphorus reserves are small and highly concentrated, and shortages could pose a severe limitation to future food production.

Box 4.3. *Experiencing landscape changes.* The changes in the natural environment influenced also the way in which it was experienced. Many authors, from ancient times onwards, have lamented the loss of once beautiful landscapes. Travellers in the 18th and 19th centuries in the Mediterranean judged the landscape in many places degraded – or at least a badland as compared to what they had imagined. Today, there are still alarming reports about degradation and desertification in the Mediterranean. However, it is a complex region with a long history of natural and anthropogenic changes, but even so ecological historians like Grove and Rackham (2001) find the evidence unconvincing and speak about the Theory of the Ruined Landscape of Lost Eden.

Just as the Greek landscape of 2000 Before Present (BP) was romanticised in some later paintings and writings, so did some authors romanticise the 19th landscape of industrialising Britain. An example is the German Prince von Pückler-Muskau who was in Yorkshire in 1827: 'With the falling dusk I reached the big factory town Leeds. The wide space which it occupies upon and amidst several hills was covered by a transparent cloud of smoke. A hundred red fires shone up from it, and as many tower-like chimneys emitting black smoke were arranged amongst them. Standing out delightfully in the scene were five-storey high huge factory buildings in which every window was illuminated by two lights, behind which the industrious worker finds himself engaged until deep in the night. In order to add a touch of romance to the bustle of enterprise, two old gothic churches arose high over the houses, on the spires of which the moon poured its golden light while, in the blue firmament, with the vivid fires of the busy people underneath, it seemed to repose in majestic rest' (Sieferle (1997) 165–166). It should be no surprise that fact, fiction and dream are often mixed in the interpretations of the causes and consequences of past and present landscape changes.

of fisheries, forests and soils exceeds in many places the natural regeneration rate. Non-renewable resources like rich metal ores are used at much higher rates than renewable substitutes are produced, although technology may provide versatility. But, crucially, fossil fuels are extracted much faster than nature accumulates them and than present-day society is producing renewable substitutes like wind and solar power. The assimilative capacity of sinks is exceeded in many places in the sense that pollutant loads are only partly and slowly broken down at rates below the inflow rates. This is sometimes obvious, as with large oil spills and aerosols. Sometimes it is known but next to invisible, as with PCBs and other toxic chemicals. All of these trends in the use of sources and sinks for human development cannot keep growing beyond the planetary scale and will inevitably stabilise in the course of this century. An important question in sustainability science is how such a process might evolve and what the options are to stay within manageable risk zones (Rockström *et al.* 2009).

4.3.3 Experiencing Change

To give numbers about the changes is one thing, to understand how the changes were experienced quite another. Landscape degradation, resource depletion and environmental pollution were observed and discussed already in antiquity. In London, air pollution from coal burning was already mentioned as a problem in the 13th century. Impending resource scarcity was another and more important reason for concern. Medieval Europe was faced with wood scarcity in several places and periods. Early in the 18th century, the threat of wood shortage in parts of Germany led to the notion of 'nachhaltende Nutzung' (sustainable use), as opposed to the common practise of clearcutting (van Zon 2002). In the same period, intensified logging forced Japanese rulers to use a mix of regenerative forestry and wood imports from tropical regions, and by the end of the century, regulation of consumption and plantation forestry were introduced (Totman 1989). Also, later concerns about depletion were heard time and again. Famous is the warning about the exhaustion of Britain's coal reserves made in 1865 by the economist Stanley Jevons. Alarms about overfishing were regularly expressed in the first half of the 20th century (Schrijver 2010). Yet, the changes during the Industrial Revolution were in the main seen as improvements: Nature was domesticated and made more productive. Wild nature was seen as evil, ruthless and selfish. The American pioneer was the symbol of entrepreneurial optimism. A telling statement of Freud made in 1929 is given by Lowenthal (1990): '. . . a country has attained a high state of civilization when we find. . . everything in it that can be helpful in exploiting the earth for man's benefit and in protecting him against nature. . . . '

It was in the second half of the 19th century that doubts arose about the acceptability of the large, ruthless exploitation of Nature such as the near-extinction of the American buffaloe and the Siberian fur-bearing animals (Scott Taylor 2011; Ponting 1991). An influential book was *Man and Nature* (1864) by the American diplomat Marsh. He deplored the harmful impacts of human activity, which were the consequences of greed and ignorance. It was the beginning of the conservationist movement. Yet, in the vein of Enlightenment and Darwinism, these first environmentalists felt that the changes were largely beneficial and that the forces of Nature exceeded those of man so much that serious harm was inconceivable. It was also, and conveniently, believed that technology and state regulation would remedy the excesses. This attitude prevailed far into the 20th century and much evidence of environmental damage was ignored. An undercurrent of ecologists developed a different view, rooted in a mixture of classical philosophy and Romanticism, which saw the equilibrium and stability of the climax ecosystem as the ideal. Nature provides a moral order and is imbued with balance, integrity, order, harmony and diversity. Advanced industrial man degrades Nature and its beneficence.

Pollution of air and water became a genuine issue when the first signs of negative impacts on human health showed up. But until late into the 20th century, action was usually only taken when the harm or damage was local and evident, as with smog or stench, and when the affected upper-class citizens could mobilise political support. This changed gradually with the rise in the standard of living. In the 1960s, the upcoming middle class in the United States and Europe expressed growing concern that pollution had negative health impacts. Also, rising noise levels, reduced visibility and nasty smells were increasingly considered unacceptable. Still, opposition and subsequent (calls for) action were largely local, where people could link the experience to the cause.

While the risks of a local and short-term nature were increasingly recognised and, though with a delay, addressed, the large-scale long-term risks remained unnoticed and ignored. It was only with reports such as the *Limits to Growth* to the Club of

Box 4.4. *Population overshoot and collapse.* How fragile the situation was in medieval times is illustrated in the historical reconstruction of population numbers in medieval France between the year 1000 CE and 1800 CE (Figure 4.8). The graph shows France's population and average Northern Hemisphere temperature relative to the period 1961–1990 (Mazoyer and Roudart 1997). Detailed analyses of agricultural practices and techniques reveal an increase in potential population density since the Carolingian era. A farmer in Mediterranean Europe needed in 900 CE about 16 hectares (ha) of which two-third were pasture to feed a five-person family (Mazoyer and Roudart 1997). This could sustain a population of 20–30 persons/km^2. In Mid- and Northern Europe, these numbers were 34 ha and 8–15 persons/km^2, respectively. This gradually rose to 30–80 persons/km^2 around 1250 CE.

In the 14th century, a series of famines and wars set in, possibly related to rather sudden drops in temperature, but there was a remarkable recovery. The fall in population liberated large amounts of productive land and better fertilisation strategies and tools, such as the use of nitrogen-fixing clover and the horse-drawn plough, led to major yields on especially the northern soils. The population quickly recovered to a new plateau with a density of up to 160 persons/km^2. This period in European history is a dramatic illustration of a cycle of expansion and decline where both natural and social forces are simultaneously at work.

Rome (Meadows *et al.* 1971) and the *Blueprint for Survival* from *The Ecologist* (1972) journal that local and regional resource depletion and environmental degradation were framed in a larger, global context and considered outside the circles of scientific experts and environmental activists. It introduced a more systemic analysis and emphasised the linkages between the various causes and consequences. One of the challenges of sustainability science is to merge these and subsequent scientific insights about the global and long-term prospects with the local potential and willingness to act in accordance with these insights.

4.4 Earth System Analysis: Regimes and Syndromes

4.4.1 Social-Ecological Regimes

The previous historical accounts provide a necessary background for an interpretation of the present situation in terms of (un)sustainability.[9] But is it possible to go beyond the anecdotical and the case studies to distinguish some larger, generic patterns? One way to look at human history is as a series of transitions from one social-ecological regime to the next (Fischer-Kowalski and Haberl 2007). A *transition* is a dynamic process in which a system changes from one relatively stable equilibrium to another (§8.6). After a take-off, important system variables first accelerate and then decelerate to a new, relatively stable state. I have already introduced the

[9] There are a vast array of books on this topic, and we refer to the suggested literature for more in-depth historical accounts.

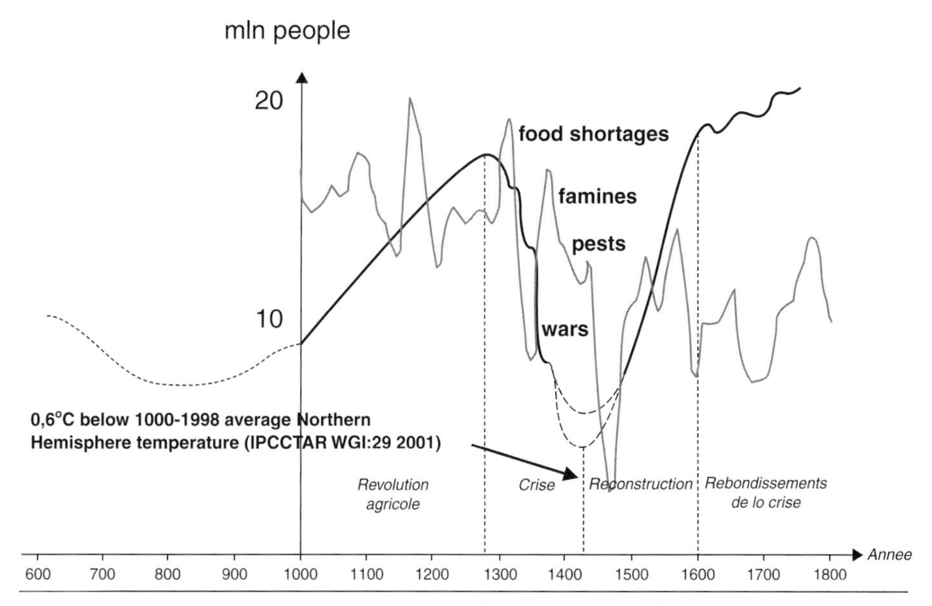

Figure 4.8. Population of France between 600 CE and 1800 CE. The population data are based on Mazoyer and Roudart (1997). The dotted curve indicates the average Northern Hemisphere temperature relative to the period 1961–1990 (IPCC 2001).

demographic transition; however, the focus is now on a transition of one *social-ecological regime*, that is, a particular pattern of society-nature interactions that is sustained in a dynamic equilibrium for a certain period, to the next (Schandl *et al.* 2009). It comprises societal (culture) as well as biophysical (nature) elements.

The transitions from one regime to another are, in a world history context, processes of deep structural change with revolutionary consequences. The old regime is more or less absorbed and transformed into the new regime. Each new regime brings an expansion of the *anthroposphere* within the biosphere. The first regime is commonly associated with the domestication of fire: People transformed a destructive natural force into a regularly available source of energy, with important consequences for culture and nature (Goudsblom 1992). The second great transformation brought about by humans was the domestication of plants and animals: the process of agrarianisation (§3.1). The resulting agriarian regime became the dominant one during the Holocene period in almost all regions of the world.

In agrarian societies, unlike in hunter-gatherer groups, people actively and deliberately intervene in the ways in which natural process convert sunlight into plant and animal biomass. They use the land to convert forest into agro-ecosystems in order to get food. The metabolism is completely dependent on incoming solar energy, at an average density of 100 W/m^2 and with strong fluctuations. There is always a delicate balance between food requirements for a growing population and food supply in a finite environment. Pre-industrial agrarian societies were thus in dynamic equilibrium, with high birth and mortality rates and life expectancy below forty years, on average. Improvements in yields and the capacity to feed people happened slowly and discontinuously and human populations were still dependent on the vagaries of the natural environment. In particular, climate change may have triggered or exacerbated existing vulnerabilities and risks and indirectly caused famines, epidemics and wars (Figure 4.8).

Around 1700, there were the first clear signs of a new social-ecological regime. It has been documented quite well for the United Kingdom of Great Britain and Northern Ireland (UK) (Schandl and Krausmann 2007). Crude death rates started to go down and life expectancy went up. To feed the growing population, agriculture had to adapt. Cropland and intensely used grassland expanded. The feudal institutional framework was restructured into a market-oriented system of landlords, tenants and agricultural labourers. This led to greatly improved yields because of increase in the size of land holdings and of innovations in strategies and tools, and until 1830, per capita food output could be sustained. Nevertheless, population started to outgrow food production capacity and shortages were relieved by importing food from the 'New World' and Russia. Around 1900, Britain imported more than 60 percent of its food, which was the equivalent of an area the size of the UK cropland area.

The expansion of the nutritional basis was one element of the transition to the new regime and the first escape from the solar-based agrarian society. Because not only land productivity but also labour productivity increased, there was a large surplus population in rural areas. Their migration to urban centres provided the labour force for industry. The appearance of transport networks (canals, trains, and so on) and the energy to operate them were an essential precondition. This, in turn, evolved on the basis of the science-technology-resource complex of coal mining, iron production and steam engines.[10] Primary energy use per person, in the range of 20–70 GJ/cap/yr and more than half biomass-based, had been more or less constant for a long period, but in the industrial era, it started to rise rapidly towards a new plateau with values in the range of up to 150 GJ/cap/yr by 1900.

The further liberation from the constraints of the agrarian regime and a new stage in the industrialisation process was the intrusion of oil and electricity into all economic and social activities. It was made possible through the internal combustion engine and the electric motor. In combination with a manifold of other inventions and innovations, not only in the United Kingdom but all over Europe and the United States, it had profound impacts on individual and societal life. Agriculture became decoupled from the solar flow constraints with the advent of road and electric power networks, the industrial production of fertiliser and the replacement of animals by tractors. It permitted a threefold increase in yields since 1950. The decentralised nature of the new energy technologies accelerated also in industry the replacement of physical labour by machines. Average working hours started to decline and human labour was increasingly engaged in operational skills and information processing (Table 4.1).[11] Urbanisation rates started to rise. At the end of this process, primary energy use per capita had made another jump, to 120–200 GJ/cap/yr or three- to fourfold the level of agrarian societies. In fact it increased much more because the

[10] It can be argued that these inventions have been induced by the wood scarcity in England – a widely used example to argue that scarcity problems generate their own solution.

[11] There are rather large discrepancies in the data on working hours per year between the data of Maddison and other sources such as the ggcd-website and other authors who use different sources and probably definitions.

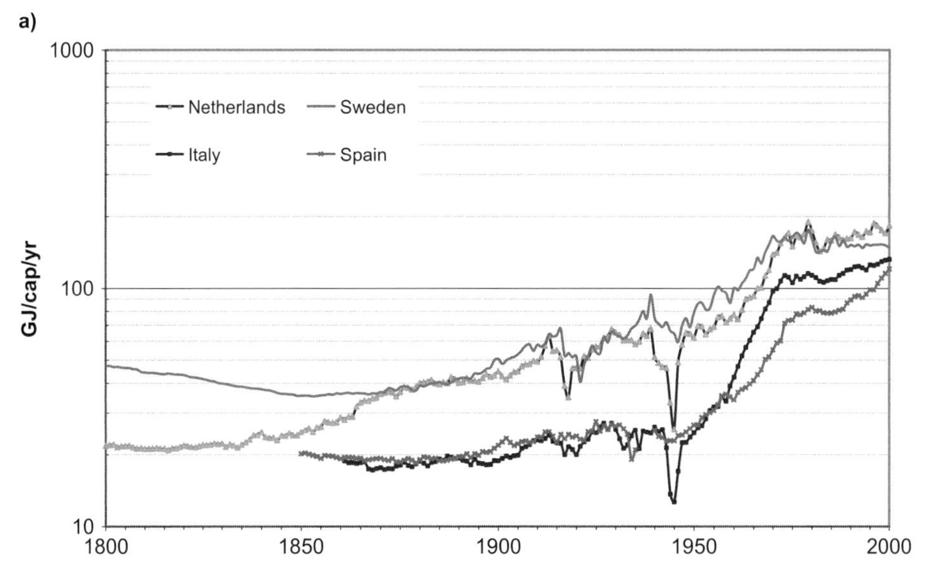

Figure 4.9a. Graphs of primary energy use per person in GJ/cap/yr since 1800 for four European countries: Netherlands, Sweden, Italy and Spain (Gales *et al.* 2007).

energy was used more efficiently (§7.3). In useful energy units, the increase was more than twentyfold!

The regime change can be seen quite well in the reconstructions by historians of the use of energy carriers in European countries. Defining food and fodder-based muscle power of men and working animals, firewood, wind and water and peat as *traditional energy carriers*, most European countries experienced a transition to less than 50 percent of primary energy carriers in the form of traditional ones between 1850 and 1950 (Gales *et al.* 2007). The relative shares of traditional energy carriers reflected local circumstances. While in Sweden, for instance, firewood dominated, muscle power prevailed in The Netherlands and Spain. After 1950, an acceleration of the transition to fossil fuels took place. The graphs of primary energy use per person per year in four European countries show this transition from an agrarian level of 20–50 GJ/cap/yr towards an industrial level of 100–200 GJ/cap/yr (Figure 4.9a).

A widely used formula to decompose primary energy use in a country is to rewrite energy use as the product of population, income (GDP/cap/yr) and energy-intensity (E/GDP):

$$E = P \cdot \frac{GDP}{P} \cdot \frac{E}{GDP} \quad \text{GJ/yr} \qquad (4.1)$$

Figure 4.9b shows per capita energy use (E/P) and energy-intensity (E/GDP) for four European countries between 1800 and 2000. There is a clear trend towards lower energy-intensity in these four European countries, which reflects primarily the net effect of improvements in the efficiency with which primary energy carriers are used in economic production.

The high-income regions entered a new, fourth social-ecological regime in the second half of the 20th century. It is exemplified by the large share of the services sector (Figure 4.3). In Northwestern Europe, the United States and Japan, the

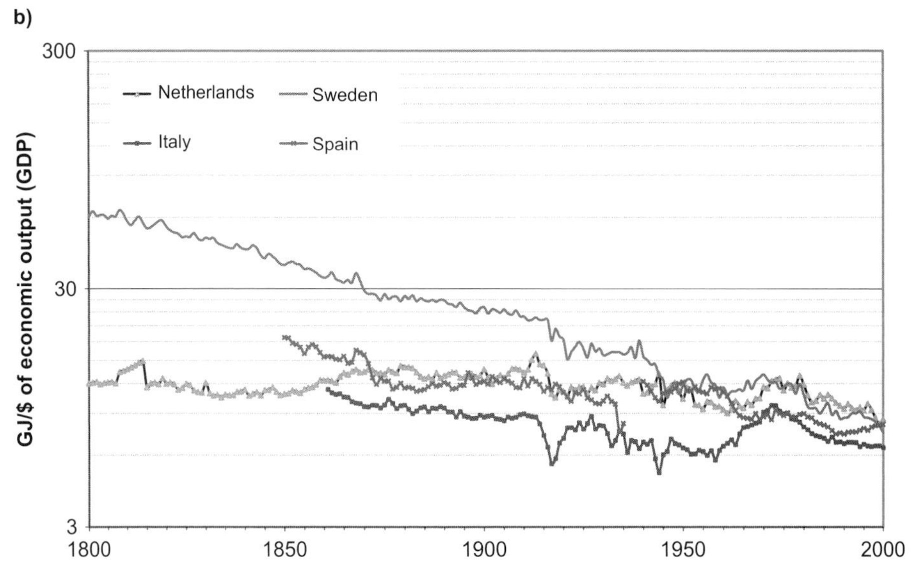

Figure 4.9b. Graphs of energy-intensity in GJ/$ since 1800 for four European countries: Netherlands, Sweden, Italy and Spain (Gales *et al.* 2007).

service sector output consisted around the year 2000 for more than one-third of public services and for more than one-third of financial services, the remainder being wholesale and retail trade, hotels and restaurants, transports and posts, and communication (Teives Henriques and Kander 2010).[12] In other words, three out of five jobs are about dealing with people, symbols and information. It is in line with a cultural change towards postindustrial self-expression values, which emphasise personal autonomy and individual choice and are conducive to more participatory forms of democracy (§4.1). Once basic needs such as food and shelter are met, the demand for mass-produced goods such as household and transport equipment start to dominate. When this becomes saturated, the richer segments of society spend an increasing part of their income on private and public services.Yet, the long-term significance of this transition is still difficult to interpret.

Evidently, the rapid development and spread of *ICT-based applications* throughout society, with the Internet as its for now supreme representation, is a key ingredient in the transition to the postindustrial world (Figure 2.7b). Measured as *useful* energy per unit of economic output, there are signs of a clear decoupling between energy use and economic output in the United States since 1970. This may be interpreted as the turning point at which information became more important than physical work as a factor of production (Ayres and Warr 2005). Some other indicators also suggest a decoupling of monetary and physical flows.

Another noteworthy phenomenon is the *productivity gap* between manufacturing and services sectors, as observed first by Baumol (1967). Those parts of the service sector where the human element is crucial – such as entertainment, educational and personal care services – cannot realise the high rates of productivity

[12] Data are for 2005 and based on constant prices.

increase that happen in manufacturing.[13] This makes economic growth in the postindustrial economy different from growth in the industrial era and is one explanation of the decline in economic growth rates with rising income. It is counteracted by replacing expensive services with do-it-yourself activities, mechanisation and robotisation. These mutually interacting and reinforcing changes influence the modern world in ways that are hard to interpret and predict.

The data on sectoral changes in emerging economies such as China and India have to be interpreted with caution, because the *informal economy*, that is, the transactions outside of organised labour unions and tax collectors, is large and growing. It consists of traditional service sector activities and may transform itself into a modern service sector without fully becoming part of official statistics. It is an important buffer against economic ups and downs for the millions of people with insuffient or inadequate skills in countries with weak governance and an increasingly global labour market.

Evidently, the contours of the new regime are not yet clearly visible, and the demarcation between industrial and service jobs and activities is rather arbitrary and vague. Are we witnessing the emergence of the *noösphere*, an expanding network of human consciousness and mental activity as the next stage in the evolution of the biosphere? Is the fulfillment of human desires gradually disconnected from material stocks and flows? Are we experiencing the first signs of a sustainability transition? Hopefully, you know at the end of this book what such a transition could be like and what your role can be.

4.4.2 Syndromes and Archetypes[14]

A fragmented, disciplinary approach does not work in sustainability science. A broader perspective transcending disciplinary boundaries is needed. One framework to investigate global change phenomena in such a broader, problem-oriented setting is the syndrome approach.[15] The word *syndrome* comes from 'running together', in particular, the simultaneous occurrence of signs of disease. Here, it refers to a typical co-occurrence of different complex natural and social dynamic phenomena or *symptoms*. They are causally connected and constitute the elements of a syndrome. The basic assumption of the syndrome approach is the existence of clusters of interrelated symptoms.

Examples of social symptoms are positive trends in individualisation and informatisation, in mobility of people and transport of goods and information, and in rural-urban migration and urban sprawl. Natural symptoms can be source-based, such as trends in use and depletion of energy and other resources and in resource use productivity. Sink-oriented natural symptoms are changes in biodiversity and in accumulation of waste and pollutants in soils, water and air. Several graphs

[13] In fact, the notion of (labour) productivity as output per unit of (labour) input is a flawed one, because it homogenises widely divergent 'outputs' on the basis of a divergent 'input', namely, wages.

[14] I would like to acknowledge the contributions of Marcel Kok and Paul Lucas, both at PBL, for this paragraph.

[15] The syndrome approach was originally proposed by the German Advisory Council on Global Change (WBGU 1994) and later conceptualised and developed at the Potsdam Institute of Climate Impact Research (PIK) (Schellnhuber *et al.* 1997; Petschel-Held *et al.* 1999).

Table 4.3. *Overview of the global change syndromes*[a]

Syndrome name	Short description of the mechanism
Utilisation syndromes	
Sahel Syndrome	Overcultivation of marginal land
Overexploitation Syndrome	Overexploitation of natural ecosystems
Rural Exodus Syndrome	Environmental degradation through abandonment of traditional agricultural practices
Dust Bowl Syndrome	Non-sustainable agro-industrial use of soils and water
Katanga Syndrome	Environmental degradation through depletion of non-renewable resources
Mass Tourism Syndrome	Development and destruction of nature for recreational ends
Scorched Earth Syndrome	Environmental destruction through war and military action
Development syndromes	
Aral Sea Syndrome	Environmental damage of natural landscapes as a result of large-scale projects
Green Revolution Syndrome	Environmental degradation through the adoption of inappropriate farming methods
Asian Tiger Syndrome	Disregard for environmental standards in the context of rapid economic growth
Favela Syndrome	Environmental degradation through uncontrolled urban growth
Urban Sprawl Syndrome	Destruction of landscapes through planned expansion of urban infrastructures
Disaster Syndrome	Singular anthropogenic environmental disasters with long-term impacts
Sink syndromes	
High Stack Syndrome	Environmental degradation through large-scale dispersion of emissions
Waste Dumping Syndrome	Environmental degradation through controlled and uncontrolled waste disposal
Contaminated Land Syndrome	Local contamination of environmental assets at industrial locations

[a] Lüdeke *et al.* 2004.

presented earlier in this chapter are representative of certain symptoms. A region may be vulnerable to a syndrome when it is exposed to a particular, causally connected set of symptoms. Its *disposition* signifies the probability that the syndrome occurs. It depends on the region's exposure and sensitivity whether and how the syndrome affects the region. *Exposure* depends on exposition factors. These are rather short-term events that may activate a syndrome if the disposition is high. They can arise from within the system (endogenous) or from the outside (exogenous) as with natural catastrophes and extreme political or economic shocks. The *sensitivity* of a region for the event and intensity of a syndrome is a complex pattern of agro-ecological, economic and socio-cultural factors. The final impact depends also on the capacity of people in the region to respond. Such a *coping capacity* or capability can prevent or mitigate the consequences of a syndrome.

Table 4.3 shows a list of situations that qualify as syndromes. Based on the dominant aspect of the human-nature interaction under consideration, a distinction can be made between *utilisation syndromes*, *development syndromes* and *sink syndromes*. Syndromes are often occurring in interaction, for instance, when an operating syndrome triggers or reinforces another syndrome. In later chapters, several stories describe some of the listed syndromes in more detail.

Building upon the syndrome approach, a similar framework has been developed under the heading of *archetypical patterns of vulnerability*. It was developed as part of the 4th Global Environmental Outlook (GEO) in response to a request from governments to show how the environment provides challenges and opportunities for development (UNEP 2007). While GEO is a global assessment, its strong regional focus bridges the gap between a coarse global overview, on the one hand, and insights from local case studies of sufficient relevance for countries and regions, on the other.

The concepts of vulnerability and risk play a key role. Similar to syndromes, vulnerability is considered a combination of exposures, sensitivities and adaptive capacities in a social-ecological system. Examination of local and regional situations and trends shows that there are certain mechanisms in human-environment systems that create specific, representative patterns of the interactions between environmental change and human well-being (Kok and Jäger 2007). These *patterns of vulnerability* tend to have characteristic space- and time-scales at which human development takes place and is confronted with constraints. They can be regarded as archetypical or generic, because they describe the common element in case studies from widely different regions of the world. The details may be quite different, but their essence is the same.

An archetypical pattern of vulnerability has two components. One is the vulnerability creating mechanisms, and the other is the spatial distribution of typical combinations of the mechanisms. Their investigation in GEO-4 is guided by five questions (Kok *et al.* 2011):

- What are the main exposures, key vulnerable groups and their sensitivities that together define the pattern of vulnerability?
- What are the basic vulnerability creating mechanisms that constitute this pattern of vulnerability?
- In what form and where does this pattern manifest itself?
- How can future changes within the human-environment system affect the human well-being situation for the vulnerable groups?
- What are the opportunities – individual responses or policy responses – to cope with and adapt to future changes?

The first two questions address a description and formalisation of a specific pattern under investigation. For instance, income, population pressure, access to markets, soil quality and water availability are identified amongst others as determinants of food security – determinants meaning rather complex elements or subsystems. Answering the third question includes quantification and statistical analysis based on the information from the answers to the previous two questions. It generates a set of proxy indicators for the most important determinants of vulnerability creating mechanisms. For instance, population pressure is measured as population density and soil quality as an index for water erosion sensitivity. The next step is a cluster analysis, that results in one functional and one spatial characterisation of a pattern of vulnerability. The functional one consists of a specific constellation of indicators that are labelled *vulnerability profiles*, while the spatial one contains the spatial

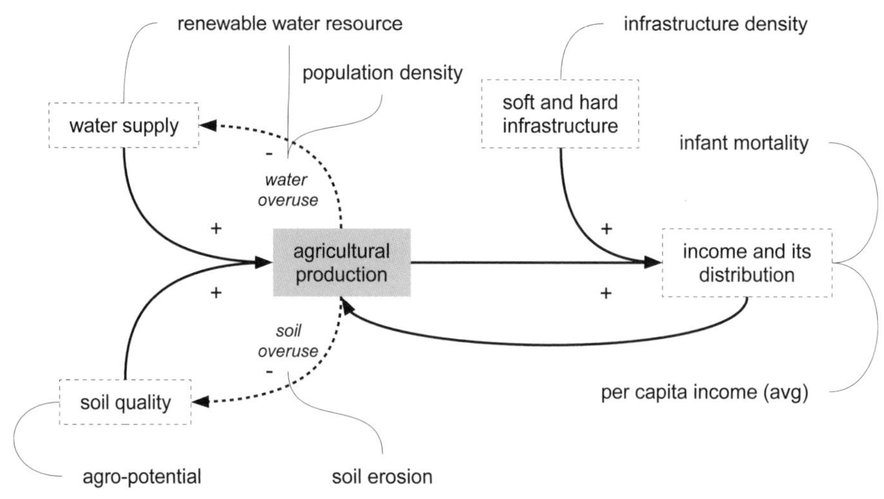

Figure 4.10. Condensed influence diagram of the pattern of vulnerability of smallholder farmers in drylands. The boxes indicate important vulnerability determinants. The arrows show hypothesised relationships and their direction of influence. The thin lines represent the indicators that are used as proxy quantification of the determinants (redrawn from Kok *et al.* 2011).

distribution of the profiles. The profiles and the differences and similarities between the profiles give insight into the factors that create the vulnerability in a specific cluster. The resulting vulnerability patterns provide an answer to the fourth and fifth question, as an entry point for identifying opportunities to reduce vulnerability and directions for policymaking.

An illustration is given for the vulnerability of small-holder farmers in dryland ecosystems, where humans have to sustain food production in an environment of erratic rainfall.[16] The hypothesised relationships between determinants of food security are represented in an influence diagram, with a focus on food production, soil and water resources, and income (Figure 4.10). For those in the dashed boxes, quantitative indicators have been collected at the grid-cell level: water availability, soil fertility and degradation, population and infrastructure density, and income. The cluster analysis is used to identify correlations in indicator space. Six of such vulnerability profiles are represented in Figure 4.11. Each profile is associated with grid-cells in space, and these are shown for each of the eight profiles in Figure 4.12. It is seen, for instance, that the resource poor regions with people in severe poverty – and characterised by high infant mortality – are largely located in a small band south of the Sahara Desert. A subsequent, more rigorous categorisation of vulnerability patterns in drylands indicates that it is indeed possible but not simple to identify generic patterns of potentially unsustainable developments in fragile regions (Sietz *et al.* 2011).

Why is it useful to identify and study such archetypical patterns? In many places on Earth, people feel powerless in the face of forces that threaten their quality of

[16] Another worked out case are forest ecosystems, which exemplify the conflicting claims on land: exploitation for human use in agriculture and for (fire)wood versus conservation to prevent loss of biodiversity.

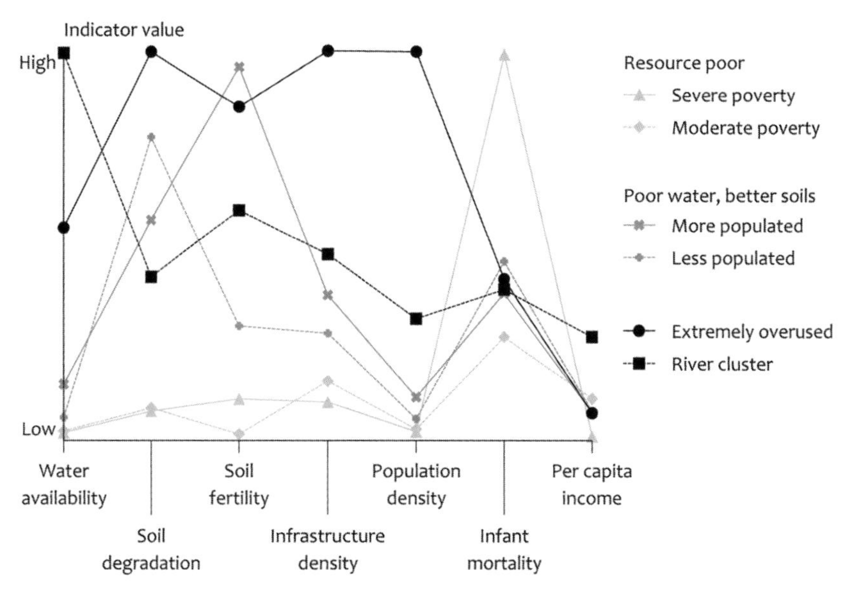

Figure 4.11. Six typical vulnerability profiles in drylands worldwide as arising from the cluster analysis. The indicators are along the x-axis. The indicator values of the respective cluster centres, from low to high and normalised (min/max), are along the y-axis (Kok *et al.* 2011).

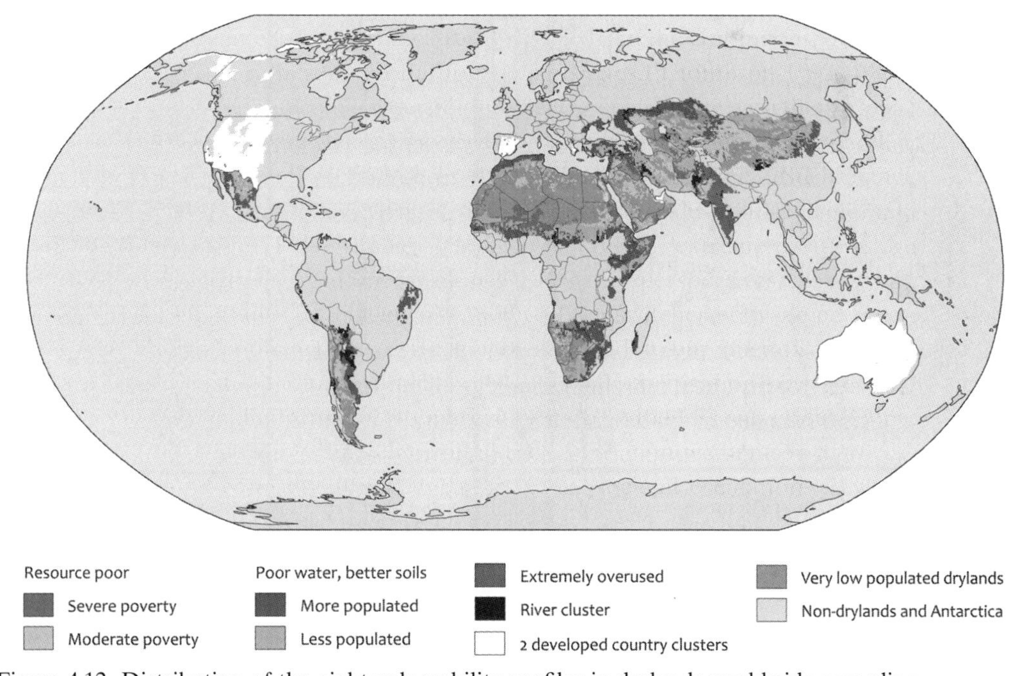

Figure 4.12. Distribution of the eight vulnerability profiles in drylands worldwide according to the vulnerability profiles in Figure 4.11 (Kok *et al.* 2011). (See color plate.)

life and appear to be inevitable. Many assessments are done all over the world to understand the causes of deteriorating quality of life and unsustainable development trends. Unfortunately, the insights gained and lessons learned often remain local or regional in their impact, due to the specifics of the situation. If archetypical patterns can be identified, such local and regional assessments can provide an analogue and a remedy for other places in the world. People in one situation may then learn how to respond, adapt and heal from insights about what happened elsewhere. It provides generic insights in the dynamics that create vulnerable situations for people and can be used as a basis for further analysis of opportunities and responses to reduce vulnerability.

4.5 Summary Points

In the past three centuries, an exponential growth has taken place in the number of human beings and their activities. The historical background to the changes since 1700 are the advent of industrial capitalism, the scientific and technological inventions and the transition to fossil fuels. Industrialisation started in Europe and its offshoots. Population growth rates started to slow down in the 20th century, as part of the demographic transition. Economic growth has been exponential throughout the period, with worldwide increases in income and a change from agricultural to manufacturing and later to service sector activities. In the second half of the 20th century, industrialism went global and expectations of hundreds of millions of people soared. There are still large income inequalities within and between countries, which is one of the threats to a sustainable future. There are a some other key points to remember:

- The changes took (and take) place in complex interaction with less easily quantified social-cultural changes. The nation-state got new roles, ranging from colonial rule to welfare provision. An extensive global trade network evolved. Values changed during the process of industrialisation, notably from traditional and survival-oriented to secular and self-expressive. The dominant ideas and values of modernity and the politics and organisation of industrial capitalism manifest themselves in the rise of science and technology and empirical rationalism, in innovations and markets as engines of economic growth, and in democratic institutions.
- Industrialisation led to massive and accelerating exploitation of natural sources and sinks, with unintended and unanticipated consequences. Past and anticipated depletion and degradation of natural resources is another, large threat to a sustainable future. The 'Western' development model itself may turn out to be unsustainable.
- As individuals, we need to interpret the situation in a long-term global perspective while appreciating its momentary acuteness and local diversity. Frameworks such as the transition to a new social-ecological regime and archetypical patterns of (un)sustainable development can provide the broader perspective that is necessary, if the transition we are in is to fulfill the promise of a good quality of life for all while staying within the planet's biophysical boundaries.

> ... until recently, ecological regimes have been losing ground vis-à-vis ... two regimes in particular: the money regime and the time regime. Both have, each in their own way, turned people's attention away from the natural environment, and from ecological issues, toward more purely social aspects of the anthroposphere.
>
> – Goudsblom, 2002: 370

> Feed the poor
> tell the truth
> make water-places
> for the thirsty
> and build tanks for a town –
> you may then go to heaven
> after death, but you'll get nowhere
> near the truth of our lord.
> And the man who knows Our Lord,
> he gets no results.
>
> From: *Speaking of Siva* (Vs. 959).
> Ed. A.K. Ramanujan. Penguin Classics 1973

> "La biosphère est tout autant (sinon davantage) la création du Soleil que la manifestation de processus terrestres. Les intuitions religieuses antiques de l'humanité qui considéraient les créatures terrestres, en particulier les hommes, comme des enfants du soleil étaient bien plus proches de la vérité que ne le pensent ceux qui voient seulement dans les êtres terrestres la création éphémères ... "
>
> – Vernadsky, *La Biosphère*, 1928:51

SUGGESTED READING

Overview of the IHOPE Dahlem conference with accounts of past civilisation developments.

Costanza, R., L., Graumlich, W. Steffen, C., Crumley, J. Dearing, K. Hibbard, R. Leemans, C. Redman, and D. Schimel. Sustainability or collapse: What can we learn from integrating the history of humans and the rest of nature? *Ambio* 36 (2007) 522–527.

Models, maps and myths about the role of the natural environment in ancient civilisations.

de Vries, B. de, and J. Goudsblom, eds. *Mappae Mundi – Humans and Their Habitats in a Long-Term Socio-ecological Perspective.* Amsterdam: Amsterdam University Press, 2002.

Economic history of the Industrial Revolution.

Heilbroner, R., and W. Milberg. *The Making of Economic Society*, 12th ed. New York: Pearson Prentice-Hall, 2005.

A natural science overview of Global Change phenomena during the last century.

IGBP Science 4 2001 www.igbp.net/documents/IGBP_ExecSummary.pdf.

One of the best documented quantitative studies on long-term trends in population and economy, with many data.

Maddison, A. *The World Economy – A Millennial Perspective.* Paris: OECD, 2001.

Although a bit outdated, this is still a landmark study about Global Change and the human-induced transformations.

Turner, B. et al. (Eds.) *The Earth as transformed by human action – Global and regional changes in the biosphere over the past 300 years.* Cambridge: Cambridge University Press, 1990.

USEFUL WEBSITES

Historical and Country Statistics

- www.pbl.nl/en/themasites/hyde/index.html is a site under the authority of the Netherlands Environmental Assessment Agency (PBL), which presents the HYDE (*History Database of the Global Environment*), that is, (gridded) historical time series of population and land use.
- www.gapminder.org is a site with large amounts of data on the world and the countries of the world, with a nice and flexible interface.
- gcmd.gsfc.nasa.gov/ is a *Global Change Master Directory* with info, data and links.
- nasadaacs.eos.nasa.gov/index.html is the Earth System Science Data and Services site, operated by NASA.
- www.ggdc.net/databases/index.htm is a database at the Groningen Growth and Development Center of the University of Groningen, with extensive data on economic history and performance.
- www.conference-board.org/data/economydatabase/index.cfm is a site of the Conference Board Total Economy Database.
- www.ilo.org is the site of the International Labour Organization (ILO), with data and information on global labour and employment related topics.
- www-histecon.kings.cam.ac.uk/history-sust/energy.htm is the site of History and Sustainability, with an overview of available historical datasets on energy use and other variables.

Reports and Analyses

- www.unep.org/publications/ is the site of UNEP, with a.o. the four *Global Environmental Outlook* (GEO) reports.
- www.igbp.net is the website of the IGBP with a.o. the Executive Summary of *Global Change and the Earth System: A Planet Under Pressure*.
- www.iiasa.ac.at is a site on various global change issues (population, land use, energy) at the International Institute for Applied Systems Analysis (IIASA).
- www.wto.org is the site of the World Trade Organization (WTO) with trade statistics, amongst others.
- www.gatt.org/ is a site where the WTO and the economic views it defends are criticised.
- www.millenniumassessment.org/en/index.aspx is a guide to the various reports published as part of the Millennium Ecosystem Assessment.
- www.ipcc.ch gives access to all reports of the Intergovernmental Panel on Climate Change (IPCC).

5 Sustainability: Concerns, Definitions, Indicators

5.1 Introduction

Historical accounts are given from a particular vantage point. Ours is the emerging field of *Sustainable Development* research in a *Global Change* context (Turner *et al.* 1990; Costanza *et al.* 2007). It explores human development and the planetary boundaries within which it unfolds (Rockström *et al.* 2009). The dominant frame is given by the natural sciences and their concepts, methods and theories. Their findings represent the prevailing outlook on the world in Western, and increasingly other, societies and are at the core of university curricula associated with *Modernism*. It is, therefore, instructive to briefly introduce the history of the ideas behind this worldview and its essential tenets. How does modern man see *earth*, *life on earth* and *man himself* after these 300 years of scientific advances and economic growth? Such a reflection also makes apparent the limitations of this worldview. Clearly, science and technology have liberated us to a large extent from the controls of Nature. There is no need anymore for Gods who have to be begged for help with prayers, virtues and sacrifices. At the same time, there is growing discontent with the scientific, modernist outlook. A primarily natural science interpretation of problems of development and sustainability and their solutions meets with opposition and critique. Chapters 6, 8 and 10 explore these critiques in more depth.

Against this background, the emergence of environmental concerns and the heightened interest in the notion of (un)sustainable development has to be understood. The second part of this chapter briefly recounts the pedigree of the concept and emphasises the historical context and changes. I abstain from a rigid definition and instead, in the last section, focus on indicators of (un)sustainability and (un)sustainable development that are proposed in order to make the concept more operational.

5.2 Global Change: The Scientific Worldview

5.2.1 The Scientific Worldview: Earth

The history of the Earth is one long process of change. It has become popular knowledge that Earth dynamics are measured in geological time-scales of hundreds of millions of years. Every year new and better reconstructions of all sorts of 'physiological variables' in the earth's system and its subspheres (atmosphere, lithosphere, cryosphere, hydrosphere) are becoming available, such as atmospheric concentration of carbon dioxide, temperature and sea level over the last hundreds of millions of years.[1] Some processes are so slow that humans hardly notice. Others move very fast, even when measured in terms of human time-scales (Table 2.3).

This scientific view of our planet is rather recent. In 1654, the Irish prelate Ussher declared that, according to extensive research, creation had taken place on 26 October in 4004 BC at 9 AM. But at the end of the 18th century, the Frenchman Buffon widened the perspective by his statement that the seven days of creation were actually seven eras of 74,832 years in total (Boorstin 1985). In his influential book *Principles of Geology*, first published in the 1830s, the Scottish geologist Lyell argued that the earth around us can be understood as the outcome of small changes of geological processes at work during very long periods of time. Whereas Newton's timeless laws of mechanics and Linnaeus' elaborate classification of plant and animal species were still consistent with traditional belief in biblical cosmology, these new discoveries and theories made such a reconciliation increasingly difficult.

Among the many discoveries and theories in the earth sciences, two 20th-century theories stand out. Wegener was, with his book *Die Entstehung der Kontinente und Ozeane* (1915), the first to propose in some detail the hypothesis that the continents were moving across the earth crust ('continental drift'). Initially dismissed on physicochemical grounds, this radical idea had developed by the 1960s into the theory of plate tectonics and is considered to be one of the important conceptual revolutions of the 20th century. It led to a whole new understanding of geological observations. For instance, what is known today as the continent of Europe has, during the last 300 million years, not only been subjected to strong temperature and sea-level fluctuations but also drifted thousands of kilometres across the earth's crust.

Milankovitch published his *Kanon der Erdbestrahlung und seine Anwendung auf das Eiszeitenproblem* in 1941. In this book, he attributed – although he was not the first one to do so – ice ages to changes in the angle between the Earth's axis and its orbit around the Sun and the shape of the Earth's orbit around the Sun. At first, meteorologists banished this idea and some even insinuated that Milankovitch had confused astrology with astronomy, which is considered a fatal reproach in the physical sciences. However, by the end of the 1960s, the results of deep-sea sediments and theoretical models of celestial mechanics and climate confirmed that these so-called Milankovitch-cycles had been one of the important causes of the change in ice cover. The Earth temperature has continued to fluctuate in the last couple of million years within a range of about 10°C.

5.2.2 The Scientific Worldview: Life

Ideas about life in ancient civilisations and in medieval times were embedded in religion, but they also reflected the practical insights from farmers and herders and from physicians and medical doctors. The developments in European science in the

[1] See the list of suggested websites for such time-series.

last few centuries have also revolutionised ideas about life. Natural history emerged as a separate branch of science in the 18th and 19th centuries, with the observations of the micro-world with the help of microscopes, the durable taxonomy of species by Linnaeus (1735) and the accurate observations framed in a romantic-holistic *Weltbeschreibung* by Von Humboldt (1814).

But ideas about life were never the same after 1859 when Darwin published his theory of evolution in the book, *On the Origin of Species by Means of Natural Selection, or the Preservation of Favoured Races in the Struggle for Life.* On the basis of his own, and other people's observations and speculations, Darwin argued that populations adapt to (changes in) the natural environment through a hereditary process in which more apt individuals have higher survival rates. Darwin's investigations were the beginning of modern biology and the idea of *natural selection* is still the essence of modern evolutionary biology. It is now scientific canon that in the past millions of years species have come and gone – the debate is about whether the causes behind the changes were evolutionary ('survival of the fittest'), revolutionary (natural catastrophes) or both.

The investigations of how species interact with their environment and with each other led to the science of *ecology*. It unravelled how plants and animals form patterns of mutual competitiveness, cooperation and parasitism, which give rise to complex behaviour of populations over time. The 20th century gave rise to a whole series of discoveries about the mechanisms behind evolutionary processes, both at the macro- and micro-level. Building upon the experiments by the Austrian monk Mendel in the 1850s, the crucial role of information (coding) and its transmission in evolutionary processes became understood. The science of genetics was born.

The life sciences have demonstrated convincingly that change and interdependence are everywhere. The idea of changeless creation has slowly but irresistibly given way to the concept of dynamic life history. With the centrality of Earth gone in the Copernican revolution, the same happened now to the human species: Human beings are no longer divine creations at the apex of the world but descendants of primitive life.[2] These changing views increasingly anchor life in the physical sciences, uncovering the material roots of life and the biological roots of human life. Philosophers such as Bergson tried to stem the tide in the early 20th century with the concept of *élan vital* (vital force), but the fence between organic and non-organic, life and non-life had been pulled down forever. In the 1960s, the chemist Lovelock proposed that organisms not only had adapted to their environment, but that life had created itself the physicochemical environment in which it could flourish (the *Gaia hypothesis*).[3] It brought down the fence between geology and biology. With the advent of complex system science and the cognitive and artificial life sciences, another fence is eroding: the one between matter and mind. For some, it reveals the unity of science. For others, it is the ultimate disenchantment of the world, to use Weber's expression.

[2] At the same time, most people in modern societies will consider human beings to be at the apex of earth history in terms of complexity and consciousness.

[3] One finding that affirms the Gaia hypothesis is that the oldest bacteria used free energy flows to build structures from hydrogen and carbon, and in this way brought oxygen into the atmosphere. The Gaia hypothesis appeared in a more esoteric and holistic gown as Mother Earth – and was vehemently rejected by many scientists (Schneider and Boston 1991).

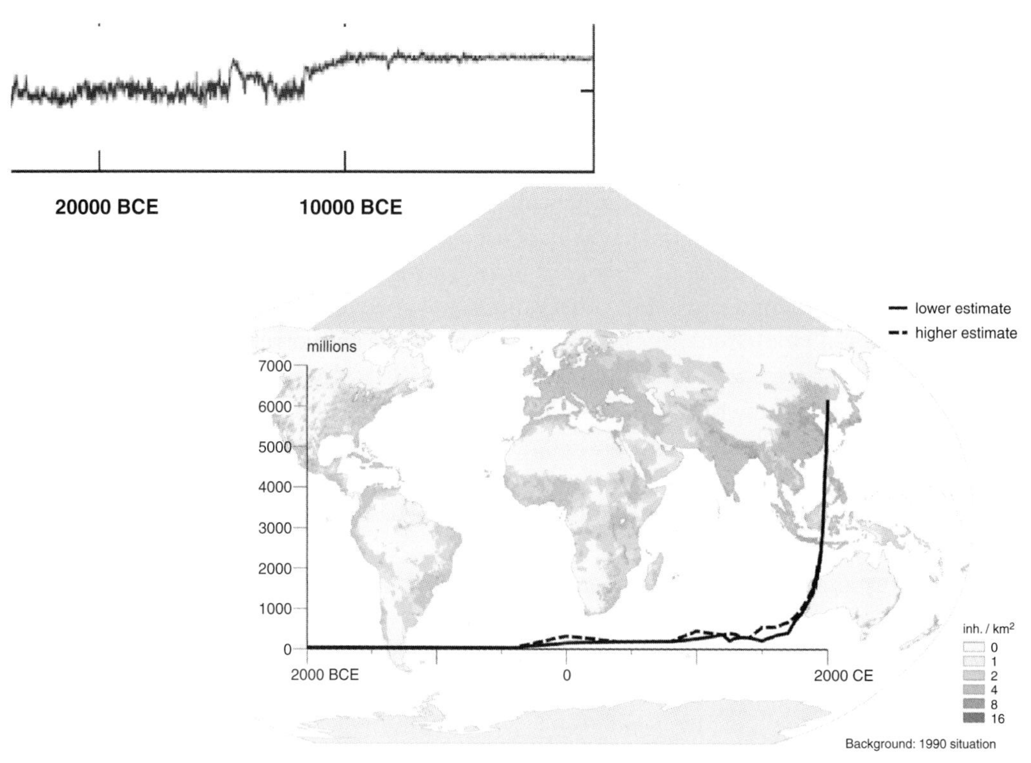

Figure 5.1. The size of the human population during the last four millennia, with an uncertainty margin. For illustrative purpose, the estimated temperature profile during the Holocene is indicated (Feynman and Uzmaikin 2007).

These changing ideas had slow but thorough implications for society. With the advent of geology, earth history was no longer the domain of theologians. With the theory of evolution, their domain further eroded and adherence to the scientific worldview became a sign of enlightenment. People extended their range of action and information beyond their own part of the world. Universalism gained an advantage over ethnocentrism. Spirits and gods no longer played a prominent role. Observations and the search for mechanisms came in its place and the 'mechanisation' and 'mathematisation' of the worldview began. In the 1960s, the full implications of the scientific paradigm became clear, with books like *Chance and Necessity* (1970) by Monod and *The Selfish Gene* (1976) by Dawkins. Witnessing the debates about Creationism and Intelligent Design, this change in orientation is still going on.

5.2.3 The Scientific Worldview: Society

Homo sapiens evolved from earlier *homines* and experienced a slow but persistent growth in numbers during most of the Holocene, which was a period with an exceptionally stable climate (Figure 5.1; Feynman and Ruzmaikin 2007). Only in the last few centuries, the human population started to grow exponentially in numbers and has become so successful that it is an identifiable force in global change: welcome in the *anthropocene* and the anthroposphere. Countless books have been written on what is the newness of *homo sapiens*, the wise and knowledgeable man. Is it his language, his religion, his tools – as *homo faber* or the working man? Or is it his playfulness – as *homo ludens* or the playing man? Or is it his consciousness, or even more: his self-consciousness, that makes the difference? Obviously, the (self) image of man has changed immensely in response to the scientific investigations of the last centuries.

The study of individual and societal dynamics has not yet delivered laws in the way they are understood in the natural sciences. The quantitative is seldom reconciled with the qualitative and formal models are at best a complement to the case studies and anecdotes that prevail in the social sciences. If one can speak of empirical laws based on observations, they are probabilistic and correlative in nature.

There are theories around which *schools* form. They adopt the name of the originator or most ardent protagonist and the essence is communicated through the 'classical works' of its greatest thinkers. In Western science, these are associated with the names of Freud, Adler, Jung and Skinner in psychology and of Comte, Durkheim and Weber in sociology. In economics, the separate schools of thought are usually indicated with the names of their proponents: Malthusianism, Walrasianism, Keynesianism, and Schumpeterianism. Or there is a historical distinction, as between the classical and the neo-classical theory. Formation of theories and schools is a normal and fruitful part of the scientific process. However, the difference is that in the natural sciences one theory survives albeit in adapted form after a period of contest and competition on the basis of evidence and the others disappear into oblivion. Natural selection seems an apt metaphor. Due to the greater complexity of the object of investigation and the more ambiguous and probabilistic nature of evidence, this process is much slower in the social and economic sciences. (In Chapters 8 and 10, I come back to this.) For this reason, it is hard to summarise how social scientists view *man on earth*.

An important question in this situation is, which social science is considered relevant and legitimate? In his book, *Le Phénomène Humain* (1955) the French priest and scientist Teilhard de Chardin advanced the idea of a cosmic unfolding towards a collective consciousness, for which he used Vernadsky's notion of the *noösphere*. In his cosmology, there are three dimensions: the infinitely large, the infinitesimally small and the infinitely complex. He viewed the increase in complexity as a process of unfolding and of interiorisation, in which each human individual partakes. Eastern philosophers had long been venturing similar ideas and these emanated again in the 1960s. For instance, Shri Aurobindo describes in his book, *Le Cycle Humain* (1972), his view of humanity's evolution towards an ever higher consciousness and Western author Capra connected 20th-century discoveries in physics with ancient Eastern philosophy in his book *The Tao of Physics* (1975).

However, such speculative and metaphysical views are not considered legitimate by most scientists. Since the Enlightenment, most social scientists have embarked on a research program and methodology along the lines of the highly successful natural sciences.[4] They favour a view of man and society that considers the experiences of the senses as the most authentic source of knowledge and falsifiability as its hallmark. Their investigations are along the lines of positivist thinkers such as Saint-Simon, Laplace, Comte and their followers. It shows up in the metaphors: The human psyche

[4] Throughout this book, the term social sciences includes economic science, unless indicated otherwise.

seen as a steam engine or a computer, society considered as a self-organising network. It is also evident in the notions borrowed from the natural sciences to describe behavioural and social phenomena: exchange processes on a market as a chemical equilibrium, innovation as evolution and political expansion as reaction-diffusion (§8.2). This book takes the position that the natural science paradigm will ever more penetrate the social sciences with the advent of new computer and communication tools and modelling applications (§10.5). This is not to say that this venture will succeed. I respect and listen to those who argue that the path of empirical reductionism, even in the new clothes of complex system science, is not solving sustainability problems. I leave this topic for further reflection in the next chapter.

From the previous summary, it can be concluded that sustainability has to be found in an ever-changing world. This paradox can be resolved by considering sustainable development as a search for the preservation of quality of life in a continuous play between change and stability. I proceed with a brief history of how the notion of sustainable development rose to prominence.

5.3 Rising Concerns

5.3.1 Early Concerns: Managing Common Inheritance[5]

Although you may think the talk about resource depletion and environmental pollution is a recent phenomenon, as previous chapters have indicated, there have been warnings of overexploitation of local and regional resources throughout the ages. When the exploitation of sources and sinks became more intense with the advent of industrialisation, it became increasingly obvious to all those involved (producers, consumers and governments) that regulations were needed to prevent outright destruction. Amongst the earliest examples of interstate cooperation to manage natural resources are the early 19th-century river commissions in Europe and the late 19th-century agreements on seal hunting and fisheries. The conventions on transboundary rivers addressed primarily the principle of free and nondiscriminatory navigation through issuing permits and enforcing navigation rules. It was in the immediate interest of industrialisation and its river- and canal-based transport needs. The seal hunting led to an arbitration between the United States and the United Kingdom about the right to unilaterally declare ownership and regulation of resources, the motive being the preservation of seals as an economic asset.[6] When steam engines led to a rapid increase in fishing capacity, genuine concerns about overharvesting some valuable fish species led to the first fisheries treaties. In 1902, the installation of the International Council for the Exploration of the Sea (ICES) was tasked to coordinate, among other things, scientific research on fisheries. It was one of the first signs that scientific information was to play a role in resource management.

After World War I, several attempts at international law-making took place under the authority of the League of Nations. One of the topics was 'the exploitation

[5] This paragraph is to a large extent based upon Schrijver (2010).

[6] Regarding the motives behind agreements, Schrijver (2010) gives the example of a 1902 convention about the protection of nature and wildlife with the telling name *The Convention for the Protection of Birds Useful to Agriculture*.

of the products of the sea'. Three considerations emerged in the debates (Schrijver 2010). First, the focus on commercial interests despite the increasing evidence that those cannot and should not be separated from biological interests. Second, the necessity to make exploitation subject to international regulation in view of the non-territorial character of fisheries, particularly of migratory species. Third, the unwillingness to accept that the principle of the 'freedom of the seas' is inadequate with respect to finite amounts of resources, more so when their exploitation intensifies.[7] All three observations are still relevant in the early 21st-century negotiations about resource management. Despite many obstacles, there were some successes during the interwar years, such as a multilateral convention on whaling (1931) and one on the preservation of fauna and flora, notably in Africa (1933).

During these years, there were also international deliberations about the access to raw materials, which became increasingly important for industrialised nations. In first instance, the focus was on commodity regulation in order to reduce the fluctuations in the prices of primary products. A more principled debate unfolded over the ownership and control of natural resources (rubber, metals, fossil fuels, phosphates, etc.). In essence, the rich countries of North America and Europe tried to enforce the rules of the 'free market' in the form of principles that forbid restriction on raw material exports, give equal rights to foreigners regarding natural resource development and ask for the prevention of excessive price increases. The selfish attitude of the then colonial powers and rich countries has not changed much since then, but the negotiating power of many resource-owning countries has increased significantly.

After World War II, the United Nations (UN) had the restoration and maintenance of peace and security as its first goal. Economic growth and expansion of trade were considered the best way to reach it. A secure resource supply was one of the preconditions. It was in this context that UN organisations got involved in debates about the timber shortages and forest restoration in Europe and about putting Middle East oil under UN control. In 1949, a conference was organised, where for the first time, scientists discussed the world resource situation and concluded that it was possible through more efficient use and new techniques 'to support a far greater population than exists today, at a much higher level of living' (Schrijver 2010). However, in the 1950s and the 1960s, the international arena was increasingly dominated by the Cold War rivalry between the (capitalist) West and the (communist) East and by the efforts of countries in the Third World to get control over their resources in a post-colonial world. Resources for development was the foremost concern, not depletion or environment.

5.3.2 The Environmental Movement

The book *Silent Spring* (1962) by Rachel Carson was a wake-up call when it appeared. It dealt with the pollution of air, water and soil in the United States as a consequence of rapidly growing industrial and consumer activities – in short, of economic growth. The population in the United States and, later, postwar Europe had become rich

[7] The principle was proposed by the Dutch jurist Grotius, with the argument that it is impossible to occupy infinite air and water spaces. It was an opposition to claims to sovereignty over the oceans by Portugal, Spain and other countries.

enough and the negative side effects had become visible enough to start worrying and questioning. In the 1960s, the destructiveness of industrialism became even more apparent, with media reporting on massive pollution of air, water and soil and on the health consequences for people. In the eyes of the protesters, environmental destruction was seen as one more manifestation, besides the repression of women and non-Europeans, of the inherently exploitative worldview of science- and technology-based industrial capitalism. Opposition was directed against the imperialist war in Vietnam, the reckless expansion of nuclear weapons and nuclear power, and the disrespectable degradation of environment and nature. Modernity had lost its innocence and industrialism its enchantment.

Many local environmental initiatives emerged, which were organised by citizen groups and gradually institutionalised into a variety of Non-Governmental Organisations (NGOs) representing civic society. In retrospect, the protests rescued the capitalist system by forcing it to adapt to the new 'environmental scarcity' – as it was rescued by the socialist movement in an earlier era by ameliorating labour conditions. The centralised communist systems suppressed all opposition, including warnings of environmental deterioration, and paid dearly for it as later developments have shown.

Scientists played an important role. The largely local actions in civil society were given a global context with two publications in the early 1970s. The first one was the report *Limits to Growth* (Meadows *et al.* 1971) to the Club of Rome, which used a computer simulation model to depict in quantitative detail the possible future trajectories of the human population, its food and industrial output and its resource base. The two main conclusions were:

> 1. If the present growth trends in world population, industrialisation, pollution, food production, and resource depletion continue unchanged, the limits to growth on this planet will be reached sometime within the next one hundred years. The most probable result will be a rather sudden and uncontrollable decline in both population and industrial capacity. 2. It is possible to alter these growth trends and to establish a condition of ecological and economic stability that is sustainable far into the future (Meadows *et al.* 1971).

The report was met with excitement in some European countries and with thorough skepticism among most scientists (Meadows 2006). The response of economists was particularly vehement. In their view, the model was nonsensical because the crucial feedback of prices in markets was not incorporated. This critique was also expressed by (economists in) organisations such as World Bank and the Organisation for Economic Co-operation and Development (OECD).

A second influential report was *The Blueprint for Survival*, originally an issue of the journal *The Ecologist* and later published as book. It had a dramatic message about the irreversible disruption of life support systems on Earth and the breakdown of society, if current trends persisted. The authors expressed the hope that 'Man will learn to live with the rest of Nature rather than against it'. In subsequent years, several more global analyses were made, some also based on computer simulation models. One of these was the *The Global 2000 Report to the President: Entering the 21st Century*, by the Millennium Institute (Barney *et al.* 1980). It informed the United States President Carter that:

> If present trends continue, the world in 2000 will be more crowded, more polluted, less stable ecologically, and more vulnerable to disruption than the world we live in now (1977)... Despite greater material output, the world's people will be poorer in many ways than they are today. For hundreds of millions of the desperately poor, the outlook for food and other necessities of life will be no better. For many it will be worse... At present and projected growth rates, the world's population would reach 10 billion by 2030 and would approach 30 billion by the end of the 21st century. These levels correspond closely to estimates... of the maximum carrying capacity of the entire earth.

It shows the danger of trend extrapolation: In 2010, most forward projections of population in 2100 were below 10 billion people.

The *Blueprint for Survival* publication preceded the first large UN Conference on the Human Environment in Stockholm with a few months. This conference was one of the first signs that environment was recognised as a global concern. At this meeting, the United Nations Environment Program (UNEP) was founded. It gave a boost to environmental science and many environmental problems were identified in scientific detail and subsequently solved or at least mitigated. In retrospect, the ozone layer depletion and the acidification by sulphur and nitrogen compounds are good examples. Apparently, environmental problems could be solved at limited cost and the industrial growth paradigm did not have to be questioned.

However, a number of environmental problems were more difficult to tackle. Examples are the depletion of open-access resources such as fisheries and groundwater reservoirs, the degradation of agricultural soils, the cutting down of tropical forests under the pressures of wood demand and the emission of greenhousegases that cause climate change. Although conferences were organised and conventions were signed about these issues, the causes are more deeply ingrained in the social and economic habits, ideas and institutions of industrial society. There was – and is – no clearcut scientific identification of and an equally straightforward policy response to these problems.

Against this background, the more comprehensive notion of sustainable development in its modern connotation became prominent in the 1980s. It reflected the growing awareness that what was at stake was more than local environmental degradation. The core of the problem went beyond the environmental sciences and had to do with the interconnectedness and speed of the changes – the very idea of development. Therefore, the quest for sustainable development became the expression of the need and desire to face the *world problématique* in a larger, holistic perspective and to reconsider the very roots of modern industrial society with its market-driven innovations, consumerism and income inequalities. This broader, systemic perspective was also the essence of the 1971 *Limits to Growth* report.

5.3.3 Our Common Future?

The launch of sustainable development as the new, overarching concept came with the publication of the report *Our Common Future* by the UN World Commission on Environment and Development (WECD), also named the Brundtland Commission after its chairperson. Its main conclusion was that 'humanity has the ability to make development sustainable to ensure that it meets the needs of the present without compromising the ability of future generations to meet their own needs'. It gave a long list of actions and policy directions in order to achieve the newly stated goal of *sustainable development*, noting that a world in which poverty is endemic will always be prone to ecological and other catastrophes. Its impact was reinforced by the two oil price crises of 1973 and 1979–1980 that confronted the industrial economies with their dependence on oil, and the Chernobyl nuclear disaster in 1986.

Upon reading this more than twenty-year-old report, one is struck by the relevance of the observations and recommendations. It has been criticised for being too optimistic in its assessment of the physical resource base and the ecological absorption capacity (Duchin and Lang 1994). In retrospect, it can also be criticised for its overconfidence in the willingness and capacities of governments to act on behalf of the poor and to cooperate internationally. The fall of communist regimes, the rise of neoliberalism and unregulated financial capitalism and the ethnic-nationalist reaction to globalism – they all significantly changed the prospects for sustainable development. None of them had been anticipated.

In the aftermath of *Our Common Future*, many meetings were organised and many reports were published on sustainable development. The UN and members of the UN system such as the United Nations Environment Programme (UNEP), United Nations Development Programme (UNDP), Food and Agricultural Organization (FAO) and the World Health Organization (WHO) took the lead. In 1992, the UN Conference on Environment and Development (UNCED) was attended by tens of thousands of people. In the outcome, the World Commission on the Environment and Development (WCED) message of balancing environmental protection and promoting development was prominently present. Two conventions were signed, one on climate change and one on biodiversity. The Commission on Sustainable Development (CSD) was installed with a broad mandate to promote and monitor sustainable development and in particular Agenda 21, the international action plan. Ever more groups from society got involved in what became a global movement. Labour unions, youth organisations, women and indigenous people's movements – they all became *stakeholders* in the Earth's future. These groups, broadly referred to as NGOs, were rather loosely organised and represented a wide variety of often local and specific issues, viewpoints and initiatives. During the 1990s, sustainable development also appeared on the agenda of other, sometimes long-established world fora such as the World Energy Council (WEC) on energy, the General Agreement on Tariffs and Trade (GATT) and later World Trade Organisation (WTO) on trade, WHO on health, UNDP on development and the United Nations Population Fund (UNFPA) on population.

By the late 1990s, global environmental change had become firmly established as agenda topic in national and international institutions and meetings. A crescendo built up towards the new millennium with the Rio+5 meeting in 1997 and the declaration of the eight Millennium Development Goals (MDGs) in 2000. One of its manifestations was the first Global Environmental Outlook (GEO), published by the United Nations Environment Program (UNEP) in 2000. It gave a voice to those within the UN who feared that development and environment were not in balance:

> Two over-riding trends characterize the beginning of the third millennium . . . the global ecosystem is threatened by grave imbalances in productivity and in the

> distribution of goods and services. A significant proportion of humanity still lives in dire poverty... if present trends in population growth, economic growth and consumption patterns continue, the natural environment will be increasingly stressed... environmental gains and improvements will probably be offset by the pace and scale of global economic growth, increased global environmental pollution and accelerated degradation of the Earth's renewable resource base (UNEP 2000).

The evaluation of the situation was influenced by, amongst other factors, the disclosure of the serious environmental degradation in the former communist Soviet-Union and Eastern European countries. UNEP'S first Global Environmental Outlook (GEO) ended politely with a message of hope: the trends towards environmental degradation can be slowed and economic activity can be shifted to a more sustainable pattern.

Some important events during the period 1972–2005:

- 1972: UN Conference on the Human Environment (Stockholm)
- 1980: The International Union for the Conservation of Nature (IUCN) presents the World Conservation Strategy
- 1982: Adoption of the UN Convention of the Law of the Sea (UNCLOS)
- 1987: The Montreal Protocol on Substances that Deplete the Ozone Layer opens for signature
- 1988: World Meteorological Organization (WMO) and UNEP set up the Intergovernmental Panel on Climate Change (IPCC)
- 1992: UN Conference on the Environment and Development (UNCED), also called the Earth Summit, leads to the Rio Declaration and Agenda 21 (Rio de Janeiro)
- 1992: Convention on Biological Diversity (CBD) and UN Framework Convention on Climate Change (UNFCCC) open for signature
- 1994: UN Convention to Combat Desertification (CCD) open for signature
- 1995: World Food Summit, organised by the UN FAO (Rome) and a 'five-year later' summit in 2000
- 1997: First World Water Forum, organised once every three years by the World Water Council
- 2000: UN Millennium Summit and the declaration of the MDGs (New York)
- 2002: World Summit on Sustainable Development (WSSD) (Johannesburg)
- 2005: Ratification of the Kyoto Protocol.

In 2002, the World Summit on Sustainable Development was organised in Johannesburg by the UN Commission on Sustainable Development (CSD) around five issues with the acronym WEHAB: Water and sanitation, Energy, Health, Agriculture, Biodiversity. In line with the trend of the 1990s to engage business, there was a growing emphasis on the implementation via partnerships in the form of agreements between governments with and between other societal stakeholders (public-private partnerships). The involvement of even more groups and interests made effective decision making more difficult. When the challenges ahead and the required costs and efforts became better understood, the divergent stakes and viewpoints of the UN member states surfaced.

Box 5.1. *Celebrations go global.* One of the many signs of globalisation, in the spirit of global villages and universal human rights, are the many UN-initiated celebrations. The more than one hundred religious festival days of medieval Europe are being replaced by days commemorating the new world culture: The UN has now twenty-six days in the year to be celebrated (www.unac.org). Quite a few are on sustainable development related aspects, amongst them World Water Day (March 22nd), World Health Day (April 7th), World Environment Day (June 5th), World Day to Combat Desertification and Drought (June 22nd), World Population Day (July 11th), International Day for the Preservation of the Ozone Layer (September 16th), World Habitat Day (1st Monday of October) and World Food Day (October 16th). Also, there is, of course, United Nations Day (October 24th) and days for many other good intentions: International Women's Day (March 8th), World Press Freedom Day (May 3rd), World Refugee Day (June 20th) and World AIDS Day (December 1st). The round of worldwide celebrations started with a citizen's initiative: In the 1970s, United States citizens started Earth Day (April 22nd) as a demonstration for a more sustainable world. The UN celebration days can help create consciousness of global issues and remind the UN every year that there is work to do.

Also, an undercurrent of different views and interests had started to surface. In the words of Bob Dylan in 1963: '*There is a battle outside and it is ragin'... For the times they are a-changin'...* ' The world followed in words, but not in actions the WCED recommendations. With the fall of the Iron Curtain and the proclamation of *The End of History*, a wave of neoliberal capitalism and Information and Communication Technology (ICT) inspired optimism set in and worries about environment and poverty became less fashionable in the postmodern world of riches and fun that emerged. In 1997, the financial crisis in Asia and the 'dotcom' stock market bubble in 2000 signalled the search for short-term gain and adventure. In 2001, after the 9/11 terrorist attack in New York, there was a sudden concern for the consequences of the global and unregulated capitalist expansion. Financial crisis management and anti-terrorism came to dominate political agendas and the willingness and trust to act for a better long-term future declined. In hindsight, in the first decade of the 21st century, the quest for sustainable development lost the momentum and coherence needed to make a sustainability transition happen.

Still, most scientists engaged in Global Change research are convinced that the accelerating changes brought upon the biosphere by the human species constitute a serious threat for the well-being of future human populations. They base their view on scientific observations and interpretations of change processes at all scales. The latest warning is from a group of ecological economists, who in 2009 concluded in a paper on planetary boundaries that:

> ...the planet's environment has been unusually stable for the past 10,000 years. This period of stability... has seen human civilizations arise, develop and thrive. Such stability may now be under threat. Since the Industrial Revolution, a new era has arisen... in which human actions have become the main driver of global environmental change. This could see human activities push the Earth system outside the

stable environmental state of the Holocene, with consequences that are detrimental or even catastrophic for large parts of the world (Rockström *et al.* 2009).

These are broad sweeping statements that are not easily validated or falsified. There are still many uncertainties about the long-term and large-scale consequences of past and present developments. Therefore, scientific findings have to be stated in probabilities and risks, not in terms of predictions and facts. One reason for this, and scientists should acknowledge it, is that a necessarily biased evaluation of the scientific facts will happen whenever there are large and as yet irreducible uncertainties and complexities. We have entered the world of *postnormal science* (Funtowicz and Ravetz 1990). It is one of the important reflections in sustainability science, and we come back to it in Chapters 6, 8 and 10. Let us now look more closely at how these concerns gradually shaped the notion of sustainable development.

5.4 The Notion of Sustainable Development

5.4.1 Prelude: Categories of Goods and Services

The previous sections teach two things. First, according to modern science, everything is connected and in change. Nothing happens in isolation, which is why we need a system's view. Second, resource management is inherently a collective affair. We interact with each other through the use of environmental sources and sinks, which are part of the public space – or, more solemnly, of the 'common heritage of mankind'. Therefore, it makes sense to introduce two notions from economic science: excludability and rivalry. When a good or service loses its use value upon 'consumption', it is called a *rival* (or rivalrous) good.[8] Food is a good example: Once consumed by me, it is no longer useful for you, at least not as food. If a good or service can be used again and again and also simultaneously by many consumers, it is called a *non-rival good*. An example is a beautiful sunset, at least as long as one does not feel bothered by other tourist-consumers. If it is possible to prevent people from consuming or enjoying a good, it is called *excludable*. Most private goods are. If one cannot exclude another person from enjoying it and the owner cannot enforce a meaningful form of payment, the good is non-excludable. The same beautiful sunset is an example. It gives aesthetic and other pleasures that are difficult to exclude people from.

Rivalry is a continuum. In principle, my enjoyment of the sunset does not interfere with yours. In practise, this is only true within certain bounds. Your wish to enjoy a natural scenery or follow a high-quality class in sustainability science may be valid and honoured if the number of visitors or students is limited. If it exceeds a certain threshold, user expectations and actual use start to interfere: The roads are clogged, so you only reach the natural park after sunset, and the classroom is overloaded with noisy students. Congestion makes the good or service rival. Excludability is also a continuum. In principle, anything that can be appropriated can be made exclusive, as some dictators and super-rich people demonstrate. But it may be difficult and not

[8] We will speak of goods but actually refer to the broad array of goods and services, which can satisfy needs and desires and for which to a smaller or larger extent material stocks and flows are involved.

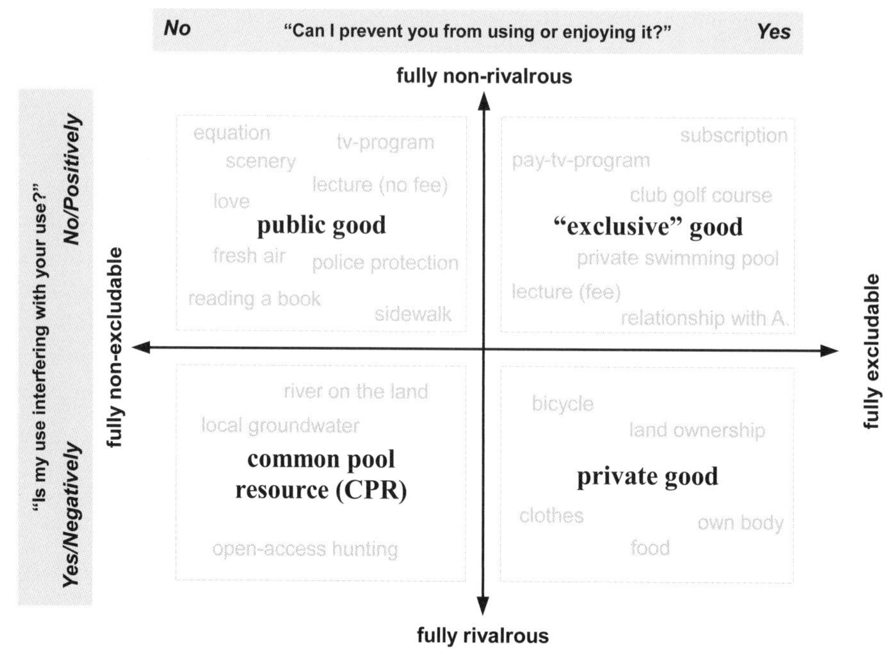

Figure 5.2. Four categories of goods and services, based on the dimensions of rivalry and excludability, with examples from the perspective of the individual in society.

worthwhile. Nobody can be excluded from fresh air near the seashore, UV-radiation from the ozone hole, or a less stable climate.

Based on these two dimensions, economists have introduced four categories (Figure 5.2). If I own a pond with fish in it, I can prevent others from catching and eating *my* fish. The fish are a rivalrous and excludable *private good*. If the pond is not yet appropriated by any individual, either by force or with societal sanctioning, its fish are called a *common pool resource* (CPR): It is rivalrous and non-excludable. The word stems from *the commons* in historical documents that refer to land that is not used by a single person or household during the whole year, but by several persons or households for parts of the year. There are equivalent terms in other European languages, and the common use was usually subject to some kind of rule. The water in a river is non-rivalrous and non-excludable: It is a *public good*.[9] Fresh air and security in the streets are public goods and services too, as it is neither possible nor necessary to prevent people from using them.

In the real world, there are hardly any absolutely non-rival or non-excludable goods and services, but some come close. Examples are given in Figure 5.2 for illustrative purposes. How to value a good or service in practice is context-dependent. Most resources have to some extent the characteristics of rivalry and excludability. This shows up in the ownership and user right arrangements of the people who use the resources. Such arrangements reflect historically grown rules and practises. In Europe and its offshoots, individual property rights have become strongly embedded

[9] Of course, this is not true for use of river water for irrigation, where the upstream user leaves less water for the downstream users. This is a well-known CPR situation for which historically solutions have been found (§ 12.3).

Table 5.1. *Five modes of provision of goods and services*[a]

Modes of provision	Frame 1: constitution of goods	Frame 2: creation rules of the provision	Frame 3: subjective dispositions
Self-provision	No exchange and no commensuration	Allocation to producer	a-personal
Informal provision	Exchange (direct) and commensuration	Social norms	personal
Market provision	Exchange (direct) and no commensuration	Price mechanism, on the basis of preferences and endowments	impersonal
Professional provision	Exchange (indirect or direct) and commensuration	Expert knowledge	impersonal
Public provision	Exchange (indirect) and commensuration	Political decision making	impersonal

[a] Based on Claassen, 2008.

in law, although other possibilities as in cooperatives exist. Many resources, in a broad sense, are appropriated as private goods, the owner claiming the right to exclude others. The owner can also be the state, and state ownership of resources is usually seen as an essential part of state sovereignty. Because resources often extend over more than one private owner or sovereign state, there are often conflicts that have to be resolved. In early times, the issue was often settled by violence – and still is in some regions. Nowadays, governments offer provision of public goods on the basis of political decision making and laws, rules and regulations. For common pool resources, governments negotiate and implement collective arrangements and regulatory frameworks. For private goods and services, the market is the dominant system of provision of goods and services, based on prices. Other modes of provision are autarchic, informal and professional and all had and have their place in the resolution of (resource-related) disputes (Claassen 2004; Table 5.1).

Regarding sustainable development, the sovereign nation-state is the most important actor and stakeholder and this explains the important role of the higher-level political entities such as the U.S. Federal Government, the EU Commission and the UN in framing policies. Many resources are shared by more than one state, and legal arrangements are needed to settle user rights – leaving aside the option of conquest. Rivers and coastal fisheries and the disputes about mineral and fossil fuel resources have already been mentioned. The situation is most problematic with regard to those resources that are outside any national jurisdiction, such as ocean fisheries, rich mineral ores on the ocean floor, and the atmosphere. The overexploitation of these *global commons* can only be prevented by treaties and conventions that are signed by enough countries to perform effective monitoring and punishment of offenders.[10] Chapters 10–12 come back to this in the context of social dilemmas. A variety of collective resources are positioned within the framework of rivalry and excludability in Table 5.2.

[10] Other 'resources' that can be considered global commons are the genetic resources of living organisms and the cultural resources of indigenous peoples, but both rivalry and excludability are more ambiguous in these cases.

Table 5.2. *Four categories of goods and services, based on the dimensions of rivalry and excludability and seen from the perspective of the nation-state*

	non-excludable outside territory	interterritorial	**excludable** territorial
non-rivalrous	**Public goods**		**'Club' goods**
	peace and security	navigable rivers	army-owned estates
	outer space		
	the atmosphere	intercountry rivers and seas	state-owned nat'l parks
	polar regions (arctic/antarctic)	wild seals (Greenland)	mineral resources (territorial)
	migratory species	marine resources (territorial sea)	forests (territorial)
	'high sea' fisheries		
	mineral resources (ocean floor)		
rivalrous	**Common pool resources (CPR)**		**Private goods**

Box 5.2. *Externalities.* Economists call the effects of economic activities that are not, or not adequately, taken into account in the market allocation *externalities.* For instance, a firm benefits from well-educated employees or an efficient infrastructure, but does not pay for it – at least not directly. Inhabitants suffer from air pollution from a local factory, although most of them will not benefit from its employment or profits – at least not directly. The externalities are often related to non-excludable goods or 'bads'. Those who benefit from positive externalities cannot be made to pay for it by those who provide it and those who suffer from negative externalities have no user right or claim which permits them to punish the polluter or be compensated. Obviously, in all these situations, the collective, a government of some sort, has to mediate. How it should do this is the topic of a large part of environmental economics. Solutions are a redistributive tax – the inclusion of the damage in the prices ('internalising'), assigning property rights and negotiating a workable arrangement among stakeholders. Each solution has its pros and cons. Sometimes there is also an issue of rivalry. For instance, if a bus lane is constructed near your house, you benefit without effort or payment – apart from tax money. It is a public service. But if many people start using it, its use becomes rivalrous – and possibly cheaper. Similarly, if a noisy and polluting factory is built near your house, you share in the 'bads' – although you may gain in other ways. It is a public disservice. You may respond to it with in-house airfilters, by moving to another place or starting a lawsuit. In the case of climate change from past carbon emissions, the problem is *de facto* because it is imposed upon people as a public bad and investigating the capability and cost of adaptation have become the dominant response.

5.4.2 Interpretations and Definitions

The list of large international meetings organised since the 1980s shows how environmental concerns gradually expanded to also include development issues and then evolved into the more comprehensive notion of sustainable development. The Rio Declaration in 1992 begins with the statement, 'Human beings are at the centre of concerns for sustainable development. They are entitled to a healthy and productive life in harmony with nature'. In 2002, the Johannesburg Declaration built on this aspiration and expressed the commitment of world leaders 'to build a humane, equitable and caring global society cognizant of the need for human dignity for all'.

Thus, sustainable development has evolved into an ethical guiding principle and leading aspiration of humankind in the 21st century, not unlike the idea of socialism in the late 19th and early 20th centuries and the 1948 Declaration of Human Rights in the late 20th century. Such principles and aspirations do not necessarily have to be defined very precisely to be effective. They should be open for reappraisal and adjustment in the light of new facts and experiences. At the same time, however, there is a clear need to operationalise the idea of sustainable development into feasible objectives and measurable targets. This demands a process of convergence towards a shared notion of what sustainable development is about and how it should effectively be implemented. This book hopes to contribute to this process. Therefore, I present a closer look at a few definitions.

In first instance, the notion of sustainability emerged in circles of nature conservationists and emphasised the aspect of human intrusions into natural ecosystems. In this case, the outlook in the *World Conservation Strategy* (1980) of the International Union of Conservation of Nature (IUCN) was outspoken bio- or eco-centric. The emphasis on the rapid degradation of biodiversity was also the major concern in the next strategy report, *Caring for the Earth* (1990). The Dutch Stichting Natuur en Milieu (1981) presented guidelines that reflect more an environmental than an ecological view:

> Requirements for a sustainable society are:
>
> - avoidance of risks at the macroscale;
> - use of renewable energy sources as much as possible;
> - limitation of the use of non-renewable resources to a minimal share of the available resources, and
> - consumption of renewable resources that does not exceed the amount that is sustainably produced by nature.

Energy was often a major element. In the late 1970s and early 1980s, it reflected the widespread concerns about air pollution, risks of nuclear power – accidents, handling and storage of radioactive waste and proliferation of material for nuclear weapons – and energy supply security. Oil prices were at an all-time high, and the Chernobyl disaster came five years later. Concern about climate change from greenhouse gas emissions only entered in the second half of the 1980s.

The definition offered by the Dutch National Institute for Public Health and the Environment (RIVM) in an influential report *Zorgen voor Morgen* (Care for Tomorrow) represented more clearly the environmental scientist:

> 'Sustainable development requires that exploitation of and pressures on reservoirs (both abiotic and biotic) is such that the powers of regeneration are not exceeded and that flows are not influenced in an undesirable manner' (Zorgen voor Morgen, RIVM 1988).

This definition used explicitly the systems terminology (§2.4). For many environmental NGOs, the ecological notion of carrying capacity was important: Sustainable development is 'improving the quality of human life while living within the carrying capacity of supporting ecosystems' (IUCN-WWF 1991).

Already early on, there was a more anthropocentric orientation on sustainable development. It emphasised unjust and unfair income and wealth distribution and was prominent among poverty and development oriented NGOs and religious institutions. An example is the statement by the World Council of Churches in 1976:

> A sustainable society requires
>
> - attention for the long term;
> - to keep open as many options for future generations as possible;
> - continuous redistribution of global energy resources; and
> - a "contract" with nature in order to reach a just and responsible use of the harvest.

With the *Our Common Future* report in 1987, the anthropocentric interpretation became even stronger. It proposed to include explicitly socio-economic aspects, as is shown in one of the most widely quoted definitions:

> Sustainable development is development that meets the needs of the present generation without compromising the ability of future generations to meet their own needs (WCED 1987).

The report argued that developments going on in the world, both in the North and the South, were environmentally, socially and economically not sustainable. Warning that behavioural changes at all levels would be needed, it described the characteristics of a sustainable society:

- a political system that ensures the effective participation of citizens in decision making;
- an economic system that generates value added in an intrinsic and sustainable way;
- a social system that is capable to resolve tensions caused by inharmonious developments;
- a production system that keeps the natural base intact;
- a technological system that searches continuously for new solutions;
- an international system that ensures sustainable structures for trade and finance; and
- a management system that is flexible and can correct itself.

This is about as broad and ambitious as the aspiration for a better world can be.

Another difference in emphasis started to influence the definition and the debate: *economy* versus *ecology*. When environmental and other economists got involved, they stressed the benefits of economic welfare that were in their view undervalued:

> '[sustainable development] requires the maximisation of the net benefits of economic development in such a way that services and the quality of natural resources are maintained' (Pearce 1990:24).

The World Bank joined with the World Development Report 1992, *Development and the Environment*, in which it observed that more than one billion people lived in acute poverty and had insufficient access to natural resources. It also advocated the theory that economic growth caused environmental problems with adverse consequences, but that they could only be solved by 'resource-efficient economic growth'. It summarised what it called the prevailing paradigm of worldwide economic growth, based on the belief that:

- the negative effects of economic growth can be reduced substantially, and
- economic development is possible without repeating the industrial countries' mistakes ('leapfrogging').

This requires:

- policy and investment to be directed towards the more efficient use of our resources;
- replacement of scarce resources; and
- application of environmentally benign technologies.

The three P's: People, Planet, Profit adage emerged in this period as a short name for the insight that neither environmental degradation nor poverty can be eliminated without profitable business. Economists, notably at the World Bank, invented in this period the broadening of the notion of capital. Besides economic capital, there is natural (or ecological) capital, human and social capital, and cultural capital to be considered. These concepts underline the durability and inertia of many processes and activities, but they are difficult to operationalise and remain rather vague. With the economist's outlook came a more positive attitude towards technological solutions and business involvement, and a more practical and action-oriented orientation.

The economy versus ecology debate has not ended. 'Mainstream' economists – many of them from the less industrialised countries – emphasise the need for economic growth and poverty reduction. Most ecological economists argue that economic growth will not relieve environmental pressure and, instead, jeopardise quality-of-life improvements and exacerbate unsustainability. The debate reflects the broader tension between anthropocentrism versus ecocentrism and between a focus on Man versus a focus on Nature. Among the followers of a more economic orientation, a distinction can be made between those who emphasise *equity* concerns and those who focus on *efficiency*. The former group stresses that more equitable access to resources and income distribution and the elimination of extreme poverty are preconditions for sustainability. Many poor people have no alternatives other than using their resources in an unsustainable way, a situation that is aggravated by exploitative trade and aid patterns. This view is prevalent among development NGOs and UN organisations like UNDP and implies that the quest for sustainable development should focus on breaking the poverty-overexploitation cycle. This asks for better governance and institutions and internationally coordinated trade and aid regulation. The latter group, with a focus on efficiency, equates sustainable development in essence with environmental protection and argues that economic growth is a precondition, in the sense of affordability, for sustainable development. Measures and policies should not impede economic growth and be done in the economically most efficient way. Prices must be market-based 'optimal'. It is the prevailing view among the business and government elites. Of course, there are multiple shades of grey in between.

5.5 An Indicator Framework for Sustainable Development

5.5.1 From Principle to Action: Indicators

We have seen how attempts to make sustainable development an operational and action-oriented concept have led to more comprehensive frames and definitions. The representation in terms of economic, natural, human and social capital, the three P's and the Nature, Society, Wellbeing, Economy in the ISIS methodology (AtKisson 2008) are examples. The evolution of definitions shows divergence in emphasis on Nature, on income and on poverty. How to connect the broad principles and goals to local and global action? One answer is: through indicators. There is vast literature on (performance) indicators in the management literature. However, I confine the discussion here to specific sustainability indicators (Kuik and Verbruggen 1991; UN 1996; Parris and Kates 2003).

Actions at the individual, group or (inter)national policy level happen with a certain degree of rationality. Figure 5.3 represents a system's perspective on the decionmaking process (§2.4; Meadows 1998). Suppose that you wish to change some situation in a desired direction and that this requires interference with some state variable X. Your perception of the situation (X_P) may differ from the actual situation (X), due to delays, filtering and so on. If the desired situation is X_G, then your action will be based upon the difference between X_G and X_P. The discrepancy between goal and perceived reality drives the desire and action to change. This simplified representation suggests that 1) a goal (target, objective or purpose) has

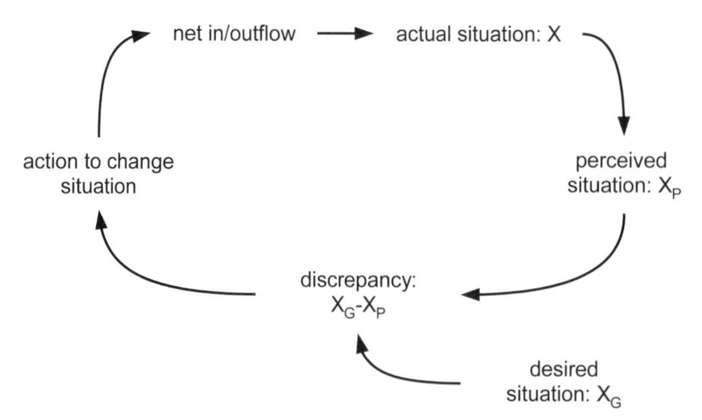

Figure 5.3. A system perspective on decision making and the role of indicators.

to be formulated, and 2) quantitative and qualitative descriptions of the goal and associated indicators should be formulated for operational action and monitoring.

First, an example at the personal level with you as house-owner. The situation: You use E units of energy per year and pay P €/year for it. You consider the situation undesirable and set the goal to reduce E by half in two years time. Your prime indicator is P, although your goal is stated in terms of E. Your perception may be somewhat wrong, for instance, because last year was exceptionally cold. Nevertheless, your task is simple: After two years, you check whether your energy use has halved as a consequence of all the actions you have undertaken.

Suppose, you live with four other people in one house together with the same energy use E and energy bill P. You have the same goal of halving energy in two years – but you want the other four inhabitants to do their share. You cannot exclude them from using energy and their energy use competes with yours in the sense that your effort to reach the common goal increases when one or more of the others don't care. It is a public good or common pool resource situation. You negotiate with the others first about your goal, because some fellow inhabitants may think it is nonsensical, impossible or both. Then, you negotiate about the distribution of the burden, as there will be different preferences regarding technical options like buying efficient lightbulbs or behavioural options such as wearing a sweater more often – each with their specific indicators such as cost.

Here is another example, but now at a town/region level. The province of Utrecht is among the most sustainable regions in the Netherlands according to the Triple P Monitor, designed and implemented by employees of a Dutch bank.[11] A score is given to each of the forty administrative regions along the dimension social (people: societal coherence), economic (profit: income generating capability) and ecological (planet: quality of nature and culture). The scores are based on a dozen properties, each in turn derived from forty-five characteristics. The results are shown in maps, one for each of the three P's and one for the aggregate.

The exercise uses statistical data in first instance. This gives it an objective appearance. However, the data have to be weighed in order to construct a single aggregate indicator for each of the three P's. This is particularly difficult when certain characteristics are ambiguous. Once, in a discussion about sustainability indicators, someone from the Netherlands suggested the number of policemen per thousand inhabitants as an indicator. She meant it as positive because the higher police count meant more safety (better). Another participant from an African country said it is negative in his country, because the higher the police count, the more trouble (worse). Another problem with approaches like the Triple P Monitor is that it omits dynamic and spatial relationships that tie regions to their past and to each other. A last and related problem is that ends and means are to a certain extent reversed: The available data are the basis for evaluation (the means), whereas the goal of the exercise, namely the promotion of sustainable activity patterns, remains largely implicit (ends). Therefore, such indicator maps are certainly useful, but they do not resolve the issues surrounding sustainability or sustainable development indicators.

There are some practical considerations about the construction and use of sustainable development indicators (SDIs). The essential role of an SDI is to guide the

[11] www.uu.nl/SiteCollectionDocuments/Corp_UU%20en%20Nieuws/triple_p_monitor.pdf.

decision-making process after a goal is agreed upon. In first instance, the SDI can be based on statistical data such as trends over time and distributions across space. This limits them to what can be and actually is measured and collected. In other words, SDIs provide quantitative information, but possibly not about what we value or what we consider important. But the advantage is that there can be precision about what is measured and about how it is measured, even though measurement errors and deliberate falsification cannot be excluded.

Information and, thus, SDI are more than statistical data. Data in time and space have *structure* that is at least partly captured in scientific models. One might say that indicators are 'on the outside' of scientific models. One disadvantage of demanding structure is that it limits SDI further to what can be and actually is modelled. It also runs the risk that an incomplete and partly value- and discipline-biased model determines the goal – or means determining ends. But an advantage is that SDI can be related to past trends and to other SDIs so that unrealistic ideas about their evolution are exposed.

How an indicator is presented is important for its (lack of) usefulness. For instance, the installed capacity of wind turbines in the world can be seen as an indicator of the progress towards a sustainable energy future (Figure 2.7b). But one can also present the same information in a different format. If windpower capacity is shown as a fraction of total installed capacity, the message is that wind turbine capacity is still only about 4 percent of the world total. If the electricity generated is used as indicator, it tells you that windturbines generate less than 2 percent of world electricity. If the subsidies for wind turbines are shown as an indicator, one realises that growth may not stop overnight if subsidies are abolished. This is not to say that installed capacity of windturbines is a bad SDI – it only warns for simplistic interpretations. A few additional model-based indicators already make it more accurate, in particular if the important causal relationships are also shown. But there is clearly a trade-off between completeness, on the one hand, and transparency and ease-of-use, on the other.

5.5.2 A Sustainable Development Indicator System (SDIS)

A satisfactory single indicator for sustainable development, as goal and performance measure, will probably never be found. A single indicator will always fall short in view of the diversity of issues and actions considered in sustainable development. Besides, its construction will always need a subjective and value-laden weighing of the components. It may be better to speak of a sustainable development information system (SDIS). One can make a list of items and conditions for such an SDIS. First, its construction is an *interdisciplinary* effort, with inputs from physics to philosophy. It must be possible to quantify or find adequate proxies and assign units to the indicators in the SDIS: physical stocks and flows to describe ecosystems and societal metabolism, monetary units to describe economic transactions and, finally, other units to describe the complexities of human society.[12] Conversions into units per capita or in units per land area or normalised with respect to a given year or reference situation can make the SDIS more informative.

[12] A widely used framework are the country-based economic input-output (I-O) tables in combination with satellite accounts for natural resource stocks and flows (§ 14.4).

Second, the SDIS should be *balanced* and address essential dimensions of sustainable development of a (sub)system such as social equity and sufficiency, economic efficiency, and environmental and ecological sustainability. In short, they are about quality of life. The SDIs in the system must be acceptable for the scientist and accessible for the politician. They communicate well, thanks to good visual and verbal representation, such as interactive views, maps, animations and so on. They are also cheap, making their construction and maintenance a relatively simple task for the statistical bureaus of countries in cooperation with scientists.

The elaboration from SDIs to SDISs can start with the inclusion of simple and transparent *models* to provide context. For example, the use of resources can be connected to the known reserves, the rate of emissions can be linked to its accumulation rate in the environment and available food per capita can be indicated in relation to a healthy nutritional diet. The system boundary, indication of mass and monetary flows and of property rights and damage across boundaries must be well defined. Distributional impacts across a population are shown in the form of distributions across income scales and location.

Clearly, the construction of an SDIS requires the art of compromise, and the quantitative has to be complemented with the qualitative. One can raise the expectations about a SDIS in two directions. The first direction is preferred by the scientist, the second direction by the politician. The first one tries to link the SDIS stronger to scientific models, preferably models with good reputation among scientists and legitimacy among politicians. This is especially true for SDIs related to the Global Change phenomena, such as the causes and effects of resource availability and depletion, land productivity and degradation, environmental quality, biodiversity and so on; such an embedding has great value added. It also gives flexibility in the sense that new insights and changing interests are easily accommodated by creating and adding new model-based indicators. As previously said, a danger is that models are often intransparent to all but the experts, and they are discredited when the integrity of the expert scientists' are in doubt or when there are more models with conflicting outcomes. The preferred solution to this is to indicate explicitly the uncertainties involved (PBL 2010a).

The second direction to go is to see the construction of an SDIS primarily as an ongoing and *participatory process* among the group of stakeholders involved (Bossel 1998; Atkisson 2008). Such direct involvement tends to focus on local situations and to bring in personal experiences and insights. This gives the SDIS more legitimacy and more support for action. Disadvantages are that the participants may lack relevant scientific knowledge, their interest and attendance may be volatile, and the focus on their own local quality of life can give a bias to the short-term and the incidental. As such, a subset of worldviews may come to dominate the process and the outcomes. The challenge is to widen the circle towards quality of life in an inclusive worldview (see Figure 1.1).

5.5.3 Quality-of-Life–Oriented and Aggregate SD-Indicators

The notion of quality of life preceded the concerns about sustainable development and there have been earlier attempts at constructing quality-of-life indicators. The underlying intention was to improve the life of human beings, with the corresponding goals being conceived in the spirit of 19th-century utilitarianism, Roosevelt's New

Deal and the UN Charter on Human Rights in the 20th century. There is now an enormous and growing amount of data available on the social and economic characteristics and trends of countries, as can be seen from the suggested websites at the end of this chapter. These are used to construct *quality-of-life indices* from national averages on infant mortality rates, life expectancy, energy use per person, household size, number of persons per physician, literacy level of women and so on. These are gradually extended to include aspects of environmental and ecological sustainability. This approach is in agreement with an emphasis on quality of life as the essential ingredient of sustainable development.

One early example of a quality-of-life index is the one introduced by Forrester (1973) in the early simulation models of world dynamics. He defined it as: 'Quality of life (QL) is a measure of performance of the world system . . . computed as a quality of life standard (QLS) multiplied by four multipliers derived from material standard of living, crowding, food, and pollution' (1973). Its value over time was calculated from the model simulation. For such aggregate, normalised indicators is the coupling to a dynamic model almost a necessity, because it is the change and not the value *per se* that matters.

A benchmark of setting targets in combination with indicators was the World Bank report, *A Better World for All* (2000). It set goals for the period 1990–2015 for a number of issues: reduce the proportion of people living in extreme poverty by half[13]; enroll all children in primary school; and make progress towards gender equality and empowering women by eliminating gender disparities in primary and secondary education by 2005. Also, the other goals focused primarily on health, notably child and maternal mortality. Only the last of the seven objectives was about the environment: implement national strategies for sustainable development by 2005 so as to reverse the loss of environmental resources by 2015. In essence, these are also the earlier mentioned goals accepted by the UN as the Millennium Development Goals (MDGs):

- Eradicate extreme poverty and hunger;
- Achieve universal primary education;
- Promote gender equality;
- Reduce child mortality;
- Improve maternal health;
- Combat major diseases; and
- Ensure environmental sustainability.

Each of these goals has its own set of indicators, ranging from the local to the global. They reflect the worldview of Modernism explicitly or implicitly and cover only part of the sustainable development spectre.

The last category worth mentioning are aggregate indicators that try to combine the three P's. The best known is the *Human Development Index* (HDI), which

[13] The goal was from 30 percent to 15 percent, and it was 25 percent in 2000. Since then, the definition of extreme poverty has been adjusted. The revised estimates suggest a drop from 41.7 percent in 1990 to 25.7 percent in 2005. See www.socialanalysis.org for a critical discussion of the measures used, for instance, extreme poverty.

was proposed in the 1990s as a complement or, for some, an alternative to the most widely used indicator of welfare, namely the Gross Domestic Product (GDP) per capita (§14.5). The HDI is the geometric mean of a life expectancy index, an education index and an income index and thus takes into account health, education and standard of living aspects of quality of life. The HDI is slowly gaining recognition as a more adequate measure than income, particularly for low-income countries.

Within a Global Change framework, there are a number of aggregate indicators that have become prominent in the sustainable development discourse. The large datasets on Global Change can best be considered as signals of the State of the World.[14] Most of these data are given as (global) maps, and I refer to this chapter's suggested websites for more information. These data are not easily aggregated – just think about how you would construct an indicator such as 'average global surface temperature' or a single index of land degradation. Many data are related to specific environmental problems, such as ozone layer depletion, land erosion and soil acidification. For most of these, one can speak of indicators, as there are regional or global targets either already accepted or under negotiation, which are part of a SDIS. There are many data available on resources, but their more regional or local features make it less amenable to aggregate indicators other than (world) market price and reserve-production ratio.

There has been one explicit attempt to include other forms of capital than economic/financial capital into the economic equation: the *Genuine Savings Rate* for countries (World Bank 1997). The rationale is that natural capital is destroyed, often reinforced by subsidies, for short-term gains and is not compensated by other forms of capital, for example, human capital. The inclusion of natural resources such as oil and forests as 'natural capital' on the balance sheet indicates that development in some countries was and often still is unsustainable. It implies a downward correction of net economic growth with several percentages, resulting in negative values. It suggests that economic growth is only sustainable when the total amount of capital is sustained – with the implicit assumption that the various forms of capital are substitutable (§14.3).

A second aggregate indicator that tries to take into account the negative parts of economic growth is the *Genuine Progress Indicator* (GPI). This later developed into the *Index of Sustainable Economic Welfare* (ISEW) for countries. Its main assumption is that GDP and income (GDP/cap) is a false measure of well-being because it counts many undesirable or harmful activities as positive. Therefore, those activities – such as pollution abatement but also car accidents and military adventures – are to be substracted from the official GDP figures. It has been found that most affluent countries are beyond an optimum income since the 1980–1990s, and that further GDP growth is no longer a net increase in well-being measured as ISEW. But this so-called 'threshold hypothesis' is controversial and the outcomes depend, of course, on the weighing factors applied.[15]

[14] Under this heading, with the subtitle *Progress Towards a Sustainable Society*, the Worldwatch Institute has published an annual report since 1984 (Brown *et al.*).

[15] It is, therefore, instructive to do an interactive exercise with the ISEW on the site of Friends of the Earth, www.foe.co.uk/community/tools/isew/templates/storyintro.html.

A third aggregate indicator is the *Ecological Footprint (EF)*.[16] It has more outspoken environmental roots. It expresses environmental impacts in terms of the equivalent space used and is calculated by adding up the cultivated land required for growing crops, pasture for grazing animals, forests for timber production, the space related to fishing grounds, the land use for the built environment (housing, transportation, infrastructure and power stations) and other space requirements. In addition, fossil fuels are converted into equivalent space via the biologically productive area needed to sequester enough CO_2 to avoid its increase in the atmosphere. It received wide publicity through a 2002 paper stating that the human population was now using 1.2 Earth – a clear case of ecological overshoot (Wackernagel *et al.* 2002). The EF is also calculated for countries. It is severely criticised by scientists for its lack of scientific consistency and the neglect of other externalities than negative environmental ones (Opschoor 2000; Grazi *et al.* 2007). However, it has acquired a solid position among large groups of people as a concept that links the individual to the state of the planet.

5.6 Summary Points

The scientific worldview teaches us that everything is in flux and that natural selection is the mechanism of change. Sustainable development is a search for the preservation of quality of life in a continuous play between change and stability. Regarding the emergence and use of the concept of sustainable development, there are a couple of points to remember:

- Since the 1950s, there have been warnings about resource supply limitations to development. In the 1970s, pollution of the environmental sinks became another concern. In the 1980s, development and environment came together in the UN report *Our Common Future* in the term *sustainable development.*
- In the course of the 1990s, the ambiguity and plurality of the concept became clear in divergent emphases on nature versus culture and on efficiency versus equity. A common denominator is to have a balance between the three P's: people, planet, profit.
- Many sustainability issues derive from the collective (public) character of resources and the interaction between resource users. This has been formalised in the notions of excludability and rivalry, and associated categories of public goods (PG) and common pool resources (CPR).
- To implement actions to reach a goal, one needs indicators. Sustainable Development Indicators (SDI) are constructed on different scales and at different aggregation levels. Some aggregate SDIs are the Human Development Index (HDI), the Genuine Savings Rate, the Index of Sustainable Economic Welfare (ISEW) and the Ecological Footprint (EF).

[16] An indicator constructed and organised along similar lines is the Water Footprint (www.waterfootprint.org). Also, the Human Appropriation of Net Primary Production, as a measure of human interference in natural photosynthesis processes, is in this category (Haberl 1997).

The optimist believes we live in the best of all worlds.
The pessimist fears she's right.

Listen, O Lord of the meeting rivers,
things standing shall fall,
but the moving ever shall stay.

Basavanna, Speaking of Shiva

SUGGESTED READING

Lumley, S., and P. Armstrong. Some of the nineteenth century origins of the sustainability concept. *Environment, Development and Sustainability* 6 (3) (2004): 367–378.

A comprehensive and accessible report on indicators, based on a 1996 workshop. The report is available via the Sustainability Institute (www.Sustainer.org) *or* www.biomimicryguild.com/alumni/documents/download/Indicators_and_information_systems_for_sustainable_develoment.pdf.

Meadows, D. H. *Indicators and Information Systems for Sustainable Development*. Report to the Balaton Group. Sustainability Institute, 1998.

Treatment of concepts of excludability, rivalry, commons and public goods.

Perman, R., Yue Ma, J. McGilvray, and M. Common, eds. *Natural Resource and Environmental Economics*. Harlow, UK: Pearson Education Ltd, 2003.

A comprehensive overview of the history of international debates and decision about resource management, and in particular the role of the United Nations.

Schrijver, N. *Development without destruction – The UN and global resource management*. Bloomington: Indiana University Press, 2010.

A conceptual natural science approach to Global Change.

Schnellnhuber, G., P. Crutzen, W. Clark, and J. Hunt. Earth system analysis for sustainability. *Environment* 47 (8) (2005) 10–27.

A good overview of some of the major issues in Global Change science from a natural science perspective. The report can be downloaded from www.igbp.net/page.php?pid=221.

Steffen, W., and P. Tyson. Global change and the Earth system: a planet under pressure. *IGBP Science* 4 (2004).

USEFUL WEBSITES

Organisations

- www.earthsummit.info/ is an excellent list of earth system science related websites.
- www.unu.edu/unupress/rio-plus-5.html is the site of the UN University (UNU).
- www.unep.org/ is the site of UNEP.
- www.undp.org/mdg/ is the site of the UNDP.
- www.fao.org/wfs/index_en.htm is the site on the 1996 World Food Summit.
- www.who.int/en/ discusses the activities of WHO.
- www.globalpolicy.org/ is the Global Policy Forum (UN-based and related to Security Council).
- www.iucn.org is the site of the International Union for Conservation of Nature (IUCN).
- www.iisd.org is the site of the International Institute for Sustainable Development (IISD).
- www.wbcsd.org is the site of the World Business Council on Sustainable Development (WBCSD).
- www.worldenergy.org/wec-geis/ discusses the activities of the World Energy Council (WEC).

- www.globelaw.com/ is about international environmental law and agreements.
- www.eoearth.org/ is an Internet encyclopedia with brief treatments of many for sustainability science relevant topics.

Reports and Conventions

- www.theecologist.info/key27.html gives access to all the papers in the Blueprint for Survival 1972 issue of *The Ecologist*.
- www.millenniuminstitute.net/resources/elibrary/index.html contains models and papers of the Millennium Institute, including the Global 2000 Report to the President published in 1980.
- www.un-documents.net/wced-ocf.htm gives online access to the 1986 report *Our Common Future* of WCED.
- www.johannesburgsummit.org/html/documents/wehab_papers.html is the site with info on the World Summit on Sustainable Development in Johannesburg in 2002.
- www.un.org/millenniumgoals/ is a site with information on the Millennium Development Goals (MDG).
- www.undp.org/mdg/basics.shtml is the site of the Millennium Development Goals (MDGs).
- publications.worldbank.org/ecommerce/ is an overview of the World Bank's publications.
- hqweb.unep.org/geo/index.asp is the site of UNEP's Global Environmental Outlooks (GEO).
- https://www.cia.gov/library/publications/the-world-factbook/appendix/appendix-c.html is the site where you can find the CIA World Factbook with a long alphabetical list of international environmental agreements.
- www.cites.org is the site of the Convention on International Trade in Endangered Species of Wild Fauna and Flora (CITES).
- www.cbd.int is the site on the Convention on Biological Diversity.
- www.unccd.int is the site of the UN Convention on Combat Desertification.

Global Change Statistics and Sustainable Development Indicators

- www.homethemovie.org/ is the site of the movie *Home* by Yann Arthus-Bertrand. It is the Story of Creation retold in the age of science, with a compelling call to stop destroying and become sustainable.
- www.iisd.org/publications/pub.aspx?id=607 is an initiative of the International Institute for Sustainable Development (IISD) to construct and maintain a database on sustainable development indicators.
- gcmd.gsfc.nasa.gov/ on global change indicators is the Global Change Master Directory, operated by the Goddard Space Flight Center, with info on Earth science data and services.
- www.ciesin.org is the site of the Center for International Earth Science Information Network with a vast amount of online available data on natural and social phenomena.
- www.lgt.lt/geoin/ is a site on geo-indicators, that is, magnitudes, frequencies, rates, trends of geological processes and phenomena.
- www.epa.gov/ is a site of the US Environmental Protection Agency (EPA) with data with brief descriptions on ozone depletion, climate change and other topics.
- www.gapminder.org is a site with vivid graphical tools to explore a large set of data. hdrstats.undp.org/en/tables/default.html is the site of the Human Development Index (HDI), operated by UNDP. On the site hdr.undp.org/en/statistics/ Human Development Reports can be downloaded and HDIs are given for countries and years.
- en.wikipedia.org/wiki/Human_Development_Index#2008_statistical_update is a wiki site that contains HDI-data for the last few years and an extensive list of other indicators.
- data.worldbank.org/data-catalog is the site with access to large datasets, for instance on the World Development Indicators (WDI).

- www.worldbank.org/html/opr/pmi/envindic.html is a site on environmental performance indicators.
- www.inequality.org/facts.html features indicators and data on income inequality.
- mdgs.un.org/unsd/mdg/ gives an overview of Millennium Development Goals (MDG) indicators.
- esl.jrc.it/dc/ is an interactive dashboard to explore the MDGs, with a data-based assessment of the MDG goals for all countries and for separate years.
- www.beyond-gdp.eu/download/bgdp-ve-gpi.pdf and dieoff.org/page11.htm report on the Genuine Progress indicator (GPI), an attempt to come up with a more adequate measure of progress and well-being.
- www.foe.co.uk/community/tools/isew/templates/storyintro.html is the interactive Friends of the Earth website on Measuring Progress, with an explanation of the Index of Sustainable Economic Welfare (ISEW) and the ISEW 1960–2000 as compared to GDP for a number of countries; it also offers a sensitivity analysis for the ISEW for the United Kingdom.
- www.isis.csuhayward.edu/alss/geography/mlee/geog2400/2400ISEW.htm is time-series for the Index for Sustainable Equitable Welfare (ISEW) for several countries.
- www.neweconomics.org/projects/happy-planet-index is a website on The Happy Planet Index. An index of human well-being and environmental impact.
- www.footprintnetwork.org/en/index.php/GFN/ is the site of the Global Footprint Network (GFN), with an extensive briefing on the footprint definition and results and the option of an interactive evaluation your own footprint at ecofoot.org/.
- www.ecologicalfootprint.com/ and www.bestfootforward.com/footprintlife.htm are sites where you can calculate your proximate ecological footprint.
- www.waterfootprint.org/?page=files/home is the site of the Water Footprint Network (WFN), along the lines of the global footprint network and definitions and results about water footprint, virtual water trade and other topics.
- wwf.panda.org/about_our_earth/all_publications/living_planet_report/living_planet_report_graphics/ is a site operated by the World Wildlife Fund (WWF), with links to GFN and WFN.
- www.uu.nl/SiteCollectionDocuments/Corp_UU%20en%20Nieuws/triple_p_monitor.pdf gives a description of the Triple P Monitor sustainability maps for the Netherlands.
- www.fwrgroup.com.au/isismethod.html is a site that introduces various methods to construct and explore interactively in a group (indicators for) sustainable development (ISIS methodology).

6 Quality of Life: On Values, Knowledge and Worldviews

6.1 Introduction

The scientific worldview cannot give meaning in itself to our lives and it cannot resolve the ethical questions surrounding sustainability issues. The scientific 'facts' about the world, important and accurate as they may be, have to be complemented with what people value and believe. It is time to have a look at the more subjective, personal side of the quest for sustainable development. Are there empirical data and theoretical concepts about the subjective side of sustainability?

Previously I have stated that sustainable development is about *quality of life*. But what is quality of life – are we merely shifting the problem? Sustainable development is to act *here and now* in such a way that the conditions for a (decent/high) quality of life *elsewhere and later* are not eroded. But for whom and for how long? Throughout history, individuals have struggled to realise their idea of 'the good life', by exploiting environmental opportunities and cooperating with and oppressing others. Since the dawn of civilisation philosophers have reflected on what 'the good life' entails. What can we learn from them?

Evidently, quality of life relates to our needs in a broad sense and to the means to satisfy those needs. What we experience as and conceive of as needs are related to our values. These, in turn, are interwoven with our beliefs. This chapter investigates the notions of needs, values and beliefs in order to discuss worldviews as a way to organise the complex field of (un)sustainable development. It is meant as a theoretical framework that is applied to a variety of issues in subsequent chapters. I do not claim scientific rigour or completeness. The objective is to provide practical context and guidelines.

6.2 Quality of Life and Values

6.2.1 Needs and the Quality of Life

The most widely used definition of sustainable development uses the word *needs* twice (§5.3). Most discourses on development are about needs, wants and desires.[1]

[1] Often, the words 'needs', 'wants' and 'desires' are used more or less as synonyms. They are indeed close in daily language. Desire is usually associated with personal impulses and longing. Wants

Quality of life results from the fulfilment of needs. It is part of *perennial wisdom* that those needs have material and immaterial aspects. In science, this insight has been rediscovered and associated with the needs hierarchy proposed by Maslow (1954). He argued, partly as a rebuttal of behaviourism, that people behave and use their resources in such a way that their physiological needs (food, shelter and so on.) are met first. When these 'basic needs' are sufficiently satisfied, people will orient themselves towards higher needs: safety, belonging and being loved. Even higher on the needs hierarchy are esteem and self-actualisation.[2] Or in Bertold Brecht's famous phrase: 'Erst kommt das Fressen, dann kommt die Moral . . . '. Maslow's theory is a holistic approach that offers the prospect of unlimited opportunities for personal growth, provided that basic needs can be fulfilled.

The different ways in which the notion of needs is approached reflect this hierarchy. Basic needs are defined almost exclusively in terms of nutritional requirements by organisations such as the Food and Agriculture Organization (FAO). Food deprivation, or in common language, hunger, are quantified in great detail and in relation to the location, gender, age and so on of the individuals concerned. Similarly, the World Health Organization (WHO) focuses on the prevalence of various diseases and its determinants. The extensive statistics on the state of health of people provide numerous 'objective' indicators such as the number of HIV/AIDS patients, access to medical services and safe water and expenditures on health education. All these data suggest the possibility to define and measure quality of life in the sense of (conditions for) the fulfilment of basic needs for food, shelter, safety and bodily health.

Yet, even the 'lower' physiological needs have less tangible and measurable aspects. Food requirements are reflecting genetic and climatic circumstances and too little or too much food can affect quality of life negatively. The prospect of starvation or disease may diminish quality of life as much as an actual food shortage or illness. Even for the 'lower' needs, then, an objective definition of needs and of their fulfilment or deprivation in relation to quality of life is difficult. For the 'higher' needs, in a complex society where needs are constructed increasingly in interaction with social exchange and technology, it is even harder.[3]

Economic science 'studies human behaviour as a relationship between ends and scarce means, which have alternative uses' in the 1935 description of Robbins. In economic textbooks, quality of life is addressed primarily as a decision problem of an individual person, who must choose between alternatives in a situation of limited resources and opportunities. Alternatives to satisfy needs – or ends – are ranked in order of their utility for the individual. The rational choice is to satisfy those needs that maximise utility/benefit or minimise regret/cost for a given income. No normative statements are made about neither ends nor outcome, because ethics is declared to be outside the domain of (welfare) economics.

There are two objections to this narrow view of quality of life. First, people's behaviour is often not according to the presumed economic rationality. It involves

are also seen as rather personal. Needs refer to more universal aspects of life. Therefore, I will consequently use the word need.

[2] How strict people adhere to such a hierarchical sequence is a matter of dispute. Immaterial needs as expressed in religion, art, culture and trade are existential, too (James 1902).

[3] The notion of mimetic desire has been introduced by Girard to point out that needs and desires are socially constructed, notably by imitation.

risk and probability assessments, contextual judgments and social imitation and comparison. Second, people's quality of life depends on more than the goods and services one can buy with money. Therefore, some economists prefer the more comprehensive notion of well-being over welfare. But usually, individual income or expenditures expressed in monetary units is used as the prime indicator of quality of life. An example is the notion of a poverty line – the $1.25 a day per person of the World Bank – which is established from the expenditures needed to buy a basket of commodities for survival. Adding indicators of work and living conditions from, for instance, the World Development Indicators gives a more complete picture.

Economists are not denying the social and other aspect, but argue that people's 'revealed preferences' as expressed in their purchases are more trustworthy than the 'stated preferences'.[4] Conversely, social scientists emphasise that quality of life is more a 'subjective' socio-cultural construct than an 'objective' state. 'Human needs and wants are generated, articulated and satisfied in an institutionalised feedback system. They do not appear from thin air but are created by the social interactions that comprise the civic community' (Douglas *et al.* 1998:259). It is, therefore, problematic to construct a quality-of-life index from statistical data at an aggregate level and to engage the individuals whose quality of life is assessed only indirectly via their economic decisions. To overcome this problem, economists are analysing people's behaviour in more depth and on the basis of surveys and experiments.[5]

A clear choice in favour of an experiential, subjective approach to quality of life is made in the expanding field of happiness research. People are asked directly whether they are happy and content with their life or not – the Subjective Well-Being (SWB) approach (Veenhoven 1991). Happiness is then described as *'the subjective enjoyment of one's life as-a-whole'* and correlated with quality of life. The previously mentioned World Values Survey (Inglehart and Welzel 2005) and the European Values Study are broader but use a similar approach.[6] This and other research shows that for the United States and the United Kingdom, the fraction of happy people did hardly change over the last half century despite a threefold increase in income. The scientific literature suggests seven factors that really matter for an individual's experience of 'being happy': family ties, financial situation, work, social environment, health, personal freedom and philosophy of life, in order of importance (Layard 2005). Income plays only a partial and indirect role. This finding, though, may be valid for European and American society only, and even there it may change because of, for instance, a growing sense of crisis or number of immigrants.

Summarising, there is evidence that well-being is a function of the 'objective' social-economic situation (income, family size, education level, health situation and so on) *but also* of subjective experiences of a person's situation (health, job, marriage, community, religion and so on). The objective approach is means-oriented and tends to focus on observable resource opportunities and constraints.

[4] An example is a construction model of health indices across countries, in which functional limitations, self-reports of health, and a physical measure are interrelated to construct health indices (Meijer *et al.* 2011). The authors find that 'health indices correlate much more strongly with income and net worth than self-reported health measures'.

[5] Much research is done in emerging fields like behavioural and experimental economics, where game theory, laboratory experiments and simulation models are combined (Gintis 2005).

[6] The results are usually presented at the country level, which may be misleading because subgroups within countries – for instance, young people or people living in large cities – may differ more from other subgroups than from similar subgroups in other countries.

Box 6.1. *Measuring utility.* Economists disagree on how to measure utility. Some consider utility to be immeasurable and an ordinal utility function the best possible way to describe the ordered preferences of an individual. Others think that a utility function is more than merely an abstract concept and that an empirical, cardinal utility function can be constructed (van Praag and Frijters 1999). The issue becomes even more difficult if one takes interactions among individuals, items without a market price, uncertainty and intergenerational aspects into account. Because there is more to life than money, economists have invented the contingent valuation method to assess the value of those parts of utility that have no market price. Presuming the existence of fixed and individual preferences, people are asked to indicate how much money they are willing to pay in order to maintain a particular (environmental) situation, or to accept as a compensation for its loss. The outcomes are called stated preferences and are different from the revealed preferences, which are derived from how people spent their money. This approach can, at least in theory, account for the non-market aspects of quality of life. One of the critiques is that respondents give answers that are strategic or in other ways biased and do not consider real-world constraints.

The subjective approach is ends-oriented with a focus on experienced quality of life. The means-oriented approach acquires a certain objectivity at the cost of excluding the experiential aspect. The ends-oriented approach has to be derived from questioning individuals with the associated methodological problems. It seems attractive to combine both.[7] Thus, quality of life can best be evaluated with objective *and* subjective indicators. Let us have a brief look at two approaches that combine both.

6.2.2 Capabilities and Satisfiers

Not only realised but also non-realised options contribute to the experience of 'the good life'. It is good to know that there is medical help in case of an accident, police in case of robbery and the countryside in case of stress. This is captured in the *Capability Approach* (Nussbaum and Sen 1993). Capabilities are what a person might wish and is capable to achieve. They represent his opportunities and the set of options from which he can choose. They reflect the potential to fulfill human needs and the freedom to choose which ones to fulfill. A person's set of capabilities depends on the means (income, resources and so on), as in a utilitarian or welfare approach, but also on personal characteristics (creativity, for example) and on social and environmental arrangements (community, access rights and so on) (Figure 6.1). A realised option is called a *functioning*: something a person manages to do or be in leading a life. The capabilities of a person thus reflect the alternative combinations of functionings that he or she can realise. Poverty is seen more as capability deprivation than income deprivation.

The capability approach connects the subjective experience of a good quality of life – *freedom to choose ends* – with the objective resource-oriented aspects – *means to realise ends*. An unstable climate, polluted drinking water or a corrupt police force

[7] Both are admittedly anthropocentric – but who can speak on behalf of (some) other living beings?

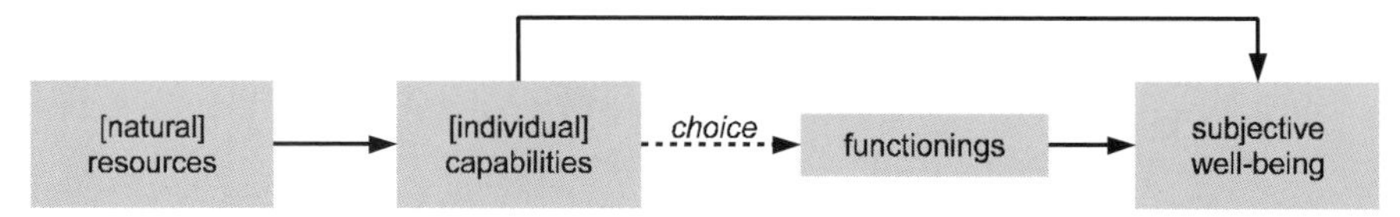

Figure 6.1. Possible causal direction from resources, capabilities and subjective well-being (Robeyns and Van der Veen 2007).

confine a person's capabilities and limit his or her freedom of choice and, therefore, the quality of life he or she can realise. It has an intuitive plausibility in concrete situations. For instance, the existence of a public transport system is a capability: Every citizen may decide to use it or not. Therefore, capabilities are connected to alternative options that can be provided in different ways and in relation to their public or private character (government, markets, and so on), as shown in Table 5.1. But operationalisation of the concepts of capabilities and functionings is difficult. An attempt at concrete implementation is the formulation of basic needs and the use of the available statistics to explore the extent of satisfaction (Nussbaum 2000; Table 6.1).

Another approach to quality of life is the *Human Scale Development* (HSD) theory (Max-Neef 1991). It emerged in the 1980s in Latin America in response to postwar 'developmentalism' and monetarist neo-liberalism. It starts from the postulate that development is about people and not about objects and that the

Table 6.1. *Quality-of-life indicators: from objective-material to subjective-immaterial. The rows indicate, in ascending order, kewords associated with the various aspects of quality of life. The lowest row indicates the source*

	Transcendence		Control over one's environment	Ethical orientation	
	Freedom		Play	Psychological needs	Political freedom
	Identity		Other species	Reproduction	Gender equality
Sahasrara (Crown)	Creation		Affiliation	Coexistence	Community life
Ajna (Throat)	Idleness		Practical reason	Adaptability	Job security
Vishuddha (Throat)	Participation		Emotions	Security	Political stability and security
Anahata (Heart)	Understanding	Interacting	Senses, imagination, thought	Freedom of action	Family life
Manipura (Solar Plexus)	Affection	Doing	Bodily integrity	Effectiveness	Material well being
Svadhisthana (Sacral)	Protection	Having	Bodily health	Subsistence	Health
Muladhara (Base)	Subsistence	Being	Life	Existence	Climate and geography
Chakra wheels	**Max-Neef** (1991)	**Max-Neef** (1991)	**Nussbaum** (2000)	**Bossel** (1998)	**Economist** (2005)

best development process is one that allows the greatest improvement in people's quality of life. Quality of life depends on the possibilities people have to adequately satisfy their fundamental human needs. These needs are understood as a system of interrelated and interactive needs without a hierarchy. There are many possible classifications, and none is final or ultimate.[8] *Existential needs* can be categorised in the four classes of being, having, doing and interacting. Along with the preceding three are *axiological needs* (Table 6.1). Human needs are met in a dynamic process of social construction of satisfiers, on the one hand, and physical means, on the other. A satisfier is anything that contributes to the actualisation of human needs. Satisfiers include space, social norms and practises, and organisational and political structures.[9] They render needs historical and cultural. Economic goods are their material manifestation.

Several classes of satisfiers can be distinguished. Pseudo-satisfiers are items or behaviours that give a false sense of satisfying an axiological need, usually via propaganda, advertisements and other forms of persuasion. Examples are certain medical treatments ('placebo') for subsistence and protection; status symbols and chauvinistic nationalism for identity; and formal democracy for participation. Inhibiting satisfiers satisfy a given need excessively and at the expense of other present and future needs satisfaction. They usually originate in customs, habits and rituals. Examples are economic competitiveness in the name of freedom, which often inhibits the needs for subsistence, protection and affection, or religious fundamentalism providing identity at the expense of understanding and freedom. Some actions destroy the possibility to satisfy needs in the name of another need. For instance, the arms race, national security doctrines and censorship and bureaucracy claim to satisfy the need for protection, but often destroy the need for subsistence, affection, participation, identity and freedom. Conversely, increasing regulation due to crowding and congestion is often experienced by the individual as a reduction in capabilities and thus in quality of life.

Social scientists have long been aware that needs and their satisfaction influence other needs in terms of effectiveness and over time. In his book *Social Limits to Growth* (1977), Hirsch observerd that the need for and value of certain items and behaviours are desired and valued for their very scarcity ('positional goods'). Economists use the notion of *reference drift* for the phenomenon that the needs of an individual influences and is influenced by the needs of others (van Praag and Frijters 1999; van Praag and Ferrer-i-Carbonell 2004; Layard 2005). Needs and their satisfaction also change over time as a consequence of habituation, new experiences and knowledge and novel goods and services. This phenomenon, called *preference drift* or hedonic treadmill, means that a person needs an increasing income in order

[8] This differs from the needs theory of Maslow in which a hierarchy is hypothesised. Max-Neef and colleagues suggest that only the need of subsistence, that is, the need to remain alive, overrules other needs (Max-Neef 1991:17). The matrix of needs should be constructed by individuals participating in a group; experiments with groups in Argentina, Bolivia, Great-Britain and Sweden have been performed and led to significant variations.

[9] There are also dissatisfiers: those rules, structures, etc., that prevent the fulfilment of needs. They may come from oppressive power structures, but they may also stem from increasing regulation because of crowding and congestion. Experienced by the individual as a reduction in capabilities they are, if well designed and implemented, the least harmful for the *total* of individuals. It explains the perennial attempts at escape from such dissatisfiers.

to remain satisfied because he is, in retrospect, less happier with the income gain than he had expected. It is a growth-promoting feedback loop. The elusive character of more income and subsequent consumption is one of the drivers of ever-increasing consumption (Jackson 2009). Table 6.1 summarises categories that have been proposed in order to come to grips with the notion of needs in relation to quality of life. They have different origins, but all contain the idea of 'lower' and 'higher' needs.

6.2.3 Values and Their Measurement

Quality of life is linked to what people *value*. The U.S. Academy of Sciences identified in its report *Our Common Journey: A Transition Toward Sustainability* (1999), four grand values in people's collective aspirations: peace, freedom, development and environment (Leiserowitz *et al.* 2006). Values about what is to be sustained and what is to be developed were listed and have been crystallised since then in more concrete form in, amongst others, the Millennum Development Goals (MDGs). But what are values? The notion of *value* is a complex one and definitions abound in the social science literature, mostly suggesting that values express a belief about a desired end, which guides individual action (Dietz *et al.* 2005; Hitlin and Piliavin 2004). This book uses the following definition (Aalbers 2006):

> a *value* is a prescriptive conviction about desirable behaviour and goals, in particular in a longer-term perspective.

Values tend to change only slowly, at both individual and societal levels, although sudden changes (catharsis, revolution) cannot be excluded. Values can develop and be expressed freely upon reflection – at least, this is the case in most societies and situations and appreciating such freedom can be regarded an important value in itself. Values are manifested in the choices people make among different alternatives and are also an expression of their moral principles.

Around both personal and societal values, there is often a tension between a desired situation and valued behaviour – an ideal – and the actual situation and behaviour. The ideal is usually a more or less shared standard or norm within a group, to which individuals may aspire and thus respond to positively, but that in practise is only partially or not at all realised. Modern-times advertising is a good example of the construction of consumerist ideals and its subsequent exploitation. Sustainable development as a collective ideal is another example. These ideals are values crystallised in ethical codes of conduct and, as such, an important cultural dimension. There is a vast social science literature on this topic. The brief discussion merely serves as a background to the conceptual framework presented later in this chapter.

There are various ways to *identify* and *measure values*. Most widely known and applied is the survey approach, in which people are questioned directly about their values. The ethnographic method makes use of in-depth interviews.[10] But because it is labour-intensive, the survey approach dominates empirical research on values. Many surveys are organised to find out what people value in life, in the public domain as well as in the private sector. The data are used for evaluation of social change and policy and for targeted marketing. One finding is that preferences for technology

[10] Laboratory experiments is a third method. Such an experimental approach gives interesting new insights, particularly with the progress in ICT, but it has limited external validity and generalizability.

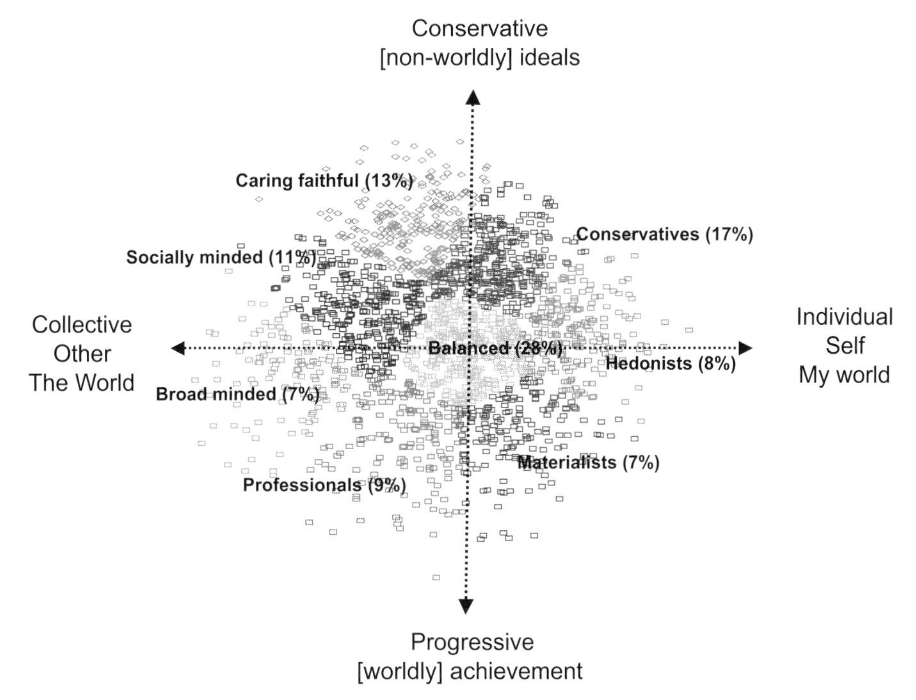

Figure 6.2a. Value orientations constructed from the answers of respondents in the Dutch population. The two major axes are identified as rather robust, but the names for the extremes are somewhat ambiguous (adapted from Vringer *et al.* 2007).

and frugality correlate well with a 'me' versus 'we' dimension. The tension between the individual and the collective manifests itself in questions about the environment as a public good. For instance, a large majority of EU citizens say to make efforts to protect the environment but doubt the effectiveness as long as others (individual and corporations) do not do the same. In the World Value Survey, a majority says that spending money on themselves and their families is one of life's greatest pleasures, but nearly as large a majority agrees that consumption threatens human cultures and the environment, that less emphasis on money is a good thing, and that gaining more time for leisure activities or family life is their biggest goal in life (Leiserowitz *et al.* 2006).

As part of the Sustainability Outlooks at the Netherlands Environmental Assessment Agency (PBL), extensive value surveys were organised between 2003 and 2006 in order to investigate the relationship between people's values, beliefs and behaviour. Data on several thousands of Dutch citizens have been analyzed over longer periods of time and ranked in a multi-dimensional value space. Using factor analysis, eight value clusters or value orientations have been identified (Aalbers 2006). On the basis of interviews and in combination with datasets on socio-economic variables (age, gender, income and so on), the value clusters have been given an 'identity' and a name: Caring (14%), Conservatives (15%), Hedonists (10%), Luxury Seekers (10%), Business-like people (8%), Cosmopolitans (9%), Engaged people (13%) and the middle group of Balanced people (21%). The last category includes people who are not aware of or do not express an outspoken position. A proximate distribution of the Dutch population over the clusters is shown in Figure 6.2a.

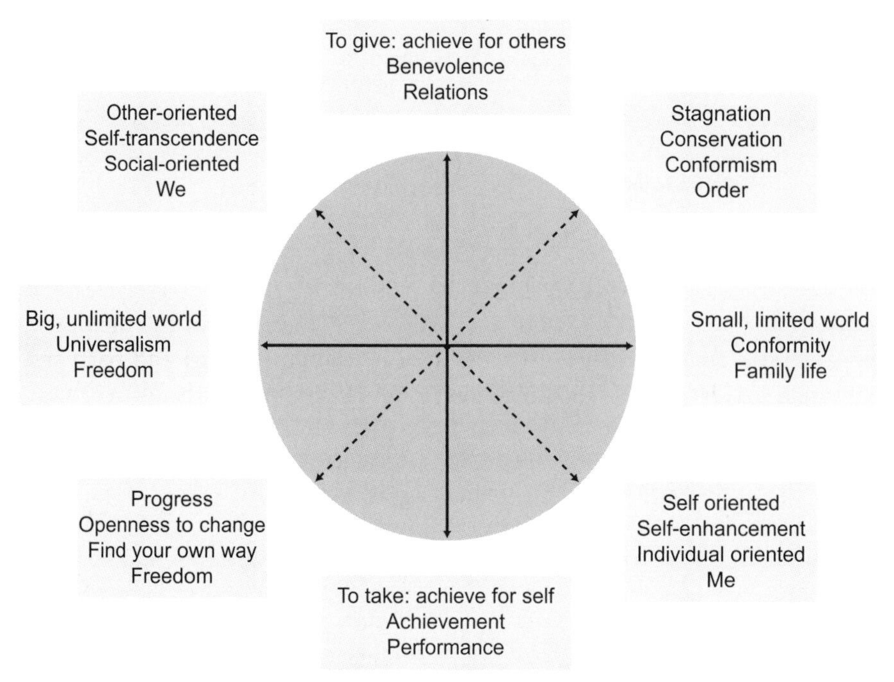

Figure 6.2b. Value orientations, based on the survey outcome in Figure 6.2a, in the broader setting of value survey outcomes.

The value space can be analyzed in terms of pairs of opposing values. A principal component analysis suggests the existence of two dimensions: the contrast *giving* (upper) versus *taking* (lower) and the contrast *small world* (right) versus *big-world* (left) (Figure 6.2b). Correlations between value positions and socio-economic variables such as age, gender, income and education turn out to be weak, albeit in the expected direction (Aalbers 2006).[11] The vertical dimension is close to the contrast *religious* versus *worldly*. The horizontal axis can be associated with an orientation on the own *local community* versus the *world* as a whole or, from a more individual stance, as the contrast between *self* and *the other(s)*. It also reflects the tension between a more egocentric and a more social attitude.

Within this framework, Caring people value immaterial aspects of life and Luxury Seekers value the material ones. Conservatives share an immaterial oriented lifestyle with the Caring people, but tend to focus more on the local 'own' situation. On the opposite side, one finds Cosmopolitans and Business-minded people, for whom the money, travel and opportunities of the Big World are important ingredients of quality of life. Hedonists are small-scale and self-oriented, in combination with a materialistic attitude. Engaged people are their opposites, with a political and idealist perspective and a focus on global-scale issues. The totality reflects the variety of value orientations existing in the population as a whole.

Comparative research on the basis of different and more extensive sets of values suggest that the two dimensions are universal in the sense of context-independent

[11] On the website of the Dutch public survey company, TNS-NIPO, you can do an interactive session that will tell you to which value orientation your worldview belongs. The questionnaire can be found at the site www.tns-nipo.com.

Box 6.2. *Characterisation of worldviews.* Since the term sustainable development came into being, people have expressed different interpretations (§5.3). Four divergent images or views of nature, man and technology can be characterised with keywords: *Technocrat-adventurer*, *Manager-engineer*, *Steward* and *Partner* (Table 6.2; de Vries 1989). There are no strict boundaries, of course. The difference between the views in essence on how one sees man's position in Nature and on the degree of control man can or may have about his own fate. Note that a person's view on science and technology is one of the crucial elements, because technology is to such a large extent a linchpin between the possible and the practical, between heaven and earth. Words like technocrat, steward and partner, have (temporarily?) gotten out of fashion. However, other words with similar connotations take their place.

and trans-situational.[12] The relative position of countries, however, is found to be different. Comparison with other empirical surveys indicate plausible correlations with properties such as lifestyle, self-control and egoism. For instance, Business-minded people are most active and in control of their lives, whereas Caring people score highest on affection and group-orientation. To find our way in this complex mix of values, beliefs and behaviours and its relation with social-cultural change, I introduce two frameworks for analysis. The first one is the Cultural Theory, a social science theory that offers a pluralist framework of individual and collective perspectives in life.

6.3 The Cultural Theory

6.3.1 Stories from the Himalayas and Bali[13]

Himalayan villagers parcel out their transactions with their physical environment to four distinct solidarities, each of which is characterised by a distinct management style. Agricultural land, for instance, is privately owned whilst grazing land and forests are communally owned. But grazing land and forests do not suffer the 'tragedy of the commons' because transactions in their products are under the control of a commons managing institution. Villagers appoint forest guardians, erect a 'social fence', that is, a declared boundary, not a physical construction, and institute a system of fines for those who allow their animals into the forest when access is forbidden or take structural timber without first obtaining permission. If the offender is also a forest guardian, the fine is doubled. If children break the rules, their parents have to pay up. There is fragmentary evidence that such regimes have occurred since at least 600 years ago.

[12] The Rokeach Value System is the most widely known (Rokeach 1973). People are asked to rank two sets of eighteen values each, one on end values and one on instrumental values. The values are portrayed along two dimensions: *self-enhancement* versus *self-transcendence* and *openness to change* versus *conservation* (or traditionalism) Schwartz and collaborators (1994) have extended the method into the Schwartz Value Survey, which lists fifty-six items to be rated by respondents on a scale between 7 and –1. The list with the ten values suggested by Schwartz to have validity across all cultures: hedonism, power, achievement, stimulation, self-direction, universalism, benevolence, conformity, tradition and security.

[13] The story on Himalayan villagers is taken from Thompson in de Vries *et al.* (2002).

Table 6.2. *Sustainable development according to four worldviews*[a]

Technocrat-Adventurer	Manager-Engineer	Steward	Partner
Computer	machine	Garden	'wilderness'
Pioneer	planner	Caretaker	participant
'frontier economy'		'mature ecosystem'	
competitve hierarchy		cooperative solidarity	
Exploitation	control	Management	harmony
courage-creativity	order	care	adaptation
struggle	planning	frugality	sprituality
Anthropocentric		ecocentric	
power-over-others		power-over-yourself	
(technological)	(material)	(just)	(spiritual)
progress	prosperity	welfare	development
science fiction		religion	
technopolis		ecotopia	
engineering sciences		'shallow' ecology	'deep' ecology
economics			

[a] de Vries 1989.

Although they may seem informal and lacking any legal status, these *commons managing arrangements* work well in the face-to-face setting of a village and its physical resources. Drawing on their 'home-made' conceptions of the natural processes that are at work – their ethnoecology – the forest guardians regulate the use of these common property resources by assessing their state of health, year by year or season by season. In other words, these transactions are regulated within a framework that assumes, first, that one can take only so much from the commons and, second, that you can assess where the line between so much and too much should be drawn. The social construction inherent to this transactional realm is that nature is bountiful within knowable limits. It is the myth of *Nature Perverse/Tolerant* (Table 6.3).

In recent years, when forests and grazing lands have suffered degradation for a variety of reasons, not the 'tragedy of the commons', villagers have responded

Table 6.3. *Characterisation of the four cultural perspectives*[a]

Worldview:	Individualist	Hierarchist	Egalitarian	Fatalist
Nature is:	Benign	perverse, tolerant	ephemeral	capricious
People are:	self-seeking	malleable: deeply flawed but redeemable	caring and sharing	inherently untrustworthy
Resources are:	infinite because they are dependent on human genius	finite and to be managed efficiently based on control and knowledge	finite and to be managed with respect and frugality	what can be grabbed
Fairness is:	who puts in most gets most out	rank-based distribution by trusted, long-lasting institutions	what matters is equality of result, not opportunity	non-existent

[a] Thompson *et al.* 1990.

by shifting some of their transactions from one realm to the other. For instance, they have allowed trees to grow on the banks between their terraced fields, thereby reducing the pressure on the village forest, and they have switched to the stall-feeding of their animals, thereby making more efficient use of the forest and grazing land *and* receiving copious amounts of manure, which they can then carry to their fields. In other words, transactions are parceled out to the management styles that seem appropriate and, if circumstances change, some of those transactions can be switched from one style to another.

Because they are subsistence farmers whose aim is to remain viable over generations – rather than to make a 'killing' in any one year and then retire to Florida – their transactions within their local environment can be characterised as *low risk, low reward*. However, during those times of the year when there is little farm work to be done, many villagers engage in trading expeditions or in migrant labour in India. Trading expeditions are family-based and family-financed, and highly speculative: *high risk, high reward*. Therefore, a farmer's individualised transactions, when added together over a full year, constitute a nicely spread risk portfolio. The attitude here, and particularly at the high-risk end of the portfolio, is that 'fortune favours the brave', 'who dares wins' and 'there are plenty more fish in the sea'. Opportunities, in other words, are there for the taking. The idea of nature here is optimistic, expansive and non-punitive: *Nature Benign*. Thus, the farmers have various ways to accommodate their interests.

On the island of Bali, there is an intricate relationship between rice cultivation and social organisation (Reader 1988). Wet rice cultivation is labour-intensive, because mechanisation is difficult to implement. It also has a high and lasting productivity, mainly because of the benefits of controlled flooding on the nitrogen balance (Lansing 2006). In this situation, a form of social complexity has arisen with a strict caste-related *hierarchy* and an important role for priests, who derive authority from high standards of behaviour. Such hierarchies can be a strong, moral organizing force in society (Dumont 1984).

Yet, Balinese society also has distinctly *egalitarian* features: 'The Balinese tend to dislike and distrust people who project themselves above the group as a whole, and where power has to be exercised they tend to disperse it very thinly... For the Balinese there is no difference between a person's spiritual and secular life... The devils and demons [in religion]... provide forceful reminders of the obligation to restrain selfish instincts for the benefit of social harmony... Rice cultivation is the ultimate expression of the Balinese readiness to follow the edicts of some greater authority' (Reader (1988) 65). The complex religion of the Balinese is an example of the functional significance of religion in human ecology. One is also reminded of the role that monks and monasteries have played in the periphery of large civilisations. The more *individualist* outlook on life is left to the trading communities, often foreigners. Nowadays, the rapidly growing tourist industry provides new opportunities for the population.

These field observations indicate that groups of people represent enough diversity to follow a variety of strategies in the face of different threats and opportunities. It is in this context that social scientists propose institutional arrangements, that do not exclude any of the voices while accepting contradictory defintions of problems and solutions and encouraging constructive argumentation. Such 'clumsy

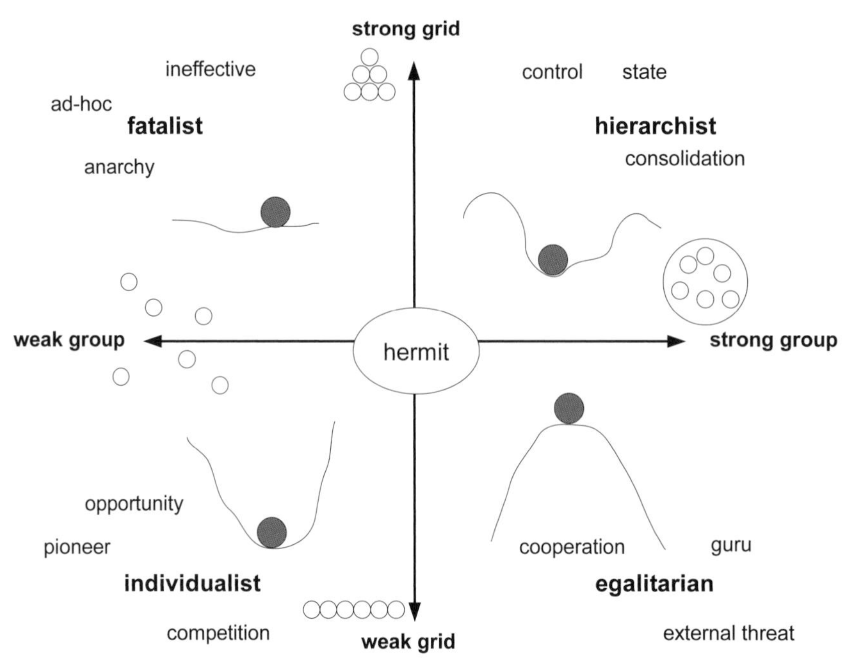

Figure 6.3. The four perspectives identified in Cultural Theory (clockwise): the hierarchist, the egalitarian, the individualist and the fatalist (based on Thompson *et al.* 1990).

institutions' is one of the solutions for sustainable resource management (Verweij and Thompson 2006).

6.3.2 The Four Perspectives of Cultural Theory

One way to understand the diversity and complexity in social-ecological systems is through the framework proposed in Cultural Theory (Thompson *et al.* 1990, 1992, 1997). It offers an interesting heuristic for an analysis of the controversies and risks about sustainable development issues such as novel technologies, climate change and biodiversity preservation (Schwartz and Thompson 1989; Adams 1995; Rayner and Malone 1998). Combining insights from cultural anthropology and ecology, Cultural Theory distinguishes four perspectives on how people experience and interpret reality, and organise themselves and manifest rationality, solidarity and fairness.[14] These perspectives – one can also speak of visions or myths – are categorised along two axis (Figure 6.3):

- the degree to which individuals behave and feel themselves part of a larger group of individuals with whom they share values and beliefs (the *group* axis), and
- the extent to which individuals are subjected to role prescriptions within a larger structural entity (the *grid* axis).

The resulting four perspectives are related to their position along these two axes: the hierarchist (high on both), the individualist (low on both), the egalitarian (high

[14] Other, similar categorisations have been proposed, and I do not claim to be complete. For instance, an investigation of business cultures across the world also identified similar distinctions (egalitarian versus hierarchical, adventurous/enterprising versus anxious/xenophobic, individualist versus collectivist, masculine versus feminine and short-term versus long-term oriented) (Hofstede 1984).

in 'group', low in 'grid') and the fatalist (low in 'group', high in 'grid').[15] Whereas the hierarchist favours control and expertise in order to guarantee stability within a world of limits, the individualist is convinced that nature has inherent stability and abundance. The egalitarian emphasises the fragility of nature and the possibility of irreversible destruction. The fatalist experiences the world as determined by pure chance. The differences are graphically depicted in Figure 6.3 with the mechanical metaphor of the rolling ball. Let us have a closer look.

Each perspective can be associated with a certain way of structuring the relationship between man and nature and between fellow men, each with characteristic judgements about quality of life and technological and environmental risks. Seen from the *individualist* perspective, the past is full of human success stories, the future full of promise. The debate about 'limits to growth' is seen as exhibiting a lack of trust in the ingenuity of humans and in the resilience of nature. It is precisely the existence of limitations that challenges people to surpass themselves, as happened so often in the past. Individualists see markets and prices as the only efficient and fair organisational scheme and have confidence in the human capacity for adaptation. Prominent and cherished values are competitiveness, success, achievement, courage, adventurism and its rewards of wealth and power. Individualists may, however, come into conflict with their own convictions. As soon as successful new areas are exploited and new markets are conquered, a loss of exclusiveness and the necessity of coordination and regulation emerge to avoid the worst excesses.

The individualist. 'False bad news about population growth, natural resources, and the environment is published widely in the face of contradictory evidence . . . Bad news sells books, newspapers, and magazines . . . the cumulative nature of exponential growth models has the power to seduce and bewitch', according to Julian Simon (*Science* 208, 1980). 'The ultimate constraint upon our capacity to enjoy unlimited raw materials at acceptable prices is knowledge. And the source of knowledge is the human mind. Ultimately, then, the key constraint is human imagination and the exercise of educated skills.'

Many administrators, managers and engineers have confidence in the results of measurements and risk calculations and have concern for as well as confidence in good economic and environmental management. They are *hierarchists*. On the basis of what experts say about the constraints and the costs and benefits, an optimum technical-economic course is determined. Seen through the eyes of the hierarchist, chaos and anarchy are just around the corner and the world is perpetually in need of management. This is the world of bureaucrats and politicians who understand the need for and subtleties of institutions, committees, and negotiations. They appreciate and support legitimacy of big government and big business and their procedural rationality. Stability, reliability, integrity and planning skills are appreciated. But they too face the shortcomings and contradictions of their view. The hierarchical organisation is intent on securing its own continuity; hence, it is not easily mobilised for policies that go beyond accepted risks or challenge the status quo. The inherent

[15] The fifth worldview, that of the hermit, is beyond the area defined by these two axes. The emphasis tends anyway to be on the three 'active' worldviews: hierarchist, individualist and egalitarian.

delays and buffers resulting from procedures and consultations may have a stabilising effect, but they also tend to paralyse commitment and creativity and cause cynicism and loss of credibility. Also, in their struggle for greater power, hierarchist organisations come across limits in the form of corruption, suspicion and indifference.

The hierarchist. The Commission on Global Governance, in their report *Our Global Neighbourhood* (1991), addresses a variety of problems related to unprecedented change on a global scale such as the New Arms Race, Widespread Violence, Persistent Poverty, and Environmental Threats. It observes that due to globalisation of capital, media and technology, 'states retain sovereignty, but governments have suffered an erosion in their authority. In the extreme case, public order may disintegrate and civil institutions collapse in the face of rampant violence'. They continue: 'The world needs to manage its activities to keep the adverse outcomes [of environmental impacts] within prudent bounds and to redress current imbalances' (p. 11). A solution is offered, 'The concept of global governance: There is no alternative to working together and using collective power to create a better world'. The United Nations should play a leading role here.

The third perspective is the *egalitarian* one which is characterised by communitarian aims and values and absence of structured authority. Nature is perceived as a fragile, connected, complex whole. Human knowledge of the earth is limited. The basic premise is equity between man and nature and among people, as expressed in the image of people as stewards or partners. A fair distribution of benefits and costs between rich and poor – the context for socialism – and between us and posterity – the context for environmentalism – is considered a precondition for sustainability, and so are justice and security. Egalitarians are risk-averse and favour the precautionary principle as management strategy. As with the other perspectives, egalitarians face limitations in maintaining theirs. They always run the risk of individualism and hierarchy arising within their ranks. The solidarity and dogmatism of their radical positions is often accompanied by a loss of creativity and effectiveness. They need a future with environmental disasters and a present peopled by acquisitive and powerful enemies in order to keep the group together.

The egalitarian. In her book *Staying Alive*, Vandana Shiva (1989) describes development as a new project of Western patriarchalism. It is rooted in reductionist science: 'As a system of knowledge about nature or life reductionist science is weak and inadequate; as a system of knowledge for the market, it is powerful and profitable'. The paradigm of Western science is itself the problem: '... development is a permanent war waged by its promoters and suffered by its victims ... and scientific knowledge, on which the development process is based, is itself a source of violence ... this transformation of nature from a living, nurturing mother to inert, dead and manipulable matter was eminently suited to the exploitation imperative of growing capitalism'. Modern science destroyed the old metaphor: 'The nurturing earth image acted [no longer] as a cultural constraint on exploitation of nature' (p. 7).

Many of the world's poor are *fatalist* in their outlook on life. They try to cope with everyday life's contingencies within the limited freedom of manoeuvre that they have. People and nature are both capricious, and all one can do is to make the best of it. Climate change and other environmental threats only get significance for them if they are actually hit by the consequences. Fatalists are passive and yet part of the fourfold scheme, because they are needed by the adherents of the other perspectives. They provide legitimacy for the hierarchists as potential members of the lower organisational ranks. For the individualist, they are the necessary complement to their success and give the American dream its appeal. And throughout history, they are again and again the potential converts and followers for egalitarian movements.

Cultural Theory claims that the interpretation of past events and the anticipation of future events are filtered through the rationalities and solidarities associated with the four perspectives. Religious and market fundamentalism, ethnic and nationalist movements, and environmental radicalism can all be understood against this background. Of course, people rarely express these perspectives in their extreme form and one should not give in to the temptation to make caricatures. In fact, one of the critiques of Cultural Theory is that real-world individuals manifest contextual behaviour and that it is unclear how adherence to a particular perspective in a particular context develops. Also, the relation between perspective and power matters. The resources at the disposal of those who adhere to and support a particular perspective influence the degree to which they can change the world in a desired direction. Thus, it is important to distinguish the perspective, that is, how one thinks the world works, and the management style, that is, how one would act if one were in power.[16]

Cultural Theory interprets social-cultural change from the continuous, dynamic interplay between the adherents of the four perspectives (Thompson 1992, 1997). Individuals alter their perspective when it is no longer reconcilable with their experience. Collective institutional change happens whenever larger groups of people start to doubt the correctness and adequateness of the dominant perspective. Such change may be triggered by surprise experiences, new information, a desire for balance and by other – for instance, demographic – changes. Therefore, extreme positions are counteracted by reactions from one or more of the other perspectives. As a result, societal development unfolds dialectically. Some of the switches between perspectives are indicated in Figure 6.4. Excessive hierarchism leads to bureaucratisation and, at some point, the system collapses and liberalisation and privatisation processes take over. Excessive individualism leads to marginalisation of the less successful people who then resort to fatalism, which in turn feeds egalitarian movements in a process of radicalisation. The social-cultural dynamics cause societal oscillations in which excessive swings to either side are corrected.

The four perspectives are a decor against which developments in the world can be interpreted, not as law-like truths but as the manifestations of human beings in their individual and social variety. As such, it is a qualitative, rich theory of complex

[16] This distinction has been applied by running computer model with different sets of assumptions on how processes are simulated and on which interventions are made (Rotmans and de Vries 1997; de Vries 1998).

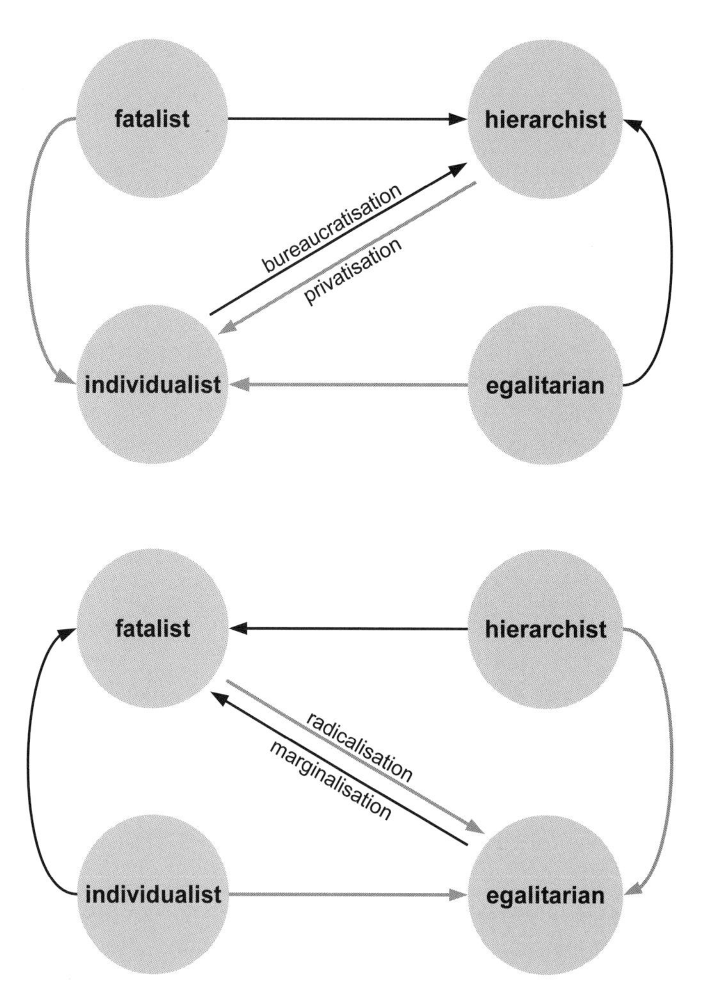

Figure 6.4. Two important mechanisms in the dynamics of social-cultural change: the interaction between hierarchist (king, state) and individualist (merchant, market) in the form of bureaucratisation and privatisation, and between egalitarian (sect) and fatalist (commoner) in the form of radicalisation and marginalisation (based on Thompson 1992).

adaptive systems. For instance, in recent times, the excessive hierarchism in the former Soviet Union ended with the collapse of the collectivist state in 1989–1991. Growing support for the individualist perspective, as proclaimed in Fukuyama's book *The End of History and the Last Man* (1992), led to a victory of neoliberalism. It accelerated financial deregulation, abolishment of trade barriers and selling out of nationalised firms to globally operating private capital funds. In combination with the Information and Communication Technology (ICT) revolution, the unfairness of global capitalism started to generate counterforces that culminated in the Seattle and Genoa revolts against globalisation and the 2001 terrorist attack on the World Trade Center in New York. In the new century, millions of less successful, excluded and marginalised individuals shed their fatalism and resort to radicalism and reform in egalitarian movements. Some are fundamentalist and (ultra)nationalist, others environmentalist or modernist in their outlook. The tensions between hierarchy and individualism and the connections between fatalism and egalitarianism are a

Box 6.3. *Mr. López the entrepreneur.* There is a story about an entrepreneur that illustrates the shift in prevailing perspective. One way to sustain an egalitarian regime is with ceremonies in which those who accumulate inordinate amounts of land or other property are compelled to give it away. However, a smart individual can always bypass such a system and create structural change.

In the 1890s, Mr. Lopez subverted the old mechanism of wealth levelling in the Mexican village of San Juan Guelavía. Traditionally, only rich people would be designated as major-domo, that is, the one who has to sponsor for the village fiestas and festivals and gains prestige from it. However, with help of the clergy, Mr. López forced the village council to designate less wealthy people – who then could not refuse – and offered the insolvent sponsors a loan with their land put up for collateral. By the eve of the Mexican Revolution, Mr. López owned most of the community's best land: By 1915, his large family owned 92.2% of the arable land and an even larger part of the irrigated land and, with the support of the church, strongly opposed any censure. Thus, equal access to strategic resources had disappeared by the perversion of a traditional regulatory mechanism – an example of the vulnerability of egalitarian cooperative regimes to individualist interference.

recurrent theme in the political-economic history of Europe and its offshoots. In the postindustrial and globalising 21st century, its cultural dynamic can be expected elsewhere, too.

As in many social science theories, the statements of Cultural Theory are hard to test and validate or falsify empirically. It is applied in energy policy, management of hazardous wastes, household consumption styles and other areas, and provides a novel and inspiring way of looking at sustainability issues. Its greatest contribution is to offer a consistent way to uncover the value-laden construction and interpretation of 'facts' and to reveal the inherent value pluralism and hence uncertainty and richness in 'solution space'. Incorporating cultural perspectives into the sustainability discourse clarifies the diversity and ambiguity in the use of the word sustainable and the clashes between people – think of the controversies about animal rights, forest clearance for new roads, genetically modified organisms (GMOs) and subsidies for renewable energy to mention a few. In the words of Thompson (1990), 'A fourfold vision is the key to understanding the contradictory visions of the future that are generated by our social interactions... Single visions – economic man, scientific rationality, any consciousness that has to insist that all others are false – are pathological'. It points at the necessity of building alliances that bridge different values and beliefs and at the merits of pragmatism: 'My view is that, if environmentalists aren't willing to engage with big business, which are among the most powerful forces in the modern world, it won't be possible to solve the world's environmental problems' (Diamond 2006: 17). In other words, we should avoid a battle between utopians and their claims to a particular view and implementation of sustainability. In the next section, I introduce another framework to operationalise such diversity and pluralism and apply it in the thematic chapters later on.

Worldview pluralism (© Stefan Verwey 1978).

6.4 Worldviews: Ways to See the World

6.4.1 Values and Beliefs: Four Worldviews

As dicussed, it is possible to get a rough idea about people's values by questioning them. A person's values in association with particular needs gives rise to certain expectations about the person's ideas and behaviour – for instance, in relation to development, environment and sustainability. If someone prefers nature walks over air travel or being employed in a multinational corporation rather than in an environmental NGO, one suspects a link with his value orientations. But the links are complex. The nature lover flies to an African wildlife park and the MNC-employee is active in a human rights NGO in the evening hours.

I do not treat the many social and cognitive science theories in this field and simply introduce our own definition of the notion of worldview:

> A *worldview* is defined as a combination of a person's value orientation and belief system, that is, of his view on how the world is best understood.

Therefore, a worldview combines a person's interpretation of and beliefs about the outer and inner world and his value orientation.[17] A worldview has a prescriptive

[17] I use the word 'beliefs' to indicate that people build up cognitive structures according to which they interpret the world around them. I use it more or less as an equivalent to interpretation, model or

and a descriptive aspect (Rokeach 1973). The prescriptive part is about the relevance of certain things and themes over others, that is, the value orientations previously discussed. They are individual, subjective and imply a certain, preferred quality of life. The descriptive part consists of the belief system about how things work. It is a mental map or model, that is, a more or less causal interpretation of the world. It can be rather objective in the sense of shared and formalised. It frames how people think they can improve or maintain an aspired quality of life. In practise, prescriptive and descriptive aspects are interrelated, as previous examples show.

One can try to identify and measure beliefs in the same way as values, by questioning people. Indeed, it is difficult to distinguish the value part from the belief part in the outcomes of social surveys because values are embedded in cognitive mental processes. Some researchers have tried to extract belief systems from answers to questionnaires or behaviour in participatory workshops. An example is an analysis of the outcomes of the World Value Survey (§4.2) over the period 1989–2004, in order to identify popular support for the transition to a market economy in India and China (Migheli 2010). They posed questions about the role of competition, government intervention, income inequality, hard work and fairness. The questions were about values (such as, do you consider competition harmful?) as well as beliefs (such as, are larger income differences an incentive for economic growth?). The answers may imply a causal relation as, for instance, in the belief that competition is good because it provides an incentive to work hard or the conviction that hard work is irrelevant because one's success depends on luck and connections.[18]

It was found that Indian people are, on average, less prone to support the characteristics of a free and competitive market and blame poverty less on laziness and lack of willpower of the poor than the Chinese. In both countries, there is a change towards preferences for a less competitive market. Perhaps, the author suggests, this is a reaction against the policies currently undertaken by the governments, because more competition means more uncertainty as compared to a strongly regulated or centralised economy. It is also found that people expect the government to intervene in order to insure a minimum level of well-being for the population and that the preference for a market economy weakens when people have confidence in the government.

A second example is about a participatory exercise organised in New Zealand, with the aim to make a shared conceptual mental model of the health system (Cavane *et al.* 1999). Such a shared model, in the form of a causal loop diagram (CLD), is supposedly helpful in finding solutions for persistent problems (Morecroft and Sterman 1992; Vennix 1999). Two groups of staff from the Ministry of Health participated in a series of workshops, during which clusters of issues were identified and subsequently translated in a conceptual model. One group consisted of policy managers, the other of clinicals (medical/health practitioners). The authors – and

mental map. Some people prefer the word mindset or, in a scientific context, paradigm. I do not use words like attitude, norms or motivation. They are close to values, but it is outside the scope of this book to treat such refinements.

[18] The author considered also the role of socio-economic variables, which are known to influence preferences: religion, education level, city size, age, gender and the state of health Besides, the different histories of India and China since the 1950s and the uncertainties of the present transition will also play a role.

most participants – found the CLDs useful in illustrating the roots of different worldviews, for instance, whether one looks at the issue from a micro- or a macro-perspective. The exercise illustrates the different ways in which individuals and groups understand their responsibility and tasks: 'The policy group looks to the effectiveness of the whole sector, while the clinicians look to the delivery of a service' (Cavane *et al.* 1999). It was evident that the values of people as expressed in their jobs are linked to their professional backgrounds and the beliefs and models acquired during education and work.

The beliefs that guide behaviour are based upon a lifelong experience with parents, neighbours, community leaders and institutions and all kinds of formal and informal education. It is a mixture of 'personal knowledge' and 'community knowledge'. The assumption of economic rationality is overly simple. Clearly, behaviour can be made more 'rational' by a more informed evaluation of the costs and benefits. But these 'costs' and 'benefits' are assessed not only in terms of money, but also of prestige, relationships, new or lost opportunities and so on. Even if we have an idea about a person's values and beliefs, or worldview – for instance, about how to achieve a more sustainable energy system or to stop the destruction of tropical forests – one can expect only a probabilistic relation with such a person's actions. Worldviews are not predictors of behaviour. Almost identical behaviour may have different backgrounds. Sometimes there are valid reasons why behaviour cannot be consistent with the worldview.[19]

Thus far, worldviews have been associated with individual persons. When a large majority of people adheres to a certain worldview, it gets social and historical significance. This is manifested in philosophical enquiries and historical events. An elementary and ancient idea is that the human mind tends to operate by thinking in opposites. Many religions and myths describe creation as the separation of primordial substance into opposites: light and darkness, heaven and earth, male and female. Modern cosmologies reiterate a similar view. The opposites are not merely mental images but are a fruitful source of psychic energy (Jung 1960). Table 6.4 summarises some of the great philosophers' investigations in terms of opposites. The two main dimensions that came out of social surveys (Figure 6.2) show a remarkable similarity with the two dimensions found in philosophical works. Combining the two gives the four worldviews that are depicted in Figure 6.5.

The vertical dimension is associated with immaterial and spiritual values on the upper side and with material and secular values on the lower side. It is the opposition religious versus worldly, giving versus taking and mind versus matter.[20] This vertical dimension can be traced back to the ancient Upanishad scriptures in India and to the writings of Plato. It reflects the controversy between realism and nominalism in medieval Europe, and it is, in the contrast between idealistic and materialistic, found in the work of the 19th-century philosopher Hegel. The sociologist Sorokin (1957) uses the words Ideational and Sensate for it. This vertical dimension is also recognised in the quality-of-life elements in Tables 6.1 and 6.2.

[19] Often, the reasons given are invented or imagined because the respondent does not give priority to the implied behavioural change or cannot resolve underlying conflicting values and ideas.

[20] It is less relevant here whether the spiritual orientation is seen as an a priori and pre-existing quality 'from above' or as a quality that is emerging 'from below' as part of human evolution. There is room for both views.

Table 6.4. *Interpretation of the vertical and horizontal axis in Figure 6.2b and 6.5*[a]

Source	universal ↔ particular	
Surveys (Fig 6.2)	Big World	Small world
	The others	The self
Weber	Collective	Individual
Sorokin	One single truth	Pluriform truth
	Universalism/Active	Singularism/Ascetic
Steiner	Objective reality	Subjective understanding
Pauli/Jung	Non-locality; Wave	Locality
		Particle
Levinas/Wilber	The Other	I
	We	
Schwartz	Group values	Individual values
Kant	Laws	Maximes
		Rules for personal behaviour
Source	**immaterial ↔ material**	
Surveys (Fig 6.2)	Giving	Taking
	Religious	Worldly
Plato	Ideas	Senses
	Mind	Body
Hegel	Idealistic	Materialistic
Sorokin	Ideational	Sensate
Steiner	Spirit	Matter
Jung	Archetype–spiritual	Instinctive-material
Wilber	Interior	Exterior
Fromm	Being	Having

[a] Van Egmond and de Vries 2011.

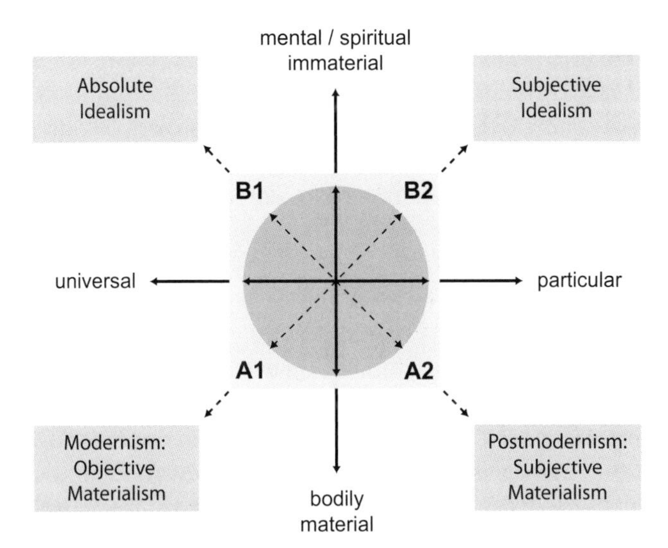

Figure 6.5. Representation of the four worldview quadrants as combinations of the value survey outcomes (Figure 6.2) and philosophical reflections (Table 6.4). The letters A1, A2, B1 and B2 refer to the names of scenarios that are discussed in Chapter 15.

Box 6.4. *Beliefs and scientific knowledge.* Beliefs about the world is knowledge. How do we acquire it and what makes it legitimate? Regarding knowledge, there is a gradual change from skills, as the ability to act according to rules, to knowing how, which implies skills-in-social-context, and from there to expertise, which includes the even higher abstraction level of not only knowing but also reflecting and adjusting the rules. Empirical-reductionist science is mostly about expertise and relies on a universal scientific truth, which can be uncovered. But in the worldview frame, science cannot claim the Truth nor the method to uncover it. Complex system science is causing a paradigm shift in the Modernist worldview, partly thanks to the new tools of computers and Internet. The path of inner reflection of, for instance, Gurdieff's self-remembering and recounted in personal accounts of the beyond are also legitimate ways to find truth. It has a long history and many appearances and leads according to its practitioners to a holistic science (Wilber 2001). It represents an aspect of the vertical dimension in Figure 6.5. Therefore, the world of humans is not only one of plural values, but also of plural beliefs about how to acquire knowledge. It explains much of the present-day debates about the legitimacy of scientific claims.

The horizontal dimensions can be given a broadly philosophical interpretation, too. Depending on the scale level, it is the opposition we versus I, collective versus individual, global versus local, universal versus particular, and whole versus fragmented. This tension between 'The World' and 'my world' is also reflected in the group axis in Cultural Theory, where the hierarchist is positioned opposite of the individualist (§6.3). Each of us has an individual existence, yet we are in innumerable ways connected to the larger material and mental world around us. The left big-world side is interpreted as shared existence in the collective realm. The right small-world side is associated with the non-shared existence in the individual realm. For a collectivity to exist, a certain level of uniformity in values and beliefs is required. In contrast, the differentiation of individuals creates and sustains diversity of value and beliefs. From an epistemological point of view, the horizontal dimension is, therefore, about whether a person adheres to an objective, universal truth or lives primarily a subjective, personal truth.

6.4.2 Worldviews in History

If worldviews can be given an historical interpretation, how would they look? Let us look again at Figure 6.5. The upper right quadrant is labelled *Subjective Idealism*, because it represents combinations of non-materialistic and individual orientations. Needs are satisfied by mastery of the sensate needs, and there is a diversity in the interpretation of truth. The material world is perceived as an incarnation one has to be liberated from. Nature is experienced as divine mystery. Big government and corporations are distrusted, science is considered unacceptably rational. In concreto, the focus is on personal growth, on the 'management of greed' and on self-reliance and conflict resolution at the community level. The Gandhian concept of *swaraj* is an expression of this worldview. Another icon of this worldview is Schumacher's book *Small is Beautiful* (1973), originally titled *Buddhist Economics*, which makes a plea for small-scale solutions and individual responsibility: 'I have no doubt that it is possible to give a new direction to technological development, a direction that shall lead it back to the real needs of man, and that also means: to the actual size of man. Man is small, and, therefore, small is beautiful' (Schumacher 1973). Its darker side are the fundamentalist sects that embrace the single subjective truth of the leader.

In the upper left quadrant, *Absolute Idealism*, a majority in society adheres to the view that there is a universal objective truth. This Truth is not to be uncovered by personal introspection or the rational mind, but should be understood from the scriptures and consolidated in churches and religious prescripts. Its essence is immaterial and collective. The institutions are hierarchist and demand order and stability, based on uniformity and obedience. Churches and their dignitaries are visible and outspoken representatives. Also, the State is an expression of this worldview inasfar as governments work for the well-being of their citizens, as reflected in, for instance, the Declaration of Human Rights. The demarcation between church and state, theocracy and democracy, is not sharp – and continuously in dispute. Widespread poverty and ecological degradation are according to this worldview caused by egoism and moral permissiveness and the remedies range from repentence to prayer and 'good works'. In a more worldly, political context, these ills are considered a consequence of oppressive structures and lack of social solidarity and are to be resolved by more adequate rules and institutions.

The diagonal opposite of the *Subjective Idealism* worldview is *Objective Materialism* or *Modernism*. It has become the dominant worldview in European and American 'Western' society, at least among the elites. It combines a belief in the existence of a universal objective truth with a material-secular value orientation and cosmology (§5.2). The world can be understood and managed according to scientific and utilitarian principles and with the societal objective of 'the greatest happiness for the greatest number'. In the Theory of Modernism, it is said that: 'At the level of the individual, modernization is characterized by an increased openness to new experiences, increased independence from authority, belief in the efficacy of science, and ambitions for oneself and one's children' (Easterlin 1983; cf. §4.3). The convergence in the science- and technology-driven modern world towards a uniform playing field and political systems is well described in Friedman's *The World Is Flat* (2005). The European welfare state but also capitalism, socialism and communism are manifestations of Modernism, although with divergent interpretations of rights and duties towards the collective. The underlying idea is that each individual person has a vote in what the collective ends should be (democracy). Academia, government organisations and corporate bureaucracies are the institutional representatives.[21] The scientific and technological success in Modernism is not only the root of much progress but also of many sustainability problems, and yet its advocates point at science- and technology-driven innovations as the main source of solutions.

The fourth worldview combines a material-secular orientation on life with a focus on the individual. It is the world of *Subjective Materialism* or *Postmodernism*. The belief in universal truths and the collective becomes a widening array of convictions, lifestyles and material manifestations of personal identity: '... in postmodern times we are moving out of an age of uniformity, collectivity and universality and

[21] Corporations wish a 'level playing field' in the form of uniformity and standardisation on the global market. Small and medium enterprises are often in a different, and dependent, position in this arena.

Box 6.5. *Universalism as a hierarchist utopia.* Universalism is an attempt of the mind to control the world. Universal knowledge, in contrast to knowledge from practical experience, tends to simplify the world in order to make it manageable. This can have dramatic consequences. 'Radically simplified designs for social organization seem to court the same risks of failure courted by radically simplified designs for natural environments. The failures and vulnerability of monocrop commercial forests and genetically engineered, mechanized monocropping mimic the failures of collective farms and planned cities' (Scott 1998:7). This observation, inspired by anarchism, reflects the evils of radicalism to the left side in Figure 6.5. The state and other hierarchical institutions can easily become obsessed with their utopias and become the enemies of human well-being. It has not disappeared with the fall of communism because 'large-scale capitalism is just as much an agency of homogenization, uniformity, grids, and heroic simplification as the state is, with the difference being that, for capitalists, simplification must pay' (Scott 1998:8). The challenge is to recognise the value in locality and diversity and to balance it with the genuine merits of universal insights, procedures and social institutions.

into one characterized by individuation, fragmentation and difference' (Hall *et al.* 1996). Postmodernism is also an expression of feelings of anxiety and insecurity: There are no anchors for meaning and consolation, neither in the collective to the left, nor in the soul above. Sustainability issues are narrowed down to issues of individual survival, with a fatalist touch and a local or national orientation. Huntington's book *Clash of Civilizations* (1997) is one of the books that identified a shift towards Postmodernism in the 1990s.

The tensions between the diagonal opposites in Figure 6.5 are reflected in philosophy and history. The opposition between upper left and lower right manifests itself in the rejection on the one side of the physical, notably the sexual, aspect of human life by the church – for instance, of Darwin's thesis that men descended from apes – and extreme forms of hedonism and body culture on the other. The diagonal from upper right to lower left mirrors the opposing views of science, with its emphasis on observation and falsification on the one side, and the religious convictions of individuals practised in esoteric and mystery schools on the other. The four quadrants together form the playing field where the prevailing values and beliefs of a people and a period are manifested. The scheme is meant to make sense out of complex phenomena, not as a comprehensive Theory of Everything.[22]

If the preceding, somewhat caricatural, positions are not only individual but also social realities, it should be possible to interpret history in terms of the four worldviews and, more importantly, to use it as a framework to explore sustainable

[22] The worldviews can be associated with the different forms of goods and services provision (§5.3). Markets claim to be a universal allocation mechanism, primarily in the material domain (A1–A2), where prices can be attached to tangible goods and services. Public and professional provision is largely oriented towards collective goals, in governments and corporations (A1–B1). Self-provision and informal provision are by choice or by necessity an orientation towards the individual needs (A2–B2). Each mode has its preferred goods and services and its specific rules and dispositions.

development pathways (Van Egmond and de Vries 2011). Certainly, it should be possible to trace the extreme manifestations in history. Let us make a small detour and examine this hypothesis for Europe. Early Christianity can be positioned in the *Subjective Idealism* upper right quadrant. It emerged in reaction to the preceding Roman sensate and 'carpe diem' culture of the lower right worldview. When more individual people got inspired by it, early Christianity got more followers and was forced to institutionalise – it became the Church of Rome. The personal and mystical was replaced by uniformity in religious beliefs. The 14th-century Arab thinker Ibn Khaldun made a similar observation about historical change (Turchin 2003; §3.5). In his view, change comes from small human groups that have a small territory but a large collective solidarity, or *asabiya* as he calls it. This gives them a large capacity for collective action and the 'ethnic momentum', shared conviction and willpower to conquer other peoples and to expand. In the process, their ideas become more widespread and are consolidated and claimed to be universal. At that point, society has arrived in the upper left quadrant.

Throughout the European Middle Ages, the worldview of *Absolute Idealism* was manifest in the dogma's of the Roman Catholic Church. The great cathedrals with their orientation towards God and made by innumerable anonymous devotees are, literally and metaphorically, its highest expression. The physical world and body were repressed, either by sanctifying ascetism or by ruthless persecution. Truth was radically monopolised and deviant convictions were forced underground. The repression led to the social discontinuities of the Reformation and the religious wars of the 16th and 17th centuries. Small groups of committed Protestants were at the root of a new inspiration, that transformed into the rationale and ethics of Euro-American capitalism according to the sociologist Weber. Other cultures and religions have experienced similar developments – for instance, the periodical repression of sufism in Middle-East Muslim countries.

The outward expansion in the direction of the upper left quadrant reached its zenith with the global exploration by the then-leading Catholic countries Portugal and Spain. However, the seeds of other orientations on the world had already entered Europe with the translation by Arab thinkers of the work of Aristoteles and other Greek philosophers and were reinforced when Europeans came in contact with previously to them unknown cultures outside Europe. The inquisitive minds of the early scientists, with their rationalism and empiricism, created scientific theories and technical inventions that initiated a shift towards a more secular universalism. In the process, it eroded the religious claims. *Modernism* slowly emerged as the new worldview that foreshadowed the great changes of globalisation and industrial revolution in the 19th and 20th centuries. It provoked a number of conflicts between Church and Science. The life of Galileo Galilei is one of the best-known illustrations. By 1900, science and technology had become successful enough to leave in its turn little or no room for diversity or dissent. It coincides with a value shift from traditional to secular/rational as part of the industrialisation process (§4.3).

This emphasis on materialistic-secular values has increasingly manifested itself as a denial of the more immaterial, spiritual-oriented worldviews. Social life became organised in the uniform way of mechanised mass production, with rigid social classes and uniform standards. Science took the place of religion as a source of authority. Liberalism and socialism were both expressions of this worldview in 19th- and

Box 6.6. *Sustainable development in Postmodernism.* The transition in the early 21st century from a science- and society-oriented, highly productive Modernist worldview to a sensate and individual oriented Postmodern worldview carries the seeds of the unfolding social-ecological crises. Science and technology can provide solutions for the great changes ahead. But a radical shift towards a Postmodernist worldview erodes the necessary legitimacy of science and technology to effectively introduce solutions. The proposed solutions reflect widely different value orientations, as, for instance, in the debate about nuclear power and GMOs, and continuous societal conflict results. Although scientists still adhere to the 'immutable laws' of nature, their recommendations and solutions succumb in ineffective and polarizing societal debates. The resulting incrementalism and localism may fail to solve the sustainability problems addressed in this book. The balance with the universal and the immaterial has to be restored if we are to find a sustainable path for the world.

20th-century Europe. Romanticism was a reaction to it, with a temporary return to the values of subjective idealism and their expression in visual arts and music, but it lost momentum. As part of the changes towards a more secular worldview and its manifestation in industrial capitalism and colonialism, there was a trend towards uniformity and collectivity at the expense of the individual. State communism and Nazi fascism in 20th-century Europe brought this worldview to its extremes, with disastrous consequences.

In recent decades, Modernism is visibly making room for *Postmodernism* of the lower right quadrant. The seeds for this change were already sown in the early 20th century, when science discovered fundamental uncertainty and complexity and the 'Brave New World' turned out less utopian than expected. Its claim to universal truth started to weaken in favour of social constructivism, a move that was reinforced by the disenchantment with the great utopias of liberalism and socialism. In the Postmodern worldview, there is room for diversity and pluralism and a shift in focus from globalisation to the smaller scale of the individual and the local group. It coincides with the transition to postindustrial values from survival to self-expression (§4.3). The Internet, with its pros and cons, is becoming the ultimate expression of this shift towards the right-hand side of Figure 6.5. Postmodernism has its dark sides, too. Extreme manifestations are the hedonism and consumerism of the shopping malls and the luxury elites, and the cynicism about other aspirations than the pursuit of individual material goals. States increasingly lack the legitimacy to formulate and implement societal goals in a complex and interconnected world. Fragmentation and disintegration are one consequence, as in the falling apart of the former USSR and former Yugoslavia and in the separatist movements in parts of Europe.

6.4.3 Mechanisms of Social Change

The historical interpretation of worldviews suggests several mechanisms of societal change along the lines discussed previously for the Cultural Theory. One mechanism of social change is a *reinforcing feedback*, which tends to drive the expression of a worldview to its extreme (Figures 6.4 and 6.5). Initially, a worldview is supported because of the original, authentic values and beliefs it proclaims. If more people adhere to it and follow its interpreters, the ascendant worldview seeks legitimacy by the introduction and consolidation of institutions that extend and rationalise its domination. It is the period of growth and is a move to the left in Figure 6.5. Authorities, representing the new worldview and its institutions, reinforce the process. People with less honorable intentions join and, gradually, other worldviews become disrespected and suppressed. Evil behaviour unfolds from good intentions. Fundamentalist rabbis, priests and imams demand the faithful to live according to The Book and avoid, convert or eliminate the non-believers. States and corporations control individuals by classification and standardisation, by administrative language and procedures, and by political and managerial censorship. At its height, the imperial powers of Rome or Britain were a dense conglomerate of vested interests – government, army and business. 'What's good for General Motors is good for America' was a characteristic statement during the United States industrial-military complex at its peak.

The power and diffusion of ideas, in the upper two quadrants, is a powerful societal force in itself. It is the force of messianic vision and tribal asabiya. Its worldly equivalent is the spread of goods and skills in the lower two quadrants. It is the force of efficiency, economies of scale and zero-sum game benefits from cooperation. Products that are successful on the market or skilled labourers who start a successful enterprise become in a positive feedback process large traders and corporations, with the same processes of expansion, consolidation and standardisation as in the formation of religions and churches.

Inevitably, the monopolisation of truth and of power creates counterforces. Although these can be suppressed for long periods, with methods ranging from torture to Internet censoring, at some moment in history they will erupt. It is the period of decline. It is an *action-reaction* mechanism that heralds large-scale social change. The more extreme the caricature of the worldview and the longer it is sustained against these forces, the more violent and catastrophic the reaction tends to be. One can distinguish two kinds of reactive forces against the legitimizing force of the dominant worldview (Castells 1997). One is conservative, as it wishes to go back to an imagined better past. The other is progressive in its embracement of a transformation into an imagined better future. Castells labels the conservative reaction as a *resistance identity*, which is taken up by those who are in devalued positions and conditions and stigmatised by the logic of domination. The progressive reaction is associated with a *project identity* and carried on by those who seek to redefine their position in society and in the process seek transformation of overall social structure. The green and feminist movements are examples of the latter, while Islamic and Christian fundamentalism and nationalism in Eastern Europe are expressions of the former.

It is individual persons who are behind the centrifugal, outward forces that turn worldviews into their own caricature and cause their collapse. Why do individuals move to one of the corners of the worldview scheme of Figure 6.5? One underlying reason is the *search for identity*. To identify oneself with certain convictions and associated social roles and material and immaterial benefits is a deep psychological need. It also serves the innate desire for control. The benefits are immaterial, as with

Box 6.7. *Values and ethics.* Values are associated with good and bad, with ethics and morals. Daly (1973) proposed a hierarchist spectrum of values, ranging from ultimate worldly means to ultimate immaterial ends. Values have been linked to 'lower' and 'higher' human needs (Maslow 1954; §6.2). In fact, all religious teachings point at a value hierarchy, often in association with Good and Evil, God and the Devil. In *La Divina Commedia*, Dante recounts his upward journey through the realm of the afterlife: Hell, Purgatory, and Paradise. In the *Bhagavad Gita*, Krishna talks about the three qualities of prakritri, the sanskrit word for nature or the basic energy from which the mental and physical worlds take shape (Easwaran (1985) 177–178):

> 'Sattva – pure, luminous, and free from sorrow – binds us with attachment
> To happiness and wisdom.
> Rajas is passion, arising from selfish desire and attachment...
> Tamas, born of ignorance, deludes all creatures
> through heedlessness, indolence and sleep...
> Those who live in sattva go upwards;
> Those in rajas remain where they are.
> But those immersed in tamas sink downwards.'

positions of status and power or getting protection and privileges, and material, in the form of money and the sensate pleasures you can buy with it. In the left quadrants, identity is strongly connected to a collective identity. To the right, identity is more individualist, less connected and more fragmented. The centrifugal move is also a flight from the center with its existential demands. The individual fears the freedom inherent to the centre, where he or she has to take full responsibility for his or her existence. Media, in its old and new appearances, often support the centrifugal forces by paying attention primarily to extreme views rather than to reconciliation. Of course, people mature during their lives and the social and cultural setting are important determinants, too.

What does this have to do with sustainable development? I put forward as a working hypothesis that *sustainability can only be found in the middle*, that is, through an acknowledgement and synthesis of worldview pluralism. Then, the conceptual framework of worldviews and their dynamics can help to identify the diverse positions with respect to sustainability issues and to resolve the conflicting tendencies. It is in line with the notion of 'clumsy institutions' advanced in Cultural Theory (§6.3). In a policy context, it implies a search for robust solutions. In the thematic chapters, this is clarified with concrete examples.

6.5 Worldviews in Action

In Chapters 7 and 9–14, I discuss a series of themes that are relevant for sustainability science: biodiversity, demography, agriculture, water and minerals, and the economy. Each chapter surveys the 'state of the world' in terms of historical trends and models. In this pivotal chapter, I argue that the search for sustainable development involves 'objective' science *and* 'subjective' perspectives and I offer a conceptual framework

Table 6.5. *Possible descriptions of extreme positions in the quadrant space of Figure 6.3*

Values	achievement, competition, adventure, efficiency, rationality, utility	income, security, health, tradition, (local) identity	individual well-being, wisdom, awareness, cooperation, compassion, respect for nature	social order, stable and fair institutions, equity, peace, human rights
Motivation	material success, worldly status	material comfort, worldly power, bodily pleasure	personal development, inner peace	immaterial success, mental and spiritual status
Beliefs	the world operates on corporate and state power hierarchies. Global 'open' markets and innovations are the way to progress and prosperity	the world consists of impulsive, emotional individuals. Be a realist; there are no universal truths and opportunism is natural	'there is more between heaven and earth . . . '. Self-sufficiency in material and spiritual affairs in a community setting	the world operates on collective and shared, religious and/or social-cultural ideals. Governments primary task is social justice and environmental integrity
Indicators	when does Chinese economy surpass the US economy? Top-500 firms . . .	personal income, (local) crime rate, tax levels	number of local shops and artisans, market share organic food	state budget for education and social security, church attendance
Policy mechanisms	stimulate international markets; create global competition on level-playing field	support regional suppliers and markets; protectionism if necessary for economic or social-cultural reasons	support and facilitate local small-scale initiatives	regulation and international (UN) binding agreements

that allows exploration of the notion of sustainable development in its full width. It is applied in the thematic chapters in the remainder of the book. At the end of each thematic chapter, I do not present a single judgment, recipe or indicator for sustainable development. Instead, I present tables with perspective-based statements on how the situation is valued and interpreted from the different worldviews. *You as reader are invited to reflect on these statements on the basis of your own values, beliefs and actions – and your own indicators.*

In practice, each chapter ends with a representation of worldviews in the quadrant space in Figure 6.5, in the form of a table (Table 6.5). The black dots in Table 6.5 indicate four extreme positions. Each position manifests itself in an individual person in the form of values, beliefs and actions. Most individuals are never fully or always in one of the extremes. Positions will depend on age, gender, income, education and so on. Therefore, the descriptions I give in the thematic chapters are best considered incomplete caricatures. If a single individual were to hold such an extreme position, it may influence the local situation, such as in a family or office, but not the outside world. Such a person would provide a mooring point in the worldview space. It is a useful exercise to find out which persons you view as representative of a particular worldview.

If many individuals adhere to such a worldview, it does have an impact on the larger world. It becomes a dominant view that is an active force in shaping human society and history. I propose that, for a sustainable development path, the dominant worldview should be a balanced 'integral' worldview. People should 'include' other worldviews in their life. In contrast, if the dominant worldview in society is on or over the edge in Figure 6.5, the situation is unsustainable. An intriguing question, then, is which are the signals ('indicators') of an extreme position ('overshoot') and which are the balancing forces to be expected ('feedback')?

The thematic chapters that follow are mostly about the material aspects of sustainability. Is nature disappearing? How and why are human populations growing in number? Are fertile soils degrading? Are groundwater and fisheries soon depleted? Will there be an oil peak within a decade? Is chemical pollution making us ill? To answer such questions requires a scientific assessment of the situation – but science cannot give definitive answers in complex situations with large uncertainties. Neither the individual, nor the collective can expect unambiguous recipes for action in a complex world. Besides, to take action implies also certain goals ('ends') and allocation of available resources ('means'), implying choices that will not be value-free. To be sure, science can give answers with near-certainty to many questions (§8.3). But there are also many questions to which science has yet no or at best probabilistic answers.[23] As to sustainability issues, the individual person has to reflect on what are for him or her the constituents of quality of life and what is valued in life. A next reflection is on what is believed about the world, from the simple to the complex. Each position suggests certain answers and behaviour. The search for a sustainable world will be a play and a battle of perspectives – and you are part of it.

6.6 Summary Points

I consider sustainable development to be about quality of life. Against this background, the following points are important:

- Quality of life has an objective side and a subjective side, that are both part of the expression and satisfaction of human needs. Food and health indicators have rather strong objective components. Income often serves as a proxy for the more subjective 'higher' needs.
- Concepts such as capabilities and satisfiers emphasise that needs, and with it quality of life, are at least partly social constructs.
- Values are important in a person's assessment of quality of life. They can be identified with surveys. A rather universal set of value clusters or orientations is found for individual persons.

[23] In some forms of Postmodernism, the ignorance of scientists is abused, using scientific controversy to defend certain interests or obstruct acquisition of relevant knowledge and transparancy See Proctor and Schiebinger's (2008) discussion on agnotology and the creation and use of ignorance.

- The Cultural Theory is one social science theory that explicitly looks at sustainable development from different cultural perspectives with their own forms of fairness, solidarity and rationality. It uses the dimensions of group and grid, rooted in ecology and anthropology, to categorise four perspectives.
- Also, beliefs about how the world 'works' are important if one wishes to understand how people see sustainable development. Values and beliefs together make up worldviews, in my definition.
- Based on value surveys, discourse analysis and historical reflections, I propose a two-dimensional framework to organise and apply worldviews in the analysis of past and present developments and their (un)sustainability. The forces that give rise to radicalisation and subsequent demise of worldviews can be understood as a search for identity, power and control and associated feedback mechanisms.
- Sustainability problems can in this broader framework be seen as the one-sided dominance of one worldview at the cost of the others. Their resolution is in balance and synthesis.

People (and not only managers) only trust their own understanding of their world as the basis for their actions.

—de Geus, in Morecroft and Sterman 1992

To be without some of the things you want is an indispensable part of happiness.

—Bertrand Russell, *The Conquest of Happiness*, 1968

...only a small portion of existing concrete reality is coloured by our value-conditioned interest and it alone is significant to us.

—Weber 1904 [1949:76] quoted in Swedberg 2003

...the irresistible tendency to account for everything on physical grounds corresponds to the horizontal development of consciousness in the last four centuries, and this horizontal perspective is a reaction against the exclusively vertical perspective of the Gothic Age. It is a manifestation of the crowd-mind, and as such it is not to be treated in terms of the consciousness of individuals... we are at first wholly unconscious of our actions, and only discover long afterwards why it was that we acted in a certain way. In the meantime, we content ourselves with all sorts of rationalized accounts of our behaviour, all equally inadequate... we overestimate physical causation and believe that it alone affords us a true explanation of life. But matter is just as inscrutable as mind.

—C.G. Jung, *Modern Man in Search of a Soul*, 1933

...belief systems are not merely beliefs – they are the home of the ego, the home of self-contraction. Even a holistic belief, like the web-of-life, always houses the ego, because beliefs are merely *mental* forms, and if the *supramental* has not been discovered, then any and all mental constructions house a tenacious ego... the web-of-life is just a concept, just a thought. Ultimate reality is not that thought, it is the Witness of that thought. Inquire into this Witness. Who is aware of both analytic and holistic concepts? Who and what in you right now is aware of all those theories?

—Wilber 2000:95/117

When all things began, the Word already was. The Word dwelt with God, and what God was, the Word was.

—John 1:1

SUGGESTED READING

An excellent comprehensive introduction into the philosophical background to this chapter.

Tarnas, R. *The Passion of the Western Mind – Understanding the Ideas That Have Shaped Our World View*. New York: Ballantine Book/Random House Publishers, 1991.

A description of and simulation experiments with a world model, exploring future pathways for the world depending on which worldview is dominant.

Rotmans, J., and B. de Vries. *Perspectives on Global Change – The TARGETS Approach*. Cambridge: Cambridge University Press, 1997.

de Vries, B., and A. Petersen. Conceptualizing sustainable development: An assessment methodology connecting values, knowledge, worldviews and scenarios. *Ecological Economics* 68 (2009) 1006–1019.

A series of short articles with opposing views on 21 environmental issues in the United States.

Easton, T. *Taking Sides: Clashing Views on Environmental Issues*. 11th ed., Contemporary Learning Series. New York: McGraw-Hill, 2006.

Thompson, M., R. Ellis, and A. Wildawsky. *Cultural Theory*. Boulder, CO: Westview Press, 1990.

van Egmond, N., and B. de Vries. Sustainability: the search for the integral worldview. *Futures* 43 (2011) 853–867.

USEFUL WEBSITES

- changingminds.org/ is a site with access to disciplines, techniques and explanations of social science topics.
- cep.lse.ac.uk/events/lectures/layard/RL030303.pdf is de weergave van de Lionel Robbins lectures on happiness door Layard in 2003.
- www.tns-nipo.com is the site of TNS-NIPO where a questionnaire can be filled in to find out about your own value orientation.
- ec.europa.eu/public_opinion/index_en.htm is the website of the European Commission on public opinion research and in particular the Eurobarometer.
- www.worldvaluessurvey.org/ on the World Value Survey with extensive information on the survey set-up, outcomes, and so on.
- www1.eur.nl/fsw/happiness/ is a database of literature on happiness research, operated by Veenhoven of Erasmus Universiteit Rotterdam.
- www.sinus-sociovision.de/ and www.motivaction.nl are two other sites on measuring values, mostly for marketing purposes.
- www.awakeningearth.org/ is the site of Duane Elgin on the role of consciousness in envisioning a sustainable future.
- www.6milliardsdautres.org/ is a site with thousands of interviews with people all over the world, who answer essential questions about their lives, such as what they value most in life and what Nature means to them.

7 Energy Fundamentals

7.1 Introduction: The Essential Resource

Energy in the form of heat and work has always been crucial for human beings. Wood usage as a source of heat goes as far back as the control of fire. Wind and water have been used for millennia, but the most important source of mechanical energy until a few centuries ago was the labour of domestic animals and human beings (peasants and slaves). This all changed with the discovery of fossil fuels as a source of high-density chemically stored energy.[1]

Coal was already known as a fuel in early dynastic China and ancient Rome. It was used for heating in medieval Europe but was considered inferior to wood because of its foul smoke. Moreover, it was harder to obtain in most places. Under the pressure of wood shortage in 17th-century England, coal became a more common fuel. Chimneys protected people from the immediate effects of smoke. By 1700, the fraction of coal in energy use in the United Kingdom exceeded already 50 percent (Figure 7.1). With the invention of the steam engine, its use got another boost. Like the use of coal, the principle of the *steam engine* had been discovered long before, in both the Chinese and Roman empires. At those early stages, no practical uses were found for it. In 18th-century England, however, circumstances were highly propitious. Initially, the steam engine was developed as a device for pumping water from the coal mine shafts, which made the coal more easily accessible. In turn, the cheaper coal was used to power other steam engines, some of them propelling the ships and locomotives that transported coal to the growing number of users. This further lowered the cost and expanded the market. In 1850, coal supplied over 90 percent of energy use in the United Kingdom. It was one of those positive reinforcing feedback processes that ignited the industrial revolution. But more was to come.

Following similar dynamics but faster, *crude oil*, and oil products like gasoline and fuel oil, and later *natural gas* entered the energy scene in a series of supply and demand innovations and expansions. These two other members of the hydrocarbon

[1] The energy use density, that is, energy flow per unit of area, in agricultural societies depended on the solar influx which is ~1320 W/m^2 at the top of the Earth's atmosphere and at full insolation ~1000 W/m^2 at the Earth's surface. Because insulation strongly fluctuates during day and year, it has, on average, an energy density of about 250 W/m^2 at the surface.

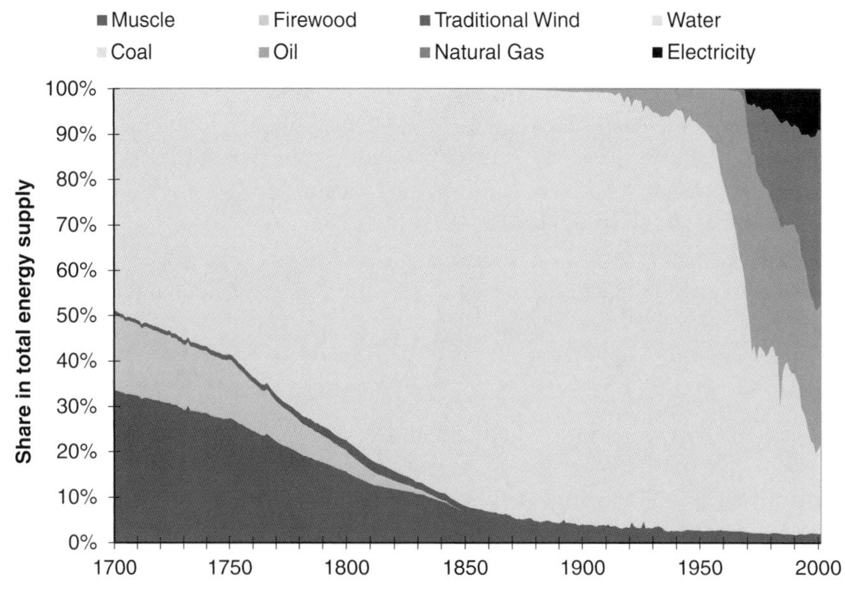

Figure 7.1. The energy transition in England/Wales 1561–1995. The graphs show the fuel shares in total primary energy use (Warde 2005). Similar transitions took place in other European countries and in the United States.

family were also known in antiquity. In the ancient Mediterranean world, bitumen pools were known, but the substance was mainly used for construction. The Chinese used bamboo to drill holes and transport gas and burn it to evaporate brine. Once its potential as a source of energy was discovered in 19th-century America, its use rapidly increased. It took oil less than eighty years to gain a 50 percent market share in the United Kingdom (Figure 7.1). Worldwide, the same happened and by 1965 oil had a greater market share than coal (Figure 7.2). The fraction of biomass-based traditional fuels, mostly wood and crop/animal residues, dropped worldwide from 90 percent to 20 percent between 1850 and 1950, although its absolute use kept increasing. These changes make up what has been called the *energy transition*. Besides, coal, and later and at much larger scale, oil and gas, were and still are used as feedstocks to produce, amongst others, cement, fertiliser and plastics. Official forecasts expect the dominant role of fossil fuels to remain for the decades to come, while traditional biomass-based fuels are increasingly replaced by modern ones. The role of renewable sources for electric power remains modest in these forecasts.

The energy transition was in essence a technological transition. On the supply side, it was mining and drilling, transporting and refining, and military support where and when needed. On the demand side, there was a non-ending series of user applications like steam ships and trains, the automobile and electric power generation and the myriad of electricity using appliances. Four out of the five long-term (Kondratiev) waves in economic activity are directly linked to energy, namely, hydropower, steam power, electrical power and motorisation based on oil.[2]

[2] The fifth and last of these waves, called 'computerisation of the economy' by Freeman and Louca (2002), also deals directly with energy, notably electricity, but at a more subtle, informational level.

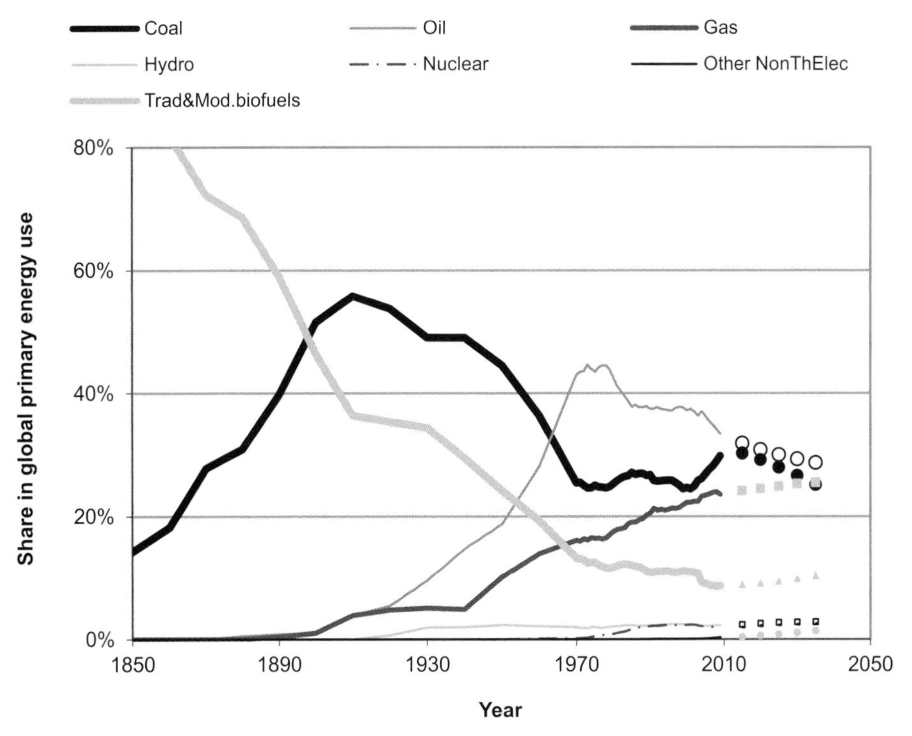

Figure 7.2. The energy transition in the world 1850–2010. The graphs show the fuel shares in global primary energy use (source: PBL). The expectations up to 2035 in the World Energy Outlook 2010 are also indicated (IEA 2010). It should be noted that nuclear and other nonthermal (NTE) options are counted in electricity equivalents.

Fossil fuel, the new source of energy, made the industrial regime fundamentally different from the earlier agricultural regime (§4.5). Fossil fuels are accumulated solar energy from a remote past, contained in geological formations and concentrated by geological processes in particular locations, from where they can be made available through human effort and ingenuity. They have a high energy density, unlike the diffuse flows of solar energy and its derived flows (waves, water, wind), and are not directly connected to the flow of solar energy, unlike plants and animals and wood. Our food is now as much or more derived from oil than from sunlight. The fossil fuel incorporated in an 'industrial' tomato is several times its calorific food content.

In advanced industrial societies, humans now have control over energy flows at levels up to 10,000 times the levels in pre-industrial societies, which is truly no small achievement. But there are several reasons why the present state is unsustainable. First, there are great disparities in energy use and access. The richest 10 percent live at levels of 20 kW/person and more, which is the equivalent of having permanently some 26 horses or some 200 slaves at one's disposal and not counting the energy needed to feed them. At the same time, a large fraction of the world population still lives in a biomass-based society at levels below 2 kW/person.[3] The distribution is

[3] This level is up to twice as high if the traditional biomass use is included. Its end-use efficiency is, however, in most cases significantly lower than for fossil fuels, such as in cooking and space heating.

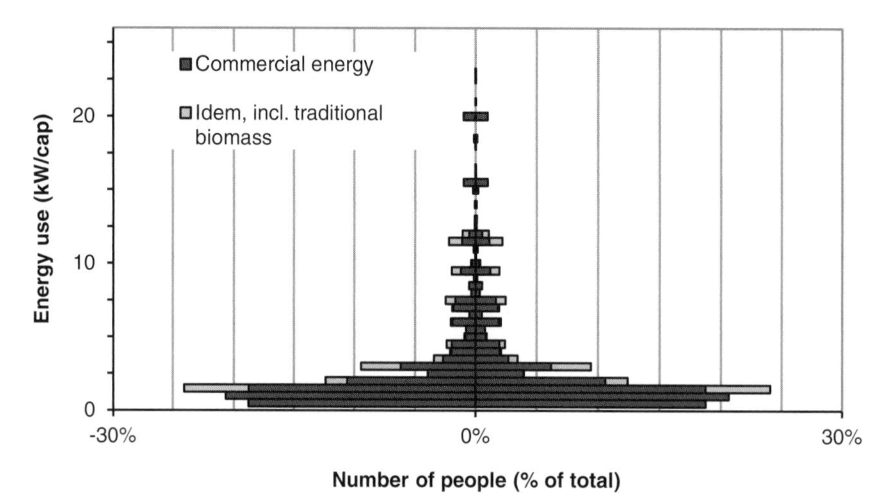

Figure 7.3. Energy pyramid: distribution of energy use in kW/cap. The width of the pyramid shows the fraction of the world population within an energy use class (kW/cap) (source: World Development Indicators 2004).

shown in the energy pyramid in Figure 7.3 and is a reflection of the similarly shaped income pyramid. It is a concern in sustainability science, because energy is crucial for the provision of basic needs like food and medical services, and a persistent shortage may cause serious social and economic imbalances.

Second, fossil fuels are *finite*: Incapable of growth or reproduction at a rate meaningful to humans, therefore, they are irreplaceable and non-renewable. They are also geographically unevenly spread. The affluent countries with high per capita energy consumption are now dependent on, although some prefer the word addicted to, finite oil and gas resources that are largely controlled by other countries. As the past century has shown, this brings serious geopolitical tensions that will aggravate when high-quality deposits get depleted and users have to switch to more expensive alternatives (Yergin 1991; §13.2). Third, the large-scale exploration, production and combustion of fossil fuels has caused and still is causing environmental damage in many places. Locally and regionally, coal mining, oil leaks and tanker accidents and emission of sulphur- and nitrogen-oxides contribute to water and air pollution and ecosystem degradation. Globally, the emission of carbon dioxide (CO_2) and methane (CH_4), two important greenhouse gases, has already led to an increase in radiative forcing and is causing, with near-certainty, changes in climate variables such as temperature and precipitation.

In daily life, most people experience energy largely in the form of payments at the petrol station and the monthly bill for electricity and gas. If one broadens the horizon, the link with the big world is evident in the news items about OPEC-revenues, Middle East oil and Russian gas power, offshore oil leakages, construction of wind turbines and nuclear power plants, agricultural crops to fuel cars and so on.

To understand the role of energy in society in more depth, an introduction into energy science is needed. In particular, two classical branches of physics, thermodynamics and mechanics, have to be studied.[4] The former teaches the basic principles of

[4] A brief introduction in the basic concepts and laws of chemistry would also be in place in a sustainability science textbook. Here it is considered outside the scope of this book.

Box 7.1. *Free energy flow density as a measure of complexity.* The astrophysicist Chaisson (2001) has proposed an interesting link between energy and complexity. Organisms can be viewed as dissipative structures: Ordered objects whose structure can be maintained thanks to a steady input of high-quality energy. The *free energy flow density* (Φ) necessary to sustain such a non-equilibrium structure is a measure of complexity. It can be expressed in erg per second per gram or, more commonly, in mW per kg (1 erg = 10^7 joule). The free energy flows are on earth supplied almost exclusively by solar radiation. It is possible to make estimates of the Φ-values for different systems on earth and the results are shown in Figure 7.4. It seems that ever since the Big Bang, which started off our story of the universe, complexity measured in this way has been continuously increasing. After early life forms like the archaebacteria in the heat gradients on the ocean floor, an important step towards higher complexity was made with the advent of photosynthesis, the biological mechanism to convert solar energy into chemically stored free energy ($\Phi \sim 0.1$ W/kg-biomass). Most animals have Φ-values in the range of 1 to 10 W/kg-biomass, depending on their activity. Birds operate at an order of magnitude higher than reptiles, which gives this measure some intuitive appeal. The human body dissipates free energy at a rate of about 1 W/kg-biomass, but if you are bicycling it is a factor ten higher – so complex is bicycling in this definition. An astounding increase in free energy density flows happened in the industrial age, with Φ-values over 100 W/kg-biomass per person. If measuring the free energy density per unit of mass of equipment, going from organic to technological systems, there is a further increase to Φ-values in the order of 1,000–100,000 W/kg. The most extreme densities are realised in military equipment and microdevices and nanodevices.

heat and work flows at the macroscale or phenomenological level. The latter explains the elementary phenomena of mass and motion. Why bother about the basic laws and models of thermodynamics and mechanics in sustainability science? First, there are some natural and technical phenomena that can only be understood on the basis of these laws and models. For example, the forces of water and wind in an erosion process or the heat exchange in heating and cooling processes and equipment. They are essential in even the simplest climate models. Understanding them facilitates an elementary assessment of natural processes and energy technologies. Second, the very notion of sustainable development is at least partly built upon concepts and models developed in the natural sciences. For instance, every process needs a source of work in order to sustain and develop against the dissipating forces of friction. The notion of energy quality is crucial here. Third, these concepts and models have often served as analogues or metaphors to investigate and discuss the behaviour of more complex systems. For instance, the notion of sustainability is often associated with equilibrium and related concepts such as stability and attractors. Elementary knowledge of mechanics is helpful to grasp their essence. Harmonic and other types of oscillators were introduced in ecology and economy as hypothetical models of population and market dynamics and such analogies had large heuristic value despite their oversimplification. Finally, sustainability science can benefit from the long

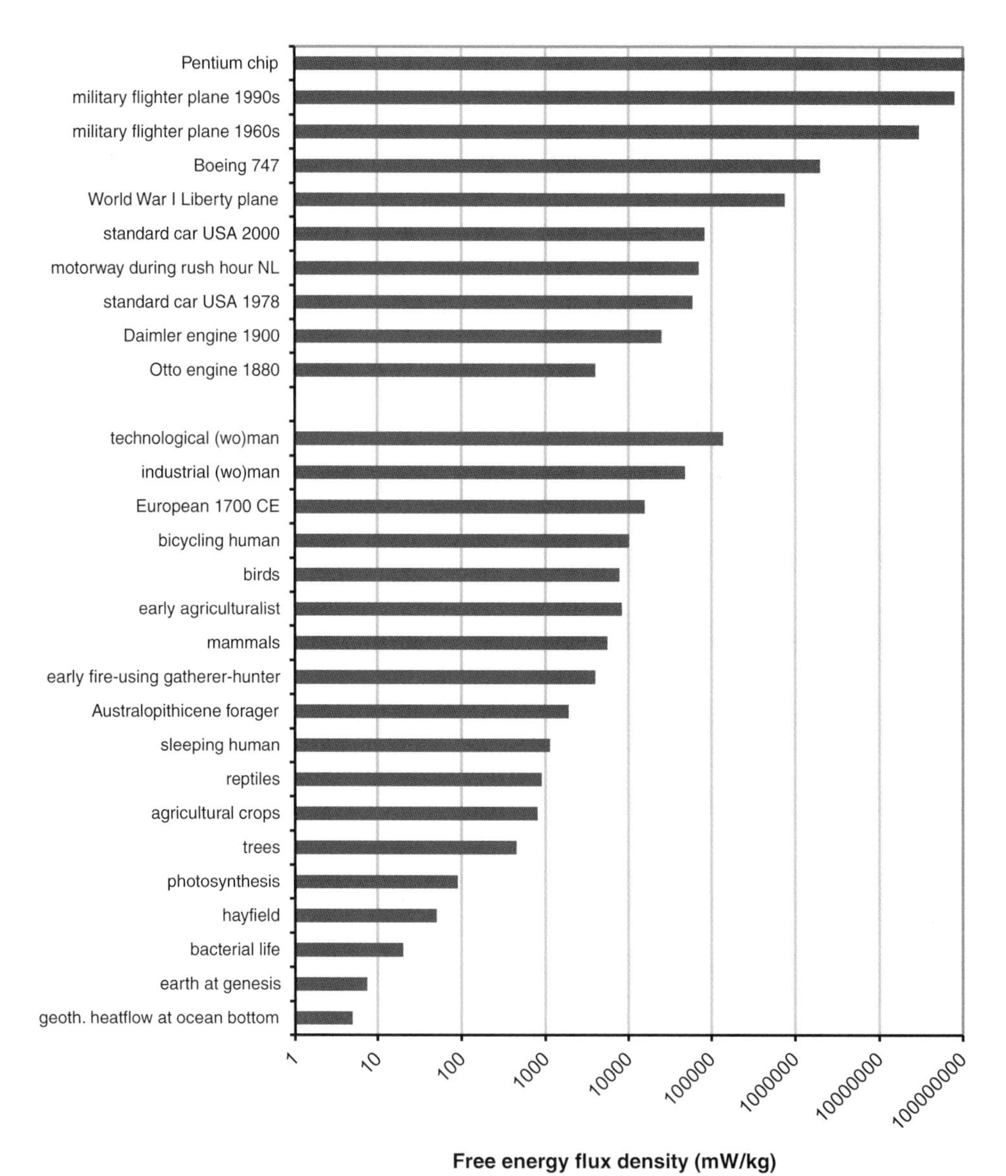

Figure 7.4. Estimated values of the free energy flow density in a number of systems (based on Chaisson 2001).

history in the natural sciences, where phenomenogical laws exist alongside more fundamental microdescriptions of dynamic processes.

7.2 Basic Energy Science: Thermodynamics

7.2.1 Classical Thermodynamics: The First and Second Law

Physics and chemistry have their own perspective on and contribution to sustainable development. That is the topic of this chapter, though oriented more towards physics than to chemistry. The word energy is derived from the Greek $\varepsilon\nu\varepsilon\rho\gamma\varepsilon\iota\alpha$ (activity), which is composed of the preposition $\varepsilon\nu$, in, and the word $\varepsilon\rho\gamma o\nu$ work. It is a fundamental quantity of nature that is transferred between systems or parts of systems. It is 'stored' in a variety of forms and associated with movement, combustion, heat and the capacity to do work. It is also widely used in more metaphorical and symbolic sense, referring to a dynamic social, psychological or spiritual force or quality and to the capacity of acting or being active.

It is important to get some familiarity with the *units* of energy stocks and flows. The energy stored in systems is a stock. Examples are the content of heat in a cup of tea or the kinetic energy in a moving car, as well as the chemical energy stored in a barrel of oil. The basic unit of an energy stock or quantity is the *joule* (J), which is about one quarter of the traditional unit of heat or the calorie. The *calorie* is the amount of heat needed to heat one gram of water one degree Celsius in temperature. The joule is such a small unit that one usually measures it in thousands: kilojoule (kJ), millions: megajoule (MJ) or billions: gigajoule (GJ). For instance, the amount of energy stored in one m^3 of natural gas is the equivalent of about 36,000.000 joule or 36 MJ.[5]

What makes the world go round are not energy stocks but energy flows. Ocean water evaporates by the heat flow of the sun, a cup of tea cools down as heat energy flows into the environment and the barrel of oil is taken out of the ground in order to burn it and use the heat flow of the combustion gases. Similarly, it is the flow movement of falling water that drives the turbine. The energy flow into a system is defined as the quantity of energy that crosses the system boundary per unit of time. The basic unit of energy flow is the joule per second (J/s) or watt (W). One hour has $60 \times 60 = 3{,}600$ seconds, so if you burn one m^3 of natural gas in the course of one hour, the energy flow equals $36{,}000{,}000/3{,}600 = 10{,}000$ W or 10 kilowatt (kW). In energy science, we do not speak of energy flows but of *power*. If the power is electrical power, the unit kWe is often used. A widely used unit for an amount of energy is the kilowatthour (kWh), which is defined as the energy content of delivering 1 kW of energy flow during one hour (1 kWh = 3,6 MJ). Table 7.1 gives some values for energy stocks and flows associated with familiar processes and systems.

Classical thermodynamics is the part of physics that deals at the macro-scale level with energy and matter and particularly with conversion of energy from one form into another.[6] Boiling water for a cup of tea, using the microwave oven, switching on the central heating system, cooling food in the refrigerator, lighting a room, riding a bicycle or driving a car are all processes in which one form of energy is converted into another. The conversion processes are described in scientific equations and models that have evolved during centuries of experimentation. This introduction touches only upon a few aspects of heat and mass transfer processes of relevance for the notion of energy and the transition to a sustainable energy system.

First, some basic concepts. Let us look at a system consisting of a given amount of a gas, a liquid (and its vapour) or a solid. The system boundaries are well defined but can move during the change processes under consideration, as is, for instance,

[5] The word 'stored' is not completely adequate, in the sense that it is stored in the 1 m^3 of natural plus the 10 m^3 of air and, as part of, the 2 m^3 of oxygen that together give on combustion a heat flow of 36 MJ.

[6] One refers to classical thermodynamics to distinguish it from the micro-approach in statistical thermodynamics.

Table 7.1. *Energy stocks and flows associated with some processes and systems. Values are approximations, because more specific system and process data are needed for precise data. See for general unit conversions www.digitaldutch.com/unitconverter/ or similar sites*

		Energy content			Energy content	
Process	System	MJ	kWh	System	MJ	kWh
Combustion	1 kg of coal	29.3	8	1 kg munic solid waste	9.4	2.6
	1 kg of crude oil	42.1	11.7	1 barrel of crude oil	5750	1597
	1 litre gasoline (petrol)	34.7	9.6	1 litre bioethanol	23.4	6.5
	1 m^3 natural gas (CH_4)	38	10.6	1 m^3 hydrogen (H_2)		
Potential (lift/fall)	1,000 kg water at 100 m	1	0.3	75 kg body 270 m up on stairway (Eiffel)	0.2	0.06
Kinetic (move)	20 ton truck at 90 km/hr	0.3	0.1	small car at 120 km/hr	0.016	0.005
Humans	One day body heat	8.6	2.4			
Animals	6 hr horse work	16.2	4.5			
Food	avg calorie intake/ adult/day	12	3.4	1 yr 1 soft drink can/day	0.23	0.06
Heat/cold	500 litre water to 50°C (bath)	105	29.2	1 kg of ice at 0°C	?	?

the case if a gas in a cylinder with a piston expands. The system boundaries may or may not be permeable for energy and matter. A system is said to be (Figure 7.5):

- *open* if the if the system boundaries are permeable for energy and matter;
- *closed* if the system boundaries are impermeable for matter but permeable for energy;
- *isolated* if the system boundaries are impermeable for energy and matter.

Isolated systems only exist in the laboratory (apart from the universe in its entirety). The Earth is a closed system, with a constant mass and a nearly constant solar energy inflow and equally large heat outflow. Other examples of (near) closed systems are underground nuclear waste deposits or the cooling circuit in a car engine. Most systems of interest are open systems. The mass of an open system changes if the inflow of matter into the system differs from the outflow. If both flows are equal, the system

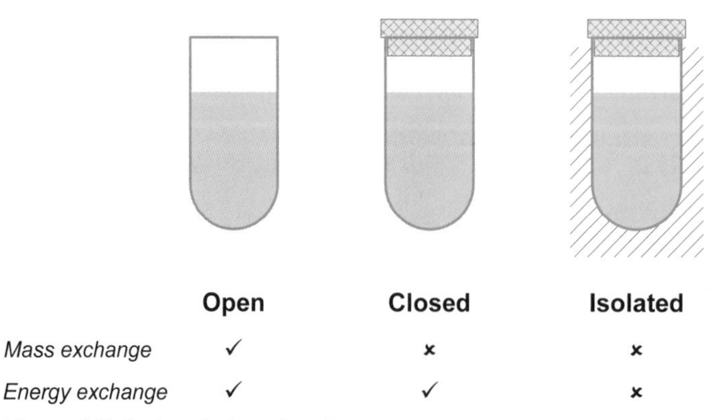

Figure 7.5. Isolated, closed and open systems.

is said to be in a *steady state* or *stationary state*, that is, in a dynamic equilibrium. Such a state has often been evoked to illustrate that a sustainable state is not a static, frozen one. In economic science, where the notion of equilibrium penetrated 19th-century thinking, this point was already made by Marshall in his book *Principles of Economics* (1890) and again by Daly in his book *Steady-State Economics* (1973).

For a (near) isolated system, there is a certain quantity that does not change, despite all kinds of changes in the system. This remarkable observation has been confirmed over and over again. The quantity is called *energy* and the observation is called the *First Law of Thermodynamics*. It expresses the conservation of energy principle. It is also stated as, energy cannot be created or destroyed (first order perpetuum mobile is impossible).[7] There is another quantity that can never decrease for an isolated system. It is called *entropy* and this remarkable observation is called the *Second Law of Thermodynamics*. It expresses the natural tendency towards disorder. It is also stated as, it is not possible to design an engine which, working in a cycle, produces no effect other than the transfer of heat from a colder to a hotter body (second order perpetuum mobile is impossible). In everyday language, you cannot extract work from a heat source without another heat sink of a lower temperature.

Since the 19th century, the two laws of thermodynamics have been formulated and reformulated. One formulation is in the form of a state equation: a system in equilibrium is characterised by the *state variables* internal energy (U), entropy (S), volume (V), masses (m_i), pressure (P), temperature (T) and chemical potentials (μ_i) for a system with $i = 1 \ldots N$ substances. The chemical potentials refer to the chemical substances and their concentrations in the system. If a state variable is proportional to the size of a system, such as U, S and V, it is an *extensive* variable. If it is independent of system size, such as P, T and μ_i, it is an *intensive* variable. Often, a system description is in specific terms, that is, the extensive variables are divided by the system size.[8] The same letter is used but in lowercase, for instance, $u = U/m$ for internal energy per unit mass. In this formulation, the two laws are:

I. The internal energy U of an *isolated* system is constant. For *closed* systems, the internal energy U of the system only changes if energy crosses the boundary and for *open* systems if energy and/or matter cross the boundary;
II. The entropy S of an *isolated* system can never decrease, tending towards a maximum in an equilibrium state.

Both laws are empirical theorems, that is, they cannot be proven deductively and they are valid only because all conclusions drawn on its basis are in agreement with experience.

Entropy is, unlike energy, *not* a conserved quantity. It is created continuously. And it is difficult to grasp. Real-world processes occur via a continuous sequence of *non-equilibrium* states. Non-equilibrium means that there are internal disturbances, for instance, a difference in temperature or pressure across the system. It generates

[7] It would be better to speak of conservation of mass-energy, in order to include the conversion of mass and energy in nuclear reactions.

[8] The measure of size is usually mass m. But it may also be volume, as in the expression for density $\rho = m/V$.

Box 7.2. *The notion of thermodynamic equilibrium.* By the end of the 19th century, the laws of thermodynamics were well formulated and confirmed by experiment. They are statements about systems in equilibrium, because only in equilibrium are the intensive variables (pressure P, temperature T or concentration μ) homogeneous in space and constant in time. As soon as this equilibrium is disturbed, for instance, by heating or pushing it from the outside, an adequate description of the system would require a huge amount of variables – for instance, the pressure and temperature differences across space and over time. Thermodynamics is actually thermostatics: The system jumps from equilibrium state X_0 at t_0 to equilibrium state X_1 at t_1.

In a *thought experiment*, one can imagine to go in steps from the initial to the final equilibrium state by briefly bringing the system in contact with the environment, then isolate it again and wait until equilibrium is reached. I will call such a process pseudo-dynamic or quasi-static. If the steps are infinitesimally small, at any given moment, an infinitesimally small gradient in an intensive variable (P, T or μ) is enough to keep the change process going. The overall process would be *reversible*, that is, one could reverse from the final to the initial state along exactly the same path. In other words, no macro-order is lost to micro-disorder. Unfortunately, it has to be infinitely slow, an unpractical condition. After all, it is only a thought experiment.

A thermodynamic equilibrium is only a static situation at the phenomenological level. At the micro-level of atoms and molecules, there is continuous movement of mass and exchange of energy. Because the number of particles is huge ($\approx 10^{23}$), statistical laws can be applied successfully, as in statistical thermodynamics and mechanics. Thermodynamic equilibrium is actually a dynamic equilibrium at micro-level.

The *Law of Chemical Equilibrium* provides an illustrative example of interference with dynamic equilibria. It was formulated in the 1860s by Guldberg and Waage and expresses the insight that chemical equilibrium is a dynamic and not a static condition because chemical reactants change in both directions at the same speed. Another important observation about chemical equilibria is the *Principle of Le Chatelier*, dating from the 1880s: If one of the factors determining the equilibrium changes, the system undergoes a change such that, if this had occurred by itself, it would have introduced a variation of the factor considered in the opposite direction. In common language, the system in equilibrium responds to an intervention with an attempt to undo it. This finding that real-world processes tend to restore the equilibrium state upon being disturbed had a large influence on the subsequent development of the life and economic sciences. Economists considered it such a basic feature of markets that it became the foundation of standard economic theory (§10.4).

internal frictions and shows up as heat. Such dissipation of energy at the microscopic level implies a loss of order. It can only be avoided when the process is infinitely slow. In statistical thermodynamics and mechanics, the second law has gotten an interpretation in terms of probability: Entropy is proportional to the probability of microstates, that is, atomic and molecular configurations (position, momentum, mass) at the microlevel.[9] I refer the interested reader to the suggested literature at the end of this chapter.

One formulation of the second law is that 'things go in a certain direction' in the universe – there is an *'arrow of time'*. Imagine a system brought in contact with the environment, such as a hot cup of coffee in a cold room or salt thrown into the water when boiling potatoes. In the state of being in contact with the environment, as in these examples, mass and energy interactions take place until equilibrium is reached and the state variables no longer change. The second law states that such a process always goes into a direction in which the differences in temperature, pressure and chemical potentials in a system tend to equilibrate. Have you ever seen a cup of hot coffee getting hotter spontaneously or the solved sugar spontaneously reappearing as a sugar cube?

How then is the order possible that we see in organisms, humans or societies? The answer is that the emergence and sustenance of such complex highly-ordered systems is only possible at the expense of an increase of disorder in the environment. Were the earth an isolated system, life would be (nearly) impossible. However, the earth is not: The solar energy influx represents an energy source from which work can be extracted before it dissipates into ambient-temperature heat – in photosynthesis and in photovoltaic cells, for instance.

If you search for it yourself, you get a big discount (© Stefan Verwey 1999).

[9] In Boltzman's formula for an ideal gas: $S = k \log W$, with k Boltzman's constant and W the number of possible microstates. Entropy has been linked to information theory by Shannon and Wiener: processes that lose information are analogous to the processes that gain energy.

7.2.2 Energy Quality: The Potential to Do Work

What does all this mean for sustainable development? If energy can neither be created nor destroyed, as the first law states, then the energy content of the universe is constant – which is, at first sight, good news. It even raises the question why mankind should economise on energy use and it suggests that we should stop speaking of *energy consumption* and *spending energy*. But there is more to it. Although human beings have always appreciated heat as a way to make their habitat more accessible and comfortable and to perform tasks like cooking and metal smelting, it really is *work* that matters most. Farming requires ploughing, irrigating, terrace building – traditionally done by human and animal muscle power, and in some places by wind or water. Manufacturing requires moving around materials and products, compressing, cutting, separating and so on – processes which could only be done at large scales and rates after the introduction of the fossil-fuel based steam engine and later steam turbine. Transport means moving people and goods around – traditionally done with animals but nowadays almost completely based on the internal combustion engine. Finally, communication increasingly means moving electrons around.

It is here that the second law comes in. It relates that not all the internal energy of a system is available as work. But classical thermodynamics is a bit of a strange science. It can only make statements about systems in equilibrium, that is, systems in which no change and no outside intervention takes place. In equilibrium, the state variables have fixed values. What we are interested in are processes from one state to another. Let there be two well-defined equilibrium states 1 and 2 of a system. Changes in the state of a systemfrom 1 to 2 are caused by interaction between the system and its environment. Such interactions are controlled interventions in experiments, like adding a certain amount of heat Q while fixing the volume V. Or everyday interactions between a system and its environment; for instance, lifting a box from the floor up on the table or boiling an egg. There are in principle infinitely many ways to bring the system from state 1 at time t to state 2 at time $t + \Delta t$. Thermodynamic analysis wants you to be precise in the description of the equilibrium states 1 and 2. The derivation of the basic equation for a steady-state or stationary state process is (Appendix 7.1):

$$Q = W_{CV} + \dot{m} \cdot \left((h_2 - h_1) + g(z_2 - z_1) + \left(\tilde{v}_2^2 - \tilde{v}_1^2 \right) /2 \right) \text{ J} \tag{7.1}$$

In this formula, a system change is considered from state 1 to state 2. During this process, Q, W and m are the heat, work and mass flows respectively across the system boundary. h is the specific enthalpy, that is, enthalpy per unit mass (in J/kg), g the earth gravitation constant, and z and v the position and the velocity with respect to a reference level. If you can apply this formula, you have a basic understanding of elementary energy conversion processes such as evapotranspiration, salinisation, steamturbines, refrigerators and hydropower plants. I refer to the suggested literature for those who accept the challenge.

This equation allows the introduction of a very important notion: *energy quality*. You probably intuitively grasp that one kWh of heat in your bathtub differs from one kWh used to power your laptop or charge your battery. So let us look again at equation 7.1 and assume that the heat flow Q is from an infinitely large reservoir of temperature T_h to another infinitely large reservoir of temperature T_l. The thermodynamic formulation of the second law in terms of entropy change is:

$$\Delta S = \left(\frac{\delta Q_T}{T}\right)_{rev} + \Delta S_{irr} \text{ J/K} \tag{7.2}$$

In words, the entropy change equals the heat flow at temperature T divided by the temperature plus the irreversible entropy production.[10] The entropy is produced whenever a process is not infinitely slow and stems from the different forms of friction if the process proceeds at finite velocity – think of the turbulence that you create stirring your cup of coffee. Because the two reservoirs are infinitely large and the only entropy change comes from irreversible entropy production, you have:

$$\Delta S_{irr} = \Delta S_{highT} - \Delta S_{lowT} = \frac{-Q_{T_h}}{T_h} - \frac{-Q_{T_l}}{T_l} = \frac{Q_{T_l}}{T_l} - \frac{Q_{T_h}}{T_h} \text{ J/K} \tag{7.3}$$

in joule per degree Kelvin (J/K). How much work can be extracted from this heat flow? The first law reflects that the sum of heat and work is conserved, so upon rewriting $Q_{Th} = Q_{Tl} + W$, you get for an infinitely slow, reversible process ($\Delta S_{irr} = 0$):

$$W_{rev} = Q_{T_h} - Q_{T_l} = Q_{T_h} - \frac{Q_{T_h}}{T_h} \cdot T_l = \left(1 - \frac{T_l}{T_h}\right) \cdot Q_{T_h} \text{ J} \tag{7.4}$$

You may wish never to read this again, but it is an extremely important result. It states that the maximum – because reversible – work that can be extracted from a heat flow between a heat source and a heat sink amounts to the fraction $(1 - T_l/T_h)$ of that heat flow. You now have an expression for the intuition that *heat has a lower quality than other forms of energy: Only part of it can be converted into work!* In qualitative terms, it expresses the fact that heat is a less ordered form of energy than potential and kinetic energy in force fields.

Now, for the last step, suppose that the low temperature reservoir is the Earth, at ambient temperature T_0 and with other intensive variables p_0 and $\mu_{i0.}$ If the heat flow happens at a finite rate and there is also a stationary mass flow in and out of the control volume, we write the equivalent of equation 7.1 for entropy:

$$\dot{S}_{irr} = \dot{m}(s_2 - s_1) + \frac{Q_{T_0}}{T_0} - \frac{Q_{T_h}}{T_h} \text{ J/K} \tag{7.5}$$

with s the specific entropy, that is, entropy per unit mass, (in J/kg/K). This can be rewritten into (Appendix 7.1):

$$\dot{S}_{irr} = \dot{m}(\varphi_2 - \varphi_1 + \Delta_{pot-en} + \Delta_{kin-en}) + \frac{Q_{T_0}}{T_0} - \frac{Q_{T_h}}{T_h} \text{ J} \tag{7.6}$$

Why is this equation so important? Because it makes it possible to calculate the maximum amount of work that can be extracted from a system that is brought reversibly into equilibrium with the Earth environment. For this, we define a new state variable: specific exergy φ. The equation tells that the maximum extractable

[10] Note that in thermodynamics, temperature T is always in degrees kelvin or K, with T = 273 + [temp in °C].

work equals the change in specific exergy and in macroscopic position and velocity, and the fraction $(1 - T_0/T_h)$ of the heat entering the system. In the real world, the work W that can be extracted in a change from state 1 to state 2 will always be less because of the irreversible entropy production $T_0 dS_{irr/dt}$. The following definition is used for the newly defined state variable:

> The *exergy* Φ of a system is defined as the maximum amount of work that can be extracted from it against a well-defined earth environment.

To calculate it, we specify the earth environment. We indicate it as (T_0, P_0, μ_{0i}), with T_0 ambient temperature (usually 273 K or 0°C), P_0 ambient pressure (usually 1 atm) and μ_{0i} the chemical potentials of the reference substances and their concentrations in the earth environment. In everyday life, you are particularly interested in two processes: fuel combustion and heat transfer. Combustion ('burning') of carbon fuels is the process in which chemically stored energy in the reactants ('fuel' and air) is released in the form of heat in the products (mostly water and carbon dioxide).[11] Heat transfer is the process in which a system changes temperature.[12] If bringing a fuel reversibly into equilibrium with (T_0, P_0, μ_{0i}), then the full exergy content can be extracted as work. But burning in an open fire, a boiler or a furnace is never reversible. The chemical energy is converted into heat of a temperature T – and only the fraction $(1 - T_0/T)$ is available as work (equation 7.6).[13]

The higher the temperature of the combustion products, the more work can be extracted. Given that work is what we want, it gives us a ranking of energy stocks according to quality. Table 7.2 lists the quality of energy and matter in a system in terms of the exergy content. The sun can be considered a black body of 6000 (degree Kelvin) K according to Planck's law. Therefore, solar radiation has in theory about 95 percent work potential and is a high-quality energy source. Nuclear energy in fission material and chemical energy in fossil fuels or biomass can be converted into a heat flux of high temperature – usually in the range of 500°C to 1500°C – and, therefore, they represent stored high-quality energy. On the other hand, using the 20°C difference in temperature between the surface of the ocean and the ocean water deep down will produce much less power per unit of heat flow. And the water of 60°C in a radiator also has a very low work potential.

It is also possible to calculate the quality of a mineral ore or metal, by calculating the exergy from the difference in chemical potential between the ore or metal and the earth environment. Each piece of metal represents work potential in an exergy analysis. Almost pure metal represents a 100 percent work potential, because most metals in nature occur only at low concentrations (<1%) (Table 7.2). Nature did work for us in accumulating certain minerals in rich ores! Vice versa, work potential is lost if pure metal is dissipated in the environment – as with lead in gasoline or titanium in paint – and it will require work to recover and concentrate it again.

[11] Chemical reaction kinetics is another branch of natural science that provides simple examples of positive ('autocatalyic') and negative feedback loops. It is also becoming a research area in complexity science (Nicolis and Prigogine 1989; Solé and Goodwin 2000).

[12] Heat transfer can also happen without temperature change, for instance, in phase transitions (evaporation of water, smelting of ice and so on).

[13] The exhaust gases drop in temperature during the heat transfer process. The logarithmic mean of entrance and exit temperature turns out to be a good approximation for use in the formula.

Table 7.2. *The quality of different forms of energy. The quality index indicates the fraction (in %) of an amount of energy that can be converted to work by bringing it reversibly into equilibrium with earth environment (after Wall 2001; www.exergy.se)*

	Form of energy/matter	Quality index (%)
Extra superior	Potential energy, e.g., high water reservoir	100
	Kinetic energy, e.g., waterfalls	100
	Electrical energy	100
	Pure element, e.g., diamond	100
Superior	Nuclear energy in nuclear fuel	almost 100
	Metals (commercial quality)	almost 100
	Sunlight	95
	Chemical energy in fossil fuels	95
	Element mixes, e.g., steel, alloys, plastics	90
	Rich mineral ore deposits	50–80
	Hot steam	60
	Poor mineral ore deposits, e.g., bauxite	20–50
	District heating	30
	Water in radiator at 50°C	15
Inferior	Waste heat	5
Valueless	Heat radiation from the earth	0

There is still one problem to be resolved. Nature does not give any clue about the absolute value of internal energy U, nor of entropy S or exergy Φ. Thus, we have to choose it ourselves, as a convention.[14] The choice is a reference situation R, where it is simply agreed that the internal energy in that situation equals zero. An obvious choice for R is the equilibrium with the ambient temperature, pressure and atmospheric concentrations (T_0, P_0, μ_{0i}). The standard or reference value for temperature is 0°C (273.15 K) and for pressure 1 atm (101.325 kPa). In chemical reactions, the standard enthalpy of formation h_{ref} ($h = u + Pv$) of a chemical compound is then defined as the change in enthalpy h when the compound is formed from the stable compounds in their natural concentrations (CO_2 and H_2O), the reactants and products all being in a given standard state The specific exergy of an important class of compounds, namely fuels, is within about 5 percent equal to the combustion enthalpy of the fuel, because the entropy term is rather small. Therefore, the fuel enthalpy is often used in exergy analysis (Table 7.1).

One last reflection on the relevance of the Second Law. Every real-world conversion process will degrade the quality of the energy in dissipative processes such as friction. Macroscopic order, in the form of, for instance, potential or kinetic energy, is converted into microscopic disorder in the form of heat. To 'sustain' such a process, there has to be a continuous supply of high-quality energy from the system's environment. This happens across gradients in the intensive variables such as temperature, pressure or concentrations. The larger the gradient, the larger the driving force, the more irreversible the process – and the larger the dissipative work. If you

[14] We do this also, in a more accessible way, for potential energy in the earth gravitation field: Without some convention of a reference height, it is not possible to calculate $E = mgh$. The same holds for the kinetic energy.

Box 7.3. *The Colosseum, slaves and containerships.* The energy needed to deliver energy has always been a concern for societies. 'All our societies require enormous flows of high-quality energy just to sustain, let alone raise, their complexity and order (to keep themselves [...] far from thermodynamic equilibrium)... [and] after a certain point in time, without dramatic new technologies for finding and using energy, a society's return on its investments to produce energy [...] starts to decline' (Homer-Dixon 2006:54–55). An example is the Colosseum in Ancient Rome (Homer-Dixon 2006). Construction began under Emperor Vespasianus between 72 and 75 CE and inaugurated in 80 CE by his son Titus, with a hundred days of games and some ten thousand beasts being killed. An estimated million ton of raw material have been moved for its construction. An estimated 185 TJ of energy was needed. Over 75 percent of it was used to feed the thousands of oxen engaged in transporting materials. The remainder powered the human labourers: Over 2,000 people working 220 days a year for 5 years. During this period, some 55 km^2 in the largely agrarian economy was needed to deliver this solar energy.

Compare these numbers with the largest container ships under construction. Such ships measure 400 m in length and over 12 m in depth and can carry 15,500 containers of 36 m^3 content or 56,2000 m^3 of freight. Their normal speed is 46 km/hr, so the kinetic energy of only the freight is equivalent to 46 GJ at an average freight density of 1,000 kg/m^3. With a 10 percent speed reduction, fuel use drops with a quarter, which means about 250,000 ton/yr less of CO_2-emissions. The next round of upscaling is to ships of 18,000 containers, which are expected to be profitable because they carry more freight but at lower speed. The builders claim a further reduction in CO_2-emission, to 3 grams per ton-km of freight. At present, Rotterdam Harbour can handle 100,000 containers/week and is among the few harbours that are deep enough for such ships (NRC 18 March 2011). The trend is an even larger scale of operations in the name of lower cost and lower emissions.

drive a distance of 110 km in a car in two hours at 55 km/hr (~15 m/s), you need about 5 kWh to overcome air resistance. If you drive the same distance in one hour at 110 km/hr (~30 m/s), you need 20 kWh or four times more (MacKay 2009). But large driving forces are desirable because they make processes go fast, and we humans are in a hurry because we do not live forever. Doing things means loss of potential work – and doing things fast means more loss of potential work. Or the other way around, doing things slowly reduces loss of potential work. Of course, doing things infinitely slowly makes no sense, so there is some optimum between what we want to accomplish and how much work potential we are able and willing to use for it.

7.2.3 Energy Forms

Energy can be categorised in several ways. One distinction is between transitional energy and stored energy. *Stored energy* is energy that is available in a mass or a position in a force field. Stored energy can be converted to another form of stored energy and the process takes place via *transitional energy*: heat Q and work W.

As discussed, energy is around in different *forms*: internal energy and potential and kinetic energy in force fields. The *internal energy* U is the energy of the constituent atoms and molecules in the form of rotational, vibrational, translational and chemical bonding energy of atoms and molecules (equation 7.2). It is manifested as sensible heat (temperature related), latent heat (phase change related, for instance, when water vapour condenses as raindrops) and chemical and nuclear energy. *Changes* in internal energy is what matters. It manifests itself in everyday experiences like burning your fingers on a stove or lighting a match. *Potential* and *kinetic energy* in a gravitational or electromagnetic field are two other energy forms (equation 7.4). A lifted arm or object represent potential energy in the earth gravitational field. Water flowing downhill, a moving bus or a metal rod spinning in a fluctuating magnetic field in an operating electromotor represent kinetic energy. Table 7.3 lists the major energy forms and their conversion processes and associated appliances.

The essential source of energy is solar electromagnetic radiation, which via photosynthesis is converted to chemically stored energy. It can be transformed in kinetic energy of movement, for instance, in the human or animal body (metabolism). The chemical energy from photosynthesis has been accumulated in geological time in fossil fuels, which upon combustion deliver thermal energy. At high temperatures, this heat is used in internal combustion engines and in steam/gas turbines to produce kinetic energy (moving piston, rotating shaft). This is either used directly for moving mass or converted into electricity in an electric power generator. If the heat comes from a geothermal or a nuclear source, it is similarly used to generate power. In these processes, part of the kinetic energy is inevitably converted into thermal energy (heat). Thermal energy is also stored or released in phase transitions (evaporation/condensation, melting/freezing). Solar radiation is directly converted into electricity, in photovoltaic (PV) solar cells, or into heat, in solar heat absorption collectors. Chemical energy is directly converted into electricity in fuel cells.

Electricity has become the dominant and most versatile form of energy available. Electric power is converted into electromagnetic radiation of different wavelengths in an increasing amount of Information and Communication Technology (ICT) appliances and other devices in household and other equipment. Electricity is also used directly in industrial processes such as electrolysis in aluminium smelters. It mostly ends up as heat at ambient temperature and is used directly for room heating. The biggest disadvantage of electricity as compared to chemical energy in fuels is that it is not easily stored. Devices like batteries and flywheels do store it in the form of chemical or kinetic energy, but they are relatively inefficient and costly.

Because *work* is the high-quality energy form of genuine interest to us, the conversion routes of energy are very important and at the core of energy technologies. Understanding the causes of exergy losses and the potential for reducing those losses, that is, increasing the exergetic efficiency by extracting work from high-temperature heat, is an important part of the search for sustainable energy systems.

7.3 Movement in Space and Time: Mechanics

Change is an elementary phenomenon in life. It is most visible in movement of matter. It is useful to look first at a simple, well observable and controllable system: a mass m at the end of a spring. It is the classical model of linear elasticity, associated

Table 7.3. *Energy conversions. Each column shows how a particular form of energy can be converted into another one that is indicated in the corresponding row to the left*

FROM:	thermal	chemical	gravitational	electric	electromagnetic	nuclear	kinetic/ mechanical
TO:							
thermal	heat flow (heat exchanger), phase transition (boiling, melting etc.)	exothermal reaction, combustion		resistance heating	solar heat collector	fission/fusion (nuclear reactor)	friction in movement, inelastic collision
chemical	endothermal reaction, thermolysis	chemical reaction		electrolysis (battery charging)	photosynthesis	ionisation	radiolysis
gravitational			mass displacement	electric elevator			lifting a mass
electric	thermoelectricity	Battery, fuel cell		transformator	solar (PV) cell	röntgen	dynamo generator
electro-magnetic	thermal radiation (light bulb), thermocouple, MHD-generator	chemolumi-nescence fire flies		electrolumi-nescence (LED-lamp), radio/tv, LCD-display etc.	photolumines cence (laser display)	radioactivity	axe on stone (sparks)
nuclear				particle accelerator	gamma reactions		charged particle reactions
kinetic/ mechanical	Expansion, heat engine, (internal combust, steam/ gas turbine)	metabolism, muscle power, fire arms	falling mass hydroturbine	electric motor, flywheel	solar sail	particle emission (nuclear bomb)	gearbox

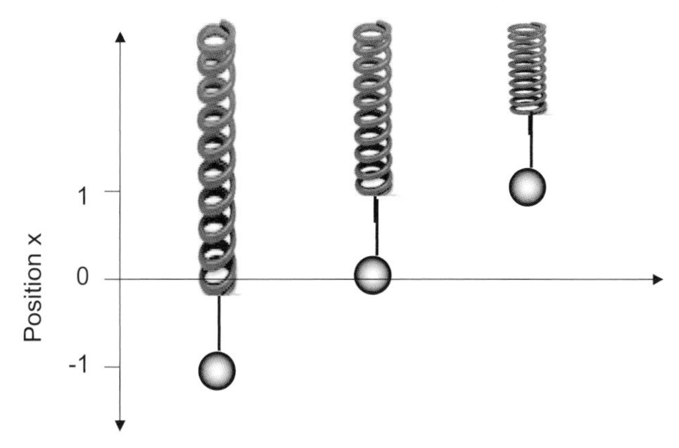

Figure 7.6. Graphical presentation of the mass on a spring. In the left, it is kept in stretched position, When left loose, it moves upwards to the middle and then the right position.

with the 17th-century British scholar Hooke. Change is about potential and kinetic energy in a gravitational field (equation 7.1).

Assume that the position of the mass is indicated with x(t) and that initially it is at rest in position x_R. If you pull the mass, it starts an oscillating movement. This happens under the influence of two forces: the gravitational force F_g and the spring force F_s. The gravitational force is given by $F_g = mg$ with g the gravitational constant in m/s^2, and the spring force is approximated by $F_s = -kx$ with k the spring constant in kg/s^2. In the rest position, the gravitational force pulls the mass a small distance down – say x_R, at which the two forces are equal: $F_g = F_s$ and $m.g = k.x_R$ or $x_R = m.g/k$. Let us normalise and define $x_R = 0$. Suppose you pull down the mass to a position – p at time t = 0. At this point the upward spring force will equal $F_s = kp$ and the potential energy in the gravitational field has dropped with an amount of mgp. The system is in equilibrium because your hand exerts a downward force equal to $F_s - F_g$. Letting loose, the mass will start to move upward because of $F_s - F_g = kp$. Its velocity v increases and the kinetic energy is $\frac{1}{2}mv^2$. The law of conservation of energy implies that the sum of potential and kinetic energy remains constant. Therefore, when the mass passes in its upward movement the point x = 0, the potential energy it had at x = p, equal to mgp, has been converted into kinetic energy. Hence, at $mgp = \frac{1}{2}mv^2$ or $v = \sqrt{2gp}$ on passing the point x_R[15]. The further you pull the mass down, the greater the speed at which the mass goes past the point x_R.

Using the formal language of mathematics, one can approximate the movement of a mass on a spring with two differential equations:

$$dx/dt = v$$
$$dv/dt = -kx \qquad (7.7a)$$

in which the small d indicates an infinitesimally small change. The first one says that the velocity v equals the rate of change of displacement and the second one that the rate of change of velocity, that is, the acceleration, is negatively proportional to the

[15] This is not completely correct, because a small part of the energy is needed to lift the spring upwards in the gravitational field and equal to mg(p-x_R).

distance from the point $x(0) = x_R = 0$. Substituting the first equation into the second yields a single second-order differential equation of the form:

$$d^2x/dt^2 = -kx \tag{7.7b}$$

which can be solved analytically[16]. I will not derive analytical solutions and instead use the simulation software package Stella® to explore the system behaviour over time. The harmonic oscillations in Figure 7.7a represent the position of a mass, x(t), over time for the reversible (ideal) frictionless spring. Plotting the velocity of the mass, v(t) = dx(t)/dt, as a function of the position x(t), one gets the phase diagram (Figure 7.7b).

Of course, you know that the mass will not keep oscillating forever. There will be friction, which shows up as a dampening force proportional to the velocity. This yields an additional term in the expression for the restoring force:

$$\begin{aligned} dx/dt &= v \\ dv/dt &= -kx - rv \end{aligned} \tag{7.8}$$

with r the friction coefficient. This description is more realistic because it incorporates the phenomenon that any process is irreversible and more ordered energy is dissipated into heat (molecular motion) in agreement with the second law (ΔS_{irr} in equation 7.2). The speed of the mass slowly diminishes over time and ends in the equilibrium point $x(0) = X_R = 0$, called the *attractor* (Figure 7.7c–d). One can also apply an external force on the mass, for instance, by pulling it at a constant frequency ω and amplitude q. Now, the natural and the forced movement interact and the position x(t) over time becomes erratic (Figure 7.7e–f). If someone would show you this graph, you would have a hard time to discover how simple the underlying system actually is!

What happens if the spring constant is itself a function of the displacement from equilibrium: k = k(x)? Suppose it is found that the spring constant behaves according to $k = (x^2 - 1)$. This implies that the system behaves like a mass at the end of a spring only for $x > 1$ or $x < -1$. For $-1 < x < 1$ the value of k is negative and the system operates in this domain not with deceleration due to friction but with acceleration. In system dynamics terms, there is a positive or amplifying feedback in this domain. Such a system is described by the two differential equations:

$$\begin{aligned} dx/dt &= v \\ dv/dt &= -(x^2 - 1)x - rv \end{aligned} \tag{7.9a}$$

The system has three attractors, two stable and one unstable. A steel spring placed between two permanent magnets is a bistable oscillator described by these equations (Bossel 1994).[17] If there is no dampening (r = 0), the system oscillates between the two attractors, as shown in the trajectory in time and in the phase diagram in Figure 7.7g–h. The damped system moves towards one of the two attractors, but it is hard to say in advance to which one. A slight uncertainty or change in the initial state, that is, x at t = 0, or in one of the parameters can make the system

[16] The general solution for a class of such equations is by separating variables (x, left; t, right) and integrating both sides. I refer to the suggested reading for details.

[17] Written as a single second-order differential equation, the model is known as the Duffing equation.

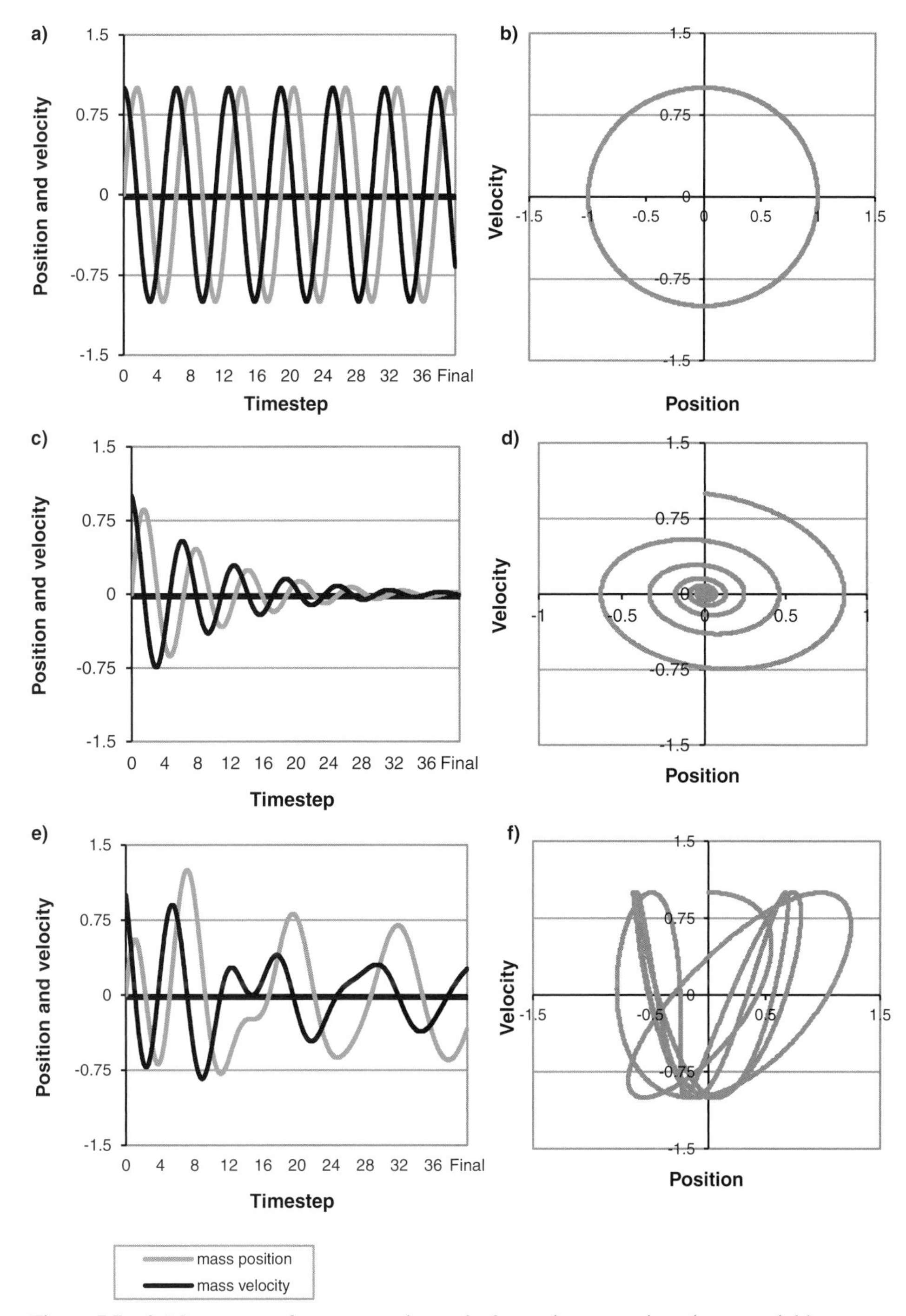

Figure 7.7a–f. Movement of a mass at the end of a spring over time (state variables: mass position x and mass velocity v; left) and in phase space (right). The middle graphs simulate a situation with friction; the lower one with friction and exogenous forcing (timestep 0,01; Runge-Kutta-4).

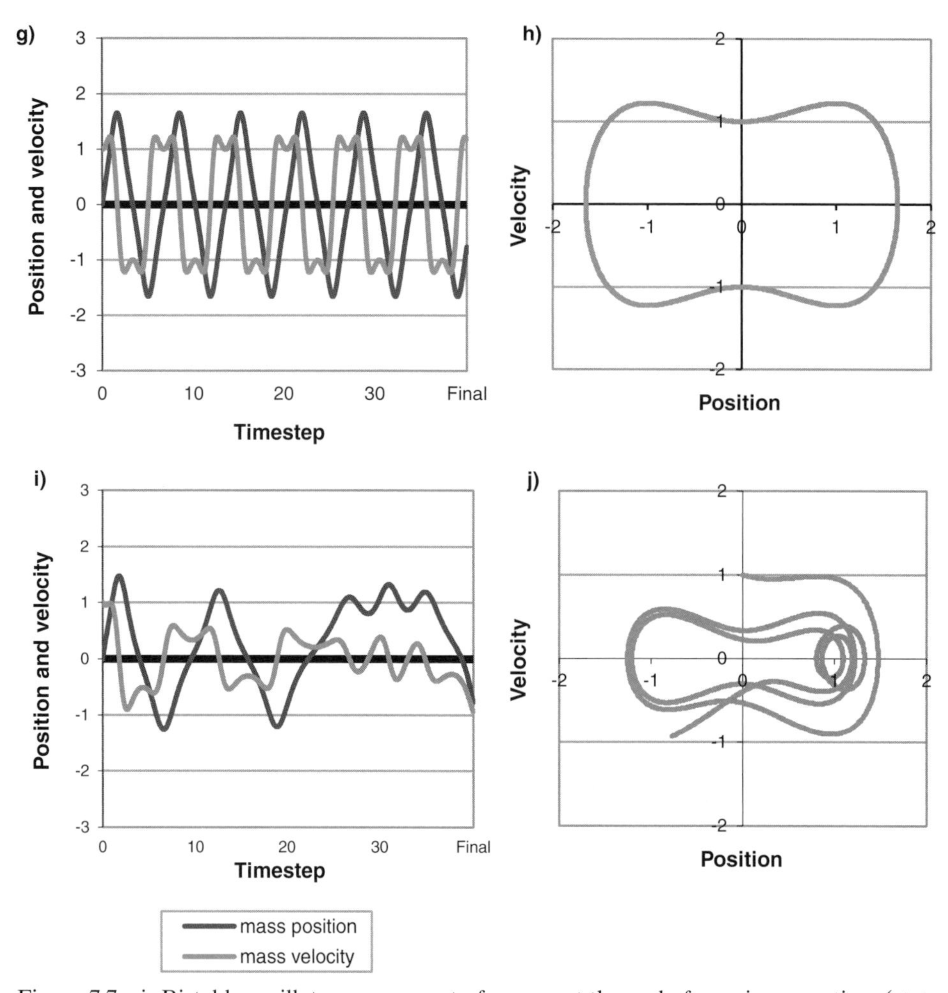

Figure 7.7g–j. Bistable oscillator: movement of a mass at the end of a spring over time (state variables: mass position x and mass velocity v; left) and in phase space (right) when the spring constant is itself a function of the displacement. The upper graphs show an undamped situation without friction; the lower graphs with an external forcing (timestep 0,01; Runge-Kutta-4).

move towards a different equilibrium state – it is extremely *sensitive to the initial conditions*. The term $(x^2 - 1)$ introduces a nonlinearity. This usually leads to multiple equilibria and critical dependence on the initial conditions. These simple technical devices show complex behaviour.

If this system is also subjected to an external force of constant frequency ω and amplitude q:

$$\begin{aligned} dx/dt &= v \\ dv/dt &= -(x^2 - 1)x - rv + q \cdot \cos(\omega t) \end{aligned} \tag{7.9b}$$

its behaviour becomes irregular and even chaotic for certain parameter domains (Figure 7.7i–j). Chaotic means in this context that the system – the same steel spring but placed now in an oscillating, not a permanent magnetic field – behaves deterministically but is nevertheless unpredictable in the sense that its future can only be known if one would know the initial state with infinite precision. Note that these

Box 7.4. *Strange attractors and butterflies.* A famous example of extreme sensitivity for initial conditions are the equations named after the meteorologist Lorenz. He applied the basic equations in fluid dynamics in order to model the movements in a closed box with smooth interior and filled with gas, which is heated at a constant rate:

$$dx/dt = a(y - x)$$
$$dy/dt = bx - y - xz$$
$$dz/dt = xy - cz \quad (7.10)$$

with x, y and z the coordinates in space. It is a simplified three-dimensional description of atmosphere processes. The behaviour of {x(t), y(t), z(t)} turns out to be highly sensitive to the initial conditions and, therefore, becomes highly unpredictable or chaotic. The system will also tend towards one of two particular patterns over time, a phenomenon called a strange attractor. The model has important implications for climate and weather prediction, because it shows that a fully deterministic system can exhibit quasi-periodic change and abrupt and apparently random change. It has become popular with the question for weather forecasters, ascribed to Lorenz, whether the flap of a butterfly's wings in Brazil can set off a tornado in Texas.

equations describe the movement of a mass in space and time (potential and kinetic energy) and not the associated heat flow (friction).

These simulations show that rather simple nonlinear equations ('models'), which may or may not be accurate descriptions of real-world systems, can exhibit complex behaviour over time. Their use as analogues is useful in the sustainability discourse, provided that one understands the idealising assumptions. In particular, the investigation of multiple equilibria in ecosystems and economic systems is a new area of research. Elementary chemical processes, for instance, chemical kinetics, provide more interesting and complex examples, but this is considered beyond the scope of this book. I refer the reader to the suggested literature.

7.4 Stories

7.4.1 Fuel Efficient Stoves for People in Darfur[18]

In the fall of 2005, Berkeley scientist Ashok Gadgil was asked by the U.S. Government to try to find a solution to a grave problem facing Darfuri families in displacement camps: Women had to walk as long as seven hours, three to five times per week to find firewood. In response, he and his colleagues and the women of Darfur designed the Berkeley-Darfur Stove (BDS). The project is a good example of sustainable livelihood improvement, as it solved several problems simultaneously. Because the BDS uses half as much firewood as traditional cooking methods, it limits harmful emissions that contribute to global warming. Users save $300 per year in

[18] This text is provided one of the project leaders, Ashok Gadgil. For more information, see www.darfurstoves.org

fuel costs. Over the five-year life span of the stove, this savings is approximately $1,500. Reduced excursions for firewood collection translate into rest and relief for the exhausted women in the displacement camps and time that could be spent with family or on income generating activities. As to safety, the stove reduces the need for women to leave the camps in search of firewood and it thereby reduces their exposure to sexual assault and violence. It also contributes to health because it limits toxic pollutants released that cause respiratory diseases such as pneumonia.

The BDS was developed by scientists and engineers at the Lawrence Berkeley National Lab, with evaluation and feedback from women in Darfur. They brought together the world's best minds in engineering to develop a simple, locally appropriate technology and to provide ongoing technical support to the field partners. The BDS is well adapted to the local situation. It is specifically tailored to the windy climate, the sandy terrain, the pot sizes and the cooking style of families living in the displacement camps in Darfur. The local participation of the women in Darfur provided feedback at every step of the process, ensuring the stove design fits their needs. The local employment and low cost due to its manufacturing process are also beneficial. The BDS starts out as sheet metal pieces stamped out in India. These 'flat-kits' are shipped to Sudan, where they are assembled by the Sustainable Action Group, a Sudanese NGO affiliated with the project partner Oxfam America. The total cost to fabricate, ship and assemble each stove is $20.

The Darfur Stoves Project and its partners produced over 15,000 stoves by the end of 2010. Each year they are in the field, these 15,000 stoves save Darfuri women as much as $2 million in fuel expenses and offset more than 20,000 tons of greenhouse gases, which is equivalent to taking >4,000 average-sized cars off the road in the United States. An estimated 300,000 families are in need of a fuel-efficient stove and the goal is to distribute a BDS to each family. This project teaches at least two lessons for a project to succeed: You have to 'walk with the people', and there are always winners and loosers and you have to identify them.

7.4.2 The South Nyírség Bioenergy Project[19]

The South Nyírség region in the northeastern part of Hungary was traditionally an agricultural area, renowned for its tobacco and apple production. Since the transition in 1990, local agriculture has gradually lost its importance. Apple export to the Soviet market has stopped, tobacco production has dropped and other agricultural activities are hindered by the relatively low quality of soil. Today, most of the landowners want to plant trees. Trees grow on lower-quality soil and there are subsidies for afforestation. There are about 80,000 hectares (ha) of privately owned forest and about 70,000 ha of state-owned traditional forests in the South Nyírség region. Most of the new plantations include fast-growing species such as black locust or poplar, and the produced wood is typically used for pellet production.

The Project was initiated in 2003 by local governments and entrepreneurs because of high unemployment and outmigration of skilled workers. A logical choice was biomass-based power generation. Hungary was already in the process of European Union accession, and it was expected that regulation would support

[19] This text has been written by Zoltán Lontay, initiator and chief engineer of the project.

renewable energy production. A company was founded that established strategic cooperation. Project development took more than four years. The local government of Szakoly village offered 20 ha, which had to be requalified from agricultural to industrial land, and dozens of permits had to be obtained. Connection to the network had to be negotiated with the owner of the local public network, the German company E.ON. The major challenge was finance. In 2007, the banks agreed to give a loan of 85 percent of the €55 million investment – thus, 15 percent equity had to be collected from elsewhere. Another company was formed to finance a complex structure with four shareholders, who each had their own interests. Given the rather unfriendly legal, regulatory and economical environment, the close relationship with the local society turned out to be crucial.

In December 2007, the first greenfield biopower plant in Hungary started commercial operation after an official ceremony and blessing by the priests of Szakoly village. The fuel is woody biomass, delivered on a contract basis with twenty to thirty small, local suppliers and the state-owned forestry company of the region. The latter covers 10–20 percent of the fuel demand of 180–200 kt/yr (wet). Thin branches, trunks, bark, vegetation from flood areas of rivers and trees from old orchards are chipped, sorted on size and used in the power plant. Research is started on how to utilise the ashes. Obstacles are the inefficiency of the state-run subsidy program (farmers receive subsidies a year after plantation) and the fluctuations in the agricultural prices (when the price of traditional agricultural products go up, the landowners switch back to them).

Economically, the situation is still precarious. The first years of production were below expectation but suppliers and operators are on a learning curve. The financial crisis in 2008 made the banks more demanding and project revenues are too small for economic operation because of a low feed-in tariff at €100/MWh, which is well below the Central European level. The government says that a higher tariff increases the electricity price and is, therefore, not possible for commercial and social reasons. External benefits of small-scale renewable projects such as lower unemployment payments, higher taxes and avoided emissions are apparently not considered.

Economic and energetic performance will be improved with the use of waste heat in a greenhouse, which is expected to produce 2,500 t/yr of vegetables that meet the European 'bio'-standard and to employ 80–100 people. The project demonstrates the opportunity for a paradigm shift in rural development. A national plan shows that the construction of 50–70 similar plants in the 5–20 MWe capacity range can help Hungary to reach the 2020 European Union targets for renewable energy production, avoid up to 3 Mt/yr of CO_2 emissions and employ over 10,000 people. The project shows that growing biomass for fuel supply is an interesting energy option for (parts of) Europe.

7.5 Energy Conversion

7.5.1 Elementary Processes

Thermodynamics can only make statements about systems in equilibrium. Therefore, many of the laws and models describing real-world near-processes or non-equilibrium processes are empirical or phenomenological laws. Here I briefly discuss a few such processes. It is not the intention to be complete, and I refer to the Suggested Reading and Websites sections for more information. The first process is *heat flow* or *heat transfer*. Heating air in a room with a stove is a heat transfer process. No work is involved: You simply produce a flow of heat Q from burning gas, oil, coal or wood. However, as soon as the air has a temperature higher than the ambient temperature, Nature strives to equilibrate the two – the consequence of the Second Law. Therefore, you not only heat the room but also the surroundings. The heat transfer from the room with an average air temperature T_{room} to the surroundings where the air has an ambient temperature T_0, is approximated by Fourier's Law:

$$\frac{\Delta Q}{\Delta t} = k \cdot A \cdot \frac{(T_{room} - T_0)}{\Delta t} \text{ W} \tag{7.11}$$

It states that the heat flow out of the room is proportional to the temperature difference between inside and outside, to the heat conductivity k (in $J/m^2/K$) and to the size A (in m^2) of the interface between room and surroundings (Δt is the time period considered).[20]

This relationship is essential knowledge if you want to insulate your house or a manufacturer wants to reduce process heat losses. The equation tells you that you can reduce the heat flow and your energy bill by:

- putting on a sweater so T_{room} and hence the difference $T_{room} - T_0$ can be lower because the difference $T_{room} - T_0$ bothers you less;
- reducing the size of the interface A – that is why living in a smaller house or having smaller windows costs less energy; and
- reducing the heat conductivity k by insulating roofs, walls, floors and windows, usually called energy conservation or efficiency improvement.

The third option is now widely applied and new buildings have to comply with ever stricter standards. With $\Delta t = 1$ second, the k-values are in the range of 0.11 $W/m^2/K$ for a 30 cm glasswool insulation layer to 2 and 6 $W/m^2/K$ for double and single glass windows, respectively. Innovations lead to ever-lower heat conductivity values, but it takes time before they are cost-competitive and accepted by house builders and owners. Insulation is very effective: Replacing a single-pane window with the latest HR-glass lowers k from 6 to 2 $W/m^2/K$ – or 2 kWh per m^2 of window on a cold day. Also, it reduces noise levels. Putting glass mineral wool under the floor or against a standard roof gives similar reductions.

There are more drastic solutions to make energy use in buildings more efficient. You may remember that one loses a lot of work potential (exergy) by burning fuel and use the heat in the exhaust gases to warm water to 60°C. If a small engine – a gas turbine, for instance – is put between the exhaust gases and the water heater, work can be extracted. Such a scheme is called *cogeneration* or Combined Heat and Power (CHP). It is one of the options for decentralised electric power supply, in combination with renewable sources like PV-solar. There are drawbacks: The small scale makes it rather expensive and the coupling with room heating confines electricity production

[20] In a more precise description, the temperature difference is indicated as a gradient across a distance d and the heat flow is given in W/m.

to cold days. However, cogeneration and CHP is already widely applied in industrial processes and increasingly in large buildings.

Another interesting option is the *heat pump*. It operates as a refrigerator: A fluid absorbs heat from an outside reservoir (river/groundwater, air) as it evaporates and delivers the heat inside your house when the vapour is compressed and condenses. In other words, it is a refrigerator, but now interest is in the hot, not the cold side. The heat exchanger in the rear of the refrigerator, or outside of the window for the air conditioner, is now the heat source (radiator). Electricity 'pumps' the outside heat to a higher temperature. The number of heat pump installations grows rapidly.

The second process to have a brief look at is the *diffusion and discharge of mass flows*. Mass flows, and their diffusion in air, water and soils are ubiquitous in biological and environmental as well as in technical processes. They are the core of micro-level understanding of processes such as changes in soil nutrients, dispersion of pollutants and (bio)accumulation. Here also there are two empirical laws that formulate the change in terms of a gradient. The first one is Fick's (first) law about the diffusion of a substance across a concentration gradient. In equation form:

$$\frac{\Delta m}{\Delta t} = -D \cdot A \cdot \frac{(c_h - c_l)}{\Delta x} \text{ kg/s} \tag{7.12}$$

with m the mass, A the diffusion surface, c_h and c_l the high and low concentration, respectively, and D the diffusivity which is the equivalent of conductivity in Fourier's law (equation 7.11). If concentration is replaced by osmotic pressure, it describes the process in which osmotic pressure difference is used to extract work from the difference in salt concentration at the outflow of river into the sea. A Norwegian pilot plant uses this salinity gradient to generate 2–4 kW of power, at 1 W/m^2 and flow rates of 10–20 litre/s. It is a truly renewable resource, but the membranes are still very expensive.

A similar relationship is used to describe the discharge of a fluid in the soil. For instance, water flows in an aquifer are determined by the pressure drop per unit length and are approximated with Darcy's law:

$$\frac{\Delta V}{\Delta t} = -\frac{k \cdot A}{\mu} \cdot \frac{(P_h - P_l)}{\Delta x} \text{ m}^3\text{/s} \tag{7.13}$$

with V the volume, k the permeability, A the discharge surface, μ the viscosity and P_h and P_l the high and low pressure levels. Here, the inverse of viscosity is the equivalent of conductivity. It is the basic equation for the calculation of the energy needed to pump up water – or oil or gas – from an underground reservoir.

These laws, and other ones like Ohm's law for electrical current and the Gaussian dispersion model for air pollutants, express the second law observation that Nature tries to reduce the difference in intensive variables (temperature, pressure, concentration) between two (parts of) systems. It is a natural negative-feedback process. These laws are never exact in the real world, their relevance is in describing the underlying process and being often a good first approximation or 'rule-of-thumb' (§8.6). Interestingly, these laws came into existence well before they were scientifically understood at the micro-level and were based on the intuition that inhomogeneities spontaneously tend to dissolve at a rate proportional to the gradient, that is, on the Second Law.

This section ends with one of the most crucial conversions in modern society: the *generation of electricity* from chemical energy (Table 7.3). It serves as an illustration of the difference between an energy and an exergy analysis. In theory, the chemical energy in a fuel can be converted into electricity with close to 100 percent exergetic efficiency.[21] In practise, most electricity is generated in a heat engine at much lower efficiency due to the Second Law. Bringing a heat reservoir of high T – for instance, exhaust gases from a furnace or engine, although this is not the infinite heat reservoir from the theory (equation 7.6) – irreversibly into equilibrium with the earth environment implies a loss in work potential (Table 7.2). But the higher the exhaust gas temperature expanded in a turbine and the lower the temperature of the at ambient river or sea water (condenser) or air (cooling tower) to which heat is exported, the lower the exergy losses. There are various technologies to extract work from a heat flow, such as the steam turbine and the gas turbine. The turbines drive a generator in an electric power plant. The system can be operated as a closed circuit as in a water-vapour Rankine-cycle steam turbine, in a closed-cycle gas turbine or in an open-circuit open-cycle gas turbine. Increasingly, large-scale electric power generation is done in Rankine-cycle operation coupled to a gasturbine in a so-called Steam And Gas Generating-Combined Cycle (STAG-CC) plant. The exergetic efficiency can be quite high. For a gas turbine, inlet temperature of 1200K or more, over 75 percent of work can in theory be extracted. Present-day STAG-CC power plants reach exergetic (chemical-to-electric) efficiencies up to 60 percent. In comparison, most power plants around 1900 worked with much lower temperatures and had efficiencies of 5–10 percent. One can increase the energetic efficiency by using the low-temperature heat to heat offices or dwellings (*district heating*) or for use in industrial processes (the previously mentioned CHP). The other option to generate shaft work are the internal combustion (Otto and diesel) engines and Stirling engines. The efficiencies are lower and they are used primarily in cars and ships and for cogeneration.

7.5.2 The Energy System

Energy features high on the sustainability agenda, because it is largely based on finite, non-renewable resources and it causes a series of environmental problems. Massive, and still increasing, amounts of fossil fuel are extracted, transported, processed and then made to react chemically with the oxygen in the air ('burnt') to generate power, heat and combustion products. In this way, one gets useful energy or *energy services* such as heating houses and offices, running vehicles and industrial ovens, and powering electricity using devices from aluminium smelters to streetlights and from vacuum cleaners to televisions and computers. Together, they constitute the 'energy chain' from source to sink.

In addition to the energy use side, the energy system comprises energy supply, transport and distribution (Figure 7.8). This is what is usually meant with the 'energy industry'. Its major elements are coal mines, oil refineries and electric power

[21] In an electrochemical or *fuel cell*, the conversion rate to electricity has in theory no inherent limitation. But there is still a gap between theory and practice, and fuel cells are not yet a reliable and cheap enough option for large-scale application.

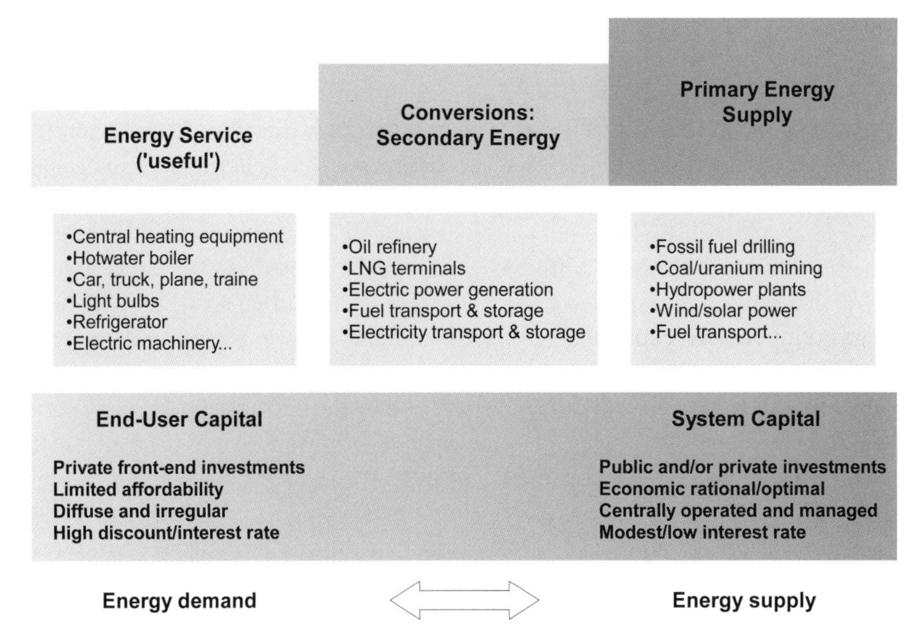

Figure 7.8. The energy system and its subsystems: energy services, energy conversion and energy supply.

plants. It is a very large and highly concentrated industry. The economic value of the transactions in the oil and gas industry amount to an estimated 2,000 billion €/yr. In the electric power industry, it is in the order of 1,000 billion €/yr. Their combined sales are about 5 percent of gross world product (GWP). The energy industry is very capital-intensive. In Europe alone, in 1971–2000, annual investments were in the order of 80–120 billion €/yr. Some of the most influential technological transitions have been in the energy field. From a decision-making and policy perspective, the large energy companies are in a more powerful position than the millions of distributed energy users. This is an important issue in the transition to non-depletable, non-carbon energy sources.

The *energy flow densities* in electric power plants and other parts of modern energy systems are extremely high. For instance, the fuel and heat flows in large power plants and the oil and gas flows in large pipelines are in the order of W/m^2 of throughput surface. This is possible because of the high energy density of fuels (20–30 GJ/m^3). Energy density is one of the big differences with renewable sources of energy. They are all based on solar energy fluxes with values of at most 1.2 kW/m^2. The consequence is that these resources are, on the one hand, abundant and 'free', apart from the equipment to harness it. On the other hand, it implies the use of space. In Europe, an average 100–200 W/m^2 can be harnessed from the sun.[22] With PV-solar installations, present-day installations produce in the order of 3–4 W/m^2. If it is converted to plant biomass, it will be difficult to produce in Europe at values above 1 W/m^2. Offshore wind turbine parks deliver at 2–3 W/m^2, and tidal power delivers in the same range. Hydroelectric dams can deliver up to 10 W/m^2 of lake area. These values are low as compared to the average oil- or gasfield or coalmine.

[22] The numbers are based on MacKay (2009) and some other sources.

For instance, the Groningen gasfield has for many years produced at a rate of 80 billion m^3/yr or 90,000 MW. If 1 percent of the province of Groningen is needed for its exploitation, the density is 3,000 W/m^2.

In electric power generation, there is a tendency towards ever larger scale in order to reduce costs – the capacity of nuclear power stations is in the order of 1,200 MWe for a single plant and possible nuclear fusion plants are envisioned to be much larger. Large-scale fossil-fuel and nuclear systems need centralised social-technical management in view of the huge upfront investments, high-tech skills, safety measures, waste disposal and so on. The capital stocks last for four or more decades and their dismantling and waste management may last many more. Responsible operation is only possible in advanced societies. A renewable energy system of spatially dispersed wind and PV-solar parks is naturally rather decentralised, although very large windturbine parks are being planned and solar power plants in American, Chinese or North African deserts could cover huge tracts of land. They face different constraints in the form of competition with other uses of land and the need for a reliable grid with storage options. Thus, the *energy transition* away from fossil fuels will rely on more efficient provision of energy services and on some mix of land-intensive and intermittent (sun/wind) renewable energy and possibly sufficiently safe forms of nuclear energy. Unless a yet-unknown, miraculous energy source is found, fossil fuels and notably natural gas will play an important role as transition fuel (§12.3).

7.6 Energy Futures

Thanks to tremendous improvements in efficiency, there has been a steady decline in the energy needed to provide one unit of energy services (Figure 4.9b). But because the number of and variety in energy services did and still does increase exponentially, total energy use keeps growing (Figures 7.2 and 13.8). As it is largely derived from burning fossil fuel, emission of CO_2 and other gases did increase as well and is expected to continue to do so (Figure 4.7). Thus, the system operates under the increasingly tight constraints of depletion ('source') and environmental pollution ('sink').

How will the future energy systems in the world unfold? Past trends can be extrapolated (Figure 7.2), but how and with which arguments? Most energy experts agree that demand for energy services will continue to increase and that fuels for transport and electricity are the two major carriers. That is where agreement stops. Table 7.4 illustrates the differences in views with a series of statements about the problems, the solutions and the mechanisms for a sustainable energy future. I apply the worldviews to contrast different possible directions, using statements from various institutions (§6.3). As the reader, you are invited to examine your own values and beliefs on the basis of these statements.

The preferred values can be summarised with the keywords availability, affordability, reliability and (environmental) sustainability. The controversies are mostly about markets and prices versus governments and standards, about the risks of energy dependence, and about the costs and risks of particular technological options. Clearly, there are serious differences in the evaluation of what is possible and desirable. If any of the four worldviews becomes dominant, the serious shortcomings

Table 7.4. *Worldviews and energy futures.*

ENERGY

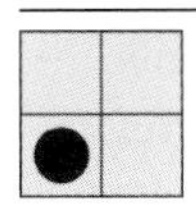	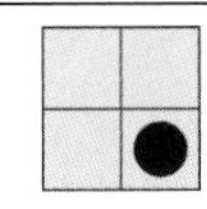	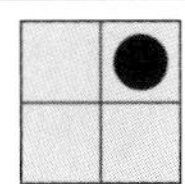	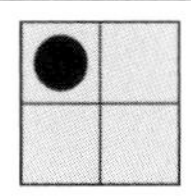
Statement 1: World energy use can and should be stabilised at the present level of about 400 EJ/yr.			
Impossible. People need and want energy. Low-income regions have legitimate aspirations for high-energy lifestyle.	I don't know and I don't care, as long as I can get my gasoline and electricity cheaply and reliably.	I don't know but I will try in my own situation. I think and feel it is necessary if we are to survive in the long term.	It is necessary in view of environmental risks. It is also possible and challenging. Poor people still need more, so the rich have to use less.
Statement 2: The world will remain heavily reliant on fossil fuels for the next forty years (CEO Shell, 2010).			
Quite possible, if political alliances are made with the oil-rich states. Climate issues can be solved with CCS, plus unconventional oil and gas and coal reserves are outside OPEC.	Why not? There are abundant fossil fuels and human-induced climate change is still unproven and maybe acceptable or even beneficial.	Irresponsible. Oil companies only think about their profits. Why do they not invest their huge profits in renewable energy?	Impossible. Shell assumes a doubling of demand between 2010 and 2050. It would mean massive coal burning – and dangerously high carbon emissions.
Statement 3: Nuclear power is neither safe nor clean. In order to save the future of our planet, we must continue to fight the expansion of nuclear power (Greenpeace 2011).			
Irresponsible. Industry needs power and nuclear power is probably the only serious, cheap and zero-carbon option available.	Nonsense. It is probably the only way out for the world energy system, given the threat of climate change and fossil fuel depletion.	Correct. The life-threatening problems of radioactive contamination, proliferation and waste disposal are not solved!	The precautionary principle should be guiding. Stricter regulatory and safety measures are needed, as the accidents make clear.
Statement 4: Governments should support energy efficiency and technologies such as the electric and hydrogen car.			
In principle, energy markets should be deregulated – energy prices should do the job. For infant high-tech support may be useful.	There is no need for these fancy options, because there is no energy shortage. It is a waste of tax money.	The technical but also the lifestyle aspects of energy efficiency should be more intensely promoted. But high-tech will never be a real solution.	There are still enormous efficiency gains possible, but price incentives should be complemented with regulation and R&D support.
Statement 5: Sufficient clean power can be generated in the world's deserts to supply mankind with enough electricity on a sustainable basis (Desertec Foundation).			
It is an ambitious project, but if it can be financed – why not? Of course, there are some technological and political challenges, but that's OK.	Is this a serious proposal? It makes us more dependent on unreliable states and in the end I will have to pay for these big projects.	Another technocratic megalomania. Renewable energy is needed, but not in this highly centralised form.	An ambitious project, that should be considered as one part of a global renewable energy system. It may also solve water problems and provide much needed employment.

of such an option would show up sooner or later – as the opponents are all too eager to point out. One can expect that different directions will be taken in different countries. For instance, France is already over, with 70 percent of its electricity generation dependent on nuclear power. The collective experiences will teach us, hopefully, about the sustainable middle road without too catastrophic and irreversible damage to future mankind's options for a good quality of life.

A value survey about the pros and cons of wind power in Denmark is an illustration of this point.[23] It revealed two very different profiles. The Profile of the Nay-sayer is characterised by statements like: Renewable energy cannot solve our energy problems; wind turbines are unreliable and dependent on the wind; wind energy is expensive; wind turbines spoil the scenery; and wind turbines are noisy. They represent the Not In My BackYard (NIMBY) attitude. The Yes-sayer, on the other hand, expresses quite different values and beliefs: Renewable energy is very much an alternative to other energy sources; climate change theory must be taken seriously; wind energy is limitless unlike fossil fuels; wind energy is non-polluting; and wind energy is safe. These people are in favour of wind farm cooperatives, small-scale and self-reliance.

Similar differences are found with respect to nuclear power. Some people do not accept the large risks and the centralised structure of a nuclear energy system, whereas others see it as a necessary step in the best of possible worlds. Also, the role of oil and gas depletion and dependence is quite differently viewed by people. Exhausting the high-quality oil and gas resources as fast as our population does now causes the risk of an irresponsible switch to coal and nuclear energy, in order to avoid disruptive energy shortages. At the same time, and accelerated by a switch to coal, there is already too much carbon released in the atmosphere to prevent at least some change in climate. Only moderation in energy use, drastic improvement in efficiency and rapid and ongoing introduction of renewable energy sources offer the prospect of a sustainable energy transition. Will this road be taken? Unlike the apparent consensus on energy futures in the 1960s, it is not clear how the energy systems in the world will look like by 2050. Because of the long lead times of oil production and conversion plants and power plants, there is quite some inertia in the system (§2.3). But for the period after 2030, it is next to impossible to make predictions about the details of the envisaged energy transition.

7.7 Summary Points

Energy is an essential aspect of development, as source of food, as driver of the industrial economy, and as cause of environmental damage. From a sustainability science perspective, the following points about energy are to be remembered:

- In agrarian societies energy use was almost exclusively based on solar energy. With industrialisation, there was a regime shift to fossil fuels;
- Classical thermodynamics gives us the equations to formulate the *energy quality* in terms of work potential and to understand energy conversion processes. It also teaches us that, although people cannot destroy energy itself, people do destroy its quality in combustion and dispersion processes;

[23] www.windwin.de/images/pdf/wc03041.pdf.

- the concept of equilibrium, important in the sciences as an 'ideal' and/or 'natural' situation, is rooted in classical thermodynamics and mechanics;
- classical mechanics describes elementary processes that are often used as metaphors and shows that apparently simple processes and models can exhibit complex behaviour;
- in the real-world of *non*-equilibrium processes, empirical/phenomenological laws are used. Their essence is that nature tries to annihilate gradients (in temperature, pressure, concentration);
- production of work (mechanical, electrical) is a crucial process in the provision of energy services and has consequences (depletion, emissions) that are at the core of some sustainable development concerns; and
- increasing the efficiency of energy conversion processes is of utmost importance to sustain or expand the energy flows required for present-day societies. The second pillar of a sustainable energy system is the introduction of renewable energy resources.

> On peut bien expliquer l'arbre par ses causes, par sa structure, par les mécanismes qu'il met en jeu, les échanges qu'il entretient avec son environnement, etc. Mais le comprendre, non : il n'y a rien à comprendre, et c'est pourquoi aucune théorie ne saurait remplacer le regard, la simplicité du regard.
>
> – André Comte-Sponville, *L'amour la solitude*, 2000:31
>
> Translation:
>
> You can very well explain a tree in terms of its causes, its structure, the mechanisms it calls into play, the exchanges with its environment, etc. But understanding it, no: there is nothing to understand, and that's why no theory can ever replace looking, simply looking.
>
> I sell here, sir, what all the world desires to have – power.
>
> – Boulton, 18th-century British industrialist, quoted in Heilbroner and Milberg, 2004:66

SUGGESTED READING

Introductory text on Life Cycle Assessment.

Baumann, H., and A.-M. Tillman. *The Hitch Hiker's Guide to LCA. An Orientation in Life Cycle Assessment Methodology and Application.* Lund, Sweden: Studentlitteratur AB, 2004.

An introductory textbook on theory and practice of energy analysis.

Blok, K. *Energy Analysis.* Amsterdam: Techne Press, 2006.

An introduction into dynamic models: a gradual build-up from elementary models to more sophisticated ones, with an emphasis on population-environment issues.

Bossel, H. *Modeling and Simulation.* Wiesbaden: AK Peters Ltd./Vieweg, 1994.

A detailed discussion of exergy and its applications.

Gong, M., and G. Wall. On exergy and sustainable development – Part 2: Indicators and methods. *Exergy International Journal* 4 (2001) 217–233.

An introduction into the scientific aspects of energy use and energy conversion technologies with much practical info and sample calculations.

MacKay, D. (2009). Sustainable energy – without the hot air. UIT Cambridge (download at www.withouthotair.com).

Introduction on sustainable development from a technical and design perspective.

Mulder, K. *Sustainable Development for Engineers.* Sheffield, UK: Greenleaf Publishing, 2006.

One of the early natural science explorations of complex non-equilibrium phenomena.
Nicolis, G. and I. Prigogine. *Exploring Complexity – An Introduction*. New York: Freeman & Company, 1989.

An instructive and extensive introduction into the history of energy in all its forms.
Smil, V. *Energy in World History*. Boulder, CO: Westview Press, 1994.

A rather advanced textbook on the mathematics of nonlinear dynamics.
Strogatz, S. *Nonlinear Dynamics and Chaos – With Applications to Physics, Biology, Chemistry, and Engineering*. Boston: Addison-Wesley Publishing Company, 1994.

A detailed discussion of exergy and its applications.
Wall, G., and M. Gong. On exergy and sustainable development – Part I: Conditions and concepts. *Exergy International Journal* 1(3) (2001): 128–145.

A comprehensive textbook on modeling environmental processes, with background and examples of natural science principles.
Wainwright, J., and M. Mulligan. *Environmental Modelling – Finding Simplicity in Complexity*. London: John Wiley& Sons, Ltd., 2004.

USEFUL WEBSITES

The Wikipedia presents for most elementary processes good descriptions. There are many sites that explain basic concepts in thermodynamics and mechanics.

Specific Techniques, Processes and Models

- www.convert-me.com/en/, www.onlineconversion.com/energy.htm and www.digitaldutch.com/unitconverter/ are three sites for unit conversion.
- www.exergy.se/ is the site where Wall offers courses on exergy analysis and exergy economics, with extensive documentation.
- www.ornl.gov/sci/roofs+walls/insulation/ins_01.html is the Oak Ridge National Lab/United States site with information and models on heat insulation.
- www.e-calculator.nl/ is a site (in Dutch) that can be used to estimate your annual energy use. There are a number of energy calculator sites in English, for instance, energyabacus.com/.
- www.carbonify.com/carbon-calculator.htm is a site that can be used to estimate annual carbon emissions (carbon footprint).
- webphysics.davidson.edu/physlet_resources/thermo_paper/ is an interactive demonstration of the expansion of a gas in a cylinder.
- www.aw-bc.com/ide/idefiles/navigation/toolindexes/9.htm#9 has several interactive models on mechanical systems, such as the harmonic oscillator, the mass-and-spring and the forced vibration, and also on electrical systems.
- www.aw-bc.com/ide/idefiles/media/JavaTools/nlhcrate.html has an interactive model on cooling.
- There are many energy models around, for instance, the Message model developed at IIASA. www.iiasa.ac.at/Research/ENE/model/extensions.html, the TIMER model developed at PBL at themasites.pbl.nl/en/themasites/image/model_details/energy_supply_demand/index.html and the GET model at Chalmers University.
- There are also many studies about the future of the energy system, for instance, the annual World Energy Outlook (WEO) of the International Energy Agency (IEA) at www.worldenergyoutlook.org/.
- www.eia.gov and www.iea.org are both sites with many data and news on energy.
- www.ewea.org/index.php?id=180 is the site of the European Wind Energy Association (EWEA).
- www.iaea.org is the site of the international Atomic Energy Association for info on nuclear energy but also other energy issues.
- www.supersmartgrid.net is the site about the supersmartgrid for Europe.

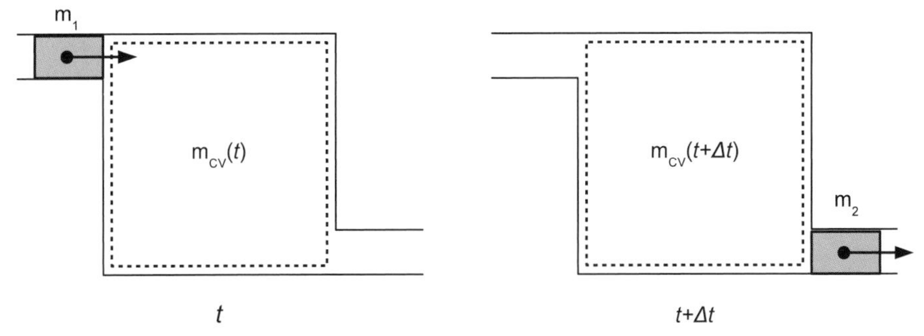

Figure 7.A1. The control volume as starting point for the steady-state energy equation.

Appendix 7.1 The Steady-State Mass-Energy Equation

We define a control system with volume CV (Figure 7.A1). Suppose that in a steady-state process a mass m_1 is entering a control volume at time t and a mass m_2 is leaving the same control volume at time t + Δt (Nieuwlaar and De Ruiter 2010). Because mass is a conserved quantity, you have:

$$m_1 + m_{CV}(t) = m_{CV}(t + \Delta t) + m_2 \text{ kg} \tag{7.A1}$$

For $\Delta t \to 0$, we get $dm_{CV}/dt = dm_1/dt - dm_2/dt$ or in physics notatio $\dot{m}_{CV} = \dot{m}_1 - \dot{m}_2$.

For this process, the first law requires that the heat exchange Q in the time interval Δt with the system must equal the change in the internal energy of the control volume E_{cv} plus the work W done. If you indicate the specific energy of a mass unit with e (in J/kg), then equation 7.1 becomes for the control volume considered:

$$Q + m_1 e_1 + E_{CV}(t) = E_{CV}(t + \Delta t) + m_2 e_2 + W \text{ J} \tag{7.A2}$$

If one considers the volume work done by the inflowing and outflowing mass, PΔV, then this equation becomes:

$$Q = E_{CV}(t + \Delta t) - E_{CV}(t) + m_2 e_2 + m_2 P_2 v_2 - m_1 e_1 - m_1 P_1 v_1 + W_{CV} \text{ J} \tag{7.A3}$$

with W_{CV} the work done *minus* the volume work done.

The specific energy e consists of the internal energy u, which is an intrinsic characteristic of mass m, plus potential and kinetic energy in macroscopic gravitational and other fields. For a system with mass m and velocity $\tilde{v}$, these are in the earth gravitation field:

$$e_{pot} = mgz \text{ and } e_{kin} = \frac{1}{2} m \tilde{v}^2 \text{ J} \tag{7.A4}$$

with z the distance above the earth surface, g the earth gravitation constant (9,8 m/s^2). Using now the formula for specific enthalpy h = u + Pv, rewrite equation 7.3:

$$\begin{aligned} Q &= [E_{CV}(t + \Delta t) - E_{CV}(t)] \\ &\quad + m_2 \left(h_2 + gz_2 + \tilde{v}_2^2/2\right) - m_1 \left(h_1 + gz_1 + \tilde{v}_1^2/2\right) + W_{CV} \text{ J} \end{aligned} \tag{7.A5}$$

You have written this balance for the change in a period Δt. As previously, it can be rewritten as a rate equation for Δt → 0:

$$Q = \dot{E}_{CV} + W_{CV} + \dot{m}_2 \cdot \left(h_2 + gz_2 + \tilde{v}_2^2/2\right) - \dot{m}_1 \cdot \left(h_1 + gz_1 + \tilde{v}_1^2/2\right) \ \mathrm{J} \qquad (7.A6)$$

The dot indicates the first derivative of time. In words, the heat transfer to the system equals the change in the internal energy of the control volume plus the work done by the control volume plus the difference in the internal energy of outflow and inflow. If you now assume that the mass flows and the state of the control volume are constant in time, you get the basic equation for a steady-state or stationary process:

$$Q = W_{CV} + \dot{m} \cdot \left((h_2 - h_1) + g(z_2 - z_1) + \left(\tilde{v}_2^2 - \tilde{v}_1^2\right)/2\right) \ \mathrm{J} \qquad (7.A7)$$

If you can apply this formula, you are able to understand elementary energy conversion appliances such as turbines and heat exchangers. I refer you to the Suggested Reading for those who accept the challenge.

The steady-state equation can now be written in an entropy form by combining equation 7.5 with equation 7.A1. It yields an expression for the work flow that can be extracted when the system is brought reversibly into equilibrium with the Earth environment:

$$W = \dot{m} \cdot \begin{bmatrix} (h_1 - T_0 s_1) - (h_2 - T_0 s_2) \\ +g(z_1 - z_2) + \left(\tilde{v}_1^2 - \tilde{v}_2^2\right)/2 \end{bmatrix} + \left(1 - \frac{T_l}{T_h}\right) \cdot Q_{T_h} - T_0 \dot{S}_{irr} \ \mathrm{J} \qquad (7.A8)$$

with $T_h = T_0$. This formula reflects that the maximum work ($\dot{S}_{irr} = 0$) you can extract on earth from the system equals the changes in the state variable $h–T_0s$, in gravitational and kinetic energy and in the reversible work from the heat flow. Call $\varphi = h–T_0s$ the specific *exergy* of the mass. In thermodynamics, it is in more general form formulated with help of a new state variable: the free enthalpy or Gibbs free energy defined as $G = U + PV – TS$ or, per unit of mass, $g = u + Pv – Ts$. The formula can be extended to include more than on substance and chemical reactions.

8 On Knowledge and Models

8.1 Introduction

Sustainability science: The word science suggests pursuit of 'scientific knowledge'. But what is scientific knowledge? Let us have a closer look at what the acquisition of scientific knowledge is in practice. Suppose you are concerned about air pollution and set up an experiment to measure the *concentration* of substance X in a well-defined area. The measuring tool is itself a specimen of scientific development. The result of your experiment is a series of concentration values at given location p and time t, c(p,t). Building upon atmospheric physics and chemistry, you interpret the results in terms of dynamic cause-and-effect processes. Such a description, framed in the formal language of mathematics, is called a scientific model.

You realise that it is actually the impact of air pollution that matters, so you decide to explore *impact on the forest* in the area. With the help of ecologists, you do additional experiments and extend the model. The concentration values c(p,t) are now inputs to descriptions of the various trees in the forest. They are a measure of the exposure of the simulated trees to external factors. Because the tree dynamics are relatively slow, longitudinal experiments have to be set up (>5 years). The result of these experiments are an indication of the sensitivity of the various trees for the particular exposure c(p,t).

Unfortunately, you cannot rely on such solid laws in this field as in atmospheric science. Estimates of tree sensitivity are based on controlled laboratory experiments and fields surveys, but they are only partly transferable to your field situation. The trees in the forest differ in age and in location-dependent parameters such as soil and water access. There is also interference with other species and other pollutants. Besides, the forest may have varied responses to different or prolonged exposure c*(p,t), which falls outside your measurement domain. It will take much effort and many years before you have valid scientific knowledge about the impact of air pollution on the forest.

One evening, you meet a friend who argues that the real issue is whether the measured air pollution has a negative effect on the *health* of the people living in the area. Taking up the challenge, you ask some medical scientists to engage in a longitudinal research project to measure the health situation of people in the area and, for comparison, of people in another area with negligible air pollution.

Although there is substantial medical knowledge about how the air pollutant affects the physiology of the human body, your long-term experiment is confronted with large uncertainties. For one thing, the group of people followed in the experiment is not constant because people move in and out of the area. There is also a large variety, both somatic and psychic, in the population samples. For instance, some individuals are more sensitive to exposure, while others are better able to cope with the effects. It is difficult and ambiguous how to deal with this variety. Often, ad-hoc research strategies have to be designed. The experience has surely made you less naïve about 'scientific knowledge'.

The ordeal is not over yet. At an environmental economics conference, an economist argues that one does not need to know the impacts of the air pollutant concentration in great detail; it is more important to know at what *cost* it can be reduced below some level, which is considered or negotiated as 'acceptable'. Recognising the appeal of this argument, you step into another research project with economists to estimate the options and costs to reduce the concentrations. You calculate the emission reductions needed for a 'safe' concentration level and some economist colleagues identify and rank the emission reduction options according to the cost per unit emission reduced. In their view, polluters respond mostly or solely to *price* incentives so a policy should focus on the proper tax levels. They quote several policy analyses and behavioural surveys to proof their point. Other economist colleagues disagree and quote scientific evidence in favour of emission *standards* for equipment. Policy should enforce stringent standards in order to induce technological options that reduce emissions at much lower cost.

During the deliberations, you discover that some of your colleagues' relationships with government officials and entrepreneurs seem to play a role in their convictions about the most successful approach. You realise that the dispute cannot be settled in the way of the natural sciences: Controlled experiments are largely excluded and all the parameters involved keep changing all the time. You realise it is time for some philosophical reflection.

As this story illustrates, if you widen the system boundary, more and more scientific disciplines get involved. Questions and answers become more complex and uncertain. People come into the picture, with their own social and cultural characteristics. You as investigator enter the scene, with your skills, limitations and biases. In this chapter, the focus is on epistemological issues and ways to handle complexity and uncertainty, as these play an important role in sustainability science.

8.2 Models in the Natural Sciences

8.2.1 The Scientific Method

The underlying paradigm in science is that there are regularities in the world, which are manifestations of dispositions and tendencies and capacities in the particulars or 'things' of the material world. The regularities result from complex interactions between these particulars, which are causal in the sense that they result from actions (Chalmers 1999).[1] To find out about the regularities, there is the scientific method.

[1] Causality is not a prerequisite: The laws of classical thermodynamics, for instance, cannot be interpreted as causal laws (§7.3).

Its essence is to generate empirical laws from observations and measurements in controlled experiments. It is an *inductive* process of inference, guided by pre-scientific intuitions and notions. It yields descriptive, phenomenological statements or 'scientific facts' and is at the core of the *empiricist* approach. Empirical reductionism or scientific materialism is the radical interpretation that this is the only valid way to acquire knowledge and it is associated with the Modernism worldview (§6.3). It is rooted in a conviction: '... all tangible phenomena, from the birth of stars to the workings of social institutions, are based on material processes that are ultimately reducible, however long and tortuous the sequences, to the laws of physics' (Wilson 1998).

Upon reflection, it becomes clear that the statements or facts cannot be derived solely from the senses: An appropriate conceptual framework and knowledge about how to apply it are needed. The most convincing proof is that what once were considered 'scientific facts' turned out to be invalid or subject to revision within a novel conceptual frame. Facts not only precede theory, theory also precedes facts. Before or with the experiments, a formal system is developed and used to generate linguistic statements and scientific theories. This is a *deductive* process during which the link with observations and measurements is less direct and the level of abstraction is high. It is the core of the *rationalist* approach. Theory is derived from facts by logical reasoning. Its essence is resounded by Simon (1969): 'The central task of natural science is to make the wonderful commonplace: to show that complexity, correctly viewed, is only a mask for simplicity; to find pattern hidden in apparent chaos'. Mathematics is a stronghold of the rationalist position because it is hard to imagine empirical progress without the concepts and methods of applied mathematics.

There are several problems with a radical empiricist approach. Given that one cannot do infinite numbers of observations, what are the rules to derive more general statements than the one that strictly follows from a few observations? In other words, when are logical deductions valid and what is, therefore, a justifiable generalisation and validity domain? Most scientists would give the pragmatic answer that a statement is true if it has a high probability to be true in the light of the evidence. Another problem is that in contemporary science, much knowledge is not accessible to direct observation; for example, think of electrons or genes.

A brief history of the ideal gas theory can illustrate the scientific process and the role of models therein. An amount of gas in a cylinder is in classical thermodynamics described on the basis of the following macroscopic observables: mass (m), temperature (T), pressure (P) and volume (V). To quantify these observables, a *reference system* is defined, which is used as the measuring device, for instance, a thermometer for T and a manometer for P[2].

If you fill your bicycle tire with air, you will notice that it takes less effort in the beginning than later on. You rediscover Boyle's law (1662): The product of pressure and volume remains constant for a given amount of gas if the temperature does not change, so more gas in a fixed volume implies higher pressure. It took over 140 years before this observation was expanded into what became known as the Boyle-Gay-Lussac law (1808): $PV \sim T$. Yet, this model was not more than an observed proportionality. With the corpuscular theory according to which gases consist of

[2] The Système International (SI) of measures and units for the sciences is based on the definition of such physical reference systems. See physics.nist.gov/cuu/Units/units.html.

Box 8.1. *Definitions of models.* There are many definitions and views of models. '*Models* and simulations of many kinds are tools for dealing with reality . . . Even thousands of years ago, buildings, boats, and machines were first tested as small models before being constructed on a large scale. Children's games have always been simulations of the world of grown-ups – using models of people, animals, objects, and vehicles. The model worlds of mythology, legends, and religions . . . have guided the behavior of generations in all cultures . . . Models range from miniaturised realistic representations of the original to technical drawings to functional diagrams. They may consist of stories, fables, and analogies or be expressed in mathematical formulae or computer programs' (Bossel 1994:1–2).

One type of model already encountered in Chapter 2 are *mental models*. Mental models are 'deeply ingrained assumptions, generalisation, or even pictures or images that influence how we understand the world and how we take action . . . The discipline of working with mental models starts with turning the mirror inwards; learning to unearth our internal pictures of the world, to bring them to the surface and hold them rigorously to scrutiny' (Senge 1990:8–9).

More formal distinctions between models are:

- deterministic vs. probabilistic or stochastic;
- analytical vs. simulation;
- continuous vs. discrete, depending on whether differential-integral calculus is used or not.

Refer to the Suggested Reading for more in-depth treatment.

small particles ('atoms'), a wide range of observations were brought together in the formula $PV = nRT$ with n the number of moles and R the gas constant.

Since then, the observed data have been refined and so have the mathematical equations. For instance, Van der Waals proposed replacing V by $V\text{-}b$ in order to correct for the non-zero volume of the gas particles. There are now a number of more advanced equations that describe the observations even better and across a larger domain of variables. But such a more refined formal description of empirical observations does not really add to our understanding. A next step was to formulate the system in terms of statistical distributions of the properties of the gas particles (speed, momentum).

The spiralling process of controlled experiments and induced hypotheses in combination with deductive formalisms has led to many hypotheses, which became accepted and mainstream – only to be later falsified and dismissed or modified when new observations were made. Often, the earlier hypothesis is not refuted but shown to have a limited validity domain. An example is Newtonian mechanics as a special case in relativistic mechanics. Or the statistical thermodynamics approach that, later combined with quantum mechanics, made it possible to relax the assumption in classical thermodynamics of equilibrium and homogeneity in the intensive variables (§7.3). A new operational model emerged, using micro-level concepts and incorporating rather than invalidating the previous model. The reverse also happens: A hypothesis is dismissed or ridiculed for lack of empirical support, only to be re-established later on when its intuitive strength is confirmed (§5.3).

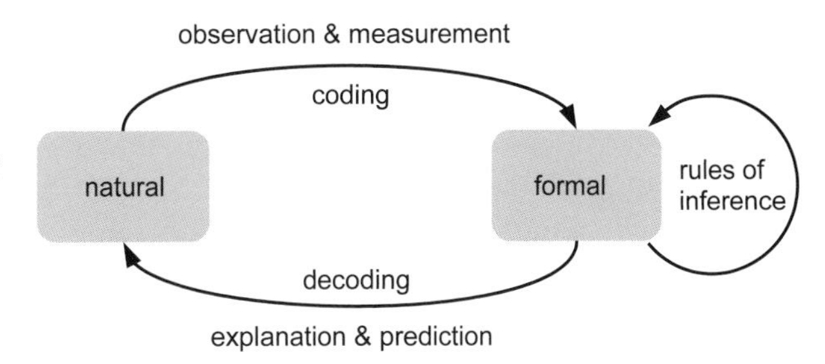

Figure 8.1. Modelling as an encoding of a natural system (after Rosen 1985).

8.2.2 Models and the Modelling Process

In a scientific context, the word *model* refers to any formal representation of observations and measurements, ranging from empirical laws to universal theories. Often, however, it is defined rather broadly and loosely. In everyday usage, a model refers to some copy or image. It may be a physical replica, usually in miniature, or an example for imitation. It can also be a description used to help visualise something that cannot be directly observed. Modelling is a way of looking at and ordering the world around us. In science, models are usually defined rather formally, for instance:

> a *model* is a representation of a part of reality, based on a system of postulates, data, and inferences presented as a mathematical description of an entity or state of affairs.

Modelling is an activity with an *objective*: to get (better) understanding of some part of reality and to directly or indirectly control or manage a system and to solve a problem. This 'part of reality' is created by explicitly and consciously drawing a boundary around what one thinks should be investigated and hence modelled: the *system boundary* (§2.3).

Modelling can be viewed as a coding process (Figure 8.1; Rosen 1985). It creates a relation between a 'natural' and a 'formal' system. The natural 'real-world' system is the *object system* or *target system* and is identified by means of observables, that is, qualities and quantities that can be observed and measured in a more or less controlled way. 'A Natural System is a collection of observables connected by mind created relations' (Rosen 1985). The formal system is the *model system*, which can be any set of mental elements (objects, symbols and so on) and relations and rules that permit the generation of statements – as in language or logic. The rules are used to generate new statements, which are decoded in order to explain and predict the behaviour of the natural system. This is why they are called rules of inference. In this sense, a model is an approximation of a part of reality. Although this view of modelling is inspired by the natural sciences, the social and psychological 'interior' reality can also be part of such a 'natural' system and investigated with participative and introspective methods (§10.4).

The process of constructing a model in interaction with the external world is sketched in Figure 8.2. The outside world (natural system) is perceived in a physiological recording through the sensory apparatus – the five senses and their technical extensions like microscope and telescope. This forms percepts in the human mind that are converted into observations. In combination with previous experiences

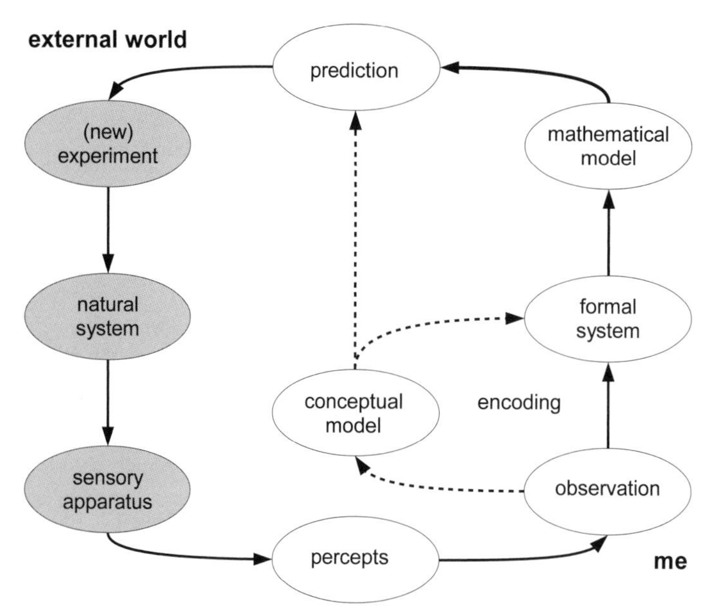

Figure 8.2. The modelling process in science.

(ideas, awareness or emotions), a conceptual model is constructed. This may directly lead to a prediction and resulting action (experiment). In a scientific process, the conceptual model is formalised and translated into a mathematical model. This permits more precise prediction and more targeted new experiments.

In principle, the modeller has total freedom in choosing the observables in the natural system. In practise, however, this freedom is constrained by the available *measuring systems*, which are part and parcel of scientific and technical advances. In fact, the simultaneous development of experiments, measuring devices and formal mathematical systems is at the root of the scientific revolution of the last couple of centuries. The measuring devices – thermometers, manometers and so on – can be considered as reference systems against which any system can be calibrated. The domain of empirical observations and measurements has expanded enormously with the development of microscopes, telescopes, echoscopes and all sorts of electromagnetic and physico-chemical measuring devices. *Mathematical techniques* have become one of the hallmarks of scientific and technical progress. As with the measuring devices in empirical analyses but more subtly, their use induces a methodological bias. In the words attributed to Mark Twain, 'If the only tool one has is a hammer, one treats everything like a nail'. In other words, in order to be analytically solvable, there is a tendency to simplify the system under study.

Sometimes one and the same natural system can be approximated by two different formal systems (Rosen 1985). This happens when some essential, though more abstract, features of two natural or formal systems are comparable.[3] In the evolution from empirical laws to universal theories, scientists construct ever more encompassing models. As long as there is no 'unified theory', two different descriptions may be accepted as complementary, as in the case of the wave and the particle

[3] In the mathematical idealisation, the two systems are said to be *isomorphic*.

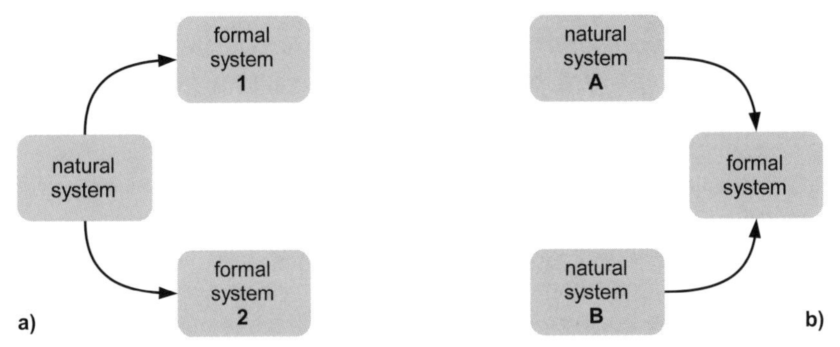

Figure 8.3. Representations of isomorphic relationships between 'natural' object and 'formal' model system: complementarity (a: left) and analogue or metaphor (b: right) (after Rosen 1985).

description of light (Figure 8.3a). Conversely, two different natural systems may apparently be so much alike that they can be described with a single model. They are analogies or, if the *isomorphism* is rather loose, *metaphors* (Figure 8.3b).

Analogies and metaphors play an important role in the construction and communication of scientific knowledge. Their use is a powerful heuristic to understand complex systems in terms of (a model of) a simpler and better understood system. It allows a connection between empirical observations on the one hand and the logic of the analog formal model on the other. The 'borrowing' of a formal system from physics and chemistry to describe observations in biology or economic science did advance science significantly. Famous examples are Hartley's hypothesis that the human heart works like a mechanical pump and Laplace's comparison of the planetary system with a mechanical clock. With the advent of the computer, electrical networks became the metaphor for the human brain and even for society. Boulding (1978) used the metaphor of the *Cowboy Economy* as an image of unlimited economic expansion. The image of *Spaceship Earth* was a symbol of enlightened engineering, often contrasted with the image of *Mother Earth*.[4] The differences between favourite metaphors reflect divergent interpretations of available knowledge and divergent value orientations (Table 6.2). Sometimes analogies or metaphors are merely evocative, such as the human psyche as a steam engine or economic processes as analogues of chemical and mechanical equilibria. However, their use reflects the search for universal principles governing the phenomenal world and helps to generate shared models in, for instance, biology and evolutionary economics and in immunology, linguistics and institutional economics (Frenken 2006; Janssen 2002).

8.3 Strong and Weak Knowledge

Classical natural science owes its success to the reduction of its domain of investigation to controlled experiments and to the combination of experiments and mathematics. The artificial constructs ('machines'), which nowadays constitute the *technosphere* and in numbers exceeds the human population, are the outcome of these

[4] Other examples are *Awakening Earth* to convey an evolution to higher collective consiousness and *Living Machines* for technologies that make use of organic processes for manufacturing and breakdown.

Box 8.2. *The mathematisation of science.* An important if not quintessential element of the scientific revolution has been the mathematisation of science. It evoked a tension with the experimental tradition, reflecting the divergence between deduction and induction. The French sociologist Bourdieu argues that 'la mathématisation est à l'origine de plusieurs phénomènes convergents qui tous tendent à renforcer l'autonomie du monde scientifique et en particulier de la physique... La mathématisation produit d'abord un effet d'exclusion: avec Newton la mathématisation de la physique tend peu à peu à instaurer un très forte coupure sociale entre les professionnels et les amateurs... Deuxième conséquence de la mathématisation [est] la transformation de l'idée de l'explication. C'est en calculant que le physicien explique le monde... Ceci entraîne un troisième effet de la mathématisation, ce qu'on peut appeler la désubstantialisation: la science moderne substitue les relations fonctionnelles, les structures, aux substances aristotéliciennes et c'est la logique de la manipulation des symboles qui guide les mains du physicien vers des conclusions nécessaires. L'usage de formulations mathématiques abstraites affaiblit la tendance à concevoir la matière en termes substantiels et conduit à mettre l'accent sur les aspects relationnels' (Bourdieu 2001:96–98).

scientific endeavours. The underlying search for more universal laws has led from coarse-grained, phenomenological descriptions of the world to fine-grained, fundamental ones (Table 2.3).

Scientific knowledge comes in grades: Not all knowledge has the same status. Can one devise a *taxonomy* of knowledge? The biologist Pantin made, in his book *The Relations between the Sciences* (1967), a distinction between the 'restricted' and the 'unrestricted' sciences, the latter being those where controlled experiments are hard or impossible – as in geology, ecology and archaeology. Along this line, we here hypothesise that a scientific theory develops by strengthening three elements (Groenewold 1981):

- logical operations (l);
- codified experiences (e); and
- hypotheses (h) which relate the experiences.

The *logical operations* constitute a more or less formalised system of concepts and rules. It can be a language syntax, differential-integral calculus or transition rules in a *cellular automata* model (§10.5). They are the rules of the formal system (Figure 8.1) and the language for conversion from conceptual to mathematical model (Figure 8.2). It is the deductive part of science. For instance, a differential equation to describe the exponential growth of a deer population is a formalisation of a particular set of observations, in particular on animal reproduction. The more it is formalised, the sharper the concepts and rules can be – but also the more the object system is simplified. The *codified experiences* refer to human experiences reframed in an experimental setting. They are the experimental set-up to interact with the natural system (Figure 8.1) and are the outcome of the conversion from percepts to observations (Figure 8.2). This is the inductive part of science. For instance, enjoying wild

Box 8.3. *On strengthening knowledge.* A recent example of how scientific knowledge is strengthened and how reference systems matter is from climate change science. From a natural science perspective, the climate change debate is full of uncertainties. This is widely acknowledged as one can read in, for instance, *The Economist* (August 13, 2005:65): 'The [climate] system itself is incredibly complex. There is only one such system, so comparative studies are impossible. And controlled experiments are equally impossible... So there will always be uncertainty and therefore room for dissent'.

For years, there has been an apparent discrepancy between what is happening to temperatures at the Earth's surface and what is happening in the troposphere. Whereas recorded surface temperatures are rising, observations on troposphere temperatures from weather balloons and satellites showed constant tropospheric temperatures since the 1970s. Are the models wrong, are the data wrong – or is there no warming at all, as climate skeptics say? Recent research suggests that the data are wrong. Weather balloons often make measurements while being exposed to sun's rays and get overheated. This was known and corrected with a correction factor. However, the correction factor has not been corrected for the design changes that have been made to reduce the problem of over-heating. A second error may have to do with the satellite measurements that are used to construct tropospheric temperatures. The constancy found since the 1970s may be caused by the change in the satellite's orbital period, which decays because of friction with the outer reaches of the atmosphere. If these corrections are introduced, the trend in tropospheric temperatures is warming and better in agreement with climate model simulations.

deer in a natural park is not codified but a well-documented count of wild deer in a certain area and period is. The *hypotheses* connect and expand the experiences via logical operations. The observation that the number of offspring of deer fluctuates with the number of wolves can induce the hypothesis that the two populations are interacting. The formalised hypotheses yield a scientific model.

A scientific theory becomes mature by gradually eliminating unnecessary hypotheses and sharpening the logical operations and the codifications of experience. Much knowledge in physics and chemistry represents what we henceforth call *strong knowledge*.[5] It is grown out of hypotheses that have been falsified and replaced by hypotheses better in agreement with observations. It is formulated in mathematical models that are thoroughly empirically validated within their domain of observation and control. Yet, there is often still much scope for improvement. An example is meteorology, where the combination of satellite data and simulation modelling has led to an enormous improvement in weather forecasting, which has still not yet ended.

A theory is strengthened by eliminating *weak* elements. Sometimes those are obvious, as in false logic, misapplied statistics or a priori judgments with a claim

[5] Often the words hard and soft are used to denote what is meant here with strong and weak. The strong-weak terminology is preferred, because hard and soft are better used to refer to the degree to which an aspect of reality can (or cannot) be influence (by men).

on absolute and universal validity. Sometimes their weakness is only obvious if one demands empirical evidence, as in statements about non-observable entities with no consequences for experience or in theological dogma's or revelations. Sometimes, statements reflect an intuition or speculation, which the *Zeitgeist* is not yet ready for or not willing to consider for cultural or political reasons.

The success of the natural sciences and the associated Modernism worldview suggest that the method of the natural sciences will penetrate all other fields of knowledge and make them in due course strong and mature (§6.4). In that sense, calling knowledge in the life and social sciences *weak(er)* sounds derogatory: It suggests that scientific knowledge about life and society will become *strong(er)* with further domination of the natural sciences. The issue, however, is complexity and there are two reasons why such domination will not happen soon. First, the method does not work for the complex systems in the real world. For a large complex system like the earth atmosphere or parts of the biosphere, it is impossible to conduct controlled experiments. At best, one can mimic the circumstances and do small-scale field and laboratory experiments. With economic and social systems, controlled experiments are hardly or not possible and the very experiment can alter the way the system functions.[6] Scientific statements about such complex systems are often highly restricted and probabilistic, as there are so many unknown system variables and interdependencies. Constraining assumptions that are helpful in physics and chemistry, such as equilibrium and linearity, are in these domains inadequate and provide pseudo-knowledge because there is no controllable reality in which they can be tested.

A second reason why the monopolising tendency of the natural sciences has its limits is that it rejects other sources of knowledge than via the scientific method. Strong scientific knowledge is active (interaction) and public (shared), but it does not imply that passive and private knowledge is weak in the sense of false or without relevant evidence. Weak statements are not 'truthless' or useless. They can have strong subjective meaning and value. Intuitive, speculative or revelatory insights can also be at the birth of scientific theories. Here the tension between worldviews is felt: to what extent should knowledge be based on empirical evidence (vertical dimension) and to what extent should it be public and universal (horizontal dimension) (Figure 6.5)? This is an important question in the search for sustainability: how to reconcile the existence of public, scientific knowledge with a *universal* claim of validity about the *material* world with the reality of millions of individuals who have *personal* knowledge with claims of validity about their exterior *and* interior world?

Therefore, statements about complex systems are usually probabilistic and non- or multicausal. One approach to strengthen them is offered by applying Bayesian conditional probabilities (Chalmers 1999). Without going into detail, the essence is that a hypothesis is becoming stronger if the prior probability of being correct increases with new evidence. The issue then is how to assign prior probabilities to hypotheses. Most people do not believe this can be done in an objective way. Prior probabilities represent the beliefs in hypotheses that scientists, as a matter of fact, state or practise. The Bayesian approach opens the way to a more constructivist,

[6] In physics, the interference between observer and observed causes an impossibility to measure certain properties with arbitrary precision. It is known as the Heisenberg uncertainty principle.

postmodern view of truth. A few examples of statements and assertions can illustrate the relevance and the problems of this approach.

Statement 1: *'A wind farm with a wind speed of 6 m/s produces a power of 2 W per m^2 of land area'* (MacKay 2009).

This statement can be tested by setting up an experiment in a windtunnel (the codified experience). The strength is less in the numbers, which are probably not exact, than in the underlying universal physical principles. Probably no one would dispute this statement, but most people would frown if you add 'yesterday' or 'particularly in the United Kingdom'. An experiment will provide new evidence, which will strengthen the statement – unless the prior probability assigned to it was already very high because the physics is known and trusted. If the new evidence does not fall narrowly within the relevant domain, the ambiguity in the word wind farm would show up.

Statement 2: *'A wide range of direct and indirect measurements confirm that the atmospheric mixing ratio of CO_2 has increased globally... from a range of 275 to 285 ppm in the pre-industrial era (AD 1000–1750) to 379 ppm in 2005'* (IPCC 2007 WGI:2.3.1).

This carefully framed statement is at the core of the hypothesis that humans influence the climate. It is based on a series of direct observations and indirect reconstructions. If one accepts the physics of gaseous diffusion and the methods used in historical reconstruction, it is a strong scientific statement. The indicated uncertainty takes into account that direct measurement of pre-industrial concentrations are not possible.

Statement 3: *'Work on simplified ecosystems in which the diversity of a single trophic level... is manipulated shows that taxonomic and functional diversity can enhance ecosystem processes such as primary productivity and nutrient retention'* (Ruiter *et al.* 2005).

A statement like this may have great importance in judging biodiversity – is it strong knowledge? The word ecosystem is first used in an experimental frame ('manipulated') and then in a broader context ('ecosystem processes'). The formulation suggests the need for new evidence from additional experiments to strengthen the assertion. The notion of diversity, even if strictly defined in the experiment, may easily be interpreted in a wider context of (bio)diversity and intensive agriculture, although such generalisations are rather weak. By implication, it becomes a matter of trust whether a statement like this is used for policy purposes.

Statement 4: *'Poor [animal] nutrition is one of the major production constraints in smallholder systems, particularly in Africa'* (Thornton 2010).

The prior probability assigned to this assertion is a difficult matter. For a local practitioner, it may be an obvious everyday experience – but not all smallholders will have a shortage of fodder and for them it is difficult to assess. Besides, what is the precise meaning of 'one of the major'? For outside experts with a broader knowledge of smallholder systems (like the author), every smallholder faced with nutrition problems for his animals will strengthen the statement. However, water experts may see lots of evidence that water shortage is a production constraints. With a statement like this, it matters already for its legitimacy who made it. If the author has a solid reputation, most people will assign a high prior probability to the statement but any experience to the contrary can weaken it.

Statement 5: '... *the individual is mainly concerned not with his absolute level of success, but rather with the difference between his success and a benchmark that changes over time*... *using economic tools, we argue that [it] can be evolutionarily advantageous in the sense of improving the individual's ability to propagate his genes*' (Rayo and Becker 2007).

The first statement is founded on 'a large body of research'. It speaks of 'an individual' and is, therefore, too general to be falsified. It is unclear how an absolute level of success and an associated benchmark can be measured. It may be assumed that such details can be found in the background literature. The association with evolution is in the form of integrating information in an individual's happiness function. It contains sophisticated mathematics and no empirical content. In their conclusion, the authors are aware of the speculative nature of their model, but do not question the scientific validity of their approach. An exercise like the one in this paper may offer insight in human behaviour, if the necessary qualifications are added. Otherwise, it is more rationalisation than insight.

Statement 6: '... *the overabundance of young people with advanced education preceded the political crises of the age of revolutions in Western Europe, in late Tokugawa Japan and in modern Iran and the Soviet-Union*' (Turchin 2008).

This statement presents a crude correlation between two observations: The age distribution of populations and political crises in history. It suggests a causal connection. Many people will feel justified to assign a prior probability to the implied hypothesis – but possibly for quite different reasons because their personal history matters for the evaluaton of statements like this one. New events like the recent political crises in African and Middle Eastern countries may strengthen the hypothesis for some people, but others remain convinced that rising food prices or ingrained corruption are the major determinants of crises. Whatever the verdict, observing correlations like these are one of many steps in understanding complex social-ecological systems.

In evaluating scientific assertions, it is also useful to occasionally shift the focus towards the person of the scientist and his psychological traits and sociological configurations. In his essay *Science de la science et réflexivité* (2002), the sociologist Bourdieu distinguishes three premises about the 'scientific enterprise' that are good to remember in judging scientific statements. The first premise is that scientists are driven by a reward system. What matters is recognition. Citation indices and networks are important. It raises also the difficult question how scientists make choices and deal with conflict. The second one is that science has largely an internal autonomous dynamic, but this is not smooth. Instead, there are periods in which 'normal science' is suddenly confronted with a revolution. Such a sudden change is called a paradigm shift by Kuhn in his book, *The Structure of Scientific Revolutions* (1962). The internal driving forces are still unclear. A third premise is that science is largely a 'contextual game' in the sense of Wittgenstein's *Philosophische Untersuchingen* (1953). The truth is not ultimate but constantly negotiated in a social domain of stakeholders. It is helpful to keep these social and psychological mechanisms

Box 8.4. *The role of disciplinary background: deforestation.* A meta-analysis of the proximate and underlying causes of tropical deforestation has been published by Geist and Lambin (2001). From a detailed literature search, they identified three clusters of proximate causes of deforestation, that is, human activities that directly affect the environment. They also listed five clusters of underlying causes of deforestation: economic, demographic, technological, policy-institutional and socio-cultural factors. For the underlying causes, they explored whether the disciplinary background of the scientific authors had an impact upon drivers and causes perceived and reported. It is concluded that there is a significant correlation for political scientists and ecologists between their disciplinary background and the main cause identified, and that research teams that combine natural and social science views show negligible bias. One may subsume that such correlations are also found in many of the questions about the evolution of complex social-natural systems. It appears that the present generation of spatial models contributes to a more thorough and integrated understanding of deforestation processes, which synthesises the various disciplinary filters.

in scientific practise in mind when dealing with (the 'production' of) knowledge in sustainability science. They incorporate the relativism of Postmodernism.

A painful illustration is the occasional fate of practical knowledge, which in certain contexts is referred to as indigenous or 'vernacular' knowledge. Scientific knowledge, even if its claims and authority are legitimate, may well be wrong, irrelevant or both. There are some stories about how mainstream scientific insights were forcefully applied in situations for which they turned out to be invalid (Earle 1988). A corollary is that indigenous knowledge acquired by practise and containing great accumulated experience and wisdom is often, as it should be, an ingredient of the scientific endeavour. The Indian activist Vandana Shiva expresses this forcefully: 'My involvement with the Chipko movement of women protecting their forests... had taught us that the powerless are not powerless due to ignorance but due to the appropriation of their resources by the powerful... literacy is not a prerequisite for knowledge... ordinary tribals, peasants and women have tremendous ecological knowledge based on their experience. They are biodiversity experts, seed experts, soil experts, water experts. The blindness of dominant systems to their knowledge and expertise is not proof of the ignorance of the poor and powerless. It is in fact proof of the ignorance of the rich and powerful' (Vandana Shiva 2009).

8.4 Complexity

Several times in this chapter we used the word *complex*. Did you ever think about what you mean when you say that a situation, a systems, a person is complex? If you buy a new mobile phone, you may find it easy and simple to use, but difficult to understand how it works. The scientist who co-designed it may consider its functioning as rather simple. The same situation is often encountered in daily life. You visit a car mechanic, a medical doctor, a lawyer, and you expect them to handle issues

Box 8.5. *Energy and complexity.* A perspective that connects energy and complexity is offered by Tainter (2000). 'Human societies often seem to become progressively more complex – this is, comprised of more parts, more kinds of parts, and greater integration of parts.... Every increase of complexity has a cost... The cost of supporting complexity is the energy, labour, time, or money needed to create, maintain, and replace a system that grows to have more and more parts and transactions, to support specialists, to regulate behaviour so that the parts of a system all work harmoniously, and to produce and control information... No society can become more complex without increasing its consumption of high-quality energy, human labour, time, or money' (Tainter 2000:6–7; §7.1).

which are (too) complex for you. You rely on an expert who is a professional. It is tempting to say then that complexity is in the eye of the beholder. In other words, saying that something is complex gives it a contextual characteristic.

In the last decades, the notion of complexity has emerged from a combination of mathematics, computer software and applications in various fields of enquiry denoted as *complexity science* or complex system science (see Appendix 8.1 for a brief historical overview and the Suggested Reading). A new language is evolving around concepts like dissipative structures, emergence and emergent properties, non-equilibrium systems, self-organisation, self-similarity, non-linearity, bifurcation, resilience, chaos, sensitivity for initial conditions, decentralised control, distributed feedbacks and others. They give new and deeper content to concepts that have become central in 20th-century science such as information, evolution, computation, order and life (Mitchell 2009).

An oscillating mass at the end of a spring is a classic example of a simple system. The standard approach works well: It can be subjected to controlled experiments and it is not needed to introduce external elements to know the movement of the spring. But, a few coupled springs can already give rise to apparently chaotic behaviour (§7.3). Conversely, the complexity of a coastline or of an ice flake turns out to be apparent because a simple equation can generate similar patterns due to the property of self-similarity (fractals). However, such *deterministic complexity* is not the common connotation of complexity. Mathematicians have proposed the notion of *algorithmic complexity*, which links complexity to information content. This is not very interesting for our discussion either. Physicists have discovered complexity in simple phenomena such as phase transitions, percolation phenomena and pattern formation in networks and chemical reactions.[7] They emphasise sensitivity for initial conditions and interdependence as key features of complex systems (Solé and Goodwin 2000; §7.3). But there is not yet an agreed-upon way to measure complexity quantitatively.

A less formal approach is to study systems experienced as complex, such as the immune system, an insect colony or a stock market, and try to list their key

[7] Nice examples are the Ising model and the percolator model, which can be inspected with quite a few other ones in the Model Library of the NetLogo software package.

characteristics (Mitchell 2009). The list of key attributes of complex systems usually comprises elements, relationships, an environment and an internal structure – as do all systems – plus learning and memory, emergent properties ('more than the sum of the parts'), self-organising and dissipative change and evolution (Manson 2001). At an abstract level, complex systems do have some intriguing properties in common. This gives rise to one possible definition (Mitchell 2009)[8]:

> A *complex system* is a system in which large networks of components with no central control and simple rules of operation give rise to complex collective behaviour, sophisticated information processing, and adaptation via learning or evolution.

This rather abstract and somewhat circular definition suggests that complexity can be defined objectively. This may be incorrect. Perhaps, as suggested before, complexity can only be defined in relation to the observer.

Two characteristics of complex systems are *distinction*, that is, variety (or heterogeneity), and *connection*, that is, constraints on parts as the result of interdependence between parts. These properties are *not* objective; they depend on what is distinguished by and how they are related to the observer. Dimensions of complexity such as identity and connection depend on what is and can be distinguished by the observer.[9] This view is expressed in most definitions proposed by social scientists, for instance, Pavard and Dugdale (www.irit.fr):

> A *complex system* is a system for which it is difficult, if not impossible to restrict its description to a limited number of parameters or characterising variables without losing [sight of] its essential global functional properties.

It is not this book's intention to dig deeper into the definition of and the debate about complex systems. Systems which apparently exhibit behaviour that is not easily understood, handled or (re)constructed by (most) observers, are complex systems. They are positioned somewhere between frozen order and chaotic anarchy where the information content is highest.[10] And, paradoxically, their micro-level complexity sometimes disappears at the phenomenological macrolevel thanks to their capacity for self-organisation. Whatever the precise definition, sustainability science is mostly about complex social-ecological systems.

In this book, I use the notion of aggregate complexity in the following way: A system A is more complex than a system B, if A has:

- more interaction with the *environment* in terms of the exchange of energy and matter and thus information;

[8] Already in 1995, over 30 definitions of complexity were given (Horgan (1995). For detailed discussions, see, for instance, the *Journal of Artificial Societies and Social Simulation* (jasss.soc.surrey.ac.uk).

[9] In anthropology the suffices '*emic*' and '*etic*' are introduced to differentiate between descriptions of a system as seen from one of its members (emic) as against a description from an outsider vantage point (etic).

[10] It is still too early for scientifically sound generalisations. For instance, complexity of a coupled system may be highest at some medium level of connectivity, where fluctuations do not yet average out but elements cannot be considered isolated either.

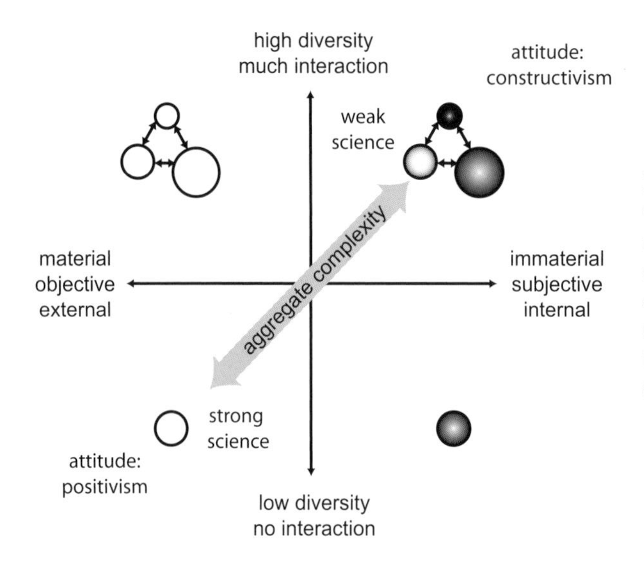

Figure 8.4. Aggregate complexity as diversity of and interaction between system elements and as degree of subjective cognition. This illustrative sketch shows also the distinction between the natural and the social sciences in the context of increasing aggregate complexity of the system under consideration.

- a larger degree of diversity and connectedness of the system *elements*, both in number and nature; and
- a larger ability of the system elements to make *representations* or models of past and future actions (memory, learning, anticipation and so on).

Following Manson (2001), the discussion focuses on such *aggregate complexity*. Most systems investigated in controlled laboratory experiments in physics and chemistry, and the derived appliances in the technosphere, are of low complexity in this sense. If experimentation is difficult or impossible or if observer (subject) and observed (object) interact, the system is of high(er) complexity. Human beings are at the highest level of complexity in terms of individual diversity, genetic and cultural codes and interconnections – at least, that is the common presumption. Knowledge about such systems tends to be weak(er).

These three features of complexity can be visualised with the scheme in Figure 8.4. It sketches two dimensions. The first one refers to complexity in terms of the number, diversity and heterogeneity of the system's elements under consideration and of their interactions – the first two of the three features previously mentioned. It is the vertical axis in Figure 8.4. Along this dimension, the difficulty to perform controlled experiments is one characteristic of complex systems and associated uncertainties. The second dimension reflects the degree of 'interiorisation' in the sense of an internal, subjective world – the last of the three features previously mentioned. It is the horizontal axis in Figure 8.4 and represents the spectrum from external/objective to internal/subjective. This axis is less common in discussions on complexity. Yet, it is here that the rift between the natural and the social sciences is felt most intensely (Hollis 2007; Döpfer 2005). It is an essential part of the artificial intelligence debate: If human agents are included in models, then which 'interior' are they given? Chapter 10 reflects on the self-image of man in more detail.

The vertical dimension in Figure 8.4 can be interpreted as describing reality *per se* (ontological) and the horizontal axis as the relation of the knower to reality (epistemological). An increase in aggregate complexity of a system can be understood as an evolution to the upper right in Figure 8.4. It can be interpreted in terms of worldviews (§6.3). In the lower left, knowledge is seen through the lense of natural science and is associated with positivism. The upper right is the domain of subjective and transcendental knowledge of cosmos, world and life. Constructivism is the dominant attitude.

With the study of real-world complexity, the role of *uncertainty* in science has become bigger and controversial: '... [the politicization of uncertainty leads to] scholastic disputations with the ferocity of sectarian politics. The scientific inputs... have the paradoxical property of promising objectivity and certainty by their form, but producing only greater contention by their substance' (Funtowicz and Ravetz 1990). Some sustainability related issues are at the core of these disputes: human-induced climate change and the risks of nuclear power and GMOs, to mention a couple. The notion of 'post-normal science' has been proposed to deal with the new situation:

> *Post-Normal Science* has been developed to deal with complex science related issues. In these, typically facts are uncertain, values in dispute, stakes high, and decisions urgent, and science is applied to them in conditions that are anything but 'normal'.

Managers and planners, whether individual or collective (firms, governments), rely on value-laden interpretations of the situation in the face of irreducible uncertainty and complexity. Faced with a complex reality, they often fall back on rather simple 'rule-of-thumb' recipes rather than on model outcomes. After all, management is often about complexity reduction, for instance, by controlling or simplifying the system and/or its environment. Therefore, in post-normal science, it is argued that for models to be relevant they need a more comprehensive assessment and participation of stakeholders. This demands a critical attitude with regard to the status of knowledge and an alertness for information quality and uncertainty. In post-normal science, 'we move from the dream of conclusive scientific demonstration to the ideal of dialogue for reconciling real antagonisms' (Ravetz 2006). It implies 'softer' forms of interference and control, for instance, in ecosystem management (Walters 1989).

In science, the empirical-reductionist mindset and paradigm are still powerful and necessary, but they can no longer be our only guide. Despite their success and a status not unlike that of rituals in pre-scientific cultures, they have to make room for new ways of observing and experimenting with the use of new and powerful tools. The availability of cheap and powerful computers, advanced software packages and communication devices in connection with Internet and social media is already changing the landscape of scientific knowledge. Table 8.1 contains a list of natural science oriented methods and tools, with applications relevant for sustainability science.[11] Their use promises a better scientific understanding of (un)sustainable development processes, because they incorporate features like non-linearity and feedbacks and engaging models users as participants and stakeholders.

[11] See Gilbert and Troitzsch (1999) and Feinstein and Thomas (2002), amongst many other books, on (quantitative) methods in social science. Table 8.1 does not contain the 'methods' in the upper right corner in Figure 6.5, which are usually considered as non-scientific: intuition, introspection, empathy and participation, and contemplation and meditation.

Table 8.1. *Methods and tools and application fields in complexity science*

Method	Fields of application
Integral-differential equations	analytical models in engineering, environmental and economic sciences and of population and evolutionary (eco)dynamics
Statistical and factor analysis	econometric models; social science relationships and 'stylised facts'
Optimisation and control theory, linear/dynamic programming	engineering sciences; optimal resource depletion; least-cost abatement strategies
Linear matrix techniques	economic input-output (I-O) theory; food webs
Systems science, systems dynamics, cybernetics	simulation models in resource and environmental economics and management
Catastrophe theory	ecosystem dynamics; social [r]evolution
Network (or graph) theory	food webs; economic, social and information networks
Game theory	social dilemmas; common pool resource (CPR) management
Evolutionary game theory	analytical models of species invasion; models in behavioural economy
Cellular Automata (CA)	simulation models of land-use land-cover change (LUCC) and of urban dynamics
Complex Adaptive Systems (CAS) and Genetic Algorithms (GA)	simulation of social-natural [co-]evolution; optimal strategy search in complex adaptive systems
Multi-Agent Simulation (MAS) or Agent-Based Modelling (ABM)	simulation models in evolutionary economics and resource use dynamics; modelling spread of innovations, diseases, information (in networks)
Participative: simulation gaming and policy exercises	simulation model of resource use management in interaction with participants; interactive simulation in social dilemma situations
Scenario analysis	connecting qualitative storytelling and quantitative modelling for resource use management

8.5 Metamodels and Organising Concepts

The acquisition of scientific knowledge – strong, public and with a claim to universality – generates expert knowledge and expert models. *Expert models* have been and still are hugely successful in understanding and manipulating the world around us, but they also demand much personal effort during construction and maintenance. An individual person can only be an expert in one or a few domains, although the options to acquire and share knowledge are increasing rapidly with the advent of ICT and Internet. Therefore, most of our knowledge and skills in pre-analytic everyday life come through observations and interactions with other people. We memorise and learn to conceptualise them into more abstract schemes, rules and theories. This 'everyday' learning and communicating process is, for most of us, in constant interaction with the findings of science via the media, colleagues and so on. We have the privilege that we do not need to discover and reinvent everything ourselves.

An expert can transform a reference or 'mother' expert model into a simplified version: a metamodel. According to Janssen *et al.* (2005):

> A *metamodel* is a reduced-form or minimal model in the sense of a simplified and coarse-grained (in space and time) version of an expert model or a relevant correlation which is probabilistic in nature.

Box 8.6. *Mental models, pants and cotton.* The Dutch author Bertus Aafjes tells a beautiful story in his travel account, *Morgen bloeien de abrikozen* (1954). He and his wife were invited to visit an English missionary, who lived in a town in southern Egypt and ran a school for the daughters of the small Christian minority. Aafjes' wife was wearing pants, which was most extraordinary in that place and time – and within hours she was known all over the village as 'omne pantalone', 'the mother of the pants'. The town lived from the cotton harvest in the province, each and every thing was experienced in terms of cotton. Unfortunately, on the morning after their arrival, the newspapers announced a further fall of the cotton price and worse was to come when it appeared that the cotton harvest in Texas was good beyond expectation.

This is dramatic news for the small community that depends on cotton sales, and the Muslim cotton traders accused Aafjes' wife of having the 'evil eye' and threatening their sustenance. The missionary was not willing to give in to this superstition and neither was Aafjes' wife. Wherever she appeared – in pants – people disappeared, as no one risked to be in contact with the evil eye. Cotton prices kept falling and the situation became precarious. Then, one morning, everything changed. Aafjes' wife made her usual promenade and, much to her surprise, she was invited by every shopkeeper and received all kinds of presents: 'please be our honourable guest, mother of the pants. . . ' She returned with lots of presents and full of surprise. The missionary went out to find the cause of this sudden change. As it turned out, an announcement had been made at the townhall that a tornado had brought irreparable damage onto the cotton fields in Texas. Cotton prices climbed and climbed, and Aafjes' wife became the most venerable person in town. Her admirers grew every day in numbers, as did the presents and invitations.

In some situations, metamodels are more adequate in communication and more useful for practical applications and policy purposes. Some people prefer the term *archetypical* or *elementary model.* Inside are the explanatory mechanism of the expert model in reduced form. In their simplest form, metamodels are like a manual or rule of thumb.

Sometimes, no reference or 'mother' expert model exists. Or there are several expert models around, each with their own expert(s) and content. In such a situation, no mature expert model may be available in the near future and one has to be satisfied with probabilistic statements and controversial schools and followers. New evidence from experiments has to be awaited before the relevant knowledge becomes stronger by refuting one or more of the hypotheses. In this situation, the available observations and measurements are often presented as *correlations*, which appear to capture salient features but for which no scientifically sound explanation can be given. The history of science has many examples. It is hoped that the correlation becomes a law – the scientists' road to fame. The scientist looking for more evidence will speak of hypotheses, the economist speaks of stylised facts, and the sceptical enquirer calls them myths or speculations, Clearly, metamodels can reflect knowledge of quite different status (strong versus weak).

Metamodels have a life cycle. After a period of excitement and controversy, either new evidence leads to qualified reformulation and acceptance, or it vanishes. A reformulation can, for instance, confine the validity of the proclaimed relationship to only a few countries or periods. It may also be that the issue was ill-phrased in the first place or that it remains controversial and has to await novel methods or concepts before any advance is being made. For a while, there can be competing models around and situations in which each individual has his or her own 'correct' model. Uncertainty is exploited to serve particular values and interests and can lead to the 'anything goes' adage of Postmodernism, examples of which are given in the thematic chapters.

Metamodels should be judged not so much for their scientific truth but for their usefulness in a given situation. They should be 'good enough' – for instance – in communicating – as basis for acting under uncertainty or as an invitation to do new experiments or collect new data. It also depends on the user. A scientist judges a situation in probabilistic terms and refines his model in scholarly dispute. A manager prefers a more pragmatic interpretation for which a metamodel can provide the best – and sometimes only – defensible strategy.

There are many relevant metamodels in sustainability science. You have already encountered several ones, for instance, Zahavi's law (§2.2). The huge amount of country data on a variety of issues give rise to coarse-grained country-level hypotheses and theories, which have the status of metamodels.[12] The chapters on demography, agriculture and economy introduce several metamodels in the proper context:

- *Population-economy:* population growth rates tend to decline with higher income levels (§10.2)[13];
- *The Engel curve:* people spend a lower fraction of their income on food at higher income levels (§11.2);
- *Economic growth and income:* GDP-growth is low at very low and very high income levels, and highest somewhere in-between (§14.1);
- *Sectoral shifts:* the activity pattern in economies tends to shift from agricultural to industrial and subsequently to service sector activities – so-called (economic) *structural change* (§4.1 and §14.1); and
- *Dematerialisation:* the use of minerals and energy per unit of GDP tends to rise and then fall, called the Intensity-of-Use (IU) hypothesis (§14.2).

Income is often the major independent variable in these correlaton-style metamodels.

The existence and use of metamodels is often a sign of complexity, uncertainty and controversy. The following example illustrates the point. Since the environment debate started in the 1960s, opponents of environmental regulation and taxes have argued that both interfere with economic growth – but how? In the industrialised countries, the argument was that *economic growth is needed to pay for environmental*

[12] An example are the empirical analysis of country GDP-growth rates versus other variables in Barro and Sala-i-Martin (2005; Chapters 13 and 14).

[13] As to income, this is usually set equal to the ratio of gross domestic product (GDP) and population of a country, in current or constant US$/cap/yr or in ppp-corrected I$/cap/yr.

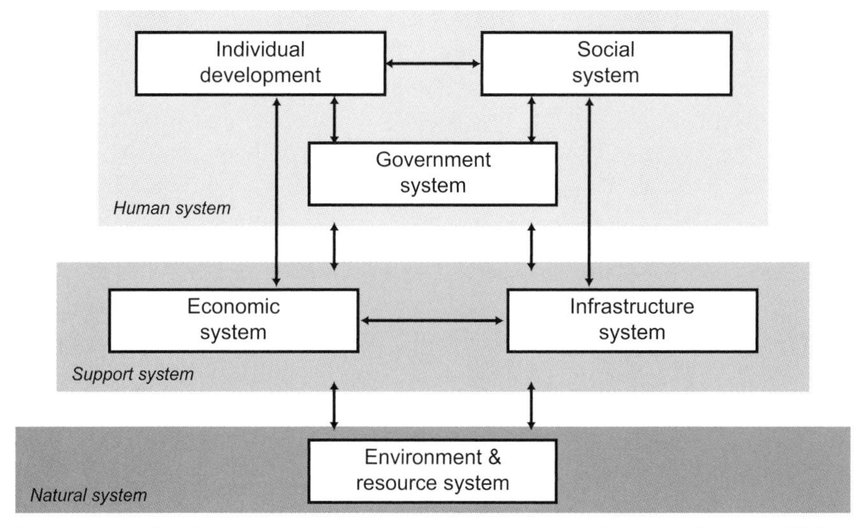

Figure 8.5a. The linkages between resources, economy and society (Bossel 1998).

measures. In less industrialised regions, the argument went differently: *We first have to become rich before we start worrying (and paying) for environmental degradation*. What is the truth about environmental expenditures and economic success? After years of research and hundreds of publications, there is no conclusive answer at the general level (Florax and De Groot 2005). The presumed negative correlation between the strength of environmental policy and intercountry trade does not occur. Some analyses use input-output tables and regression analysis, others use abstract theories such as neoclassical trade theory or gravity models. The heterogeneity in the different analyses is too large to provide a meaningful metamodel. Perhaps, more data and more variables – think of innovation dynamics – will make it stronger over time.

A less formal and more abstract way to deal with ill-structured problems in sustainability science is the use of *organising concepts* – but people also use the words conceptual model, frame or template. It is a way to frame the issues and express and communicate ideas and values. Sometimes, organising concepts are close to an analogue or metaphor and their adequacy and usefulness are valued differently in different worldviews. The previously introduced regimes and syndromes (§4.5) and the scheme of aggregate complexity (§8.5) are both examples.

Organising schemes and concepts can be used to analyse change in a structured way and at different spatial and time scales. Figure 8.5 presents two such schemes. The first one is *ecosystems as life-support systems* (Figure 8.5a) and represents the relationship between 'nature' and 'culture' and between ecology and economy (§10.1). Physical nature is seen as the material substratum – or life-support system – upon which the use of food, water, energy and materials and, ultimately, all human life is based. At the top are the social and cultural developments of individuals and society. The institutional arrangements (government) and the economic and infrastructure systems connect top and bottom at the intermediate level. It is a common scheme in the analysis of global change phenomena and the twin

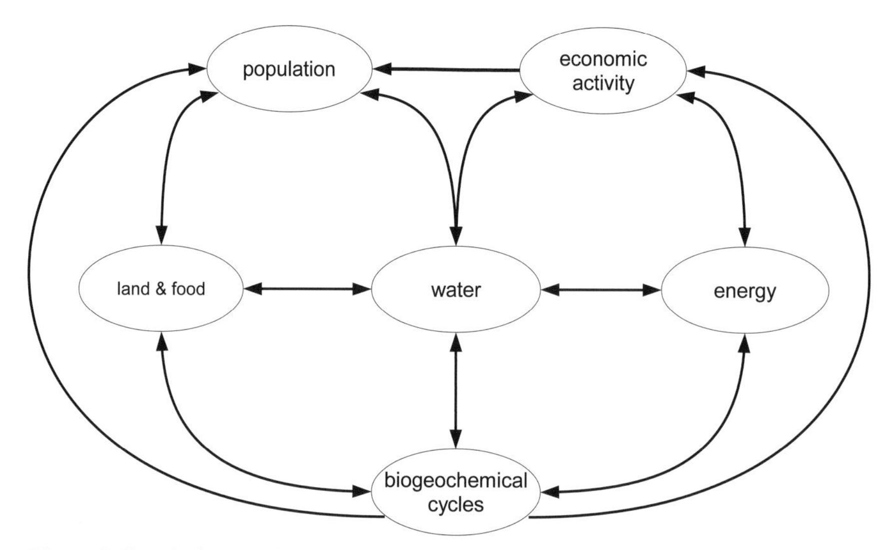

Figure 8.5b. The interactions between population and economy on the one hand and biogeochemical cycles on the other (Rotmans and de Vries 1997).

issues of development and sustainability.[14] Another scheme considers population and economic activities as the drivers behind the need for and supply and use of food, energy and water, which in turn impacts natural processes (Figure 8.5b). Such a *Pressure-State-Impact-Response (PSIR) scheme* is one of the representations of the 'nature-culture' nexus (§10.1). Another example of an organising scheme are the *ecocycles*, which indicate how (eco)systems may evolve according to a sequel of states (§9.7).

An important organising concept in sustainability science is *transition*. In Chapter 4, it has already been discussed in the context of the agrarian and (post)industrial regimes. Here, it is introduced in order to structure complex phenomena at a high level of abstraction:

> A *transition* is an evolutionary process in which social, economic, institutional and technological structures develop in mutual interaction and change drastically in the long run.

A formal representation of a transition is the logistic growth process dealt with in Chapters 2 and 9. Amongst the first transitions identified was the *demographic transition* followed by the *epidemiological transition*. Other transitions are the *health transition*, the food or *nutrition transition*, the *land use and forest transitions* and the *energy transition* and *transport transition*.

A generic process that is part of all these transitions are *technological transitions*. These have been defined as processes of major and long-term changes in technologies to fulfill societal functions such as transportation, housing, communication and food provision. They not only involve technological changes but also changes in user practises, regulation, industrial networks, infrastructure and symbolic meaning.

[14] The scheme is reminiscent of Marx' view of a material base or substructure and an ideological superstructure. It is also reflected in the 'value pyramid' proposed by Daly, which arranges knowledge and actions from ultimate means at the bottom to ultimate ends at the top (Meadows 1998). Both are expressions of the axis materialist versus mental/spiritual (Table 6.1).

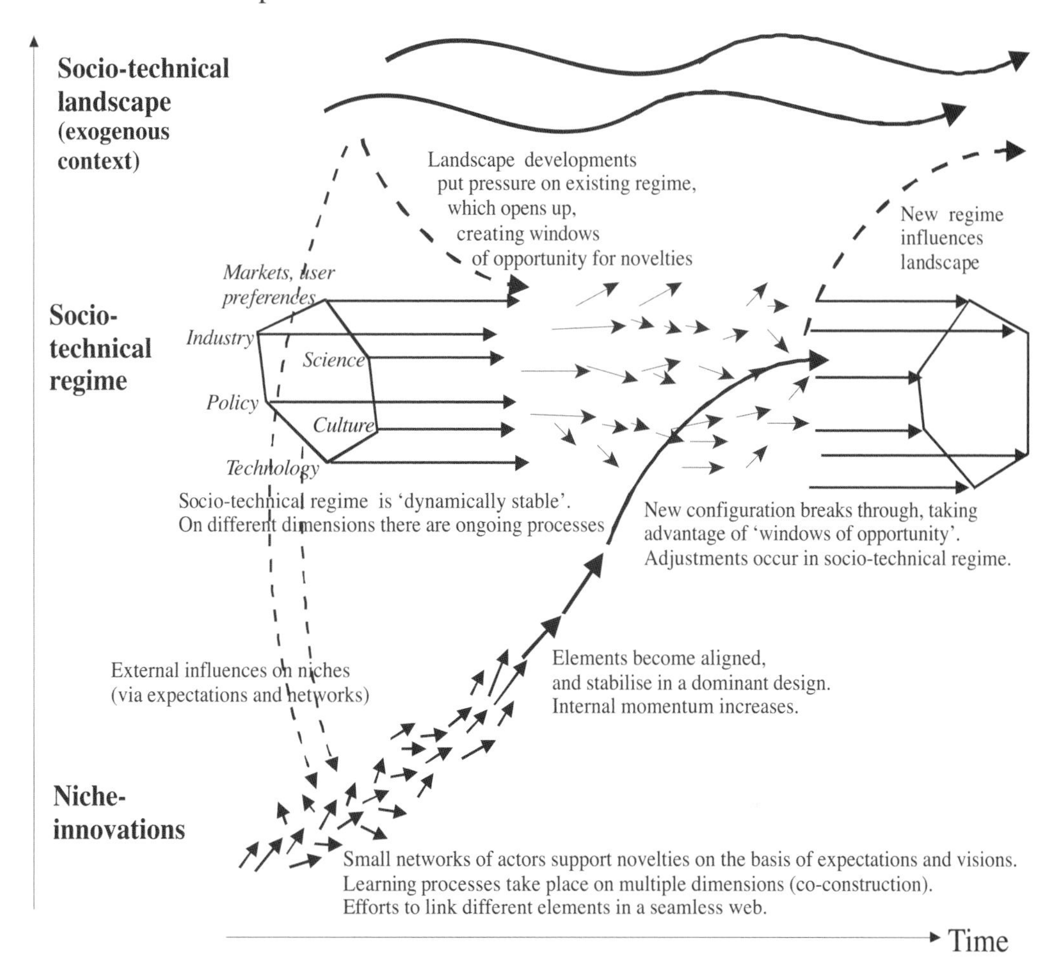

Figure 8.6. Multi-level perspective on transitions (Geels and Schot 2007, adapted from Geels, 2002:1263).

In economic science, such changes have been studied in the context of capital stock and factor substitution dynamics, for instance, the gradual change in steelmaking from the Bessemer to the basic oxygen and later the electric arc process. One of the first models of technological transitions was the logistic substitution model (Grübler 1999; §14.3). It is based on the simple predator-prey model and shows that there is a remarkable robustness in such long-term substitution of processes. A broader conceptual framework is being developed and applied at the interface of sociology, innovation science and evolutionary economics. It distinguishes a micro-, meso- and macro-level (Geels 2002; Geels and Schot 2007). At the micro-level, there are protective niches in which novel technologies ('configurations that work') can emerge. The meso-level of socio-technical regimes is where technological trajectories are consolidated and incremental innovations occur. The macro-level is the slowly evolving landscape of societal change (Figure 8.6).

Box 8.7. *Hierarchy and consciousness.* Possibly because it is in the nature of humans to think in terms of hierarchies, many forms of hierarchical order have been postulated in human history. There are many authors who have suggested a hierarchy from a worldly, humanist orientation. Besides Marx's distinction between infrastructure and superstructure, one can think of Maslow's spectrum of needs (§6.1). Jantsch (1980) sees hierarchy in the unfolding of 'The Self-organizing Universe'. In the sense of the material-immaterial dimension in Figure 6.5, this insight appears to be part of perennial wisdom (§6.4). Schumacher (1977) introduced levels of being, with reference to Thomas of Aquino, in his book *A Guide for the Perplexed.* Teilhard de Chardin (1959) advanced complexity and *interiorisation* as a third dimension alongside the infinitely large and the infinitesimally small, and the *noösphere* as part of the process of human evolution. Aurobindo (1955/1998), Elgin (1993) and others also see an evolution towards an ever higher consciousness. The Eastern chakra doctrine expresses a similar insight.

One might postulate an 'ascending order of complexity' and associate the levels along the complexity dimension or axis with an ascending order of intentionality and consciousness. Whether this is read as an unfolding or push from below as in the materialist-empiricist orientation, or as a teleological drift upwards as in the metaphysical spiritual orientations, is as yet an open question for each individual human – as well as for human society at large.

A series of historical change processes is investigated within this framework, such as the gradual displacement of sailing ships by steamships and of the horse-drawn carriage to automobiles. The previously mentioned transitions are all characterised by coevolution of social and technological forces. An illustration is the gradual shift from surface water use to piped water and personal hygiene, where engineers and users (or technology and culture) are both intrinsic to the change process (Geels 2005).

The transition concept is used to guide policies. *Transition management* emphasises the opportunities to influence technological innovations in a desired direction. Not unexpectedly, the applications still struggle with some pressing underlying questions such as: What is the end-point? Is it worldview-dependent? Can we steer towards it, and if so, how? But the concept is useful in its attempt to transcend partial and formal (meta)models and frame the broader, integrated perspective. It also frames a prime aspiration in sustainability science: a *sustainability transition* in which society moves towards high-efficiency, low-emission technological systems and associated economic institutions and social lifestyles (Meadows *et al.* 1991).

8.6 Science in the Age of Complexity

It is one thing to have reflections on and make models of complex systems, but quite another to manage such systems. Because they are complex, there are lots of uncertainties about how they function and respond to interventions. Because there are uncertainties, there are also controversies and every intervention will be

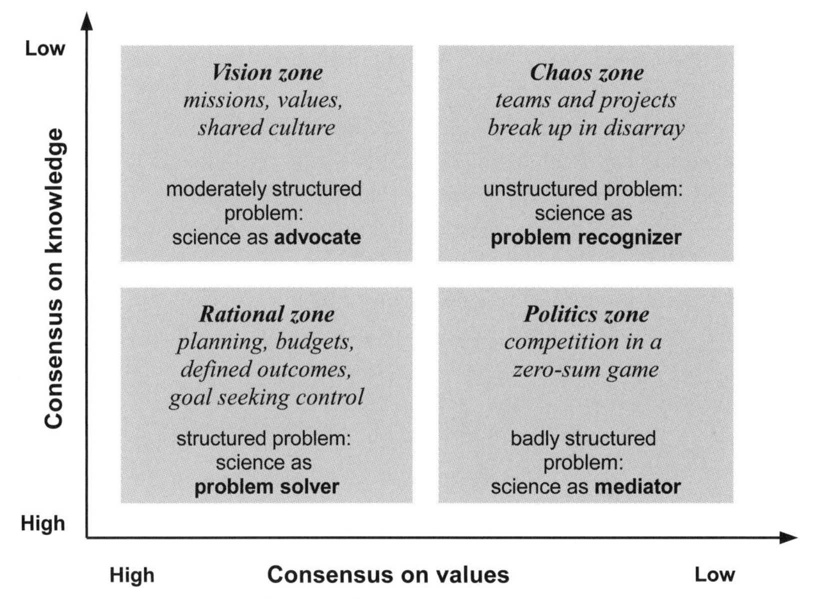

Figure 8.7. The role of (scientific) knowledge in situations of low versus high consensus on values and on knowledge, and the corresponding management regimes (de Vries 2006). It illustrates also a way to connect the science and policy domains.

rationalised within the frame of a worldview. Therefore, we must explicitly consider the role of values and knowledge in decision making with regard to sustainable development. This is also the key message of the aforementioned post-normal science.

A useful metamodel is shown in Figure 8.7 (de Vries 2006; Hisschemöller *et al.* 2009). The problem under consideration is evaluated in terms of (lack of) consensus on knowledge and in terms of (lack of) consensus on values. The stronger the knowledge, the more frequent there is a knowledge consensus. The more dominant the worldview, the more frequent there is a consensus on values.[15] If there is a large consensus on knowledge *and* on values (lower left corner), science can be invoked as a *problem solver*. It is the domain of applied science. Many environmental models have served this purpose in the past – think of air pollutant dispersion or process models for water cleaning. In management science, this is called the rational zone. The buzzwords are planning, budgets, defined outcomes, and goal-seeking control. It naturally fits the Modernist worldview. The usual way to deal with uncertainty is statistics.

If strong knowledge is not (yet) available, but agreement on values prevail, there is room for alternative models and scientists become *advocates* in the decision-making arena (upper left corner). Some elements of the debate on human-induced climate change are illustrative, for instance, the mechanisms of cloud formation in climate change or the technical potential of renewable energy sources (Hulme 2009). The expectation is that, with more research, such controversies will be solved and

[15] This is of course a simplification. Often, the power relationships dictate a degree of agreement that is as much the reflection of compromise and tactics as of genuine value consensus.

action can be taken. However, as long as the situation persists, vocal individual scientists can play an influential role, and in business this is the corner for inspiring motivators and visionary leaders.

If there is consensus on knowledge, *but not* on values, the situation is one of *mediation*, offering rule systems and evaluation methods (lower right corner). 'Science' offers practical assistance to find 'rational' ways of managing complexity and uncertainty. Professional consultancy offers value explication and decision support with tools such as resource accounting, cost-benefit analysis and risk assessment. Also, participatory methods such as simulation gaming and policy exercises can serve this purpose (§10.7). These methods primarily explain, clarify and communicate in order to teach *'how to play the game'*. An example are the various ways to bring about a reduction in greenhouse gas emissions: Should the focus be on a well-designed emission trading system or should some kind of equity-based allocation scheme be negotiated (den Elzen *et al.* 2008)? Mission statements and partisan positions are prominent tools to negotiate that which is known into action. This zone is associated with politics: 'We know damned well what the problem is, but how can we best protect our interests?'

The fourth and last position is when there is lack of consensus in *both* knowledge and values. The primary role of science here is *problem recognition*. It appears that major global change issues such as the risks and consequences of a sudden collapse of the West Antarctic Ice Sheet (WAIS) or dramatic decline in biodiversity are in this quadrant. In the eyes of the planner/manager, this is the dramatic chaos zone (upper right corner). In this corner of controversy on knowledge as well as on values, the arguments, claims, interests and values on which policy options are based cost much energy to organise. Here, too, novel interactive and participatory approaches are useful.

A similar scheme has been introduced by Pielke (2007). It uses two axes: One is the *view of science* associated with knowledge con/dissensus, and the other is the *view of democracy* akin to value con/dissensus. This leads to four roles for the scientist: the pure scientist (upper right), the issue advocate (lower right), the science arbiter (upper left) and the honest broker of policy options (lower left). The importance of the various schemes is more their usefulness in alerting us to the different role of science than to their scientific 'correctness'. Media play an important role in the dynamics of the discourse.

8.7 Summary Points

Given the interdisciplinary and complex character of sustainability problems, one must reflect on the nature of (scientific) knowledge. This chapter introduces some epistemological considerations and basic ideas about models, uncertainty and complexity. Key points to remember are:

- the scientific method is a combination of inductive empiricism and deductive rationalism – the one cannot be without the other;
- a *model* maps an object or item in the 'real world' into a formal system with formal (mathematical) rules;

- analogues and metaphors are complementary maps or representations;
- knowledge can be qualified on a scale from strong to weak, on the basis of an assessment of logical operations, codified experiences and hypotheses;
- *complexity* can be defined in various ways, but aggregate complexity is primarily an increase in the diversity of and interactions between system elements;
- metamodels are a way to represent and communicate knowledge about complex systems (in expert models or as black box correlations); and
- organising concepts, such as *transition*, play a useful role in structuring the investigation of complex systems.

> One of the impulses behind science is the desire to gain reliable knowledge about the world so that we can control it.
>
> – Solé and Goodwin, 2000:1

> For the Chinese the bamboo expresses the will to survive, the spirit to endure under adverse circumstances. Great trees in their strength resist the winds and are broken; but the pliant, yielding bamboo, twirled and tossed about madly in the storm, bends and bows unresting, and survives.
>
> – Diana Kan, Chinese Painting, 1979:45

> Fire can burn but cannot move.
> Wind can move but cannot
> burn. Do men know it's like
> that with knowing and doing?
> Till fire joins wind it cannot take a step.
>
> – Devara Dasimayya, in: Speaking of Siva (Vs. 127).
> Ed. A.K. Ramanujan. Penguin Classics 1973

> . . . each of us knows exactly one mind from the inside, and no two of us know the same mind from the inside. No other kind of thing is known about in that way.
>
> – Dennett, *Kinds of Minds: The Origins of Consciousness*, 2001:2

SUGGESTED READING

A helpful guide about the status of scientific knowledge and the philosophy of science from a natural science perspective.

Chalmers, R. *What Is this Thing Called Science?*, 3rd. ed. Queensland University Press/Open University Press, 1999.

An interesting collection of controversial views of U.S. authors, clarifying divergent worldviews.

Easton, T. *Taking Sides – Clashing Views on Environmental Issues*, 11th ed. Contemporary Learning Series. New York: McGraw-Hill, 2006.

This paper gives sixteen reasons, other than prediction, to build models.

Epstein, J. Why Model? (2008). jasss.soc.surrey.ac.uk/11/4/12.html.

A thorough introduction into the philosophy of science from a social science perspective.

Hollis, M. *The Philosophy of Science – An introduction*. New York: Cambridge University Press 2007.

A systematic introduction into what complexity is on the basis of a set of concepts and theories in 20th-century science.

Mitchell, M. *Complexity – A Guided Tour*. New York: Oxford University Press, 2009.

Contributions and applications in the field of sustainable development, notably ecology and anthropology.
Norberg, J., and G. Cummings, eds. *Complexity Theory for a Sustainable Future.* New York: Columbia University Press, 2008.

A not uncontroversial but concise introduction in the rationale for a post-normal science.
Ravetz, J. *The No-Nonsense Guide to Science.* Oxford: New Internationalist, 2006.

An inspiring exploration of complexity in a variety of models and systems.
Solé, R., and B. Goodwin. *Signs of Life – How Complexity Pervades Biology.* New York: Basic Books, 2000.

An introduction into environmental modelling with a stepwise introduction of complex system approaches.
Wainwright, J., and M. Mulligan, eds. *Environmental Modelling – Finding Simplicity in Complexity.* London: John Wiley & Sons, Ltd., 2004.

USEFUL WEBSITES

- www.bayesian.org/bayesexp/bayesexp.html is the website of the International Society of Bayesian Statistics and applications.
- www.nusap.net/ is the NUSAP website on uncertainty in post-normal science.
- math.rice.edu/~lanius/frac/ offers a website with fractal geometry examples; there are many more on this and other complex system science theories and applications.
- www.inclusivescience.org/ presents ideas and principles regarding an 'inclusive science' in a/the New Age.
- www.image.nl/fair is the website of the FAIR emission allocation model.

Appendix 8.1 A Brief History of Complex Systems Science

Complex systems science (CSS) emerged in the 1970s and 1980s. In the wake of research by early 20th-century mathematicians like Poincaré and Julia, rigorous experimentation and mathematical analysis in combination with rapidly advancing computing power resulted in several novel approaches to the study of real-world complexity. In the 1960s and 1970s, Thom and Zeeman developed catastrophe theory, showing how the qualitative nature of system behaviour depends on the equation parameters and introducing nonlinearities in a variety of models with broad applicability. In the 1970s, Mandelbrot discovered fractal geometry while studying complexity and chaos, of stock markets, amongst other phenomena, and increasing computing power permitted to draw fascinating graphs. In 1976, May published his discovery that simple discrete mathematical models could generate chaotic behaviour.

In 1980, Prigogine published the book *From Being to Becoming*, followed by the book *Order Out of Chaos*, written with his colleague Stengers and based on earlier work on irreversible thermodynamics. In this and his 1987 book, *Die Erforschung des Komplexen*, with Nicolis, he announced the end of the Cartesian-Newtonian paradigm. In retrospect, it was only one of the 20th-century events that signalled the end of 'modern' science and the advent of 'post-modern' science.

Since the 1990s, the pathbreaking research has occurred in the life and social sciences. Kauffman proposed ideas about how life could have evolved from self-organising chemical reactions in his book, *At Home in the Universe* (1995). An early application of game theory to animal behaviour introduced a completely new understanding of complex dynamics in the book, *Evolution and Theory of Games* (Maynard-Smith 1982), which has further been developed into the subdiscipline of *Evolutionary Dynamics* (Nowak 2006). Watts and colleagues attracted attention with the book *Small Worlds* (1999), which explores the science behind the phenomenon that any two people, selected randomly from almost anywhere, are 'connected' via a chain of only a few, six or seven, intermediate acquaintances. Graph theory got a further boost when computing power permitted network analyses of large networks such as citations, the World Wide Web and metabolic reaction schemes.

In the 1980s and 1990s, complexity science has become popular by some events and books. Several books popularised the ideas of chaos and complexity and the scientists around it, notably at the Santa Fe Institute of Complexity (www.santafe.edu). To mention a few, Gleick's *Chaos* (1988) made the new science the scene of enthralling and personal theatre; Waldrop popularised notions such as the edge of chaos in his book *Complexity* (1992); Gladwell wrote his book *The Tipping Point* (2000) 'on how little things can make a big difference' after being inspired by the nonlinearities and shocks of infectious disease epidemics.

9 Land and Nature

9.1 Introduction

At the dawn of the environmental movement in the 1960s and 1970s, there was increasing concern about the damage done by humans to *nature* – and nature protection and conservation were the stated goals. Ecology and the emerging environmental sciences were at the forefront. With the advent of the idea of sustainable development in the 1980s, more emphasis was put on the legitimate aspirations of many people to (material) well-being, that is, development. There was also an increasing realisation that natural systems are always changing and evolving, and that not much 'undisturbed nature' had been left after millennia of human evolution.

Nevertheless, ecology is still considered by many the core of sustainability science. The word is derived from the Greek οικοσ, house, and λογοσ, reason or idea, and it is, in a broad sense, the art and science of seeing things as a whole. As such, it has its formal scientific offshoots in theoretical and systems ecology, and its more social and transcendent expressions in social and human ecology. Its core idea has also shaped new bridging disciplines like landscape ecology and environmental and ecological economics. Ecology supports a rich interpretation of sustainability and a broad view of the environment-development nexus. It should not be confounded with environmental science, which has branched out in more practical and applied forms across various disciplines (chemistry, economics and others). Ecology is crucial for understanding the theory and practise of sustainable development, and more than one chapter should be devoted to it. This is not possible so I refer to some textbooks in the Suggested Reading.

Another core science in sustainability science is geography or, more broadly, the earth sciences. Life has evolved in the biosphere in interaction with geological change processes (§5.2). The resulting *landscapes* have been classified in terms of soils, vegetation and other characteristic variables. The dynamic changes in landscapes and the general principles and mechanisms behind them are the topic of land change science (Lambin and Geist 2006). Large parts of the natural world can apparently be understood on the basis of a rather small set of physical-chemical principles and exploited and (re)designed with help of the engineering (mineral, hydrological and civic) sciences. Amplified by the exponential growth in population and productivity, hardly any landscape on earth has remained untouched by human activity. This chapter investigates the natural and anthropogenic changes in landscapes and ecosystems.

9.2 Earth: Soil Climate Vegetation Maps

The interplay between geological features and forces and life as it manifests itself in ecosystems is a necessary part of understanding (un)sustainable development. In this section, we briefly introduce a few classifications and change processes in the geosciences. Soils are the outcome of aeons of geological change processes and the processes of life, including human life. The different soil types reflect differences in chemical composition, topsoil layer thickness, degree of weathering and so on. In combination with the incoming solar energy and its derivatives (such as winds, clouds, precipitation and vegetation), soils are an integral part of landscapes and climate. Figure 9.1 shows the FAO world *soil map*. Figure 9.2 shows the maps for two important climate parameters: average annual surface *temperature* and average annual *precipitation*. One should be aware that averages are constructed from a – sometimes sparse – network of observation points. They conceal the large differences in space and time, and local heterogeneities and monthly or even daily or hourly fluctuations are often of great importance from a sustainable development perspective. Increasingly, with the development of geographical information systems (GIS) and associated software, global and regional maps and time-series are available on the Internet on geology, soils and climate. The interested reader can look up the suggested websites.

The three maps in Figures 9.1 and 9.2 are a representation of the abiotic environment on which life can flourish in all its forms. It is the basis for the *land cover change (LCC) classification*. Oldest and still widely used is the life zone classification system proposed by Holdridge in 1947. It distinguishes thirty-three classes on the basis of two dimensions: annual precipitation rate (rain in mm/yr) and potential evapotranspiration (ratio of water loss in evaporation/transpiration and precipitation) (Figure 9.3). If precipitation is high but the potential evapotranspiration is low, you are in the humid tropical forests (lower right corner). At low latitudes and high temperatures, these ecosystems fix large amounts of water in soil and vegetation. If temperatures are high but there is not much rainfall, the evaporation rate is high and you are in a (semi-)arid desert (lower left corner). The little water there is in the system is not easily contained. Little rainfall and low temperatures are characteristic of the polar deserts (upper corner). The various classes coincide with two features mentioned before: geographical location (latitude) and elevation (altitude), both are shown in Figure 9.3. A bioclimatic classification like this is useful to investigate *potential vegetation*, that is, the vegetation that one would expect without human impact. Of course, there will always be local variety due to the natural gradients in altitude, temperature, soils and so on.

The Holdridge classification is used in several global change models to perform integrated assessment of atmosphere-biosphere interactions. For instance, the IMAGE-model maps land cover and use changes in combination with greenhouse gas emissions and climate change (Figure 9.3; Bouwman *et al.* 2006). Assigning to

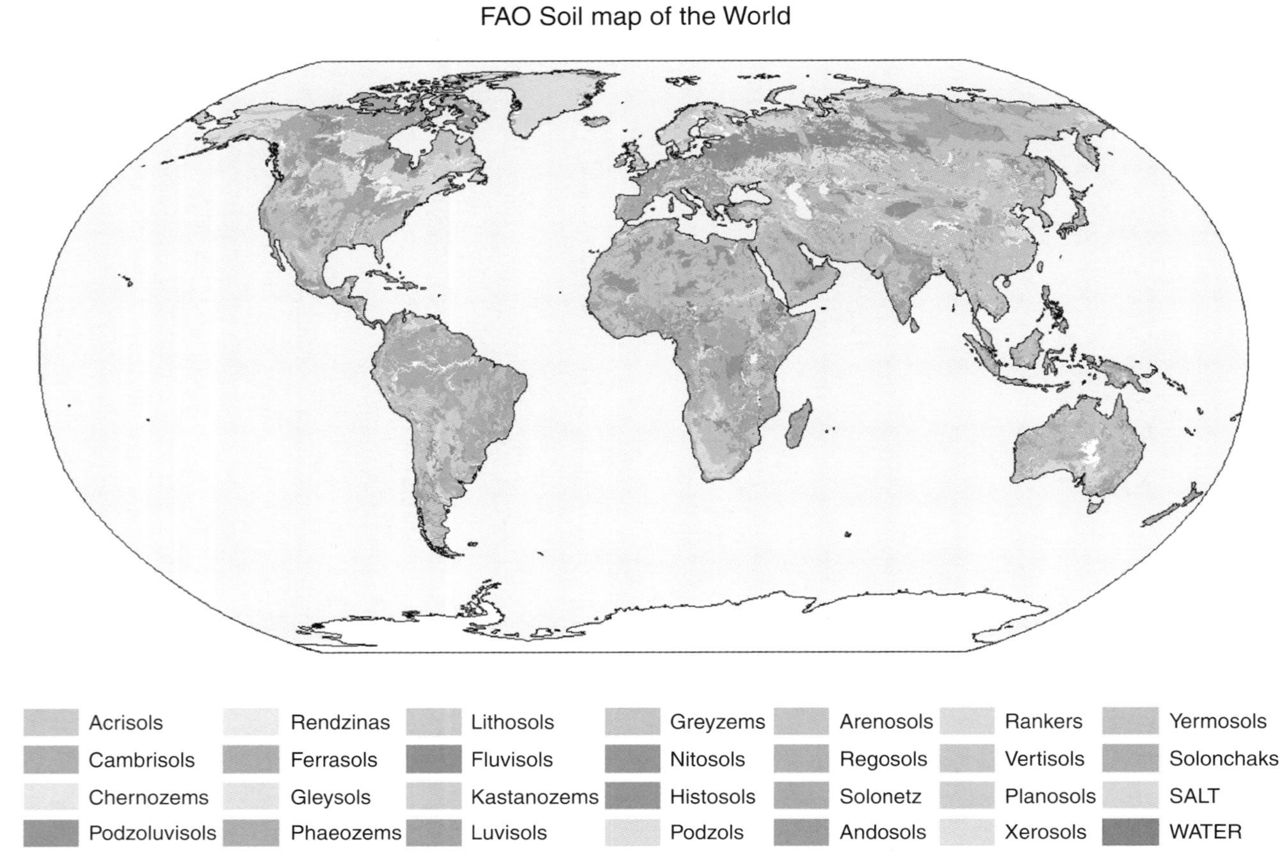

Figure 9.1. Soil map of Earth (source of map: FAO). (See color plate.)

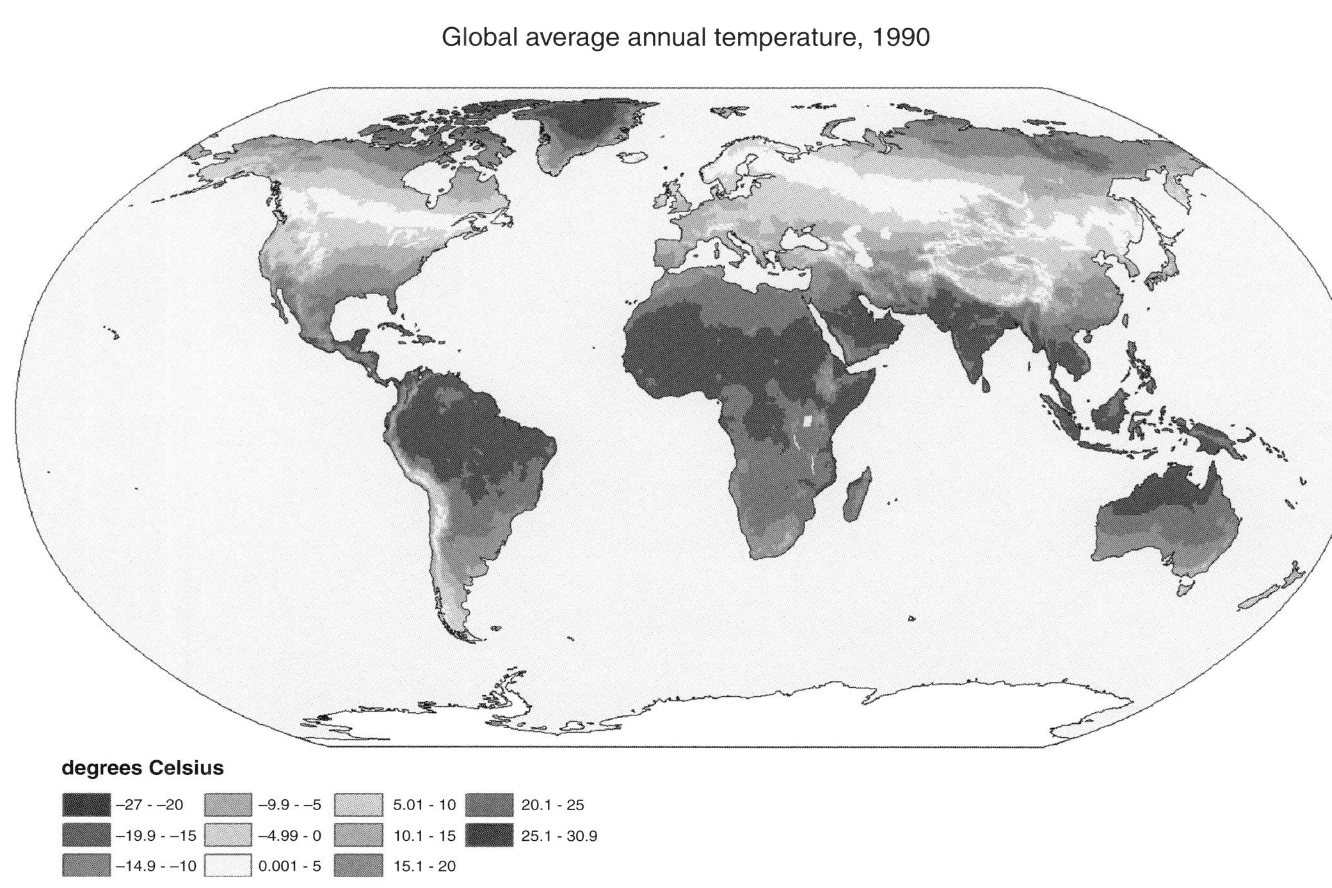

Figure 9.2a. Climate map of Earth: annual average temperature and rainfall (source of maps: PBL). (See color plate.)

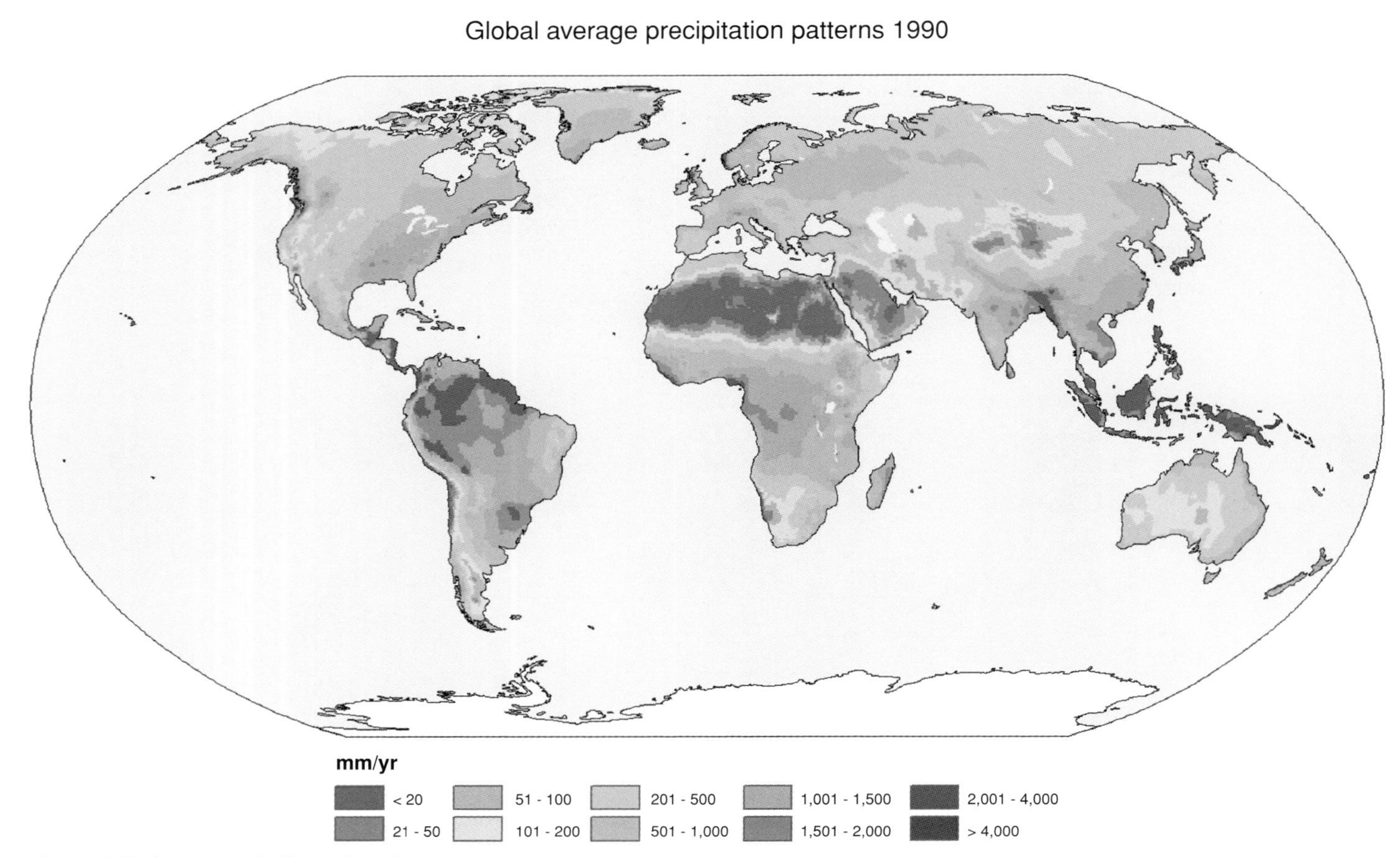

Figure 9.2b (*continued*) (See color plate.)

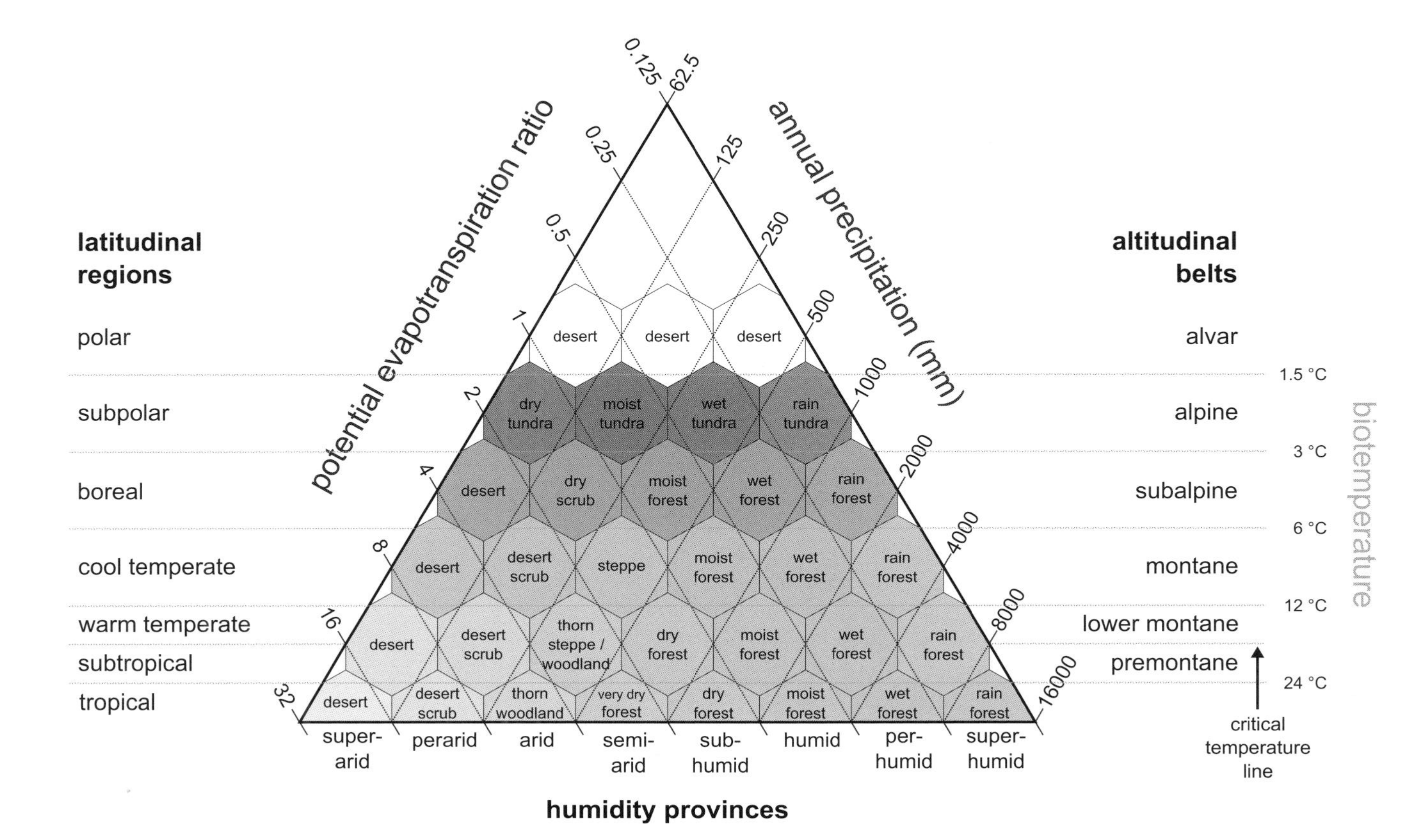

Figure 9.3. Life zone classification on the basis of annual precipitation rate and potential evapotranspiration ratio (source: Peter Haiasz, creatice commons).

the crude LCC-classes average characteristic properties such as carbon uptake, it allows an assessment of the role of the biosphere in climate processes. Figure 9.4 is a more empirical and refined way to construct such a map from the growing number of satellite data. It shows a potential vegetation map constructed from satellite and census data and after eliminating human influences (Ramankutty and Foley 1999).

Scientists attempt to simulate the mechanisms of land and soil change processes in order to understand and predict their dynamic evolution in global change integrated assessment models (GC-IAMs). The models are improved continuously by adding more detail such as plant types, and by increasing resolution, for example, from monthly to hourly or even shorter timesteps and to smaller spatial units. This permits a more detailed investigation of the complex interactions between vegetation, climate and climate change–inducing greenhouse gas concentrations. This topic could fill a whole book in itself. In this chapter the discussion is confined to a few aspects of undisturbed 'nature': what kind of processes occur and how do human interventions affect ecosystems and their 'services'? But, first, a few narratives from the real world.

Box 9.1. *Universal and global processes.* Natural science principles have led to an abstract, formal understanding of real-world phenomena at the micro-level, in the form of a limited number of generic processes, which are widely taught and used in environmental science, engineering and management. Such processes can occur in many places on Earth, but they are local/regional in scale and appearance. They are called *universal* to distinguish them from processes that involve the earth system as a whole. The latter obey the same laws but happen on a much larger scale, which is why they are called *global*. The distinction is not sharp, but scale does matter.

Soil erosion due to wind and water, evapotranspiration in vegetation and the dispersion of combustion products are ubiquitous and thus universal. In the 1960s, it was discovered that the emission of sulphur and nitrogen oxides in countries like the United Kingdom and The Netherlands acidified the lakes in Sweden. It was a regional-scale phenomenon. Similarly, the dust storms in China and Africa have reached regional-scale proportions and now also affect California and Latin America. One of their causes may be the clearing of tropical forests. These processes are (or are becoming) global (change) processes. Widespread but local emission of chlorofluorocarbon (CFC) compounds penetrated the stratosphere and broke down ozone molecules, causing the global phenomenon of the 'ozone hole'. When emitted from fossil fuel combustion and land use activities, greenhouse gases like carbon dioxide spread quickly through the atmosphere and thus cause a global increase in concentration, which in turn affects the radiation balance of the Earth – creating a global phenomenon. In turn, the resulting changes in temperature and precipitation affect local conditions. The existence of universal change processes as part of global change phenomena is a characteristic setting for many sustainability science issues. This micro-macro aspect is also referred to as 'nested dynamics'.

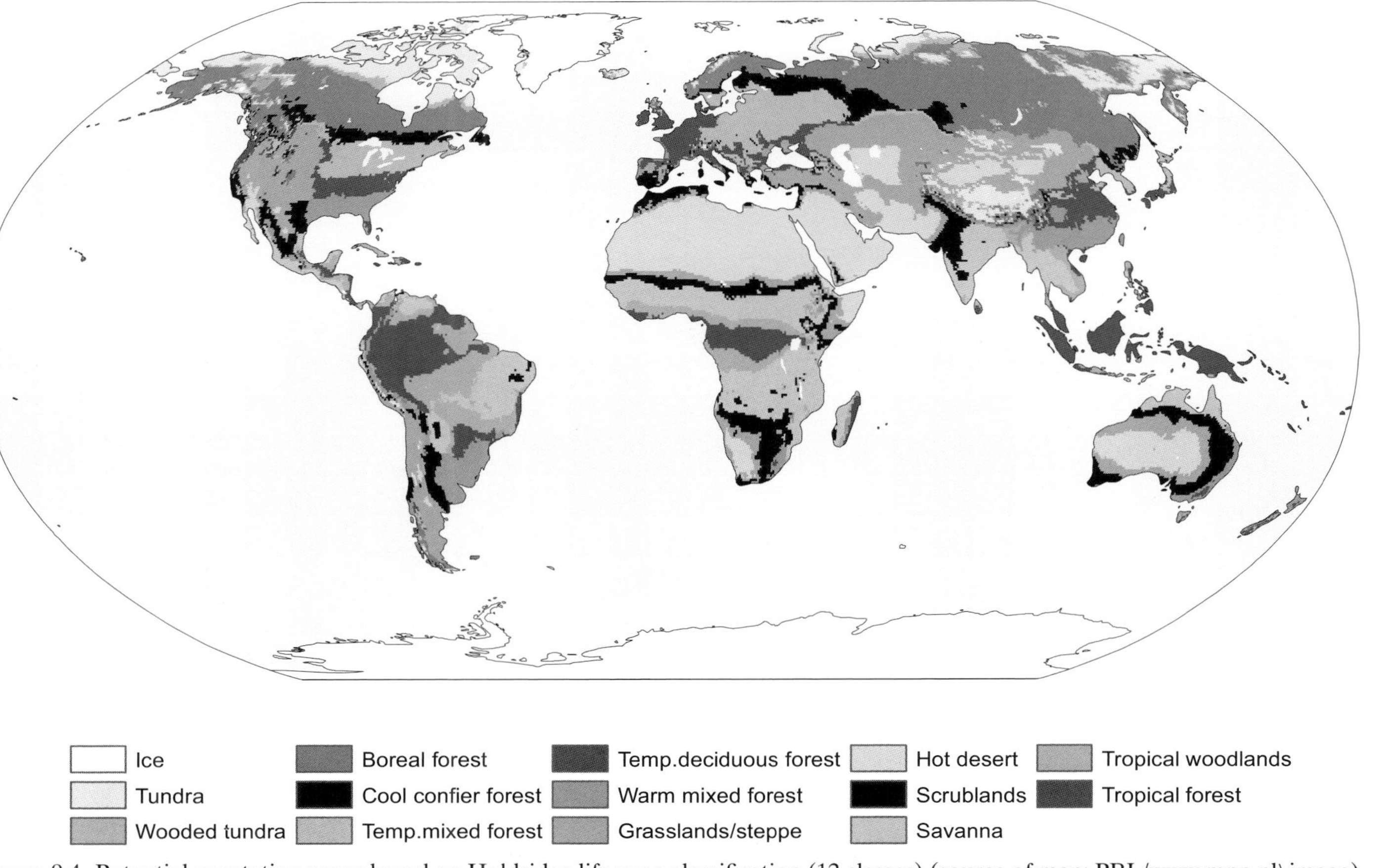

Figure 9.4. Potential vegetation cover based on Holdridge life zone classification (12 classes) (source of map: PBL/www.mnp.nl\image). (See color plate.)

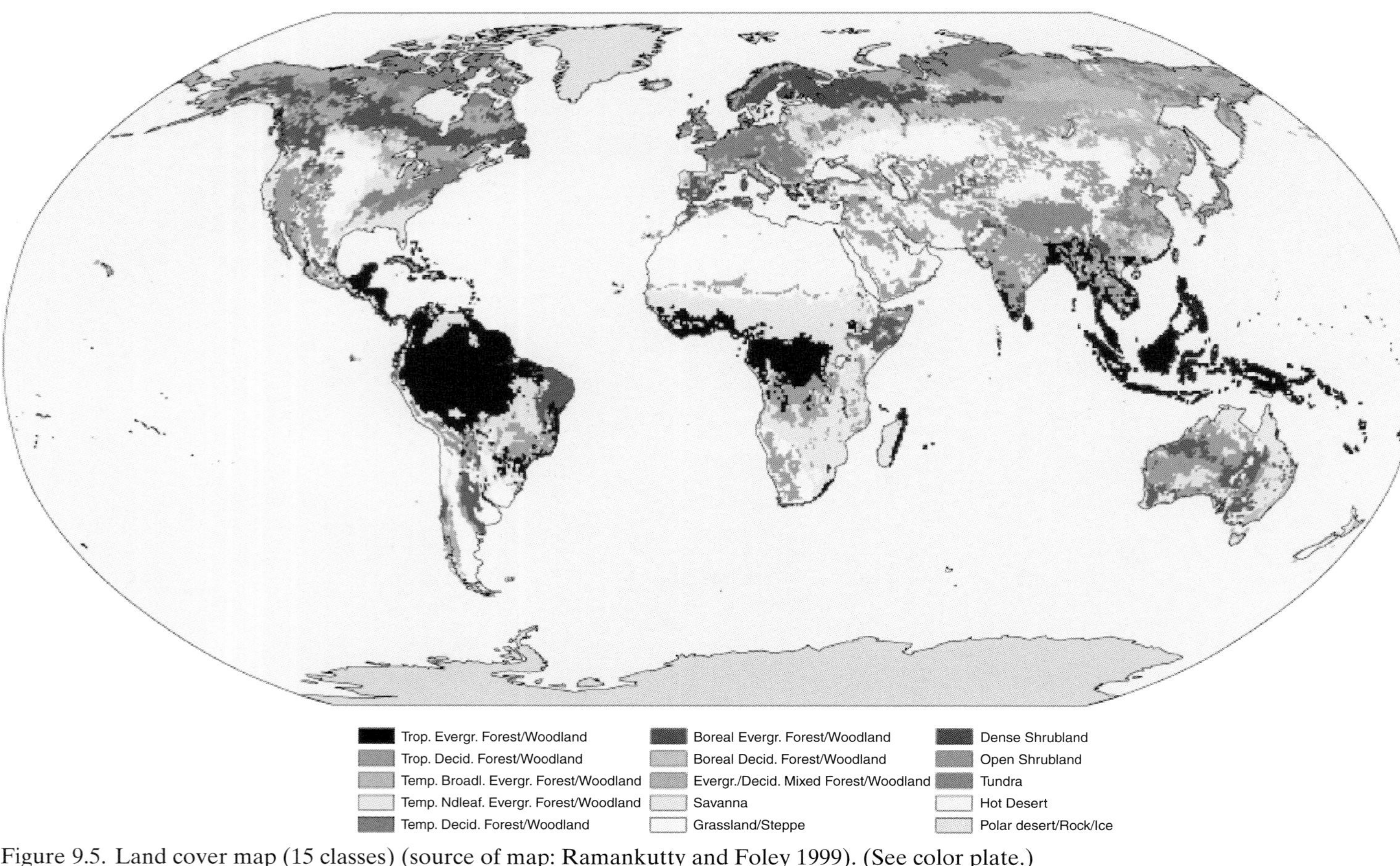

Figure 9.5. Land cover map (15 classes) (source of map: Ramankutty and Foley 1999). (See color plate.)

9.3 Stories

9.3.1 Forest Fires[1]

One of the causes of forest destruction is *fire*. It is a natural phenomenon, and it is instrumental in regulating ecosystems and rejuvenates their biodiversity. Natural fires are mostly triggered by lightning, other fires are caused by agricultural practises and other human activities. What are the impacts of fire? In Mediterranean Europe, the annual area burnt and the average fire size has roughly been constant since 1990 and the number of fires is declining since 2000. However, they can still have devastating impacts.

A group of French researchers has investigated representative Mediterranean forest areas in the Provence (oak and pine trees). They compared the vegetation, soil and biodiversity in areas where no fire had occurred for fifty years or more with the same variables in areas where up to five fires had taken place in rather short periods. It turns out that after some fifty undisturbed years, the forest has built up some resilience against disturbances. The structure of the original forest has only come back after 150 to 200 years. During those first fifty years of recovery, multiple fire events can permanently degrade the potential of the ecosystem. The risk of permanent damage is increased by the more frequent droughts as a consequence of, most likely, climate change. There is a threshold effect: 'A forest can recover from a succession of three fires in 50 years (one every 25 years on average), but the fourth one is critical... The combined effects of recurrent fires and droughts are devastating', according to the study coordinator Vennetier. What to do? Protecting with priority the oldest forests (150–200 years), which are rare and contain species that might get extinct, is one option. Better protection against fires is another one. If unmanaged, the fires will transform the landscape into one of bushes (*maquis* and *garrigue*) with considerable loss of flora and fauna.

It is difficult to get strong knowledge about the relationship between size and frequency of forest fires and climate change. In June 2010, a 'blocking anticyclone' caused a heat wave of unprecedented intensity and duration in the southern part of Russia and in eastern Ukraine. It is still unclear whether there is relation with climate change. Summer temperature in large cities such as Moscow and Kazan had reached 39°C. In August, temperatures started to drop and the month was not the hottest ever, but very dry. Human activity and natural processes, possibly self-ignition in the extremely dry and hot conditions, caused huge fires in the forests and the peat-bogs. The data are controversial, because of different definitions, measuring methods and possibly distortion of information. Official data speak of an affected area of 1 to 1.5 million hectare (Mha), whereas satellite data from scientific organisations give estimates of 5 to 12 Mha. Hundreds of thousands of professionals and volunteers were involved in combating the fires and rescuing people. Liquidation of state protection of forest and peat bogs, termination of work to prevent fires and abolishment of monitoring and suppression of fires at early stages became the reason of catastrophic

[1] Le Hir in *Le Monde* 4 août 2009. See also effis.jrc.ec.europa.eu/. The text on forest fires in Russia is written by Sergey Minosyants, Ph.D. student at Mendeleyev University of Chemical Technology of Russia and based on, *The Conclusion of the Public commission on investigation of the reasons and consequences of natural fires in Russia in 2010* (September 30, 2010).

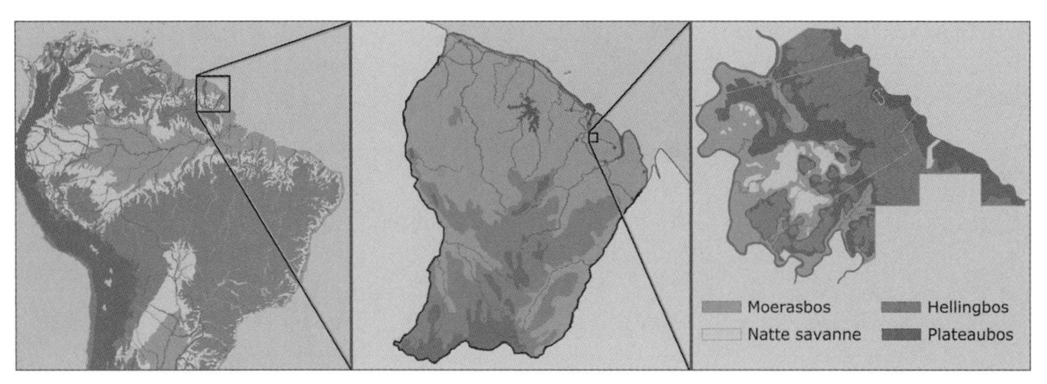

Figure 9.6. Nature reserves in North French-Guiana. The Trésor Reserve borders the twenty-five times bigger Kaw-Roura (moerasbos: swamp forest; hellingbos: slope forest; natte savanne: wet savanna; plateaubos: plateau forest). (See color plate.)

fires of forest and peat bogs. Preliminary estimates of the damage indicate that the economic costs might exceed U.S. $375 billion or in the order of 25 percent of annual GDP. Besides, an estimated 30–100 Mt CO_2 were emitted and hospitalisation and mortality rates increased significantly during the heat wave and the fires.

9.3.2 Biodiversity in South America: The Trésor Project[2]

The *Trésor Foundation* was founded in 1995 by enthusiastic and committed individuals at the University of Utrecht, the Dutch, French and U.S. branches of the World Wildlife Fund (WWF) and the Dutch business community. The Foundation's mission is: to conserve biodiversity and develop sustainable land use in a way that is equally beneficial to all the parties directly and indirectly involved, by means of cooperation, consultation and local participation. The Foundation bought 2,600 hectare (ha) of rainforest from the French-Guianese diocese in the northeast of French-Guiana and later expanded with another 1,400 ha (Figure 9.6). The university made an assessment of the area's carbon storage capacity, which motivated companies to participate with an investment to compensate part of their carbon emissions and as a way of branding.

A 'commercial' nature reserve was born with three objectives. The first one is protection of the biological diversity and the existing ecosystems in and nearby the park. The second one is stimulation and support of scientific research in tropical systems. The third one is providing botanic and environmental education by starting or contributing to educational programmes for schools and visitors of the Trésor Reserve. The park management is done by a voluntary local board (Association Trésor) with three full-time employees at their service. Because the park is being granted an official national park status (Réserve Naturelle Regionale) as a token of social appreciation, their salaries are now paid for by the regional government.

Some original fundraising methods were created to finance the park and the activities. A rainforest adoption system was created that makes it possible for private

[2] This story is written by Joeri Zwerts and Vijko Lukkien at Utrecht University. See also www.iucn.nl/about_us/members_1/trsor_foundation/.

contributors to adopt square meters of the reserve for a certain amount of money and a certificate in return. In this way, donating money becomes a more personal affair, even though the certificate gives no further rights to its owner, and the management of the reserve can remain simple. Compensating for greenhouse gas emissions is another way of fundraising. Several university departments compensate their CO_2 emissions from flight trips with donations to the Trésor Foundation. For the future, management seeks to facilitate corporate research for bioprospecting, as relevant knowledge of the area can significantly reduce a company's investment costs prior to the research. In return, an agreement will be made to finance further development of the area.

The project has large positive effects for the people in the area and as a result receives much local support. The area gains much (inter)national interest and a local ecotourism industry is developing. Investments create revenues and employment for the local population. One example is the production of durable wooden signposts for alongside the walking trails: The company that was asked to make them now also provides them to other reserves across the continent. It illustrates the positive link between local economic growth and nature preservation when park management involves the local population. The neighbouring, and twenty-five times larger, Kaw-Roura Reserve now also obtained official protected status, as a result of the example set by Trésor.

The project illustrates what can be achieved by a number of committed and knowledgeable individuals and by collaboration between local communities, environmentalists, governments, science and business. The central message is that in order to accomplish your objectives, intensive, innovative and adaptable management is a must. Poverty alleviation and education play a central role and respect for the local situation is key because no two forests and no two cultures of the people that inhabit them are alike. For the Foundation itself, after fifteen years, the future perspectives are upscaling the reserve and disseminating knowledge to serve the general cause of nature conservation.

9.3.3 Mining in Papua New Guinea[3]

Papua New Guinea is one of the world's largest islands. Called the 'Last Great Place', it is home to hundreds of unique species of animals and plants as well as to an upward of 820 languages. The Porgera gold mine is situated in the highlands. It produced around 18 tons of gold per year and more than U.S. $1 billion of profits in 2006, according to Barrick Gold, a Canadian corporation that assumed a majority share of the mine. Barrick operates twenty-six mines worldwide and boasts of having the industry's largest reserves.

Porgera, New Guinea's biggest gold mine, accounts for 72 percent of the country's export earnings. It utilises the most advanced extraction technologies and helicopters fly people and gold in and out on a daily basis. Its extraction process creates cyanide-laced wastewater that the company discharges directly into the local river system. One millionth of a gram of cyanide in a litre of water is enough to kill the fish. Because Porgera is literally at the top of the country, the streams into which

[3] Part of this information is from the site http://www.corpwatch.org/article.php?id=14381.

the toxins are dumped flow into many other tributaries before they reach the sea. After fourteen years, the mine waste has slowly torn the hills from under the local inhabitants and turned the small valley below into a choked river of dirt creeping toward the Gulf of Papua 200 miles away. The large rainforests and its inhabitants are also under threat: Half of the forests will disappear by 2020, if present deforestation rates continue. The causes of deforestation are exploitation for timber and clearing by small farmers.

Since the 1970s, in the Indonesian part of New Guinea, the province Irian Jaya, the American mining company Freeport has been exploiting some of the world's largest reserves of copper and gold. Here, too, huge profits are made while ecosystems are destroyed. Massive deforestation for infrastructure took place. Almost all processed material is released and fills up several square kilometres at the mouth of the nearby river. These mines are part of what the United Nations Industrial Development Organizations (UNIDO) calls a 'gold rush in the Third World' that began in the 1980s and spread to Tanzania, Surinam and many other countries. Mining for gold is one of the world's most grotesque industries, consuming vast resources and producing mountains of waste to produce a small amount of soft, pliable metal with few practical uses. To make one gold wedding band, at least twenty tons of earth must be excavated.

9.4 Land Cover Change and Degradation

Two key ingredients of ecosystems are soils and vegetation. *Soils* are the outcome of very complex processes over long periods of time. They are at the interface of the lithosphere, biosphere, hydrosphere and atmosphere and are under natural conditions in dynamic equilibrium with climate and vegetation. Their classification is based on a series of constituent characteristics, such as parent material, chemical composition and physical features, topography and regional climate and vegetation (Figure 9.1). *Vegetation* is the cover of plants and trees on the Earth's surface. It ranges from a small number of species in simple arrangements to very diverse and complex ecosystems, depending on the prevailing conditions. Vegetation is characterised by different spatial features and changes over different time scales. The gradients in climate variables such as temperature and precipitation and in elevation are the major determinants of spatial heterogeneity. Together, soils and vegetation constitute the larger part of the biospheric stocks and flows of mass and energy.

An essential factor in soil-vegetation systems is *water*. Indeed, the role of water in all its forms is so diverse and pervasive that it is hard to give a concise description. The system of water stocks and flows is called the *hydrological cycle* (Figure 9.7). By far, the largest amount of water on Earth is stored in the oceans. Other important stocks are ice and groundwater. The major flows are evaporation and subsequent transport and precipitation. With temporary storage in snow, soils and vegetation (the equivalent of delays in a system dynamics context), water flows back to the oceans via the several hundreds of river basins. These flows are large in absolute amounts and in a human context – up to hundreds of cubic km per year (km^3/yr) – but tiny compared to the three aforementioned stocks ($<$2 percent). I refer the reader to the Suggested Reading and Useful Websites for more detailed information. Human use of water is dealt with in §12.3.

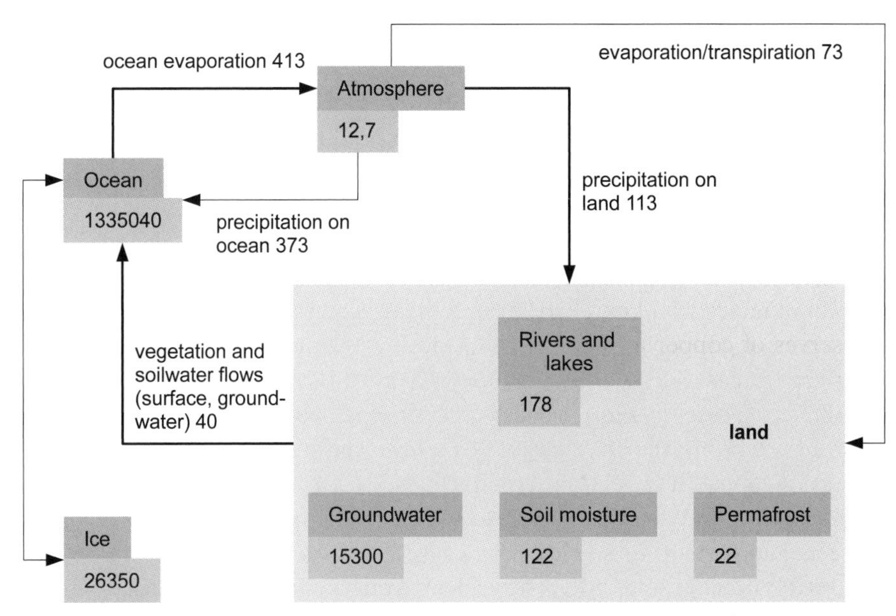

Figure 9.7. The hydrological cycle (source of data: www.cgd.ucar.edu). Units are thousand (10^3) km^3.

A major sustainability concern is the degradation of landscapes and, in particular, soils, with important consequences for ecosystem functioning. Human interference with soils, vegetation and water has occurred for millennia. With the ongoing population growth and the advent of industrial agriculture, interventions have increased in extent and intensity. Soils are said to degrade. But what is land degradation and to what extent is it human-induced? Also, how will it evolve under the increasing pressures of human activity and climate change?

In many landscapes *erosion* by the forces of water, wind and gravitation is at work as a perennial force behind changes in land cover. There has always been erosion, as it is one of the geomorphological forces. It is a fundamental and complex natural process. Physically speaking, erosion is the detachment of particles of soil and surface sediments and rocks. The ultimate driving force of erosion is tectonic action, which lifts the Earth's surface from which material can then be eroded. It is the basic formation mechanism of sediment and is, therefore, at the origin of much cultivable land and fertile deltas. A second influence comes from climate, in particular, rainfall intensity and rainfall pattern in interaction with vegetation. The third factor – which is usually the focus in sustainable development contexts – is erosion induced by human activities such as burning forest, ploughing and animal grazing. These are in some areas like the Mediterranean and Asian river deltas age-old forces shaping the landscape, with overgrazing and fire amongst the most influential processes. It is difficult, however, to separate human-induced erosion from natural erosion.

Erosion and its impacts tend to be greatest in the so-called drylands (hyper-arid, arid, semi-arid and dry subhumid), as these are most vulnerable with their poorly developed soils and little vegetation. These drylands make up some 25 to 40 percent of the earth surface and are habitat to one-fifth of the world population. They are prone to *desertification*, which narrowly defined is a process of increasing aridity and gradual dominance of abiotic over biotic processes (Okin 2002). Desertification has three dimensions: meteorological, ecological and human (Reynolds and Stafford Smith 2002). The seasonality and frequency of (extreme) rain events are an important factor in fragile drylands, and there are still large uncertainties in the causes of variability in precipitation over the years. The natural vegetation in drylands is a mixture of grasslands, shrublands and savannas, and the relatively unprotected soils are poor in organic matter. Their use as rainfed or irrigated cropland and, much more important, as rangeland can easily degrade soils by tillage and grazing. The human or socio-economic dimension, therefore, often plays a large role, but in different forms and at different scales. Debt and market failure are direct and immediate causes of degradation, for instance, in the form of local overgrazing. Tourism and trade are indirect driving forces.

One of the processes leading to desertification is *salinisation*. Semi-arid and arid regions tend to be naturally salty, because evapotranspiration exceeds precipitation and there is not enough water to leach out the soluble soil material. Salts from rivers and sea intrusion and atmospheric deposition are not carried away to rivers and seas and accumulates in the soil. This is exacerbated in closed drainage basins. Human activity, notably water diversion for irrigation and dam construction, can intensify salt accumulation. Improper management of irrigated soils may accelerate it. Human-induced salinisation happened already in the early Mesopotamian civilisations (§3.3). An estimated 10 percent of the world's irrigated area for crops is severely degraded from salinisation.

One of the many definitions of land degradation is given in the UN Convention to Combat Desertification and Land Degradation (UNCCD), often referred to as the 'desertification convention' (Okin 2002):

> [*land degradation*] is the reduction or loss of the biological and economic productivity and complexity of terrestrial ecosystems, including soils, vegetation, other biota, and the ecological, biogeochemical and hydrological processes that operate therein, in arid and semi-arid lands, resulting from various factors including climatic variations and human activities.

There is still little reliable information on land degradation globally. A map of human-induced land degradation was made in the late 1980s by Oldeman *et al.* (1990) at the International Soil Reference and Information Centre (ISRIC) as part of the Global Assessment of Soil Degradation (GLASOD) project. It was based on surveys amongst national experts on type (water erosion, wind erosion, chemical deterioration, physical deterioration and so on), degree (light, moderate, strong or extreme), relative extent and causative factors of degradation. The latter comprised deforestation, overgrazing, agricultural activities (improper agricultural management), overexploitation of vegetation (cutting for fuelwood and so on) and industrial activities (pollution). The GLASOD-analysis indicated that almost 2 Gha or 15 percent of the land surface was degraded, of which four-fifths was due to erosion.[4] There are only a few other data sources on land degradation. One study estimates the area affected by water and soil erosion at 1,64 Gha or 12 percent of the land area (Leemans and Kleidon 2002). The Millennium Ecosystem Assessment (MEA 2005) gives similar qualitative and quantitative estimates in its investigation of ecosystem

[4] Land area is indicated in million hectares (Mha) or billion hectares (Gha). Sometimes km^2 are used, 1 km^2 = 100 ha.

services. In 2000, a FAO-based assessment of human-induced degradation found severe degradation from deforestation, agriculture and overgrazing in countries such as Iceland, Cameroon, Ethiopia, Nigeria and Turkey (20 to 30 percent) and very severe in countries like Costa Rica (68 percent) (Lambin and Geist 2006).

In 2008, ISRIC published an updated 'proxy global assessment of land degradation' (Bai *et al.* 2008). In the new assessment, degradation is (re)defined as a long-term decline in ecosystem function and productivity. The so-called normalized difference vegetation index (NDVI) is taken as a proxy-indicator of ecosystem function and productivity, which is based on satellite observations of net primary production (NPP) corrected for changes in rain-use efficiency (ratio of NPP to precipitation). Trends in the NDVI map over time are an indication of land degradation or improvement, but can unfortunately not tell anything about the kind of change happening. The results show that 24 percent of the global land area had suffered from a declining rain-use efficiency adjusted NDVI during the period 1981–2003 (Figure 9.8). Severely affected areas are found in Africa south of the equator, Southeast Asia, Australia, the South American pampas and the high-latitude forests in Canada and Siberia. An estimated 24 percent of the world population, or more than 1.5 billion people, are affected. If overlaid with land use/cover maps, 19 percent of degradation in this twenty-two-year time period took place on croplands and 43 percent on forested lands. More rigorous analysis is needed to understand the relationships with, for instance, population density and poverty.

The results differ substantially from the earlier GLASOD estimate, because the latter did not distinguish between historical and current rates of land degradation.[5] Historically, degraded lands have mostly become stable landscapes with low productivity. In the new estimate of changes between 1981 and 2003, they hardly contribute to the rate of degradation. In other words, the earlier legacy of humanity is not recorded in the new data. It also explains why notoriously degraded areas in the Nile Delta and the Iraqi marshes, amongst others, do not show up and why current degradation is not primarily a dryland issue. The primary productivity-based NDVI approach has some advantage over the earlier GLASOD approach, which is based on expert opinion. It is globally consistent, repeatable, quantitative and measures actual degradation. There are, however, problems of interpretation because other effects cannot easily be separated from the data. A mix of both approaches, with a focus on abandoned farmlands, may be the best strategy for the future.

Can we now answer the question whether land degradation is caused by humans? Undoubtedly, erosion processes can be strongly modified – and usually accelerated – by human activities such as forest clearing, agriculture (ploughing, irrigation), pastoralism (grazing), dam and road construction, surface mining and urbanisation. Erosion of arable land is a serious issue. In some places, erosion rates are significantly above the natural 'undisturbed' rates, which can cause accelerated river sedimentation and hill denudation. The rather scarce field experiments confirm that intense rainfall storms and large-scale land clearing (bulldozing) are the largest contributors. According to the aforementioned GLASOD inventory, the partly overlapping

[5] The earlier GLASOD assessment was 'a map based on perceptions, not a measure of land degradation; its qualitative judgments... have proven inconsistent and hardly reproducible, relationships between land degradation and policy-pertinent criteria were unverified... although, to be fair, its authors were the first to point out the limitations' (Bai *et al.* 2008:223).

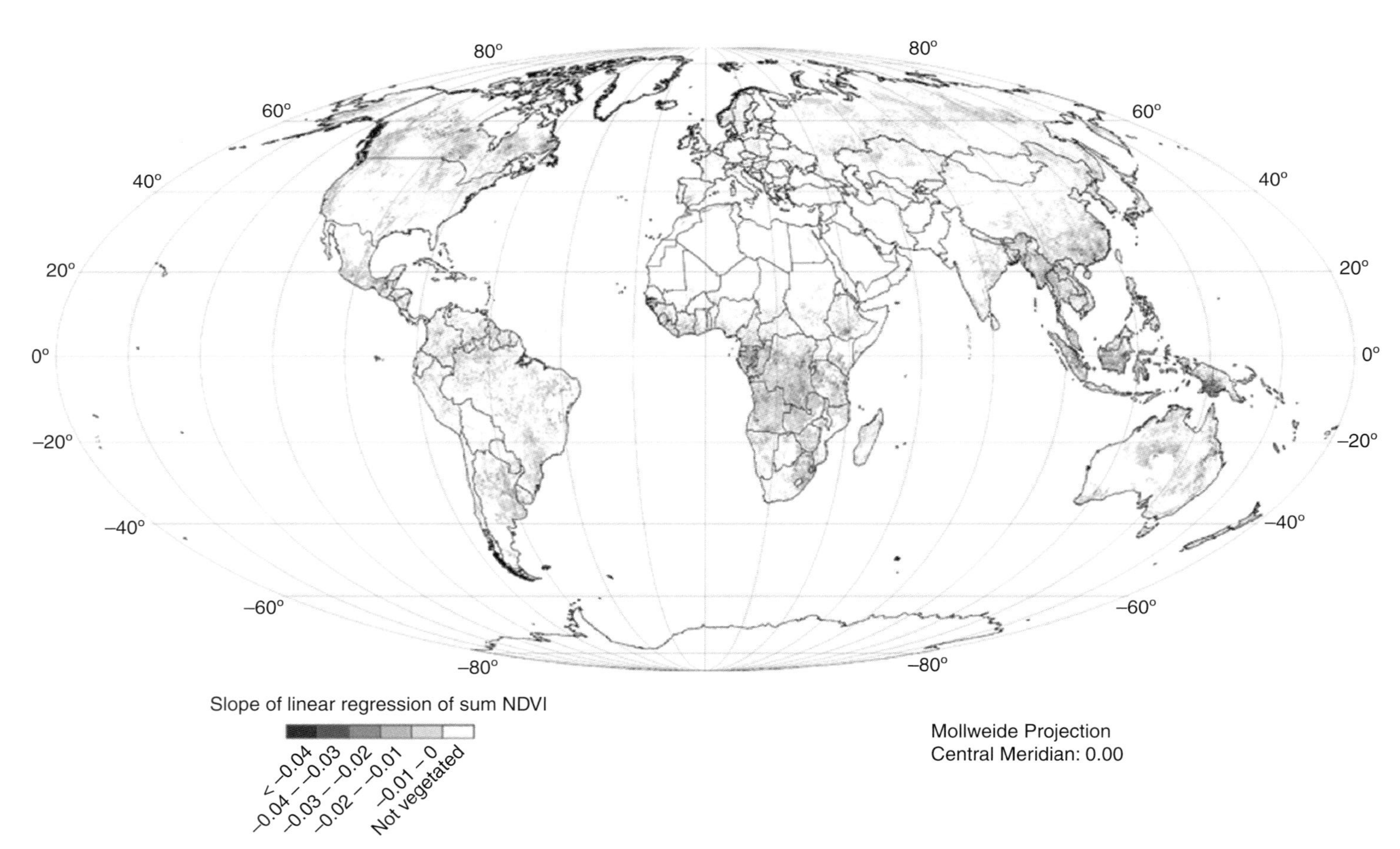

Figure 9.8. Global negative trend in rain-use efficiency-adjusted normalised difference vegetation index, 1981–2003 (source of map: Bai et al. 2008). (See color plate.)

Box 9.2. *USLE, erosion and worldviews.* Scientific knowledge of universal processes is instrumental in understanding the processes that concern us from a sustainability perspective. It is instructive to do simple erosion modelling exercises with software packages such as NetLogo 4.1 or PCRaster, in order to get a feel for process dynamics. Some of the suggested websites at the end of the chapter have interactive models that use basic equations parameterised for specific situations (Wainwright and Mulligen (2004)). Erosion models often use the Universal Soil Loss Equation (USLE), because it is a simple and widely used way to assess erosion processes.

Ecological historians like Grove and Rackham are critical of the use of the USLE-equation: 'The "Universal Soil Loss Equation"... is not really an equation, but an empirical relationship between observed rates of erosion and various parameters of environment and land use... Collection of data began in the 1920s and led to a number of regional soil loss "equations", which were combined in the USLE. The data that generated that Equation were gathered east of the Rocky Mountains... the term "universal" means universal within the United States... Empirical models will work only within the range of conditions over which the original data were collected... The Mediterranean lies outside the range of climates, soils, crops and cropping practices for which the USLE was constructed. North American soils were generally not cultivated until the 19th century, whereas Mediterranean soils have been cultivated and eroded for centuries... The USLE fails to predict rates of sheet and rill erosion in the Mediterranean. It is even worse at predicting total erosion, which it was never meant to do... '. (Grove and Rackham (2001).

Environmental modellers often have a different and more practical view: '... erosion prediction is still dominated by the use of one model – the USLE. It provides the user with a simple model which can be executed using nothing more sophisticated than a pocket calculator, yet in model comparison studies predicts annual erosion as well as any model. This may not reflect any superiority in terms of the model, but simply the fact that a five-parameter empirical model can do as well as a multi-parameter model in capturing the extremely variable response of a hillside system' (Quinton, in Wainwright and Mulligan (2004) 192).

activities of overgrazing (35 percent), agriculture (28 percent) and deforestation (30 percent) are the major causes. Such results have to be interpreted with caution, however, because of inconsistencies, limited reliability and reproducibility and divergent class definitions and interpretations. Besides, there is a difficult to decipher the legacy of the past. Some experts believe that published data and maps grossly overestimate the extent of land degradation and that awaiting better observations in combination with clearer definitions is needed.

Indeed, controversy about the causes and the consequences of desertification has increased in the last decades (van der Leeuw 1998; Grove and Rackham 2001; Reynolds and Stafford Smith 2002, Bai *et al.* 2008). Some areas have been impacted by humans for millennia and humans and their animal herds are as much part of the resulting social-ecological system as other species. In a long-term and larger-scale perspective, events are also more ambiguous in their effects. Eroded hills are the remnants of sedimentation processes that created fertile valleys. Damming and diverting large rivers such as the Nile, Yellow and Colorado significantly reduces the sediment flows and thus block the nutrient inflow in the estuaries (MEA 2005). Tectonic uplift and climate change processes are occurring slowly and over long periods, and human populations may too easily and incorrectly be blamed for erosion of (top)soil, particularly in fragile areas such as the Mediterranean. Landslides and the various forces of water are sometimes neglected or misinterpreted. The role of trees may be overestimated, the role of other vegetation underestimated. 'Ruined landscape theory is [thus] deeply pessimistic. If it is true, soils are unconservable except by growing trees. All cultivation – even orchards – is ultimately self-destructive; farmers can survive only by going out of business...' (Grove and Rackham 2001).

So what about sustainability? The degradation processes occur at rather large time-scales and are partially or wholly irreversible. For instance, organic matter decomposition happens over five to twenty years, accelerated erosion over ten to fifty years, and accelerated mineral weathering over centuries (Leemans and Kleidon 2002). This is often outside the time span considered in human resource management and, therefore, often pernicious – remember the shift in risk spectrum as part of societal collapse (§3.5). Land overexploitation and degradation should, therefore, be viewed as part of the long-term evolution of landscapes and sustainable management of soil vegetation systems as context-dependent. But the basic sustainability principle remains: the (top)soil removal rate should not exceed the rate at which topsoil accumulates from natural processes.

9.5 Ecosystem Dynamics: Population Ecology

9.5.1 Population Ecology: Logistic Growth

As you may recall from §2.3, exponential growth can never go on indefinitely. At some point, growth constraining/stabilising forces start to work. This can be introduced in a model by making the growth rate a function of the available resource (Edelstein-Keshet 1988; Case 2000). It can be done in several ways, but, historically, the first and most simple method stands out. In the 19th century, Verhulst had observed processes of population growth of organisms under limiting environmental constraints and proposed to describe these with an equation of the form:

$$dX/dt = \alpha X\,(1 - X/K)\ (\mathrm{a} > 0) \tag{9.1a}$$

with X the population size (§2.4).[6] This logistic growth equation can be rewritten in the form of the population growth rate:

$$\frac{1}{X}\frac{dX}{dt} = \alpha\left(1 - \frac{X}{K}\right) \tag{9.1b}$$

This shows that the growth rate of the population declines with population size.

The shape of the population over time is an S-curve. The reference behaviour is initially fast growth, then a smooth transition towards the attractor or equilibrium

[6] See Appendix 2.1 for mathematical details.

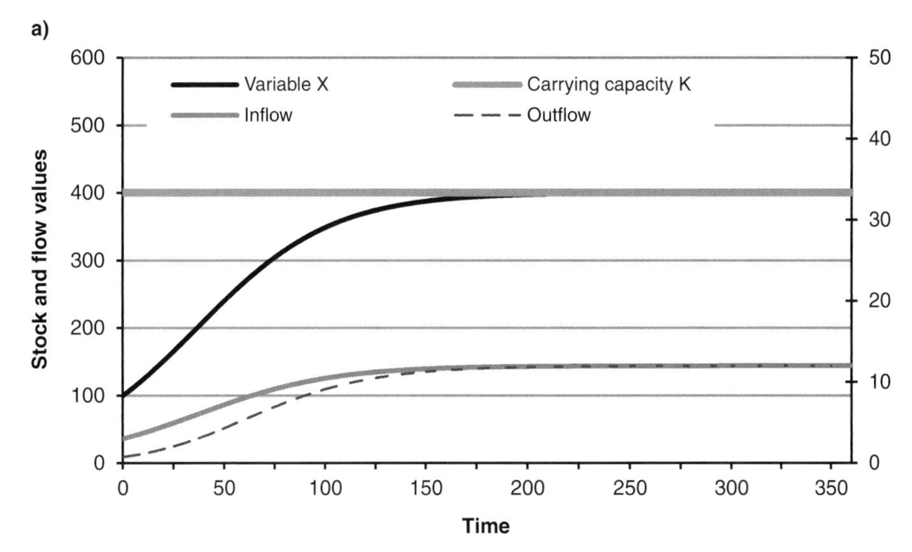

Figure 9.9a. Characteristic logistic growth path of the state variable Population (1), approaching the carrying capacity value (2) of 400 (population) units after about 200 timesteps ((α = 0.03, $\Delta t = 0.25$, Runge-Kutta4). The lower curves indicate the in- and outflux (3,4), with the influx initially larger than the outflux.

value K (Figures 2.10 and 9.9a). It is easily seen that for $X << \frac{1}{2}K$ the equation becomes the one for exponential growth ($dX/dt \sim \alpha X$) and that for $X >> \frac{1}{2}K$, it approaches towards an attractor of value K ($dX/dt \sim \alpha(X\text{-}K)$). Setting $dX/dt = 0$ shows that there is one equilibrium state at $X = K$ (saturation) and one at $X = 0$ (extinction). At $X = K$, the system is in dynamic equilibrium: outflow equals inflow. K is associated with the *carrying capacity* because it is the maximum value to which the population size X ($0 < X < K$) can grow in the specific environment.

The logistic growth equation or 'model' is a combination of a positive and a negative feedback loop. It is used widely to describe growth processes, which are confronted with external limits to growth. These limits constrain further growth and can be food scarcity, density-related diseases, overcrowding and so on. A well-known empirical case is bacterial growth in a resource-limited environment. In ecology, a broader interpretation is that it describes the succession dynamics from a simple pioneer to a mature climax ecosystem. In system dynamics, it is known as the *Limits-to-Growth* archetype (§2.4).

In our simulated model world, birth and death are deterministic processes: At a given population size, they are exactly a particular fraction of the population size. In the real world, such processes are often *discrete* events – a mother gives birth to one or two children but not to 1.5 child, and there is no law regulating its frequency. Only the assumption of large numbers justifies the approximation. The discrete character of (birth and death) events can be addressed by formulating the logistic equation in discrete form.[7] It gives some surprising insights (May 1976).

[7] See §2.3. Applying simulation software to explore the model behaviour over time is always done with a discrete timestep. It is important to test results for the sensitivity for timestep and integration method. Seppelt and Richter (2005) investigated an ecosystem model constructed with different modelling tools such as STELLA® and MatLab. They found it hard to generate the same outcome with different software tools and warn for the differences from uncertainty in integration methods.

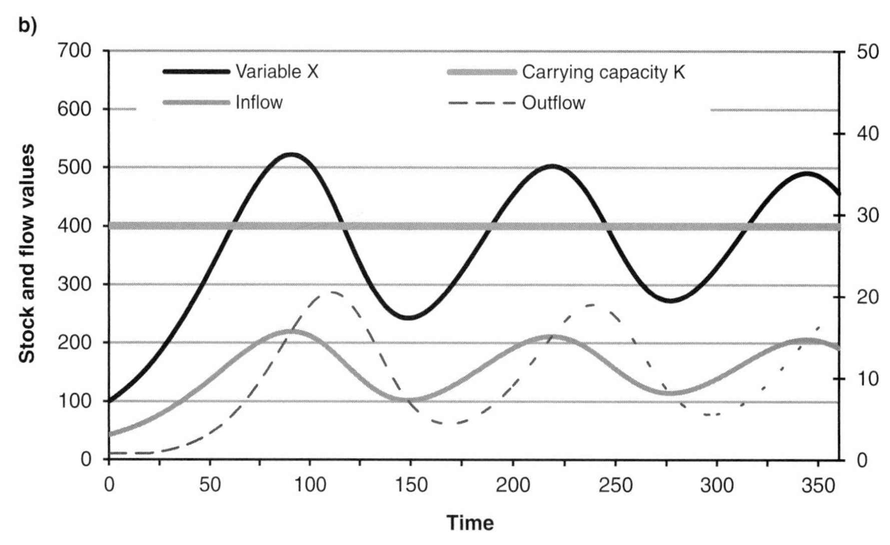

Figure 9.9b. Logistic growth in an oscillatory dampening mode ($\alpha = 0.03$, $\Delta t = 0.25$, Runge-Kutta4, delay period $= 20\ \Delta t$).

Animals, and often people too, probably do not anticipate that the environment has a maximum amount of food it can provide. The forces behind the (exponential) growth process do not 'see' the environmental constraint posed by the carrying capacity, because the relevant information is not available and there cannot be an immediate feedback in time and space. Such a delay as consequence of weak/incomplete signals and absence of anticipation cause 'overshoot': The population grows above the level that can naturally be sustained. The process can be simulated by adjusting the equation in such a way that the outflow (death rate, the negative term in equation 9.1a) responds to a change in the population size with a delay of one or more timesteps. The result of a twenty-period delay is shown in Figure 9.9b. Instead of a smooth approach towards the carrying capacity, the system gets in an oscillatory 'overshoot-and-collapse' mode.

In real-world processes, events are not only discrete but also *stochastic*. Birth and death rates, as well as all kinds of phenomena in the environment, show up as statistical variations in experimental data on population sizes. The stochasticity – that is, randomness or noise – originates in processes that are not considered explicitly in the model but influence the dynamics, such as fluctuating precipitation, an extremely hot summer, a sudden disease or invasion, a social revolt, novel inventions and so on (Turchin 2003). For instance, rain variability is amongst the factors that make erosion a complex process and disease epidemics can cause erratic death rates in populations.

A more general growth equation can be expressed as $dX/dt = g(X) \cdot X$. The function $g(X)$ is called the intrinsic growth rate. Its logistic growth form $g(X) = \alpha(1–X/K)$ can be extended to, in first instance:

$$dX/dt = g(X) \cdot X = (\alpha_0 + \alpha_1 X + \alpha_2 X^2) \cdot X \qquad (9.1c)$$

The growth rate is now a parabolic function of the population size and has for $\alpha_1 > 0$ and $\alpha_2 < 0$ a maximum at some population size X_{maxgr}. The birth rate is now proportional to population size at very low population levels, unlike in the logistic growth (equation 9.1b). A biological explanation of this so-called Allee effect, that has empirically been observed, is that mating is more difficult at low population density.[8] Many other growth functions have been proposed for a variety of ecological and biological processes (Case 2000).

As nice as the notion of carrying capacity may be from a mathematical point of view, its real-world interpretation is not without problems. Having studied the millennial history of social-ecological systems in the Mediterranean, Grove and Rackham (2001) conclude that carrying capacity is not a precise measurement: 'Is [carrying capacity] to be based on an average year, or the worst year in ten? Or a hundred? Are the animals attended and moved around, or can they go where they will? Even in a stable environment the ecological carrying capacity... is typically at least 50% more than the economic carrying capacity.' A more dynamic perspective on carrying capacity, in niche construction theory, is discussed in an evolutionary context in Chapter 10. The next section discusses an obvious refinement: interacting populations of more than one species.

9.5.2 Population Models: Prey-Predator Dynamics[9]

The logistic growth equation describes the dynamics of a single population, such as bacteria, algae, deer, lynxes – or humans. However, populations never live in isolation, so what happens if more than one species is introduced in the model? This has been a research area for more than a century in ecology and biology. It had already been noticed that animal populations sometimes show cyclical behaviour. For instance, there are records dating back to the 1840s kept by the Hudson Bay Company in Canada and their trade in pelts of the snowshoe hare and its predator the lynx reveals that the relative abundance of the two species undergoes dramatic cycles. Also, other 'ecohistories' show the importance of species interactions, for instance, the Kaibab and Pribiloff Island narratives.

In ecology, three fundamental *types of species interaction* are posited: antagonistic interactions between two competitors; consumer-resource interactions between a predator and its prey; and indirect cooperative interactions or mutualism in chains of species (Roos and Persson 2005). The following discussion focuses on the best known one: *predator-prey*. For one predator population of Y and one prey population of X individuals, you get the classical formulation given by Lotka and Volterra in the 1920s as a set of two linear differential equations:

$$dX/dt = aX - bYX = X(a - bY)$$
$$dY/dt = -cY + dXY = Y(dX - c) \quad (9.3)$$

The equations reflect two common-sense interpretations (Figure 9.10):

- because the prey is food for the predator (X in dXY), it is part of the carrying capacity of the predator; and
- more predators increase the prey death rate (Y in –bYX).

[8] There may also be a maximum growth rate for human populations, when crowding causes the net population growth to decline due to emigration and wars or a declining desire for children (§8.4).

[9] See Appendix 9.1 for some mathematical details.

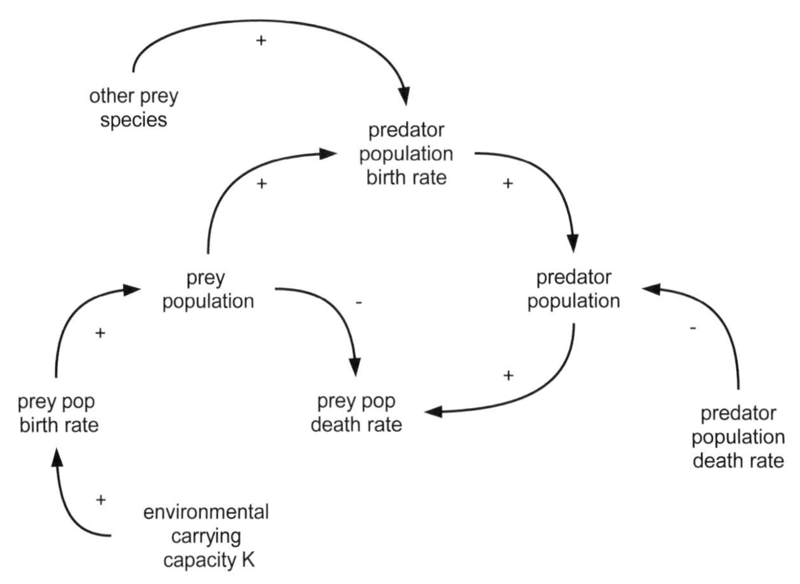

Figure 9.10. The interaction between a species and its environment and between two species, represented in a causal loop diagram.

Note that the equations represent exponential growth with a net growth rate dependent on the population of the other species (equation 9.1b). By setting $dX/dt = 0$ and $dY/dt = 0$, one finds the attractors or equilibrium states, meaning the state in which both populations do not change. One stable state is the trivial (0,0) and the other is (c/d,a/b). The coefficients a, b, c and d make up the so-called ecological *community matrix*.[10]

A typical pattern over time of a system described by the Lotka-Volterra equations is shown in Figure 9.11a. When there is prey abundance, the predator population increases. The prey population starts declining due to predator abundance, after which predator population will go down again. The populations will oscillate over time, but these oscillations are rather ephemeral because the actual path is sensitive to the initial populations, as is seen in the phase diagram with trajectories for three different initial predator populations (Figure 9.11b). The system is neutrally stable: a delicate balance between stability and instability. If prey is harvested, the oscillations are dampenend or even disappear altogether. Invasion of prey causes the oscillations to be amplified (Figure 9.11c,d). Invasion of predators also stabilises the system, but harvesting them tends to destabilise it. In other words, population sizes easily crash or explode under small perturbations. This is one reason why you should not take the model too seriously. As with the logistic equation, it is a stylised description of a phenomenon, but it does convey the key point that species interaction can cause oscillations in population sizes.

In the course of the years, numerous multi-population models have been proposed and analysed – from simple refinements of the Lotka-Volterra model to ones

[10] The model has actually been borrowed from chemical kinetics and provides in turn a metaphor for more complex processes, for example, in economic and social systems – remember the arms race model.

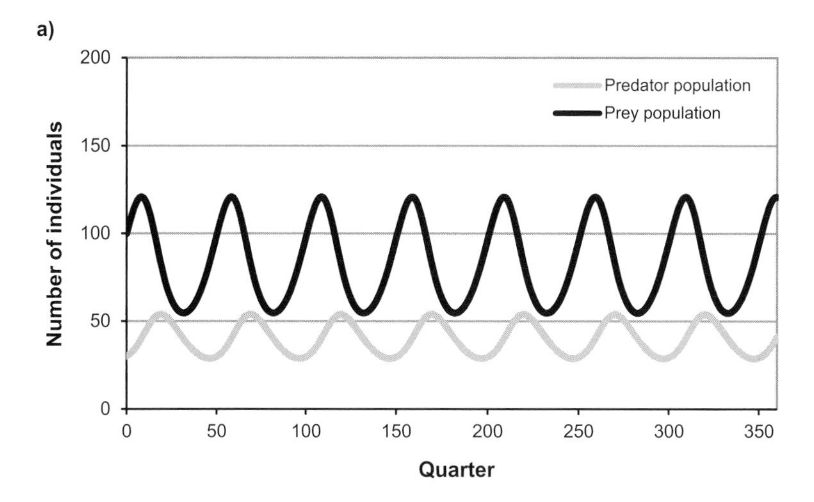

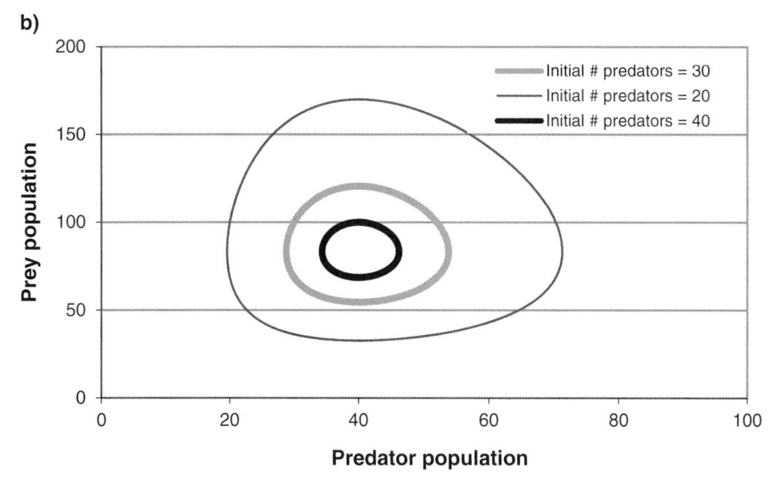

Figure 9.11a,b. Simulated populations of prey (1) and predator (2) for the system description of equation 7.5b ($a = 0{,}004$; $b = 0{,}16$; $d = 0{,}1$; $k = 0{,}3$; $\Delta t = 0.25$, Runge-Kutta4). The populations oscillate permanently around the equilibrium values (3,4).

with many species and classes. One extension is to introduce a finite carrying capacity for the prey by replacing the term aX in equation 9.3 by aX(1–X/K). This changes the steady state from a neutral stability to a stable spiral. If prey is harvested, the system remains in a stable oscillation within the attractor basin for not too large disturbances. Further extensions are, for instance, to add vegetation as food for the prey and to introduce competition between species for the same resource. Such models have been examined in great detail (Case 2000). In some cases, they provide robust predictions, as in the case of the arctic lemming (Gilg *et al.* 2003). It is one of the simplest real-world ecosystems, with one prey (the lemming) and four predators. Four years of empirical data are nicely reproduced with a straightforward predator-prey model. Interestingly, no finite carrying capacity for the prey has to be introduced – stabilisation comes from predator interaction. Nevertheless, the models are criticised by empirically oriented ecologists because the relationships are often

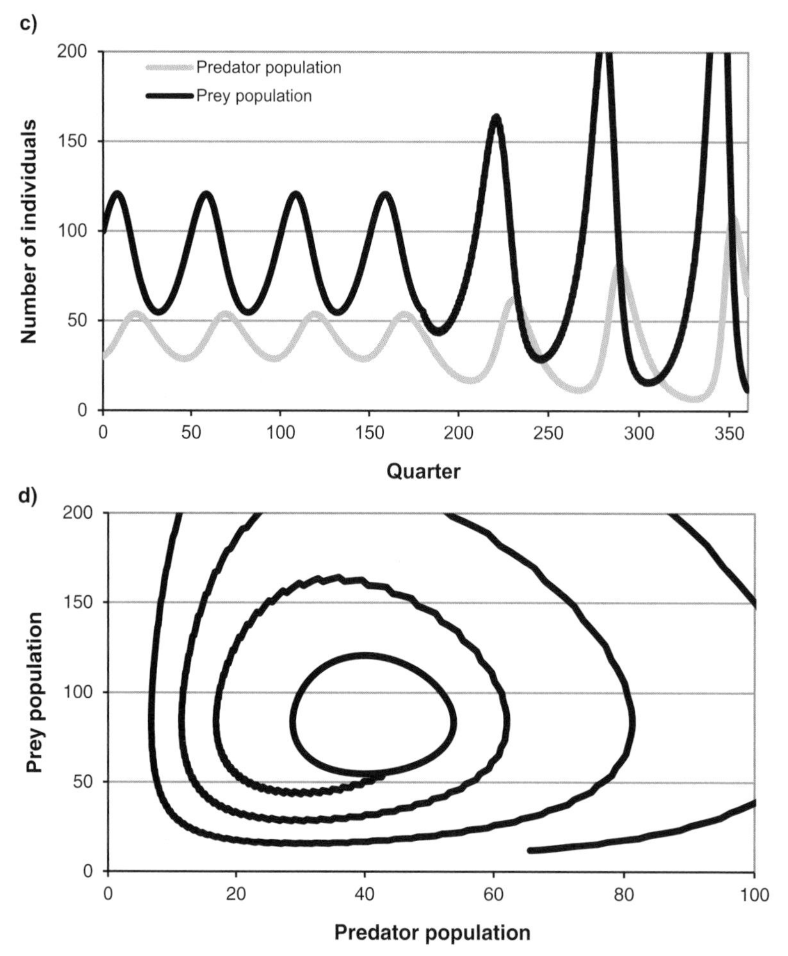

Figure 9.11c,d. Simulated populations of prey (1) and predator (2) for the system description of eqn. 7.5b (a = 0,004; b = 0,16; d = 0,1; k = 0,3; $\Delta t = 0.25$, Runge-Kutta4). The populations oscillate permanently around the equilibrium values (3,4) but switch to a new trajectory after a permanent disturbance (harvesting 3 prey, that is, 3 percent) is introduced in quarter 180.

weakly linked to field observations. It turns out that oscillations in real-world ecological communities are more exception than rule because of other interactions, such as positive feedbacks, and because evolutionary processes are omitted in population models.

In this respect, an interesting modelling exercise has been done by Fay and Greeff (2006). They built, starting from empirical data on lion, wildebeest and zebra populations in Kruger Park in the period 1968–1994, a stepwise more comprehensive three-species population model. In each step, they try to reproduce statistical time-series with the model. The first one is the simplest predator-prey set of equations (equation 9.2). Next, mutualism between wildebeest and zebra is introduced; it turns out that historical data are badly reproduced. It becomes hardly better if a carrying capacity is introduced for the prey species on the assumption that even with predators there may be overpopulation. The resulting spiral point attractor gives no good fit. In a subsequent step, a limiting factor to predation is introduced to reflect that only

Box 9.3. *The Kaibab narrative: management on ill-understood systems.* Prior to 1907, the deer herd on the Kaibab plateau, which consists of some 727,000 acres and is on the north side of the Grand Canyon in Arizona, numbered about 4,000. In 1907, a bounty was placed on cougars, wolves and coyotes – all natural predators of deer. Within fifteen to twenty years, the larger part of these predators (over 8,000) was wiped out and a consequent and immediate irruption of the deer population followed (Figure 9.12). By 1918, the deer population had increased more than tenfold. The evident overbrowsing of the area brought the first of a series of warnings by competent investigators, none of which produced a much-needed quick change in either the bounty policy or the policy dealing with deer removal. In the absence of predation by its natural predators (such as cougars, wolves and coyotes) or by man as a hunter, the herd reached 100,000 in 1924; in the absence of sufficient food, 60 percent of the herd died off in two successive winters. By then, the girdling of so much of the vegetation through browsing precluded recovery of the food reserve to such an extent that subsequent die-off and reduced natality yielded a population about half that which could theoretically have previously maintained (Roberts *et al.* (1983); Meadows (1986)).

In the article *The Rise and Fall of a Reindeer Herd*, Scheffer (1951) tells a similar story about forty reindeer placed in 1911 on the two Pribiloff Islands, St. Paul and St. George, in Alaska to provide the native residents with a sustained source of fresh meat. There were no predators except humans. The herd on St. George grew quickly to 222 animals, after which it declined to a stable level of forty to fifty animals. The herd on the larger St. Paul increased to more than 2,000 animals by the late 1930s, after which it collapsed to only eight animals in 1950. It turned out that the reindeer survived in winter by eating certain shrublike lichen. When the herd size increased, the reindeer consumed the lichen faster than their regeneration rate and it needed only one severe winter to exterminate most animals on St. Paul.

hungry lions kill. This dampens the fluctuations in population sizes but still leaves some historical trends unexplained. Finally, two more assumptions were put into the formal model: wildebeest calving in a six-week spring period instead of spread out over the year, and the 'cropping of lions' in the period 1976–1984. This gives a satisfactory resemblance between the simulated and historical data, considering the errors in observations. Only after including historical contingencies, such as a human intervention, did they get it right. This illustrates the necessity to combine quantitative, mathematical models with qualitative, historical case studies.

An objection against the general population models is the root *assumption of homogeneity in space and behaviour.* The isomorphism with the kinetics of chemical reactions suggests that ecological interactions are basically an encounter of two or more individuals. However, the interaction mechanisms are usually much more complex and are part of a whole food web with space- and scale-dependent patterns. The assumption of identical individuals is also unrealistic, in view of the consequences of different life histories of individuals and the variance in their traits. This is not unimportant, because if the models of ecosystem dynamics suggest generality but

a)

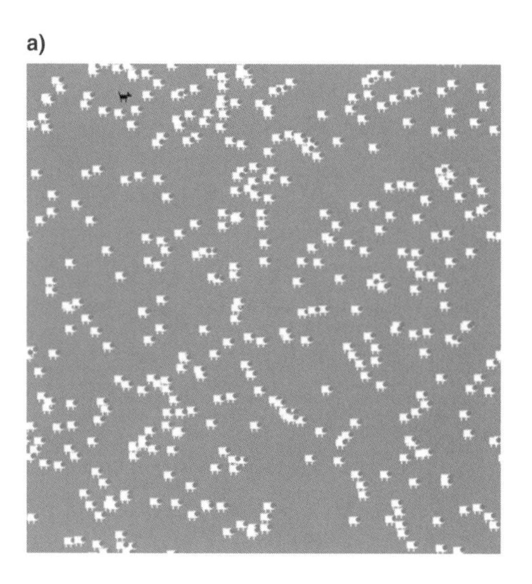

b)

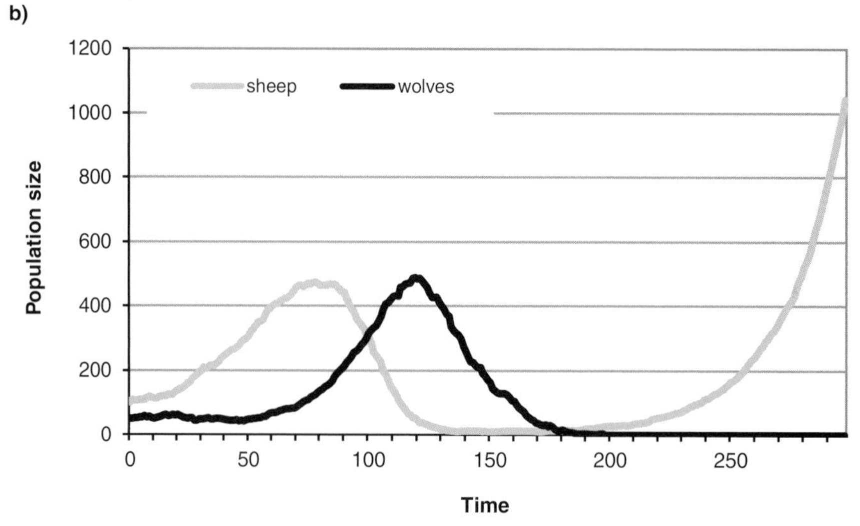

Figure 9.12a,b. Spatial dynamics of a simple, CA-based wolf-sheep predation model, with grass 'off'. It is almost impossible to find a limit cycle path. The lower left graph shows the population trajectories (NetLogo 4.1).

fail in particular situations, ecosystem preservation and restoration policies may be inadequate or even catastrophically wrong. Again, there is the tension between the universal claims of theory and the contingency of practise (§6.3).

A logical next improvement is, therefore, to include spatial aspects and behavioural rules, in combination with disaggregation into age, gender and other classes. Introducing behavioural variety has been pioneered by evolutionary game theorists like Maynard Smith (1982) and elaborated by Nowak and colleagues (2006). It is investigated with cellular automata (CA) algorithms and GIS-based software packages (§10.4; Engelen *et al.* 1993) and with agent-based models (§10.5; Janssen 2002). Some aspects of real-world complexity are introduced in this way and point at the possibly stabilising role of spatial heterogeneity and behavioural variety.

c)

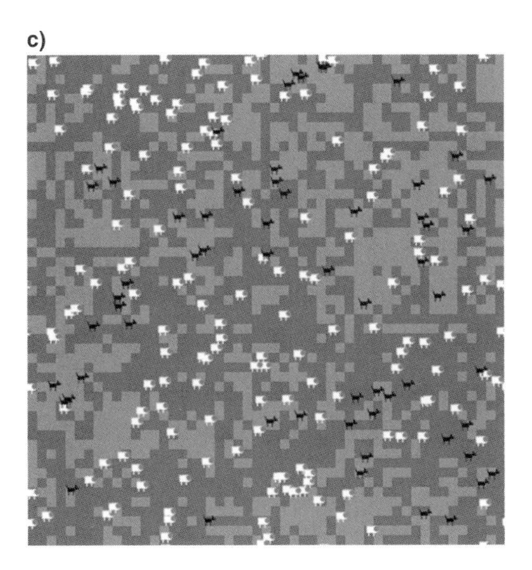

d)

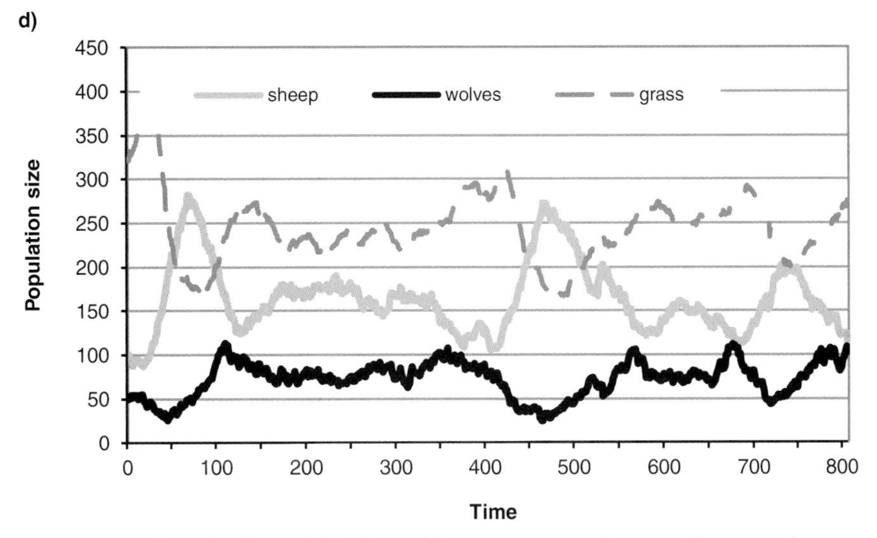

Figure 9.12c,d. As in Figure 9.12a, but with grass regenerating at a 30 percent/year rate (grass 'on'). The lower left graph shows the population trajectories (NetLogo 4.1).

The discussion about the stabilising feedbacks in predator-prey systems is nicely illustrated with a spatially explicit simulation model. The predator (wolf) and prey (sheep) move in space according to elementary rules such as 'wolf-eats-sheep if they happen to move into the same spot'.[11] Figure 9.12a,b shows the square area populated by wolves chasing and eating sheep and the population sizes over time. It is a spatial representation of the predator-prey model (equation 9.2). The simulation ends for almost any parameterisation in an unrealistic situation: either both wolves and sheep extinct or sheep population growing unchecked to infinity. Do you see the last wolf and the many sheep that still do not know their fate? The reason for

[11] The model can be downloaded from the NetLogo website as a demo model under the name Predator-Prey.

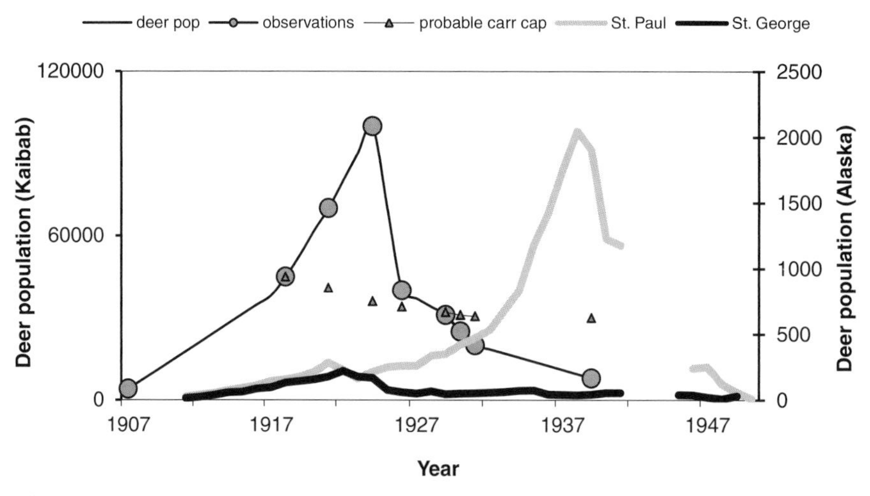

Figure 9.13. The deer population on the Kaibab plateau population and the reindeer population on two Alaskan islands. An interactive simulation model is available on the site forio.com/simulate/e.pruyt/ex4–6-the-kaibab-case-2/model/explore/ to explore the dynamics.

this system behaviour is clear: The sheep are not confronted with resource scarcity. If grass is introduced into the model and it is allowed to regenerate towards its carrying capacity (equation 9.1), the system exhibits oscillations in all of its three state variables (wolves, sheep and grass) (Figure 9.12c,d). It is now a stable oscillation, in agreement with the previous analysis. Thus, an elegant but simple mathematical description may be faulty because of its artificial system boundary; in this case, the lack of food for the sheep.

9.6 Food Webs: The Stocks and Flows in Ecosystems

9.6.1 Food Webs and Their Representations

A first step to deal with ecological complexity is setting up a good classification system. The Swedish botanist Linnaeus was given the name Father of Taxonomy for his *Systema Naturae* (1735) and *Species Plantarum* (1753), a plant classification system. It is still widely used in botany. In subsequent centuries, the static classification was extended with a dynamic representation of trophic systems and food chains. In ecology, populations *per se* are understood on the basis of individuals with characteristics and interactions with individuals of the same and other populations, as members of an ecological *community*. Plants and animals, constituting the elements of the system, are linked to each other through eating and being eaten, and, therefore, several forms of ecological succession processes can happen. Such interactions – predator-prey as well as competition, mutualism and parasitism – give the community a dynamic structure and cause emergent properties which make it a 'sum more than its parts'. They form a food chain, in which producer-consumer-decomposer functions and relationships determine trophic levels. Currently, most ecologists prefer to speak of *food webs*. If one also includes the physical environment, the system is called an *ecosystem*. This is a core concept in ecology.

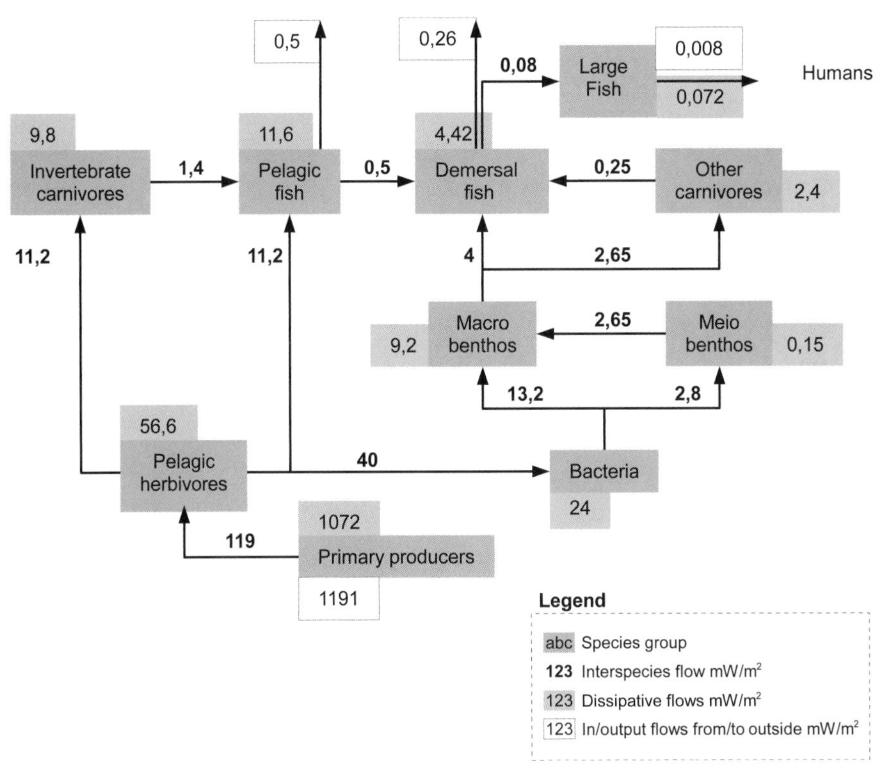

Figure 9.14. Estimated flows in mW/m^2 through the food web of the North Sea around 1970 (source of data: Ulanowicz 1986:50, after Steele 1974).

For a given classification, the species populations and their interactions can be sketched in a box-and-arrow scheme or graph.[12] The populations in a food web sustain energy and material flows of carbon (C), nitrogen (N) and phosphorus (P) amongst others through the system. The basis is formed by biomass from sunlight via photosynthesis. Every component of the web uses part of the energy flow to sustain itself. How much energy from the sun can be used depends on the location, soil, climate and so on. A common measure for biomass via photosynthesis is net primary production (NPP), usually expressed in ton per hectare (t/ha).

The actual measurement of the mass and energy flows making up a food web is far from simple. Odum and colleagues (1976) have formulated and applied principles of energetics in theoretical/systems ecology for the description of food webs. Many ecosystems have been conceptualised and empirically represented in terms of energy and mass stocks and flows – with many results of food web analyses having been published. Figure 9.14 gives an example of the food web in the North Sea around 1970. The boxes indicate the species. The linkages represent species interactions expressed as energy flows. The total system throughput, or the sum of all the

[12] Equivalently, they can be represented in an *input-output (I-O) matrix*, with the species classes arranged in a square matrix and the coefficients indicating linkage strength. The I-O formalism is widely applied in economics since Leontief proposed it in the 1940s and is used also in ecological phenomenology (Ulanowicz 1986; §14.3).

system flows, is 2.6 W/m^2. There are several things to note from such a food web analysis:

- the system is driven by the solar energy inflow, but only a tiny fraction of it (1.2 W/m^2) is used (solar radiation is in the order of an average 100 W/m^2);
- the larger part of this energy flow is already dissipated, like that used for metabolism, at the lowest level of the primary producers; and
- human preference for large fish taps into a very small energy flow at the very top.

A food web representation suggests also a clear difference between *growth*, which is an increase in the total system throughput, and *development*, which is a rise in the average mutual information of the network flow structure.

9.6.2 Stability and Resilience

Since the concept of food webs gained prominence in the 1970s, researchers have investigated the relationship between complexity and stability in ecosystem models. Connectivity and stability, it turns out, are not simply linked (May 1976; Pimm 1979). More realistic descriptions than a set of coupled differential equations are needed. Community complexity does apparently buffer against perturbations, when environmental heterogeneity creates subsystems (compartments) that permit higher-level species to do adaptive foraging and in that way stabilise the system. Introducing spatial and temporal variation and heterogeneity amongst the individuals in population models, which has become possible with advances in computational speed and methods, gives new insights in the stability, resilience and persistence of biotic communities. The application of network theory is also generating new insights into the structure and evolution of ecosystems (§10.4).

Perhaps you remember a chemistry experiment from high school: You dissolve a salt in hot water, let the solution cool down, put a small crystal in the now oversaturated solution – then all of a sudden crystals form. Or you may have read about landslides in countries such as Brazil or the Philippines where it rains for a long time and, suddenly, a large mass of earth starts to slide and whole villages are covered under mud. These phenomena are nonlinear, which means that the effect of a change is not proportional to the change but, at a certain point, superproportional or catastrophic. The system jumps from one equilibrium to another one; for example, from an oversaturated solution to crystals or from a stable layer of soil to a huge mass of earth down in the valley. Usually, positive or reinforcing feedback loops are involved. The word *tipping point* has become popular to describe such a sudden change. It's not evolution, it's revolution. It has to do with stability, resilience and multiple equilibria.

What, then, is (eco)system stability? Already in the mid-1990s, there were no less than 163 definitions of 70 different stability concepts in ecological literature (Grimm and Wissel 1997). In general,

> *stability* of a system is the condition that it a) stays essentially unchanged (constancy); b) persists through time (persistence); and c) returns to the reference state (or dynamic) after a temporary disturbance (reversibility).

The property of constancy is most simple but begs the question: *What* in the system is constant? The other two properties refer to the ability of a system to absorb *disturbances* in its environment. A system is in a stable state when it develops forces or movements that *restore the original condition* upon being perturbed from a condition of static or dynamic equilibrium (Giampietro 2004). An indicator of stability is the degree to which a system is amplified upon perturbations or disturbances. A mass at the end of a spring or a pendulum in a rest position are both typical examples of a stable system. If you pull the spring or push the pendulum, it will tend to restore its initial stable state, which is no movement (Figure 7.6). Similarly, a ball in a bowl will tend to come at rest at the point where the potential and kinetic energy are smallest, which is on the bottom (§7.4). A system described by the predator-prey system of equation 9.2 is stable in certain parameter domains, in the sense that a disturbance will cause it to switch to a new oscillation but not make it explode or die out. As long as a system can be plausibly described by a set of differential equations, stability can mathematically be investigated (Appendix 2.1).

Many real-world systems do not behave according to the stylised linear models that were thought to be 'reasonable approximations' of real-world phenomena. Instead, many ecological as well as economic and social systems exhibit non-linear behaviour outside a certain domain. For example, if extreme rainfall causes a landslide or more intense hunting causes populations to collapse, the system is unstable. The apparently stable climate system may exhibit amplifying feedbacks, such as the mechanism that global warming accelerates the release of methane from permafrost areas, which in turn accelerates temperature rise. If a shortage of a commodity leads to rapid and large price fluctuations or one month of rioting brings down a government, the underlying system is in an unstable state. Such systems may have an attractor, which is itself an oscillation – a so-called limit cycle – or have multiple attractors and sometimes strange attractors.

To account for more complex system properties, the concept of *resilience* has been introduced (Holling 1973). Here, again, there are several definitions, reflecting the author's background. A widely used definition is:

> *resilience* is the ability to preserve identity under external perturbations or disturbances and to return to the original structure when the perturbation or disturbance is over.

A measure of the resilience of a system is the full range of perturbations over which the system can maintain itself. For instance, the maximum wind speed at which a tree branch can still remain unbroken by bending. Or the human body, viewed as a system trying to maintain body temperature around 37°C, has a rather large domain of environmental situations in which it can function. This condition has been called *homeostasis*.[13] The notion of identity makes the definition of resilience somewhat ambiguous and a well-defined system boundary is important. For example, clothing and heating equipment increase the domain of circumstances under which the human body is resilient provided they are included in a newly defined, larger system. Their provision, however, may lead to less resilience in the even larger system due to new

[13] The essence of the Gaia-theory is that System Earth is in homeostatis, such as it has its own internal feedback, which keep essential state variables within bounds (Schneider and Boston 1991).

dependencies, for instance, on wool and fossil fuels (§3.4). Think of the 200,000-inhabitant city of Yakutsk in Siberia, where −25°C is considered mild and people die if fuel supplies fail because of rivers being frozen longer than anticipated.

Historically, there has grown a distinction between engineering resilience and ecological resilience, one characteristic for the technosphere and the other for the ecosphere. 'One focuses on maintaining efficiency of function (engineering resilience); the other focuses on maintaining existence of function (ecological resilience). Those contrasts... can become alternative paradigms whose devotees reflect traditions of a discipline or of an attitude more than of a reality of nature' (Gunderson and Holling 2002).

Engineering resilience focuses on efficiency, control, constancy, and predictability – all attributes at the core of desires for fail-safe design and optimal performance.[14] Such a desire is appropriate for systems with low uncertainty in a largely controlled environment. It concentrates on stability near an equilibrium state and is measured as resistance to disturbance and speed of return to equilibrium. For instance, the capability of a strained entity to recover its original condition (such as size, shape or structural characteristics) after a deformation caused by stress can be considered as system resilience (Giampietro 2004).

Ecological resilience focuses on persistence, adaptiveness, variability, and unpredictability – attributes embraced from an evolutionary or developmental perspective. The emphasis is on the instabilities associated with multiple equilibria and far from equilibrium states. Indeed, it is better to think of basins of attraction or, broader even, regimes instead of (point) attractors. Resilience is now measured by the magnitude of disturbance that can be absorbed before the system undergoes a *structural* change, which is in a mathematical sense the shift from one attractor basin to another ('regime shift'). Consequently, ecologists consider the maximum amplitude of disturbance that still allows the system to return to the attractor basin as an indicator. This notion of resilience is also penetrating the social sciences, where it is associated with the capacity of institutions to adjust to perturbations (Scheffer 2009). The more resilient an institution, the less vulnerable it is to changes or perturbations from the outside world. This aspect has recently become prominent with regard to financial institutions and political regimes in an increasingly globalising and volatile world.

Engineering resilience is an inadequate concept for dynamic, evolving 'living' systems with high uncertainty, more than one attractor basin and a 'noisy' environment. Ecosystems are such systems, with a bewildering richness and complexity: 'Indeed, when considered in detail, many fluctuations in the populations of many animals and plants show erratic patterns that are likely to be the joint result of fluctuating external conditions and interacting intrinsic cycles in the ecosystem' (Scheffer *et al.* 2001). If drawn to the outer bounds of the attractor basin, there is an increasing probability to transit to another attractor, depending on the magnitude of the disturbance *and* on the underlying resilience of the current state. There is a loss of 'identity'. Nonlinear thresholds, random fluctuations in the environment and

[14] In a technology context, the German word *Fehlerfreundlichkeit* expresses this feature to indicate that machinery should be designed in such a way that (human) errors do not cause irreversible damage.

complex couplings across space and in time make it hard to predict their evolution. It is also possible that 'latent' energy is built up endogenously in the system, which is suddenly released.[15] Human interventions further complexify the situation.

From a management perspective, the question is also what is desirable – an issue at the heart of many sustainability debates. Often, humans intend to avoid dangerous catastrophic shifts or at least moderate their extent and speed in order to make adaptation more feasible. For instance, if the frequency of extreme floods or droughts because of climate change were to increase, this may in many ecosystems become a trigger for further and accelerated change. Risk analysis based on such models can bring some clarity in possible management options. But from a techno-economic or social-political point of view, the intention is often to move a system out of an undesirable basin of attraction – an issue taken up in Chapter 14 under the heading of 'lock-in'. In any event, a better understanding is crucial: 'The central theme of food web research is the understanding of structure, function, dynamics, and complexity... this task will determine in part the management of our ecosystems towards sustainability' (de Ruiter *et al.* 2005).

9.7 Catastrophic Change in Ecosystems

Real-world ecological communities are networks of highly structured, often dense interactions in three dimensions and with individual trait variance. Their dynamics has, traditionally, been understood as *ecological succession*, a process thought to be controlled by some form of exploitation. A stage of rapid colonisation of open or disturbed areas is succeeded by a stage of conservation, during which the emphasis is on slow accumulation and storage of energy and material. So-called *r-species* (or r-strategists) are associated with the first, exploitative phase: They have extensive dispersal ability and rapid growth in a 'winner-takes-it-all' competitive environment. In the second, conservative phase the so-called *K-species* (or K-strategists) prevail, with slower growth rates and more elaborate forms of competition.

A more comprehensive scheme of ecological change processes is proposed by Holling, in the context of adaptive resource use and management (Holling 1986, Gunderson and Holling 2002; Figure 9.15).[16] The process of transformation from an initially unstructured state of *r*-strategists into a structured and stratified climax community of *K*-strategists is still the basis. But observations suggest that there is more: The climax community eventually becomes so complex that its own stability is undermined and collapse becomes an inevitable consequence, as in the aforementioned self-organised criticality and societal collapse mechanisms (§3.5). At such a catastrophic moment, all the energy that is tied up in the niches and interdependencies of the climax community is released. This energy may be used again by

[15] Such events – for instance, the sudden outburst of catastrophic forest fires – have a power law size distribution and are named self-organised criticality. The power law size distribution implies that there are many events of small size but a few of (very) large size. Natural phenomena such as forest fires and earthquakes exhibit such distributions.

[16] The authors usually represent the changes in the form of a lemniscaat, in which the r-K change is one towards more potential and more connectedness. I use the more familiar system dynamics representation, which suggests a resemblance with the worldview framework and dynamics introduced in §6.5.

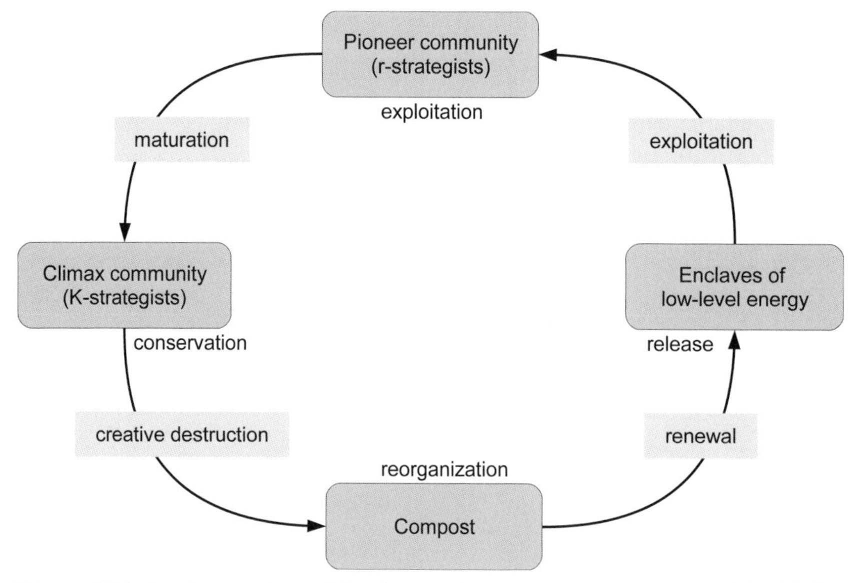

Figure 9.15. A scheme of possible changes in ecosystems, as proposed by Holling and co-workers.

opportunistic, fast-breeding and highly competitive r-species, who in the process prepare the road for the energy-conserving strategies that characterise the K-species. There is as yet only limited evidence to validate the scheme. This is not surprising because in most systems an even more complex reality operates at multiple scales in space and time. The adaptive 'ecocycle' dynamics has been applied as an organising framework to deal with the complexity of ecosystems and, by analogy, of social and economic systems (Holling 1986; Gunderson and Holling 2002).[17] The scheme has heuristic value, as an extended analogy of economic and social systems with divergent cultural perspectives as agents of change.

It has been known for quite some time that ecosystems occasionally experience sudden, discontinuous change as, for instance, in the collapse of a climax community. What is behind such 'catastrophic change'? Throughout the last decades, nonlinear behaviour has been studied with the help of formal models in order to understand it and use it for adaptive resource management. Catastrophic change can happen when the underlying system has more than one equilibrium state or basin of attraction. It is a sudden shift in system structure and behaviour, a *regime shift*. A simple nonlinear equation that represents this phenomenon is a third order differential equation of the form[18]:

$$dX/dt = -X(X^2 - \alpha) \qquad (9.4)$$

Developed in the 1960s and 1970s by Thom and Zeeman, this was at the basis of catastrophe theory. The essential point is that a change in the slow parameter α influences the behaviour of the fast changing state variable X.

[17] Holling (1986), well aware of the parallel with Schumpeter's (1942) theory of economic maturity, collapse and renewal, calls this transition from climax community to compost 'creative destruction'.

[18] See Appendix 9.2 for some mathematical details.

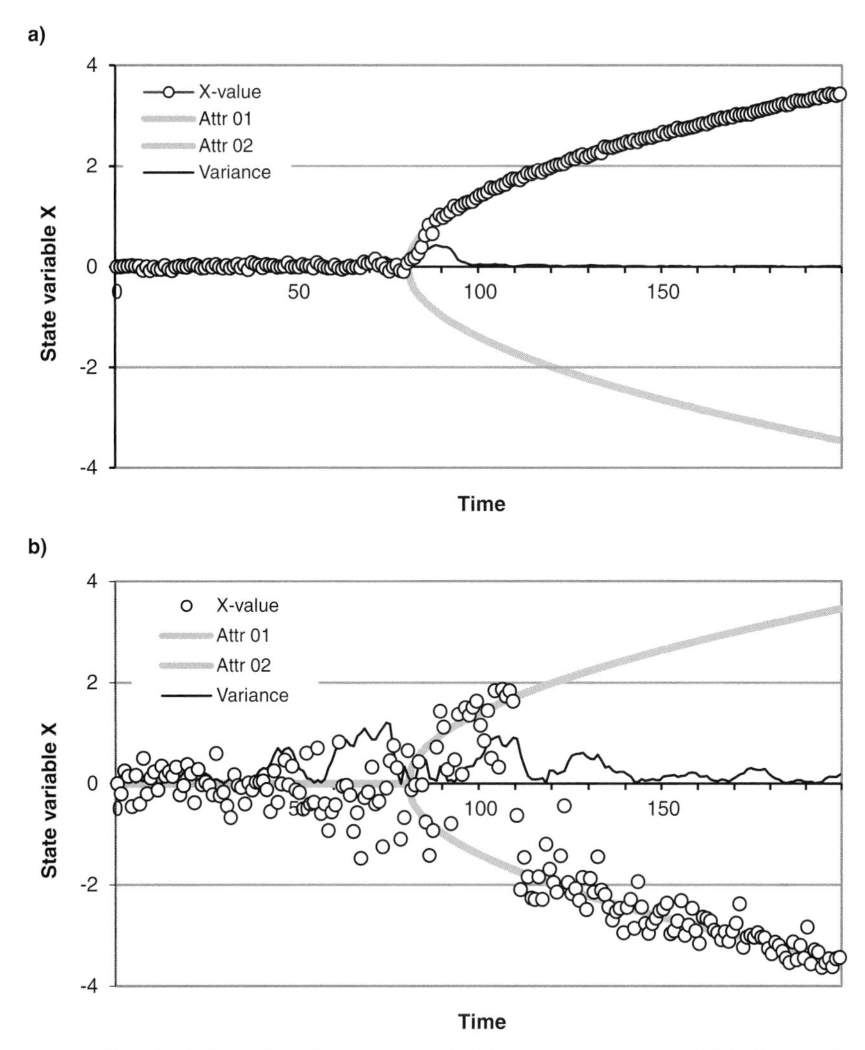

Figure 9.16a,b. Bifurcation diagram. The dots indicate the state variable X according to equation 9.4, for a change of the slow variable α from -5 to $+5$ ($\Delta t = 0{,}01$, Runge-Kutta4). The grey curves are the attractor curves. In the upper graph (a), the random fluctuation equals 2. In the lower graph (b), it is 20. The thin lines are a measure of the variance.

A simple simulation can clarify the model. Suppose that a slow variable, represented by the value of α in equation 9.3, changes in the course of a period of 100 quarters (25 year) from –5 to +5. During this change, the system will shift from one attractor to three (one unstable, two stable) attractors (Appendix 9.2). This is indicated in Figure 9.16a with the grey curves. Each point on the curve is a steady-state and the emergence of the fork at $\alpha = 0$ is called a bifurcation. The state variable X is the fast-changing variable and remains, in the absence of environmental fluctuations, at the attractor $X = 0$. However, because this attractor is unstable for $\alpha > 0$, the smallest perturbation will make it switch to one of the stable attractors (grey curves).

If there are such fluctuations, X will for $\alpha < 0$, simply fluctuate around the then stable equilibrium at $X = 0$. As soon as the bifurcation emerges, at $\alpha = 0$, it depends on chance which of the two pathways is chosen (Figure 9.16a). If the fluctuation

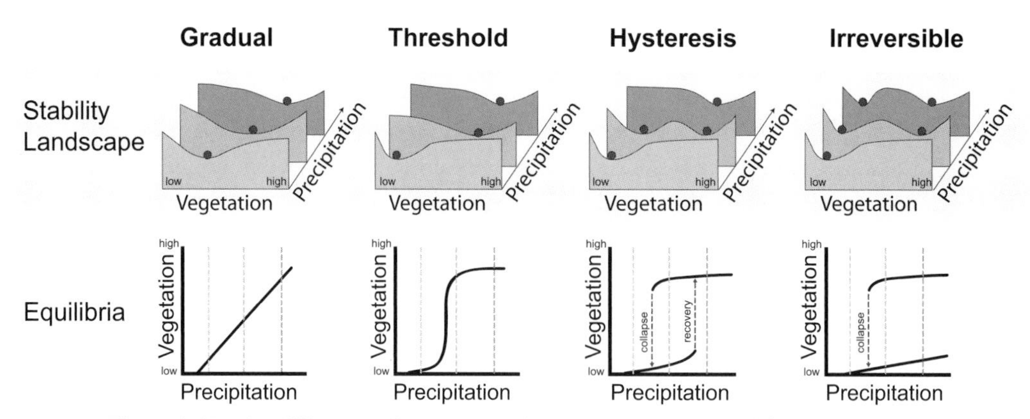

Figure 9.17. The differences between gradual ecological change (a) and three different types of regime shifts (b–d) using precipitation–vegetation interactions as an example (Gordon *et al.* 2008).

is large relative to the separation between the two branches, the system may even switch from one branch to the other (Figure 9.16b). It is hard in this situation to derive firm statements about the state of the system, because in the neighbourhood of the bifurcation point, the stability landscape is rather flat. One consequence is that environmental noise can have a relatively large impact on X around this point. This is seen in the value of the variance during the simulation. Recent investigations suggest that careful analysis of the patterns of fluctuations over time may be an indicator of a critical transition (Scheffer *et al.* 2009). Such 'before-the-collapse' signals can serve as an early warning indicator.

It is instructive, again, to illustrate regime shift graphically with the ball-in-the-landscape metaphor (Figure A2.1). The ball indicates the position of the system in the 'stability landscape'. Let the x-axis represent a state variable X and the y-axis a response variable Y, then the question is how the two are related (Figure 9.17). If their relationship is linear, a change in X will cause a proportionate change in Y. The relation becomes nonlinear when a threshold develops. A next step is the emergence of a 'fold': The system has one unstable and two stable attractors and gives rise to a phenomenon known in physics as hysteresis. This happens for $\alpha > 0$. The system can follow either of two trajectories, depending on the initial state (Figure 7.7). Fluctuations ('noise') in the environment may cause the system to flip from one attractor basin to the next. For large α, the system enters irreversibly the one or the other attractor basin. As discussed, the width of the attractor basin is a defining characteristic of resilience.

For three categories of *agriculture water* interactions, regime shifts can be envisaged as is shown in the graphs in Figure 9.17 (Gordon *et al.* 2008). The first group is agriculture aquatic systems, where changes in runoff quality and quantity cause regime shifts in downstream aquatic systems. The second one is agriculture soil systems, in which changes in infiltration and soil moisture result in terrestrial regime shifts. The third one is the agriculture atmosphere system, in which changes in evapotranspiration result in regime shifts in terrestrial ecosystems and the climatic system. The graphs illustrating these situations are for precipitation as state variable and vegetation as response variable. The system under stress is characterised by two branches of equilibria (b) and sometimes an area in which these two overlap (c,d) and sudden change can occur. Note that each curve indicates values of the two interacting variables (vegetation and precipitation) at which the system is in equilibrium.

There are several real-world examples that suggest the existence of the kind of dynamics previously described. A first classical example is the *Spruce budworm and forest*, which is about the interactive dynamics between the spruce budworm, its predators and the boreal forest (Holling 1986; Meadows 2008). For centuries, the spruce budworm has been killing spruce and fir trees periodically in North America. When the virgin pine forests became depleted and the lumber industry became interested in the spruce and fir, the budworm became a 'pest' and northern forests were sprayed with the insecticide known as DDT to control it. Later, other insecticides had to be used and spraying costs kept rising, but success in killing the budworm was meagre. It was only after Holling and co-workers looked at it as a system that they started to understand why.

It turns out that the budworms are crucial in maintaining forest diversity. They are part of a natural stable ecological cycle, during which their favourite food, the balsam fir tree, increases until with the help of a few hot summers the budworm population explodes. Even its natural predators are not able to control it, and the outbreak only stops when the balsam fir population is drastically reduced. The budworm dies off from the combined forces of balancing loops: starvation and predators. In mathematical terms, there are two equilibrium branches with regular catastrophic change. Interference by spraying insecticides kept the budworm population under control, but it also kept the budworm food stock (balsam fir) at a high level and killed off the natural predators. In consequence, the forest managers set up a situation of 'persistent semi-outbreak', as Holling called it. A similar cycle has been observed with regard to budworms, tree foliage and predators (birds). Two equilibrium branches exist: one with low budworm populations and young, growing trees; the other with high budworm populations and mature trees. The essential element is a nonlinear relation in predator (bird) effectiveness: As foliage increases, the budworms become less visible and their population is no longer controlled. A similar 'near-outbreak' situation can occur in forests where effective fire protection prevents natural fires from happening.

A second and well-researched case is *Eutrophication of shallow lakes* (Scheffer *et al.* 2001). It has become another archetype of ecosystems under stress of a disturbance such as a pollutant. The storyline is that an influx of nutrients from inflowing fertiliser and wastewater and industrial effluent cause a growth of phytoplankton. This causes bottom plants to disappear as they get less or no light, which makes the lake look greenish and turbid. Also, small animals living on the bottom vegetation die off, several fish species disappear and a monotonous community is what remains. Bird populations visiting the lake drop by an order of magnitude. These observations can be formalised in terms of a catastrophic change model (equation 9.4). Turbidity is the slow-changing variable, nutrient concentration the fast variable and there are two branches of stable equilibria: One with low and one with high turbidity-nutrient levels that partly overlap (Figure 9.17c). A clear lake with bottom vegetation will become more turbid as nutrient concentration rises, until a certain threshold is reached – the critical turbidity. At this point, the bottom vegetation collapses and the system jumps to the upper branch without vegetation and high turbidity. 'Overall, the diversity of animal and plant communities of shallow lakes

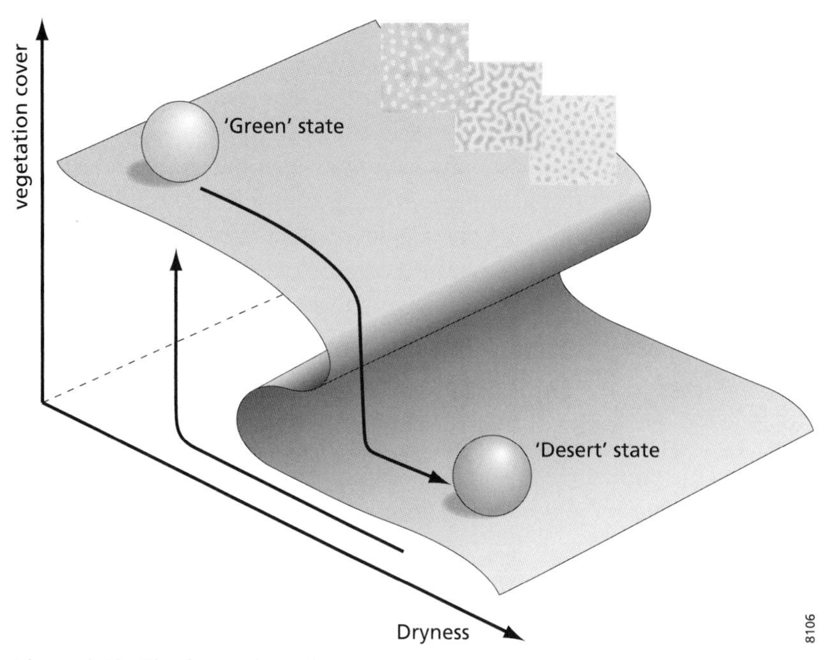

Figure 9.18. The figure shows how ecosystems may undergo a predictable sequence of emerging self-organised patchiness as resource input decreases or increases. Thick solid lines represent mean equilibrium densities of consumers functioning as 'ecosystem engineers'. The area between the dotted arrows is the area of two attractors where catastrophic shifts between self-organised patchy and homogeneous states, and vice versa can occur. The squares indicate the associated vegetation patterns, dark colours representing high density (Rietkerk *et al.* 2004).

in the turbid state is strikingly lower than that of lakes in the clear state' (Scheffer *et al.* 2001).

The model indicates an important point for management: To bring the system back to the original state along the lower branch, one has to go back to much lower nutrient concentrations than when the switch to the upper branch occurred. A detailed analysis of the shallow lake area De Wieden in The Netherlands has shown that understanding such complexity is useful indeed. The official norm of 0.05 mg total-P/litre, inspired by the aim of nature protection, requires significant reduction of phosphorus (P) emissions. Much effort and money was spent on reversing the trend by reducing the nutrient influx, mostly without success. The catastrophic shift model indicates that the goal of clear water can more effectively and at lower cost be realised by biomanipulation (removal of the benthivorous fish) (Hein 2005). It has also consequences for the more optimal use of ecosystems services (Mäler *et al.* 2003).

A third example of nonlinear regime shifts in response to external and associated internal ecosystem variables is when there is a positive feedback between consumers (such as plants) in combination with limiting resources (such as water or nutrients). This is quite common: Fine-scale interactions lead to spatial resource concentration and self-organised patchiness because of endogenous causes. In such situations – for instance, in *semi-arid regions* – a regime shift can happen if grazing pressure exceeds a critical threshold, with possibly catastrophic consequences (Rietkerk *et al.* 2004;

Figure 9.18). This phenomenon has been investigated for heterogeneous landscapes using a stochastic cellular automata model and field data from three regions in the Mediterranean (Kefi *et al.* 2007). It is found that the patch-size distribution of the vegetation follows a power law and that this can be explained from local feedback interactions amongst plants. When the model is used to simulate the effects of increasing grazing pressure, it turns out that the deviations from power laws seen in the field data also emerge in the model simulations and, importantly, that they always and only occur close to a transition into a desert. The researchers suggest that patch-size distributions may be a warning signal for the onset of desertification, a spatial equivalent of the previously discussed time-signal of a critical transition (Figure 9.16). Imagine that our complex social-economic-cultural systems also have such thresholds beyond which sudden catastrophic change might occur – quite a different metaphor from the one of smooth ongoing growth in material welfare usually presented.

MDG Goal 7: Ensure environmental sustainability.
TARGETS

1. Integrate the principles of sustainable development into country policies and programmes and reverse the loss of environmental resources.
2. Reduce biodiversity loss, achieving, by 2010, a significant reduction in the rate of loss.

'Global deforestation – mainly the conversion of tropical forests to agricultural land – is slowing, but continues at a high rate in many countries... Though some success in biodiversity conservation has been achieved, and the situation may well have been worse without the 2010 target, the loss of biodiversity continues unrelentingly... '

– The Millennium Development Goals Report 2010, New York

9.8 Biodiversity and Ecosystem Services

Nature has become increasingly an input for the production of marketable goods and services of the world economy. Sometimes this is a very tangible input, such as food and timber from the land and fossil fuel and mineral ore. There are also less tangible forms, such as regulating water flows or satisfying the longing for beauty. Ideas about and action against the diverse ways in which nature is threatened by the advance of civilisation have a long history but are nowadays mostly associated with nature conservation and protection of biodiversity. It is evaluated scientifically and at a global scale in a number of large-scale assessments. According to the UN Convention on Biological Diversity (CBD):

> *biodiversity* is... the variability among living organisms from all sources including, inter alia, terrestrial, marine and other aquatic systems and the ecological complexes of which they are part; this includes diversity within species, between species and of ecosystems.

As with other definitions, defining something is one thing and operationalising it is another. Several indicators are used to represent the state of biodiversity and its change (Alkemade *et al.* 2009; PBL 2010b):

- the biome extent: the biome area according to climatic and geographical potential minus agri/urban areas and forest plantations (Figure 9.4);
- the *mean species abundance* (MSA): the ratio of actual and potential species composition and abundance, with 100 percent being undisturbed nature by definition; and
- the status of threatened species and the genetic diversity of domesticated species: this is a rather diverse set comprising, for instance, the Red List Index with a summary of threats for species.

Other indicators of biodiversity are coverage of protected or wilderness areas (Table 11.1) and food web–based trophic indices.

The MSA can be estimated on the basis of biomes and from trends in the driving forces of biodiversity loss such as land use for crops and grazing, forestry, infrastructure and others (Dirzo and Raven 2003, Benítez-López et al. 2010). Figure 9.19 gives an impression of the MSA in the regions of the world in the years 1970 and 2000, and one possible future trajectory for the year 2050. In the outlook for 2050, biodiversity loss as measured in the MSA is assumed to continue at the rate of loss during the 20th century, a plausible assumption unless explicit and effective policies are implemented (PBL 2010b). Not surprisingly, in view of the still limited understanding of food web dynamics, there is controversy amongst scientists about the consequences of further loss in biodiversity. However, there are good reasons to think that biodiversity is important for the stability of ecosystems by offering complementarity through diversity in species functions and insurance through diversity in responses to external disturbances. The rate of biodiversity loss is already, together with the nitrogen cycle disturbances and climate change, considered to exceed acceptable levels from an integrated planetary boundary perspective (Rockström *et al.* 2009).

The CBD has now 193 signatory nations. The process shows, however, the familiar features of global negotiations: There is general agreement about the urgent need to protect biodiversity, but there are widely different views on issues such as access to genetic resources for medical purposes. It is also unclear how initiatives are to be funded. It is proposed, for instance, to phase out part of the global subsidies to marine fisheries – estimated at U.S. $15 to 35 billion per year (*Nature*, 18 October 2010) – in order to restore and protect nature, similar to the idea of abolishing subsidies to fossil fuels in order to combat climate change. However, powerful interest groups will not easily give in to such measures.

As in debates on other sustainability issues, the question of why to care about biodiversity has been raised. Economists have pointed out that there are trade-offs to be considered if certain levels of biodiversity are to be sustained. The argument that many citizens consider it a necessity is countered with the observation that there is a discrepancy between what people say and what they do. Therefore, what is the economic benefit of preserving biodiversity?

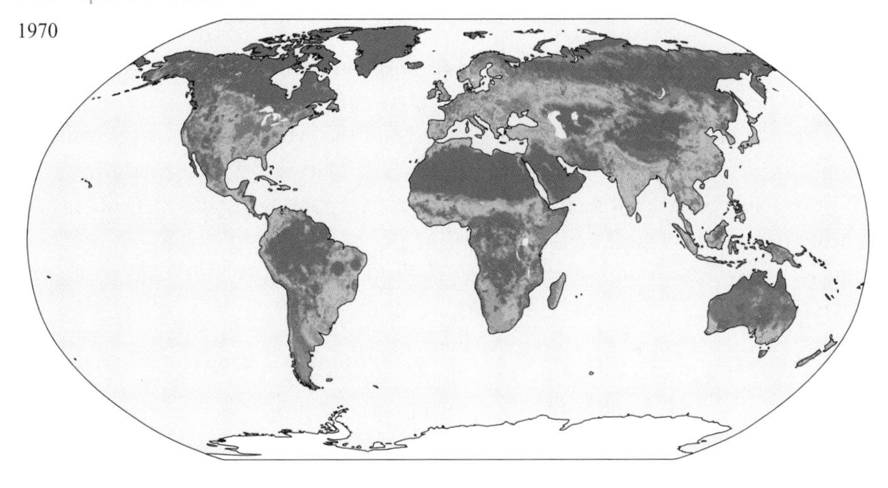

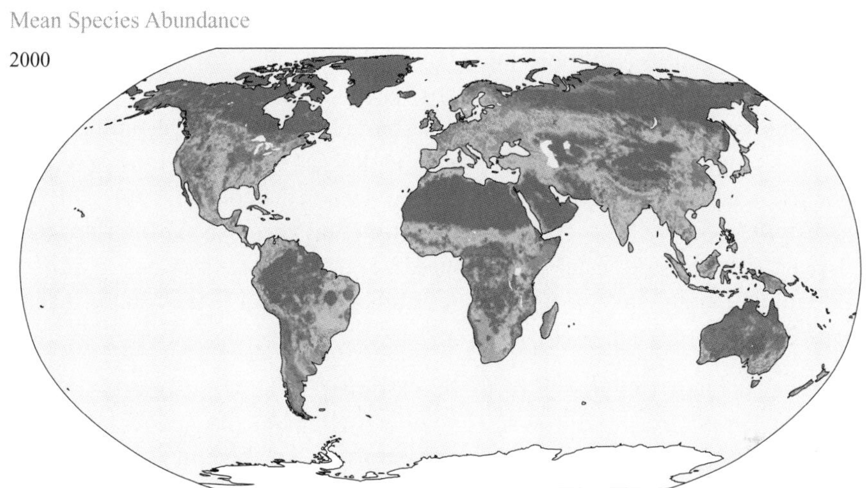

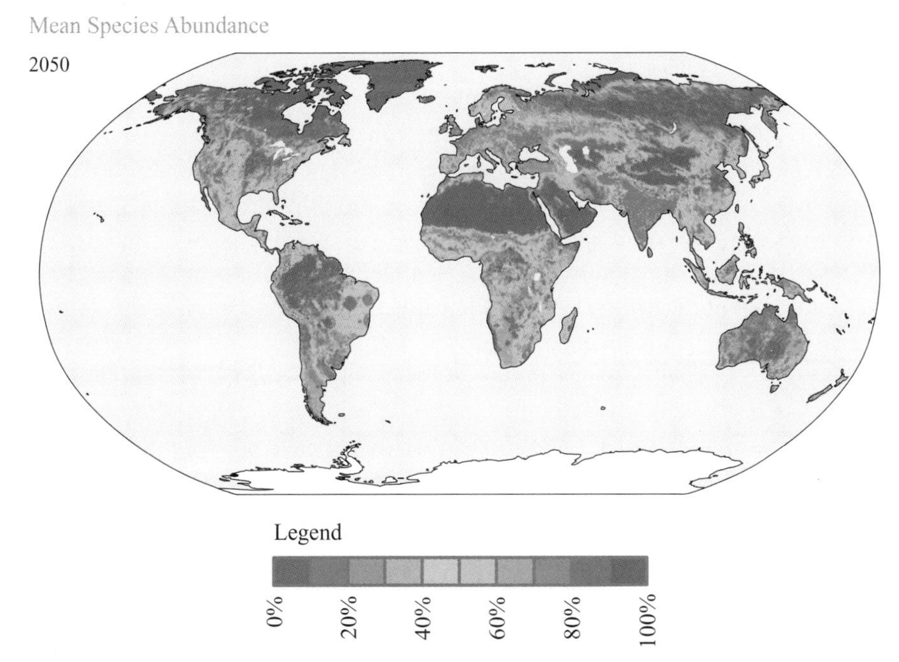

Figure 9.19. Map of terrestrial mean species abundance (MSA) in 1970 and 2000, and a projection for 2050, based on calculations with the Globio model (Alkemade et al. 2009, PBL 2010b). (See color plate.)

Already in the 1990s, estimates were made of the economic value of the services nature is providing 'for free' (Hueting 1980; Daly and Cobb 1989; Limburg *et al.* 2002). The *Nature* issue on *Pricing the Planet* (15 May 1997) came up with an estimate surprisingly close to the gross world product (GWP). Two-thirds of it was related to marine biomes. Since then, more in-depth analyses are made of what has since then become known as *ecosystem services*. The notion of ecosystem services was preceded by the idea of ecosystem functions, in the sense that it is the function that results in the supply of a service. A general definition is that:

> an *ecosystem service* is the capacity of an ecosystem to provide goods and services that satisfy human needs.

A telling example of ecosystem services is the honeybee. The bee is by far the most important animal pollinator and crucial for an estimated one-third of global food production. The decline in bee colonies is only partly understood and may 'cost' dozens of billion of dollars (Ratnieck and Carreck 2010).

The most comprehensive and authoritative assessments of the world's ecosystems and their services is the Millennium Ecosystem Assessment (MA) published in 2005 after four years of researching, assessing and describing by hundreds of scientists all over the world. It has chosen ecosystem services as the core concept, with the leading question: How well can ecosystems (continue to) provide the services that people depend on but so often take for granted? The ecosystem services identified in the Millennium Ecosystem Assessment are listed in Table 9.1 (MEA 2005). These services can be grouped in a few classes:

- *supporting*: nutrient cycling, soil formation, primary production and so on;
- *provisioning*: food, fresh water, wood and fiber, fuel, medicine, genetic material;
- *regulating*: climate regulation, carbon sequestration, erosion and sedimentation control, pollination, biological nitrogen fixation, disease regulation, water purification, storm and flood protection; and
- *cultural*: aesthetic, spiritual, educational, recreational.

All four are intricately linked to aspects of human well-being, such as material quality of life, security, health and social relations. The four categories are widely accepted, although names may differ, such as production instead of provisioning services. Yet, it may be argued that the supporting services should not be listed separately because, first, given their extent and complexity, there is no solid basis for their inclusion and, second, their inclusion in valuation may lead to double counting as their value is already part of the other services (Hein 2005). Ecosystem services in all four categories are under various forms of stress: '... *humans have changed* ecosystems faster and more extensively than in any period in human history... The changes made to ecosystems have contributed to *substantial gains in human* well-being and economic development, but these gains have been achieved *at growing costs*. These costs include the degradation of many ecosystem services... This *[...] could get significantly worse* during the next 50 years... *Reversing [...]* while meeting increasing demands for their services is a challenge' (MEA 2005).

Many ecosystem functions have been an everyday experience for people since the dawn of humanity, in particular the provisioning and cultural services. But the scale of human interference and its consequences has instigated two changes:

Table 9.1. *Ecosystem services (based on MEA 2005)*

Ecosystem service	Subcategory
Provisioning services	
Food	Crops
	Livestock
	Capture fisheries
	Aquaculture
	Wild plant and animal food products
Fiber	Timber
	Cotton, hemp, silk
	Wood fuel
Biotic	Genetic resources
	Biochemicals
	Natural medicines
	Pharmaceuticals
Fresh water	
Regulating services	
Atmospheric and climate regulation	Composition, temperature, precipitation, etc.
Water- and wind-related regulation	Air quality
	Water
	Erosion
	Water purification
	Water treatment (removal, breakdown, etc.)
Biotic	Disease and pest regulation
	Pollination
	Refugia
Natural hazard regulation	
Cultural services	
Values	Cultural diversity
	Spiritual and religious values
	Educational values
	Aesthetic values
	Inspiration, sense of place
	Cultural heritage
Social and economic	Recreation and ecotourism
	Social relations
Supporting services	
Soil formation	
Photosynthesis and primary production	
Nutrient and water cycling	

1. ever more people become aware, in the context of the modern scientific world-view, how intricate the 'web of life' is and how dependent humans are for many of its functions; and
2. as part of the modern capitalist market economy, the idea has spread that only by accounting for such functions in economic/monetary terms will induce the necessary prudence.

As to the first point, many so-called indigenous people still have a huge reservoir of pre-modern experiences and insights about the functioning of their environment.

Box 9.4. *Attitudes towards biodiversity.* On request of EU Directorate-General Environment, Eurobarometer has asked more than 27,000 European Union citizens in spring 2010 how familiar they are with the term *biodiversity* and with the concept of *biodiversity loss* (EC 2010). Two-thirds of these citizens were familiar with the term biodiversity in the sense of knowing the meaning (38 percent) or having heard of it (28 percent). Thus, more than one-third (34 percent) have never heard of the term and, unsurprisingly, respondents with low-level education, manual workers and non-working respondents are significantly overrepresented in this group. There are significant differences between countries. As to biodiversity loss, 43 percent says that it is about certain animals and plants disappearing and 19 percent that certain animals and plants are endangered. Therefore, it is primarily associated with species, although a minority (13 to 18 percent) mentions habitat change, declining forest and loss of natural heritage. Most knowledgeable and best informed are the citizens of Germany and Austria.

The survey also posed questions about the threats to biodiversity and the importance of protecting biodiversity and natural areas. More than a quarter of the interviewees mention air and water pollution as the greatest threat to biodiversity. Next come intensive farming, deforestation and overfishing (19 percent), climate change (13 percent) and the creation of more roads, houses or industrial sites and land use changes (9 percent). The perception of the seriousness of biodiversity loss varies widely. Whereas in Finland only 9 percent consider biodiversity loss a *very serious domestic* problem, it is about two-thirds in Italy, Greece, Romania and Portugal. A large majority consider it a *very serious global* problem. Over three-quarters of citizens thinks that biodiversity should be protected for moral reasons but also because it is essential for economic prosperity. One-third of the respondents opt for stricter regulation, about one-fifth prefer a focus on better information. The results also indicate that there is relatively low familiarity with the *Natura 2000* European Union-wide network of nature protection areas.

This has often been replaced and lost. Yet, it can still be a valuable source of knowledge (Berkes *et al.* 2003). Regarding the second point, the argument is that market prices in a private property setting are the most effective signals to secure sustainable development. By ascribing 'services' to ecosystems in monetary terms, one makes the average citizen aware of these processes and their precious role for many societal processes – or so it is thought. But such 'monetisation' of nature is still controversial, if only because it is hard to define and value those services adequately.

9.9 Nature and Sustainable Development

Clearly, people have quite different values, feelings and experiences with what we call nature.[19] For poor farmers and herders in the world's villages, it is a source

[19] The site www.6billionothers.org/ gives a fascinating account of the diversity in people's views of nature – as well as of other aspects of life.

of food and beauty but also a place of danger filled with wild animals, diseases and floods. For all of us, the forces of Nature are feared in earthquakes and tsunamis. Some people express fear of nature and its uncontrollable forces; for others it is a source of spirituality and a way to connect to life. Some people and scientists fear that the relation between man and nature is out of balance; for others the urban jungle is as livable as the natural jungle. Underneath are the local and global changes that people experience in their material life: less trees, less water, less wild nature, more urban environments, more agriculture and livestock, and more roads. For most urban populations, nature has become a rather far and away experience. Children in towns and cities have less and less direct contact with animals other than with pets, in zoos and in animations. As one young urbanite expresses it: 'I'm happy in the urban jungle.' It brings new forms of economic activity: In 2007, the pet food market for dogs and cats amounted to more than U.S. $45 billion and was dominated by only four global corporations. Other people worry and battle. According to Greenpeace, 'Without healthy, thriving forests, planet Earth cannot sustain life. As much as eighty per cent of the world's forests have been degraded or destroyed. Greenpeace is campaigning for zero deforestation by 2020 to protect what is left of these extraordinary ecosystems' (www.greenpeace.org).

According to cultural theory, one may expect a limited set of perspectives on nature (§6.3). An investigation of Australian attitudes towards wildlife, for instance, shows a distinct difference between aboriginals and Anglo-Australians (Aslin and Bennett 2000). Whereas aboriginal society shared a substantial consensus about what is acceptable behaviour towards animals, the present Anglo-Australian population has a more fragmented view on and less personal connection to or responsibilities for wildlife. An analysis of various global change scenarios inspired by cultural theory reveals a difference in emphasis regarding biodiversity loss: For some it is an environmental problem; for others primarily a social-cultural problem and for a third group an ethical-cultural issue (Beumer and Martens 2010).

It is not my intention to make any prediction about the future evolution of ecosystems and, closely related, social-ecological systems. But it is possible to make some worldview-based explorations. Table 9.2 illustrates the differences in worldview with a couple of questions and answers. In the Modernist paradigm, an evolution of man is perceived towards ever greater control and beneficial use of nature. In its utopia, the loss of the natural world is successfully compensated by a combination of strictly managed nature reserves, three-dimensional (3-D) documentaries on the television or Internet and biotechnological imitations and improvements. In the eyes of the opposite worldview, those people with a more local and spiritually inspired belief system, the Modernist paradigm is an attitude of hybris and flatness. The deep ecology movement and its teaching of ecosophy is one of its manifestations. One should look for strategies that sustain all the diverse qualities of life associated with nature. They certainly require rational decisions and practical policies, as well as a broad understanding and appreciation of diversity, a moral conviction about man's relation to Nature and a deep respect for Life in all its appearances.

Table 9.2. *Worldviews and ecosystem and biodiversity futures*

ECOSYSTEMS AND BIODIVERSITY			
Statement 1: No single species of living being has more [of this particular] right to live and unfold than any other species. (Naess/Deep ecology)			
In theory, yes. In practise, the human species is most successful, but we do have to use our power sensibly.	Incorrect. We have to eat. Besides, it is God who created the world and put us on top.	A valuable insight, that may bring us the respect for other sentient beings that many indigenous people still have.	In theory, yes. In practise, it is difficult in view of the large populations. We certainly should reserve some place for other species.
Statement 2: Recreational tourism and partnerships with farmers are the most promising options for Nature conservation.			
This is a good example of public-private partnership success. It can help landowners and farmers to preserve biodiversity.	Someone has to pay for all those natural parks, so these options are an interesting source of revenues. Win-win, I would say.	It is marketed as Nature conservation, but in fact it may become the largest threat to nature. It is already in many low-income regions.	Provided it is well regulated on scientific grounds, this seems one of the options to get funding.
Statement 3: Pressures leading to biodiversity loss are, in some cases, intensifying the consequences of this collective failure, if it is not quickly corrected, will be severe for us all. (GBC 2010)			
It is unclear how serious the consequences will be. A genebank and agreements on intellectual property rights can reduce the risks.	I don't think the consequences for me are serious. As long as I can watch beautiful nature movies on television...	I know this to be true in my own environment, so I assume that it is true also globally. Extinction of species is anyway terrible.	Correct. Most scientists agree that it is a loss of valuable ecosystem services – water, medicinal plants and so on. Time for strong international action.
Statement 4: Without healthy, thriving forests, planet Earth cannot sustain life. (Greenpeace 2010)			
Can this scientifically be proven? Nature turns out to be quite robust... Maybe, with climate change, forests will conquer us.	This is an abstract and pretentious statement, typical for a radical environmental NGO.	This resonates with my own feelings and intuitions about Life.	The importance of the role of forests cannot be overestimated. We should do whatever possible to preserve them.
Statement 5: Trees are my connection to the spiritual world.			
Trees are respected in many cultures, but they are also commercially useful...	Sorry, I am not sure what you mean. I don't understand New Age stuff.	Trees, the sky, the ocean – they all make me so peaceful inside. They are love to me in the proper sense of the word.	Right, it is an important ecosystem service and we should be grateful for it, but as a religion, it is rather primitive.

9.10 Summary Points

Land and nature are the substrate of human life. Geography and ecology are two core disciplines in sustainability science and their subjects of investigation are landscapes and ecosystems. Points to remember from this chapter are:

- Soil and vegetation constitute the life support system. Science has developed classifications of life zones that are associated with soils and vegetation and with climate. An essential part is water. Its role is represented in the hydrological cycle.
- Nature is under the pressures from accelerating changes in land cover, and the consequences such as soil erosion and overexploitation of water, fish and forest. It is difficult, however, to assess degradation and to separate the natural and the human causes.
- Erosion is an illustrative and important universal process of change, and one of the processes at the root of declining resource quality (soil).
- Population ecology is about the species in ecosystems: how they reproduce, eat, mate, die. Simple population models, such as the logistic growth and predator-prey models, represent an elementary understanding of real-world ecosystem dynamics. They also provide useful metaphors, for instance, for overshoot-and-collapse and competition processes.
- Food webs provide insights in interactions and connections, experimental as well as theoretical. They are complex, have multiple attractors and can exhibit catastrophic change. Stability and resilience are important properties of ecosystems and food webs, but there are as yet no general theorems. The precautionary principle and adaptive management, that works along with the systemic forces instead of overlooking or fighting them, seem a good guide here.
- Ecosystems provide many less visible, but important services (supporting, provisioning, regulating or cultural). Loss of biodiversity may destroy or degrade ecosystem services. Their value should be incorporated in development plans and management strategies.

> The best people are like water.
> They benefit all things,
> And do not compete with them.
> They settle in low places,
> One with nature, one with Tao.
>
> — Lao Tzu, Tao Te Jing

> Furcht, nämlich – das ist des Menschen Erb- und Grundgefühl; aus der Furcht erklärt sich jegliches, Erbsünde und Erbtugend. Aus der Furcht wuchs auch meine Tugend, die heist: Wissenschaft.
>
> Die Furcht nämlich vor wildem Getier – die wurde dem Menschen am längsten angezüchtet, einschliesslich das Tier, das er in sich selber birgt und fürchtet: – Zarathustra heisst es "das innere Vieh".
>
> Solche lange alte Furcht, endlich fein geworden, geistlich, geistig – heute, dünkt mich, heisst sie: Wissenschaft.
>
> — Friedrich Nietzsche, Also sprach Zarathustra

Translation:

Fear, you see – fear is a basic, inherited feeling of man; original sin and original virtue can be explained from this fear. My virtue grew also out of this fear, that is called: science. That is to say: the fear of wild animals – it was bred in man over a long period of time, including the animal that he hides within himself and of which he is afraid: – Zarathustra calls it "the inner cow". Such a long and ancient fear, finally become subtle, sacred, intellectual – now, it seems to me, it is called: science.

You are the forest
you are all the great trees
in the forest
you are bird and beast
playing in and out
of all the trees
o lord white as jasmine
filling and filled b y all
why don't you
show me your face?

From: Speaking of Siva (Vs. 75).
Ed. A.K. Ramanujan, Penguin Classics 1973

SUGGESTED READING

A dozen story-research chapters on social-ecological systems around the world.

Berkes, F., and C. Folke, eds. *Linking Social and Ecological Systems – Management Practices and Social Mechanisms for Building Resilience.* Cambridge: Cambridge University Press, 1998.

A collection of chapters on the philosophy and background disciplines constituting the environmental sciences.

Boersema, J., and L. Reijnders, eds. *Principles of Environmental Science.* New York: Springer Science and Business Media B.V., 2009.

A thorough introduction into mathematical models in ecology.

Case, T. *An Illustrated Guide to Theoretical Ecology.* Oxford: Oxford University Press, 2000.

A broad and solid textbook introduction on ecosystems and environmental systems.

Jarvis, P. *Ecological Principles and Environmental Issues.* Essex, United Kingdom: Prentice Hall/Pearson Education, 2000.

A comprehensive and beautifully illustrated overview of biodiversity and the possible impacts of climate change.

Lovejoy, T., and L. Hannah, eds. *Climate Change and Biodiversity*, 2005, New Haven/ London: Yale University Press.

Freely downloadable assessment of the state of ecosystems and their services as of 2001–2005.

Millennium Ecosystem Assessment. *Ecosystems and Human Well-Being – Our Human Planet. Synthesis and Summary for Decision Makers.* Washington: Island Press, 2005.

Thorough introduction on critical transitions in complex dynamical systems, mostly from an ecological background and with models presented in appendices.

Scheffer, M. *Critical Transitions in Nature and Society.* Princeton: Princeton University, 2009.

An extensive introduction in models of environmental system and the various modelling methods and techniques.

Wainwright, J., and M. Mulligan. *Environmental Modelling – Finding Simplicity in Complexity*. London: John Wiley & Sons, Ltd., 2004.

USEFUL WEBSITES

- earthtrends.wri.org/index.php# is a site operated by the World Resources Institute with data and maps on ecosystems and biodiversity.
- www.isric.org/UK/About+Soils/Soil+data/Thematic+data/Soil+Geographic+Data/ is the site of ISRIC on World Soil Information, with definitions, data and maps.
- soils.usda.gov/use/worldsoils/mapindex/ has more than a dozen world maps on topics such as soils, climate and biomes.
- www.globalsoilmap.net/ is a site for maps and contains a legend for a worldwide project on constructing soil maps.
- www.fao.org/ag/agl/agll/wrb/soilres.stm has information and data on natural, land and water resources.
- modis-land.gsfc.nasa.gov/ is the site of the MODIS team at the NASA/Goddard Space Flight Center on global change research and resource management.
- www.resalliance.org is the website of the Resilience Alliance, which provides papers and models about research of socio-ecological systems.
- www.ecologyandsociety.org/ is the site of the Journal Ecology and Society, on integrated science for resilience and sustainability.
- www.millenniumassessment.org/ is the website of the Millennium Ecosystem Assessment (MEA 2005)
- www.cbd.int/ is the site of the Convention on Biological Diversity (CBD) and gbo3.cbd.int/ is the site of the Global Biodiversity Outlook, the flagship publication of the CBD.
- www.iucnredlist.org/ is the IUCN-site on the Red List index of threatened species.
- wwf.panda.org/is the site of WWF with, amongst other items, the Living Planet index.

Modelling and software

- www.globio.info/ is the site of the Globio model, a tool to assess human activities on biodiversity and developed and operated by PBL.
- www-binf.bio.uu.nl/rdb/books/mpd.pdf is a coursebook on population modelling in biology/ecology and the GRIND software, written by Rob de Boer of Utrecht University.
- www.aw-bc.com/ide/idefiles/navigation/main.html is a site that permits exploration of differential equations interactively, for instance, logistic growth, predator-prey and spruce-budworm bifurcation models.
- www.itc.nl/ is the site of Geo-Information Science and Earth Observation of Twente University in the Netherlands, with courses and research on geographical and resource modelling.

Appendix 9.1 Prey-Predator Models and Stability Analysis

Appendix 2.1 gives the description of a system of first-order ordinary differential equations. In vector notation:

$$d\vec{X}_i/dt = \underline{A} \cdot \vec{X} \tag{A9.1}$$

with X_i the state variables and $\boldsymbol{A}$ the Jacobian matrix. In an ecological context, the matrix $\boldsymbol{A}$ is the *community matrix*. It describes the interactions between the species in an ecological community. Setting $dX_i/dt = 0$ for $i = 1,2$, one finds the attractor $\{X_1^*, X_2^*\}$. A stability analysis reveals how the system responds to perturbations of the equilibrium. In a prey-predator system (equation 9.3), for X_1 = prey (hare, deer, daphnia . . .) and X_2 = predator (lynx, wolves, fish . . .), $a_{12} < 0$ and $a_{21} > 0$. The a_{11} and a_{22} are the species-specific net growth rates. Stability analysis shows that under certain conditions the system may exhibit a neutral oscillation, well-known from mechanical analogues, or an unstable spiral (Figure 7.5). Of course, this model is an extreme simplification of real-world ecosystems.

An example of a more extended and rather general predator-prey model is given by:

$$\begin{aligned} \frac{dN}{dt} &= r \cdot N \cdot \left(1 - \frac{\lambda_N N}{D_N + K_N} - \kappa \cdot \frac{P}{N+E}\right) \\ \frac{dP}{dt} &= s \cdot P \cdot \left(1 - \frac{\lambda_P P}{D_P + K_P} - \frac{hP}{N}\right) \end{aligned} \tag{A9.2}$$

with P the size of the predator population and N of the prey population. The parameters s and r are the respective natural growth rates of the predator and prey populations. The equations look impressive, yet they can be understood rather easily. First, for a vanishing predator population ($P \to 0$), the first equation describes a logistic growth process of the prey. Its dynamics then depends on the regeneration of its food base. The second term in the first equation denotes the additional mortality of prey because of the presence of predators, with $\kappa/(N + E)$ the catch effectiveness of the predator. In other words, $a_{12} = r\,\kappa\, N/(N + E) > 0$. Now the role of the parameter E is seen: For $E \to \infty$, this term and, therefore, catch effectiveness becomes zero as the prey has infinite space. Third, if the prey population becomes very large, the second equation turns into the logistic growth process of the predator. h/N is the additional mortality of the predator because its food, the prey, has become scarce. For $\kappa = h = 0$ all interactions vanish. Therefore, this extended model incorporates a number of 'sensible' assumptions about a predator-prey system. It can be shown that such systems can under certain conditions show rather complex but stable oscillatory behaviour around an attractor, a so-called *limit cycle*.

Appendix 9.2 Catastrophic Change and Bifurcations

When a parameter change in a system of differential equations causes the stability of the equilibrium to change, a (local) bifurcation can happen: The simplest equation to explain bifurcations is of the form:

$$dX/dt = \beta - X^2 \tag{A9.3}$$

If one solves for the attractor, such as for $dX/dt = 0$, there are two roots: $X^* = \pm\sqrt{\beta}$. In the phase plane of dX/dt as function of X, for $\beta > 0$, the two attractors are at the intersections with the x-axis (Figure A9.1). The positive one is locally stable, and the negative one is locally unstable. For $\beta = 0$, the two coincide. For $\beta < 0$, the roots are complex numbers and there are oscillations. If a system has a fast variable X and a slow variable β, then a gradual increase in β moves the system to either the branch with the stable or the branch with the then unstable attractor.

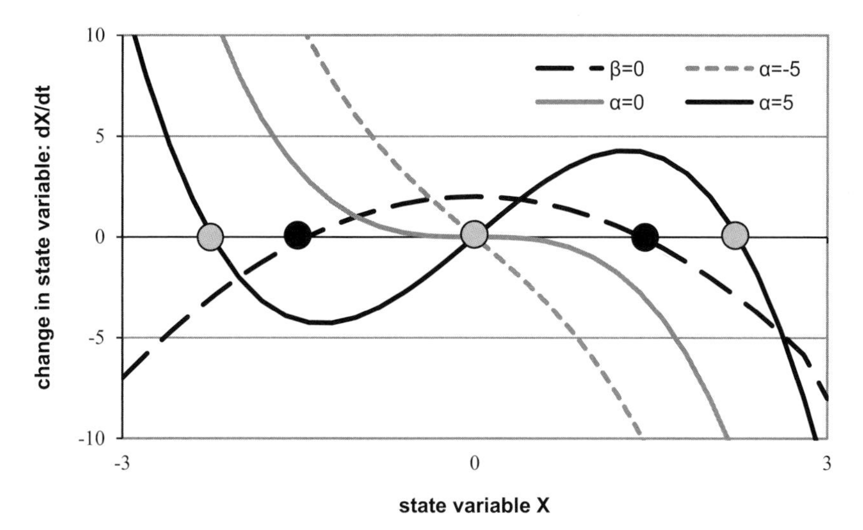

Figure A9.1. Phase space diagram for a second order (equation A9.3) and third order (equation A9.4) differential equation. The dashed curve and the two black dots are the curve and the attractors for equation A9.3. The grey curves are the curve for equation A9.4 for α-values –5, 0 and 5. The three grey dots are the attractors for the curve of $\alpha = 5$.

Similar bifurcation behaviour happens in a system described by equation 9.4:

$$dX/dt = -X(X^2 - \alpha) \qquad \text{(A9.4)}$$

The attractors occur for dX/dt = 0. One root is X = 0, and it represents a globally unstable attractor. The other two roots are for $X^* = \pm\sqrt{\alpha}$. Unlike in the previous case, for $\alpha > 0$, the two non-zero attractors are both locally stable ($X = \pm\sqrt{\alpha}$). This is seen in the phase plane for $\alpha = 5$ (Figure A9.1). For $\alpha \leq 0$, there is one globally stable attractor. A change in the parameter α leads to a structural change in behaviour. If, again, X is the fast variable, for instance, the size of a population, and α is the slow variable, for instance, sensitivity to climate change, the time-path of the state variable X will for slowly rising α gradually move from one with a globally stable attractor to one which has two locally stable attractors and, therefore, can undergo a rather sudden switch from one equilibrium state to another.

10 Human Populations and Human Behaviour

10.1 Introduction: The Image of Man

The previous chapter looked at 'nature' through the lens of geography and ecology. But almost everywhere, 'nature' has been influenced by 'culture'. The word culture is rooted in the Latin word *colere*, which means to tend or till the land but is also related to cultus and cultivate in the sense of care, labour and worship. The relation between nature and culture is an ancient philosophical question. How deeply are you rooted in 'nature' and how much and what kind of freedom can be realised in 'culture'? The narratives about past societies give provisional answers. This chapter focuses more explicitly on the species *Homo sapiens* in nature.

One can see nature as fundamentally subservient to human interests, as it is phrased in some biblical texts and entrenched in modernity, or as a cosmos in which humans participate as it is experienced in most pre-modern religions and modern forms of ecospirituality (Figure 8.5). Clearly, there is more than one answer to the question of how we see ourselves in nature. Are people not much more than instinct-driven animals? Are human beings the pinnacle of evolution, on the verge of a revolutionary jump in cosmic consciousness? Are we, basically, rational survival-oriented individuals with an ever larger array of tools to control our destiny? Or are we all of this? There is a huge perennial philosophy at our disposal as source of reflection and insight. Philosophers like Wilber (2001) build an integrated psychological-spiritual view in the footsteps of ancient traditions, incorporating novel scientific insights. Social scientists have developed grand theories from which some major characteristics of men as social beings emerge (§5.2; Table 6.3). Biologists and economists have their own particular image of man. The latter prefer to see men as rational beings; the former emphasise the biological roots in an evolutionary framework.

An important development in this respect is the advance in understanding complex systems (§8.5). We have associated complexity with an increase in behavioural diversity of and interactions between system elements and with an increasing 'interiority'. Which survival strategies can viruses and bacteria develop is a crucial question in understanding the spread of diseases. How do sheep behave when grazing – one aspect of overgrazing. What are the rules and strategies of humans in converting

Internal-Collective (I-C) We → Inter-Subjectivity [shared/collective knowledge, invisible social codes and implicit ontologies, informal norms and conventions] **"Noösphere"**	**Internal-Individual (I-I)** I → Subjectivity [mental states, emotions, desires, intentions, cognition] **"Interiority"**
External-Collective (E-C) Them, All this → Inter-Objectivity [reified social facts and structures, organizations, institutions] **"Sociosphere"**	**External-Individual (E-I)** It, This → Objectivity [agent behaviour, object process, physical entities] **"Observables, Exteriority"**

Figure 10.1. (Human) agent representations within the horizontal dimension of internal-external and the vertical dimension of collective-individual. Complexity science approaches are implicitly or explicitly using such representations (based on Ferber 2007).

land, catching fish, producing ores and fuels? What is the role of interaction between people in choosing or not choosing organic food or favouring the car over public transport? And, the most difficult one, how do people self-organise in multiscale governance arrangements and institutions?

Many of the issues raised in sustainability science are so difficult to tackle because of the role of humans. Increasingly, such questions are addressed with the novel methods of complex system science (CSS). There are two reasons to consider these methods and their applications in sustainability science. First, I introduce quality of life as a largely subjective yet integral part of the search for sustainable development (§6.2–4). Therefore, I must reflect on the psychological and social dimensions of individuals and societies: values, beliefs and perceptions, expectations and attitudes, mental constructs and contemplative practises.[1] Such a reflection points at meaning, at transcendence. Second, the methods and techniques of CSS allow experiments *in silico*, to test hypotheses about people's mental and inner functionings and about their interactions with others in complement to laboratory and field experiments.

This chapter focuses on the 4-Quadrant framework originating in the field of artificial intelligence (Figure 10.1). Two dimensions are distinguished: external versus

[1] The magical rituals and mythical accounts of the members of civilisations – their Great Stories – can also play an important role in understanding the deeper, cultural drives, even in apparently rationalist, modern societies.

internal and individual versus collective (see also Table 6.5). The horizontal dimension is about whether a person is primarily seen and modelled as an individual in (relative) isolation, or as part of a larger group with mutual interactions (tribe, network or society). The vertical dimension represents the degree to which the focus of description is on external observable behaviour or on internal mental processes (memory, imagination or anticipation). Of course, these are not black-and-white distinctions.[2]

Mainstream science, operating within the paradigm of empirical reductionism, has the tendency to focus on the E-I domain (lower right corner) where laboratory experiments with animals or people give 'objective' scientific results and 'strong' knowledge. It is dominant in biology and medical science, and it is an interpretation and aspiration for most scientists in the economic and social sciences. But how do emotions, values, intentions determine and change individual behaviour? For that, one has to consider the interior state of a person – the I-I domain (upper right corner). This is considered the domain of psychology as well as of the esoteric and spiritual. And how do the connections between people come into the picture? Individual animals and human beings are never isolated and this recognition brings people into the domain of the other social sciences – the E-C domain (lower left corner). Because human beings are both (self-)conscious individuals and members of collectives, such structures become internalised in the form of collectively shared social codes and ideas – the I-C domain (upper left corner). It is here that the intangible and complex elements of cultures, religions and ethics reside. In the university, philosophy and the humanities are closest to this quadrant. As to methods, network and agent-based models study emergent phenomena and evolutionary dynamics (E-C corner), while investigations of cognitive science and artificial life complement the 'natural' introspective techniques (I-I corner).

This brief philosophical and methodological detour provides also a bridge between the natural and the social sciences. It structures this chapter. The first sections describe some important issues in demography and, in particular, population models. Next, the chapter briefly explores the evolutionary perspective in biology and discusses the image of man as seen in mainstream economic science and recent extensions. I briefly introduce the methods, techniques and tools of CSS and illustrate with a few applications how it can help to better frame and examine core issues in sustainability science, among them the discourse on the relation between science and religion and on the role of (self-)awareness. Thus, sustainability science can become part of a personal journey through the diverse manifestations of human life, towards a [r]evolution of consciousness. This is what an integral approach aims for, ultimately.

World Population

6,890,729,773 on February 4, 2011 8:00 A.M.
world-population.com/

[2] The horizontal dimension is similar to the group dimension in cultural theory and the individual-collective axis in the worldview framework; the vertical dimension reflects partly the material-immaterial axis in the worldview framework (§6.3–4).

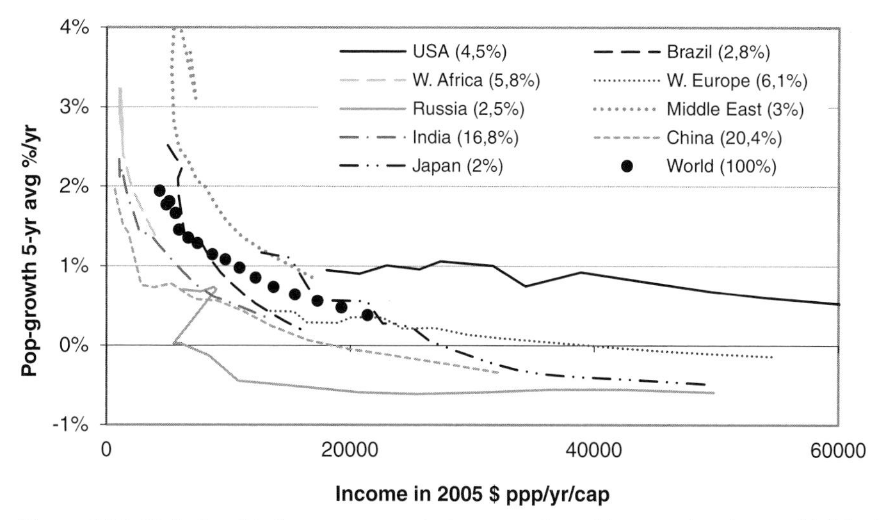

Figure 10.2. Relationship between population growth and income for the world (black dots) and nine world regions in the period 1970–2005, and their model-based extrapolation onto 2050 (source of data: PBL/World Bank).

10.2 Demography: Human Population Dynamics

10.2.1 Modelling Population[3]

The human population in the world has grown from less than one million at the start of the Holocene to somewhere around 50 million individuals 4,000 years ago and to more than 6.5 million in 2005 (Figure 5.1). The rate of growth has been declining since the 1980s, with a consistent fall in the population growth rate in nearly all countries. Income, measured as gross domestic product (GDP) per capita, is a major driving force through a variety of mechanisms (§4.2). At the regional level, the net population change rate declines exponentially with income – one of the metamodels or 'stylised facts' presented in Chapter 8[4] (Figure 10.2). The former USSR republics, including Russia, and Central Europe experienced negative population growth rates for the period after the fall of the Iron Curtain in 1989–1990. The growth rate of the United States' population declined, but at a significantly higher level than that of the population in Europe and Japan throughout the period. Probably, the flow of interregional migration is a factor here. There are significant exceptions, each with their explanations on closer inspection. For instance, South Africa experienced a period of large socio-political change and a rise in the death rate because of HIV/AIDS. Saudi Arabia's combination of high population and high income growth reflect its traditional culture and its oil wealth and immigrant inflow.

Understanding population dynamics is important because population size is an important determinant of the prospects for quality of life and sustainability in a direct

[3] I want to acknowledge the contribution of Henk Hilderink, at PBL, to this section.

[4] Figure 10.2 is based on the world data of income as GDP/cap/yr (Y) and of population growth (PopGR) (Bakkes *et al.* 2008, OECD 2008). For most regions, there is a good fit with a power curve PopGR = aY^{-b} (Figure 10.3). Also, an exponential decline curve gives a good fit. For low-income regions, there is a widespread, but the power curve remains a good fit. The high-income regions follow a logarithmic decline of the form PopGR = a-b LN(Y).

way, such as per capita resource use, and via indirect ways, such as health costs and immigration pressure. Some people even insist that the present size of the human population, having reached 7 billion in 2011, prevents any form of sustainability – but this is impossible to prove without an idea about their quality of life. The organisation and rising costs of health services – in rich countries as part of aging and advancing medical technology and in poor countries as part of the intense desire for adequate health care – are one of the population-economy links that determine future quality of life for many. Another important link are the provision and costs of education. I refer the reader to the Suggest Readings for more specific discussion on these important topics.

Throughout history, people have tried to forecast the size of the population. At the end of the 17th century, one estimate even went up to the year 20,000. The early long-range projections were based on mathematical extrapolation of total numbers or growth rates. In system's terminology, the growth of a population (stock) is the result of the difference between the total number of births (inflow), deaths (outflow) and the net migration flow. The *crude birth rate* (CBR) is defined as the number of births per 1,000 persons in a given year. Generally, it declines as a result of a decline in the fertility rate, which is defined as the expected number of children born per woman during her lifetime. The *crude death rate* (CDR) is similarly defined as the number of deaths per 1,000 persons in a year. The death or mortality rates can be specified for different ages of the deceased individuals. The resulting graph is an asymmetric U-shaped function: Mortality is high for the children up to five years, relatively low between five and twenty-five years and then steadily increasing for higher ages. There is a non-negligible distinction between the two sexes, for example, women have lower mortality rates, especially at higher ages. A more elaborate description considers population age structure and income trends and inequalities (Lutz 2001, 2003).

A country's population also changes because of *international migration*, while within a country the population density changes due to rural-urban migration. Usually, migration trends are simply extrapolated from historical data. Net migration is the outcome of two processes with different drivers: emigration and immigration. Determinants of migration are economic opportunities, conflict situations and extreme environmental events. They operate at local, regional and global levels. Long-term environmental change such as more intense droughts or sea level rise, often in combination with overpopulation and conflicts, are becoming a more frequent cause of migration. The number of environmental refugees is estimated in the tens of millions and increasing.

The profound historical changes in fertility and mortality patterns is generally referred to as *demographic transition* (§4.2; Figure 10.3). It is the transition from an equilibrium situation with high birth and death rates (stage I) to another equilibrium situation with low birth and death rates (stage III). The period in between (stage II) is one where mortality is falling, sometimes rapidly, but the birth rate is still high because people respond to the declining mortality with a delay that causes a temporary difference between death and birth rates and hence fast population growth. The transition process is visible in the age distribution of the population. Figure 10.4 shows age and gender classes for the population in four countries calculated for the year 2010. One can clearly see the different structure of the population pyramid.

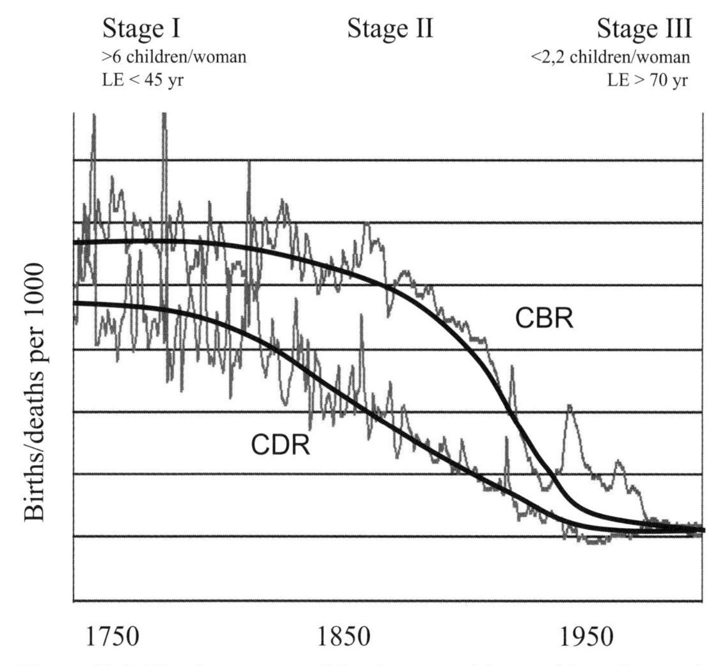

Figure 10.3. The three stages of the demographic transition: changes in crude birth rate (CBR) and crude death rate (CDR) and in life expectancy (LE). The historical data on crude birth and crude death rate are for Sweden (source of data: en.wikipedia.org/wiki/File:Demographic_change_in_Sweden_1735–2000.gif under Creative Commons).

West Africa had the highest growth rates and now, as a consequence, has the largest fraction of young people. Because the young people will also get children, such broad-base pyramid structure represents a large population momentum – the notion of inertia discussed with respect to stocks (§2.3). Western Europe and, to a lesser extent, India are past stage II and the age cohorts of the high-growth periods are past childhood. Japan is a clear example of an aging population.

The demographic transition is not an isolated process but part of a broader transformation observed in a growing number of countries since the 18th century that is associated with 'modernisation'. It leads to considerable variety in fertility decline between and within populations, as they are in different stages of the transition. Yet, the pattern of the transition is remarkably consistent all over the world and historical data indicate that fertility and mortality rates change only gradually. The transition can be disrupted in some of the poorest countries or regions of the world, if a combination of natural and societal forces cause collapse and people do not have the money or connections to import food and/or are unable or unwilling to move elsewhere.[5] In essence, population models are not different from the models in population ecology (§9.5) – with one important exception: They do not consider a (fixed) carrying capacity. The reason is that human populations have been quite successful so far to overcome external constraints through outmigration and ingenuity-driven intensification.

[5] There are also countries where rigorous population policies have had a significant influence on population structure and dynamics, for instance, Rumania and China. Also, the consequences of large wars are seen, for instance, in the population pyramid of Japan (Figure 10.4).

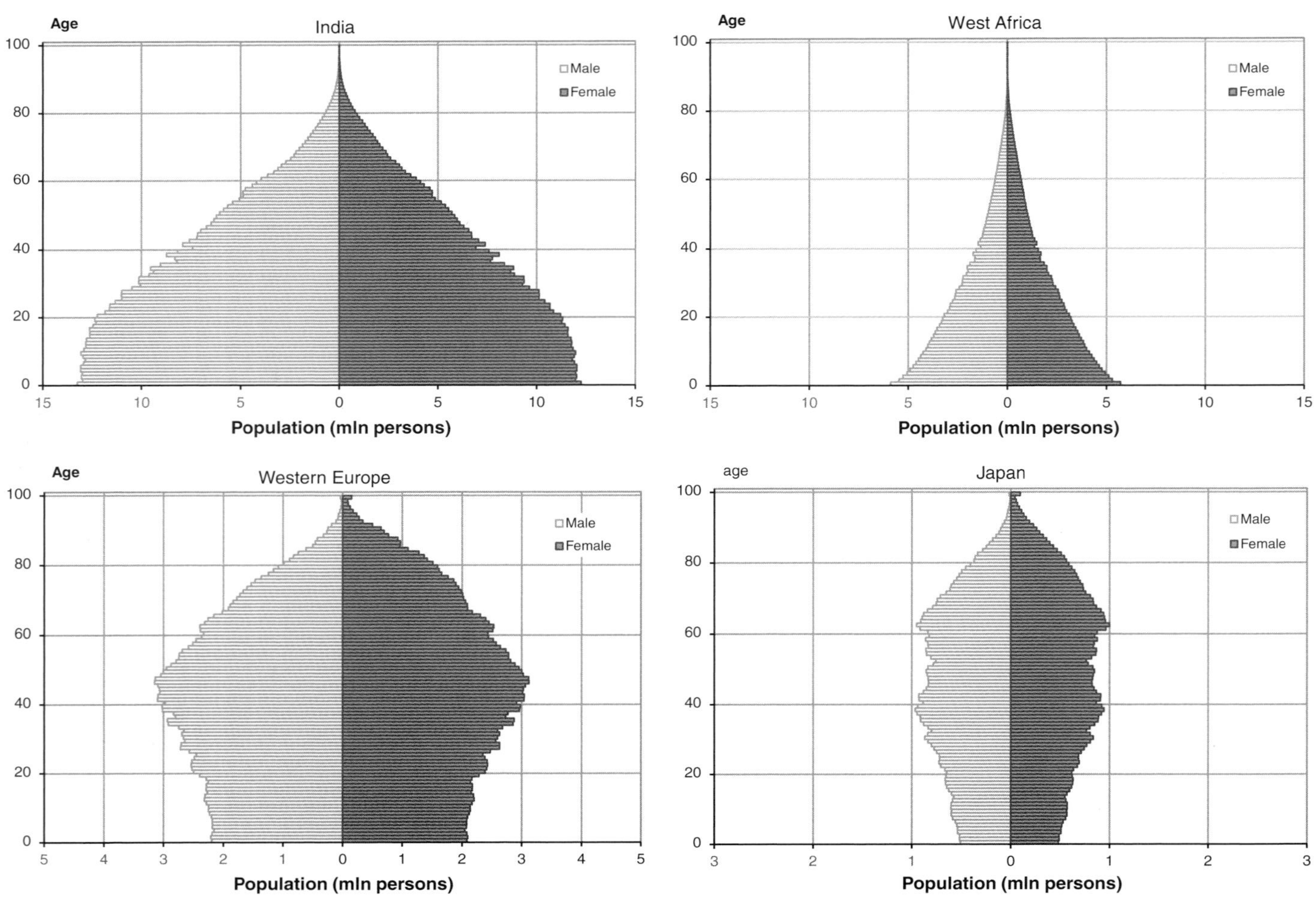

Figure 10.4. Age structure of the population for four countries in different stages of the demographic transition, based on simulation with Phoenix model (Hilderink *et al.* 2009). The horizontal bars indicate the number of people (male/female) in a particular age group or cohort, in millions.

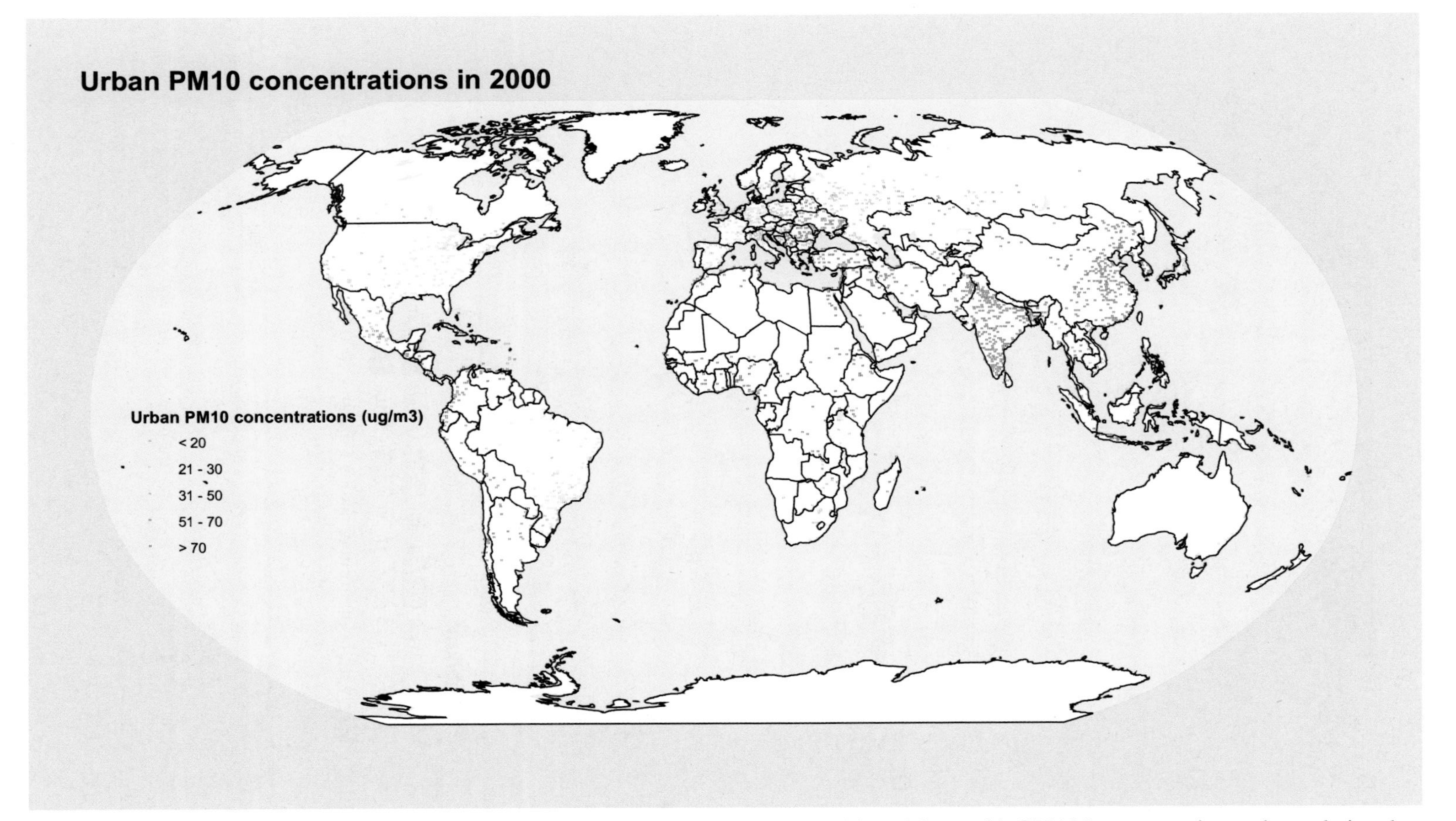

Figure 2.3b. Air quality in terms of Particulate Matter (PM10) concentrations in the cities of the world. GUAM concentration and population data are coupled to city locations (x,y). The values are aggregated to an average value for IMAGE-model $0.5^{\circ} \times 0.5^{\circ}$ grid cells. The cells are converted into square points. The darker the colour, the worse urban air quality (source of data: PBL).

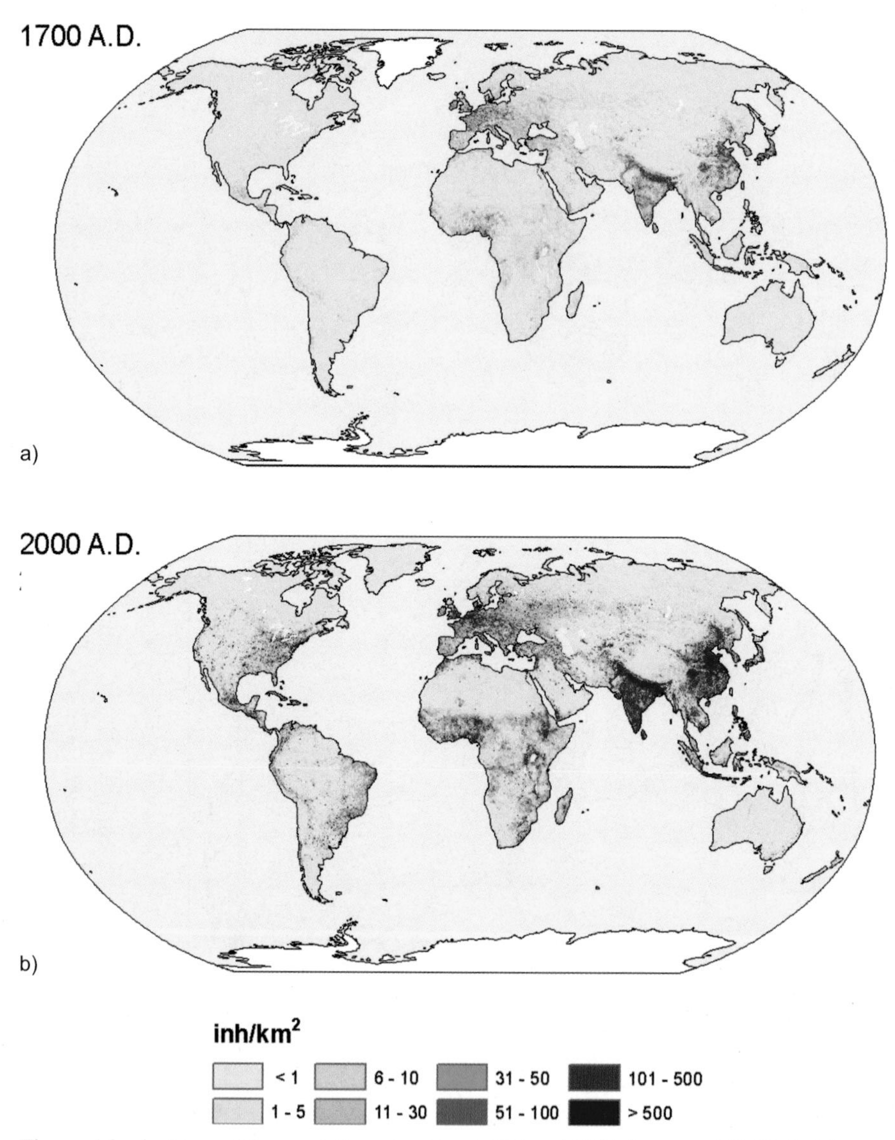

Figure 4.2.a,b. Map with population densities 1700 and 2000 (Source: HYDE/PBL).

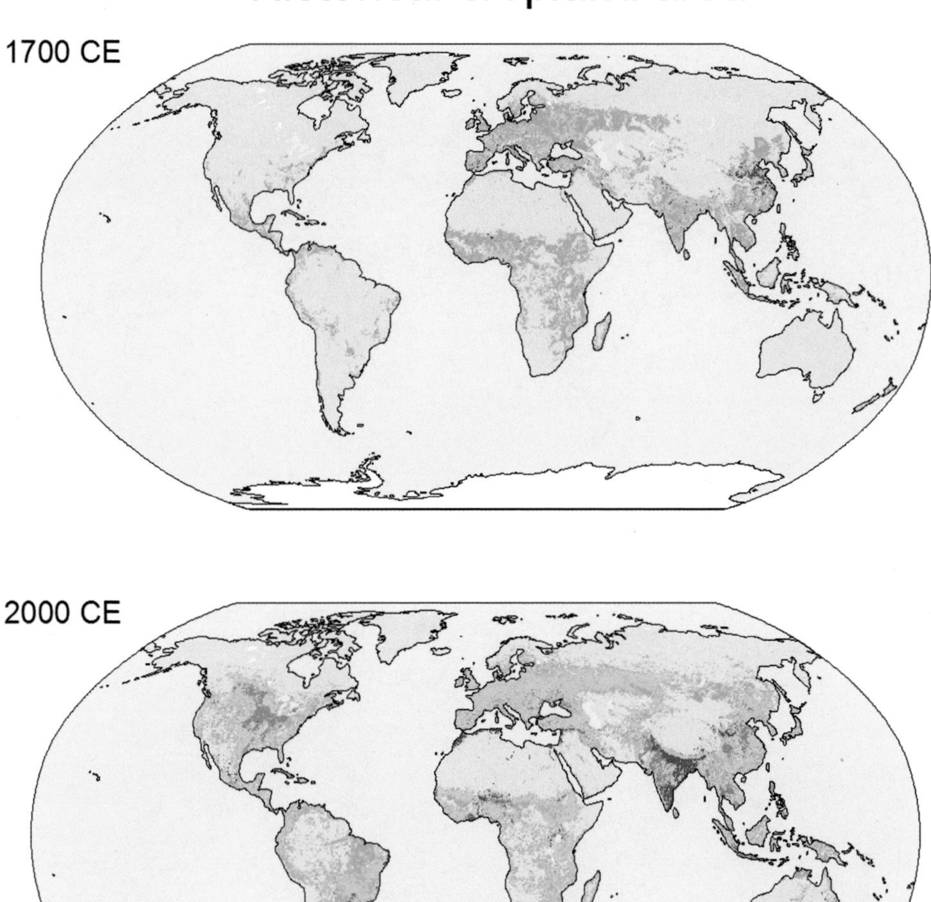

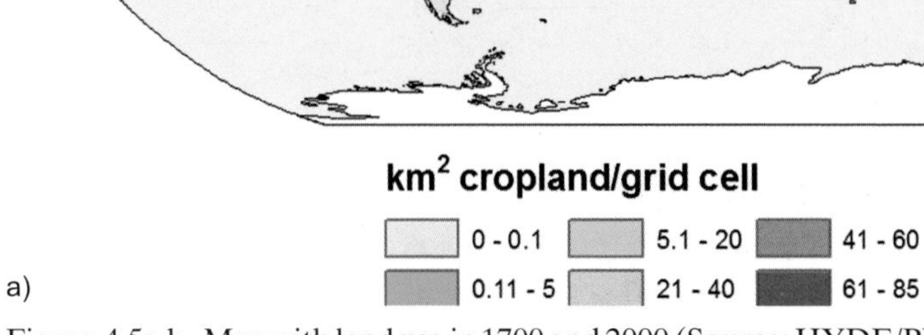

Figure 4.5a,b. Map with land use in 1700 and 2000 (Source: HYDE/PBL). The upper (a) shows the fraction of a gridcell used as cropland; the lower (b) shows the fraction used as pasture.

Historical pasture area

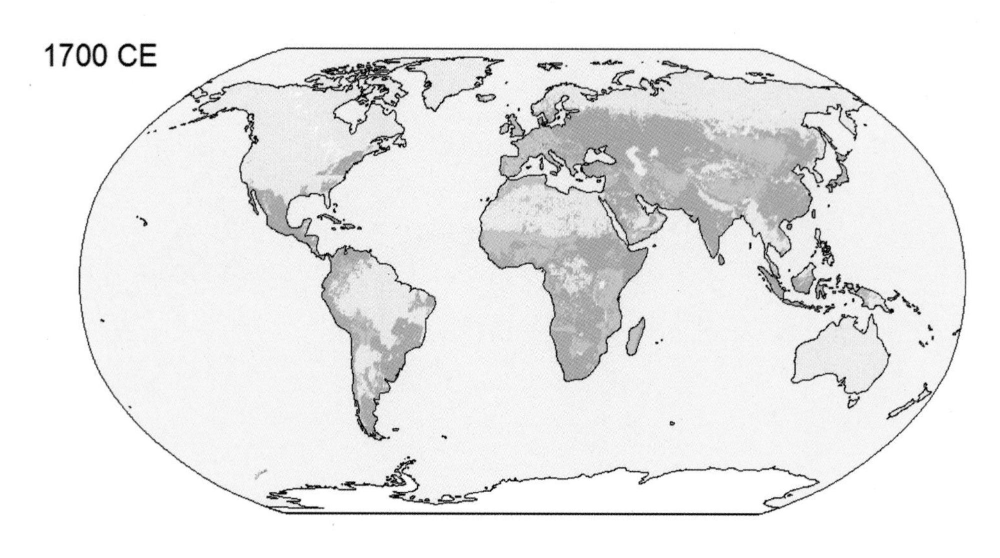

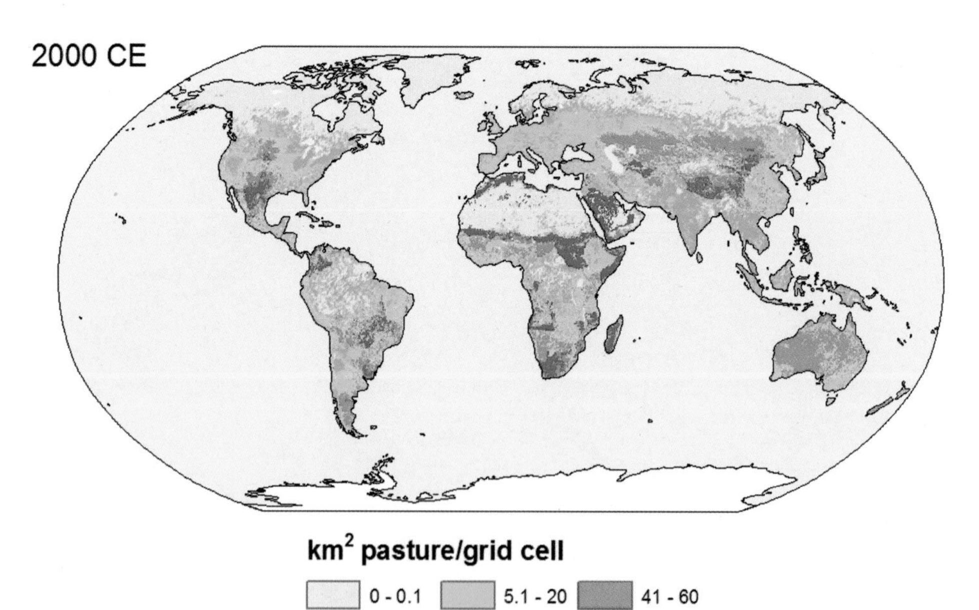

Figure 4.5a,b (*continued*)

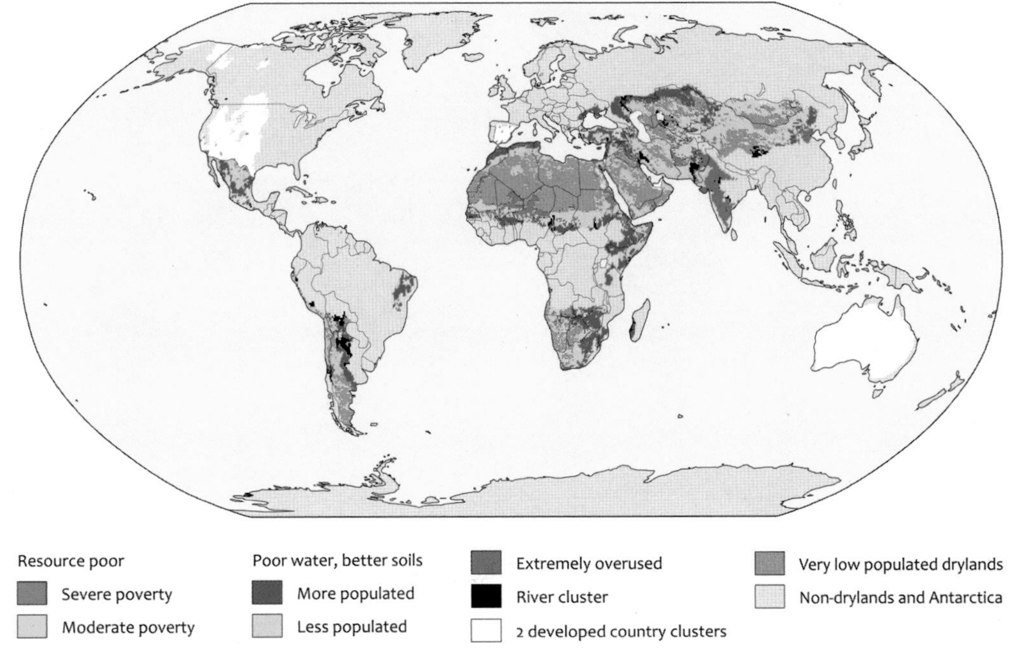

Figure 4.12. Distribution of the eight vulnerability profiles in drylands worldwide according to the vulnerability profiles in Figure 4.11 (Kok *et al.* 2011).

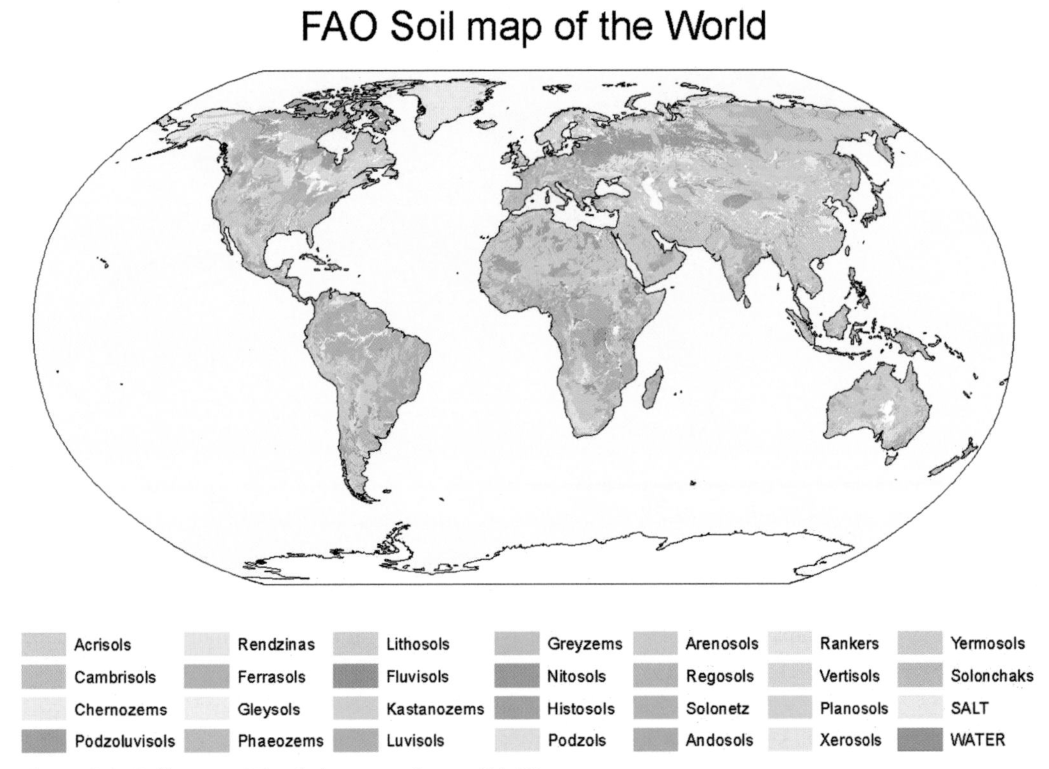

Figure 9.1. Soil map of Earth (source of map: FAO).

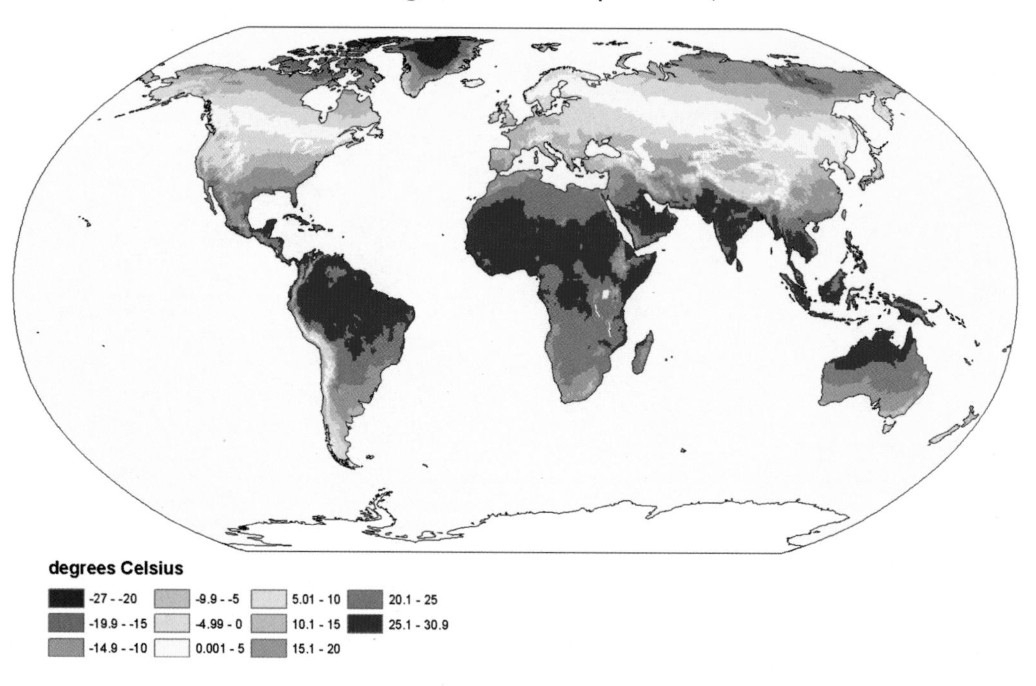

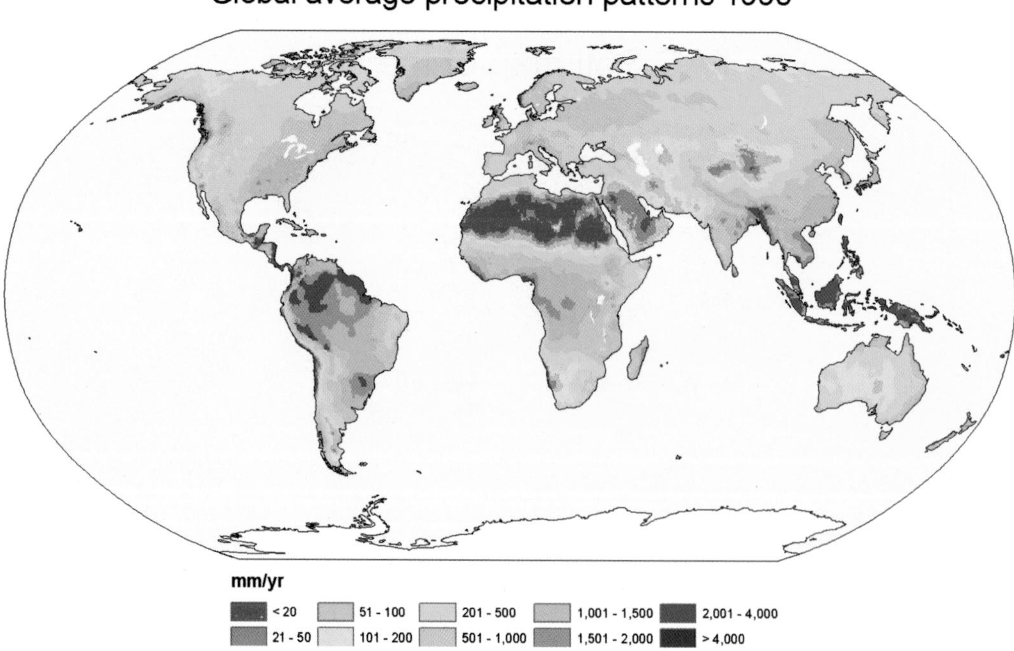

Figure 9.2. Climate map of Earth: annual average temperature and rainfall (source of maps: PBL).

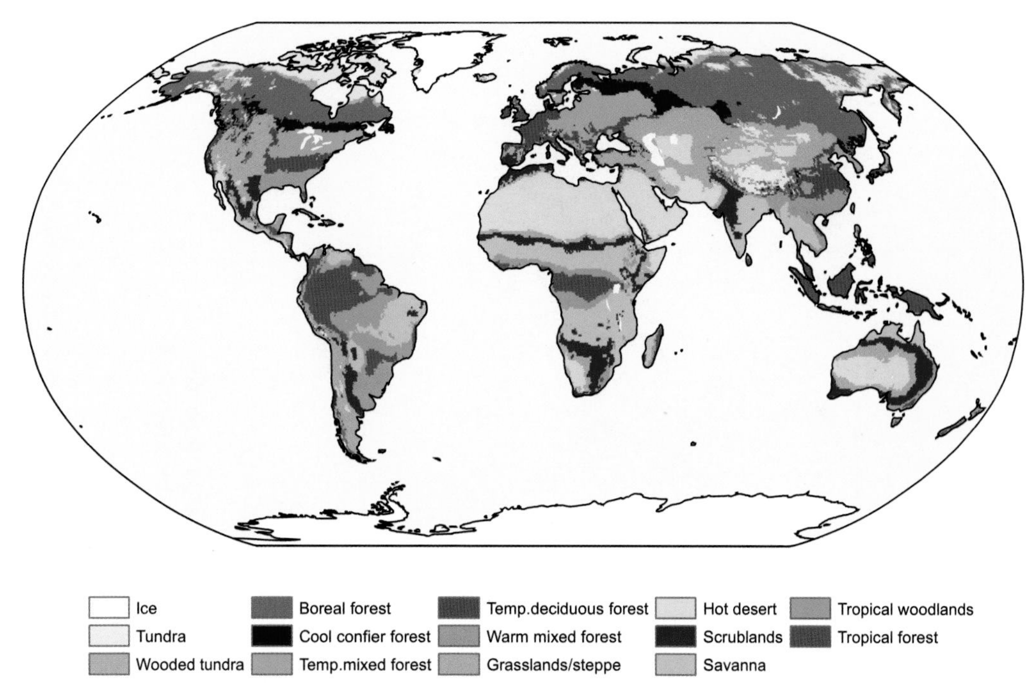

Figure 9.4. Potential vegetation cover based on Holdridge life zone classification (12 classes) (source of map: PBL/www.mnp.nl\image).

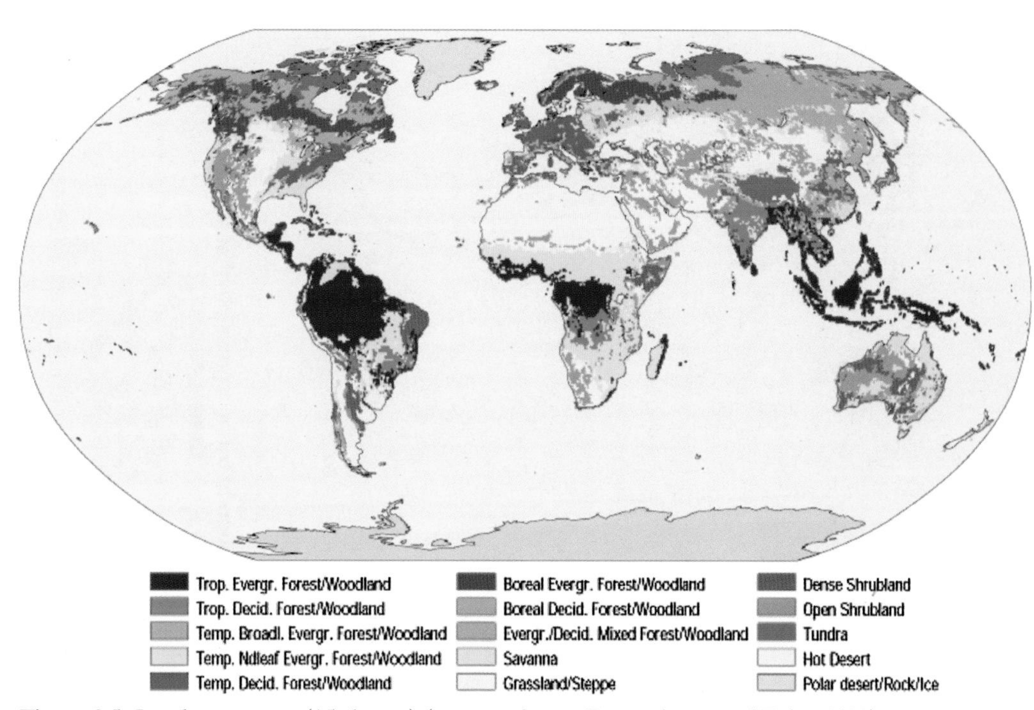

Figure 9.5. Land cover map (15 classes) (source of map: Ramankutty and Foley 1999).

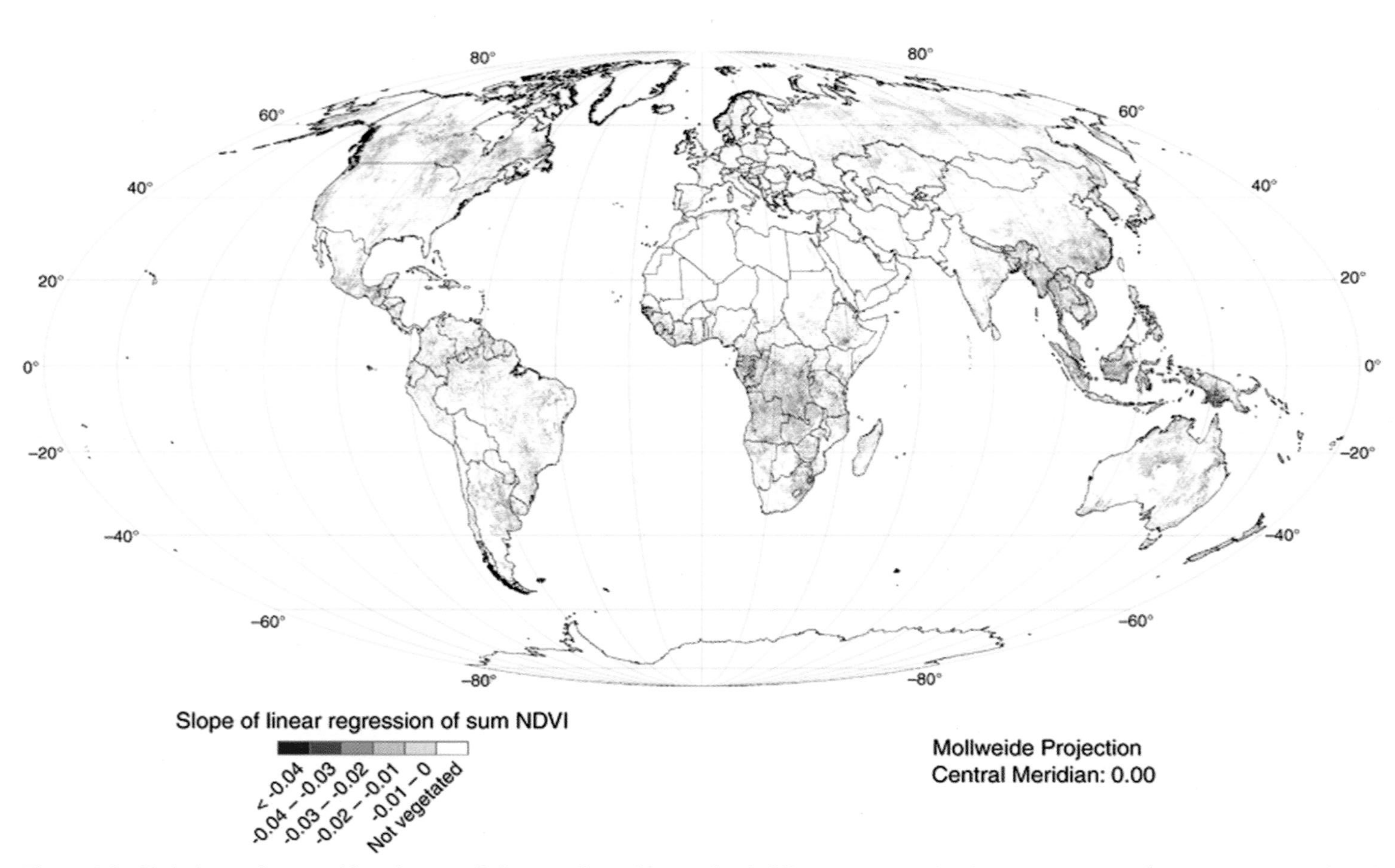

Figure 9.8. Global negative trend in rain-use efficiency-adjusted normalised difference vegetation index, 1981–2003 (source of map: Bai *et al.* 2008).

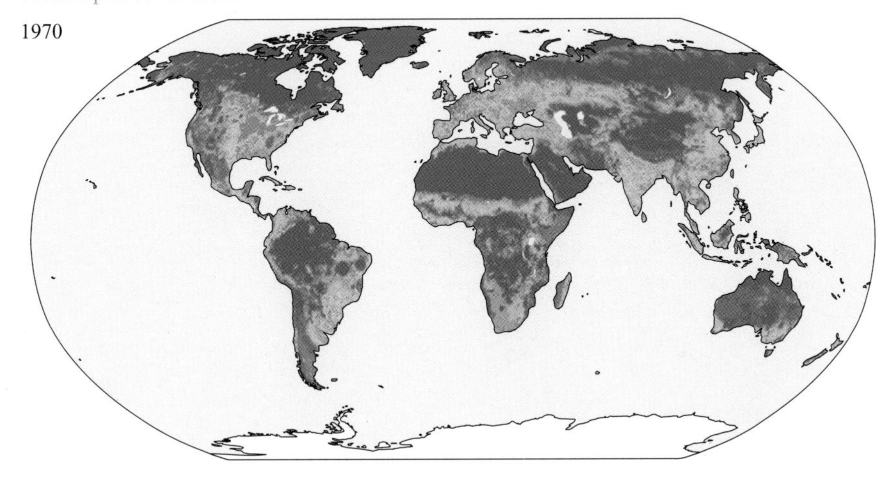

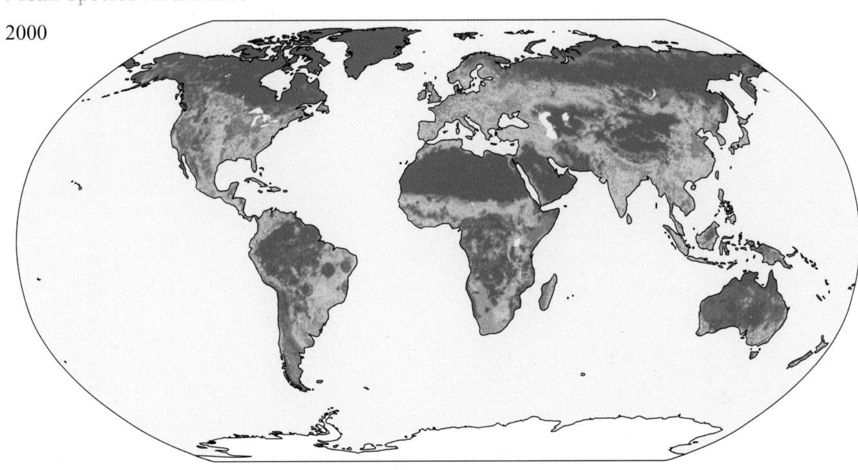

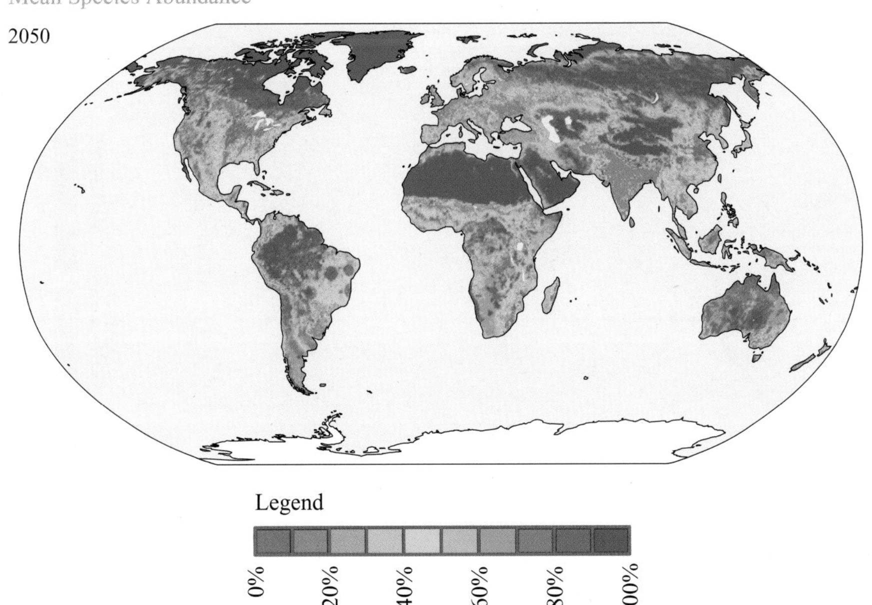

Figure 9.19. Map of terrestrial Mean Species Abundance (MSA) in 1970 and 2000, and a projection for 2050, based on calculations with the Globio-model (Alkemade *et al.* 2009, PBL 2010b).

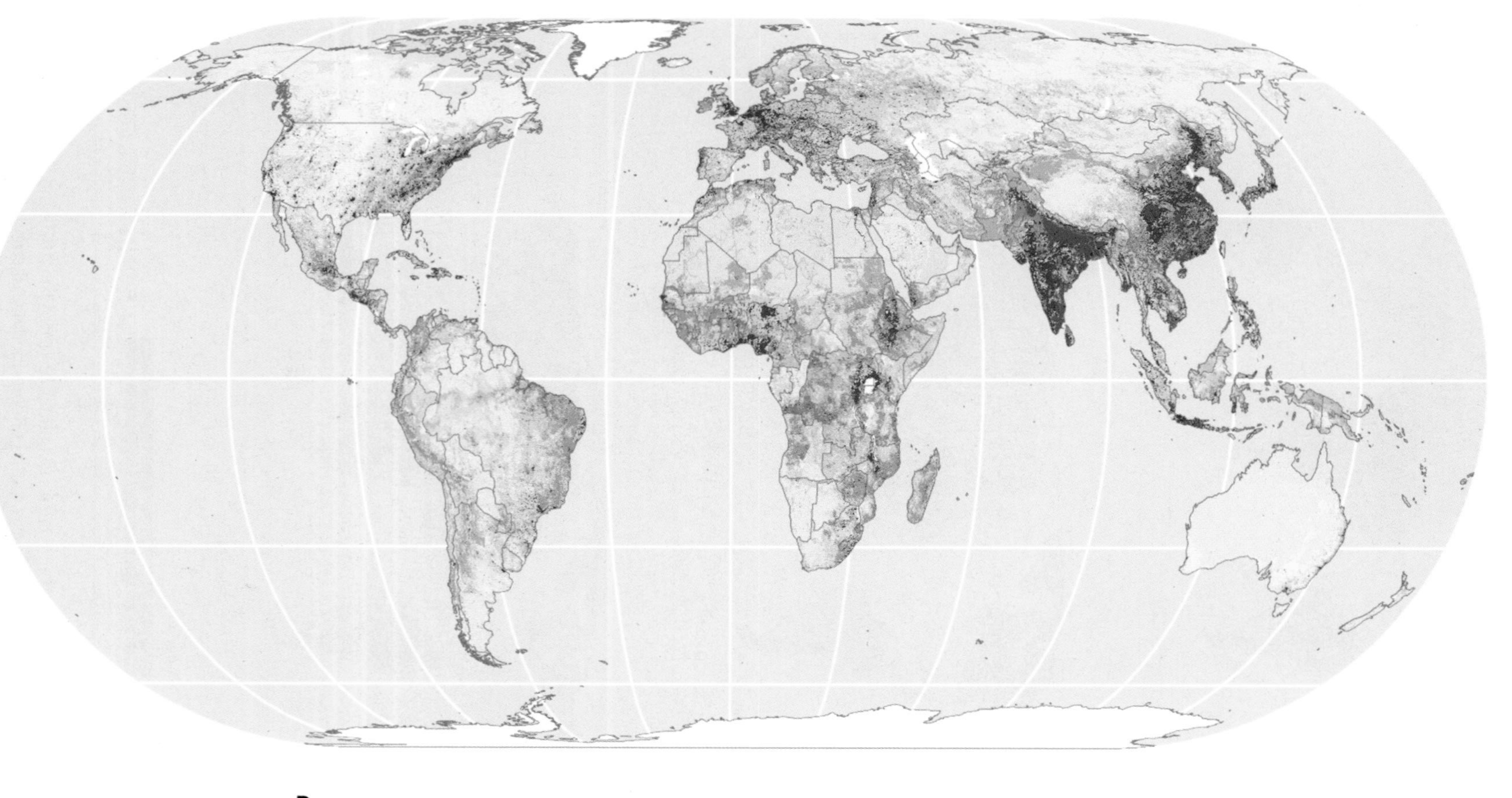

Figure 11.2. Map of anthropogenic biomes or anthromes (Ellis *et al.* 2010).

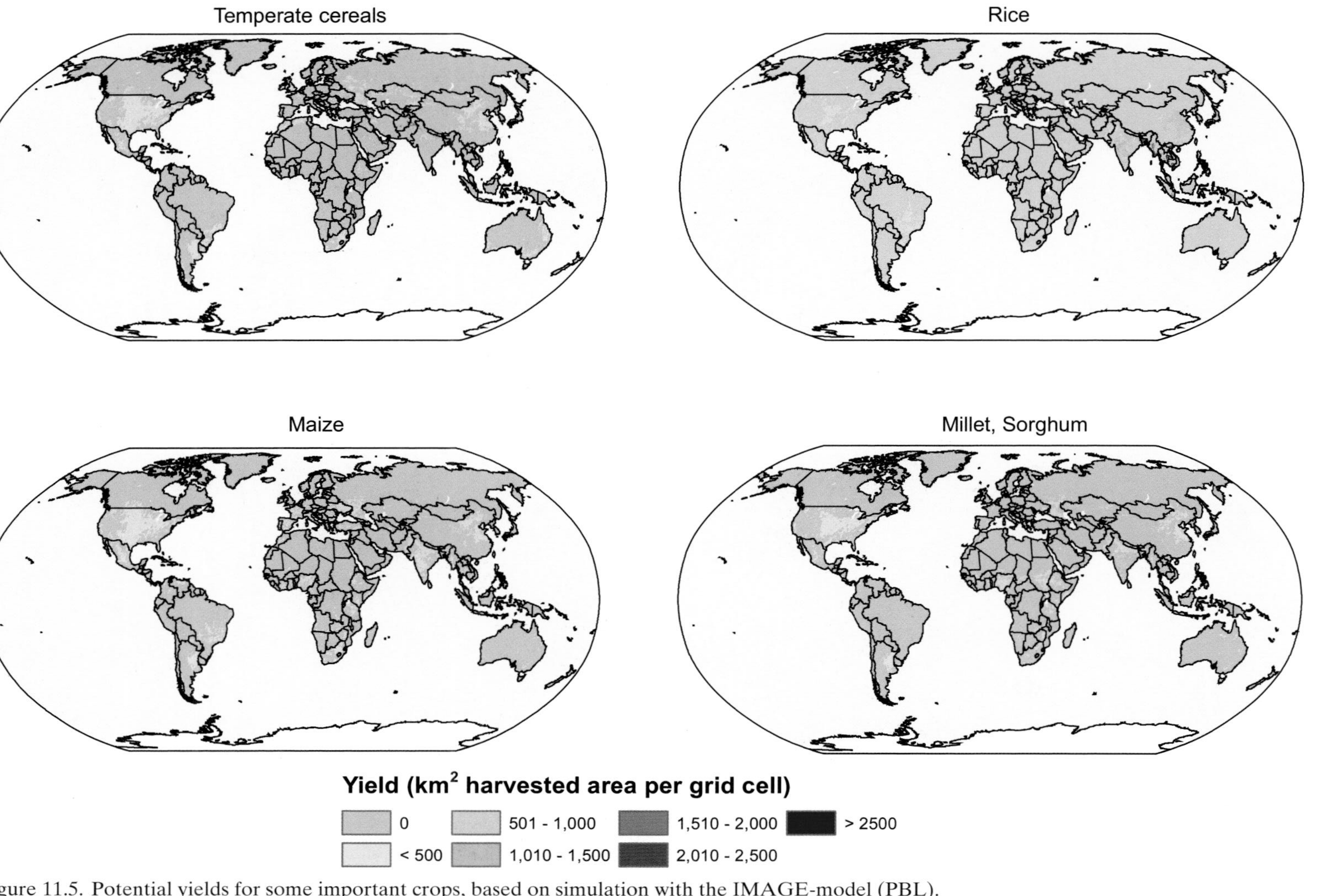

Figure 11.5. Potential yields for some important crops, based on simulation with the IMAGE-model (PBL).

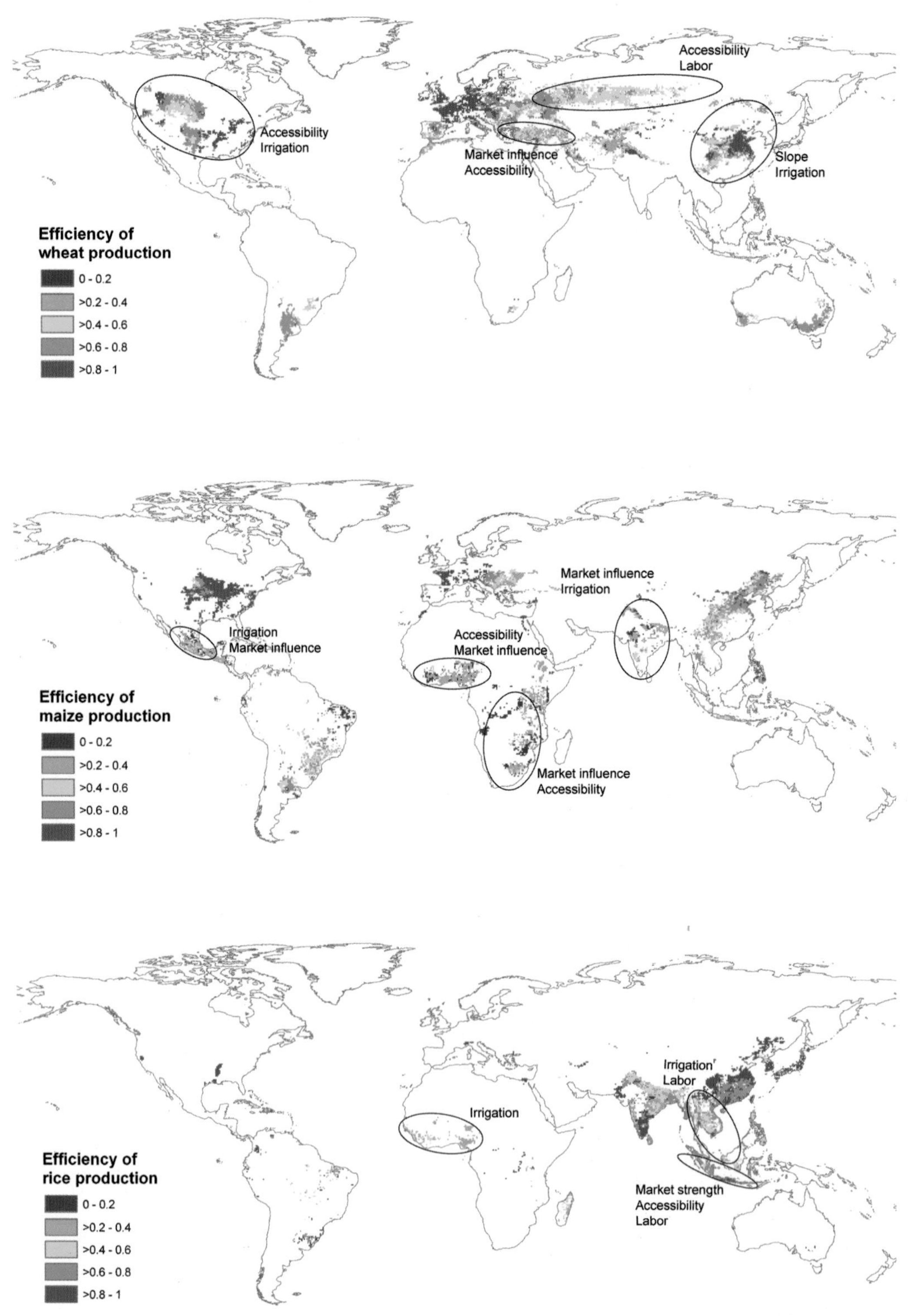

Figure 11.6. Estimate of the efficiency in wheat production (Neumann *et al.* 2010). Efficiency is defined as the ratio of the observed output to the corresponding frontier output.

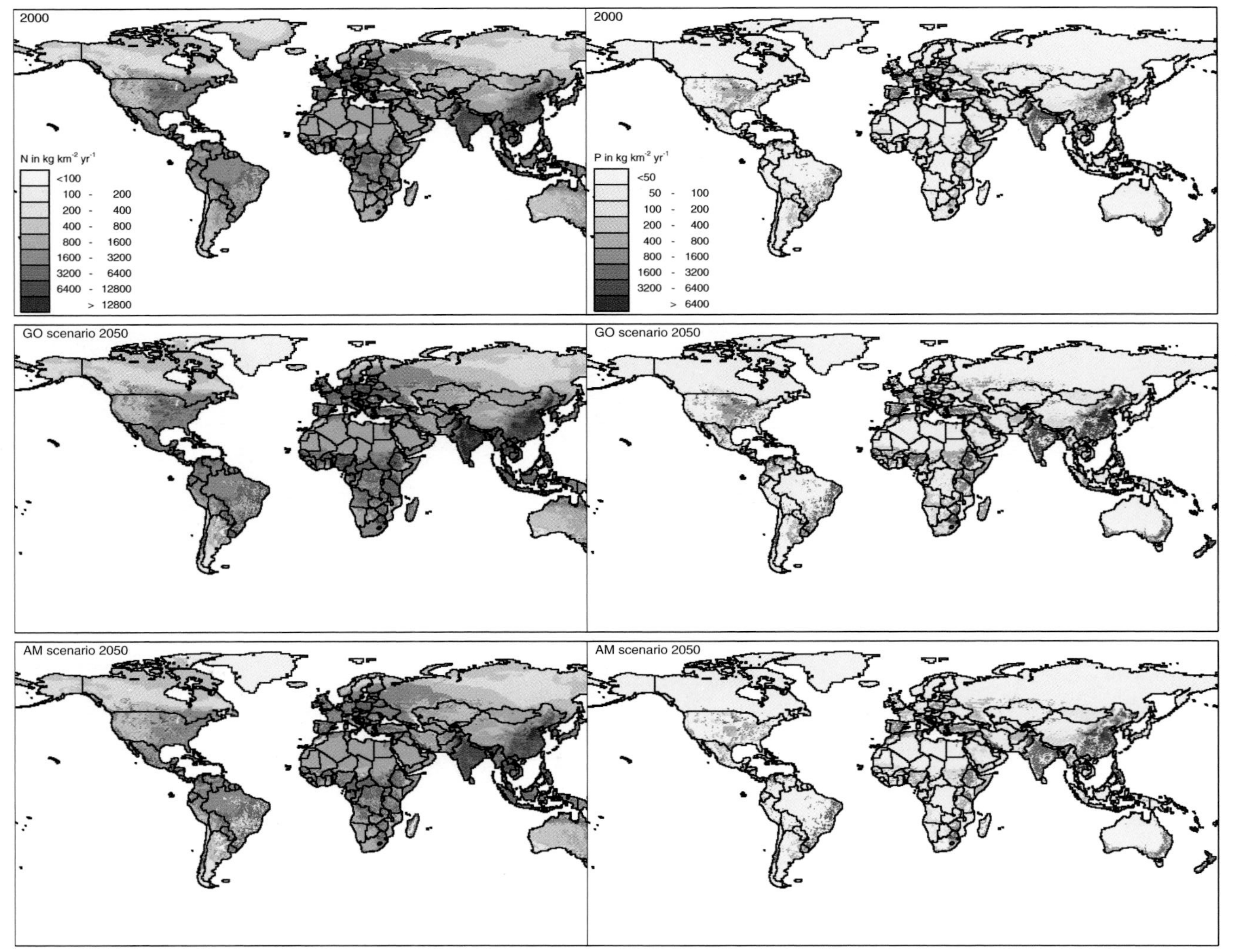

Figure 11.8. Total balance for (left) N and (right) P for natural ecosystems and agriculture for 2000 (Bouwman *et al.* 2009).

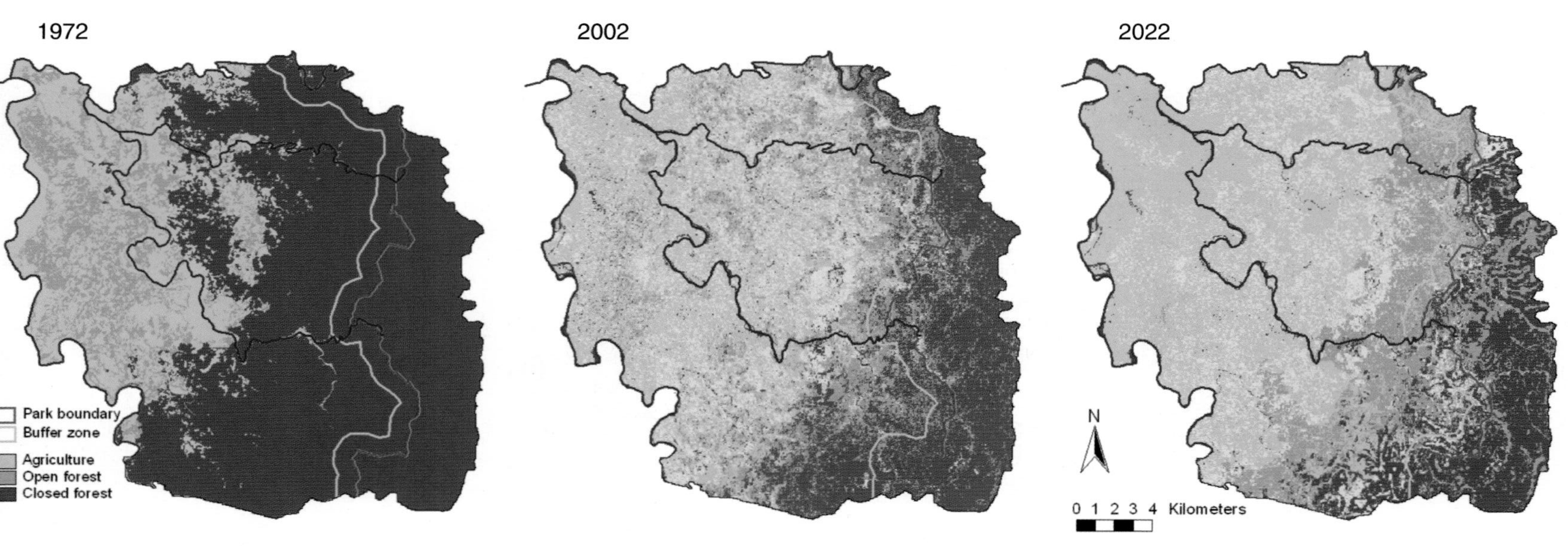

Figure 11.9. Generalised land use in San Mariano municipality in the northern Philippines: 1972 and 2002 based on interpretation of respectively aerial photographs and SPOT images; 2022 simulated by the spatial landscape-level model (Verburg *et al.* 2006).

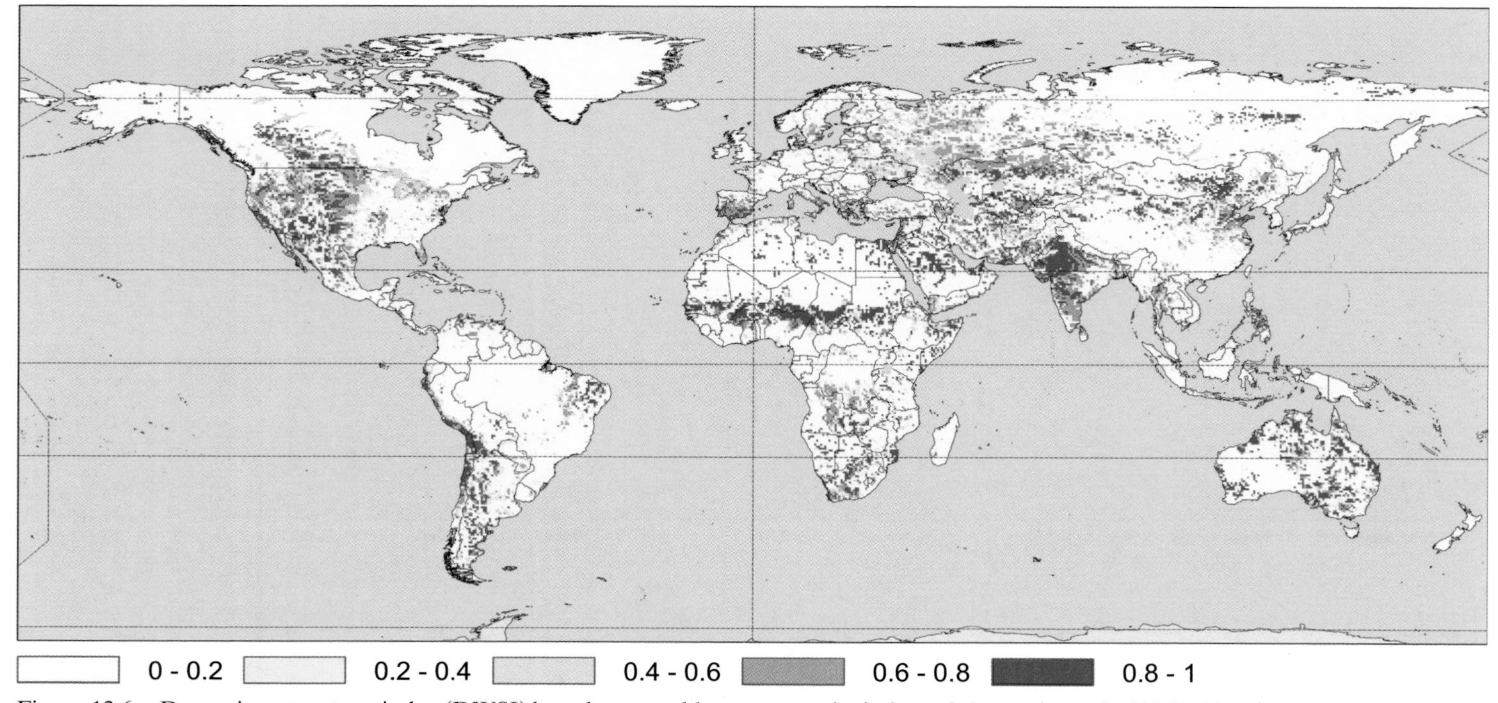

Figure 12.6a. Dynamic water stress index (DWSI) based on monthly water scarcity index ≥ 0.4 over the period 1958–2001 (Low: 0–0.2, Moderate: 0.2–0.4, High: 0.4–0.6, Very high: 0.6–0.8, Extremely high: 0.8–1.0) (Wada *et al.* 2011).

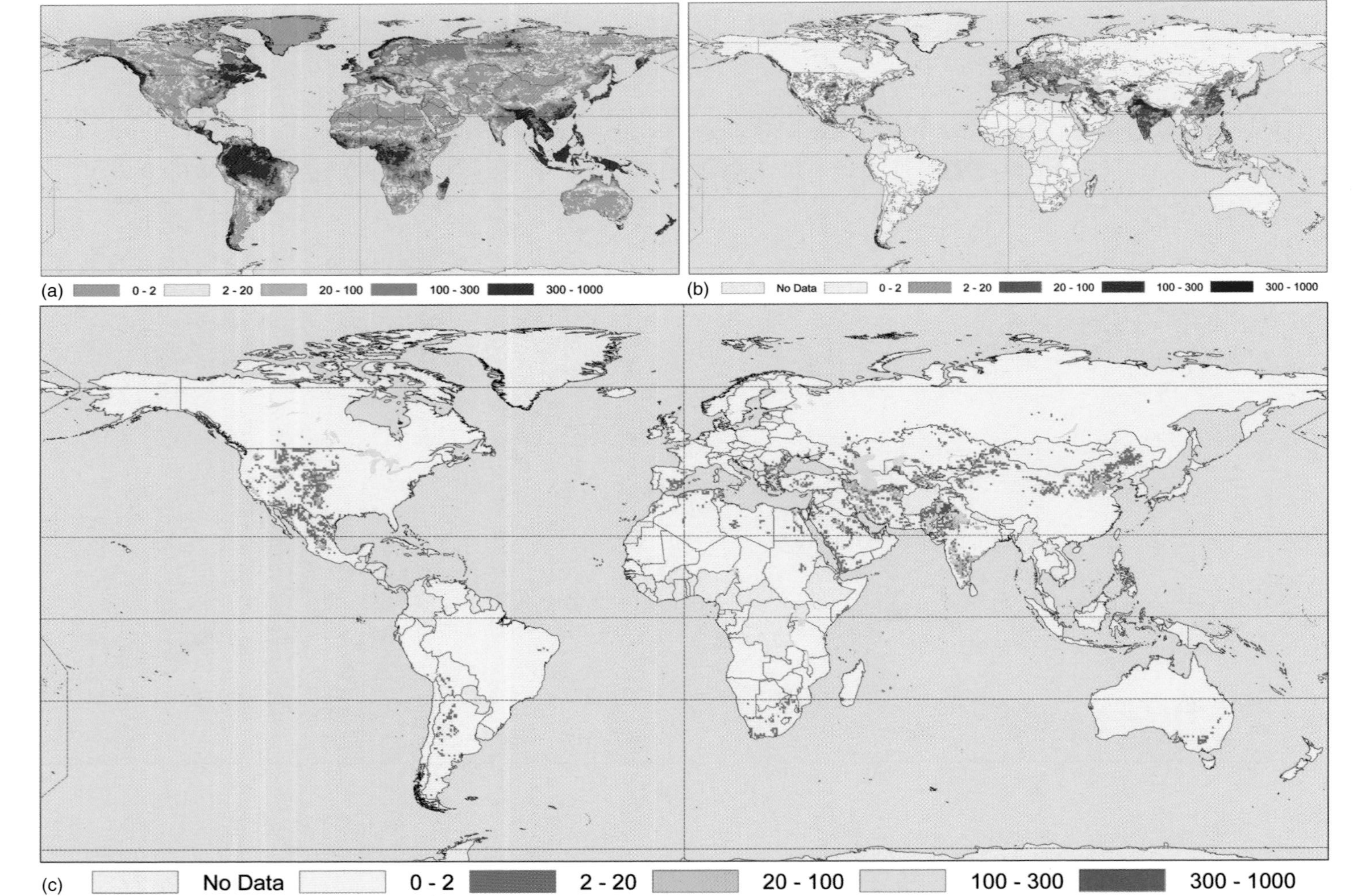

Figure 12.6b. Groundwater depletion in the regions of the United States, Europe, China and India and the Middle East for the year 2000 (mm · a−1; clockwise from top left). (Wada *et al.* 2010; reproduced/modified by permission of American Geophysical Union).

10.2.2 Driving Forces

Using historical data and trends for CBR and CDR, one can construct a simple stock-flow model and make population projections. But demographers try to understand in more detail the determinants of birth and death rate and more refined concepts and age-specific data have been developed (Lutz 2001, Hilderink *et al.* 2009). A first step is to distinguish separate classes (cohorts) in the population, notably age group and gender (Figures 2.10b and 10.4). The age structure has implications for future population growth via the number of fertile women in the population. A second step is to investigate the biological, socio-economic and behavioural determinants of fertility and morbidity/mortality rates in each age/gender group. Figure 10.5 is a diagram that portrays the most important determinants and connections. There is a clear correlation between income and fertility/mortality rates, directly through socio-economic and cultural factors and indirectly via state provision of education and health services – another metamodel or 'stylised fact'.

Human *fertility* is a biological process in which social, economic, cultural and environmental variables play a role. The number of births mainly depends on the potential number of children women can have (total fecundity), the rate of marriage, and the degree to which contraception and induced abortion are applied. Fecundity is related to nutrition, amongst other factors. Model analysis usually focuses on the average *total fertility rate* (TFR), which is the number of children a woman would, on average, have given birth to at the end of her reproductive life span if current age-specific fertility rates would prevail.

Proximate drivers are people's desire to have less children as incomes rise because child mortality drops and old-age provisions improve, and the increase in opportunities for women to use contraceptives and to make a career – and a decision – of their own. Other determinants are largely socio-cultural and vary amongst societies and over time, sometimes even as part of explicit population policy. Some of these relationships can be estimated from correlations with a woman's social-economic status. Women's levels of educational attainment turn out to have an especially strong effect on fertility in particular, more even than men's education levels or other characteristics of households such as wealth. Education levels are, not surprisingly, correlated with income.

In many countries, there is a clear preference for male children (*son preference*). This gives an upward pressure on fertility levels because many couples continue to have children until a son is born, resulting in a higher TFR than without gender preference. At the global level, this effect is rather small, but it may be significant at the country level. Data from six countries with strong son preference – Bangladesh, Egypt, India, Nepal, Pakistan and Turkey – show that the number of women pregnant at the time of the survey would be lower by 9 percent to 21 percent in the absence of son preference. The effect becomes more evident with a rise in contraceptive prevalence and a decline in fertility level (Nag 1991). In Sub-Saharan Africa, Southeast Asia, Latin America and the Caribbean, fertility levels are relatively unaffected by couples' desire to have a child of a specific sex (INFO Project 2003). With increasing education and continuing modernisation, son preference seems to fade away gradually.

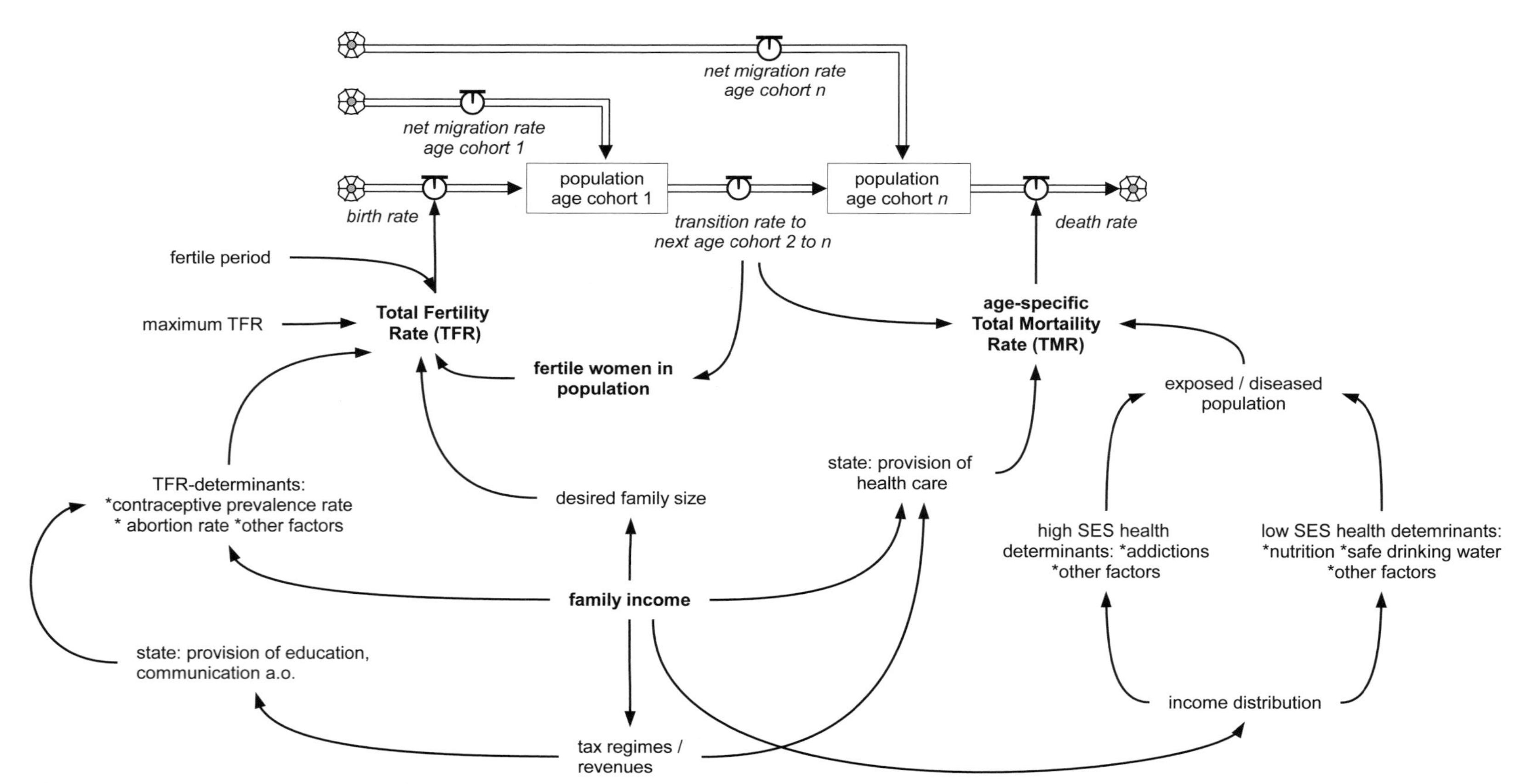

Figure 10.5 Causal loop diagram (CLD) with determinants of population. The important determinants of the flows (birth and death rate) are indicated.

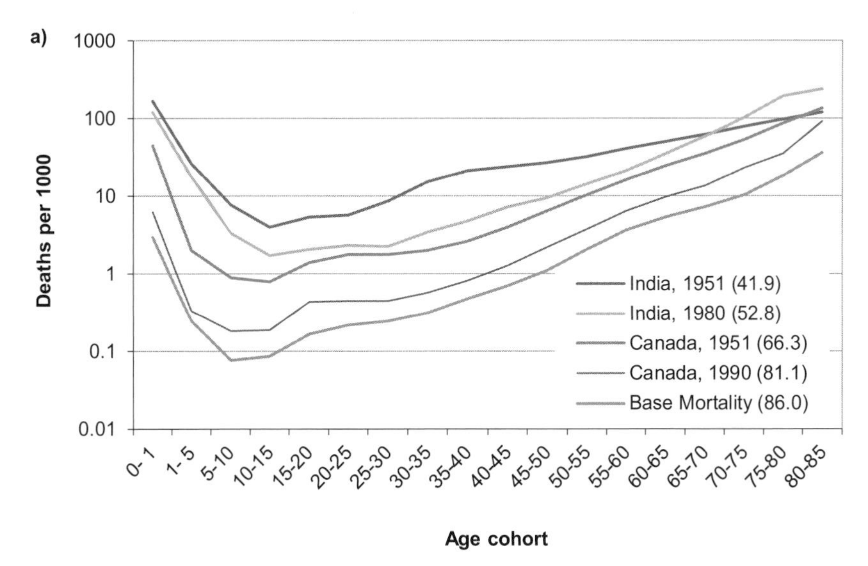

Figure 10.6a. Base mortality rates for the five-year age cohorts for Canada and India in the years 1951, 1980 and 1990. The lowest curve is the base mortality rate. The corresponding life expectancy is indicated between brackets (Source of data: Hilderink *et al.* 2009).

The causes behind human *mortality* are more varied than for fertility. The common approach is to start with an age-specific base mortality rate, such as the lowest possible. An estimation for different age classes is shown in Figure 10.6a. Mortality from explicitly listed health risks (diseases, accidents, war and so on) are estimated from exposure to these health risks and added to the base mortality. Exposure is estimated from socio-economic and environmental variables such as malnutrition and deficient water supply and sanitation. These tend to correlate with income, education, climate and other indirect variables. An additional factor to be considered is the level and effectiveness of health services, which tend to be correlated with health expenditures and, again, with income. To calculate the total mortality, an excess mortality is added from unexplained and non-health risk–related ones. The *life expectancy* (LE) can be calculated from the age-specific mortality rates: It is the number of years a certain age group may expect to live given the present mortality experience of a population (Table 4.2).

From these correlations, one can postulate causative factors that allow an estimation of mortality rates as a consequence of health risks and as a function of age and gender. Simulating the changes in the determinants of mortality, particularly income, it is possible to reconstruct the death rate and its causes over time. There are still large differences amongst the world regions (Figure 10.6b-e). The trends in the causes of death in India changed significantly since 1950, one of the clear marks of modernisation. In Western Europe, the major causes are the 'welfare'-related obesity, smoking and blood pressure, and the changes are much smaller. For both regions, an extrapolation is shown, based on an officially expected income growth path. In India, one can expect a significant increase in welfare-related diseases.

From the probabilities of certain risk-disease combinations, you can calculate the number of years lived with a disease and of years lost by premature death.

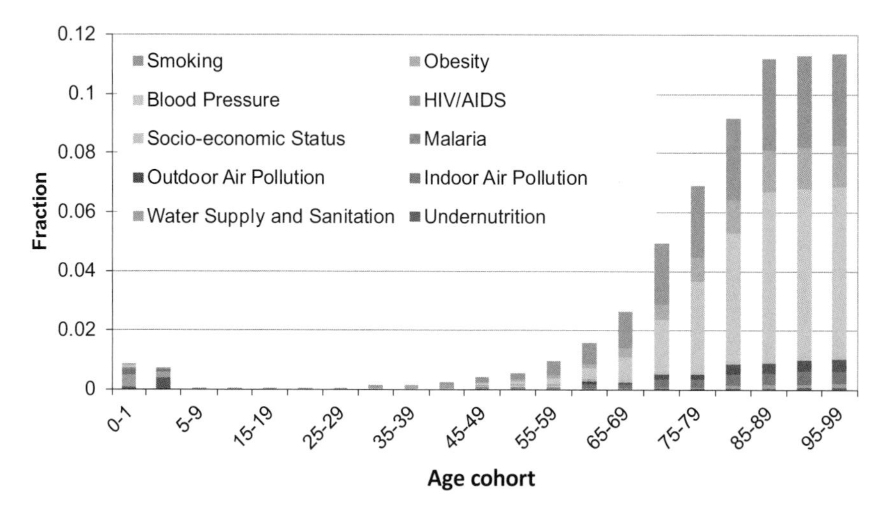

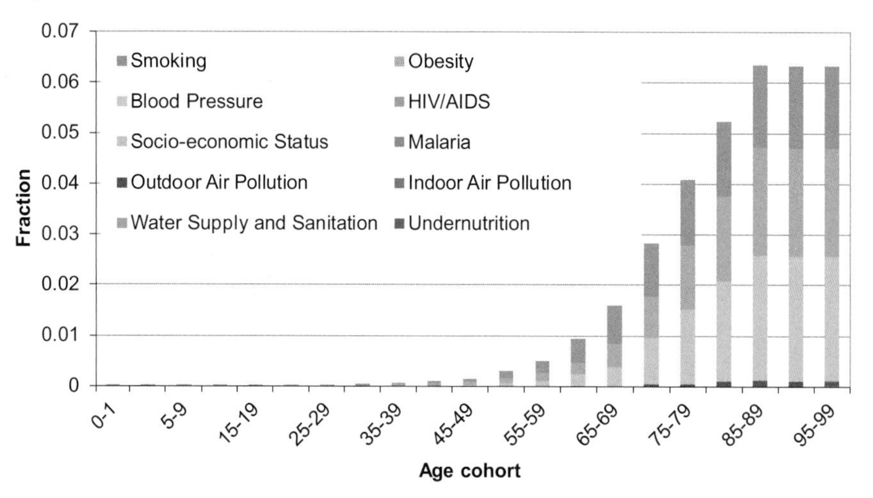

Figure 10.6b,c. Attributable mortality by exposure for ten health risks in Western Europe and India in 2010, based on simulation with the Phoenix model (Hilderink *et al.* 2009).

The sum of these two is the *disability adjusted life years* (DALY). The DALY has become the measure of the burden of disease in a population. The main risk factors from which the DALYs are estimated are poverty- and environment-related risks in poor countries and welfare-related risks in rich countries (Figure 10.7). Poverty (childhood and maternal undernutrition) and environment (notably lack of adequate sanitation and safe water) dominate risks in poor countries. For instance, in Africa, childhood and maternal undernutrition account for more than 40 percent of all attributable DALYs. In rich countries, welfare-related risks prevail. In Europe, addictive substances (such as tobacco and alcohol) and diet-related and physical inactivity factors (such as blood pressure and obesity) are responsible for the larger part of health losses. These estimates point at the areas where policy interventions are most effective in terms of gains in healthy life years.

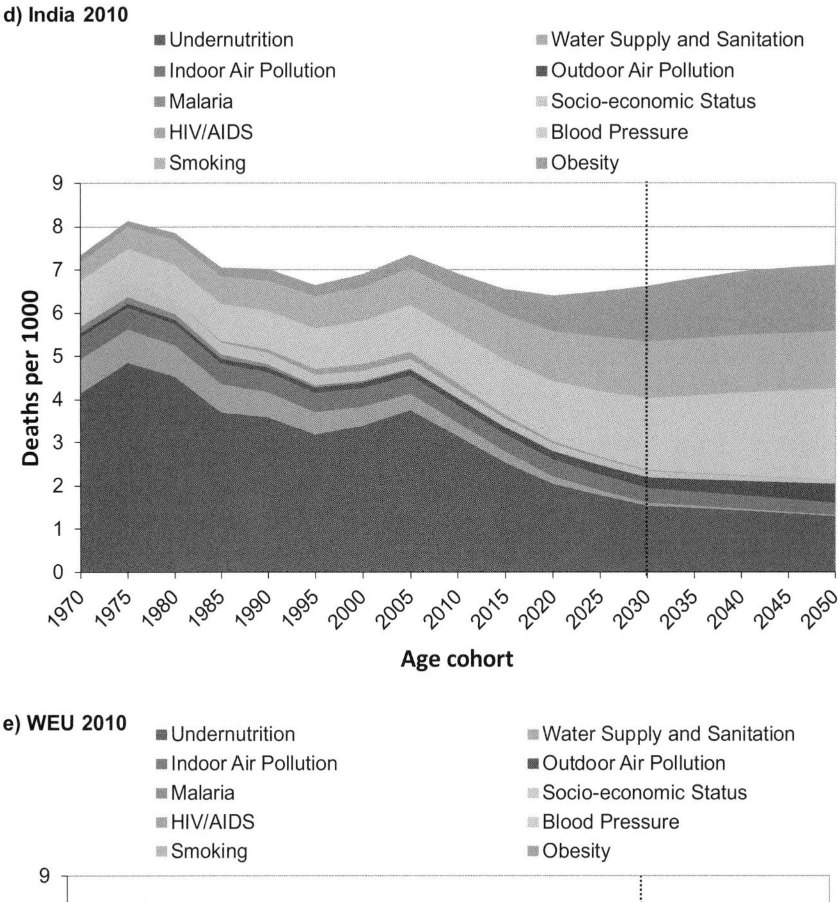

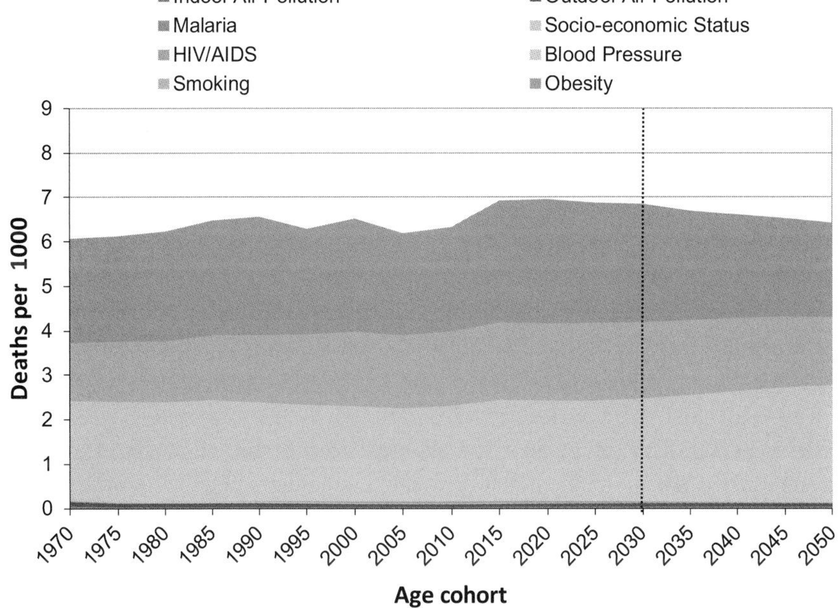

Figure 10.6d,e. The epidemiological or health transition: attributable mortality by exposure (per 1000 persons) in Western Europe and India in the period 1970–2050, based on simulation with the Phoenix model (Hilderink *et al.* 2009).

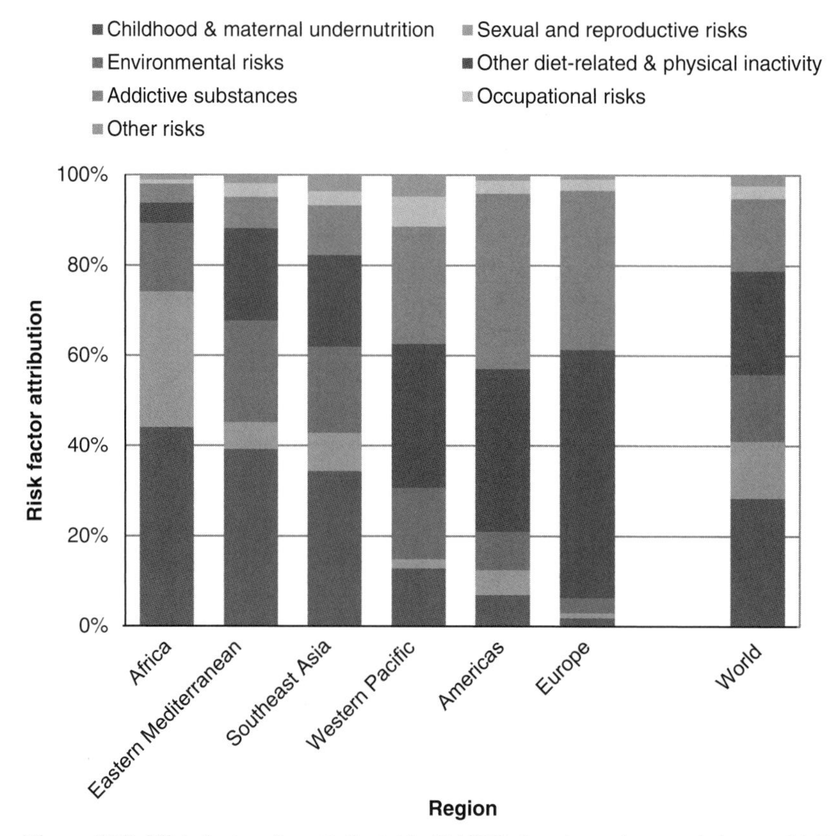

Figure 10.7. Risk factors for attributable DALYs for six regions and the world. The percentages indicate which part of the total disease burden (DALY) can be attributed to each of the seven risk factors (Source: WHO 2002).

The gradual shift in risks with rising incomes is called the *epidemiological* or *health transition* (Martens 2002). Currently, one of the clear manifestations of this phenomenon is the rapid rise of obesity and, as a consequence, diabetes and heart diseases amongst the middle classes in countries such as India and China. The main reason is the change in lifestyle, from traditional to fast food. Environmental risks are present in all regions but are far more important in poor countries and cause about 15 percent of the DALY loss worldwide. The effect of sexual and reproductive risks, mainly HIV/AIDS, is clearly visible in Africa but forms a growing threat in other regions such as Asia.

10.3 Evolution: Our Biological Roots

Human development has for a long time been primarily a co-evolution with plants, trees and animals. What insights on sustainable development are received from a biological perspective? The central concept is evolution, primarily in a Darwinian sense. The two essential principles of evolution theory are (Mitchell 2009):

- natural selection on randomly generated, small variations in individuals is the major mechanism of evolutionary change and adaptation, and
- macroscopic phenomena are understood by microscopic processes of gene variation and natural selection.

Box 10.1. *Population regulation amongst the !Kung San.* The !Kung San people live in small bands in the hostile environment of the Kalahari desert. Studies by anthropologists give insights in the interaction between humans and their environment. A simple model has been constructed by Read (1998). It assumes the resource base is fixed regardless of the intensity of foraging and hunting, and that population growth is determined by the 'textbook' logistic growth equation with a population density-dependent fertility rate. A behavioural element is added to the model: !Kung San women want as many children as possible, but they are concerned for the well-being of their family and balance the time and energy spent on children and on family support–related activities. How much time they can spend on the family depends on the resource characteristics, and so does their desired family size. The fertility rate depends on the resource via the energy expenditure of a !Kung San woman. If the energy expenditure exceeds a threshold value, the fertility rate drops to zero. This causes a crash-and-boom periodicity in the population size for a low energy expenditure per child – a pattern that may have characterised the Netsilik Eskimo with periodic starvation from unexpected and substantial changes in their resource base. If the energy expenditure per child is high and women therefore have to take into account the time and energy requirements for caring for offspring, there is still boom-and-bust behaviour, but of a much lower amplitude.

Ethnographers have found that !Kung San women did not want additional children *unless* they could care for them properly. Therefore, birth spacing decreased with lower resource procurement efforts. This means that a woman's decision on how to allocate her energy between the well-being of the family, such as searching for food, and the desire for more children is further tilted towards the former. In a subsequent extension of the model, the threshold value for the energy expenditure is, therefore, assumed to increase with population size in order to reflect resource scarcity. This gives an even more adequate tracking of historical population size and resource adequacy. Instead of overshoot-and-collapse, there is stabilisation. What is the lesson? First, it matters whether females place high priority on the well-being of the family or instead focus on their desire for more children. Resource abundance is usually not a control option, but birth spacing is! Second, culturally mediated individual behaviour introduces anthropological insights with possible population projections that are very different from what (over)simplistic models tell us.

The evolutionary perspective on human nature emphasises the ability to change and adapt, to adopt new behaviours and to produce social norms and culture. Human nature is seen as adaptive and innovative, more than as a limitation or something to be subdued. This view, with its idea of progress and its competitive search for efficiency, is an important part of Modernism. 'The theory of evolution by cumulative natural selection is the only theory we know of that is in principle capable of explaining the existence of organized complexity' (Mitchell 2009).

The evolutionary perspective can offer unexpected and new vistas. The advances in the life sciences (biology, cognitive and neuroscience) shed new light on the question how malleable and diverse 'human nature' is. Some biologists claim that genetics are the key to understanding individual and social behaviour. This view is part of a long-standing thread in the scientific discourse, with Dawkins' books, *The Selfish Gene* (1976) and *The God Delusion* (2006), amongst the recent expressions. Its rigorous application to social organisation of not only ants and bees but also of humans has led to the controversial branch of sociobiology as described first by Wilson in his book *Sociobiology – The New Synthesis* (1975). It emphasises the role of natural selection in social behaviours such as mating and hunting. It borders on biological determinism and has been used ideologically to justify social inequity and racism. This variant of *social Darwinism* is unacceptable to those who cherish a less materialist-empiricist worldview. Similarly, some neuroscientists are also making claims such as 'we are our brain', on the basis of brainscans and experiments. Strong statements about physico-chemical processes and behaviour under simplified controlled conditions are broadened to weak statements about the free will, responsibility, the soul and other complex notions, with sometimes strong ideological impact. It reflects the tendencies of the (post)modern worldview.

The arguments that evolutionary biology provides the central organising principles for understanding the behaviour of humans as well as other animals concentrates on a few observations and interpretations. First, human behaviour with respect to offspring and consumption has biological roots, as is seen from the similarity with animal populations. Why do people want offspring and under which conditions do they reduce the birth rate? Reproductive success is in many, and not only traditional, societies positively associated with great wealth and high position. Biological research suggests biological roots of this phenomeon: 'Costly exaggerated displays enable high quality males to honestly advertise their quality to potential mates and rivals because only high quality individuals can bear the costs of the display' (Penn 2003). People acquire display items to demonstrate group membership and they need items worn by high-status individuals to rise in social status. The economist Veblen (1899) introduced the notion of 'conspicuous consumption' and suggested why people pursue extravagant goods such as fashionable clothes, luxurious cars, and massive homes: 'Consumption of... excellent goods is evidence of wealth... the failure to consume in due quantity and quality is a mark of inferiority and demerit.'

Biology also sheds light on competitive versus cooperative behaviour amongst humans. There is a fierce debate going on about the balance between selfishness and altruism. 'It is human nature to want more than what is necessary to survive and reproduce – more resources, more social status, more mates – but it is also human nature to want fairness and to shame individuals that behave selfishly' (Penn 2003). The study of primate behaviour points at the existence of altruism in animal populations, such as behaviour that is beneficial for the receiver and has a cost for the giver irrespective of motive (de Waal 2007). The rationality of this can be understood in a context of reciprocity and interdependence, including a sense of empathy, retribution and justice. There is also *in silico* evidence that higher forms of cooperation and organisation can arise out of individual self-interest (Nowak and Sigmund 2004). But the more complex and consistent moral considerations are probably uniquely human outcomes of social interactions. Even apart from the ethical implications, these issues matter in sustainability science because most

Box 10.2. *Rationality in economic science.* Economists usually interpret *rationality* as self-regarding materialism and not as consistency. The British economist Kay argues in his book *Culture and Prosperity* (2001) that this occurs to keep the analysis simple: 'Economists insist on rationality because they do not like the alternatives... [this is why] economists adopt a concept of rationality that reduces to self-interest. It seems to offer an anchor in an ocean of otherwise unpredictable human behaviour. The assumption of rationality gives economics rigor that distinguishes it from other social sciences... self-regarding, materialistic behaviour would be the norm because no other behaviour could persist in a market economy' (Kay 2001: 212).

History provides interesting lessons about behavioural motives and social context. During the Cold War, Intelligence Services from the United States, CIA, and the former Soviet Union, KGB, actively recruited spies in Western Europe. They realised that persons qualifying as a spy would have different motives. The FBI distinguished four possible motives, indicated by the acronym MICE: 'M' for Money, 'I' for Ideology, 'C' for being prone to be compromised (financially or sexually) and 'E' for Ego. This is a summary of why individuals may become free riders and break with group solidarity and cooperation.

environmental issues are collective-action problems (§5.4). To solve such problems, social pressure and coercion are supposedly effective because people care about their reputation: shame works, and public shaming in combination with other incentives may force cooperation (Axelrod 1997).

Evolutionary biology indicates that people have certain traits that result in maladaptation to changing circumstances. Cognitive research shows that human beings can solve extraordinarily complex problems and yet fail to do simpler tasks because 'human cognition and behaviour are designed for ecological rationality and we rely on simple heuristics rather than complex algorithms to solve problems faced by our ancestors' (Penn 2003). Perhaps, the computers of the information age will help to overcome these shortcomings. A related question is why people prefer the short term over the long term. Evolutionary game theory suggests that it is the outcome of weighing expected present and future fitness pay-offs under uncertainty in the quest for survival. In a sustainability science context, an interesting development on the interface between life and social sciences is niche construction theory (Odding-Smee *et al.* 2003). Its main tenet is that organisms change their environment and, in the process, modify the selection pressures of their own and other species. Such a co-evolution of populations and their environments is backed up by field and simulation experiments and makes the carrying capacity a dynamic element in population ecology. The idea that species 'engineer' and control their environment is another bridge with humans, who can be seen as the ultimate niche constructors. An empirical illustration are the historical population estimates for England 1545–1975 and Japan 1100–1992, which clearly show jumps in the carrying capacity as a consequence of human-environment-technology interactions (Meyer and Ausübel 1999).

10.4 *Homo Economicus* and Its Critics

10.4.1 Consumers and Producers

In most of the earth and life sciences, there is only a descriptive and no prescriptive component. Intentions and goals may be assigned to plants and animals, but there is no 'should' or 'ought to' involved. The rain just falls in a gravitational field. Plants and trees occupy their place ('niche') within soil and climate constraints. The predator kills the prey and is not judged for it. Optimality and efficiency in natural processes are evolutionary effective but without normative content. But inevitably, with the introduction of human beings, normative and ideological elements enter. This is the essence of human freedom and responsibility. Therefore, observations on economic behaviour are always somewhere between 'what is' and 'what-should-be-done', fact and value, science and ethics or means and ends. As part of sustainability science, a reflection on the image of man in economic science is a necessity. Again, this book merely scratches the surface of the vast literature, so review the Suggested Readings.

In the 20th century, economic theory has become dominated by the neoclassical economic school with a particular image of man. Humans – or *agents* using modelling parlance – in economic decision processes are rational and choose with perfect information and foresight in a perfectly competitive market. As consumers, they are in isolation maximising their utility within a budget constraint. As producers, they are constantly in competition on 'free' markets. This *Homo economicus* species is rooted in the Modernist interpretation of science (rational) and evolutionary theory (survival and competition). Sophisticated theories and models have been constructed in (mathematical) decision analysis to extend and refine this image of man, but without changing the essence. In Chapter 6, the discussion of worldviews made clear the inherent reductionism in this standard image of man in (micro)economic science.

Consumer theory presumes a concave *utility function* $u = u(c)$ with u per capita utility and c per capita consumption. This theoretical construct represents a set of ordered, fixed preferences of an individual person (§6.1). A lot of attention has been given to theories and practises of strategies to maximise utility in a narrow (consumers) and broad (nations) setting. In most macro-economic models, an aggregate 'national' utility function U is maximised across generations in order to deal with, for instance, climate change (de Vries 2010). In formula:

$$\max \int_{t=1}^{t=T} U(C) \cdot e^{-rt} dt \tag{10.1}$$

with C the national consumption and T the time horizon over which the decision matters. The term e^{-rt} is the discount factor. It converts the annual income streams into the discounted sumtotal in order to take into account that people value the present more than the future (time preference).[6] Such a formulation reflects a utilitarian ethic and a belief in perfect markets. The dominant criticism focuses on the presumed rationality and the lack of interaction with others and attempts are made

[6] The sum total is called the *present value* of the income stream. The underlying behaviour is referred to as *time preference*.

to introduce more realistic behaviour such as cooperation and altruism (Nowak *et al.* 1990, Brock and Durlauf 2001).

Already in the 1960s, in his book *Models of Man*, Simon (1957) criticised absence of cognitive and social capabilities in economic models and searched for a genuine theory of human motivation to understand individuals in organisations better than prevailing market theories. He suggested that people are *satisficing*, that is: looking for adequate and not for optimal strategies in everyday decision making. Besides, it has been empirically confirmed that uncertainty and risk perception play an important role (Kahneman *et al.* 1999; van den Bergh *et al.* 2000) and that context and availability of and differences in information influence people's decisions (Gintis 2000, 2005). Status, experience and novelty in postmodern consumer culture are to be considered in a larger system perspective (Hirsch 1977; Pine and Gilmore 1999). Consumption of goods and services is much more than satisfying basic material needs for food, shelter and so on. 'Material things offer the ability to facilitate our participation in the life of society... our attachment to material things can sometimes be so strong that we even feel a sense of bereavement and loss when they are taken from us... Novelty plays an absolutely central role in all this' (Jackson 2009). Consumption is partly a substitute for religious consolation, a filling up of the 'empty self' and combating a sense of meaninglessness (Handy 1998; Figure 6.2b). In recent times, the resulting restless desire of the consumer is merging with the restless innovation drive of the entrepreneur.

Traditional economics fails to consider the extent to which people are guided by noneconomic motivations. The theory ignores animal spirits: 'The economics of the textbooks seeks to minimise as much as possible departures from pure economic motivation and from rationality' (Akerlof and Shiller 2009). A possible remedy consists of theoretical fragments dealing with real-world phenomena such as confidence, fairness, corruption and bad faith and money illusion. In recent years, significant refinements are introduced in economic theory, and connections are made with empirical findings in psychology and sociology. At the interface, new disciplines emerge such as evolutionary economics and behavioural and experimental economics.

Production theory assumes *profit-maximising entrepreneurs* who enter markets until the last entrant no longer sees an opportunity to gain a profit. It implies a constant search for cost-minimisation. In its simplest form, it is written as:

$$\min C = \min \sum_{i=1..m} p_i X_i \text{ under the condition: } P = P_0 \cdot X_1^{\alpha} \cdots X_m^{\mu} \qquad (10.2)$$

with C the production costs, X_i the production factor i = 1..m, p_i the price of production factor X_i and P the output. Primarily, the factors are in neoclassical theory labour L and capital K. The condition represents the economic *production function*, such as the possible combinations of inputs with which the output P can be realised. Applying mathematics produces elegant formulas. But the assumption that an entrepreneur or corporation is rational and that the market is without oligopolism, trade barriers and institutional failure is often invalidated. The real-world economy is never in equilibrium, innovation is an evolutionary process and there are inherent tendencies toward concentration and centralisation away from the

Box 10.3. *Competition in a globalising world.* In Friedman's *The World is Flat* (2005), the following anecdote appears about the competitive world economy. A Chinese manager posts an African proverb, translated into Mandarin, on his factory floor:

> Every morning in Africa, a gazelle wakes up
> It knows it must run faster than the fastest lion or it will be killed.
> Every morning a lion wakes up.
> It knows it must outrun the slowest gazelle or it will starve to death.
> It doesn't matter whether you are a lion or a gazelle,
> When the sun comes up, you better start running.

ideal market.[7] In the 1980s, the first attempts were made for a new foundation, which is rooted in evolutionary mechanisms (Nelson and Winter 1982). In the new branch of evolutionary economics, the *Homo economicus* is replaced by a pragmatic and adaptive organisation that is in constant search for bettering its goals in interaction with other organisations and the environment. Behaviour is self-sustaining and self-reinforcing and not necessarily (cost-)efficient. Its metaphor is selection and adaptation in an evolutionary process, not utility maximisation or cost-minimisation in a mechanical world (Döpfer 2005; Frenken 2006). A similar critique and alternative descriptions of economic production processes are developed in system dynamics (Sterman 2000).

The simplifications in economic theory, many of them fashioned after classical physics, are understandable from a technical and a psychological vantage point. For early agricultural societies, it may be possible to give a description of human behaviour that is largely correct and permits formal modelling. For a modern, (post-) industrial society, one has to bring in more variety in motives and behaviour, and formal mathematical analysis becomes difficult. An example is the penetration of new products in the market, where the usual assumption of no returns to scale turns out to be invalid in many situations (Appendix 10.1). I refer the interested reader to the Suggested Reading, and I come back to it in Chapter 14.

10.4.2 Games, Dilemmas and Cooperation

An interesting and growing field is *game theory*.[8] It analyses the strategic interactions between individual agents: How do people interact? The most elementary device in game theory is the *pay-off matrix*. It gives the costs and benefits for an agent of two or more possible actions in an interaction with one or more other agents. Examples are the hawk-dove game, where two animals upon encounter can either

[7] The shortcomings of conventional economic theory and the historical explanation in terms of the aspiration to imitate the natural sciences and of the limitations of available analytical techniques are addressed in, for instance, Kirman (1993), Ormerod (1998) and Kay (2004). The analogy with 19th-century equilibrium thermodynamics and chemistry is a notorious feature (Döpfer 2005).

[8] Its beginning is usually associated with the book *Theory of Games and Economic Behaviour* (1944), by Von Neumann and Morgenstern. With the book *Evolution and the Theory of Games* (1972) by Maynard Smith, biology, ecology and ethology have also become areas of application.

Table 10.1. *Application fields of game theory*

Situation	Kind of strategies/dynamics	Discipline
Gene selection	Evolutionary strategies	Genetic biology
Animal behaviour	Hawk-dove game	Etology/ecology
	Competition-cooperation in reproduction	
Consumers	Social limits to growth	Economics
	Solid waste disposal	
Wars	Arms race: Security dilemma	Sociology
Crime	Prisoner's dilemma	
Common Pool Resources (CPR)	Monitoring and sanction regime	
Relationships	Duel: Game of chicken	Psychology
	Stay/divorce: Battle of the sexes	

attack (hawk) or retreat (dove); the big and the small monkey, collecting coconuts with or without cooperation; the two car drivers that compete on the issue of who will first slow down in the face of an abyss. Table 10.1 lists some of these situations and contexts of application.

One of the widely studied and better known games is the so-called *prisoner's dilemma*. This and similar dilemma situations (chicken game or ultimatum game) can be represented with a pay-off matrix as shown in Table 10.2. Each of the persons involved can either act in his or her own selfish interest (defect) or in the collective interest (cooperate). Interest is indicated with a benefit (reward or gain) that can be expressed in money but can also involve recognition, love and so on. In the prisoner's dilemma, the prisoners – who are not allowed to communicate – are told that if both plead not guilty, they both get a short-term sentence because of lack of evidence. However, if both plead guilty, they get a medium-term sentence. If one of them pleads guilty and the other does not, he is set free and the other gets a long-term sentence. If the pay-off is interpreted as the number of years in prison: $c = q > a = b > r = s > d = p$. Clearly, it is best for both people to plead not guilty. But in the absence of communication and trust, each considers the other person's

Table 10.2a. *Pay-off matrix of a dilemma in a two-person game*

		Agent B / Your adversary follows strategy:	
Agent A / You follow strategy:		1. cooperate	2. defect
	1. cooperate	(a,b)	(p,q)
	2. defect	(c,d)	(r,s)

Table 10.2b. *Pay-off matrix in a social dilemma*

		Everyone else follows strategy:	
You follow strategy:		1. cooperate	2. defect
	1. cooperate	(a,b) Win-win solution	(p,q) You are the sucker
	2. defect	(c,d) You are the free rider	(r,s) Commons tragedy

reasoning and concludes that the most rational strategy is to plead guilty! In game theory jargon, the Nash equilibrium strategy.

It matters, of course, whether you play the game once, twice or the rest of your life – it is all the difference between a purchase in Timbuktu, doing an exam or stepping into a marriage. There are many similar dilemma situations that can be represented by a pay-off matrix and agent strategies (Barash 2003). For instance, if you play as agent A for the first time and distrust your adversary, you may defect and be proven correct (r,s). If you wish to do business with the other person and build up a good reputation, you may first cooperate, until the other person defects once or twice and you feel abused (p,q). It may also be you who first cheats (c,d). To figure out which strategies are optimal is the essence of game theory – but reflecting upon such situations is millennia old. The 'players' are presumed to be competitive and antagonistic in the pursuit of their goals, but the resulting behaviour depends crucially on the perceived costs and benefits of each strategy. If one of the prisoner's feels that prison is the best place to be for a couple of years, the whole game changes. Being an experienced criminal also makes a difference. Game theory can clarify thinking and puts complex social interactions in a new perspective – and may be fun, too.

The common view amongst economists is that rational agents are *not* likely to cooperate in certain settings, even when such cooperation would be to their mutual benefit. Based on considerations such as those in Olson's book *The Logic of Collective Action* (1965), it is asserted that rational, self-interested individuals will not act to achieve their common or group interests unless the number of individuals in a group is quite small or there is coercion or another way to make individuals act in their common interest. Elementary game theory seems to confirm this, but more advanced computer-based simulations show much richer possibilities. An early and famous iterative game experiment was the tournament organised by Axelrod in the 1980s around the question: What is the best strategy in a repeated infinite prisoner's dilemma (Axelrod 1984)? It turned out that the tit-for-tat (TFT) strategy – always respond with the same strategy as your opponent – was almost always the winning strategy in terms of overall score. This tournament set the stage for many explorations into the complexities of cooperative behaviour (Axelrod 1997). The resulting evolutionary game theory suggests that in iterative games cooperative behaviour is also not rational, because the argument of the prisoners can be extended backwards to the very first encounter – the Nash equilibrium strategy. But *in silico* experiments show that patches of cooperative agents do survive and the conditions under which this happens are intensely investigated (Nowak and Sigmund 2004; Helbing and Yu 2009).

Often and more relevant in the present context are the situations in which a single person 'plays' against a large group of others with similar ends and means. These situations are called *social dilemmas*. They represent rather complex situations in which the gain or advantage being sought is limited by the fact that others are also reaching for the same goal. A social dilemma is intricately related to the context of a decision in space and time. The dilemma is that each player is in a narrow context best off if he acts according to his own individual interest, but in a larger context, he is better off if he acts according to collective interest. 'Once you start identifying social dilemmas, it's difficult not to see them everywhere, whether in public affairs or private matters' (Barash 2003).

Individual rationality will not induce people to follow strategies in their own long-term self-interest without externally enforced rules. But the assumption of individual selfishness or rational egoism in economic theory has limited validity in civic society. Experimental economics has established that the rational egoist assumption does indeed predict behaviour in auctions and in competitive market situations quite well, but that it is less general than often thought and confined to particular spots in societies, such as economic science students. Empirical fieldwork and simulation experiments confirm the everyday experience that individuals in all walks of life and all parts of the world voluntarily organise themselves for a collective purpose. Evolutionary game theory is increasing the understanding of this presence or absence of cooperation and the conditions for and dynamics of its emergence (Wright 2000; Nowak and Sigmund 2004).

Theoretical and empirical analysis of resource-related social dilemmas and the establishment of design rules for sustainable resource use are a great challenge for sustainability science. Resource allocation issues are often social dilemma's about allocation in a public goods (PG) or common pool resource (CPR) setting (§5.3). Think of the following examples in a physical, economic or social scarcity frame:

- watering your lawn during a drought;
- littering in your neighbourhood;
- crowding in a natural park area;
- diverting your income to tax havens in order to evade taxes;
- membership of a trade union that fights for higher wages – also for you;
- protesting in a dictatorship with secret police;
- emitting carbon dioxide that causes global warming;
- admission to a class on sustainability science.

These are all situations that can be framed in terms of a pay-off matrix, with different strategies and outcomes. As discussed in Chapter 12, the sustainable use of resources such as land, water and fish are indeed dependent on our creativity and willingness to to deal with these situations.

The key is how to induce cooperation and, relatedly, coordination. Historical analyses indicate that there are more options than state- or market-based governance regimes, and laboratory experiments begin to generate relevant insights about human behaviour (Ostrom 2000; Ostrom *et al.* 2002, 2008):

- individuals contribute significantly of their own endowments for a public good in a PG game; their contribution declines as the game progresses and/or the last round is announced;
- face-to-face communication produces substantial increases in cooperation;
- individuals who believe that others will cooperate in social dilemmas are more likely to cooperate themselves; learning to play the game also matters; and
- the availability of a sanctioning mechanism (punishment or reward) tends to change individuals who are initially the least trusting into strong cooperators.

Human behaviour is rich and heterogeneous, with some people more willing to initiate and sustain reciprocal and cooperative behaviour than others. More research

Table 10.3. *Worldviews, population futures and ideas about human behaviour*

Population and Human Behaviour

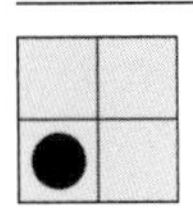	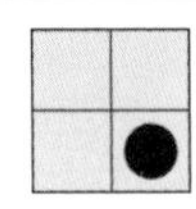	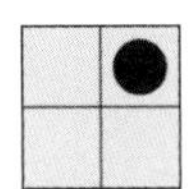	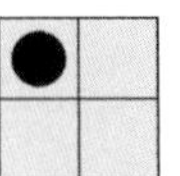
Statement 1: Population growth is still the greatest threat to sustainable development.			
This is incorrect. Indeed, the ingenuity and labour of people is the very source of societal wealth and progress.	Unfortunately, this may be true. The poor should quickly stop reproducing, for their and our welfare. Provide contraceptives.	Every human being deserves respect and every soul on earth has potential and purpose. It is inner development, not number of people that counts.	We have already overshot the Earth carrying capacity. Birth control should be a high priority and education of women is the best approach to it.
Statement 2: Aging populations in the rich countries will become an ever larger economic burden because of high-tech medical treatments.			
It is indeed one of the problems we are facing. The rich will insist on costly high-tech medication, but the poor will benefit later, too.	This fear is justified. It is one more argument against immigrants who put pressure on the necessary social security arrangements.	This is typically the preoccupation of a wealthy and materialist society. Personal well-being is more than individual health.	It is indeed one of the problems we are facing. It is important to prepare for a decent health care for and availability of medical services for all.
Statement 3: Humans are selfish by nature.			
This is the safest bet in everyday life, but it is not necessarily bad for society!	Correct. Life is about survival.	Incorrect. Human beings have the unique capacity for empathy and altruism, as part of their spiritual development.	This is true for the larger part of the human population. Precisely for that reason, government is needed.
Statement 4: Most people let their actions, not their words, be determined by money.			
Correct. It can be seen all over the world. As the saying goes, blood is thicker than water.	Of course. If someone tells you he's different, don't believe him!	It is true for many people – but what else can one expect in a materialistic society?	One only has to meet all the committed people in NGOs and governments to know that this is not true . . .

is needed to establish firmer rules and construct a more adequate and integrative theory than the prevailing one in economic science. There are probably no universal solutions based on simple models, and we should look 'beyond panaceas' in the search for sustainable management of social-ecological systems (Ostrom 2009).

We have explored in this section the image of man as seen in demography, biology and economic science. In Table 10.3, some sustainability-related issues that pertain to these disciplines are addressed from different perspectives. Clearly, the ways in which these sciences see man are not unambiguous 'facts' but instead reflect differences in worldview. Following the framework presented in Chapter 6, we invite you to trace some of the differences and find your own position.

Box 10.4. *Corporations.* Agents are often thought of as individual people. But institutions such as governments and businesses are the higher-level and more powerful agents. The most important actors in the modern industrial economy are the multinational corporations (MNCs). If corporate sales and country GDPs are considered equivalent measures of economic importance, 51 out of the 100 largest economic systems were corporations in the year 2000 (www.ips-dc.org). In resource-related sectors (fossil fuels, metals, chemicals and food processing), a dozen MNCs create the larger portion of value added.

The corporation was invented in 18th-century Britain and 'separated ownership from management – one group of people, directors and managers, ran the firm, while another group, shareholders, owned it... [Its] genius as a business form was – and is – its capacity to combine the capital, and thus the economic power, of unlimited numbers of people' (Bakan (2004) 6–8). Early 20th century, the Anglo-American corporation became a legal 'person' and has become a remarkably efficient wealth-creating machine. Their operation reflects the principles of laissez-faire capitalism.

The legal status and power of corporations has dark sides. Bakan, a Canadian Law professor, warns that social and environmental values are for corporations not ends in themselves but strategic resources to enhance business performance. 'Greed and moral indifference define the corporate world's culture [in the USA]... as pressure builds on CEOs to increase shareholder value, corporations do anything to be competitive... the managers may be kind and caring people but are allowed, often compelled, by the corporation's culture to disassociate themselves from their own values' (Bakan (2004) 9). He characterises the (American) corporation as psychopathic: It is irresponsible, manipulates, has a lack of empathy, has asocial tendencies and is unable to feel remorse. Its common features are: obsession with profits and share prices, greed, lack of concern for others, and a penchant for breaking rules. It is 'an externalising machine', with predatory instincts and a built-in compulsion to externalise its costs. Although this may be a value-laden caricature, the unethical and amoral behaviour on the part of big multinational corporations is at the roots of many of the world's social and environmental ills.

Corporate social responsibility (CSR) is for this reason one of the contentious sustainability related issues. Within the business community, CSR is considered a sign of good entrepreneurship and part of the solution for sustainable development. Opponents see it largely as window dressing and argue that 'the people' have to reconceive the corporation as it was originally intended: a public institution whose purpose it is to serve national interests and advance the public good. This is *a forteriori* argued with respect to banks. Such a transformation will not be easy, but within certain worldviews, it is an essential part of the sustainability transition. One implication is that the state reclaims its responsibility in high-income countries. In low-income countries with a weak state, corporate wealth easily mingles with national and local elites at the expense of the majority of citisens. Here, NGOs should oppose such practises.

10.5 Simulating Human Behaviour

10.5.1 Introduction

If the wish is to improve sustainable development models and introduce human behaviour, which way should we go? There are two strands of scientific research that seem to come together. On the one hand, there are the traditional, qualitative social science theories, and on the other hand, there are the formal modelling approaches of CSS. Increasingly, bridges are being built between the two. But for both, it is not possible to study human behaviour without some 'image of man' and the tension between a bottom-up and a top-down view of the world (§10.1). In the bottom-up view, humans act individually in interaction with their direct neighbours – direct not only in physical but also in affective, mental and social sense. In contrast, in the top-down perspective, planning and order from above dominate individual human behaviour. It is the tension between the 'invisible hand' of the market and the 'Big Brother' eye of the state or corporation, represented by the horizontal axis in the worldview scheme of Figure 6.5. Of course, the distinction is not sharp. This chapter focuses on three approaches in CSS and gives some illustrative applications, to illustrate the bottom-up simulation model approach. It can, at least in principle, give an idea about when to oppose and when to give in to, or even use, the forces 'from below' with intelligent regulation and institutions 'from above'.

CSS methods are complementing, not replacing calculus. Classical mathematics is extremely useful for a variety of analyses, but often the analytical models cannot be solved, even for a skilled mathematician, unless the system is drastically simplified. The models become quickly intractable if discrete heterogeneous objects, informational delays and feedbacks and many nonlinear interactions are to be included (§2.4).[9] Moreover, the interpretation becomes often too strenuous to be successfully communicable. Most of the more interesting and useful models in sustainability science, therefore, rely on simulation techniques and software. The advances in this area are inextricably linked to the digital, discrete world of computers.

Let us take the notion of agent as starting point. It is the equivalent of element in system dynamics and defined as:

> an *agent* is a model entity which can represent animals, people or organizations; can be reactive or proactive; may sensor the environment; communicate with other agents; learn, remember, move and have emotions (Janssen 2002).

From a cognitive science and artificial intelligence point of view, the question is how to construct an adequate representation of a (human) agent with perception, beliefs and goals, which in combination with intentions and motivations lead to (inter)action (Gilbert and Troitzsch 1999; Phan and Amblard 2007). One way is to refine the internal representation of simulated agents, in the direction of the I-I quadrant (Figure 10.1). Another direction is to improve the interactions between simulated agents (their *topology*), in the direction of the I-E quadrant (Figure 10.1). The two approaches are interrelated, as indicated in the scheme of Figure 10.8. There are no sharp boundaries in the wealth of recent applications.

[9] This is even more true if space is explicitly introduced in the form of partial differential-integral equations.

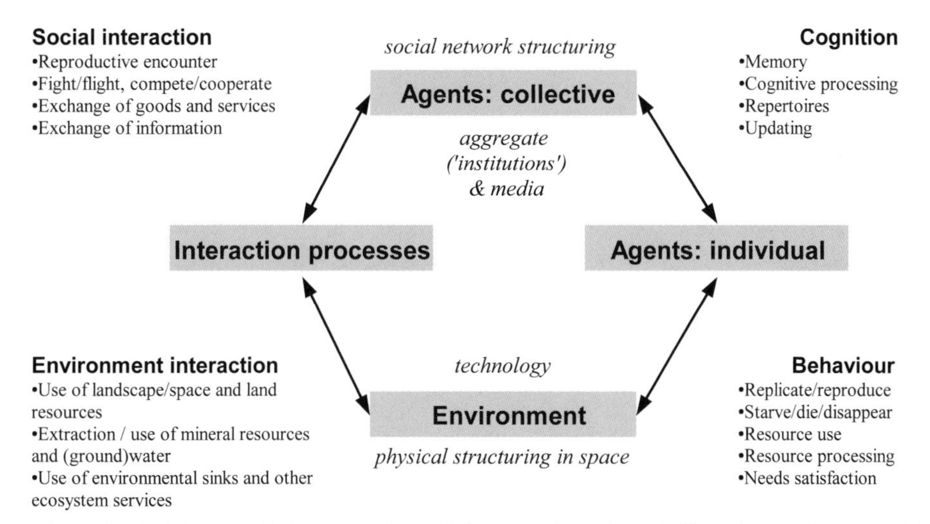

Figure 10.8. Scheme of the ways in which agent-based modelling is applied: cognition, behaviour and interaction in a dynamic environment.

10.5.2 Cellular Automata Models

Have you ever seen fireflies flashing on and off above the seawater? It turns out that the behaviour of a firefly can be understood by the interactions with the neighbouring fireflies. This is the idea behind the *cellular automata* (CA) approach. The CA framework assumes a number N of ordered cells. The ordering can be in one dimension (in sequence on a line or circle), in two dimensions (in a 2-D grid on a surface) or in three dimensions (in a 3-D grid in a cube) or in other not necessarily spatial configurations. Which cells are considered to be neighbours has to be defined explicitly; for instance, a neighbourhood can contain four or eight cells around a cell in a 2-D surface (Batty 2005). Usually, the cells are adjacent squares and represent a part of an imaginary or real surface. Each cell works as a kind of minicomputer: It is in a certain state $S(t)$ at time t and changes into a subsequent state $S(t+1)$ at time $t+1$. The cell states are updated simultaneously according to some transformation rule, or algorithm, that depends on the state of the cell and its neighbours. The rule can be, for instance, to switch state if and only if all neighbouring cells are in a different state.

There are many real-world phenomena that can adequately be described on the basis of CA, for instance, the already mentioned fireflies, chemical oscillators such as the famous Belousov-Zhabotinski reaction and computing devices. CA models simulate behaviour of individual elements (bottom-up) without coordination by a king, a government or a CPU in a computer (top-down), although some coordination can be introduced with synchronous updating. Macroscopic order can emerge from *local* transition rules.[10]

[10] There are several software packages available to construct CA models. A user-friendly package is NetLogo 4.1 that simulates agents on a grid. The predator-prey and the firefly model are both examples. Another package, specifically for geographical applications, is PCRASTER. It has been developed at Utrecht University and provides a high-level simulation environment for cartographic and dynamic modelling. In more advanced software tools, it is also possible to relax the assumptions and have, for instance, cells of different sizes and shapes and non-identical rules.

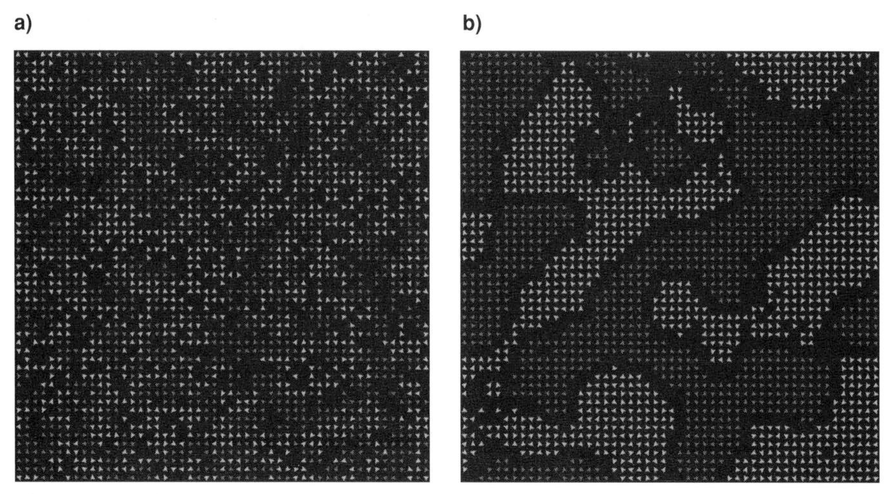

Figure 10.9. Simulation of a segregation process amongst 2,000 people with a similar-wanted threshold of 30 percent (a) and 70 percent (b). In the 30 percent situation, people are happy at much higher levels of integration than in the 70 percent situation (Courtesy: Netlogo 4.1 Segregation model). (See color plate.)

One of the simplest, yet evocative CA examples is the *Game of Life*, constructed by Conway in 1970. Cells can be 'alive' or 'dead', and their state in every subsequent step depends on the state of neighbouring cells in processes of birth, survival, loneliness and crowding. For instance, sudden catastrophic forest fire can be simulated and understood in a simple CA-based model as self-organised criticality (§9.6). Another interesting application is a model of the mechanism of neighbourhood segregation proposed by the economist Schelling in his book *Micromotives and Macrobehavior* (1978). He argued that the segregation of people from different ethnic backgrounds (a macro-phenomenon) could be understood from people's individual behaviour (a micro-foundation). This is an example of a threshold or tipping-point process. In the Netlogo implementation, agents – known in the NetLogo jargon as turtles – have preference for a certain percentage of people in their direct neighbourhood being 'similar' (ethnicity, profession or whatever is seen as defining identity). At any moment, an agent looks around and 'sees' how many agents in the neighbourhood are similar and decides on the basis of this information whether to move to another location or not. If the number is above a certain threshold (percent similar-wanted), he or she feels happy and will not move. Because agents are initially randomly distributed, a number of people will for a given average threshold still have less similar people as neighbours than they would like and they decide to move. For low thresholds, after a few rounds, a stable pattern emerges with everyone being satisfied. If the average threshold goes up, more and more people start to feel unhappy and to move around. After some dozens of rounds, a stable and rather segregated pattern consolidates the situation (Figure 10.9). For even higher thresholds (>80 percent), the system cannot find a stable configuration and keeps oscillating indefinitely. The model gives a vivid demonstration of the formation of spatially segregated patterns from a single simple rule. The approach has been used to explain and predict locations of ethnic conflict and to suggest mechanisms to promote peace in regions with ethnic or cultural segregation tendencies (Lim *et al.* 2007).

Urban and economic geography and the environmental sciences are probably the disciplines where CA models are most widely created and used (White and Engelen 1997). CA models in urban and rural geography use GIS-data, which characterise the grid cells (elevation, temperature and so on). They represent a potential field with attractors or repellers that direct the activities through, for instance, the relative attractiveness of an area for rice growing, tourism or office space.[11] Elaborate transformation rules on neighbourhood, diffusion and noise have been investigated to simulate the growth of cities (Batty 2005). Also, in the natural and engineering sciences, CA models are becoming available and applied, for instance, on water erosion and pollutant dispersal (Wainwright and Mulligan 2004). Detailed simulations of land use dynamics on the basis of a variety of transformation rules are used to understand processes such as deforestation and overgrazing (Verburg *et al.* 2003; Bouwman *et al.* 2006). A model-based exploration of land degradation risks from tourism in Crete is amongst the early CA models to explore sustainable development issues (Clark *et al.* 1995). CA models are used in many other settings, for instance, to understand land price developments and to investigate the impacts and adaptation options in a situation of rising sea levels. Indeed, it is hard to imagine how the inherently local and intricate dynamics of sustainability-related issues can be understood without spatially explicit CA models. In Chapters 11–13, some applications are presented in more detail.

A more theoretical application of CA models is the simulation of the evolution of agent behaviour on a grid in a repeated prisoner's dilemma (Lindgren 1991). Each agent has at any timestep an encounter with one of its neighbours. When agents meet each other, they can apply one out of four strategies in their encounter (always defect DD, always cooperate CC, reply tit-for-tat DC and reply anti–tit-for-tat CD, with the action on the left the response to the action on the right) and each strategy has an outcome that is determined by the pay-off matrix A (Figure 10.10a). Agents memorise the outcome of encounters and adjust their strategy according to the replicator equation, which is a core equation in evolutionary dynamics (Appendix 10.2). The evolution of simple strategies in a population can be simulated and visualised on a grid and over time.

In a repeated prisoner's dilemma with 1,000 players interacting with everyone else and the occurrence of mistakes, an evolutionary path unfolds in the course of 600 generations, with different strategies emerging and disappearing (Figure 10.10b). There are, for instance, long periods in which tit-for-tat (DC) is dominant, but cooperative strategies (CC) are sometimes significant for considerable periods of time. The introduction of mistakes makes the strategies more sophisticated but also more cooperative. If mutation, duplication, innovations or a stochastic environments are introduced, even more complex patterns emerge (Eriksson and Lindgren 2002).

In silico simulations such as these alert to the richness of real-world interactions and warn against general statements about human nature on the basis of simplistic

[11] Geographical information systems (GIS) are data representations in metric space. Note that spatial information on economic-geographic systems is in discrete form, which makes the CA approach more amenable to simulate change in space than the analytical equivalent of partial differential equations.

(a)

Pay-off matrix:

		Action for player B	
		Cooperate	Defect
Action for player A	Cooperate	(3,3)	(0,5)
	Defect	(5,0)	(1,1)

(b)

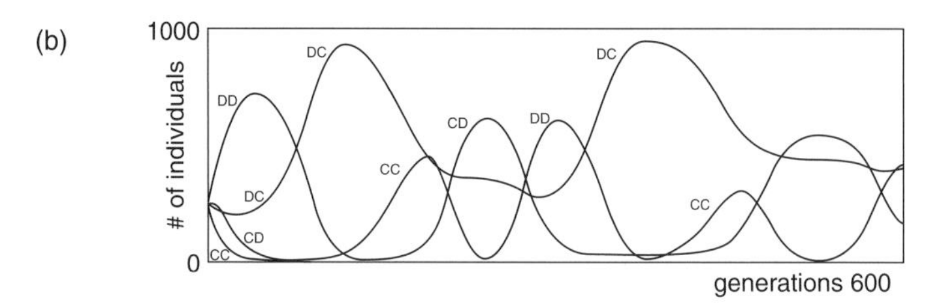

Initial population, 250 each of memory 1 strategies:
CC – Always Cooperate
DD – Always Defect
DC – Tit-for-tat
CD – Anti–Tit-for-tat

Figure 10.10. (a) The pay-off matrix for cooperation or defection (b) a trajectory of sub-populations in a repeated prisoner's dilemma with mistakes (Lindgren 1991).

theoretical models. These models are also simplifications, but they complement the outcomes of even simpler game-theoretic analyses. They instill amongst some advocates a postmodern enthusiasm: 'In an evolutionary system . . . there is no single winner, no optimal, no best strategy. Rather, anyone who is alive at a particular point in time, is in effect a winner, because everyone else is dead. To be alive at all, an agent must have a strategy with something going for it, some way of making a living, defending against competitors, and dealing with the vagaries of its environment' (Beinhocker 2005). These models are like mirrors with which we uncover our own image. There is still a long way to go, but there is the promise that, one day, they will be of help in designing and implementing effective strategies and policies for sustainable development.

Box 10.5. *Small worlds.* In 1967, the psychologist Milgram sent out a number of packets to people who had agreed to participate in an experiment, namely that the packets had to be sent to someone in Massachusetts, but the participants were only allowed to send their packets to someone they knew by first name. It turned out that only a median of five intermediaries were needed to get the package from the participants to the person in Massachusetts. This has been called the 'small-world phenomenon': Each of us is only six steps away from every other person on earth. Network theory has shown that in a network where every vertex is connected to its two neighbours, only a few additional random connections bring forth the 'small-world phenomenon' (Watts (1999)). A few shortcuts do shrink the world dramatically. In other words, a few species or persons can make a huge difference in overall connectivity and hence function and performance of a system.

10.5.3 Interaction: Networks

The properties of systems with many individual elements can be understood from the interactions between the elements and from their individual behaviour. *Network theory* or *mathematical graph theory* is a generic way to look at the behaviour and structure that emerge from interactions. It started centuries ago but has only recently moved into the analysis of large networks with the availability of large datasets and the enormous computing power of today. The emphasis in this brief introduction is on relevant examples from various disciplines. It is not the intention to go into mathematical details. For some basic definitions, I refer to Appendix 10.3 and the Suggested Reading.

A network consists of a number of *vertices* (or nodes) and *edges* (or links) connecting nodes. In first instance, the analysis of a network is a description of the vertices and edges and an examination of the properties of the network. A fundamental property is the *degree* of a vertex. It is the number of edges to or from that vertex. A plot of the fraction of vertices of a certain degree as a function of the degree is called the degree distribution. Some networks have a degree distribution, which can be approximated with a *binomial distribution*. These are often random networks because their structure is what one would get if the edges are randomly distributed amongst the vertices. The degrees of nodes in random networks do not differ strongly: All nodes have roughly the same degree. In contrast, many real-world networks have a degree distribution that can be approximated with a power law. Such networks have many weakly connected and a few strongly connected vertices ('hubs').

The availability of large computing power makes it possible not only to analyse a particular network at a given moment in time but also to simulate the *evolution of networks* over time. It is one of the fascinating applications in modern network theory. The construction of a network *in silico* is done by setting up a number of vertices and connecting them by adding edges according to certain rules.[12] The construction rule for a *random network* is very simple: Add edges between randomly chosen pairs of vertices in a network of fixed size and see what happens. With this rule, the probability p that any two edges are connected increases and one expects a binomial degree distribution. It turns out, upon adding edges and thus increasing p, the relative size of the largest connected (or giant) component ($n_{giant\ comnponent}/n$) jumps discontinuously from small (~ 0) to large (~ 1). The network has no particular structure.

A more interesting rule to construct a network is to connect vertices with already many edges with a higher probability than those with few edges, such as the most connected vertices that are preferred in the growth process. Therefore, the mechanism is called *preferential attachment*. It generates in a growing network scale-free structure. Figure 10.11a shows the dendritic configuration typical of such networks. The degree distribution is approximated with a power-law function. In system terms, it is a reinforcing feedback. In system archetype terms: *Success breeds success, the rich get richer.*

[12] The first attempts to examine the growth of networks were by Erdös, Rapoport and others in the 1950s.

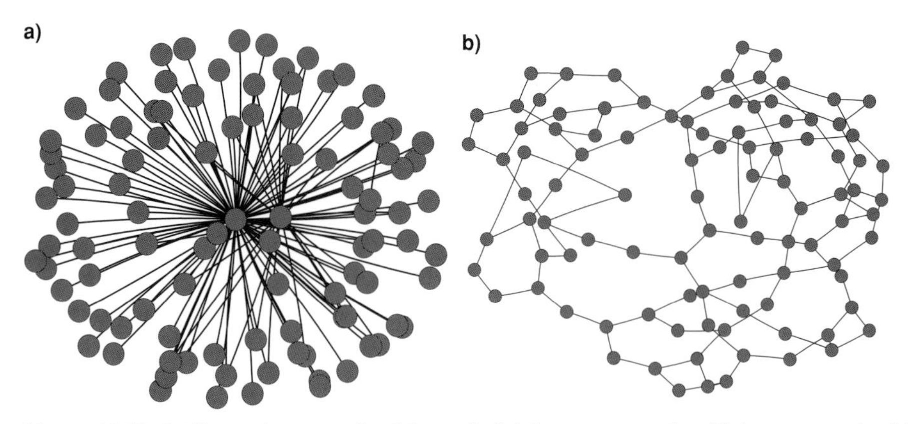

Figure 10.11a,b. Example networks. Network (a) is an extremely efficient network, (b) an extremely resilient one. Examples are all networks with 100 vertices and 250 edges (Brede and de Vries 2009a).

There is an interesting link between efficiency and resilience that can be examined with network analysis (Brede and de Vries 2009a). Efficiency of, for instance, transport and communication networks can be related to the average shortest pathlength. Efficient networks are star-shaped and consist of a periphery made up of a highly connected core and outward branching nodes. Resilient networks are associated with the stability around a presumed existing stationary state and have the property of redundancy (§9.6).[13] One extreme is an efficient network with a single 'super-hub' (Figure 10.11a). The other extreme is a network with long loops (Figure 10.11b). Networks that realise a trade-off between both extremes exhibit core-periphery structures, where the average degree of core nodes decreases but core size increases as the weight is gradually shifted from a strong requirement for efficiency and limited resilience towards a smaller requirement for efficiency and a strong demand for resilience. Both efficiency and resilience are important requirements for network design and a balance between the two may be a principle of system robustness and resilience.

There is a rapidly increasing number of analyses of real-world networks and *in silico* produced networks. Network analyses are done in different fields of science, such as ecology (food web stability), transport and other infrastructure (congestion or emergency), epidemiology (spread of disease) and the social and economic sciences (information, power or money) (Appendix 10.3). Most investigations of networks focus on the statistical properties of real-world networks. The rapid advance of Internet and social media and the instabilities of financial systems accelerate the interest in networks.

The networks in the different domains exhibit large structural differences. Many *metabolic networks* appear to be scale-free networks, with a few highly connected nodes (Jeong *et al.* 2000). There are a few hubs that dominate the overall connectivity and are, therefore, the system's most sensitive elements. 'Living

[13] Mathematically, the smaller the largest eigenvalue of the adjacency matrix of the network, the more stable and thus resilient it is.

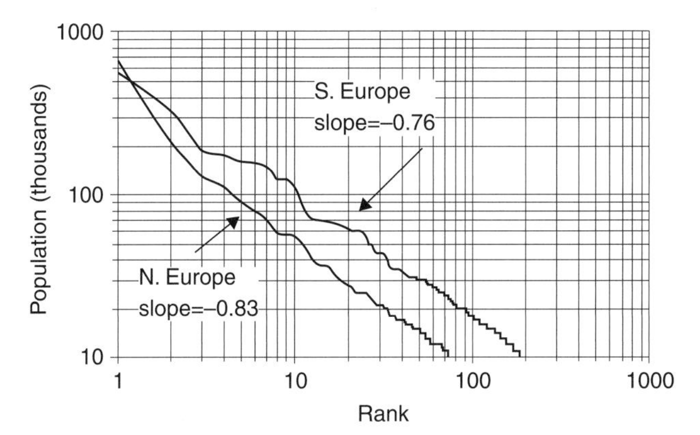

Figure 10.12. Rank-size distribution for European cities in 1750 (Blum and Dudley 2001). The lower curve is for northern European cities, and the upper curve for southern European cities.

networks' that grow by accretion do often have the dendritic shape, with telling examples in biology and geography. But the structure of 16 high-quality *food webs* from a variety of ecosystems in the United States differ from what one would expect in random networks and in scale-free networks in terms of characteristic path length, clustering coefficient and degree distribution (Dunne *et al.* 2002). Instead, the sixteen food webs show a variety of functional forms and their size plays a role in the network degree distribution. Investigations such as these enhance understanding of structural complexity and help to design better ecosystem management strategies (Ruiter *et al.* 2005).

There are also interesting applications in *geography*. The structure of river catchments can be reproduced in *in silico* experiments with rules based on erosion dynamics (Rodriguez-Iturbe and Rinaldo 1997; Buchanan 2002). These rules are also characteristic of the percolator process in physics. In urban geography, the frequency versus size histogram of cities and towns exhibit a scale-free structure: a few very large cities, many towns and many more small villages.[14] The urban population distribution in northern and southern Europe around 1750 are examples of such a rank-size distribution (Blum and Dudley 2001; Figure 10.12). They can be produced by adding edges randomly *and* multiplicatively, that is, proportional to existing density (Anderson 2005; Batty 2005). Is the mechanism really understood? Early theories focused on the *rich get richer* preferential attachment mechanism that is also at work in scale-free networks. It is not difficult to imagine how it might work: Once a village becomes a town, it starts to attract more people than neighbouring villages because of all kinds of advantages (trade, scale and so on), and this further reinforces its growth. But this is probably not the whole truth, and a more accurate representation of cities as entities in an evolving network is needed.

Increasingly, the social sciences use network analysis to explore structure of financial, economic and social systems. An analysis of interbank payments between

[14] It is called the rank-size distribution of city sizes and its power law representation is also called Zipf's law.

Box 10.6. *Hierarchy in networks.* Hierarchical networks with a modular structure are found in urban systems and organisations (Pumain (2006)). Intrinsic hierarchy is found in a variety of communication networks, whereas networks with a geographic organisation, such as power grid networks and Internet router nodes, do not apparently have such a topology. In archaeology and anthropology, the observation of hierarchy in (past) human groups has been questioned. There are reasons to assume that the ways in which human groups have organised themselves are much richer, but the ones with a high degree of centralisation and concentration may have left the most clear traces.

The sociologist Simon offers a rationale for the prevalence of hierarchical structures: 'Complex systems will evolve from simple systems much more rapidly if there are stable intermediate forms than if there are not. The resulting complex forms in the former case will be hierarchic. We have only to turn the argument around to explain the observed predominance of hierarchies amongst the complex systems nature presents to us. Amongst possible complex forms, hierarchies are the ones that have the time to evolve' (Simon (1969) 90–91/98–99).

commercial banks pointed at a network with both a low average path length and low connectivity (Soramäki *et al.* 2007). There is a tightly connected core of banks to which most other banks connect. The degree distribution is scale-free over a substantial range. A recent analysis points at the high connectivity amongst the financial institutions of the world (Schweitzer *et al.* 2009). Mutual shareholdings and closed loops are a sign that the financial sector is strongly interdependent. It can make the network unstable, obstruct market competition and pose systemic risks.

Another social science application is interpersonal networks of friendship/influence. Over thirty years old, there is the example of the analysis of networks of influence regarding nuclear power policy in The Netherlands (de Vries *et al.* 1977). The vertices are the members of governing boards, directorates and advisory councils of all the institutions (companies, ministries and councils) involved in the nuclear policy process, or the stakeholders. Whenever a person has positions in two or more of these institutions, he is assumed to be a channel for the exchange of information and the exertion of influence. These 'double-functions' make up the network edges. The analysis revealed that only a handful of people (<20) carried a rather dense network of relationships that connected construction firms, (public) electricity generating companies, research institutions and ministries. Such organisational networks around large-scale ventures can promote effective decision making, but they also tend to exclude or ignore alternative viewpoints. Social network theory has been greatly refined in the last decades, but the applications in sustainability related areas are still rare.

Are network theory and applications relevant for sustainability science? Every complex system that consists of many interacting units can be described and understood as a network. A striking number of complex systems fit into a few, broad classes. If rather rigid laws determine the development pattern of such systems, it constrains the options for growth and planning. For instance, systems with scale-free architectures are quite susceptible to targeted interventions – both positive and negative ones (Figure 10.11). The removal of one or a few 'hub' species from a food web with a scale-free–like structure can cause its collapse. Targeted vaccination in a population suffering from an epidemic outbreak can contain the spread of the disease. As a corollary, scale-free network systems are remarkably robust against random errors because most of the vertices can be removed without noticeable impact on its structure. Therefore, the network structure may explain the robustness of some ecosystems against random forms of interference (pollution or invasion) and the vulnerability of centralised systems to (terrorist) attack. Understanding such generic mechanisms and 'laws' of system structure and evolution can teach us about a system's boundaries and permissible trajectories and about its development potential.

10.5.4 Multi-Agent Simulation: Behavioural Variety

When the emphasis is more on the representation of agent behaviour (such as memory, choice criteria, and so on) than agent linkages, one tends to speak of agent-based modelling (ABM) or multi-agent simulation (MAS).[15] With the advent of object-oriented simulation languages and platforms, it is increasingly feasible to incorporate agents in simulation models. They are 'intelligent agents' in the sense that they are autonomous entities in a virtual world, with a model of their environment, rules and possibly a model of themselves. Their 'intelligence' ranges from response behaviour to information signals to anticipating the intentions of other agents. In these simulated worlds, memory and adaptation are introduced by acting on the basis of past experiences and discovering new strategies or rules. Agent-based models have an inherent tendency to focus on heterogeneity and interdisciplinarity. The self-organisation exhibited in ABM suggests that order emerges 'from below', such as from rather simple rules at the individual level (Kauffman 1995; Holland 1996; Beinhocker 2005). According to some practitioners, it supports the market ideology – an interesting link between worldview and scientific method.

Do you remember the story of Easter Island (§3.2)? A human population can build a society and destroy it. *The Lord of the Flies* (1954) by Golding is another famous story about an island. In this book, primitive rituals underneath a layer of civilisation take over in a group of boys on an isolated island after an imaginary plane crash. The story suggests that destruction does not have to come from cutting all trees or polluting all water – it may come from a loss of 'civilised' behaviour. In Chapter 3, other evidence of collapse because of human behaviour is mentioned (§3.4). But how does such a process really work? ABM may help to understand such processes with *in silico* experiments.

The first people to make and use ABM were probably the makers of computer games (SimAnt, Civilization, SimCity and so on). A first, more rigorous model of how individual agents can evolve is *Sugarscape*. It was presented in 1996 by Epstein

[15] Also in differential equations and system dynamics models, one can interpret model behaviour in terms of agents. For instance, the interaction between a prey and a predator (equation 9.5) is a condensed representation of random encounters in a homogeneous environment, but the simulation of behaviour is too restricted to be acceptable for social scientists.

and Axtell in their book with the telling title, *Growing Artificial Societies: Social Science from the Bottom Up*. Again, the story happens on an island – the 'ideal gas' of the social scientist.

Imagine an island of 2,500 km^2 divided in 50 × 50 squares (cells) of 1 km^2. The cells have only one characteristic: the amount of sugar, which differs across cells. At the start of the imaginary simulated world, there are agents with a limited behavioural repertoire: look around in neighbouring cells for sugar (inspection), move and eat sugar (metabolism). They differ in two respects: the inspection capability (how many cells can be inspected) and metabolic rate (sugar needs for survival). The sugar is unevenly distributed across the island, as with real resources. What will happen in this simulated world? The agents – with different capabilities and positions – start off and look for sugar. When they move around and hit more sugar than they need, they accumulate it, which extends their life. If they cannot satisfy their needs, they starve. Thus, agent A and agent B in Figure 10.13 develop in different directions because of their different initial positions, inspection capability and metabolic rate. Whereas agent A can collect hardly enough sugar to survive, agent B accumulates with a bit of luck enough sugar to live on for a long time. Notice that there is no direct interaction, unlike in the evolutionary game presented earlier.

What makes this simple model interesting is that, when run with many agents and many times, it is possible to construct the statistics on the system in the form of probability distribution functions of key variables. It turns out that the evolution of this sugar economy shows some remarkable similarities with real-world observations. Although agents start from a random distribution (location and metabolism), the simulation experiments generate a shift from an egalitarian situation with an uniformly distributed wealth (accumulated sugar) to a highly skewed distribution in which only a few agents own most of the wealth and a large group is near starvation. This feature is consistent with the Pareto distribution found for income. It is an *emergent property* in the sense that it is not the consequence of any one cause in particular, but instead the outcome of the mix of initial distributions, rules of the game and luck. Macroscopic order emerging from microscopic behaviour.

Much behaviour is mediated by *information exchange* and a number of ABMs have been made to explore the penetration of innovations. An example is the simulation of the spread of environmentally friendly innovations in a population of farmers (Weisbuch and Boudjema 1999). Individual agents are given partial knowledge of their environment and are endowed with different preferences and motivations and an internal representation including the ability to learn and adapt. Encounters with other agents, structured according to an imposed social network, can 'infect' a potential adopter with the innovation. With the assumption of full rationality of economic theory, an exact prediction of the fraction of farmers F_{eq} that will adopt the environmentally friendly option is possible and can have any value between 0 and 1. But with social networks and agent heterogeneity, a form of *bounded rationality* is introduced. Herd behaviour occurs and prediction is difficult. There is a tendency for F_{eq} to be near 1 or 0, not unlike what happens in the Polya model (Appendix 10.1). Another area of ABM-applications is the spread of diseases such as AIDS and mad cow disease, inspired by epidemiological models (Dunham 2005). In Chapters 11–13, I present some examples.

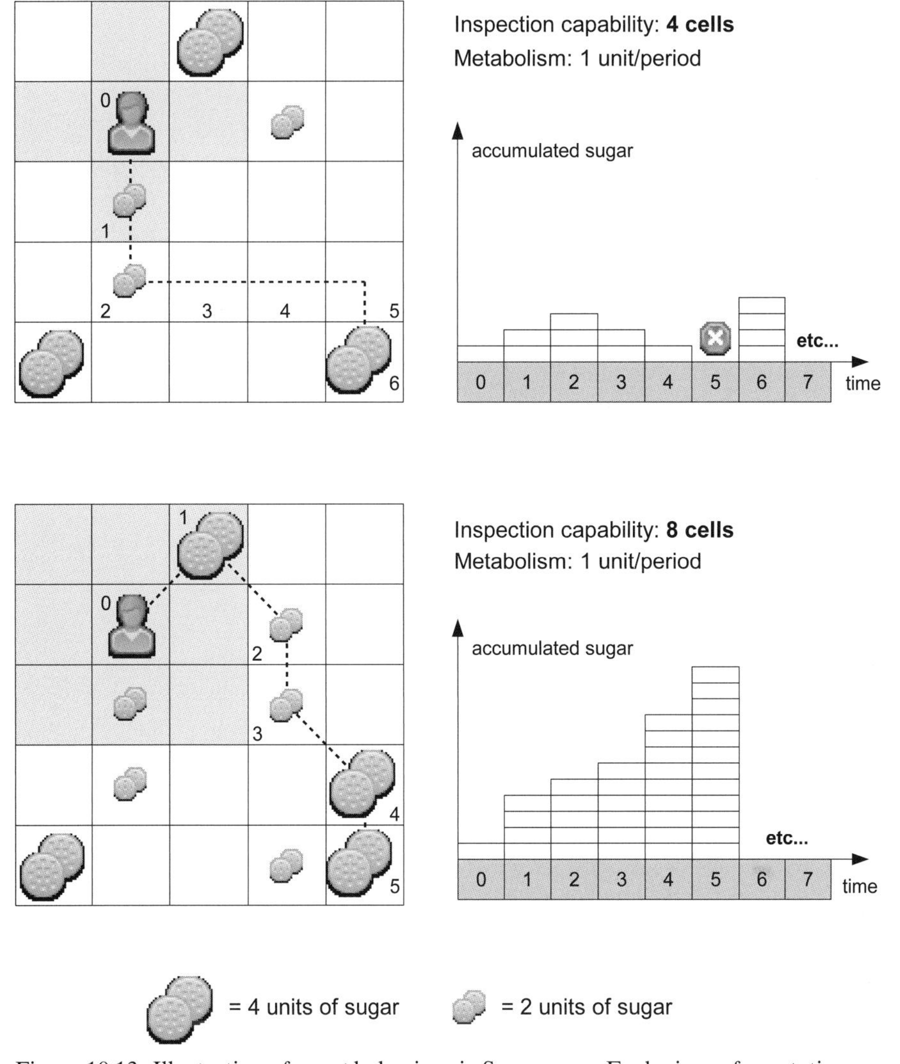

Figure 10.13. Illustration of agent behaviour in Sugarscape. Each piece of vegetation accounts for one units of resource (sugar).

Random events and rule-guided behaviour are standard features in ABM. Order is seen to come from below and not as a result of a deterministic process – more like a flight of starlings than an externally forced pendulum. The formulation in terms of discrete events and objects makes it easier to link them to everyday experiences and to cognitive maps. A possible drawback is that an ABM is more difficult to control and validate. A fancy interface graphics may hide the internal complexity. Their relevance in sustainability science is that it permits an experimental investigation of the interactions between humans and their environment (Jager and Mosler 2007). Computer simulations are the 'laboratory' in which social science hypotheses about human behaviour can be tested and intervention policies can be explored. Given the large and increasing complexity of sustainability issues, this may prove to be a decisive factor in the design and implementation of adequate policies.

Box 10.7. *Do ABMs have added value?* Modelling human behaviour, including motives and habits, leads to a more in-depth understanding of the forces behind (un)sustainable development and resource exploitation and provides more meaningful levers for policy experiments. But is it really giving new insights? Janssen (2002) has explored the advantages of adding spatial heterogeneity to models of range-land management in Australia. Experimenting with various types of sheep behaviour using a relatively simple model – including fire and water-points – it is found that non-uniform grazing due to different rules regarding biomass density, waterpoint locations, and other sheep, may cause system behaviour to differ significantly from uniform grazing assumed in non-spatial models. This could influence notions of ecosystem resilience and indicate the need for more prudent exploitation strategies. A cautious conclusion is that 'understanding the implications of spatial processes is vital in the large open paddocks of range-lands; however, it is likely that these processes are also important, perhaps in a more subtle way, in other environments' (Janssen 2002:123).

10.6 Summary Points

If designing policies for a sustainable future, one must ask the questions: Who am I and who is the other? Or broader: What is the prevailing image of man? A brief exploration into biology, demography and economics gives some indicative answers. Here are some general insights:

- The image of man in science is largely focused on the individual and the outside, as part of Modernism. The methods of complex system science (CSS) can complement the traditional social science approaches in building stronger knowledge about interactions between and cognitive aspects of individuals.
- Population dynamics (demography) and phenomena such as the demographic transition are important for an understanding of (un)sustainable development. The biological, notably the evolutionary, perspective on human behaviour has deeply influenced the prevailing self-image in industrial societies.
- The image of man in economic science, known as *Homo economicus*, is biased towards rationality and competition. The variety of human needs and behaviours is explored explicitly in new areas such as behavioural and experimental economics and evolutionary game theory.
- Their relevance for sustainability is, amongst others, the study of strategic behaviour in a public goods (PG) or common pool resource (CPR) setting. It gives more realistic ideas about competition versus cooperation and the individual vs. the collective and associated worldviews.

Novel methods to introduce heterogeneity in behaviour are cellular automata (CA), network dynamics and agent-based models (ABM). They offer the prospect to better understand human behaviour in complex social-economic systems, notably at the micro-level.

> Death is very likely the single best invention of Life. It is Life's change agent. It clears out the old to make way for the new.
>
> – Steve Jobs, former CEO Apple

> I suspect that the fate of all complex adaptive systems in the biosphere – from single cells to economies – is to evolve to a natural state between order and chaos, a grand compromise between structure and surprise... [from the complex system point of view] we cannot know the true consequences of our best actions. All we players can do is be locally wise, not globally wise.
>
> – Kaufmann 1994:15/29

> To the eternal triple question which has always remained unanswered, Who are we? Where do we come from? Where are we going? I reply: As far as I, personally, am concerned, I am me; I come from just down the road; and I am now going home.
>
> – Pierre Dirac, quoted in *Zeldin, The French*

> Il y a dans tout homme, à toute heure, deux postulations simultanées, l'une vers Dieu, l'autre vers Satan. L'invocation à Dieu, ou spiritualité, est un désir de monter en rade; celle de Satan, ou animalité, est une joie de descendre.
>
> – Baudelaire

> Translation:
>
> There are in every man, at all times, two simultaneous tendencies, one toward God, the other toward Satan. The invocation of God, or spirituality, is a desire to follow the road upward; to invoke Satan, or animality, is taking delight in falling down.

SUGGESTED READING

An entertaining yet thorough introduction to game theory and its applications.

Barash, D. *The Survival Game: How Game Theory Explains the Biology of Cooperation and Competition*. New York: Times Books, 2003.

A thorough introduction in the theory and applications of cellular automata, with emphasis on geographical applications.

Batty, M. *Cities and Complexity – Understanding Cities with Cellular Automata, Agent-Based Models, and Fractals*. Cambridge, MA: MIT Press, 2005.

An overview of environmental science with contributions from a variety of disciplines.

Boersema, J., and L. Reijnders. *Principles of Environmental Sciences*. New York: Springer Science and Business Media B.V., 2009.

A popular, well-written account of network theory and its applications.

Buchanan, M. *Nexus/Small Worlds and the Groundbreaking Science of Networks*. New York: Norton Company, 2002.

One of the early introductions on agent-based simulation models.

Ferber, J. *Multi-Agent Systems – An Introduction to Distributed Artificial Intelligence*. London: Addison-Wesley, 2000.

A lucid and thorough introduction into evolutionary game theory and its applications.

Nowak, M. *Evolutionary Dynamics – Exploring the Equations of Life*. Cambridge, MA: Harvard University Press, 2006.

Introduction into complexity science with applications in various fields.

Perez, P., and D. Batten, eds. *Complex Science for a Complex World – Exploring Human Ecosystems with Agents*. Canberra: ANU Press, 2003.

Series of papers on causes and consequences, such as aging, urbanisation, life stages, kinship networks and so on.

Demeny, P., and G. McNicoll. *Population and Development Review 37*. Special Volume on Demographic Transition, 2011.
A practical guide on the technique of agent-based modelling.
Railsback, S., and V. Grimm. *Agent-Based and Individual-Based Modeling*. Princeton, NJ: Princeton University Press, 2011.
Advanced introduction into nonlinear dynamic models and their applications.
Strogatz, S. *Nonlinear Dynamics and Chaos – With Applications to Physics, Biology, Chemistry, and Engineering*. Boston: Addison-Wesley Publishing Company, 1994.

USEFUL WEBSITES

- www.nicheconstruction.com/ is a site that gives a broad and well-referenced overview of niche construction theory.
- www.unfpa.org/swp/swpmain.htm provides info on the State of World Population reports by UNFPA.
- sedac.ciesin.columbia.edu/gpw/index.jsp is a site operated by CIESIN at Columbia University with many data and maps on population density, urban-rural and so on.
- www.pbl.nl/en/themasites/phoenix/index.html is the website of the demographic model PHOENIX of the Netherlands Environmental Assessment Agency (PBL).
- www.ifs.du.edu/ is a site showing an interactive exploration of population-economic-environmental futures of the world's countries. The International Futures (IFs) model can also be downloaded.
- www.who.int/en/ is the site of the World Health Organization (WHO) with the World Health Report and data on health.
- www1.fee.uva.nl/cendef/ is the site of the Center for Nonlinear Dynamics in Economics and Finance at the University of Amsterdam.
- www.bitstorm.org/gameoflife/ is a clear illustration of the Game of Life. Many more versions, including 3-D ones, can be found on the Internet.
- jasss.soc.surrey.ac.uk/12/1/6/appendixB/EpsteinAxtell1996.html is a description of the Sugarscape model.
- www.youtube.com/watch?v=rN8DzlgMt3M&feature=relatedshows a predator-prey simulation in 3-D, with bottom-up rules creating macroscopic patterns.

Wikipedia has some excellent entries on complex system methods and applications.

SOFTWARE

- ccl.northwestern.edu/netlogo/ is the website of the Netlogo agent-based simulation platform. Netlogo 4.1 can be downloaded for free.
- www.pcraster.nl is a GIS-MAS package developed at Utrecht University, with diverse applications in geography and geophysics.
- www.pajek.org is the website of the network construction software package, freely downloadable.
- www.openabm.org/ is the OpenABM site with tutorials, a model library and discussion forum on computational modelling
- www.railsback-grimm-abm-book.com/index.html is an introduction to agent-based modelling procedures using NetLogo (Railsback and Grimm 2010).

Appendix 10.1 Models of Economic Decision Making

A simple model to allocate fractional shares amongst competing options is *the multinomial logit* (MNL) model, a member of the logit/probit models. Suppose a company or a consumer must make a decision between two alternative options – such as purchasing equipment or a product of brand A or brand B. First, define a broadly defined 'cost' of each option: price, reputation, environmental impacts, fuel use and recyclability. The higher the cost, the less attractive the option. There is a large group of potential buyers and the number of times that option A is chosen in a certain period or number of decisions is called its market share μ_A. The MNL-formula states that the market share of option i converges to:

$$\mu_A = \frac{c_A^{-\lambda}}{\sum_{i=1}^{2} c_i^{-\lambda}} \quad \text{or} \quad \mu_A = \frac{1}{1+(c_B/c_A)^{-\lambda}} \qquad \text{(A10.1)}$$

with μ_i the market share of option i. The competing options are assumed to be fully substitutable and the parameter λ is a measure of the substitution elasticity. A small λ-value implies that cost differences hardly affect the choice (inelastic substitution). Taxes or subsidies to affect choice, therefore, do not work. For a high λ-value, it is the opposite.

The model is consistent with economic production theory, but for some products, it is incorrect. Notorious examples are where the apparently most attractive option lost are the QWERTY keyboard and the VHS video system. The mechanism is that, once a product or process has reached a certain market share, economies of scale and scope, access and inertia of infrastructure and other factors operate to give it an even larger market share. Reinforcing feedbacks create *one winner takes it all.* Economists call it positive or increasing returns to scale.

It can be simulated with a different model: the Polya generator (Arthur 1994). The choice process is seen as a series of random choice events that are *not independent.* A large number of producer or consumers choose in a sequence of n events either option A or option B and option A has been chosen n_A times. Let the probability that during the next event option A is chosen be equal to $P(A) = F(n_A)$, such as the probability depends on how often option A has been chosen before. If you normalise $F(n_A)$ into $G(\mu_A)$ with $\mu_A = n_A/n$, it depends on the function G what will happen. If G declines with increasing μ_A, there is one stable attractor and no returns to scale. This is the usual assumption in economics. But if G is proportional to P(A), there can be multiple attractors and positive or negative returns to scale and the actual development becomes sensitive to what is chosen during the first few events. Processes become path-dependent and inventions can get *'locked-in'*, or become dominant even in the face of clear disadvantages. The objective may be to get the system out of the attractor basin, instead of containing it within (§9.6). History matters![16]

An example of real-world processes with positive returns to scale are industrial location: The probability that a new firm is established in a region increases with the number of firms already there. It is known as *first mover advantage* and operates in many infrastructure systems. An example is the choice of standards in ICT-infrastructure/networks: If the task to be performed requires connections with other users, each next user has an advantage in choosing the one with most users connected. Other examples are alternating current (AC) versus direct current (DC) electricity, gasoline versus biofuels, and private transport by car versus public transport. Reputation is an example that illustrates the importance of information:

[16] A vivid illustration of its consequences in economic decision making is given by Kirman (1993) with his ant model.

Table A10.1. *Descriptions of networks in different disciplines*

Network theory	Vertex	Edge
Physics	Site	Bond
Computer science	Node	Link
Ecology/economy	Point, site, place	Link, connection, bridge
Sociology	Actor	Tie

The more an option earns a good reputation (quality, reliability and so on), the more often it is chosen. This Polya generator model leads to a better understanding of how technologies penetrate markets. It is important for the rapid, continued and widespread introduction of more resource-efficient ways of producing and consuming.

Appendix 10.2 Replicator Dynamics

Let the pay-off for an interaction of strategy i with strategy j be given by $\{a_{ij}\}$ then one can define the *fitness* of a strategy by the product of its frequency and its pay-off. For instance, if a strategy CC has a certain pay-off a_{CC} and its frequency in a particular round is x_{CC}, then the expected pay-off in the agent population is:

$$f_{CC} = \sum_{j=1}^{n} x_{CC} a_{CC} \tag{A10.2}$$

If pay-off is equated with fitness, the replicator equation determines the rate of change in the frequency of the strategy:

$$\frac{dx_{CC}}{dt} = x_{CC}(f_{CC} - \varphi) \tag{A10.3}$$

with φ the average fitness across all agents and strategies (Nowak 2006). In this way, behavioural strategies can invade the system. I refer the interested reader to the Suggested Reading for more details.

Appendix 10.3 Network or Graph Theory

A network is defined as a set of system elements. The elements are called *vertices* and the links are called *edges*, but there are different names in different fields (Table A10.1). The number of vertices, n, is called the *size* of the network. The edges represent some kind of exchange (goods, influence, progeny and so on) and may be assigned directions, weights or sizes, as, for instance, in transport, trade or communication channels. The vertices may have a variety of properties, such as position in space, membership of a class (plants, animals, peoples, cities and so on) or characteristics of such membership (weight, age, metabolic rate, communication frequency, number of inhabitants and so on). If the vertices are used to formulate a square matrix and the value of edges is indicated in the cells, the matrix is called an *adjacency* matrix.

The *degree* of a vertex is the number of edges incident on, or connected to that vertex. The shape of the *degree distribution* is an important network property. It can

Table A10.2. *Overview of systems that have been studied as networks*[a]

Network	Type	No. of vertices n	No. of edges m	Mean degree z	Mean vertex-vertex distance l	Clustering coefficient C	If power-law distribution: exponent α
Metabolic	Undirected	765	3,686	9.64	2.56	0.09	2.2
Marine food web	Directed	135	598	4.43	2.05	0.16	n.a.
Neural	Directed	307	2,359	7.68	3.97	0.18	n.a.
Train routes	Undirected	587	19,603	66.79	2.16		n.a.
Power grid	Undirected	4,941	6,594	2.67	18.99	0.10	n.a.
Internet	Undirected	10,697	31,992	5.98	3.31	0.035	2.5
Citation	Directed	783,339	6,716,198	8.57			3.0
Sexual contacts	Undirected	2,810					3.2
Physics coauthorship	Undirected	52,909	245,300	9.27	6.19	0.45	n.a.
Company directors	Undirected	7,673	55,392	14.44	4.60	0.59	n.a.

[a] *Source*: Newman 2003.

be approximated with an analytic function P_k, which should be interpreted as the probability that a randomly chosen vertex has k connections. Strictly speaking, one should distinguish a histogram, that is based on empirical data, from its mathematical abstraction, that is called a probability distribution. When the distribution follows a power law: $P_k = k^{-\alpha}$, one gets a straight line for a plot of P_k as function of k on a double logarithmic or log-log scale. They are called *power-law networks*. They are also called *scale-free networks* because they remain unchanged under rescaling of the independent variable k with a multiplication factor m, as is seen upon introduction of $\kappa = mk$ (then $p_{k'} \sim \kappa^{-\alpha} = (mk)^{-\alpha} = m'k^{-\alpha} = m'\, p_k$).

Another interesting large-scale property of networks is the *mean degree z*, which is defined as the average number of edges connected to a vertex. It equals the ratio of the number of edges and the maximum possible number of edges. The *transitivity* or clustering coefficient C indicates the extent to which 'the friend of your friend turns out to be also my friend'. The *community structure* of a network is a measure of the presence of subgroups of vertices with a high density of edges within them and a lower density of edges between them. The *resilience* of a network is an indication of a network's response to the removal of vertices. If vertices are removed, paths between connected vertices will become longer, and at some point, the vertices will become disconnected.

Newman (2003) collected data on a variety of networks and characterises four loosely connected categories (Table A10.2):

1. *biochemical and bio/ecological networks*: metabolic pathways, protein interaction networks, genetic regulatory networks, food webs, neural networks;
2. *technical and infrastructure networks*: river and canal networks, road and railway networks and airline routes, electric power grids and telephone and Internet communication networks. Many of these networks are related to space- and geography-governed infrastructure and designed for distribution;

3. *information* (or knowledge) *networks*: a classic example is the network of citations, more recent ones are the World Wide Web and networks of people's preferences for objects used in targeted advertising; and
4. *social networks*: intermarriages, sexual contacts, mail contacts, friendship relations, business and collaboration relationships, and influence networks of executives and politicians. The vertices are (groups of) people that are in some forms linked to each other. They overlap with communication and information networks.

11 Agro-Food Systems

11.1 Introduction: The Human Habitat

The topic of a previous chapter is 'pristine' nature, but pristine nature is becoming scarce. Human populations have been very successful as a species and now have impact upon one-third or more of the biosphere. Humans have always interfered with the natural environment: Agriculture and pastoralism are sustained forms of manipulation of plants and animals. During the process of agrarianisation, they have learned to exploit soils, fish and forests for the provision of food, fiber and fuel. The larger part of the world population is still engaged in agriculture, and food production is still the most direct and largest interface between humans and nature. In monetary terms, agricultural products make up only a few percent of gross world product (GWP), partly because subsistence agriculture is outside conventional economic statistics. If food processing, transporting and retailing are considered, the economic role is much larger and in the order of 8 percent. But in physical terms, agriculture and the larger food system are of enormous importance. This chapter explores sustainability issues in present-day agricultural populations and systems. Agriculture is understood here in the broad sense of agroecology and the agro-food systems in the world, from local production to global trading and processing. It connects the Latin words *colere* and *ager*, the latter meaning field.

There are many ways in which human populations interfere with the 'natural' world of biogeochemical cycles and ecosystems. Although one can discern some universal tendencies in human-environment interactions, farming and animal husbandry are intrinsically local and the natural cycles still dictate the human activities in space and time. Land is locality *par excellence.* It is only logical, then, to focus first on the *human habitat.* The relevant branch of science is geography, along with ecology a core discipline in sustainability science.

11.1.1 Land and People

The natural energy flows are an essential part of the biosphere dynamics and the mean net primary production (NPP) is a good measure of it. The NPP reflects the growth of vegetation, which depends on location, soil quality, elevation, precipitation, temperature and other factors. There is as of yet no internationally accepted

Table 11.1. *Characteristic data for different land use land cover categories around the year 2000. Ten to 20 percent of the forested and mountainous regions are under some kind of protection*[a]

					Population density		
	Unit	Area (mln km²)	%Frac Land (%)	Mean NPP kgC/m²	Rural persons/ km²	Urban persons/ km²	Income ($/yr/cap)
Water	Marine	349.3	(–)	0.15	0	0	0
	Coastal	17.2	7.4	0.52	70	1,105	8,960
	Inland water	10.3	4.4	0.36	26	817	7,300
Forest/woodland	Forest: (sub)tropical	23.3	10.0	0.95	14	565	6,854
	Forest: temperate	6.2	2.6	0.45	7	320	17,109
	Forest: boreal	12.4	5.3	0.29	0.1	114	13,142
Dryland	Dryland: hyperarid	9.6	4.1	0.01	1	1,061	5,930
	Dryland: arid	15.3	6.5	0.12	3	568	4,680
	Dryland: semiarid	22.3	9.5	0.34	10	643	5,580
	Dryland: dry subhumid	12.7	5.4	0.49	25	711	4,270
Island	Outside island states	2.4	1.0	0.72	82	1,220	12,162
	Island states	4.7	2.0	0.45	14	918	11,148
Mountain	Mountain 300–1,000 m	13	5.6	0.47	3	58	7,815
	Mountain 1,000–2,500 m	11.3	4.8	0.45	3	69	5,080
	Mountain 2,500–4,500 m	9.6	4.1	0.28	2	90	4,144
	Moutain >4,500 m	1.8	0.8	0.06	0	104	3,663
Polar	Polar	23	9.8	0.06	0.06	161	15,401
Cultivated	Cultivated: pasture	0.1	0.0	0.64	10	419	15,790
	Cultivated: cropland	8.3	3.5	0.49	118	1,014	4,430
	Cultivated: mixed (crop and other)	26.9	11.5	0.6	22	575	11,060
Urban	Urban	3.6	1.5	0.47	0	681	12,057

[a] Millennium Ecosystem Assessment 2005.

land use/cover classification system, but several United Nations–led initiatives are underway.[1] Quantitative assessments have, therefore, still uncertainties and ambiguities. Marine ecosystems (oceans with >50 metres (m) water depth) are dominant with more than two-thirds of the Earth's surface. Much of the terrestrial surface is covered by drylands: hyper-arid deserts (6.5 percent) and (semi-)arid and dry sub-humid regions (34.3 percent) (Reynolds and Stafford Smith 2002). The remainder is forest/woodland (28.4 percent), mountain areas (24.3 percent) and cultivated land (23.9 percent). Due to definitions, overlaps occur and the sum does exceed 100 percent.

Data on land cover, NPP, population and income densities from the Millennium Ecosystem Assessment (MEA 2005) are listed in Table 11.1. Not surprisingly, the highest rural population densities tend to be at medium values of the mean NPP levels. There is more to eat than at the low NPP values of the tundras, and there is, or at least was, less competition with other species than in the tropical forests with their

[1] *Land cover* refers to the physical and biological cover over the surface of land. *Land use* is defined by natural scientists in terms human activities and by social scientists and land managers more broadly, including the social and economic circumstances. Land cover can be observed in the field or by remote sensing. Land use is more difficult to establish and needs an integrated approach.

high NPP-values. Probably the most thorough dataset of *croplands and pastures* so far has been made by Ramankutty *et al.* (2008). They report 15 million square kilometre (km^2) or 1.5 billion hectare (Gha^2) of cropland (90 percent confidence interval 12.2–17.1) and 28 million km^2 or 2.8 Gha of pasture (90 percent confidence interval 23.6–30). This is 12 percent and 22 percent respectively of the total ice-free land surface. Using a broader definition of agriculture, namely those areas where at least 30 percent of the land is used for cropland or highly managed pasture, agroecosystems cover 28 percent of total land area excluding Greenland and Antarctica (WRI 2000–2001).

Income (GDP/cap) as a measure of activity correlates positively with NPP, with the exception of the desert and polar regions where the high income is from fossil fuels or other resources, and the boreal forests and (sub)tropical forest/woodland areas where climate and ecology suit other species better than humans. The economic activity per unit area on the basis of population density data and economic data at lower than national (state or province) level can be estimated (Nordhaus 2005). The resulting gross cell product (GCP, or GDP per unit of cell area), on a $1° \times 1°$ spatial grid, shows that income tends to increase with distance from the equator, while GCP has a clear maximum in the temperate zones (5°C to 20°C temperature zone).

It is also possible to map the distribution of human individuals with respect to four geographical and climate parameters: altitude, nearness to permanent rivers and sea coasts, temperature and precipitation (Small and Cohen 2004). The overlay maps and their statistical representations clearly show the importance of climate, vegetation and geography as determinants of human habitat (Figure 11.1). The data indicate that human beings have been quite successful in adapting over a wide range of temperature and precipitation, with the preferred niche largely between 10°C and 20°C and between 500 to 2,000 millimetres per year (mm/yr). It suggests a reinforcing loop of success breeding success 'The talent for ingenious adaptation... has always provoked a need of yet more ingenuity... the very success of an ecological adaptation creates a need to develop some means of keeping population growth under control' (Reader 1988). A second observation is that human populations tend to live overwhelmingly (>70 percent) in areas less than 100 kilometres (km) from large rivers, less than 1,000 km from a sea coast and below 1,000 m elevation. They overlap with the large river plains in the world, and it points at a solid biogeographical foundation for human evolution and development.

11.1.2 Anthromes

In §9.2, maps were presented of potential 'natural' vegetation. But the activities of human populations are influencing directly and indirectly the larger part of the biosphere, so it is more relevant and interesting to construct actual 'humans-in-nature' land use/cover maps – maps of humanscapes.[3] Ellis and Ramankutty (2008) and

[2] Land area is indicated in million hectares (Mha) or billion hectares (Gha). Sometimes km^2 are used, 1 km^2 = 100 ha.

[3] Romanova and colleagues of Lomonosov State University in Moscow have reconstructed such maps for the pre-Roman and Roman era in Europe and other regions and times – see de Vries and Goudsblom (2002).

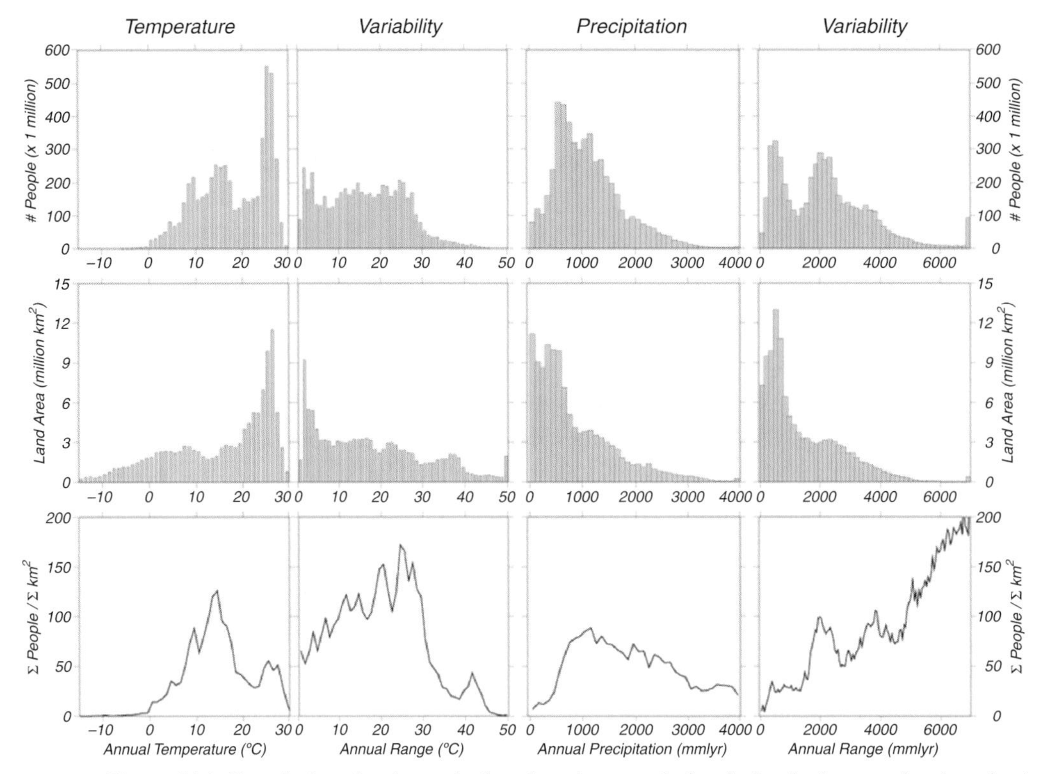

Figure 11.1. Population density, calculated as the population in land of a certain elevation/distance divided by its corresponding area (thick curves) (Small and Cohen 2004). The thinner curves are a rescaling; the peak in this curve at 2,300 m is the Mexico Plateau.

colleagues (Ellis *et al.* 2010) have constructed maps of what they call *anthropogenic biomes* or *anthromes*. It is based on the definition that:

> a *biome* is a unit of species composition and ecological processes, and a function of climate and climate-derived parameters.

The authors propose a total of twenty-one categories: dense settlements (two), villages (six), croplands (four), rangelands (three), seminatural (four – mostly woodlands) and wildlands (two).[4] The anthrome map is shown in Figure 11.2.

The anthrome map clearly shows the large differences with the potential vegetation map (Figure 9.5). The wildlands and, to a lesser extent, the seminatural woodlands and rangelands cover almost 50 percent of the world terrestrial surface. Human settlements and land use processes increasingly dominate the other 50 percent. The maps for the year 2000 confirm in more quantitative detail other statistical estimates (Table 11.1), notably that:

- more than 80 percent of all people live in densely populated urban and village biomes;

[4] The categories have been slightly updated since the first anthrome map because of new insights – for instance, one cropland and one wildland category are put under the new category seminatural in view of rather ambiguous demarcations.

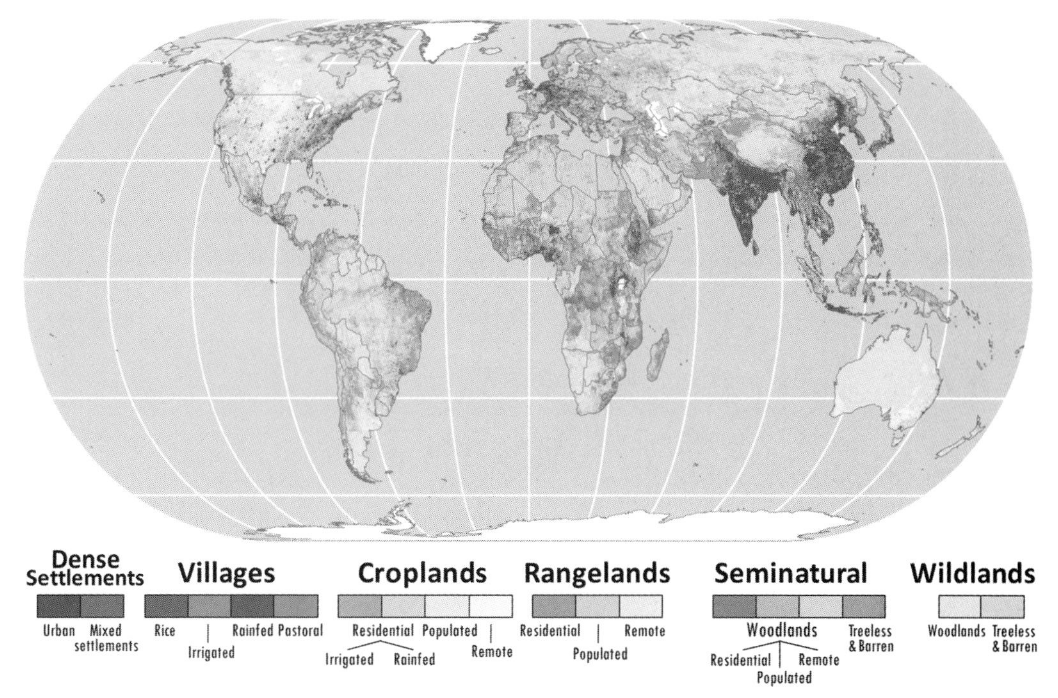

Figure 11.2. Map of anthropogenic biomes or anthromes (Ellis *et al.* 2010). (See color plate.)

- agricultural villages are the most widespread form of densely populated biomes: one in four people lives in them; and
- most of the terrestrial biosphere has been altered by human residence and agriculture, to the extent that now less than a quarter of Earth's ice-free land is 'wild', with only 20 percent of it forested and more than 36 percent of it barren.

The anthrome maps have also been constructed for the years 1700, 1800 and 1900 (Ellis *et al.* 2010). The reconstruction and other data sources indicate that the era of massive expansion into new lands (extensification) is over. Total cultivated area in this classification has hardly changed since 1980 and amounts to about 700 million hectare (Mha). There are still hundreds of millions of ha of cultivable land, but their conversion – also for non-agricultural purposes – would have serious social-economic and ecological impacts because many of these lands are in the tropical forests of Latin America, Africa and East Asia.

The anthrome maps give an integrated view of both ecosystems and the human activities within these systems, and they make Social-Ecological Systems (SES) the natural unit of investigation. An SES is defined as:

> an integrated system of ecosystems and human society with reciprocal feedback and interdependence. The concept emphasizes the 'humans-in-nature' perspective (www.resalliance.org).

Subsequent chapters often speak of SES because they offer the best way to identify and appreciate local variety and contingency and to explore sustainable development pathways that are based upon it.

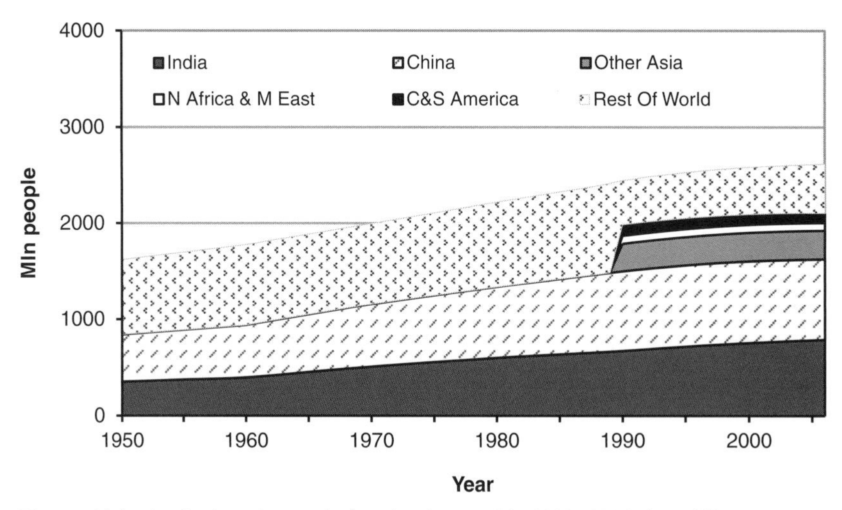

Figure 11.3. Agricultural population in the world, 1990–2006, in million persons (source of data: faostat.fao.org).

11.2 Agricultural Systems

11.2.1 Agro-Food Systems in the World

Many people in industrialised high-income countries tend to associate the natural environment with rural areas and natural parks. Where they live, intensive agriculture is concentrated in a few highly productive regions and dominates the food system. Abandoned rural areas are re-inhabited by rich urbanites. People enjoy forests and lakes in natural parks for recreation and travel as tourists and/or adventurers in the remaining wild nature abroad. But for nearly 3 billion people, the natural environment is an everyday experience, for sustenance, enjoyment and survival. In the 1960s, about 80 percent of the world population was engaged in agriculture and in 2005 still more than one-third of the world's people are classified as *agricultural population*: persons depending for their livelihood on agriculture, hunting, fishing or forestry[5] (Figure 11.3). They live in the agricultural villages and towns of the world, much as their ancestors did for millennia. Essentially, they have a biomass economy and their knowledge of local 'ecosystem services' is their most valuable and necessary asset for their livelihood.

Figure 11.4 shows in a simplified way routes and endpoints of human populations in their food procurement. Several evolutionary pathways have led towards the systematic and intensive interference with natural processes in modern agriculture. Categories and transition processes are not sharply defined and estimated populations and areas are rather uncertain, but the scheme in Figure 11.4 can guide you, with a few exceptions, through the various agroecosystems in the present-day world.

Very small numbers of humans are still living from *hunting and gathering* and survive in the periphery of the anthroposphere – a couple of hundreds of thousands of

[5] According to the FAO definition and data. The estimate comprises all persons actively engaged in agriculture and their non-working dependants. Also, an estimated 800 million people are actively engaged in urban agriculture (WRI 2000–2001).

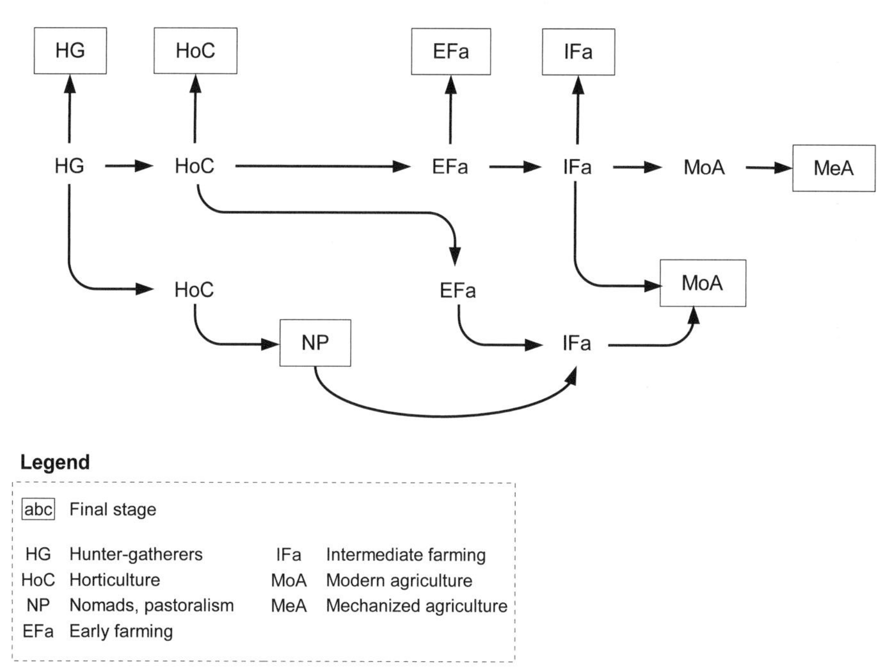

Figure 11.4. Various routes from hunter-gatherers to forms of agriculture (de Vries and Goudsblom 2002).

indigenous inhabitants of tropical forests and of the Arctic. Millions of herders live a (semi-)nomadic life and practise *pastoralism* in landscapes with short-lived grasses as the dominant vegetation: treeless temperate and tropical grasslands, savannas with woody overstory, and arid shrubby grasslands. The area under permanent pasture is huge but unevenly spread across the globe. While more than 80 percent in Oceania, Sub-Saharan Africa, South America and Southeast Asia is considered pastureland, the fraction is less than 15 percent in South and Southeast Asia. In the 1970s, an estimated 10 to 20 million people lived more or less as *nomadic herders*, mostly in the belt from the Sahara to Mongolia, in an area of about 2.5 Gha or 20 percent of the total ice-free land surface (Grigg 1974). Their number has increased – one estimate suggests 20 million to 25 million herders in the Sahel region alone (Niamir-Fuller and Turner 1999).[6] A recent estimate considers pastoral areas the home to more than 200 million people who are distributed over a grassland area of 5.25 Gha (~0,04 person/ha). They own some 35 million cattle, two-thirds in the African continent, and 9.5 million sheep and goats (Disperati *et al.* 2009). The numbers are imprecise because of different definitions and classifications and lack of data. They also change over time because grasslands spontaneously transit to forest and, reversely, forest is cleared for pasture.

Nomadic herders are vulnerable for external disturbances such as a severe drought. Therefore, and also to control and tax them, governments and aid agencies encourage them to switch or diversify to subsistence farming. The success of such a switch is unclear because mobility is their best strategy to cope with disturbances

[6] The estimate refers to households in which 50 percent of gross revenue comes from livestock or livestock-related activities.

(Pedersen and Benjaminsen 2008). Horticulture and agroforestry can be complementary sources of subsistence for these populations and thus enhance food security.

An outgrow of pastoral nomadism is *extensive commercial grazing* or ranching, practised in large-scale, low-productivity farms in parts of the Americas, Australia and New Zealand for the production of beef and wool. Where land was or became scarce, it was followed by intensification and introduction of novel breeding and other techniques. Cattle, pigs and poultry stocks are grown and processed at high densities in large manufacturing plants ('bio-industry'). It allows the optimisation procedures and practises of engineering science and the large-scale output and sales of animal-based products in present-day urban supermarkets. Evidently, the view of nature of those involved has undergone a transformation, too.

Shifting cultivation, the most prominent and highly varied form of early farming, still exists in parts of Southeast Asia, Latin America and Africa. Around 1970, an estimated 50 million people were living from shifting cultivation, cropping 10 to 18 Mha in a total area of 100 to 110 Mha (~0,5 person/ha) (Grigg 1974). More recent estimates of the number of people involved indicate a range from 40 to 500 million people, and the area may be a significant fraction of the 850 Mha of secondary forest (Mertz *et al.* 2008). Depending on available nitrogen and duration of fallow period, the system has yields in the order of 1 ton/ha (feeding ~0,5 person/ha) with low labour input and hardly any capital input. It is a useful system in tropical forest regions. Yet, it is declining under the pressure of higher population density, globalisation and rising food demand, which forces a transformation to permanent agriculture and cash crop plantations.

The intermediate, modern and mechanised forms of agriculture can be divided into two main groups: wet-rice culture and mixed farming.[7] The *wet-rice system* requires intensive land use with high labour inputs. It has several advantages, notably high energetic yield and minor soil degradation (Lansing 2006).[8] It has developed mainly in Southeast Asia in the large river basins because much water is needed. Natural floods are the most important irrigation form. Livestock is unimportant in wet-rice regions. Because up to half of the rice harvested used to be consumed on the farm, it is often considered subsistence farming. The so-called Green Revolution has given a boost to rice yields to values of 3 to 4 ton/ha (feeding >2 person/ha), with the use of hybrid varieties as well as with a need for abundant irrigation and fertiliser. It also led to a rapid increase of production for the market, although the fraction of rice traded internationally is still only about 7 percent of production.

The agricultural system, which has developed in Europe and spread to North America and other temperate areas of European settlement, is known as the *mixed farming system*. Three trends since the early Middle Ages determined present-day mixed farming in Europe: reduction of the fallow period, the extinction of common rights (enclosure) and increasing importance of livestock (Grigg 1974, 1987). Other elements in its evolution were the introduction of the heavy plough and the three-field system, the cultivation of roots and grasses as fodder, more and more intense

[7] Mediterranean agriculture and tropical crop plantations are two other, more or less, separate systems.

[8] Possibly, the success of rice cultivation allowed populations to grow to such densities that there was no place for animals and that vegetarianism became the dominant food practice. It is an example of a connection between cultural-religious practices and agricultural circumstances.

Box 11.1. *Science for agriculture: Liebig's Law.* One of the early 'laws' about agricultural systems is *Liebig's Law of the Minimum*. It is a principle stating that growth is controlled by the scarcest resource and not by the total of resources available. The principle is founded in the observation that increasing the nutrient flows does not lead to more plant growth – growth only increases if a nutrient is applied, which is most scarce of all the nutrients needed by the plant. This nutrient is called the limiting factor. This insight was essential for the development and application of fertilisers. In chemical kinetics, the limiting factor is known as the rate determining step.

Liebig's Law has been extended to biological populations and natural resources. In a logistic growth process, it is the scarcest resource in the environment that starts to slow down growth (§9.2). Scarcity often has a spatial and temporal dimension. For example, the growth of a biological population may not be limited by the total amount of resources available throughout the year, but by the minimum amount of resources available to that population at the time of year of greatest scarcity. The principle is also used in the formulation of economic production functions and market substitution processes.

use of inputs in farming and a subsequent shift to more high-value products such as livestock and vegetables. When real incomes started to rise in the 19th century, a further shift from cereals to high-value products took place. With Europe's mild winters, fertile soils, long days in summer and reliable rainfall, European wheat farmers nowadays attain amongst the highest yields in the world (6 to 8 ton/ha).

In the process, agriculture has become more and more dependent on science and manufacturing for inputs such as industrial fertiliser, tractors, transport and refrigeration equipment and fuel and electricity to run them. It also has to comply with ever stricter hygiene, environmental and of late animal welfare standards, partly as a result of the complexity and associated vulnerability of the system as a whole. These developments were interwoven with trends in globalisation – which in the 19th century was tantamount to European colonialism – and commercialisation. Modern mixed farming as practised in Europe and North America is now a highly commercialised form of near-industrial activity with high yields and high levels of non-labour inputs.

11.2.2 Food: Needs and Consumption

Evidently, any exploration of sustainable development has to acknowledge the great variety and diversity of agro-food systems in the world. Farmers tend to simplify and control their environment, but the combination of inherent variety of the land and growing connectivity of food producers make the global agro-food system very complex. The first question is: What are the food needs in the world? Often, one speaks of food demand but the nutritional food requirement for a healthy person can be determined quite well so the link between needs and demand is relatively straightforward. Food demand and, if available, food consumption is determined by the number of people and their nutritional needs and habits. This can be measured not only in physical units but also as the part of income spent on food. In a 1895 article, the Belgian economist Engel analysed the cost of living of worker families and observed that food expenditures as a proportion of income decreased with increasing household income. In Belgium, for instance, the share of food expenditures in total household expenditures was around 60 to 65 percent between 1860 and 1920, then declined slowly to 50 percent by 1950 and dropped precipitously to less than 18 percent by 1980 (Swinnen *et al.* 2001). Economists speak about *Engel's law*, which is confirmed by more recent trends in emerging economies.

Economists define in the analysis of food demand the notion of *income elasticity* as the percentage increase in food expenditures per percent increase in income (Appendix 11.1). If the elastitiy is less than one, the product for which this happens is considered a necessity. Examples are potatoes and other starchy staple food items. If the elasticity exceeds one, it is called a luxury good. Income elasticities as well as similarly defined *price-elasticities* are estimated from statistical data. The results confirm that the fraction of starchy staple food in total food consumption decreases with rising income, but that consumption of protein-rich food items like fish and meat – the luxury items for the poor – increases.[9] It implies that demand for food depends not only on average income, but also on income distribution. It also points at the high vulnerability of poor people for rising food prices. Throughout history, rising and volatile prices of staple food have hit the poor segments of inegalitarian societies the hardest and caused riots and revolts. It will not be different in the 21st century, as the social unrest in African countries in the first decade of the century shows.

These trends are part of the food or *nutrition transition* (Kearney 2010). There are other trends that further complicate the analysis. The food industry endeavours to upgrade and process food, which keeps the food expenditures higher than it would otherwise be. In growing economies, food preparation at home is replaced by instant meals and outside snacks and dining, as part of modernity. These trends show up in macro-economic data as an increasing fraction of value-added in the food processing and hotel and restaurant sectors.

11.2.3 Food: Resource and Potential Supply

A second look at the system is the biophysical resource perspective: How much food can be supplied? *Soils* provide the substrate for vegetation – and thus for food, fodder and fiber. Critical environmental variables are the climate, notably temperature and precipitation, and derivatives such as length of growing season and growing degree days. Soil properties such as organic matter and moisture content are other critical variables. Based on these variables, classification in agro-ecological zones and potential vegetation are made (Figures 9.2 to 9.5). It is possible to use an extended set of variables to determine the *potential yield* of specific crops (wheat, rice, maize and so on) for specific locations. It sounds straightforward, but it is not.

[9] For instance, in 1996–1997 the fraction of income spent on food in Argentina, Brazil and Mexico was more than 50 percent amongst the poorest 10 percent of the population and less than 20 percent for the richest 10 percent (Sabates *et al.* 2001). Incorporation of income distribution may cause a 5 to 10 percent difference in food demand estimates (Cirera and Masset 2010).

For instance, one has to define precisely for which crop and in which situation, because genetic potential, landscape and other factors also play a role. In a couple of regional and global models, potential yields are calculated across large areas. Figure 11.5 shows potential yields for the three most important grains (wheat, rice and maize) around the year 2000 on a 0.5° × 0.5° (about 50 × 50 km) resolution as simulated with the IMAGE2.2 model (Bouwman *et al.* 2006).

Of course, these potential yields are not realised in practise because of natural variations and suboptimal management. The difference between the potential and the actual yield is referred to as the *yield gap*. Licker *et al.* (2010) assign crop yield data on a 10′ × 5′ grid for eighteen crops and separate them in 100 different climate zones. Using global maps of growing degree days and soil moisture availability, they analyse the role of other factors such as management practises across regions with similar climates. The potential yield in a climate zone is in their analysis defined as the lowest level of the best 10 percent of the grid cells (90th percentile). Therefore, it is not a biophysical potential, but today's near-maximum achievable yield. There is a significant spread in the potential yields thus defined, even within similar climate zones. Comparison between this potential yield and the actual yields indicate room for improvement. The results point at possible improvements of 60 percent for wheat, nearly 40 percent for rice and 50 percent for maize with practises now adopted by the most productive farmers.

Neumann *et al.* (2010) follow a somewhat different agro-economic method. They specify an agro-engineering production frontier, which relates the yield to growth-defining, growth-limiting and growth-reducing factors. The first group consists of climate-related variables such as photosynthetically active radiation (PAR), temperature, CO_2 concentration and crop characteristics.[10] Taking into account precipitation and soil fertility constraints, they determine the production frontier. The role of the growth-defining (frontier) and growth-limiting/-reducing (inefficiency) factors is explored by estimating a log-linear relationship between yield and a set of variables that are considered proxies for these growth determinants. Aggregated across regions, the results for 26 regions suggest similar yield gaps as in the Licker *et al.* (2010) study, namely 43 percent, 47 percent and 60 percent for wheat, rice and maize, respectively. Defining efficiency as the ratio between the observed yield and the production frontier for that crop in that region, it is possible to identify the extent and determinants of yield gaps (Figure 11.6). Therefore, it seems that there is ample room for yield improvements, but the researchers are cautious with firm conclusions because what really can reduce the yield gap requires a better understanding of the region-specific socio-economic factors.

11.2.4 More Food: Can It Be Supplied Sustainably?

Most of the increase in world food production since 1980 came not from more cropland, but from yield improvements. The agricultural land area increased with less than 10 percent since the 1960s. Globally averaged yields have jumped between 1980 and 2000 from 3.3 ton/ha to 5.7 ton/ha (Lambin *et al.* 2003). The average price of

[10] PAR indicates the fraction of sunlight that can be used by organisms in photosynthesis. It is radiation in the 400 to 900 nanometres realm.

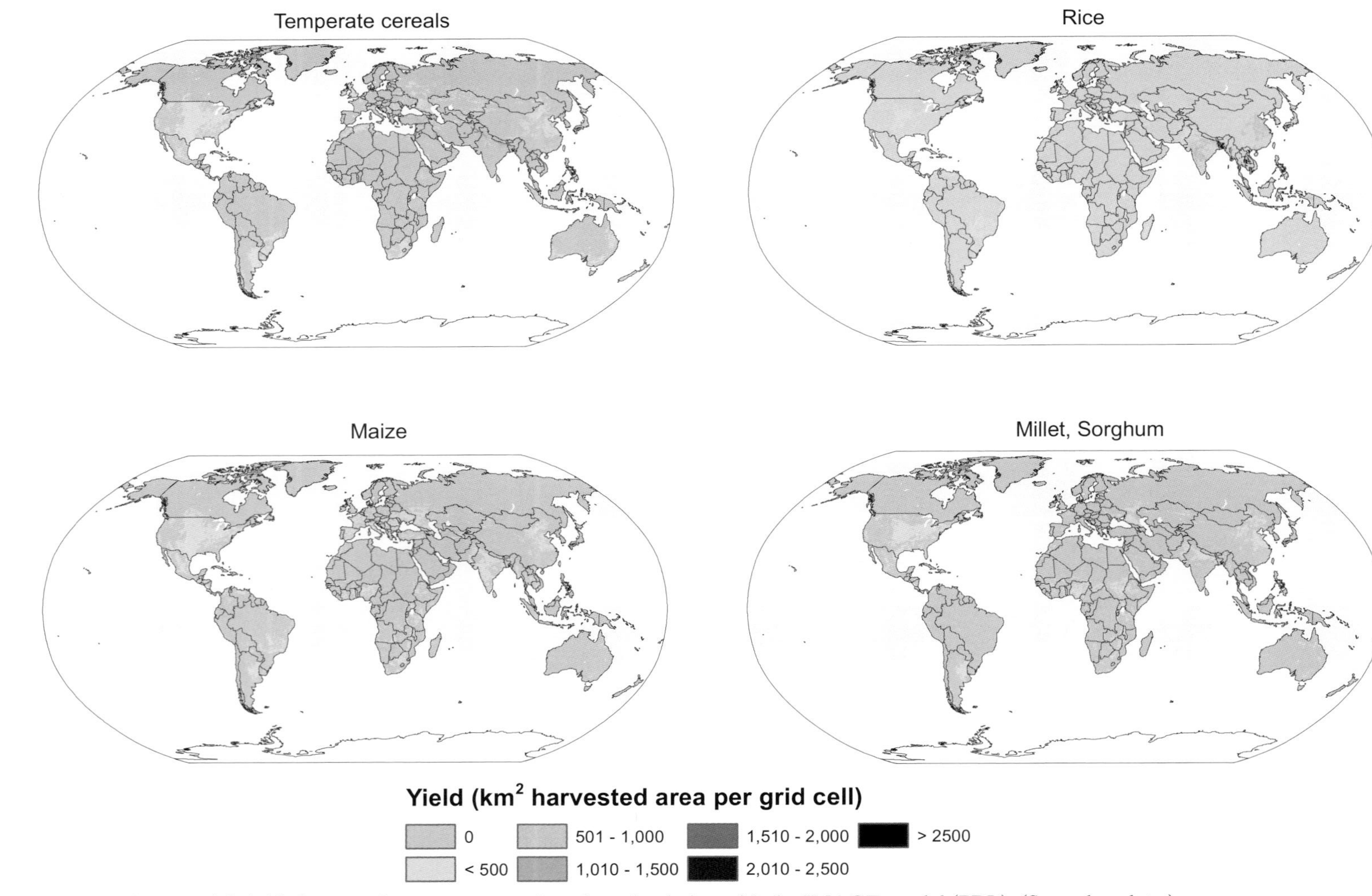

Figure 11.5. Potential yields for some important crops, based on simulation with the IMAGE-model (PBL). (See color plate.)

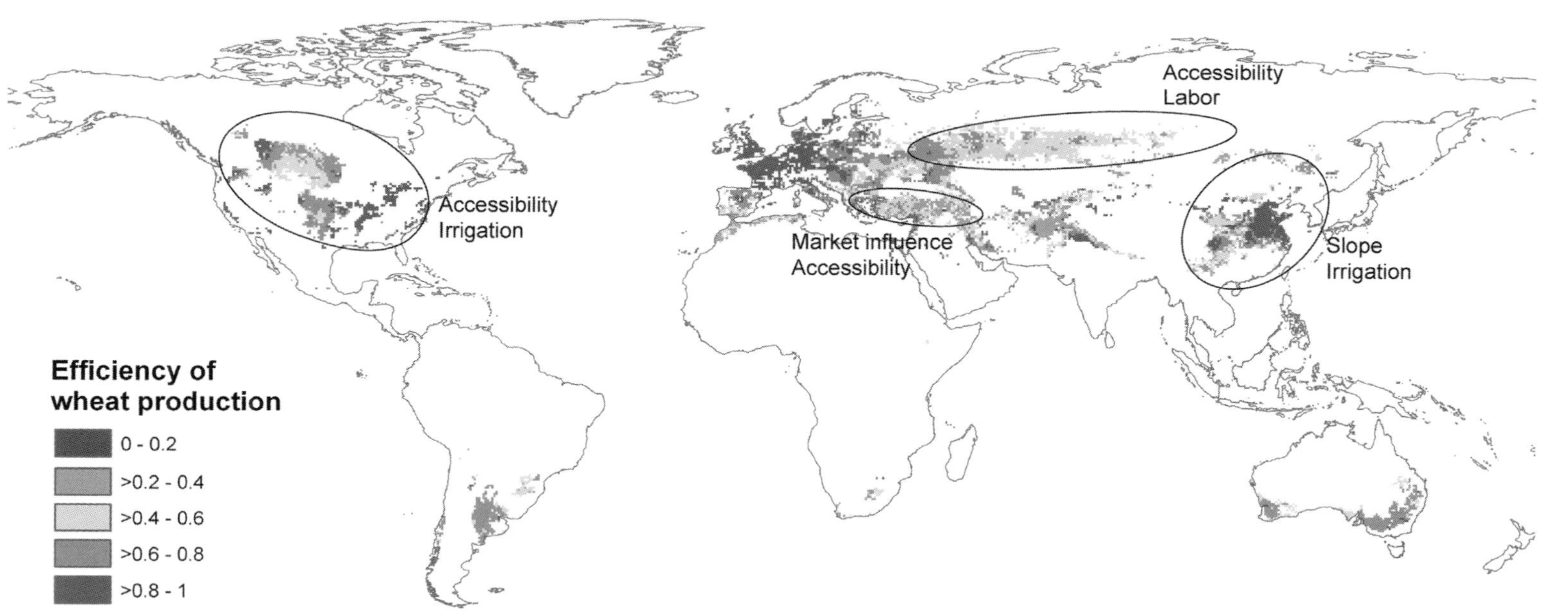

Figure 11.6. Estimate of the efficiency in wheat production (Neumann *et al.* 2010). Efficiency is defined as the ratio of the observed output to the corresponding frontier output. (See color plate.)

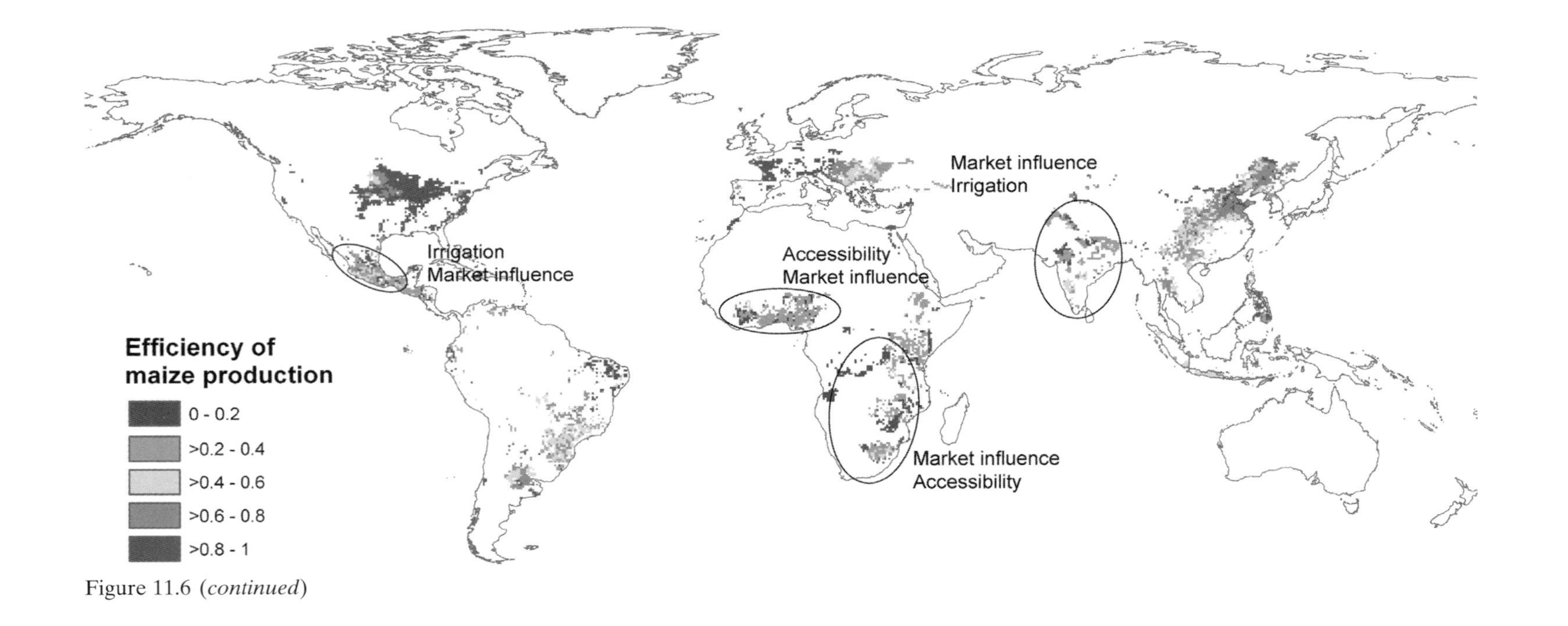

Figure 11.6 (*continued*)

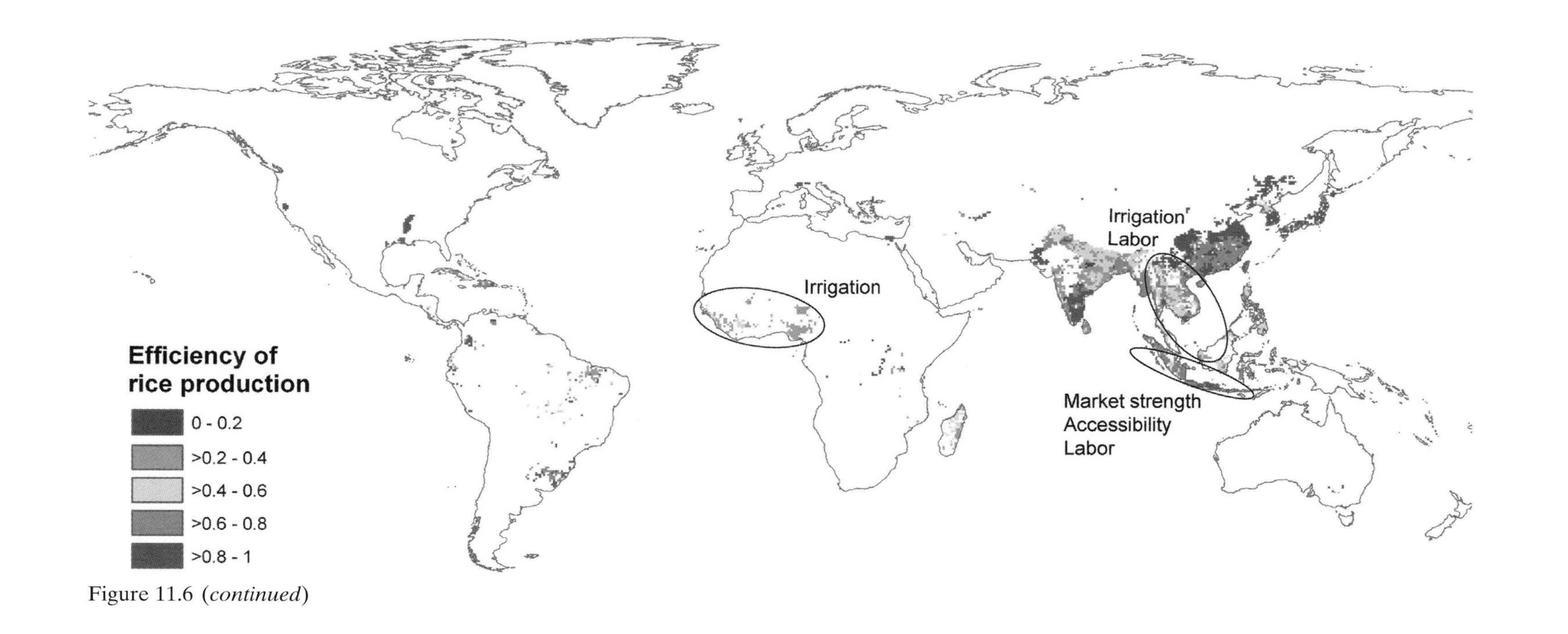

Figure 11.6 (*continued*)

ten major crops more than halved in this period. This successful intensification has been possible thanks to a series of changes: irrigation; multiple cropping; application of more and more specific fertilisers, herbicides and pesticides; improved land management (optimal planting, pest controls and so on); and genetic advances (raising the seed/biomass ratio or harvest index and new breeding techniques). More innovations in the agrosciences are expected and some people talk about a second Green Revolution and a livestock revolution (Jaggard *et al.* 2010; Thornton 2010). It is unclear to which extent this is a solid expectation or wishful thinking. In any event, it is widely believed that the increase in food production necessary to relieve hunger and feed the 2 to 3 billion additional people is only possible with further intensification. As the calculated yield gaps presented in the previous section indicate, there is still room for improvement with existent practises and technologies. But will it aggravate existing problems of erosion, salinisation, pollution and other forms of degradation? Will it make farming less sustainable?

Soils are the crucial resource in this respect. A straightforward definition of sustainable soil management is to preserve its *inherent fertility*. The inherent fertility is in essence a legacy from the past: Some areas in the world are endowed with very fertile soils of metres thickness, whereas in other places there is only weathered-down rock. The natural processes that created the legacy are still at work, but they are mostly slow in comparison with the rate at which humans are now intervening. And not only the rate but also the extent and variety of human interventions have become a matter of great importance. A number of soil characteristics determine the suitability for food production. Sustainable soil management thus implies from a system dynamics point of view (§2.3) that:

- the rate of outflow of nutrients does not exceed the rate of inflow by natural and human processes; and
- the rate of inflow of fertility-reducing substances ('pollutants') does not exceed the rate at which they flow out or are broken down to non-damaging substances.

The scheme in Figure 11.7 shows the most important processes in soil formation and losses and in soil fertility maintenance.

There was widespread decline in ecosystem functions and productivity in the last decades of the 20th century. It prohibits certain forms of land use and affects hundreds of millions of people (Figure 9.7). There is controversy about causes and consequences of such *land/soil degradation* (§ 9.4). Many soils in the world have been degraded by a mix of natural and man-induced processes, notably wind and water erosion. In fragile drylands with large natural variations in rain, human activities have accelerated the degradation and the inherent soil fertility has declined because of loss of organic matter and salinisation. Such irreversible processes on a time-scales of several decades are already impacting on food yield and output in several regions in the world (Brown 2004). The areas of most concern are the densely populated regions in the tropical zones, where cropping and foraging intensity are already above sustainable levels and where demand for food and feed is expected to increase further.

Ongoing intensification requires more inputs and creates new and sometimes harmful side effects. An important intensification measure is irrigation. *Water* is a crucial input, but water use is unsustainable in many places, as discussed in the

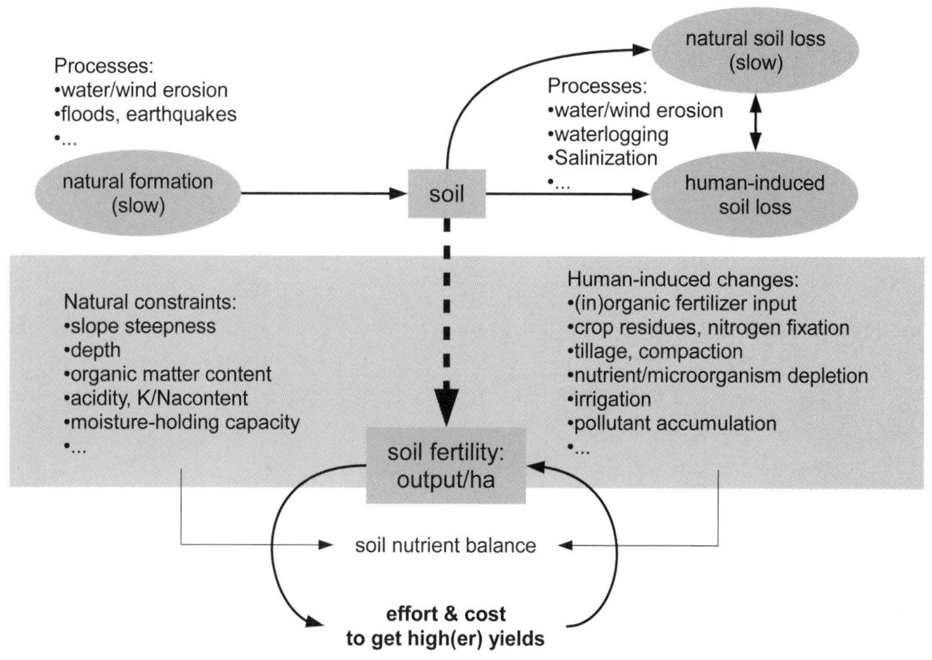

Figure 11.7. Scheme of the factors involved in (un)sustainable soil use.

next chapter. More efficient water use is a necessity. There is also use of more chemical and machinery inputs and a continuous flow of genetic innovations with only partly foreseeable impacts. One of the consequences is a massive and still increasing interference with the natural nitrogen (N) and phosphorous (P) cycles. Figure 11.8 is a map of the N- and P-balances for natural ecosystems and agriculture. In the densely populated regions, the emission densities (in kg/km^2/yr) are very high. Excess values contribute to *eutrophication* of water and soils with deleterious concequences in the long run.

Further intensification thus aggravates some of the undesirable trends that have come along with the successes:

- the widespread and, in many places, still increasing use of inputs pollute soils, rivers, lakes and seas, notably eutrophication from N- and P-compounds;
- increasing health risks at several levels, from the individual farmer who uses herbicides and pesticides to the system-large disruptions because of disease outbreaks or the impacts of using genetically modified organisms (GMOs);
- larger pressure on (ground)water resources for irrigation, contributing to further loss of biodiversity and ecosystem services; and
- in a larger system context, the agro-food system becomes increasingly dependent on finite resources, notably oil and phosphorous.

One can, therefore, expect that to supply more food will bring new risks, with sometimes dramatically different consequences in different locations.

Parts of the agro-food system, notably meat production, contributes to greenhouse gas emissions and the subsequent climate change. Direct effects from rising

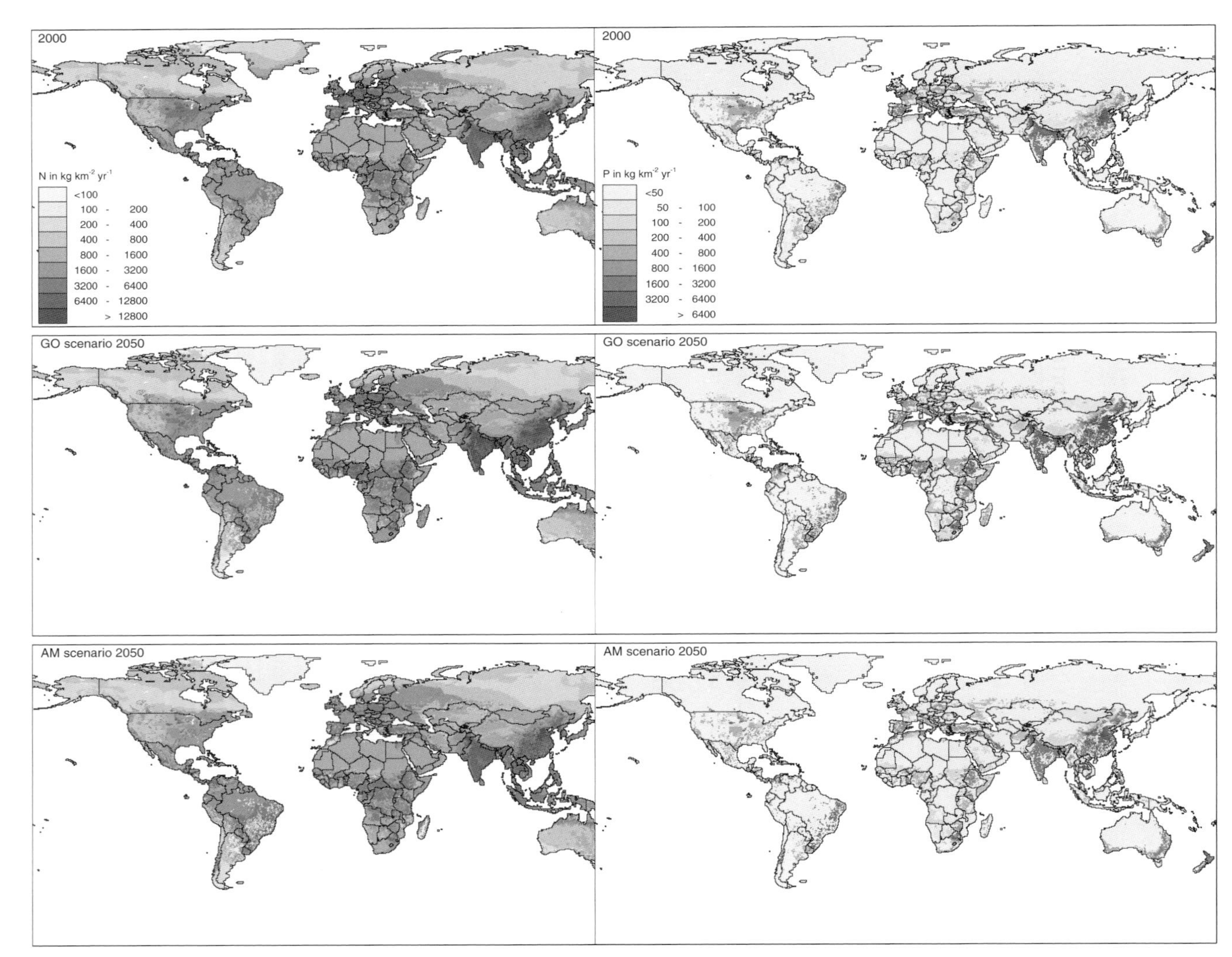

Figure 11.8. Total balance for (left) N and (right) P for natural ecosystems and agriculture for 2000 and for 2050 according to two scenarios (Bouwman *et al.* 2009). (See color plate.)

atmospheric carbon dioxide (CO_2) concentrations have in the theory and the laboratory a positive impact on yields through a higher rate of photosynthesis and a higher water use efficiency. It can be beneficial for some regions. Generally speaking, however, the consequences of *higher surface temperatures and changes in precipitation patterns* from climate change as a consequence of rising atmospheric concentrations of CO_2 and other greenhouse gases can be severe: extreme rainfall or drought, heat stress during heat waves, and large-scale adaptations and relocations. And despite massive field tests and model simulations, it is still uncertain how yields will respond in reality to changes in temperature and precipitation (IPCC 2007; PBL 2010a). Rising concentrations of surface ozone (O_3) in the Northern Hemisphere could annihilate any yield increase from CO_2 fertilisation (Jaggard *et al.* 2010). Whatever the scientific details, there is widespread concern about the possible consequences of climate change for agriculture and much research is done to identify the vulnerable populations and to strengthen the capabilities for adaptation (UNEP 2007).

How all these changes will impact food production and food security is very difficult to say. The human response in the form of innovations and lifestyle changes will be varied and make conditional forecasts even more uncertain. A natural next step is to investigate in more depth the social-economic and cultural processes: the facts, mechanisms and models of land use and land cover changes. But first, I present some stories from the real world in order to ensure appreciation of the human factor and illustrate both the locality and the universality of the quest for a sustainable food provision.

11.3 Stories from the Real World

11.3.1 Nomads in Mongolia[11]

In April 1998, the U.S. Embassy in Beijing reported on the tension between the nomadic herders in Inner Mongolia and nearby farming people, who are invading in large numbers and cause environmental havoc. It shows how closely natural and human factors are intertwined. 'Two hundred thousand impoverished farmers, mostly Hui minority people from the Ningxia Hui Autonomous Region, devastate the grasslands of Inner Mongolia each summer. They go there to harvest facai grass. Land per farmer in Ningxia fell by one-third during the 1980s.... [and] Mongol minority herders are no match for gangs of 20 or more Hui nationality farmers, sometimes armed, who often poison wells to drive the herders away. (Chinese) desertification experts say facai harvesting is the chief cause of desertification in Inner Mongolia. A 1997 expert report... recommends a ban on facai harvesting and commerce combined with poverty alleviation programs for the 200.000 poor farmers involved. Human activities, rather than long range climate change, appear to be the main causes of desertification in Inner Mongolia. Just as in other parts of China, the map of severe poverty largely coincides with that of rapidly growing minority populations, desertification and environmental devastation'.

[11] The quotes in this story are from www.usembassy-china.org.cn/sandt/desmngca.htm and www.reuters.com/article/idUST3183820071008.

According to a 2007 report from China, the large grassland steppes of Inner Mongolia are turning into sand deserts, 'The wild grass reached up to my knees in the past,' according to a 40-year-old herdsman. 'But there's very little grass now. It hasn't rained here in six years and we have to buy fertilizers and feed for our livestock. We never needed these before.' Desert area in China is estimated to have increased from 17.6 percent of total land area in 1994 to about 27.5 percent in 2006. According to the report, many homes in Inner Mongolia and other western provinces have been swallowed up by sand, dumping sand in springtime, not only on Beijing but also sending dust particles as far away as Korea, Japan and even the United States. It is causing respiratory problems, especially for children and the elderly. 'Eye infections are getting more serious and common because of the sandstorms', according to the chief of the Xilinhot City Peoples' Hospital in Inner Mongolia. A 'Green Great Wall' of 700 km barrier of shrubs and trees has slowed down the desertification, but has not stopped it completely.

11.3.2 Can and Should Rural France Be Saved?[12]

'When you understand that the take-it-or-leave-it prices now being offered [by wholesalers] mean that you'll pay more to produce crops than you'll get back in proceeds, you're left with the choice of either becoming a slave to this impossible system or find a niche to begin other activities.' This statement is from a French farmer who took over her parents' farm and diversified from crops to horseback excursions. She expresses a general trend in French agriculture. Since 1960, the number of farms in France decreased from 2 million to 0.66 million in 2010. They employ less than 3 percent of the workforce. A further decline to 0.32 million by 2020 is expected. A fourfold increase in labour productivity in the last 40 years has transformed many farms into mega-operations run with an industrial mindset. Prices paid for crops have fallen 60 percent in this period – and still the gains in efficiency are insufficient to get a decent income because the wholesalers who supply food processing companies have gotten access to even cheaper alternatives, such as wheat from Ukraine and strawberries from Morocco.

It is a downward spiral of competition and globalisation, which, in combination with elimination and a 'green' re-orientation of European Union subsidies, forces farmers to look for alternatives. One avenue is bio-agriculture in which quality, flavour and local tradition is emphasised. Another option is agritourism, which benefits from the infatuation of urban French citizens with the countryside. A third option, widely used also in other parts of Europe, is to let the farmer play a key role in landscape and nature management in its various forms. It appears improbable, however, that these trends will make major inroads in mainstream European agriculture.

Can the farmer be saved? In the summer of 2010, a group of farmers persuaded a fastfood chain to give them permission to sell their fruit and vegetables in the parking place. There was widespread media attention: 'The raison d' être of farmers is to feed people and manage the land, not the market... we work with a loss and consumers cannot pay our products... why do I get 17 cents for a kilo of apples and does the

[12] The first two lines are based on an article by Crumley, How to Save Rural France? in *TIME* August 2, 2010.

consumer pay 1,70 to 3 Euro'? Should the farmer be saved? Consumers benefit from lower prices and global competition forces farmers to be more competitive or leave it to – often poor – farmers elsewhere. From an equity and environmental perspective, that is not necessarily worse than subsidising European, or American, farmers. Inevitably, worldviews and painful trade-offs enter the evaluation.

11.4 Land Use and Cover Change

11.4.1 Land Use Changes and Its Causes

Land use and the associated changes in land cover or *land use and cover change* (LUCC) are an important cause and consequence of global environmental changes (§4.5). This is particularly true for the low-income regions in the world where the already large population density is expected to increase even more, and for the regions that are industrialising rapidly under the opportunities offered by global markets. How and why did and does land cover and land use change happen? Despite advances in data availability and modelling, understanding the causal mechanisms behind LUCC is fragmented. The complexity of meteorology and ecology is compounded with the human dimension.

Evidently, the quest for food, fodder and fiber is at the root of LUCC as well as the – often related – construction of dams and roads and mining and other activities have played and do play an important role. The major process is the still ongoing transition from local and small-scale subsistence farming to large-scale industrial farming for an increasingly global market. Ongoing *deforestation* is part of this process (Figure 4.4). In the last three centuries, massive conversion of forested/woodland area and savanna/steppe/grassland area into cropland (from 300 to 400 and from 1,500 to 1,800 Mha) and pastureland (from 400 to 500 and from 3,200 to 3,300 Mha) has taken place (Lambin *et al.* 2003; Figure 4.4). In the 18th century, conversion accelerated in Europe, the northern part of the Indian subcontinent and in China. This was followed in the 19th century by rapid cropland expansion in the newly developed regions of North America and the former Soviet Union. The greatest cropland expansion in the 20th century took place in South America and Southeast Asia. Much of the remaining forest in these regions and in Africa is under pressure of deforestation, despite impressive forest regrowth and afforestation rates in some parts of the world. On the other hand, tens of millions of hectares have been taken out of production in United States and Western Europe in recent decades as a consequence of rationalisation in agriculture.

Following the concept of transitions applied in demography (§10.3), the notion of a land use/cover or *LUCC transition* is used to describe the manifold and multicausal processes involved. As explained before, transitions are not deterministic trajectories but rather co-evolutionary and reversible change processes with a variety of social, economic and ecological feedbacks. A recent inventory of changes during the last few decades within this framework indicates that (Lambin and Geist 2006):

- different parts of the world are in different phases of LUCC transition – for instance, cropland is decreasing in the temperate and increasing in the tropical zones;

- large-scale deforestation takes place in the Amazon Basin, in Southeast Asia (rapid cropland expansion) and in boreal Eurasia (logging and fire); and
- there are still large gaps, inconsistencies and uncertainties in the existing databases on LUCC (extent and rate of change).

The mechanisms behind LUCC are complex and multicausal. Often, population pressure and poverty dominate the debate but one should also consider people's responses to economic opportunities within the local, national and global institutional frames. Proximate (or direct) causes have to be distinguished from underlying (or indirect or root) causes. In a survey of a series of case studies, it was found that five fundamental underlying causes drive LUCC (Lambin *et al.* 2003):

- resource scarcity leading to an increase in the pressure of production on resources;
- changing opportunities created by markets;
- outside policy intervention;
- loss of adaptive capacity and increased vulnerability; and
- changes in social organisation, in resource access, and in attitudes.

The most important determinant of what actually happens is probably the behavioural responses of people to the perceived opportunities and constraints of markets and policies. Increasingly, this is influenced by forces that operate at regional and global scale and amplify or attenuate local factors.

In the context of worldviews, one recognises in LUCC transitions the tension between the local biogeography and the associated customs – the small world – on the one hand, and modern science and national/global government and business interests – the big world – on the other (§6.3). This multi-scale character is often a challenge in the search for sustainability. For instance, Raquez and Lambin (2006) conclude from meta-analysis of forty-six case studies on factors behind sustainable land use that 'in general, governments seem to be reluctant to take into account the diversity of environmental contexts to which local communities have adapted to properly manage their natural resources. They tend to dictate land use management strategies and deal with environmental problems in a top down manner'.

11.4.2 Modelling Land Use and Cover Change

LUCC processes are often conceptualised in the form of a land use change or transition matrix in order to simulate changes amongst land categories over time (Geist 2006). The basis is the collection, storage and analysis of data in a geographical information system (GIS). Land conversion processes are also simulated at a rather aggregate level in 'top-down' global change integrated assessment models (GC-IAMs). They are the tools *par excellence* to study how the local biophysical and socioeconomic changes have wider consequences through complex feedback linkages. In some of these models, the emphasis is on the biophysical processes, as in the Integrated Model to Assess the Global Environment (IMAGE) model (Bouwman *et al.* 2006) and the Global AgroEcological Zones (GAEZ) model (Fischer *et al.* 2001). Other models focus on aspects of the regional agro-food systems and the agro-economic dynamics and links (IFPRI 2008).

Box 11.2. *Modelling ecosystem dynamics and services: rangeland overgrazing in Senegal.* One of the challenges in (resource-)economic models is to incorporate the use of ecosystem services in a broader, complex dynamical context. An example is the model by Hein *et al.* (2005) on rangeland management in northern Senegal. It is an area where livestock grazing is the main source of income. Sustained, heavy grazing pressures have an impact on the vegetation which depends to a large extent on the annual rainfall. The data are available to test different models. First, a simple model was constructed to assess the optimal grazing pressure in animals/ha. Basic assumptions are a logistic growth process of livestock growth, no feedback from grazing upon the pasture forage production and hence upon livestock carrying capacity, given annual rainfall and fixed prices. In subsequent steps, this benchmark model was refined with stochastic variation in rainfall, a negative feedback of grazing upon rain-use efficiency, and price fluctuations in a situation of drought.

The inclusion of the loss of resilience for drought because of high grazing pressure indicates that an optimal long-term stocking rate should be 10 to 20 percent below the benchmark model value. The current stocking rates are in the order of 50 percent *above* the indicated optima. There are clear warnings that sudden ecosystem change might happen, with catastrophic results for the inhabitants (§9.7). An important task is to implement ecosystem dynamics and services in simulation models of long-term climate-induced changes in the (semi-)arid regions of the world. Only then can these stories *and* these models prepare for such events and their consequences. Increasingly, agent-based models (ABM) are developed to investigate situations like these and suggest strategies for sustainable management (§10.5).

However, the uniqueness of local situations make it necessary to analyse LUCC processes in a 'bottom-up' way, not only to perform local assessments but also to identify and validate universal mechanisms used in GC-IAMs. It requires a rather explicit and accurate image of 'agricultural man' (Verburg *et al.* 2003). At the lowest scale is the *unit of production*, usually the family farm with its family structure, labour division, wages and values, against the background of environmental variables such as microclimate, soil characteristics and seasonal fluctuations. The actual processes: clearing, burning, sowing, harvesting, soil erosion and so on, take place here. One scale level higher, the *landscape* sets the social-economic context: How is the community organised; how are property rights organised and secured; how is access to technology; and so on. Its biophysical counterparts are the landscape features like altitude, topography and soil and water basin characteristics. Irrigation schemes and erosion and sedimentation processes take place at this level. The *region* provides the social, economic and cultural as well as biophysical embedding of farmers and communities. The relevant macro-processes are colonisation and migration, large-scale infrastructural developments, world market food prices and trade opportunities and the like. Usually, assessments of food production are also made at this level on the basis of climate parameters (temperature and rainfall patterns) and vegetation- and soil-based crop potential (§11.2).

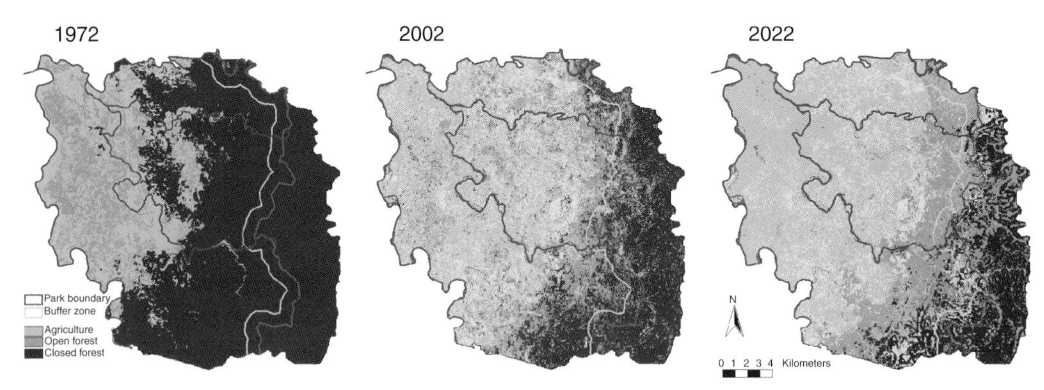

Figure 11.9. Generalised land use in San Mariano municipality in the northern Philippines: 1972 and 2002 based on interpretation of respectively aerial photographs and SPOT images; 2022 simulated by the spatial landscape-level model (Verburg *et al.* 2006). (See color plate.)

An example of a simulation model focussing on the local and regional scale is the CLUE (Conversion of Land Use and its Effects) modelling framework (Verburg *et al.* 2006). It uses a Cellular Automata (CA) approach and simulates land use transitions driven by human and biophysical factors. For each land type, partly overlapping suitability classes are assigned to the cells in a grid. These suitabilities may change over time as the result of feedback mechanisms such as land degradation. Regionally available food from the cells is calculated by incorporating the effects of past land use, pests or diseases, annual fluctuations with existing technology levels, economic values and two years of food/money reserves. In a second step, the regional land-use demand is calculated from population and variables such as management level, urban expansion rate, yield level and land use strategy (both a function of [inter]national technology level). People's values and preferences are considered, too, in particular concerning cattle and natural vegetation. In a third step, a land use allocation model translates regional demand into actual land use changes. In this way, the model can be used to explore agricultural transitions, taking into account food security and technological adaptation. The CLUE-model has been applied in regions in China, the Philippines and Central and South America, where traditional farming methods are gradually replaced or complemented by cash crops and other forms of commercial farming and where pressure on the forest and soils leads to land degradation and loss of biodiversity (Verburg *et al.* 2003; Wassenaar *et al.* 2007). It is also used in the EUruralis project on the future of rural Europe (Westhoek *et al.* 2006; Verburg *et al.* 2010).

One illustrative application of the CLUE model at the local scale focuses on a major biodiversity hotspot in the world: the Philippines, and in particular on the San Mariano municipality in the north (Verburg *et al.* 2006). It is an area of 48,500 ha with twenty villages and bordering one of the largest protected areas in the Philippines, the Northern Sierra Madre National Park. The still growing population largely are or descend from migrants. Logging companies deforested a large part of the area between 1960 and 1990. Since the logging moratorium in 1989–1992, employment declined and people switched to agriculture. A landscape-level analysis zoomed in on an area at the border of the National Park. Aerial photographs and remote sensing are used to present the changes between 1972 and 2002 (Figure 11.9a,b). It shows the impact of large-scale logging, with road construction and farmer immigration – a process that happened in many places in the world in the past centuries at an accelerating pace. Using the most recent land use maps (1990), the CLUE model was applied to forecast land use changes in the Philippines for the period until 2022 with an imposed deforestation rate based on trend extrapolation and for different policies and scenarios (Figure 11.9c). Given the need for more agricultural land and the resulting deforestation pressure in the Philippines, deforestation is expected to continue at a rate of one-third of the rate in the 1972–2002 period. To design and implement adequate policies to prevent ongoing illegal logging and conversion to agricultural land, the models need to address the different scale levels of local, regional and global forces in an integrated fashion.

A rapidly growing number of model-based analyses like these are made nowadays. They can help to understand the seemingly relentless drive behind deforestation at the different scale levels and help to formulate and implement feasible preservation strategies. The simulations of biophysical potential and socio-geographic change form a useful background to explore the prospects for 'sustainable food for all' and for the effort to eradicate extreme poverty and hunger – Millennium Development (MDG) Goal 1.

> MDG Goal 1: Eradicate extreme poverty and hunger.
> **TARGET**
>
> *Halve, between 1990 and 2015, the proportion of people whose income is less than $1 a day.*
>
> According to the United Nations *Millennium Development Goals Report* (2010), 'The number of people who are undernourished has continued to grow, while slow progress in reducing the prevalence of hunger stalled – or even reversed itself – in some regions between 2000–2002 and 2005–2007. About one in four children under the age of five are underweight, mainly because of lack of food and quality food, inadequate water, sanitation and health services, and poor care and feeding practices.'

11.5 Towards a Global Industrial Agro-Food System

11.5.1 Diversity in Transition

Although there are undesirable trends such as overexploitation and pollution of soils and water, the general view is that modern farmers have overcome ancient environmental constraints. Expansion into new agricultural lands has almost come to a halt. Intensification has led to a steady rise in yields and a declining trend in food prices, on an almost constant cultivated area. The quantity and diversity of food products offered on the local and regional market has increased. There are different judgments about the quality. There is a shift in diets towards high-value food, but junk food, obesity and diabetes are also part of the story. Improved logistics give a better match between demand and supply, which reduces the vulnerability of human

Table 11.2. *Agro-food systems: key indicators for some countries/regions in the world (2009)*

	Unit	USA	Japan	EU[a]	Korea, Rep	S. Arabia	RussianFed
Population (2009)	mln	307	128	499	49	25	142
Income (2009)	PPP US$/cap	**46.437**	**32.433**	**30.543**	**27.195**	**23.421**	**18.938**
HDI (2010)		0.902	0.884	0.854± 0.080	0.921	0.752	0.719
Ratio food import and exports (2008)	$/$	0.7	20.7	1.2	6.4	9.3	4
Cereals harvested area (2009)	ha/cap	0.18	002	0.11	0.02	0.02	0.3
Cereals production (2009)	kg/cap/yr	1335	90	820	150	95	675
Fertiliser use (2008)	kg/ha/yr	290	620	295	730	555	45
Capital stock (2003)	1995US k$/cap	10.8	35.7	10.3	15.7	29.6	2.7
	Unit	**Mexico**	**Brazil**	**China**	**India**	**Vietnam**	**Nigeria**
Population (2009)	Mln	107	194	1332	1155	87	155
Income (2009)	PPP US$/cap	**14.341**	**10.429**	**6.838**	**3.275**	**2.956**	**2.149**
HDI (2010)		0.75	0.699	0.663	0.519	0.572	0.423
Ratio food import and exports (2008)	$/$	1.5	0.15	2.2	0.55	0.7	4
Cereals harvested area (2009)	ha/cap	0.09	0.1	0.07	0.08	0.1	0.12
Cereals production (2009)	kg/cap/yr	290	370	360	205	490	195
Fertiliser use (2008)	kg/ha/yr	105	485	585	240	210	25
Capital stock (2003)	1995US k$/cap	7.7	9.0	7.4	2.9	4.3	1.5

[a] Population weighted average of France, Germany and United Kingdom.

Source: FAO; for HDI: en.wikipedia.org/wiki/List_of_countries_by_Human_Development_Index#Very_high_human_development_.28developed_countries.29.

populations to famine. Are we witnessing an irreversible change towards a global agro-food system that fulfills the promise of eradicating hunger, even in remote pockets? It is difficult to say, partly because there are still such large differences between and within the usual unit of measurement: the country (Table 11.2). The countries that produce food can roughly be divided into three groups: subsistence biomass-based, agro-industrial and transition.

One-third to one-half of the world population still lives in a *biomass economy*, with natural energy flows for their sustenance. They are located on the periphery of the world system, across all continents but especially in Asia, Africa and Latin America (Figure 11.3). Their livelihood is from subsistence farming and herding. They are often poor in both income and wealth and are faced with increasing land, water and energy constraints under the pressure of population growth, resource depletion and globalisation in all its appearances. Most have low levels of education and health and lack access to capital. Many still adhere to a mythical worldview, with gods and rituals playing a direct role in their affairs. Often, they live in a society with a rigid social hierarchy, dominance of male values and enforcement of dogmatic beliefs. For them, survival values are still dominant. Food is more than just calories: Its production and consumption is rooted in ancient traditions and has profound social and cultural meaning. Their knowledge of local ecosystems is a source of resilience. Some typical characteristics of these social-ecological systems (SES) are:

- regional biogeography is a key defining factor, that leads to inherent diversity, locality and plurality of human cultures and institutions;
- part of their activities and transactions is not monetised; therefore, it is in-house, subsistence farming/herding and *in natura* exchange;
- livelihood is sustained on the basis of local resources which are exploited as common pool resources (CPR) with a mix of competitive and cooperative arrangements;
- social stratification is often strong and the resource-based surplus is exploited by local landlords or national government and industrial elites (cash crop or mining);
- people are vulnerable for natural and human-made perturbations because of limited resource and adaptation options.

Sustainable development for them is about the risks and challenges posed by a combination of overpopulation and local resource overexploitation. Trade and out-migration bring relief and opportunity, as well as dependence and new risks.

On the other extreme are the farmers in the advanced industrial economies in the centre of the world-system. They are a mere 3 to 4 percent of the world's agricultural population, but their products represented more than two-thirds of the market value of world agricultural output in 1995–1997 (Figure 11.3; WRI 2000–2001).[13] Their farming activities have gradually been industrialised to the extent that their relationship with natural ecosystems has largely vanished. Their business is high-input industrial agriculture, as part of the increasingly global agro-food chain of food producers, processors, traders and retailers. Their farming is large-scale growing of staple crops such as wheat and maize, production of high value-added fruits and vegetables, and large-scale animal husbandry for dairy and meat products. Competition forces them to operate at ever larger scale and to aim for ever higher yields with the latest biotechnological techniques. They also face risks: those associated with financing the large capital requirements and trading on global markets (Roberts 2008). Increasing globalisation tends to erode their income security. Postindustrial concern about environment, ecological restoration and animal welfare confront them with new and additional constraints and risks.

In between these two extremes is a large and growing number of farming families who try to adapt to the rapid regional and global changes they are faced with. They are in the midst of the nutrition and land use transition. In some regions, the Green Revolution offered the opportunity for large-scale wheat farming on irrigated lands,

[13] WRI (2000) gives an estimate of 80 percent for 1995–1997. It has declined since then, but reliable recent data have not been found. The large share is partly because part of the food production does not enter into the formal market accounts.

Box 11.3. *Entering the cash economy.* 'Chotu Ram knew that "development" was not necessarily to the advantage of everyone... Chotu Ram was a Kumhar, or potter, by caste, but nowadays the farmers were cash-conscious and unwilling to spare any of their land for his caste to dig clay. What's more, clay cups and pots were today considered inferior – stainless steel, plastic and, among the wealthier, china had taken over. When the service of potters had been as essential to [the village] Thakurdwara as that of barbers or washermen, Chotu Ram's family had been reasonably well looked after. They had received grain and other gifts that at least provided their basic requirements. Now Chotu Ram had become a modern man, he'd entered the cash economy, but the wages he received barely allowed him to buy anything beyond the grain his family ate, and they certainly did not cover the emergencies that every family in the village had to face from time to time' (Tully *The Heart of India* (1995) 22–23).

as in northern India. Others switched to high value-added fruits and vegetables for the local market, with rising incomes and demand, as in China. Cooperatives have been founded to market the milk from the small herds and satisfy the growing demand for dairy products, as in villages of rural India. Many try to survive on a mixture of subsistence farming, cash crop production and migrant work in urban centres, as in Kenya and Nigeria. Many also become labourers on large-scale plantations owned by foreign investors and producing cash crops for the global market, as in Indonesia and Malaysia (palm oil), Ecuador (bananas) and Vietnam (coffee). Rising food prices and prospects for biofuel have stimulated the interest of speculative foreign investors to buy land from these people – million of hectares are bought in notably African countries with weak legal and government institutions ('land grabbing').[14] This may intensify if governments of land- and water-scarce countries face an insecure food supply and enter the global food market for imports. Sustainable development for these people means job and income opportunities in order to share in the basic amenities of modern life. In the process, many will end up in the (mega)cities.

11.5.2 The Global Agro-Food System

Trade can diversify the food supply and thus reduce the vulnerability to large-scale famine – as does conquest (§3.3). Both are an important ingredient of food security since the earliest civilisations. In economic terms, *trade* was driven largely by a comparative cost advantage: Some places are more suitable than others for food production. The appearance of novel products (potatoes, sugar, tropical spices and fruits), the use of military force by the colonial powers and the advances in the agrosciences were also part of the story. In the second half of the 20th century, world

[14] With economic, not military means, some Asian countries are repeating the history of colonialism. Up to 50 percent of agricultural land in Madagascar is said to be sold to foreigners from Korea and also in other African countries farmers are apparently selling their land to foreigners from China, India and Saudi Arabia.

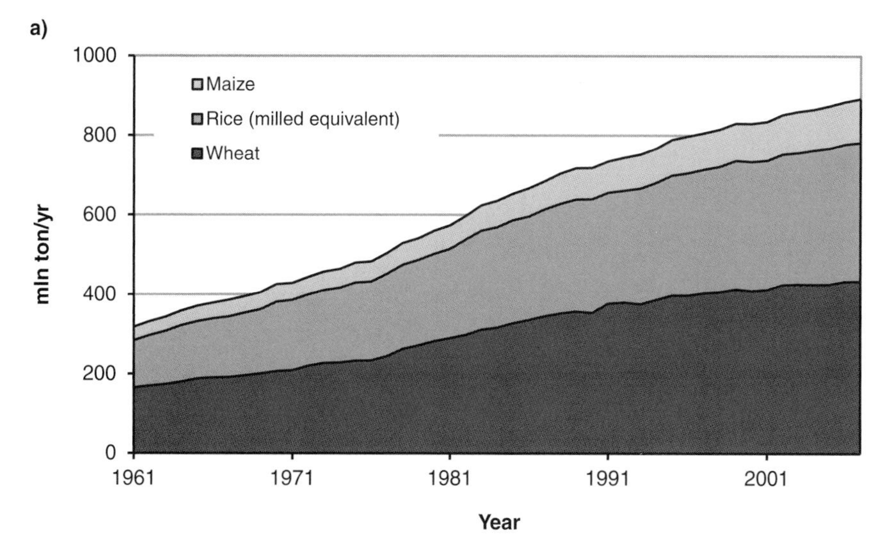

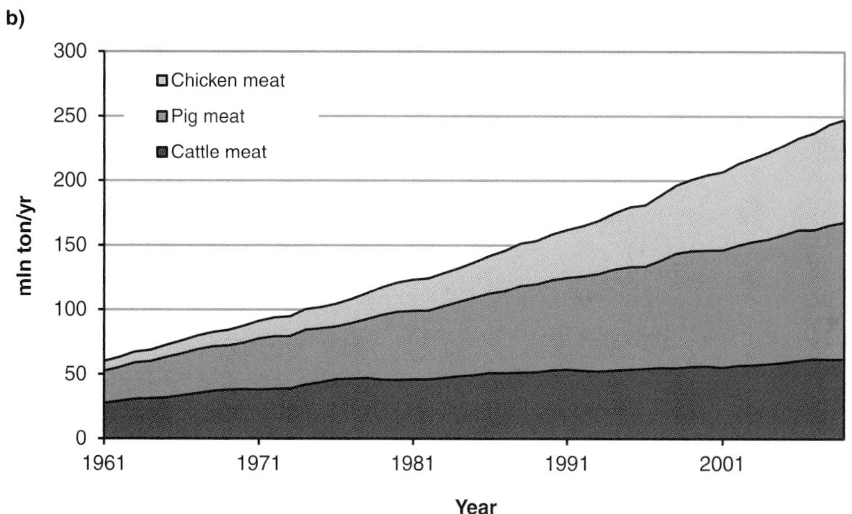

Figure 11.10a,b. World output of three dominant cereals (maize, wheat and rice) and of the most important sources of animal protein, 1960–2008 (source of data: faostat.fao.org).

food production started to increase rapidly and so did food trade. Annual output of the dominant cereals – or grains, notably wheat, rice and maize – almost tripled since 1960 (Figure 11.10a). Besides, there are other staple food crops (potato, millet or sorghum) and vegetables and (sub)tropical plantation produce (sugar, rubber, coconut, palm oil, cacao, coffee, tea and many other crops). Extensive grazing and dairy farming – notably for milk, butter and cheese – became large-scale and annual production of meat from cattle, pigs and chickens increased more than fourfold in the last five decades (Figure 11.10b).

Food production is the upstream source part of the *global food chains*. It represents an economic output in the order of 5 percent of GWP.[15] But the larger part

[15] The estimate of 1,300 billion US$/yr is for the years 1995–1997 and based on WRI 2000. More recent data are difficult to find.

of value-added is in the downstream part of the chains. Total monetary transactions in growing, processing, trading and retailing the ever larger array of food products amounted in 2007 to an estimated US$3,500 billion or more than 8 percent of GWP. About 80 percent had to do with food and beverages manufacturing and grocery retail (www.etcgroup.org). These activities become ever more *concentrated* in the hands of a dozen globally operating companies, who account for more than one-third of the total world transactions. Food processor Nestlé and supermarket chain Walmart are the respective leaders. The remaining 20 percent consist of the manufacturing of seeds, agrochemicals, pharmaceuticals, fertiliser production and biotech activities. These activities are even more concentrated in the hands of a few multinational corporations: The top ten producers have 65 to 90 percent of the market, with Syngenta, Monsanto and Bayer as the largest ones in 2007 (www.etcgroup.org). A relentless competitive struggle for market share and expansion is the driving force behind this concentration and the production and control of scientific and technological knowledge. It is one of the forces behind and outcomes of global capitalism.

Globalisation is a mixed blessing. The growth of financial institutions and the movement towards open trade increasingly forces farmers worldwide to compete with large-scale intensive farming and low labour-cost areas. Globally operating agro-food producers and traders exploit differences in labour costs. In combination with storage strategies and speculation, the increasingly connected system exhibits *food price instabilities*. The clearest sign of failure is *hunger*. There are still more than one billion undernourished people in 2010. There is enough food for them, and probably also for the 8 to 10 billion people expected to live on the planet by 2050. But those who need it cannot afford it or have no access. Relative abundance and improved logistics have made *food aid* a common practise. But it has also encouraged populist policies to appease the urban poor – the *panem et circenses* in ancient Rome. It is even argued that food aid aggravates the very problems it is supposed to solve, because they make urban populations dependent on food from elsewhere and discourage local farmers to make the necessary investments (Moyo 2009). Forward projections for some countries read like a Malthusian doomsday: Investments in local farming fall short, traditional employment declines, fertility rates remain high because education of women stalls, rural areas get depopulated and city slums swell to ever larger proportions. The activities of NGOs for human rights, poverty relief and environmental standards gain momentum, but still only slowly redirect some of the basic trends.

In view of the growth in human populations and the trend towards more meat, food demand is expected to increase. On the supply side is still considerable room for expansion (§11.2). There is the prospect of *innovations* at an unprecedented scale and in various directions, from high-tech GMOs to organic farming, from reducing food chain losses to dietary changes and from ever more industrialised farming to community-based local schemes. But these bring new risks and challenges. Can a further increase in fertiliser and water use and in pesticides and herbicides use be part of sustainable development? Is dependence on fossil fuels and control of the food chain by a small group of powerful corporations compatible with sustainability? The long-term future of agro-food systems are hard to imagine let alone predict. But it is a near-certainty that a decline in soil fertility, water availability and waste absorption

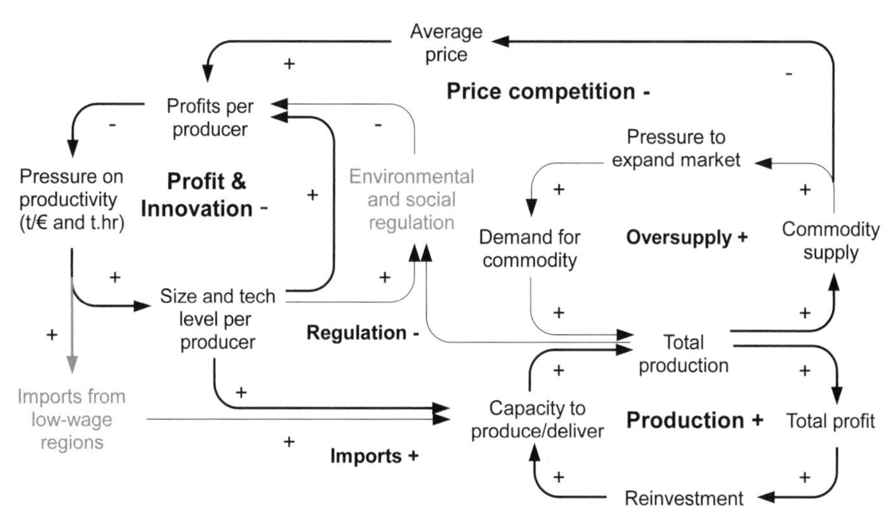

Figure 11.11. Causal loop diagram of feedback processes in agro-food systems (partly based on Sustainability Institute 2003).

capacity and an ongoing economic, social and cultural globalisation will make the world agro-food system more complex and more vulnerable.

11.5.3 Markets, Scale and Innovations As Driving Forces

Many different mechanisms and forces are at work in the global agro-food system. Some generic insights can be formalised in a system dynamics model in order to get a better grasp of the options for a sustainable path.[16] What are the mechanisms behind the trends? One can distinguish several loops (Figure 11.11). The first one is the *production* loop. It operates in every capitalist market system: The farmer-entrepreneur makes an investment in food producing capital and, with other inputs like labour and land, the production capacity generates output and revenues/profits. Part of the profits are reinvested in order to further increase producing capital. It is a reinforcing growth loop as long as profits are sufficiently high, which depends amongst others on demand. Otherwise, capital owners will look for other opportunities.

Second, a balancing *price competition* loop is operating. Providing there is open access to the market, competition amongst existing and new producers drives down the average price. This, in turn, tends to increase demand until the profit margin has declined again to the extent that farmer-entrepreneurs and capital owners put their effort and money somewhere else. Producers can respond to stagnating sales and profits with an attempt to find new markets. A subsidiary loop is *marketing/upgrading* in which demand is driven up by processing and advertising. A more effective and important way to boost profits is to bring down the cost of production by increasing the scale of production and by technical innovations. Thus, a third loop

[16] This analysis is inspired by the report *Commodity Systems Challenges* by the Sustainability Institute (2003). The starting point was the observation that output and prices of agricultural products between 1950 and 2000 in the U.S. food system all show an upward trend in output and a downward trend in price. The search for generic patterns is also the essence of the syndrome/archetype approach (§4.5). See also Roberts (2008).

of *upscaling/innovation* ('rationalisation') leads to increasing productivity of labour and capital and declining product cost. The prospect of rising profits will provide an incentive to invest in new capacity, which in turn reinforces production growth.

These three feedback processes have given a boost to productivity in food provision and greatly benefited populations in the United States and Europe. Government regulation for the sake of food security and consumer protection played an important role, although there were different views on and forms of agricultural policies. But in many places elsewhere, weak property rights and no or weak rule of law permit parasitical politicians and vested business interests to reap short-term profits and impede productivity growth. Market incentives that could respond to the pressures of rising fertiliser cost and large losses during transport are blocked. These are causes of hunger.

There are two more loops in a globalising and postindustrial world. First, despite scale and innovations, the product may still be more expensive than equivalent imports. In a world committed to open trade, producers are confronted with competitive imports. This *import* loop adds to the capacity to deliver and puts a downward pressure on price and an upward pressure on demand. Protective measures for reasons of food security and employment were and are a natural response of governments. Second, with rising income people become more demanding with respect to the environmental and social aspects of food quality and production. They also become more aware of environmental, health and social impacts of the industrialising and globalising agro-food system. It manifests itself as a *regulation* loop: More stringent health, safety, environmental and animal welfare standards tend to increase costs. This trend towards postindustrial values contributes to what is perceived as a good quality of life, but puts also new strains on the system.

The United States and European agro-business model has with the fall of communism become the dominant vision for the future of agriculture. Many consumers will perceive it as a virtuous cycle of ever more food available at ever lower prices. From the perspective of a farmer, it can also be experienced as a vicious cycle of income erosion and increasing dependence. In an environmental resource context, it poses serious long-term dangers. The system grows in response to increasing consumer demand and shifting preferences, driven by increase in scale and competitive innovations (Figure 11.11). Against such a larger system background, a number of mechanisms and interventions determine its path:

- *food surplus and trade*: the scale/innovation loop may be so successful that a food surplus results. Well-known examples are the European Union 'milk and butter mountain' in the 1990s. The income 'race to the bottom' of United States and European farmers in their relentless search for the cheapest supplies is one consequence. Food exports in poor countries, promoted by governments and business against the interests of their own population, is one implication of global trade;
- *food shortages and food security*: like any commodity system, the agro-food system experiences fluctuations in outputs and prices. Rapidly rising demand and globalisation intensify this phenomenon. It is reinforced by short-term responses such as speculation, social unrest and government protectionism and aggravated by expansion of biofuel plantations, possibly climate change–related harvest

Box 11.4. *Global food.* 'Paprika is the latest recruit to a revolution in Peruvian farming... the country has added almost 400 different export crops to its traditional staples of coffee, cotton and sugar... Farm exports totalled $1,3 billion in 2004, up a third on the previous year.' (The Economist, September 7, 2005). Ukrainian soils are the most fertile in the world, and Ukrainian wheat exports has risen tenfold between 1995 and 2010, but the farmers do not benefit. Instead, the introduction of export permits brought great fortunes to a small elite of entrepreneurs and bureaucrats. Because banks ask 20 percent interest to small farmers, international capital funds invest in Ukranian agriculture. Rationalisation and upscaling increase yields (NRC January 4, 2011). The harvest was good in southern Africa in 2008 – but people cannot afford to buy the food. They spend more than half of their income on food. The FAO estimates that low-income food-deficit countries will import 2 percent less cereals but pay 35 percent more.

In the world food system, events unfold as a series of connected causes and consequences too complex for a simple single story. Rich countries buy Chinese goods while building up huge debts. The money is used by a Chinese middle class to eat more meat. Chinese corn and wheat production fall short of local needs. Simultaneously, the U.S. government decides to reduce oil dependence by subsidising ethanol from corn. Exports start to fall and cause a 50 percent rise in the corn price in Mexico, where it is a staple food. Worldwide droughts and floods cause harvests to be less than normal in a number of countries. Argentina and Brazil respond with rapid and large-scale expansion of soybean production for export, partly on deforested lands. Although less than 10 percent of food is traded internationally, markets do work and the additional demand causes food prices to rise. The Russian government responds with a ban on exports, causing food riots in Egypt with political consequences. And so on...

failures and natural catastrophes and famines. In the background are the 'slow' dynamics of resource degradation and environmental change;

- *externalising environmental and social consequences*: the push to bring production costs down causes serious environmental damage and social destruction, that is passed on to poor populations and future generations. This process, too, has been intensified by globalisation. Competition has not brought innovations throughout the food chain but also caused havoc to many poor farmers and their communities (income disparities, debts, starvation and loss of local skills and resilience);
- *environmental and social regulation*: increasing awareness and demands of consumers in rich countries force governments to introduce more stringent standards and regulations, with diverse impacts throughout the food chain. Food chain lobbies tend to resist, arguing that it makes them less competitive vis-à-vis foreign producers who are not faced with such standards. Governments of low-income, food exporting countries consider such measures a form of protectionism and cannot or do not comply. On this not-so-level playing field, an increasing number of trade disputes take place with the World Trade Organization (WTO), governments, corporations and a multitude of NGOs as participants.

Although this analysis pertains primarily to important, globally traded crops such as wheat, rice, maize and soybean, similar forces are at work in other parts of the agro-food system. High-value commodities such as cotton, palm oil and tea, which are cultivated on large-scale plantations, and fresh fruit and vegetables are increasingly embedded in global corporate and trading networks. The Big Food and Big Retail business conglomerates and sectors and the allied government departments operate in a sense as effective top predators.

11.6 Perspectives on Food and Agriculture

There are two long-term 'objective' constraints in the global agro-food system: Humans need a minimum amount of food for physical survival, and the food-providing resource base is finite in size and in capacity to supply and adapt. Within these bounds, the directions and options for a sustainable future are shaped largely from the *status quo* framework of forces and powers. But quite different pathways are still possible and the trajectory is a continuous balancing act between utopian expectations and dystopian tendencies. Some pathways are more desirable, feasible or possible than others. Some are more sustainable than others in terms of adequacy, affordability, security and safety for the majority of present and future people and for other species on the planet.

The system has for some decades evolved according to a neoliberal 'free' market–oriented worldview. In theory, this leads to improvements with the side effects remaining manageable through standards, labels and so on, under the guidance of the United Nations and enlightened and proactive corporations. In practise, it is in quite some places a postmodern nightmare in which private gain, speculation and exploitation dominate people's lives and in which prospects for local small-scale farming erode because of lack of investment and infrastructure. Are major disruptions inevitable when food prices go up and become more volatile? Is more technology the solution, and, if so, what kind of technology? Will resource overexploitation outstrip yield improvements? Can small-scale, ecological and community-oriented forms of agro-forestry and animal husbandry contribute to food security and quality? Is a new solidarity and ethic needed to bring an end to hunger? The last section of this chapter examines some prospects for the future with explicit use of divergent worldviews.

Many models are around that make conditional forecasts of future food and land developments. Some trends are unmistakable. The human population will grow and so will its needs and desires for food. Intensification and upscaling are inevitable. But how, how fast and with what consequences is unclear. Many of the seeds of change are hidden in small and fragmented events. Look at the following items in the newspapers:

- The 'light-green' customer is discovered by supermarkets. Their concerns are personal health, low prices and supermarket convenience, and they are less 'fundamentalist' than the original organic food market client. But Walmart, where one out of every ten shopping dollars in the United States is spent, has announced to compete more intensely on prices. And some nutrition specialists claim that fast-food-plus-additives is the most healthy meal, despite evidence that cheap food is bad for health.

- ‘Without us food would be even more expensive’, says a commodity trader in a 2008 interview. Author of *How To Make Big Money Investing In Metals, Food And Energy* (1998), he has great expectations of the second-generation biofuels, notably jatropha curcas. It grows in semi-arid regions, does not need much water and its nuts have to be harvested manually and are used to make biofuel. ‘The commercial world jumps on it . . . the financial world is still asleep’. But according to UN specialists, food trade is increasingly disturbed by investment banks and pension funds, with profits as their sole aim. Their activities should be regulated in order to prevent food crises.
- NGOs say that genetic technologies are pushed by an unholy alliance of science, corporations and governments. A company spokesperson says that ‘they are ready to be applied, but the fear of the label ‘genetically manipulated’ has to be overcome’.
- Using satellite photos, Dutch farmers can do ‘precision agriculture’ and significantly reduce fertiliser and pesticide inputs. ‘Comfort-class’ livestock buildings are highly mechanised and animal-friendly. Farmers earn additional income from recreational and educational services and renewable energy production.
- Eighty percent of Indian farmers has not more than 2 ha farmland. Yields are rather low, inputs such as electricity and water are subsidised and rural employment is high. What is needed are investments in infrastructure and mechanisation – but such a long-term policy does not deliver votes.
- Some 800 farmers in the Ferghana valley in Kirgizistan are producing biocotton with support of development NGOs. Certification helps them to get a higher price. A Bio Farmers Public Union helps with access to the international market. The goal is to double production by 2012.
- Pesticides that are forbidden nearly everywhere in the world are still used on banana plantations in Nicaragua. Numerous children have distorted limbs, psychic disturbances and heart defects.
- Coastal seas can provide endless amounts of algae and weeds, sources of protein rich food. The first farms in nutrient-rich waters are established. It reduces eutrophication and may also become a source of biofuels, according to researchers.
- ‘The world is at its end’, laments an old herder. ‘We knew how to deal with drought, but something in the world has changed’. The latest of a series of droughts is about to give the death blow to the Samburu nomads in northern Kenya. Like so many nomads, they will end up in the city slums.
- The Chinese government invests more than €100 million for equipment to create artificial rain in some of its driest regions, in response to severe water problems. A recent operation led to 17 percent more rain, it is claimed.

Each fragment reveals a particular perception and interpretation of the situation. Each fragment also expresses hopes and fears about the ‘good life’. Each can be assigned a position in the worldview space of Figure 6.5. And they are partly inconsistent, contradictory and incompatible, as is the world.

Table 11.3 is an attempt to express the different worldviews in terms of what a person desires and expects and the policies he or she thinks are most effective in reaching a world in which he or she would like to live. In the Modernist worldview,

Table 11.3. *Worldviews and agro-food futures*

Agro-Food Systems			
(dot lower left)	(dot lower right)	(dot upper right)	(dot upper left)
Statement 1: Agricultural trade liberalisation should be promoted in order to improve food security and boost incomes.			
Correct. It is through food trade that famines in many low-income regions have been prevented. More food security for all!	We should be very cautious, because it increases our dependence in the critical domain of food.	Local food production is to be preferred – more independence, less environmental cost of transport. If trade, then Fair Trade!	The advantages of food trade liberalisation are at least partly undone by speculation and food price volatility.
Statement 2: Sustainable organic farming is not only beneficial to us, and to animals and crops, but rather, much more to our environment (www.theorganicfoods.net/).			
This may be correct, but these claims should be as rigorously validated as for non-organic food	I am only interested in the health claims, and those are still controversial. Besides, organic food is still too expensive.	Correct. In a larger system perspective, organic food brings many benefits. Unfortunately, many of these are not recognised on commercial markets.	Correct. However, it is still a luxury option for most people. Globally, it offers no solution because it would need too much land.
Statement 3: The environmental movement has done much harm with its opposition to genetic engineering.			
This is probably correct. The fear and resistance about GMOs seems unwarranted. The risks are smaller than many other risks in industrial society.	Correct. It has delayed the necessary innovations to keep food cheap. It is also responsible for unnecessary hunger.	Unfair. The risks of genetic engineering are still unknown, and we should be very careful. It is mostly pushed by greedy corporations.	There are strict regulations regarding GMOs and the global food situation can probably not without GMOs. Opposition may no longer be a responsible strategy.
Statement 4: Further intensification of agriculture is only possible with stringent environmental measures to limit eutrophication.			
Correct. Technically, it is already possible to reduce the use of inputs. Prices and innovations will solve the problem. No regulation needed.	I don't know. There is already too much regulation for farmers – is it really a problem?	Intensive agriculture is in many ways problematic. It is absolutely necessary to reduce emissions – and organic farming is one of the solutions.	Essential for sustainable agriculture. In low-income regions, the impacts are rapidly becoming serious. International standards should be introduced.
Statement 5: The dramatic rise in private investment into agriculture has led to neglect of small-scale farming and 'land grabbing'.			
Unfair. Upscaling and private investments are the most effective way to increase productivity and create employment in poor countries.	If the poor want to catch up with the rich, this seems one of the few ways that work.	Wrong direction. This is what the 'free' market does: economic colonialism (slavery and land grabbing) by rich countries.	This is an unintended consequence of globalisation. In the long term, it harms local food security and should be regulated.

the problem is primarily seen as lack of efficiency and ingenuity. The solutions are to increase scale, eliminate trade barriers, reduce losses, stimulate innovations and ensure market-based incentives. Governments should not interfere and subsidies should be abolished. In a more postmodern reality, corporations and governments try to steer the world amidst the dangers of hunger, social unrest and environmental degradation, and to sustain the existing power structure – as befits hierarchist institutions (§6.2).

The opponents will attack them from various sides. Markets only favour the rich, technology brings new problems, corporate responsibility is window-dressing. 'We' should introduce fair trade and green label certificates, eat more vegetarian, forbid child labour, support local cooperatives and farming communities, establish a fund for climate adaptation, and value again local food and traditions. And 'we' should avoid paternalism and neo-colonialism: 'Are African pastoralists the panda bear and ice bear of global agriculture?' One direction is voiced by Wangari Maathai's 'challenge for Africa' Green Belt movement and the Chipko movement in India: fight forest destruction and promote small-scale production for the local market, establish community-based cooperatives, practise organic agriculture and create local employment. Another direction is to focus on UN-based regulation, with global targets for poverty reduction and greenhouse gas emissions and with strict standards for child labour, fair trade and sustainability. Somehow a balance must be found.

11.7 Summary Points

Food is an essential need of people. It is produced in the agro-food systems of the world, which are diverse, local and increasingly connected. In combination with population growth and environmental constraints, its adequate provision is one of the great challenges. Some important points to remember are:

- The basis of agro-food systems is the human habitat, that can be categorised in anthromes. Social-ecological systems (SES) are the natural unit of investigation.
- The present-day agro-food systems are the outcome of a millennial evolution and the human population lived and still lives in quite different anthromes and stages.
- Since the 1950s, world food production is a success story. With little more land, much more food is produced. This intensification has unintended and undesired side effects, such as input dependence and environmentally harmful outputs. There is room for another 40 percent to 60 percent increase in food production, but it may further endanger long-term sustainability.
- The causes of land use/cover change (LUCC) are diverse and local. Important drivers are the opportunities for the poor and the conversion for urban areas and infrastructure.
- The global food system is driven by a few causal loops, notably cost competition, upscaling and innovation. Governments and multinational corporations have become the major actors. Trade liberalisation and consumer demands for social, environmental and health regulations are important co-drivers of development. Both are important but only partly understood ingredients of a transition to a sustainable food system.

The Months
Januar: By this fire I warme my handes,
Februar: And with my spade I delfe my landes.
Marche: Here I sette my nthinge to springe,
Aprile: And here I heer the fowlès singe.
Maii: I am as light as birde in bow,
Junii: And I weede my corne well ynow.
Julii: With my sithe my mede I mowe,
Auguste: And here I shere my corne full lowe.
September: With my flail I erne my bred,
October: And here I sowe my whete so red.
November: At Marinesmasse I kille my swine,
December: And at Christesmasse I drinke red wine.

15th century poem. The Oxford Book of
Medieval English Verse, Eds. Sisam,
– Clarendon Press, Oxford

Ït's the land that feeds our children, you cannot own the land it's the land that owns you.

– Dolores Keane

The counter-seasonality of production in the Southern Hemisphere for Northern Hemisphere consumers has created what has been termed 'permanent global summertime'.

– Dicken, *Global Shift*, 2009:376

SUGGESTED READING

A number of case studies and models about resource management in local communities.

Berkes, F., and C. Folke, eds. *Linking Social and Ecological Systems – Management Practices and Social Mechanisms for Building Resilience*. New York: Cambridge University Press, 1998.

Although from the 1980s, this book is still a good, systematic overview of elementary interactions between man and the biosphere.

Goudie, A. *The Human Impact – Man's Role in Environmental Change*. Cambridge: MIT Press, 1981.

A comprehensive overview of LUCC-related topics.

Lambin, E., and H. Geist. *Land-Use and Land-Cover Change – Local Processes and Global Impacts*. The IGBP Series. Berlin: Springer, 2006.

A thorough introduction in economic aspects of resource and environment issues, partly on food.

Perman, R., Yue Ma, J. McGilvray and M. Common, eds. *Natural Resource and Environmental Economics*. Harlow: Pearson Education Ltd., 2003.

Twelve fascinating anthropological essays about peoples all over the world in their relationships with the Earth.

Reader, J. *Man on Earth: A Celebration of Mankind*. New York: Perennial Library, 1988.

An historical account of the meaning of food in society.

Tannahill, R. *Food in History*. New York: Three Rivers Press, 1973.

USEFUL WEBSITES

Wikipedia presents good descriptions of resource and environmental economics definitions and concepts. There are also sites on specific models and applications:

- glossary.eea.europa.eu/ is a general site of the European Environmental Agency with brief explanation of terms in the area of environmental economics and policy.
- themasites.pbl.nl/en/themasites/image/index.html is the site of the Integrated Model to Assess the Global Environment (IMAGE) model. Also, the HYDE project on historical land use, the PHOENIX-model on population and some other models and databases can be accessed via this site.
- www.iiasa.ac.at/Research/LUC/GAEZ/ is the IIASA-operated site where the data on global agro-ecological zones and on the Harmonized World Soil Database (HWSD) can be accessed.
- www.fao.org/ of the UN FAO with many news items and downloadable papers and statistical data on food and agriculture; a subsite is on sustainable development, in particular: *The State of Food and Agriculture* (SOFA), *The State of the World Food Insecurity* 2010 (SOFI) and *The State of Agricultural Commodity Markets* (SOCO).
- www.grida.no/ is a site with extensive data, graphs and maps on Environmental Knowledge for Change, operated by UNEP/GRID Arendal.
- www.itc.nl/ is the site of the ITC, a research and education institute on GIS-based data collection and processing.
- www.cluemodel.nl is the site of the Conversion of Land Use and its Effects (CLUE) model.
- www.wri.org/project/ is the site of the World Resources Institute (WRI) with data and reports on major global environment and development issues.
- www.maweb.org/en/index.aspx offers an extensive guide to the Millennium Ecosystem Assessment published in 2005.
- www.ifpri.org/ is the site of the International Food Policy Research Institute (IFPRI) with an overview of their reports and projects.
- www.sustainablefoodlab.org/ is an organisation and is the site of a global partnership for innovative solutions towards a sustainable food system.
- www.eururalis.eu is the site of *Eururalis* project, an extensive scenario analysis of futures and policies regarding the rural areas in Europe.
- www.avsf.org/ is the site of the Agronomes sans Frontières, an NGO for assistance to poor farmers worldwide.

Appendix 11.1 Income and Price Elasticity

The income elasticity can be expressed in formula for food expenditures F and income I:

$$\varepsilon = \frac{\Delta F/F}{\Delta I/I} \tag{A11.1}$$

Δ is the change during the period of observation, usually one year. If the ratio of food expenditure and income in year t equals R_t and the income growth rate is α, then it can be shown that:

$$\frac{R_{t+1}}{R_t} = \frac{F+\Delta F}{I+\Delta I} = \frac{1+\varepsilon\alpha}{1+\alpha} \tag{A11.2}$$

The part of income spent on food will decline if $\varepsilon < 1$. The price-elasticity π is defined in similar fashion:

$$\pi = -\frac{\Delta F/F}{\Delta p/p}$$

with p the price of the food commodity. Because it is assumed that amounts decrease for increasing price, it is defined with a minus sign.

12 Renewable Resources: Water, Fish and Forest

12.1 Introduction: Lakeland

There are other renewable resources than soil such as water, fish and forest produce. Let us make a trip to the imaginary country Lakeland to illustrate the nexus between population and resource exploitation. The country consists of two subsystems: a lake or coastal area with fish and their food – say, daphnia, a kind of water flea – and the economy of people, fishing boats and a reserve of mineral ore with high gold content (Figure 12.1). The Lakelanders get their food from fishing in the local lake and make their boats and nets from local materials. There are only two capital stocks: boats and nets, and mining equipment. They change because of investment and depreciation, with the latter assumed to be inversely proportional to the economic lifetime. The demand for fish is growing over time because of population growth. This means more boats and nets. Therefore, you can make three reasonable assumptions about the fishermen behaviour and the ecosystem:

- fishermen invest in new boats and nets in order to satisfy expected growing demand;
- there is an upper bound to the annual fish catch; and
- gold mining causes pollution, which negatively affects the daphnia.

Each season the fishermen go out harvesting fish, which is sold on the local market at the end of the season. Assuming a population growth of 1 percent per year (%/yr), the Lakeland fishing grounds are sustainably exploited over the chosen simulation period of about twenty years (1,100 weeks; Figure 12.2a).[1] As long as the population and fish demand hardly change, the fleet size remains about constant. Such a sustainable state has existed in many places and for long periods, with different institutional arrangements. Note that the model is meant to be an illustration and has not been implemented for a real-world situation.

The islanders can 'develop' and enter into the large 'world' economy by opening up the gold mine for exploitation and by allowing foreign fishing vessels to enter the lake. Often, the population does not really have a choice and is simply exposed

[1] The Stella® version of the model is available upon request.

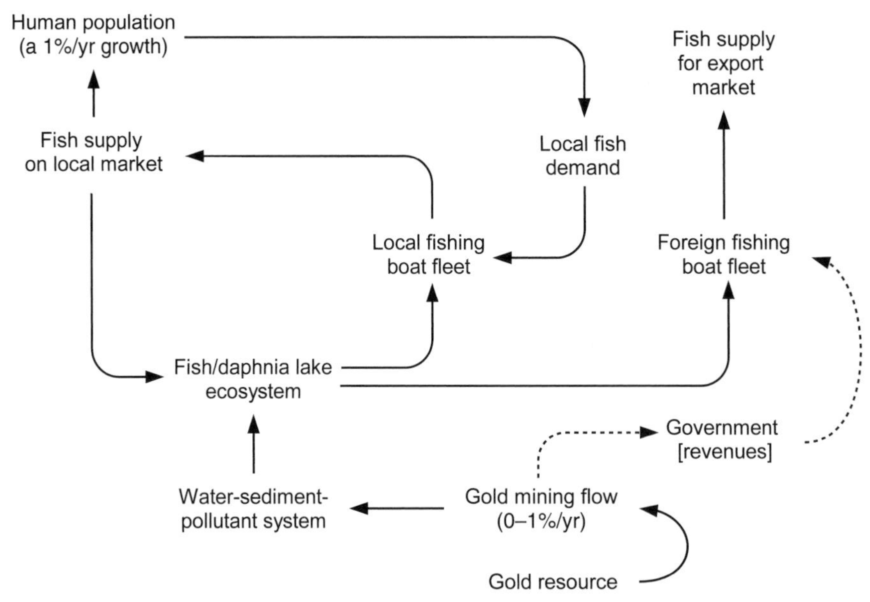

Figure 12.1. The Lakeland model world: a human population that sustains itself by fishing and exploiting a gold mine, and an ecosystem with a prey-predator dynamic and pollution of water and sediment.

to such developments by foreigners, either as military or commercial force. Aldous Huxley gives a compelling description in his book *Island* (1962) how such a process can unfold.

Exploiting the gold deposit can become the first and most important, export-oriented sector in the economy. It not only provides revenues to the Lakeland community or government, but it also generates mining waste that flows into the lake and pollutes the water. The pollutant concentration influences the reproduction rate of the daphnia, which in turn affects the fish population that predates on the daphnia. The mining waste slowly accumulates in the lake sediment, building up a

Box 12.1. *Lakeland – a model world.* Lakeland is an island with a large and beautiful lake that has been populated by fish as long as the Lakeland people can remember. The Lakelanders loved to eat the large fish which swam around so abundantly. The human population was in equilibrium with the ecosystem – a sustainable state as long as population growth remained small. However, one day, gold was discovered on the island, and it underwent 'development' in the form of a gold mine. The mining caused water pollution. Then another 'development' event took place: Foreign fishing companies got permission to catch and export fish. After a few years, the fishermen noticed that it was becoming harder and harder to catch fish. Fish on the local market was becoming more expensive, and, to their distress, the Lakelanders discovered that the water quality in the lake was declining. Not long after that, the gold resources also became depleted.

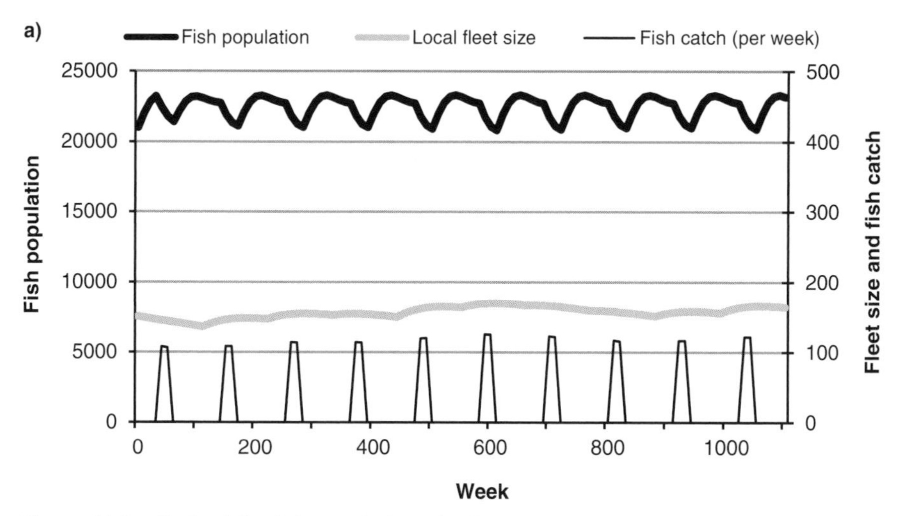

Figure 12.2a. Path of the fish population, the fleet and the weekly catch during the fishing season (note: time axis is in weeks). The catch changes the fish-daphnia system only temporarily and insignificantly ($\Delta t = 1$, Euler).

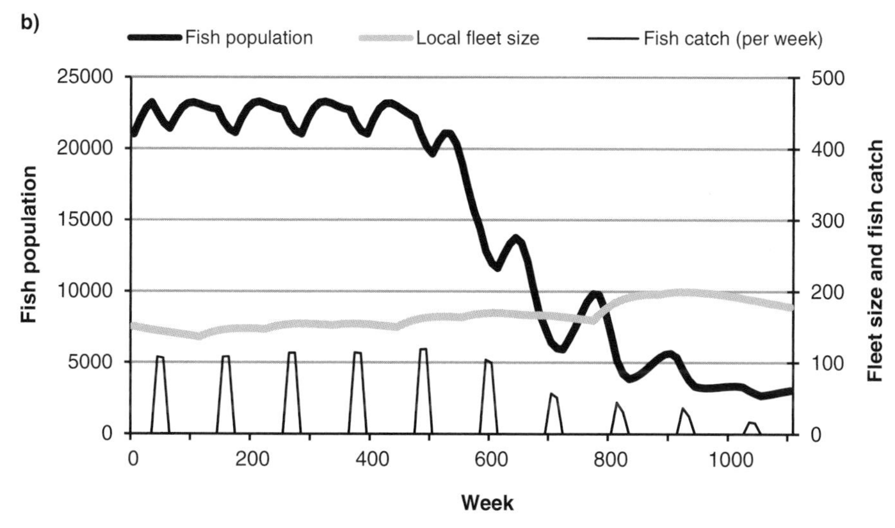

Figure 12.2b. As in (a), and with pollution from gold mining at 1%/yr, affecting the daphnia population and, subsequently, the fish population and fish catch ($\Delta t = 1$, Euler).

stock of pollutant. As a consequence, even if pollution stops the concentration in the water declines only slowly because the pollutant in the sediment dissolves again.[2]

A possible outcome is shown in Figure 12.2b. The growing pollutant concentration in the lake reduces the fishing productivity, defined as catch per unit effort. Fishermen respond by building and sending out more boats. But the mining waste flow continues to harm the ecosystem, and after about ten years, the fish population collapses (Figure 12.2b). The Lakelanders have to import fish – for which they can

[2] In environmental science, this phenomenon is known in its possibly catastrophic form as *chemical timebomb*. See Carpenter and Brock (2006) for the analysis of such a system for phosphorous in a lake-sediment system and the detection of an early warning indicator of regime shift (§9.5).

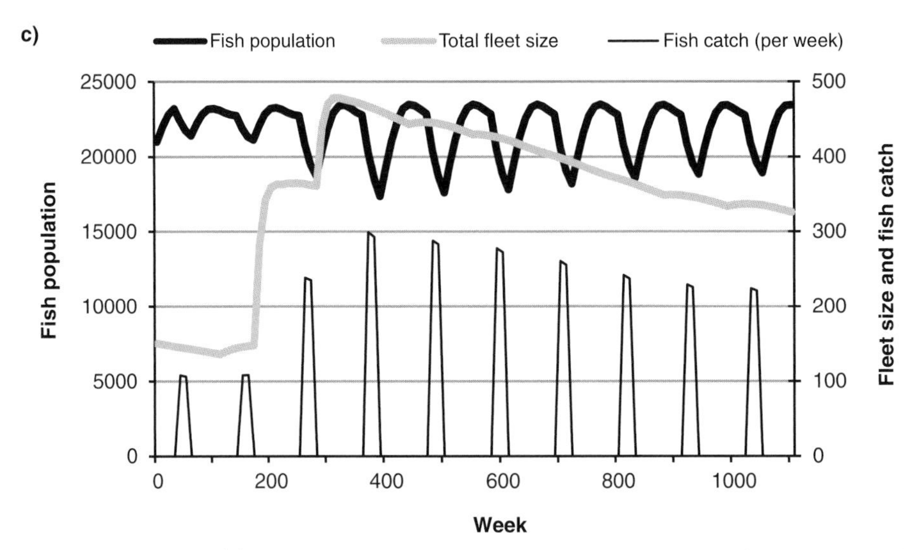

Figure 12.2c. As in (a), and with permission for a maximum of 500 foreign boats to fish for export ($\Delta t = 1$, Euler).

use the revenues from gold mining, provided these are not spent on luxury items or transferred to a foreign bank. The lake is no longer a food resource and the sediment has become loaded with pollutant (Figure 12.3a). The phase diagram plot in Figure 12.3b shows how fish population (x-axis) and daphnia population (y-axis) move in time. The gold resource gets depleted and the ecosystem degrades to a much lower ecosystem biomass. The system has shifted to a new equilibrium with a significantly lower quality of life for a human population.

The other option for development in Lakeland is to let a *foreign fishing fleet* enter the fishing grounds. The foreign boats fish for export to the world market

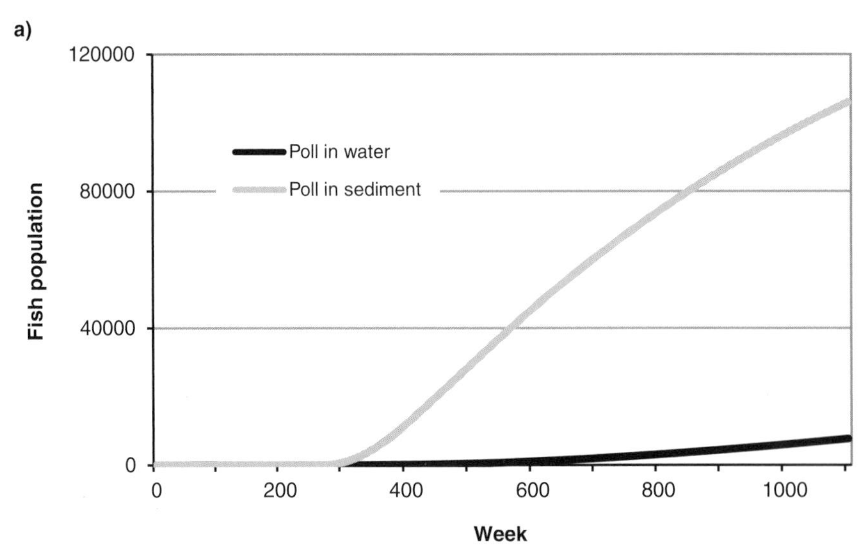

Figure 12.3a. The pollution level in water and sediment for the 1%/yr gold mining ($\Delta t = 1$, Euler).

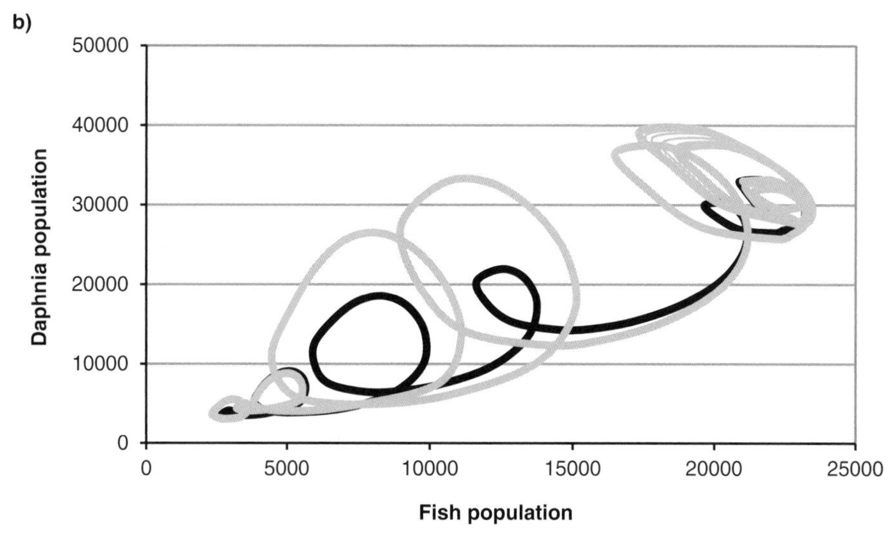

Figure 12.3b. Phase space diagram for the simulation with gold exploitation (black curve; Figure 12.2b) and for the simulation with gold exploitation and foreign fishing boats as discussed in the text (grey curve) ($\Delta t = 1$, Euler).

where prices are assumed to be constant. Whether and how many foreign fishing boats will come fishing depends on the profit ratio. They will go elsewhere if the net income, equal to fish sales minus operational cost, is less than the capital cost of the vessels. The foreign fleet is in fact a top predator, with the financial sector with its return on capital criteria as the ultimate top predator. The typical pattern is that the foreign fleet rapidly enters the area, increases in size towards the maximum catch permitted and declines in response to declining fishing productivity from overfishing.

Assuming that the government issues permits from the third season onwards for a total of 500 ships, a possible outcome is shown in Figure 12.2c. The story is that foreign boats capture the opportunity of high returns on capital investment, but start leaving the area once the high return on capital can no longer be sustained. The total fleet (local plus foreign) then reaches a sustainable exploitation level. The damage is not irreversible, but government revenues from fishing permits and local fish availability both decline. The situation may get worse if the government gives in to the temptation to ignore pollution and overfishing and, instead, decides to ward off local opposition and subsidise local fishermen for as long as it has revenues to do so. Such a policy actually accelerates the collapse and is part of what is discussed in Chapter 13 as the *resource curse* (Ross 2001). The combination of gold mining and foreign fleet fishing spells disaster (Figure 12.3b).

More sustainable pathways are possible. The government can, for instance, enforce environmental regulation on the mining waste flow or spend part of the revenues on pollution abatement. It can also be more stringent with permits for foreign fishing companies and tax them more heavily, particularly if mining is also permitted. From a policy point of view, the trade-off for the Lakeland community is between a variety of assets, each with its own indicators: ecosystem health, gold resource, local food supply and supply security and, last but not least, government income from taxing gold mining and fishing activities.

Table 12.1. *A classification of renewable (bio)resources*

Resource	Stock	Inflow	Outflow
Soils (Chapter 11)	Soils area/thickness Organic content	Soil formation (sedimentation, vegetation, etc.)	Soil erosion (dust storms etc.)
Rivers	(Flow resource)	Water flow downstream	Natural outflow in sea Abstraction (dams)
Groundwater	Water reservoir content	Natural recharge	Natural discharge Abstraction (pumps)
Fisheries	Fish population	Natural birth	Natural death Harvest/catch
Fish ponds (aquaculture)	Fish population	Organised growth (farming)	Harvest
Forests	Tree biomass	Regrowth, afforestation	Natural death Cutting/deforestation
Forest produce	Game; fruit, nut etc. biomass	Reproduction Produce (re)growth	Natural death Hunting Harvest/catch
Wild animals (game)	Animal population (deer, swine and so on)	Natural birth	Natural death Hunting
Sunlight, wind	(Flow resource)	Electromagnetic radiation Moving air mass	Same (collector, panel, turbine)

12.2 Renewable Resources

12.2.1 Renewable Resource Use: An Archetypical Model

Resources that can be harvested are called *renewable (bio)resources* when their regrowth rate as part of natural processes is in the same order of magnitude as the harvest rate. The regenerative and absorption/degradation functions of the environment ('sink') can also be considered a renewable resource, because they provide ecosystem services and have limited regrowth rates. If the regrowth or regeneration rate is happening on much longer timescales, as with fossil fuel and mineral ore formation, resources are categorised as *non-renewable* or *finite resources*, which is the topic of next chapter.[3] If harvesting relates to the tapping from a flow, as with river water or wind and solar energy, it is best to speak of *flow resources*. Their prime characteristic is that they are, for all practical purposes, infinite in time as well as local and fluctuating. They are 'free' but their conversion into useful energy requires capital investments as with other resources. Table 12.1 gives a list of bioresources.[4]

Soils are an essential renewable resource in the agro-food system. Other such resources are water, the fish in seas, rivers and lakes, timber, wild animals and other produce in forests. To a large extent, they were and still are exploited 'in

[3] Also, fossil groundwater reservoirs can be considered finite, as their formation rate is very slow.

[4] What actually constitutes a resource is not as objective as one may think. A resource is a part of the natural environment that provides human beings (the opportunity of) satisfying needs and desires. As such, it reflects prevailing needs and desires as well as technical skills and organisational capabilities (§6.1).

the wild', as open access resources, because people are less easily excluded from using them and they are less easily appropriated (§5.2). Their overexploitation and mismanagement affect ever more people and managing them sustainably becomes an urgent task. Science contributes with research in scientific disciplines such as fisheries and forestry economics and human ecology, and numerous models have been formulated and applied. I refer the reader to the Suggested Reading.

The archetypical or generic population-resource models simulates a hunter-gatherer or horticultural/agricultural population of size P, that exploits a local renewable resource R. The population growth rate is a function of the number of children per woman during her fertile period and on the life expectancy at birth. These in turn depend on the resource, which in this case is the available food. The resource has a finite capacity to regenerate itself from harvest or other disturbances. In theory, it can be used infinitely if it is harvested below the natural regeneration rate. Questions to explore are:

- how will resource size and quality constrain the harvest rate, for instance, through lower animal density, soil erosion and the like;
- how will this in turn affect the population dynamics, for instance, through increased mortality and subsequent starvation or violent deaths from mutual competition; and
- how will the human population respond, for instance, by outward migration, adjustments of the number of births, technical innovations or concentration in certain places where chances for survival are highest.

These questions are also addressed in the discussion on past civilisations and their collapse in Chapter 4 as well as in the syndrome/archetype analysis in Chapter 4. A simple mathematical description is the following model (Appendix 12.1):

$$dP/dt = bP - P/L \tag{12.1a}$$

$$dR/dt = \alpha R(1 - R/K) - \beta P \tag{12.1b}$$

where P is population (stock), b is the birth rate parameter and L is average life expectancy. R signifies the resource size (stock) and dR/dt its change (rate), α its regeneration rate parameter and K its carrying capacity in the area under consideration.[5] The harvest rate H is set proportional to the population, with the parameter β the resource use per person. Let us assume that the population does not grow ($dP/dt = 0 \rightarrow b = 1/L$). Given the harvest rate $H = \beta P$, sustainable exploitation happens for $dR/dt = 0$. This condition is fulfilled if:

$$\alpha R(1 - R/K) = \beta P \tag{12.2}$$

It is useful to plot the exploitation rate dR/dt as a function of the resource R in the phase plane (Figure 12.4a). The *maximum sustainable yield* (MSY) happens for $H_{max} = 1/4\,\alpha K$ at which the resource regenerates at its maximum rate (Appendix 12.1). Exploitation is sustainable as long as $H \leq H_{max}$, which is possible for all

[5] 1/L is used for the mortality rate, which implies an exponential decay of the population if no children are born (§2.3). We assume that the resource is homogenously distributed and follows a logistic growth equation (§9.2).

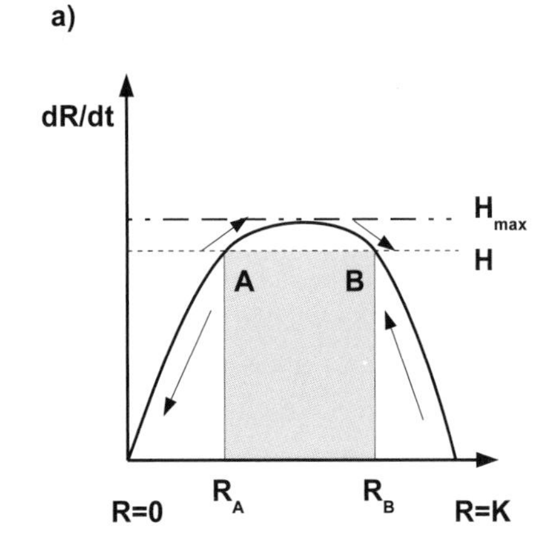

Figure 12.4a. Phase plane diagram of the resource depletion according to equation 12.2a. There are two attractors, one stable and one unstable. The arrows indicate the direction in which the fish population (stock/state variable) will move.

combinations of P and β for which $P\beta \leq H_{max}$. The carrying capacity for people in this situation is $P_{max} = 1/4\ \alpha K/\beta$. For $P > P_{max}$ exploitation exceeds the maximum rate at which the resource can regenerate and the system is exploited unsustainably.

This extremely simple model shows that every *growing* population will at some point be confronted with exploitation of its resource beyond the sustainable level. The situation is worse if the resource regeneration rate is diminishing as a consequence of human interference, for instance, as with soil erosion or age-specific fish removal. This is tantamount to a decrease of the carrying capacity K. It is also worse in the presence of random fluctuations in environmental parameters (May 1977). If the carrying capacity is exceeded for prolonged periods and resource productivity starts falling, the first option is to move to another area with or discover new unexploited resources (extensify). A second option is to reduce demand for the resource by more sober and efficient use or use of substitutes, for instance, fish from ponds (reduce/substitute). A third option is to make the resource more productive by applying innovations, for instance, feeding the fish with agricultural remains (intensify). These options, if successful, are effectively increasing the carrying capacity K for the population.

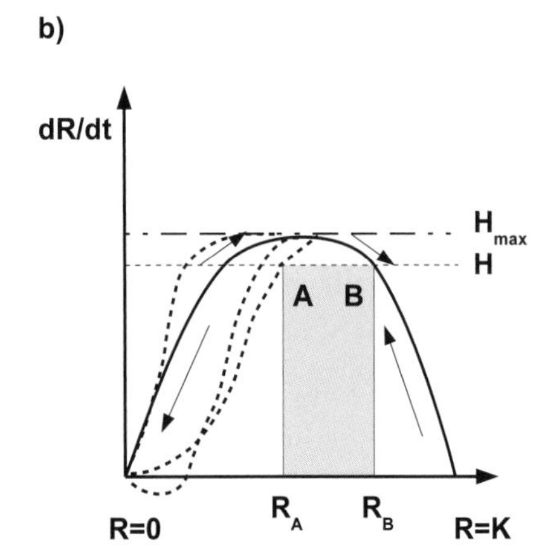

Figure 12.4b. As Figure 12.4a, but with non-logistic dynamics.

Box 12.2. *Archetypical human-environment models.* The simple mathematical population-resource model suggests four typical population trajectories (Figure 12.5). Curve (1) upper left is an exponential population growth (0.6%/yr) leading to a growth from 10 to more than 400 individuals within 600 timesteps. Without an upper bound, it leads to infinity and hence has to stop (§2.3). Curve (2) upper right shows what happens if there is a limit: a smooth approach towards the carrying capacity of 400 individuals, which may, however, easily become oscillatory (overshoot-and-collapse) (§9.3). Curve (3) lower left shows what happens when the limit is lifted due to technical ingenuity or behavioural changes. This is the optimists' world. The limit may also be brought down by natural processes or mismanagement. Curve (4) lower right shows this pessimists' pathway.

The use of the logistic growth model to forecast growth and stabilisation of human populations has led to notoriously wrong results. Some claim that the carrying capacity is already exceeded with the present nearly 7 billion people. Others have calculated that, from a food supply point of view, the carrying capacity for the human world population is 30 billion individuals or more. There is not much empirical basis for extrapolation, so the value of K has to be chosen from other considerations. This is difficult because human societies are complex adaptive systems with continuous interaction between expansion and constraint, opportunity and threat. They coevolve with their environment, as explained in niche construction theory (§10.3).

12.2.2 Model Extensions and Management Principles

The model previously outlined is far too simple to be used for management. A first objection, from the side of biologists and ecologists, is that the logistic growth equation for R may be a good approximation for certain non-migratory animals and biomass stocks, but a more complex description is needed to represent real-world dynamics (§9.5). A possible consequence for resource management is shown in the phase plane plot of dR/dt against R (Figure 12.4). In the logistic model, the area $R_A < R < R_B$ is the 'safe' region at a harvest rate H (Figure 12.4a). However, if the natural regeneration rate is smaller or even below a certain threshold value of the stock – because of density-effects, as shown in Figure 12.4b – the 'safe' region is significantly smaller (Allee effect). The system has now three attractors. The reproduction rate can at (very) low population densities also be higher than in the logistic growth model. For instance, there is a possibility that reproduction rates go up at low density as a consequence of genotypic adaptation, such as evolutionary trait adjustment via earlier maturation, and not of phenotypic adaptation, which results in less crowding and less competition for food. The 'safe' region is larger. What is sustainable depends on what is proven to be the 'correct' model. Introducing features such as seasonality, age-structure of populations, energy metabolism, trophic levels, spatial interactions and genetic adaptation make the models more complex and introduce almost certainly more equilibrium states and the possibility of sudden change (§9.5).

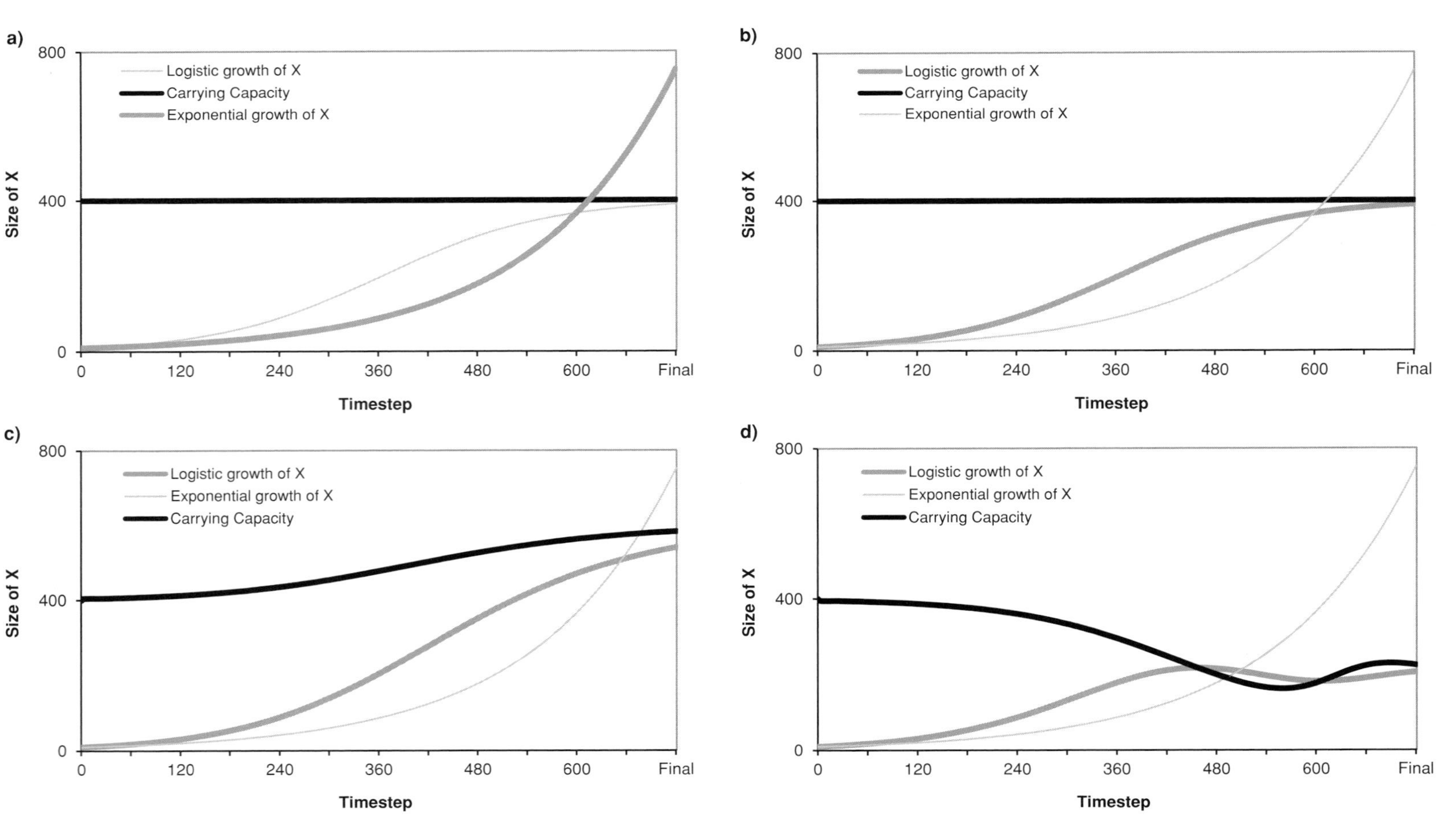

Figure 12.5. Population-resource dynamics: four possible interactions.

A second objection comes from economists. They wish to know the least-cost extraction rate so as to replace the simple extraction rate H $= \beta$P with a more realistic harvesting function. In an open-access situation, everyone has the right to fish, hunt, let animals graze or use upstream river water. The usual assumption is that the population will exploit the resource up to the point where one more exploiter has (perceived) gains that no longer outweigh (perceived) efforts. The following introduces some simplifying assumptions not unlike those in the Lakeland world:

- the harvest H is a function of the harvesting effort E, measured, for instance, as the number of boats times the fishing period;[6]
- the harvest also depends on the size of the stock R. The more resource, the higher the harvest for a given effort level E;
- both the resource and the firm are experiencing diminishing returns, that is, the additional harvest ΔH per unit of additional effort ΔE and of resource ΔR respectively is declining.

In an open-access situation, in theory, entry and exit are free and without cost; therefore, new exploiters keep entering the business until for the last entrant the profit is zero. If this is introduced as a behavioural rule in a competitive open-access situation, the steady-state effort level will rise for higher resource price and for lower resource cost (Appendix 12.2). Thus, if consumers are willing to pay higher prices and/or firms can bring down costs by innovating, the resource will more intensely be exploited. This is precisely the mechanism behind the near-extinction of several fish populations and some other resources in the world.

The exploiter of the resource is expected to be primarily interested in maximising profits along a particular resource extraction path. Already in the 1960s, mathematical models have been proposed in order to formulate *optimal* resource extraction paths (Dasgupta and Heal 1979). An optimal harvest rate maximises over time the profit (price minus cost). In formula:

$$\max \int_{t=1}^{t=T} [(p-c)\,H] \cdot e^{-rt}\,dt \tag{12.3}$$

with T the planning or time horizon in years, p the price and c the cost in money per unit of catch. The term e^{-rt} is the discount factor (§10.4). The variable r can be interpreted as the interest rate on capital. The mathematics to solve this maximisation problem has been dealt with extensively since the first models appeared in the 1960s (Clark 1990; Perman *et al.* 2003). The agent in these models can be an oligopolistic firm or a government planning agency. With a non-zero discount rate, the resource is always exploited until depletion or extinction. More advanced models have been constructed, but the basic economic rationale remains the same and so do the long-run outcomes. I refer the reader to the Suggested Reading for more details. And remember: *All models are false, but some are useful.* In other words, one has to strike

[6] Often, a simple function of the form $H = \varepsilon E^a R^b$ with a, b < 1 is used. Such a function is what economists call an economic or engineering production function. This form is known as the Cobb–Douglas production function (§14.2). An even simpler assumptions is that H is proportional to E (a = 1, b = 0).

the balance between adding more scientifically relevant and interesting detail and ensuring that the model is still transparent and useful for practitioners.

Is it possible to manage renewable (bio)resources sustainably? They used to be open access resources that everyone could use and from which people could not easily be excluded. If such a resource becomes less productive when it is more deeply exploited, it is not only non-excludable but also becomes rivalrous. It is called a *common pool resource* (CPR) (§5.3). In a widely cited paper entitled *The Tragedy of the Commons* (1968), the biologist Hardin suggested that there is an inherent tendency amongst humans to overexploit such a shared, common (or collective) resource: 'Picture a pasture open to all. It is to be expected that each herdsman will try to keep as many cattle as possible on the commons... As a rational being, each herdsman seeks to maximize his gain [and] concludes that the only sensible course for him to pursue is to add another animal to his herd. And another; and another... But this is the conclusion reached by each and every rational herdsman sharing a commons. Therein is the tragedy' (Hardin 1968). This is what the previous simple model showed: Open-access CPRs are exploited until the very last unit as long as someone pays for it. It was concluded that only less people or less freedom, under one private or public centralised owner/manager, can avoid tragedy. The situation that economically rational behaviour of individuals causes an undesirable outcome for the collective is known in social science as a *social dilemma* (§10.4).

There have been many responses to Hardin's paper, and a whole branch of science has evolved around CPR-management. Currently, a widely shared view is that a tragedy of the commons is not inevitable. Both markets and centralised control can cause degradation and collapse. But there are other management regimes, besides the options to limit access by private ownership or government-issued permits and quotas. In her book *Governing the Commons* (1990), Ostrom has shown that people all over the world have developed institutions to harvest the CPRs on their lands. Some of these have lasted over 1,000 years and have survived droughts, floods, diseases and economic and political turmoil. These management regimes operated often in highly variable and uncertain environments. The appropriators, that is, the individuals who use the CPR, have 'designed basic operational rules, created organisations to undertake the operational management of their common property regimes, and modified their rules over time in light of past experience according to their own collective-choice and constitutional-choice rules' (Ostrom 1990).[7] They solved the problem of commitment and mutual monitoring – amongst them the avoidance of free-riders and effective sanctioning – without resorting to either centralised power exercised by external agents or competitive market institutions.

Based on historical analyses, fieldwork and outcomes of behavioural experiments, a provisional list of design principles for sustainable management of and institutions for CPR can be constructed (§10.4; Ostrom 2000; Ostrom *et al.* 2008):

- clear boundary rules: who is in and who is not;
- local rules-in-use should restrict amount, timing and technology of harvesting, relate benefits to inputs, and take local conditions into account;

[7] The term common property regimes (CPR) has been replaced, after careful consideration, by the term common pool resources (CPR) management (Ostrom *et al.* 2002).

- individuals affected by the resource regime can participate in making and modifying their rules;
- users select their own monitors, who are accountable to the users or are users themselves and who keeps an eye on resource conditions as well as on user behaviour;
- sanctions should be graduated, depending on the seriousness and context of the offence.

Often, resource exploitation occurs under a regime that violates most, if not all of, these rules and in which an urban business or government elite is the primary external force. The real threat are then loss of community and the anonimity that facilitates overexploitation and shifting-the-burden mechanisms. Proper mixes of centralised, top-down regulatory and local, bottom-up experiences and rules appear most promising and should be experimented with. But it is a complex matter and there are no one-for-all solutions.

MDG Goal 7: *Ensure environmental sustainability.*
TARGETS
C. Halve, by 2015, the proportion of the population without sustainable access to safe drinking water and basic sanitation.
"If current trends continue, the world will meet or even exceed the MDG drinking water target by 2015. By that time, an estimated 86 per cent of the population in developing regions will have gained access to improved sources of drinking water . . . ".
– The Millennium Development Goals Report 2010, New York

12.3 Water Resources

12.3.1 Water Availability and Use

Water is essential for human life, as drinking water, for crops and animals, and for many industrial processes. The hydrological cycle plays a crucial role in biospheric processes and in human affairs, and water is with carbon dioxide (CO_2) the building block for organic compounds via photosynthesis (Figure 9.7). Hydrologists make a distinction between freshwater in rivers, lakes and aquifers called 'blue water' resources and moisture in the soil and vegetation called 'green water' resources. *Blue water* availability is the total river discharge, which equals precipitation minus evapotranspiration or the sum of surface runoff and groundwater. It depends strongly on seasonality and interannual variability in the flow. Ultimately, the water recharges aquifers and flows off to the oceans. The larger part of rainfall infiltrates into the soil and gets stored as *green water* on its way to the oceans and the atmosphere. Its availability depends on soil water holding capacity and other soil characteristics. It is moisture or water stored in the soil and productively used for transpiration by vegetation. It is important for irrigated and rainfed crop growing and animal husbandry, much less for industrial and domestic use.

Water availability varies in amount and time of year with the local situation. Water use is diverse: A city using nearby river water for drinking water, for cooling of electric power condensers and for diluting urban waste flows is very different from a village dependent on rainwater collection or a groundwater reservoir for irrigating crops. What is sustainable water use? As with all bioresource exploitation, the answer has to consider the availability at a given time and place and of a given quality with respect to the demand. Demand is related to needs and functions, but, as with food, it is usually equated to water use for the past and projected use for the future. The resulting supply-demand match has usually grown out of a long history. Water engineering and management for drinking and irrigation goes back thousands of years in the Middle East and Asia. Populations in arid and semi-arid regions have developed all kinds of ingenious solutions and schemes, aware of the common pool aspect of water. Famous are the *qanats* in Iran, constructions of shafts and tunnels, which can deliver water without pumping. In Spanish Andalucia, the Moors are remembered for their intricate irrigation canals, constructed during the 10th and 11th century, and the sophisticated water allocation schemes some of which still survive. But in the last half century, the traditional solutions are in many places overwhelmed by growing demand and changing lifestyles.

With the accelerating growth in human numbers and activities, the use of the freshwater stocks (lakes, soil/groundwater) and flows (rivers) has increased in extent and intensity. Water is no longer an open access 'free' resource, for which one does not have to pay. It has become a potential source of profit and water-related manufacturing sales (pumps, pipes, filtration, desalination equipment and so on) amounting already to more than €300 billion per year. Because water has deeply social and cultural connotations, like food, its commodification as part of modernisation is still resisted in many places (Robert 1994).

One estimate of global water use is that food production around the year 2000 consumed 6,800 cubic kilometres per year (km^3/yr) worldwide, of which 1,800 km^3/yr are from blue water sources and the remaining 5,000 km^3/yr are green water flows in evapotranspiration processes in rainfed agriculture (Falkenmark and Rockström 2006). The majority of assessments have focussed on blue water use by human activities. An estimate of total global water withdrawal rates for human purposes in the 1990s is in the range of 2,500 km^3/yr, if fossil groundwater and non-local blue water are included (Hanjra and Qureshi 2010). The total withdrawal rate is estimated to have more than doubled since the 1960s (Wada *et al.* 2011; Figure 12.6). Over 40 percent is in China and India. There are large differences in per capita availability and withdrawal rates (Table 12.2). Significant fractions of rural populations have, at least according to official estimates, no access to safe water for drinking and sanitation, which impacts life expectancy and other quality-of-life indicators (§10.3).

In order to meet water demand, people have interfered in many ways with the natural stocks and flows, notably by the construction of dams, embankments and dikes to contain and use river flows and by construction of open and groundwater reservoirs for irrigation. Dams to store water are an ancient solution to control the fluctuating supply and demand, later in combination with power generation. Worldwide the number of dams increased from less than 1,000 in the year 1900 to over 30,000 in 2000, almost half of them being constructed for irrigation purposes (Haddeland 2006). Besides direct interference in the water cycle, the natural water

Table 12.2. *Indicators about worldwide water withdrawal, supply and access*[a]

	Total freshwater withdrawal rate	Idem, per person	Water use for agriculture	Water use for industry	Water use for domestic	Access to water (urban/rural) supply	sanitation
Unit:	km^3/yr	m^3/cap/yr		% of total		% population serves	
Africa	213.2	260	83.1	4.3	12.6	85/47	85/45
Asia	3,900.8	1,060	84.9	7.2	7.9	93/74	78/31
Europe	728	998	29.3	48.5	22.2	99/88	99/74
N America	510.9	1,073	44.1	33.9	22	100/100	100/100
S America	375	1,099	84.8	6.4	8.8	93/62	87/48
Oceania	31.4	1,094	64.9	10.4	24.7	100/67	100/78

[a] *Source:* Wada *et al.* (2011).

stocks and flows have been affected as a consequence of changes in land cover and land use. In retrospect, the issue is always the tension between human activities and natural possibilities. In other words: scarcity.

The first attempts to assess *water scarcity* at the global level were made by Falkenmark and Lindh (1976), who classified national per capita water use from no stress (>1,700 m^3/cap/yr)[8] to absolute scarcity (<500 m^3/cap/yr). Later analyses have refined the notion by including local features down to the scale of water basins. Other indicators of scarcity have been proposed, such as the ratio of annual withdrawals and availability rates (criticality ratio) and the ratio of total water demand and river discharge. The higher their value, the higher the water stress. Usually, a threshold value of 0.4 is used, above which there is high water stress. Based on indicators like these, it is estimated that some 20 percent of the world population is confronted with water scarcity, defined as a situation in which the amount of water use is close to or exceeding that of water availability. In these areas, water is only available a few hours per day or has to be bought at high prices from tank wagon or ship supply. The annual averages conceal the large local differences and the role of seasonal and interannual fluctuations. An additional complication in water assessments is *water quality*. With increasing population density and agricultural and industrial activity, water pollution is in many places a serious concern with important consequences for people's health (§10.3).

In the last decade, assessments are made on a spatially explicit river basin or even more detailed level (Alcamo *et al.* 2003). One recent assessment is based on the global hydrological model PCR-GLOBWB, which is used to calculate the water stress on a 0.5° × 0.5° spatial resolution (van Beek and Bierkens 2008; van Beek *et al.* 2011, Wada *et al.* 2011). The model considers soil and vegetation characteristics and runoff and evaporation flows, amongst others. An average climate is constructed from climatological data for the period 1958–2001. Monthly water use and water availability are calculated and converted into a distributed *water stress index* WSI:

$$WSI_{blue/green} = W_{demand,blue/green}/W_{available,blue/green} \text{mln m}^3\text{/month} \qquad (12.4)$$

[8] Common units in water use are litre per person per day and cubic metre per person per day (m^3/cap/yr). The conversion factor is 1 m^3/cap/yr = 1,000/365 ~ 2.74 litre/cap/day.

It is done separately for blue and green water because the latter is only available for crops. Next, a dynamic water stress index, DWSI, is calculated that quantifies the severity of water stress by incorporating information on the exceedance of a water-stress threshold value and the duration and frequency of the periods of exceedance.[9] The dynamic water stress map for blue and green water is shown in Figure 12.6a. There are spots of severe water stress throughout the world, but it is worst in the densely populated and partly arid parts of India and China, in a Sub-Saharan belt and in (semi-)arid regions of North America.

An important blue water source are *groundwater aquifers*. Their exploitation has rapidly increased since the 1960s, particularly for agricultural and urban use. It can at low cost be extracted with easily applicable water pumps, has excellent quality and can be used with low transport losses in irrigation. In urban areas, clean groundwater provides a substitute for polluted surface water. The largest withdrawals are in India, the United States, Pakistan, China and Iran, together an estimated 70 percent of global withdrawals (Giordano 2009). From a sustainable management perspective, the ratio between the actual abstraction rate and the renewable recharge rate is the relevant indicator and should be less than one. This ratio is alarmingly high in (semi-) arid countries such as Saudi Arabia (9.5), Libya (8) and Egypt (3.5) and is increasing rapidly in densely populated countries like Pakistan (1.1), Iran (1), Bangladesh (0.5) and India (0.45) (www.worldwater.org). Globally, only 6 percent of the recharge is abstracted – but global aggregates are misleading. Using the system principle that outflow should not exceed inflow, groundwater use is utterly unsustainable in quite a few places on earth.

Wada *et al.* (2010) assessed the average yearly *groundwater* recharge flows on a $0.5° \times 0.5°$ spatial grid and downscaled country-based groundwater abstraction rates to the same spatial resolution by using water demand as a proxy. The resulting maps give an indication of the extent to which groundwater depletion, defined as abstraction minus recharge rate, occurs. It is a hot-spot map that coincides quite well with the information from local qualitative and quantitative accounts. According to this analysis, in the order of 40 percent of groundwater abstraction contributes to depletion (Figure 12.6b). There is, for instance, severe groundwater depletion in northern India. Naturally, there is a relation with the extent and type of agriculture and pastoralism in view of the large water demand for crops and animals (Ramankutty *et al.* 2008).[10]

What are the prospects for sufficient and sustainable water supply for human populations?[11] Blue water river sources are now drained to such an extent that some major rivers – Colorado, Nile, Ganges, Indus and Yellow – are almost dried up for parts of the year upon reaching the sea. Some of the large inland lakes in the world are shrinking because the inflowing river flows have declined to a trickle – the

[9] The complex nature of water stocks and flows make the definition and interpretation of indices such as the WSI and the DWSI a difficult topic. We refer to Wada *et al.* (2011) for the details.

[10] In the analysis, only water demand for livestock is considered and it turns out to be only 1–2 percent of total water demand. It, therefore, has supposedly only a minor impact.

[11] Some water-scarce countries are diminishing the problem by importing food and thus, indirectly, water. It is called 'virtual' water and is calculated in the form of the water footprint. It is only a solution for countries that have the means to pay for the food imports, such as the oil-rich Middle East countries.

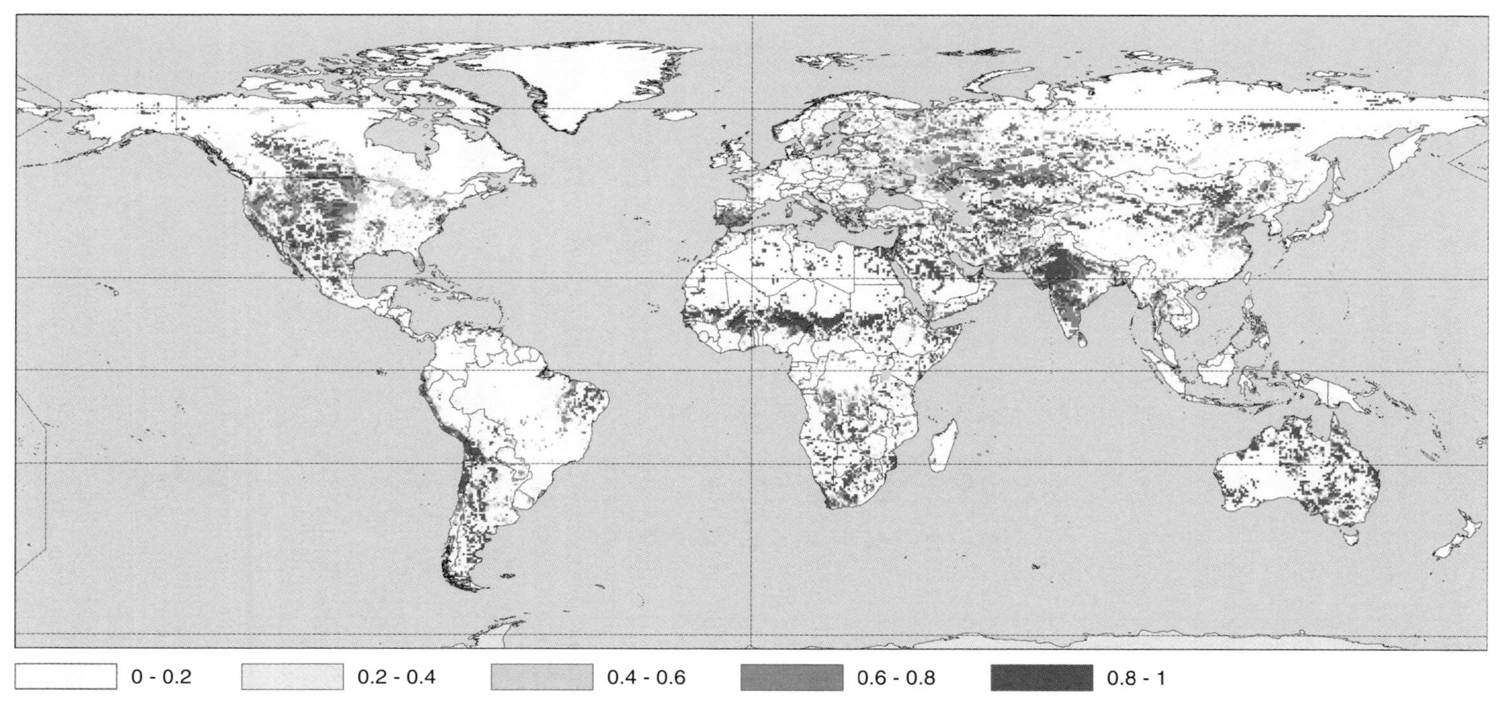

Figure 12.6a. Dynamic water stress index (DWSI) based on monthly water scarcity index ≥0.4 over the period 1958–2001 (Low: 0–0.2, Moderate: 0.2–0.4, High: 0.4–0.6, Very high: 0.6–0.8, Extremely high: 0.8–1.0) (Wada *et al.* 2011). (See color plate.)

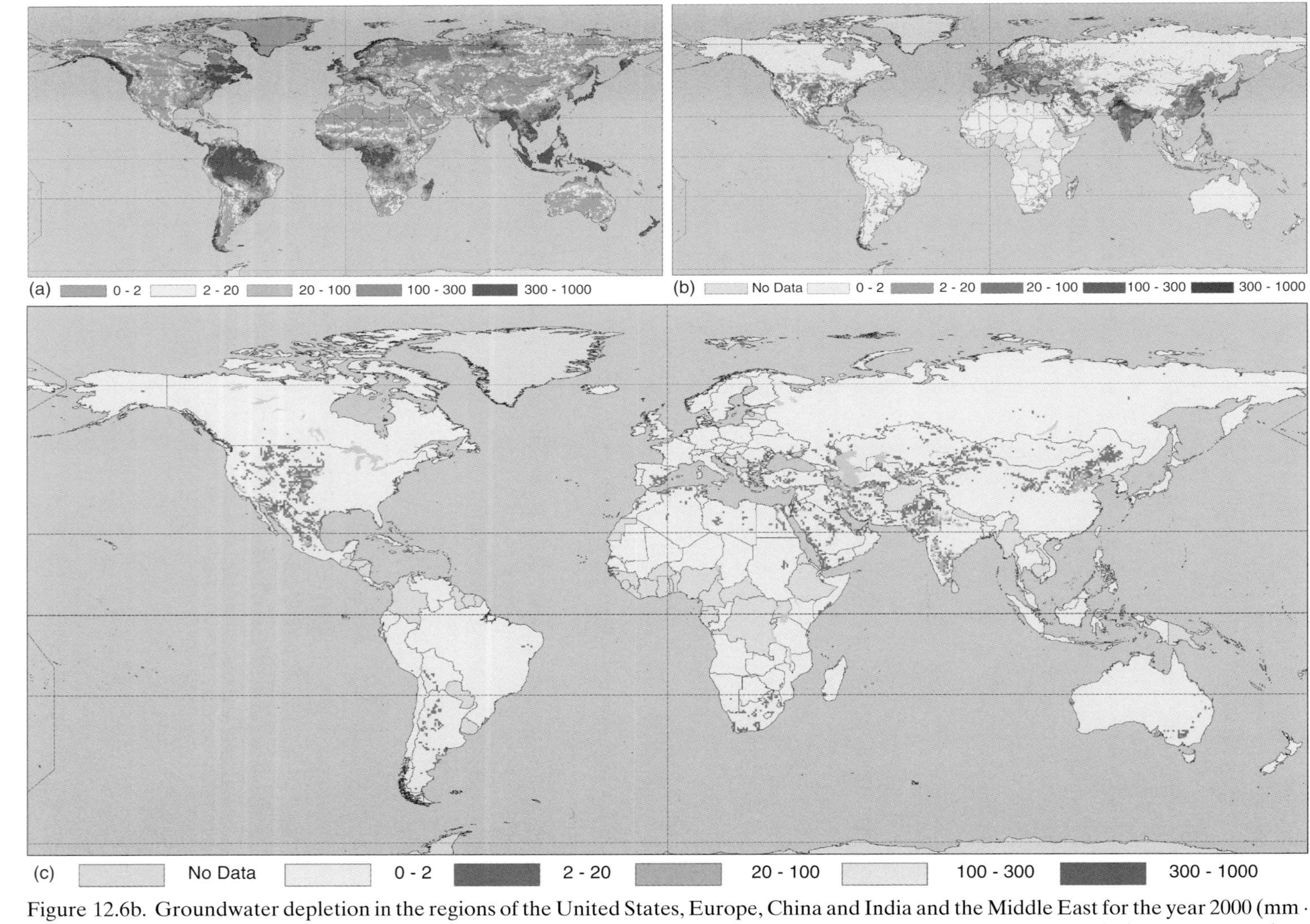

Figure 12.6b. Groundwater depletion in the regions of the United States, Europe, China and India and the Middle East for the year 2000 (mm · a−1; clockwise from top left). (Wada *et al.* 2010; reproduced/modified by permission of American Geophysical Union). (See color plate.)

Box 12.3. *Groundwater resources.* Groundwater stems from precipitation, which infiltrates the land surface and then percolates through a zone of unsaturated water towards deeper, water-saturated layers. The two are separated by the water table. Soil moisture in the unsaturated zone is important in the transport, conversion and storage of dissolved substances and is one of the determinants of water quality. Use of agrochemicals, for instance, can cause water pollution that takes sometimes up to a century to fall to original levels. If the rock pores in the layer below, the water table is saturated with water and the water can flow under natural or applied pressure; it is a groundwater reservoir or *aquifer*. There are different kinds of aquifers and they are usually exploited by drilling wells. If the water has been sealed for very long periods, the aquifer contains fossil groundwater and is effectively a non-renewable resource.

Development and use of groundwater is sustainable when abstraction can be maintained indefinitely without causing unacceptable environmental, economic or social consequences and risks. Stated in system terms, the groundwater outflow rate (natural drainage and withdrawal by humans) should not exceed the inflow rate (rainfall, rivers or glaciers). Often, only the groundwater *level* in various places can be directly measured and water managers rely on 'black box' approaches that correlate water levels to rain patterns and withdrawal rates.

satellite photos of the Aral Sea and Lake Chad tell a compelling story. With sovereign nation-states still the major political unit, the use of river water in river basins such as the Tigris-Euphrates, the Mekong, the Nile and the Colorado is bound to cause allocation disputes (§5.3).

Although water withdrawal rates are relatively small in comparison with the total water stocks and flows in the biosphere, abstraction and diversions cause changes in the local and regional landscape and ecosystems. In combination with other changes such as deforestation, increasing population densities, urbanisation and climate change, it can have serious consequences for human habitation. Not surprisingly, water is a central concern in the assessment of climate change impacts and adaptation measures (IPCC 2007). Many regional water systems are exploited unsustainably. It compromises 'the ability of future generations to meet their own needs', in the WCED-definition of sustainable development.

There is a variety of solutions that are available and are increasingly implemented (Stikker 2007; Gleick and associates 2011). First, there is a large potential to use water more efficiently, with techniques like drip irrigation, water-saving toilets and showers, and recycling. Rainwater collection can bring relief in semi-arid regions – installation is already obligatory, for instance, in new housing in parts of India. Another option is to expand supply by seawater desalination. It is still energy- and capital-intensive, but innovative designs are being tested and it seems eminently suitable for use of solar energy. It will not bring relief everywhere, because water transport over long distances is rather expensive and impractical. The implementation of these options will be accelerated when water is priced more correctly, but regulation and allocation schemes must be part of the solution as well. With growing demand, water as a commons will thus slowly disappear and become part of the market system.

12.3.2 Water for Irrigation: A Case Study and a Model

Because the water situations in the world are so diverse, there are lots of anecdotes and stories but only few commonalities. For further inspection, the following is an analysis of groundwater use in Spain (Fernández and Selma 2004). Many landscapes in the Mediterranean are characterised by traditional irrigation systems along river valleys, from which they take surface water. Since the opening of European and global markets for Spanish producers, large-scale intensive agriculture has been expanding. In the newly irrigated lands in the Segura basin in southeastern Spain, the area of irrigated lands has increased between 1960 and 2000 from 1,200 hectares (ha) to over 17,000 ha. Groundwater levels have on average fallen 15 to 20 metres in this period. Overabstraction has resulted in seawater intrusion and water salinisation and in outflow of nitrates and phosphates.[12] The outflow from natural springs in these four decades is reduced with a factor of ten, causing a decline in biodiversity. Valuable habitats as well as future profitability are threatened. 'In south-east Spain, most of the aquifers are over-exploited... The whole process clearly fits the desertification syndrome... and constitutes a paradigm of environmental and socio-economic unsustainability... the present total water demand, mostly originating from agriculture, amounts to 228% of the internal renewable resources of the Segura basin' (Fernández and Selma 2004). In short, it is unsustainable.

The decline in size and quality of the local groundwater reserves from over-exploitation is being countered by importing water from neighbouring groundwater reservoirs, re-use of wastewater flows and seawater desalination plants. The hydraulic infrastructure (dams or river diversion) is extended and innovations (better crop varieties, computer-aided fertilisation and irrigation) are introduced, but these responses and the associated policies are controversial. A system dynamics model has been constructed and implemented to examine the (ground)water policies for the area (Fernández and Selma 2004; Figure 12.7). It has five sectors: irrigated lands, profitability, available area, water resources and pollution. The crucial state variables are the area of irrigated lands (tree crops, open air and greenhouse horticultural crops) and the local groundwater resources (thirty-nine small aquifers). The most important development loop is the conversion of drylands and shrublands into irrigated lands (Figure 12.7 – bold arrows). This was considered profitable when it began in the 1960s. Water demand started growing and groundwater withdrawal caused declining water levels and rising pumping costs. It tended to decrease profitability, in turn slowing down new irrigation projects. Later, also the decrease in suitable lands and the environmental impacts put a brake on further development.

The model has been calibrated with historical data and 'tells' the water history of the area. It is applied to explore three scenarios for the future: one scenario of trend extrapolation, one with no increase in external water resources, and one with some increase and technological intensification. The researchers conclude that the currently implemented plans may well aggravate the unsustainability of large-scale irrigated agri/horticulture, with a significant drop in profitability over the next two decades. It is the lack of a systemic approach of the underlying driving factors that leads to misconceived policies. The construction, application and lessons of models

[12] In the Segura basin, soil salinisation is not affecting crop yields and profitability, because 80 percent of irrigated lands use artificial substrata which are replaced every two years. Like the nitrates and phosphates, the salt ends up outside the system – an externalisation of potential costs.

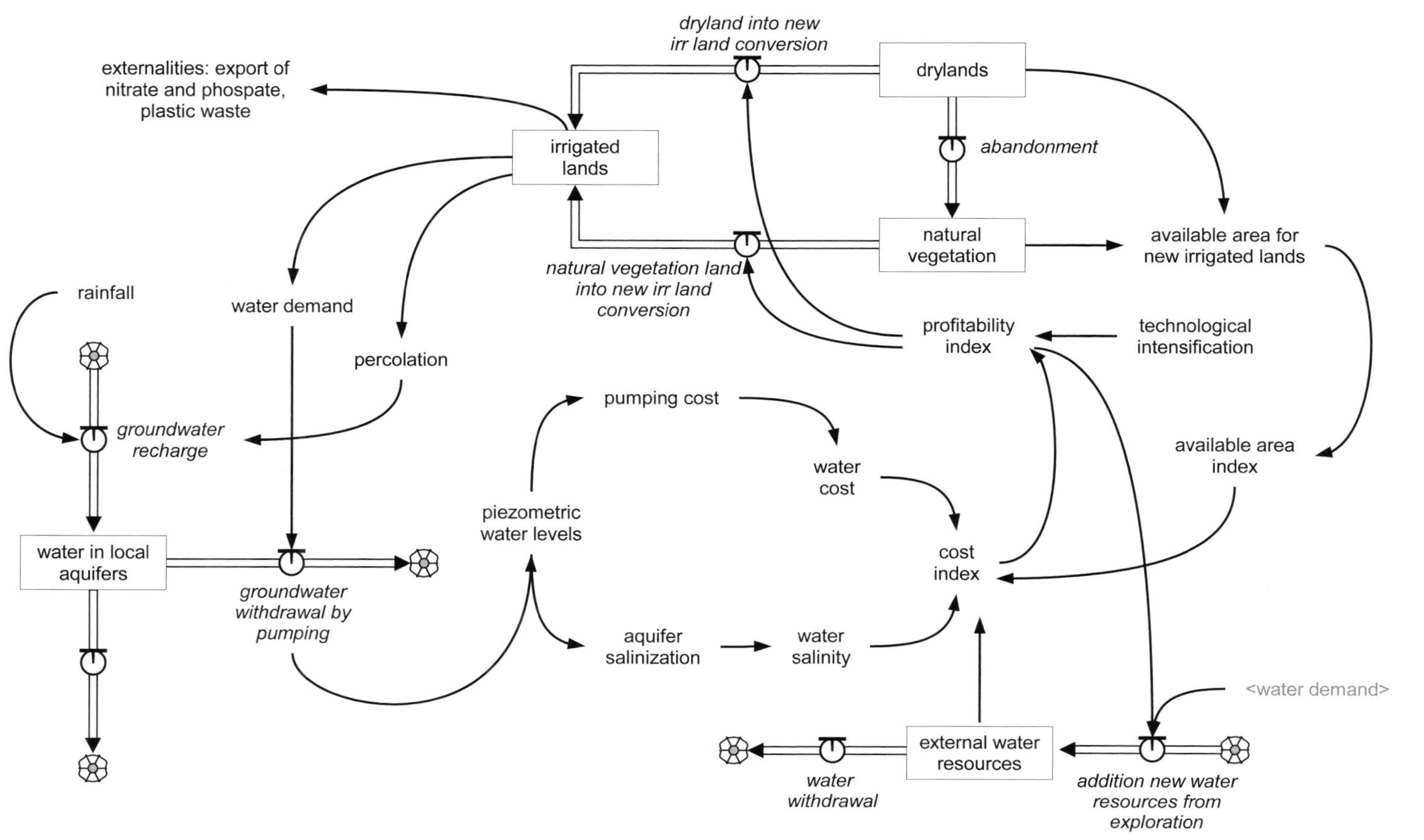

Figure 12.7. Simplified causal loop diagram of the large-scale projects on new irrigated lands in southeastern Spain. State variables are grey (based on Fernández and Selma 2004).

such as this one is a necessary step in development projects if the scarce water resources are to be used sustainably.

12.4 Stories

12.4.1 The Canadian Fish Drama

Northeastern Canadian fisheries – cod, haddock and other species – have been in serious crisis since the 1990s. It is 'the classic case of the failure of conventional science-based fisheries management: the collapse of the northern cod of Newfoundland and Labrador' (Finlayson and McCay 1998). For centuries, the extraordinary abundance of cod has been exploited by European settlers. It was done in the summer months – storms and ice made it impossible in the winter months – with relatively simple and inexpensive techniques. Cod fishery was the focal point of the small coastal communities, which tended to have egalitarian social relations.

In the 1960s, the situation changed with the arrival of foreign trawlers, which itself was driven by hungry post-war European and Russian populations and new, cheap techniques such as the devastatingly effective factory freezer trawler. Fishing now also happened during winter months by fleets that stayed all year round at sea and were directed by their corporate or state owners wherever catch rates were highest: 'By the mid-1960s their numbers were so great that the Newfoundland fishing banks at night were described as a "city of light"' (Finlayson and McCay 1998). The total catch went up from 300 kilotons per year (kton/yr) in 1956 to 783 kton/yr in 1968. Distant water fishing increased most rapidly. During these years, there was minor and ineffective regulation because of a 'lowest common denominator' policy and restrained enforcement power. It led to massive overfishing and collapse – total catch precipitously declined to 288 kton/yr in 1975. A pragmatic sustainable yield is estimated to be 200 kton/yr.

Local communities declined rapidly affecting some 35,000 people and inducing large recompensation sums, and massive government funds went into restructuring the industry after the collapse. This tragedy of an international, open-access commons unfolded until the Canadian government closed the northern cod fisheries in 1995 for probably several decades. In retrospect, the scientific assessments of the resource turned out to be wrong. Knowledge of the local fishermen was neglected. Political and ideological factors played a role: 'Scientists knew the truth but were not heard or not allowed to speak because those charged with making fisheries policy had reasons to favour more generous assessments' (Finlayson and McCay 1998). Bureaucratic inflexibility, public relations, a managerial industry perspective and awkward relations between scientists and policymakers all were other factors. Did anyone learn lessons? Possibly not, because most attention has gone to the crises, not to the underlying causes.

12.4.2 European Union Fisheries Policy in Senegal[13]

'For us, it has no sense or benefit because the industrial fishing boats don't leave us any chance of survival. They fish right up to the coast without being stopped and

[13] This story is based on UNDP Human Development Report 2005 Occasional Paper (hdr.undp.org).

the government doesn't have the means to control their activities. If the government would listen to us, we wouldn't sign an agreement with people who catch everything, even the small fish,' says Mario Alberto Da Silva, a West African artisanal fisherman. The fisheries sector is an essential component in Senegal's development: It provides 75 percent of local protein needs and generates about 100,000 direct jobs for Senegalese nationals, of which more than 90 percent are in small scale (artisanal) fishing. Another 600,000 people or 15 percent of the working Senegalese population are employed in related industries. The rising global demand for fish combined with pressures on world supplies means that Senegal's 'blue gold' is an increasingly valuable resource.

The UN Convention on the Law of the Sea (UNCLOS), signed in 1982 and in force since 1994, provides for an Exclusive Economic Zone (EEZ) of 200 nautical miles. This places 95 percent of the world's fish stocks and 35 percent of the oceans under the jurisdiction of coastal nations. The open access previously enjoyed by the long distance fishing fleets was lost, so the European Union (EU) had to negotiate access through fisheries agreements or through private license and joint venture arrangements. Already in the 1980s, the Senegalese government first reduced support for the fisheries sector and then introduced export-stimulating mechanisms as part of an agreement with IMF-World Bank. Since then, it has concluded a number of *fishing access agreements*, principally with the EU, that have a significant contribution to government revenues. Like many African governments, short-term financial compensation from fisheries access agreements was favoured over a thriving informal domestic sector from which it is difficult to extract revenue. Fish exports have grown and the revenues have been an important source for debt repayments.

However, the value of this sector is being eroded by multilateral trade liberalisation. The pressure on West African fish stocks increased six-fold between the 1960s and the 1990s, mainly as a result of fishing by the EU, Russian and Asian fleets. The small-scale fishing industry of Senegal is now in direct competition with the fishing fleets of the EU to supply both the local and the EU market. The risk of supply shortages and price increases on local markets looms ahead as fishing efforts shift from locally consumed to export-oriented species. The EU has been strongly criticised for its role in the depletion. It 'exports' EU fishing fleets to already resource-scarce regions and partly subsidises these vessels. It concludes intransparent agreements with signatory states who have quite limited capability to monitor or control the EU fleet activities. Local interests suffer from its lobbying power. The fisheries policies are in conflict with EU development policies in West Africa. To address these criticisms, the EU has developed a new approach to Fisheries Agreements. The latest agreement with Senegal (2002–2006) made some improvements but is still an old-style 'cash for access' agreement.

12.5 Fisheries and Forests

12.5.1 World Fisheries

Fisheries and fishery management are amongst the best researched renewable resource topics, which does not prevent their near-depletion. *Fish* is an important source of high-quality food for millions of people and has been so for a long time. Originally, most fish was harvested from inland surface waters (rivers, lakes, floodplains or lagoons). In high-income regions, fishing 'in the wild' is only for recreation, but elsewhere it is still of direct importance for people's livelihood. Inland fisheries still provide an estimated 10 million ton/yr and possibly more because of underreporting and illegal catch. The catch is a valuable source of protein-rich food and employment. Governments tend to stimulate a transition to aquaculture in private ownership because 'governments and resource developers see inland fisheries as an impediment to their desires to expropriate the wealth of the rivers – the transfer of generalized wealth (nutritional security, livelihoods) from powerless people into focused income streams that benefit powerful people' (Welcomme *et al.* 2010). As with pastoralists, there is a tendency of the state or those representing it to 'tame the wild' and exert control according to universal principles (Scott 1998). It is a process in which a common pool resource gradually comes under the influence of government institutions and market incentives, in order to satisfy a growing food demand through intensification.[14] But a transition is not without problems because aquafarming requires inputs and physical and institutional infrastructure, produces different species and demands a different lifestyle of the farmer.

By far the larger part of aquatic food is nowadays coming from marine and coastal fisheries. World capture of fish increased from about 20 million ton per year (Mt/yr) around 1950 to 144 Mt/yr in 2006, including the 10 Mt/yr from inland fisheries (Figures 4.6b and 12.8a). Increasingly, fish is produced from the high seas and from coastal and inland aquaculture. The latter has spectacularly risen from a few Mt/yr in 1985 to 52 Mt/yr in 2006, with a dominant role for Chinese aquaculture. Besides, there is an estimated 11–26 Mt/yr illegal, unreported and unregulated catch and 9.5 Mt/yr discarded catch. Global fisheries contribute only a few percent to gross world product (GWP), but directly and indirectly (vessels, processing and so on) they provide employment and income to an estimated 42 million people (Garcia and Rosenberg 2010). The sector as a whole is said to support more than 500 million livelihoods. According to the FAO, fish contributes almost 20 percent to average per capita intake of animal protein for more than 1.5 billion people, many of them in low-income food-deficit countries (WRI 2000–2001). In some countries, fisheries contribute more than 7 percent to GDP and are essential for poor segments of the population.

Fisheries exploitation happens in stages, from undeveloped to developing to mature and finally to senescent which is a state of overexploitation, depletion or recovery. The statistics for 169 national fishing areas indicate that 27 percent is overexploited or depleted with respect to the state in which maximum sustainable yield is supported (Garcia and Rosenberg 2010). Nearly 60 percent of fish landings are estimated to come from fully/overexploited stocks (Figure 12.8b). There is widespread agreement amongst experts that marine fisheries are at or near the upper production limit of an estimated 80–125 Mt/yr (Garcia and Grainger 2005). This level can probably not be sustained. To increase aquatic food output, the rapid growth in fish from aquafarming and aquaculture must continue (Figure 4.6b). But overall system efficiency is low, because the fish has to be fed, usually with seafish. The use of chemicals to fertilise the ponds and protect the fish against diseases has already led

[14] Intensive aquafarming in ponds can deliver up to 1,000 times higher yields than in natural production situations (Welcomme *et al.* 2010).

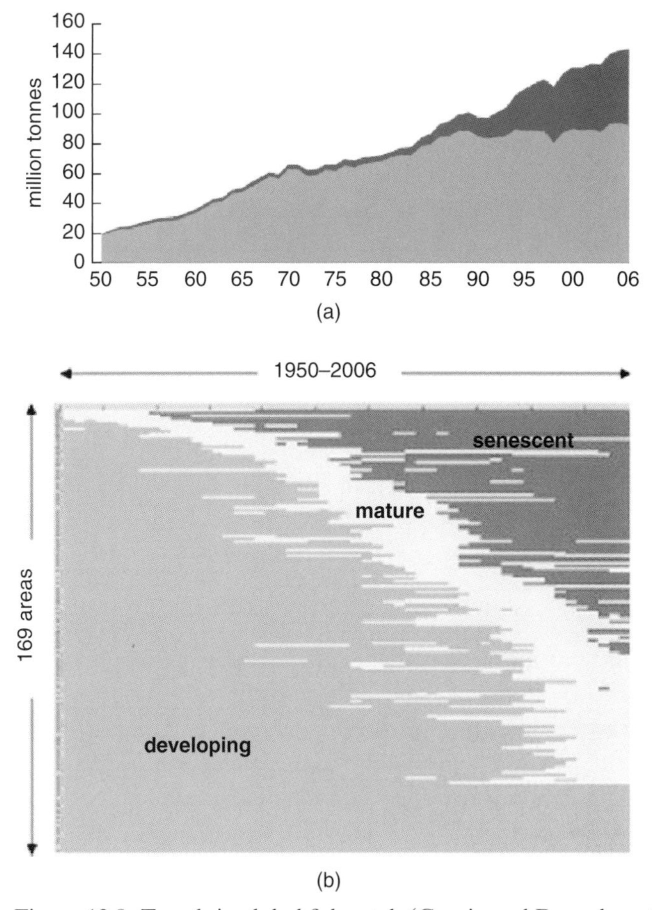

Figure 12.8. Trends in global fish catch (Garcia and Rosenberg 2010). (See color plate.)

to health and pollution risks and epidemic contagion of wild fish (Ford and Myers 2008; Arquitt *et al.* 2005). Sustainable aquafarming will be one of the challenges for the future.

Fisheries depletion is not only a matter of quantity but also of quality. Food web analysis of fisheries suggests that fishing grounds are 'fished down' in a double sense. First, average depth of fish catches has increased from 150 to 200 metres (m) to about 250 m with the shift to deep sea fishing. Second, the mean trophic level has decreased in most coastal areas and the North Atlantic ocean (MEA 2005). Analysis of catch data for the western central and north Atlantic and the southeastern and northwestern Pacific reveal that since 1970 the mean trophic level of fish catch has been declining in all of these fishing areas. This fishing down of marine food webs is the result of sequential depletion of a fishing ground by catching species lower down the food web (intensive or 'deeper'), not by moving to new fishing grounds (extensive or 'broader') (Pauly and Palomares 2005). But the empirical evidence for this is still contested and more research is needed.

The exploitation mechanism in fisheries is not different from other resource systems. Fishing power, defined as the product of number of ships and potential catch per ship, has increased sixfold between 1970 and 2005 and, in combination with

Box 12.4. *Depletion and technology: Long-term trends.* In a recent analysis of UK fisheries, Thurstan *et al.* (2010) have compiled data on bottom-living fish species landings into the United Kingdon from 1889 to 2007. They also estimated total fishing effort, taking the innovation waves from sail ships to steam trawlers and then, from 1950 onwards, to motor trawlers into account. The data show a rapid increase in landings between 1890 and 1915, a further increase until 1950 followed by an accelerating decline. Around the year 2000, landings decreased to half the 1890 level.

After factoring out the innovations introduced to increase catching power, such as better gear and nets, it is possible to construct a reliable, long-term indexed productivity of fishing activity measured as catch per unit of effort. It turns out that between 1890 and 1920 productivity in ton per unit of fishing power dropped dramatically, but the increased landings masked the steep decline in fish stocks. Productivity doubled again between 1920 and 1960, as fishing vessels exploited new grounds in the Arctic and West Africa. After 1960, a precipitous decline in productivity set in and reached unprecedentedly low levels by 1980 – a thirtyfold decline since 1889 (Figure 12.9).

Thus, these long-term empirical data confirm theoretical analyses that predatory fish biomass is currently at most 10 percent of pre-industrial levels, and that productivity has gone down so much that 'for every unit of fishing power expended today, bottom trawlers land little more than one-seventeenth of the catches in the late 19th century' (Thurstan *et al.* 2010). It is a classical tragedy of the commons story and urgent action is needed to eliminate overexploitation of European fisheries and rebuild fish stocks.

increasing demand, led to a relentless search for new fishing grounds and to their subsequent depletion. The tendency to overinvest in capacity to exploit renewable resources is found time and again.[15] One probable cause is the misperception of feedbacks and delays in the system (Moxnes 1998). In a broader context, the system is seen to be fatefully drawn into the capitalist logic of catching as much fish as fast as possible and putting the money into a bank account or, better even, investing it in still profitable fisheries or other resource extraction business elsewhere. Are scientific models useful for sustainable management? Not that much, it seems. Already for decades, science-based regulation with quota and other measures are introduced and implemented, but the negotiations and enforcements are slow and tedious and measures are often too late. Yet, coordinated regulation along the previously mentioned CPR management principles is the only road towards sustainable exploitation.

12.5.2 Fisheries Models: Strategies and Interactions

The past century has taught that scientific advise in fisheries management has not led to sustainable harvesting. Apart from the difficulty to adequately measure fish

[15] The game Fish Banks (Meadows 1996) convincingly makes the point that sustainable exploitation of open-access fisheries need some kind of coordination.

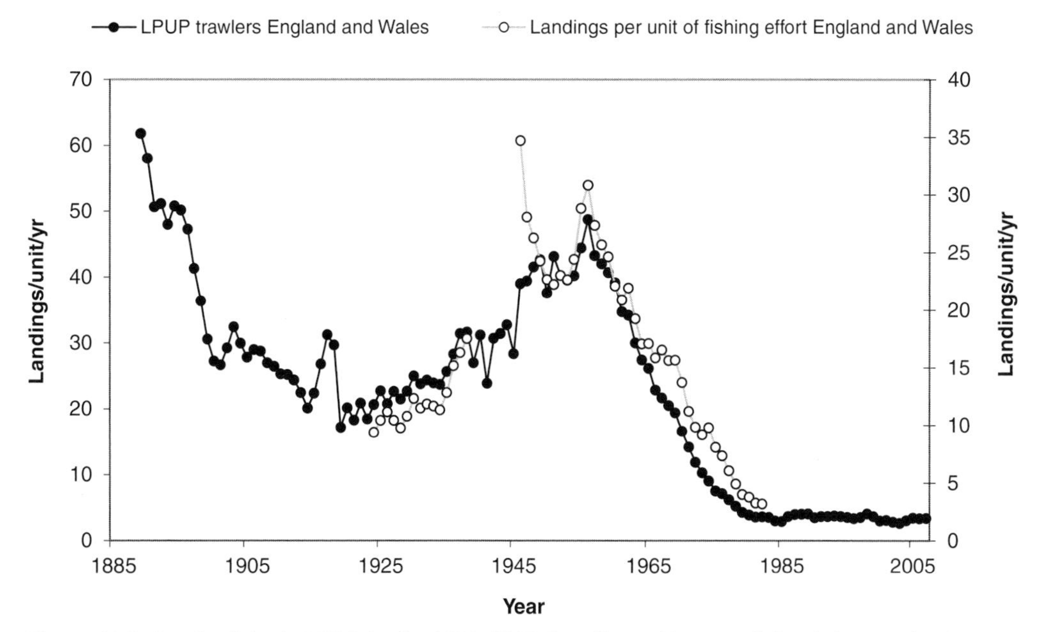

Figure 12.9. Productivity in UK fisheries 1889–2007. Landings of bottom-living fish per unit of fishing power of large British trawlers. Closed circles show landings per unit of fishing power (LPUP) into England and Wales, open circles show landings per unit of fishing effort of large British trawlers (corrected for changes in fishing power) into England and Wales (Thurstan *et al.* 2010).

stocks, there is the problem that most scientific models tend to simulate the ecosystem and then tell what is rational and optimal to do. But humans are as much part of the resource system as the fish, and what is rational and optimal for the individual fishermen tends to be irrational and suboptimal for the system at large – the essence of a social dilemma. For instance, in the Nova Scotia Groundfish Fisheries, only for one of three fish species – haddock – has a relation between catch and effort has been observed (equation A12.2). Why? It turns out that including the well-known fact that the birth rate of haddock is extremely variable, fish populations can suddenly start to fluctuate wildly instead of the smooth equilibrium behaviour in a deterministic simulation (May 1977; Allen and McGlade 1987). Apparently, human exploitation can amplify the natural fast fluctuations and, therefore, resource systems with a natural background noise can probably never yield a constant and satisfactory economic return.

One way to improve the models is to introduce and simulate explicitly the *behavioural strategies of resource exploiters*. An early example of such an agent-based model (ABM; §10.5) is the investigation of two possible fishing strategies in an open-access, spatially distributed resource area (Allen and McGlade 1987). Let us suppose that there are two strategies, one that disregards catch data from boats and chooses randomly where to fish, and the other that decides where to fish on the basis of differences in (expected) profitability and can be considered an optimising 'ultra rationalist' skipper. The latter are rational optimisers cartesians. The former are opportunists stochasts. Using an evolutionary selection mechanism (Appendix 10.2), the risk-averse *cartesians* go fishing where they know it to be profitable and then deplete the zone and move on, whereas the risk-taking *stochasts* exploit in more or less random fashion also new territory. The two strategies represent a trade-off between efficiency and innovation, hence their names. Normally, the stochasts outcompete the cartesians because they map the resource more quickly and completely, and find more fish although they use information less efficiently. But the cartesians can develop novel strategies in response and complex patterns may evolve.

From a sustainable development point of view, the lesson is that in real-world situations resource use strategies can be expected to fluctuate between random behaviour with long-term benefits (knowledge or innovations) and efficient behaviour with short-term gains (profits or stability). The resulting exploitation cycles can be interpreted in favour of stochasts: 'The model shows the opposition between short and long term decision making... and the importance of management principles which would maintain diversity... If we are to avoid a future of ferocious and ever growing competition, in a shrinking world with a single perspective and the common values of a single culture, then we must encourage "stochasts", and the diversity and expansion which only they can bring' (Allen and McGlade 1987). The difference between stochasts and cartesians echos the difference between the individualist and the hierarchist cultural perspective (§6.3) and shows that multiple interpretations of depletion and, therefore, divergent recommendations for management can be valid.

Significant advances in resource modelling have been made since this seminal paper (Garcia and Charles 2007; Little and McDonald 2007). One extension of the behavioural fisheries model is to introduce different fishing zones with different carrying capacities and growth rates and several fishing strategies (Brede and de Vries 2009b). What happens if a crowd of agents or fleets fish around in a few separate, open-access fishing grounds *without* centralised coordination? Suppose that you are a fisherman, then you can behave as a stochast, but there are other strategies. For instance, if many people go fishing in the high-yield area A, you imitate and join. If everyone does the same, A will soon be overexploited and yields will decline. Some people start moving to area B, and the same will happen there. This kind of crowd behaviour is known as the minority game (MG).[16]

An ABM has been constructed with this minority game and three other strategies that can be followed by individual fishermen in order to explore the role of different strategies for the exploitation history (Appendix 12.3). A series of model simulations has been performed to see under which conditions strategies survive in an evolutionary competition. If the harvesting pressure increases, the average catch per vessel declines (Figure 12.10a) and the number of vessels following a particular strategy changes (Figure 12.10b).[17] Which strategy becomes dominant depends on the harvesting pressure. If harvesting pressure is low and resources are abundant, the best performing agents are the stochasts, such as the ones that follow an uncoordinated chance-based harvesting schedule (RAND agents; Figure 12.10a). When the harvesting pressure increases, the the more cooperative COINS-strategy starts to dominate in the agent population and the MGS-strategy also becomes more

[16] The minority game was originally proposed to simulate behaviour in stock markets.

[17] Harvesting pressure is calculated as the maximum possible catch for the given number of vessels divided by the recovery period. It increases for shorter recovery periods. It is an equivalent of resource scarcity or abundance.

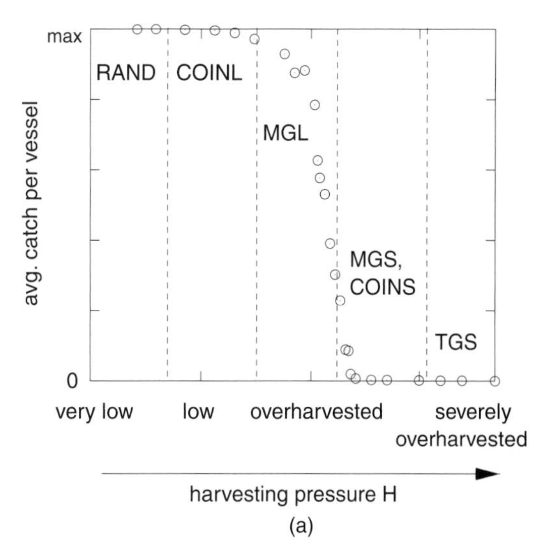

Figure 12.10a. The average catch per vessel declines with increasing harvesting pressure. Long-term (L) strategies perform better than short-term (S) strategies (Brede and de Vries 2009b).

effective. In severely overharvested situations, the team game strategy (TGS) gives the best results. The agents are assumed to have a short time horizon (S). A refinement of the model is to introduce agents who make a long-term (L) projection on the basis of their knowledge of the resource. The long-term strategies perform consistently better than the equivalent short-term strategies (Figure 12.10a). The COINL-strategy is now superior as long as the resource is not overexploited, but beyond a certain harvesting pressure, some of the distributed resources are occasionally overharvested and the minority game (MGL) strategy rises to dominance (Figure 12.10b).

The simulation results suggest that the exploitation strategy co-evolves with the intensity of exploitation. They also point at the inevitability of coordinated monitoring and regulation if resource use becomes more intense. A cooperative and long-term oriented strategy such as COINL emerges as the most sustainable

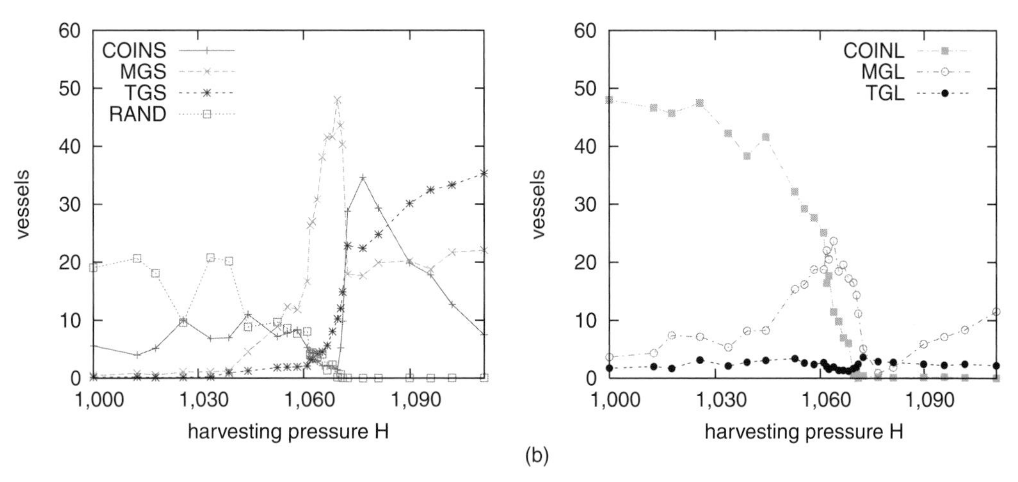

Figure 12.10b. The average number of vessels that follow one of the strategies, as a function of the harvesting pressure. The coordinated long-term (COINL) strategy is the most attractive one as long as the resource is not overexploited (Brede and de Vries 2009b).

strategy: it relieves the pressure on the resource, reduces fluctuations and diminishes the risk of overharvesting. Therefore, it can reconcile group interests and individual interests. An interesting question is whether a population discovers such a COIN strategy in time to prevent the collapse into the selfish (MG) and poverty (TG) strategies. This brings in the question of regulation and policy.

Government policies have long influenced fishermen's strategies. A comprehensive fishery management system dynamics model has been constructed to explore the role of government interference in the United States (Dudley 2008). It has three sub-models. The first one is the ecosystem model based on a somewhat refined logistic harvest function (equation A12.2).[18] The second one is the investment model. If the actual productivity, defined as catch per unit of effort, exceeds the desired or target value, new ships are ordered. Otherwise, ships will leave the area in search for more profitable fishing grounds but, before leaving, fishermen will attempt to increase catch efficiency in order to sustain productivity. The third model simulates lobbying and regulatory policy that affects the incentives and strategies chosen by the fisheries managers.

The most important finding is that, in an open-access non-managed situation, an oscillation in stock and catch ('overshoot and collapse') occurs with a frequency of three cycles per century for a reasonable model parameterisation (Figure 12.11). In the low-level periods, a large part of the stock addition consists of newborn fish and this makes the system vulnerable to ecological and climatic fluctuations. This result resembles the outcome for a minority game (MG) strategy and reflects the empirical findings for the United Kingdom.

Second, full implementation of the management goal to regulate the number of vessels on the basis of productivity gives in the long term the highest fish biomass stock and income per unit of fishing effort. This more rational strategy resembles the COIN strategy and is also the more sustainable strategy. It does not exhibit boom-and-bust oscillations.[19] However, the catch value in $/yr is lower than for a less well-managed, short-term–oriented strategy. A worrying result of the model experiments is that stock levels tend to decline to much lower levels than necessary, if lobbying and political influences disrupt full implementation of management goals. Unfortunately, this is what may have happened, at least in the United States: 'Interpretation of what's going on in fisheries has become very complex, not in the least due to management failures, and, in response, more and more differentiated regulation as well as litigation and non-compliance' (Dudley 2008).

12.5.3 Fisheries Models: Behavioural Variety

Some researchers have introduced more complex agent behaviour than in the previous models. An example is the *consumat model* (Jager *et al.* 2000). The décor is Lakeland, the imaginary island in an earlier section. The Lakelanders are endowed

[18] The reproductive rate α is assumed to be a function of the biomass R and biomass growth is separated into growth of existing stock and of new fish (recruitment). Besides, the ecosystem carrying capacity K will drop in proportion with the cumulative fishing effort so the ecosystem will recover, but slowly and non-linearly.

[19] The insertion of random fluctuations in the model causes boom-and-bust behaviour even in a fully managed fishery, in line with previous findings.

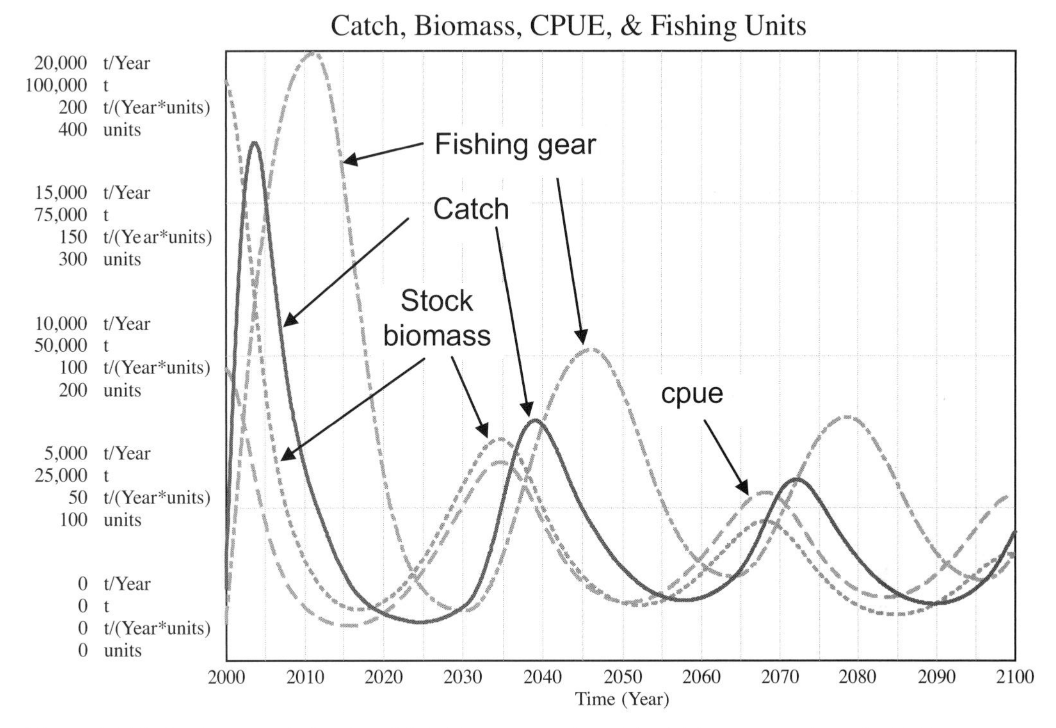

Figure 12.11. Trajectory of simulated fish biomass stock, catch, fishing gear and productivity, calculated as catch per unit effort (cpue), for an unmanaged fishery when the cpue is sufficiently low to cause overfishing to happen (Dudley 2008).

with different preferences and behaviours and are called consumats. They have a number of stylised characteristics that are based on findings in economics and psychology. In summary form, these characteristics are:

- they are equipped with certain *abilities* for fishing and mining and they allocate their time to leisure, fishing and mining according to their individual preferences;
- they spend the money earned from gold mining on fish imports and on non-fish consumption 'luxury' goods – the *opportunities*;
- their behaviour as producer (ability) and consumer (opportunity) depends on their level of satisfaction and level of uncertainty.

The last point is important: Each consumat only behaves economically rational if he is not satisfied and certain. This behaviour is called *deliberation* (reasoned or individual) and is characteristic of the homo economicus (§10.4). But agents can also behave differently: *imitation* (automated and socially determined) if satisfied and uncertain, *social comparison* (reasoned, socially determined) if unsatisfied and uncertain, and *repetition* (automated, individual) if satisfied and certain (Figure 12.12).

A consumat updates every timestep his memory with information on his own and other agents' performance. When he decides about job and spending according to the decision scheme in Figure 12.12, it results in certain levels of *needs satisfaction* and *uncertainty*. The needs considered are proxies for leisure, identity, subsistence

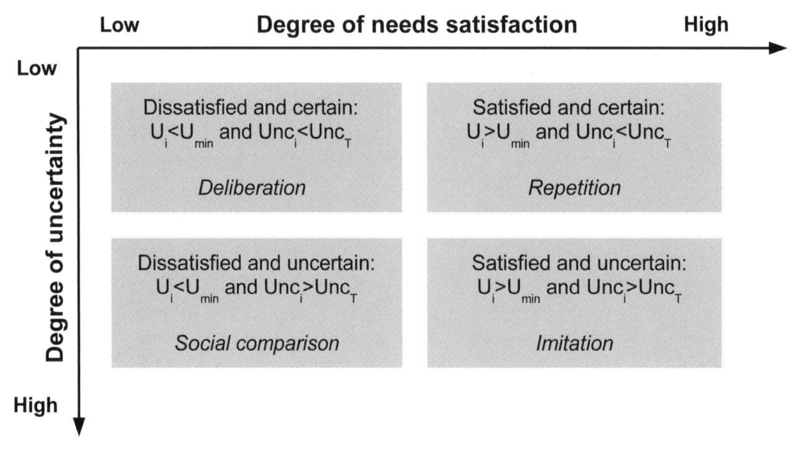

Figure 12.12. The agent model for consumats (after Jager *et al.* 2000). The level of needs satisfaction and of uncertainty are important contextual determinants of people's behaviour.

and freedom.[20] The satisfaction of the need for leisure is related to the share of the time spent on leisure and for identity to the amount of money the consumat owns in comparison to other consumats with similar abilities. Needs satisfaction for subsistence depends on fish consumption; for freedom, it is a function of the available money. What can we learn from this model about sustainable development?

Sixteen consumats can satisfy their personal needs by exploiting two natural resources: fish and gold (Figure 12.1). Which strategy a consumat chooses depends on present and past levels of need satisfaction and uncertainty. If not satisfied and certain, he or she aspires for more consumption and is certain enough to show reasoned behaviour. If fully satisfied and highly uncertain, he or she will resort to automated socially determined behaviour. This is similar for the two other cognitive processes. Development in Lakeland happens because agents switch strategy if their performance falls below the expectation they build up in their memory. The consumats are in a social dilemma situation: Each consumat is inclined to get as much income at the lowest effort possible, but in doing so they may collectively destroy the option to sustain income and quality of life for the long term.

Let us look at some illustrative outcomes of this model if fishing is the only option. To keep the analysis transparant, two archetypical consumats are defined: the *Homo economicus (He)* and the *Homo psychologicus (Hp)*. The former favours deliberation, because he operates with high levels of need satisfaction and uncertainty reduction. The latter is quickly satisfied and an uncertain imitator. If fishing is the only option, the *He* use deliberate rationality to increase their income and as a result the fish stock is depleted before year forty (Figure 12.13a). The *Hp*, on the other hand, are quickly satisfied and do not fish more than needed. Most of them spend their time as happy, lazy, poor repetitors and imitators, until the initial budget has been fully spent, in year eighteen, and many of them become unsatisfied and switch to deliberate behaviour in order to restore income (Figure 12.13b). As a consequence, the fish stock remains at a high level (Figure 12.13a,b). The two

[20] These needs are supposed to play an important role and are selected out of a larger set of needs, as described by Max-Neef (1991) (§6.2).

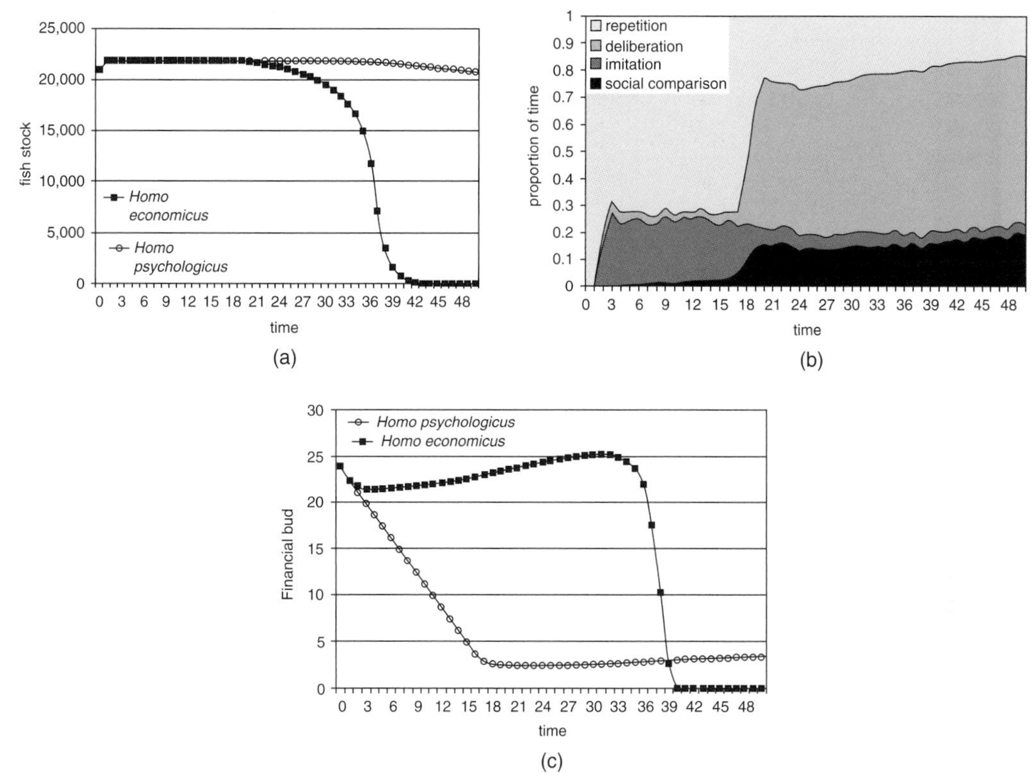

Figure 12.13a-c. Illustrative pathways in a fishing-only no-gold-mining consumat world for the *Homo economicus* (*He*) and *Homo psychologicus* (*Hp*) in the first 50 timesteps (averages for 100 simulation runs) (Jager *et al.* 2000). (a) The fish stock, (b) the proportion of time spent in one of the four strategies (c), the financial budget.

consumat-types have a very different financial budget profile and in the *He*-collapse period there are large income differences amongst the agents (Figure 12.13c).[21] If uncertainty is completely removed, the *He* does not overharvest and the *Hp* spends even less time fishing. This is in agreement with the finding that uncertainty tends to cause resource overexpoitation (May 1977).

If the consumats can engage in gold mining as an additional source of income, a transition happens from a fishing into a mining community for both *He* and *Hp*. Both do pollute the lake, with a negative impact on the fish stock (Figure 12.13d). The *He* are keen to grasp the new opportunity and the fish stock is less exploited. The *Hp* switch at a slower rate to mining, suffer more from pollution impacts and are forced into an accelerated transition to mining. One sees here two different development patterns with the same outcome: A growth in income (from mining) takes place at the expense of the environment (pollution or depletion).

Discovering and implementing labour-saving opportunities is what innovation and economic development are about. So what happens if a deliberating consumat discovers a new fishing opportunity and increases his productivity (fish caught per hour)? Nothing happens in the *Hp world*, because he or she is easily satisfied (is

[21] Income differences are explored in more depth by giving the agents different abilities.

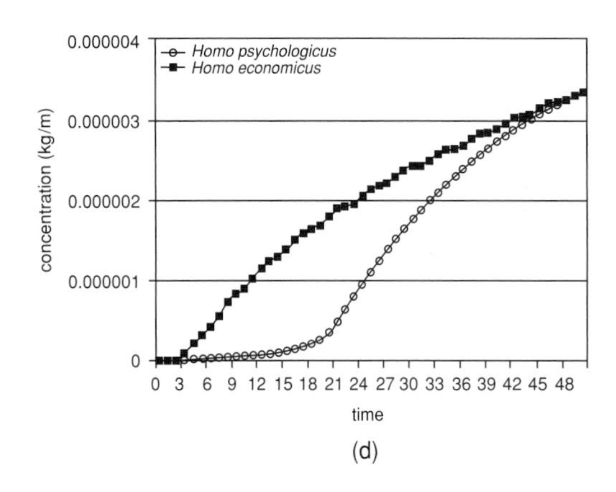

Figure 12.13d. Illustrative pathway of the pollutant concentration in the lake in a fishing and gold-mining consumat world for the *Homo economicus* and *Homo psychologicus* in the first 50 timesteps (averages for 100 simulation runs) (Jager *et al.* 2000).

she lazy?) and does not engage in deliberation (is he stupid?). In a world populated by *He*, the innovation makes one individual more productive than others and it rapidly spreads in this competitive and entrepreneurial world through imitation. As a consequence, the fish stock is depleted faster than without the innovation. This is precisely what happened when fishing boats became more effective. It illustrates the ambiguity of technology: Innovations in resource exploitation are both cause and solution of sustainability problems. Of course, an illustrative model like this has to be backed up by empirical fieldwork and to be extended with more realistic behaviour (Janssen 2002; Perez and Batten 2003; Walker *et al.* 2006).

12.5.4 World Forests

Reconstructed maps of Europe show the enormous decline of forested areas since medieval times. The more recent large-scale deforestation in parts of Canada and the United States, in Argentina and Costa Rica, in Indonesia and in other parts of the world has also been reconstructed and shows an acceleration in the rate at which human populations deforest the planet (Turner *et al.* 1990; Williams 2008). But in North America and Western Europe, the forested area is again increasing for decades, as part of what some call the *forest transition* (§11.4; Satake and Rudel 2007).

Forests provide already for millennia important products such as fuelwood for cooking and heating, timber for construction and charcoal for industrial processes. Currently, wood is an important commodity and traded worldwide. Of late, global wood use appears to stabilise for various reasons (Figure 4.5). Besides, forest produce is a very diverse source of food and materials for forest dwelling populations. Forest exploitation has been one of the first situations where the notion of sustainable resource use emerged (van Zon 2002). It was also one of the first areas in which enlightened management techniques have been introduced.

The assessment of forest resources, both quantity and quality, is even more difficult and fraught with definition issues and uncertainties than of water and fish resources. The major reason is the huge diversity and the complex dynamics of forests, notably mature tropical forests (Bossel and Krieger 1991). Global estimates of forest cover suggest an ongoing decline in forested area worldwide. Amongst the driving forces is the large-scale clearing of forest for the production of soybeans and

Box 12.5. *Forests in Japan.* Forests in Japan have been under pressure already for millennia (Totman 1989). More than 2000 years ago, rice culture caused the first dramatic modifications of woodlands and bronze and iron smelting started to put pressure on the forest due to the demand for high-quality charcoal. Metallurgy led to new, more powerful tools and the assault on the woodlands intensified. The warrior castes needed timber for ships, stockades and large residences and for coffins buried in huge mounds rivalling the Egyptian pyramids in size. Wood was also used as fuelwood for pottery and charcoal for weapons. Farmers also needed wood for fuel but also for fodder and, most importantly, for green fertiliser material which sometimes relieved cutting pressure. There was also a large demand for the construction of monasteries, shrines and temples and, owing to termites and rot, most wooden buildings had to be rebuilt every 20 years.

The deforestation had all kinds of consequences such as wildfire, flooding and erosion, often forcing people to move. Kings and emperors often had to move, possibly because local wood supplies dwindled. There were occasional attempts to control the use of woodland on the part of governments and monasteries. Ruling warriors tightened control to assure themselves of resources for military use. It was a history of outright exploitation without concern for preservation or reforestation.

When political struggles subsided in the 17th century, population and construction rose rapidly and the demand for timber soared. The demands of the peasant families led to widespread but less intensive use than the more concentrated demands from the rulers in the cities. Logging expanded and intensified, erosion denuded mountains and damaged lowlands, and Japanese rulers were forced into a combination of regenerative forestry and imports from the tropical rain forest in nearby regions, particularly in Malaysia and Indonesia. Like with food in 19th-century Britain, the pressure on scarce land resources was relieved by importing resources from abroad (§4.3). Regulation to restrain consumption was introduced, plantation forestry emerged by the late 18th century and most forested areas came under some sort of management. It is one of the reasons that Japan now remains more forested than nearly any other country in the temperate zone.

for plantations of palm trees (for palm oil) and eucalyptus (for paper). But one has to be careful with interpretations and conclusions. For instance, there is a rather vague distinction between old (primary) and young (secondary) forest.

For forests, too, models have been constructed to understand the basic processes and to design harvesting strategies (Acevedo *et al.* 1995; Bossel 1996). Of course, given the variety in forests, the models have different characteristics and aims. They used to focus on optimal harvesting strategies in the context of forest economics (Clark 1990). Important variables are the rotation period and, as with fisheries models, the discount rate. Such a framework had its origin in the hierarchist and rationalist approach of the Modernist worldview, where the forester tries to shape the trees according to the desire for control and efficiency. German foresters and mathematicians even specified a standardised tree (*Normalbaum*) in an utilitarian attempt to narrow the view of the forest down to commercial wood (Scott 1998). The underlying worldview of Modernism fits in with the practise of maximising monetary benefits from resource exploitation (Figure 6.5).

An example of an economic optimisation model is a global timber market model by Sohngen *et al.* (1999) (equation 12.3). The authors calculated the optimal management of the world's forests over the period 1995 to 2135. After parameterising the model for nine world regions, the main conclusion is that only inaccessibility of forests can protect biodiversity in a globalising market-driven world. Although of little practical relevance, the model points out that economic rationality in open-access resource exploitation tends to preserve resources only if the harvesting costs are infinitely high and exploiters are forced to look for substitutes. Such high cost occur when a resource is taxed very heavily or inaccessible. (However, what is inaccessible for modern technology?) It should be added that neither demand nor substitutes are considered in this model.

The reductionist optimisation approach has less appeal nowadays, with the discovery of the rich biodiversity of tropical forests and the recognition of the wealth of forest ecosystem services. The existence and complexity of ecosystem services is only slowly incorporated into management models and practise. A simple way to account for (the loss of) environmental and other services is to include externalities in the harvesting model. A combination of consumer initiatives such as ecolabels and producer considerations such as relative benefits for landowners of forestry and agriculture can guide exploitation into more sustainable directions (Nilsson 2005; Satake and Rudel 2007). A more integrated approach is to include topsoils, water and ecosystems services such as erosion control and the possible occurrence of irreversible change due to thresholds explicitly in the model (Mäler *et al.* 2003; Hein and Van Ierland 2006). This gives a natural link to biodiversity. Last but not least, it is important to engage local stakeholders and concerns into the models and the decisions, because forest exploitation is quite local in its causes and consequences. This point deserves some special attention and is the topic of the next section.

12.6 Interactive Modelling for Sustainable Livelihood

Many applications in environmental science aim at models being used as decision support systems (DSS). It can be combined with techniques such as scenario writing and backcasting, as is done in policy exercises (§8.6; Toth 1988). The first attempts to construct simulation games in environmental education and policy date from the 1980s. The rapid development in computer technology and software applications has given a boost to the interactive formulation and use of models in close interaction with stakeholders. The latest development is to run games and models through the Internet. Modellers and experts interact with participants and stakeholders in order to formulate a shared problem perception, to understand and communicate the dynamics of the relevant processes, and to design feasible, mutually beneficial strategies. The objective is a combination of teaching skills, conveying information, increasing motivation and interest, changing attitudes and inducing evaluation (Greenblat 1988; Bousquet and Voinov 2010). As discussed in §10.4, interactive modelling can bring together the worlds of science, policymaking and computer games. There are many initiatives, projects and platforms in simulation gaming and decision making on environmental and sustainability issues, so I refer the interested reader to the Useful Websites.

A group with longstanding experience is the CORMAS team in France (www.cormas.fr). The CORMAS platform allows construction of spatially explicit multi-agent simulation (MAS) or ABMs. This is done in interaction with stakeholders and in combination with other participatory methods like role-playing games. It has led to their own brand of *companion modelling* called ComMod (Daré *et al.* 2009). The basic element in the ComMod approach is experiential learning through collaboratively tackling a question or an issue, co-constructing a shared representation (mental map or conceptual model), implementing it in a computer model and visualising the dynamics, and, finally, collaboratively designing scenarios and intervention strategies.

It is a learning and communication process between scientists and stakeholders, during which new forms of knowledge and knowledge gaps come to the fore. The participants progress from individual to collective mental models and are learning for action and empowerment, both important aspects in a CPR context (§8.5). There have been many and varied projects and workshops with the ComMod approach, for instance, on integrated and shared water management (Brazil); water availability, migration and (shortage of) farm labour (Laos); participation in how to manage and develop multiple use regions in wetlands (France); and learning about negotiation mechanisms (Brazil). It is as yet unclear to what extent people do change behaviour, once they are back in everyday reality. The experiments with ComMod have not yet led to general principles of sustainable management, but many data and insights have been accumulated about the *role of context* in what people believe and how they behave (§6.4).

An illustrative application of the ComMod approach concerns a clarification and search for solutions in a conflict between sheep farmers, foresters and environmentalists in the Causse ('plateau') Méjean near Montpellier (Etienne *et al.* 2003). The problem is that Scots pines and Austrian black pines are encroaching into a natural ecosystem of high biological diversity. This affects the development objectives of the stakeholders (sheep production, timber production or nature conservation) in different ways. An ABM has been developed, using a 200 m × 200 m square grid of 5.726 cells (4 ha each). Biological characteristics are introduced for the initialisation, formulated in the form of 'spatial entities' such as management units making sense according to some specific perception of the ecosystem, such as already invaded ridges, farming and woodland areas. Then, three sets of actors are defined, each with their own strategy vis-à-vis the spatial entities: thirty-seven sheep farmers, two foresters (one dealing with afforestation, one with the exploitation of native woodlands), and a national park ranger (the 'conservationist').

The model has been used in an interactive decision setup, with an approach based on four concepts:

- *The point of view*: the specific way in which each stakeholder perceives resources and identifies management entities – not unlike the worldviews presented in §6.3;
- *The viewpoint*: a spatial representation of a point of view, such as a vegetation map, a map of tree ages, a map of tree stands or a map of landscape units;

- *Indicators*: a set of markers selected by the stakeholders to monitor the dynamics of the system; and
- *Scenarios*: prospective management rules to tackle the problem at hand (such as pine encroachment of the Causse).

During these exercises, it became evident that two aspects were particularly effective in structuring the problem and looking for solutions: sharing experiences and views of the underlying processes and comparing contrasting scenarios. A major conclusion was that 'it helped the participants to understand that all their views were legitimate, but also subjective and partial'. Another example of an interactive modelling tool is the *Geonamica Dynamic Landuse Planning* software package (Appendix 12.4).

12.7 Perspectives on Water, Fish and Forest

Just as with agro-food systems, the interpretation of what is going on and what should happen with the world's renewable resources diverge. Listen to the stories from NGOs:

- the Mau Forest, Kenya's largest area for water withdrawal, is under threat. One-third of the 0.4 Mha forest has been cleared in the last decade because of politicians giving out favours and illegal occupation;
- megacities are badly in need of water management. Yet, in megacities such as Mexico City and Mumbai, there is hardly any long-term strategy for supply, reuse and conservation of water;
- the largest rubber plantation, with 8 million rubber trees, is in Liberia. It is operated by the Japanese/U.S. tire manufacturer Firestone/Bridgestone. Labour conditions are bad: $4/day payment, use of pesticides (www.stopfirestone.org).

The following are the voices from the pragmatic world of engineers and politicians:

- 'As long as the farm breaks even on a cash-flow business I'm OK', the farmer in Arizona said. 'But my retirement is not such that I could suck up the water bill if it keeps rising, and we still have a loan on the land. Wouldn't take long to suck all the retirement into that just to keep it'.
- 'If we don't do something, the most expensive part of Jakarta will disappear into the sea', says a director at the Agency for Development and Planning. 'We investigate a 30-km dike in the sea, but it is the most ambitious option'. Jakarta sinks every year about 10 centimetres (cm), mostly due to groundwater pumping. The nineteen-year La Niña will, in combination with sea level rise, hit hard again in 2025.
- 'The economy [in Liberia] has always been associated with forest exploitation. I do not say that REDD (Reducing Emissions from Deforestation and Degradation) is not a good strategy, but when you have to survive from day to day, your first thoughts are not about climate change'. Can a carbon tax help to preserve the forest in Liberia?

But there is always opportunity and hope:

- In 2009, a large investment fund advertised: 'Water will become scarce. So it will give many opportunities all over the world. Invest with Robeco'. Water is a 'growth market' and investors and speculators are becoming more interested.

There are many water funds, and some note returns on investment of more than 30 percent:

- The World Wildlife Fund (www.wwf.org) works with corporations such as Unilever in several countries on the introduction of ecolabels in order to promote sustainable fishing.
- 'L'Europe est sur le point de fermer ses portes aux entreprises qui prospèrent sur la destruction des forêts', says a Greenpeace activist in Brussels. Between 1995 and 2009, the trade volume in primary tropical wood has fallen 40 percent due to ecolabels as well as depletion and rising domestic use in Indonesia and Malaysia.
- Drip irrigation is a water conservation option with a large potential. Besides saving water, it reduces weed problems, soil erosion, electricity use and helps to reduce overexploitation of groundwater. In poor regions, it may mean survival for farmers.
- Today's scientific advances in water desalination dramatically increase our ability to transform sea water into fresh water and quench the thirst of 1.2 billion people facing shortages of water. In 1956, Israel's Prime Minister Ben Gurian said it: 'Die Wüste mit entsalztem Meerwasser zu bewässern mag für viele volkommen unrealistisch klingen, aber Israel sollte keine Angst vor scheinbar unrealistischen Ideeen haben, die durch die Macht von Visionen, Wissenschaft und Pioniergeist die vorgegebene Ordnung ändern'.

These statements make it clear that the problems and solutions regarding renewable resource use are diverse and are viewed differently. In a security-oriented world, the greatest risk is that conflicts about resources intensify – and the solution may be more armaments. Advocates of the Modernist worldview have high expectations of new technologies and rational property and pricing arrangements. The idealists who voice other worldviews frame the issue in terms of equal access, of frugal use and of vulnerability. As in Chapters 7 and 9–11, I have attempted to link more specifically certain renewable resource issues to the different worldviews, with Table 12.3 as the result. It is an invitation to the reader to explore his or her own position. The challenge is to listen to each other, respect the unity in diversity and find sustainable resource management somewhere in the centre.

12.8 Summary Points

This chapter used an imaginary country to illustrate pathways of (un)sustainable resource management. Many models of renewable resource use have been constructed and applied, with modest success: overexploitation happened and still happens in many situations. The key insights to remember from the models and stories are the following:

- Renewable or (bio)resources have a finite regeneration rate and can, therefore, only be harvested sustainably below a certain harvesting rate;
- The driving force behind (over)exploitation is the size and productivity of the resource-exploiting capital stock (groundwater pumps, fishing boats, tree-cutting machinery and so on). Open-access (bio)resources will be overexploited without some form of regulation;

Table 12.3. *Worldviews and renewable resource futures*

Renewable Resources 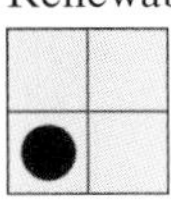	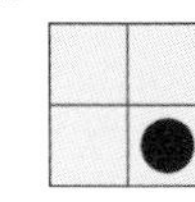	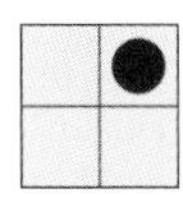	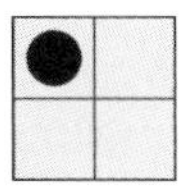
Statement 1: As a result of overfishing, overall ecological unity of our oceans are under stress and at risk of collapse (www.overfishing.org).			
Admittedly, fisheries are overexploited. But what is the ecological unity of our oceans and how can it collapse? Besides, Nature is more robust than we think.	This is an alarmist statement. I don't understand and I don't believe it.	The web of life is very complex and intricate, so if you destroy a part, you may destroy the whole.	Scientists have found evidence that overfishing is affecting whole food webs, not just single species.
Statement 2: The rapid cost reduction in seawater desalination make this technology an important ingredient of the transition to a sustainable water future.			
This is certainly true for many of the semi-arid regions in the world. The oil-rich countries can already afford it, and their use makes further innovations possible.	It sounds like this is how they use their oil revenues in the Middle East. Not bad, water is their scarcest resource.	Maybe. These centralised, capital-intensive options tend to discourage small-scale options such as more efficient use and rainwater collection.	In semi-arid regions, it is often the only option and relieves the pressure on groundwater. It should be in balance with other options.
Statement 3: Certificates such as FSC for tropical timber will save the tropical forests.			
Indeed, the introduction of green certificates is a very effective way to sustainable resource use. Consumers want it, and industry does it.	Possible but improbable: There is so much corruption in those countries. Anyway, as long as the green certificate does not increase the price.	It is good that people who buy such timber can make a 'green' choice. Still, they may consider to use Europe's own timber – with treatment it is equally durable.	Green certificates are working well for imports in Europe. Still, illegal logging, population pressure and growing local demand pose severe threats to the tropical forests.
Statement 4: The world community should implement and enforce criteria for the ecologically sustainable, socially appropriate and economically viable production and use of biomass (IUCN 2010).			
A typical NGO-statement: A political compromise and impossible promise...	Biomass use may be needed in our energy system. Its use should primarily be justified on economic grounds.	Worldwide biomass production to satisfy the energy addiction of the rich is not a solution. Criteria such as these may at least avoid serous damage.	Biomass plantations can have negative impacts for indigenous people, local food supply and the environment. Regulation is critically needed.

- Historical experiences with and recent investigations of common pool resource (CPR) management suggest clear sets of rules to avoid the government planning versus private property dichotomy and to overcome the 'tragedy of the commons';
- Water is a key resource with diverse and local characteristics. Duration and frequency of mismatch between water demand and (blue and green) water availability are already significant in several regions and are expected to increase. Groundwater resources are overexploited in many places. Increasing water stress can have significant consequences for quality of life of many people;
- Global fisheries exploitation is occurring above sustainable yields and the carrying capacity is eroded. The switch to large-scale aquaculture raises new concerns. Scientific models must be improved and include human behaviour and (the ambiguities of) technological change to be valid and useful. The same is true for global forestries.
- Interactive simulation gaming and participatory modelling are useful, new tools to engage larger groups of people, as stakeholders and citizens, into the theory and practise of more sustainable resource use.

> The highest form of goodness is like water
> Water knows how to benefit all things without striving with them
> It stays in places loathed by all men
> Therefore, it comes near the Tao.
> Nothing in the world is softer and weaker than water;
> But, for attacking the hard and strong,
> There is nothing like it!
> For nothing can take its place.
> That the weak overcomes the strong, and the soft overcomes the hard.
> This is something known by all, but practices by none.
> – Lao Tze, quoted in Allerd Stikker, *Water: The Blood of the Earth*

> Then they saw the Cedar mountain, the Dwelling of the Gods,
> The throne dais of Imini.
> Across the face of the mountain the Cedar brought forth luxurious foliage,
> Its shade was good.
> – Gilgamesh and Enkidu arriving at the sacred grove – Epic of Gilgamesh

SUGGESTED READING

A textbook with several chapters on renewable resources.

Common, M., and S. Stagl. *Ecological Economics – An Introduction.* Cambridge: Cambridge University Press, 2005.

A special issue with a series of articles on system dynamic models in theory and practise.

Cavana, R., and A. Ford. Environmental and resource systems, *System Dynamics Review (Special Issue)* 20 (2004): 2.

Thorough, mathematical introduction into renewable resource economics.

Clark, C. *Mathematical Bioeconomics – The Optimal Management of Renewable Resources.* New York: Wiley Interscience, 1990.

This book has several system dynamics models in Stella® and with diagrams about fisheries and irrigation, with emphasis on spatial and commons aspects.
Costanza, R., B. Low, E. Ostrom, and J. Wilson, eds. *Institutions, Ecosystems, and Sustainability*. Ecological Economics Series. London: CRC Press, 2001.
A textbook with several chapters on renewable resources.
Folmer, H., H. Landis Gabel, and H. Opschoor. *Principles of Environmental and Resource Economics – A Guide for Students and Decision-Makers*. Cheltenham, UK: Edward Elgar, 1995.
A textbook with several chapters on renewable resources.
Perman, R., Yue Ma, J. McGilvray and M. Common, eds. *Natural Resource and Environmental Economics*. Boston: Pearson Education Ltd., (2003).
An introduction into resource system dynamics models in Stella®, with diagrams about fisheries, forests and so on.
Ruth, M., and B. Hannon. *Modeling Dynamic Economic Systems*. New York: Springer, 1997.

USEFUL WEBSITES

- glossary.eea.europa.eu/ is a general site of the European Environmental Agency with brief explanation of terms in the area of environmental economics and policy.
- www.worldwater.org/ is an informative site on water in a series of reports *The World's Water* published since 1998, operated by Gleick.
- www.waterfootprint.org is the site of the waterfootprint, with definitions and calculations.
- ec.europa.eu/fisheries/cfp/fisheries_sector_en.htm is a rather complete and official account of fisheries in the EU.
- www.worldfishcenter.org is a site on fisheries. It started as a centre in 1977 in the Philippines. The website states as its mission 'to reduce poverty and hunger by improving fisheries and aquaculture'.
- maps.grida.no/go/graphicslib have maps of global fisheries and global forests on the site.
- www.fao.org has maps, statistics and documents such as the *State of the World's Forests*.
- www.worldagroforestry.org/ is a site with info and links on world agroforestry events and issues.
- www.sciencemag.org/site/feature/data/deutschman/forest_model.htm is a site on forest dynamics models.

Websites on interactive models and games

- ivem.eldoc.ub.rug.nl/ivempubs/Software/Stratagemmanual/ is a description of the interactive game Stratagem (Meadows 1990), about running a country and experiencing that stabilisation of population growth is a sine qua non for long-term sustainability.
- earthednet.org/Support/materials/FishBanks/fishbanks1.htm is one of the sites about the game Fish Banks Ltd. (Meadows 1996), in which up to six teams get the opportunity to experience, and prevent, the tragedy of the commons.
- www.rug.nl/fmns-research/ivem/publications/software/index has an overview of resource models, among these the interactive model for electric power planning PowerPlan (de Vries and Benders 1989).
- CORMAS (cormas.cirad.fr/) is the site on the CORMAS interactive simulation platform, with practical recommendations for companion modelling (Daré *et al.* 2009). It is also published in English.
- www.iemss.org is a site of the International Environmental Modelling & Software Society (IEMSS), with information on a.o. the GEONAMICA® software environment for integrated environmental spatial modelling.

Appendix 12.1 The Simple Population and Renewable Resource Model

A general description of a resource or environmental sink exploitation model is:

$$\frac{dR(t)}{dt} = F[R(t); G; t] - H(R) \qquad \text{(A12.1)}$$

with dR/dt the rate of change of the resource or environment stock *R* and F[R(t);G;t] a function expressing its degradation and regeneration. It accounts for depletion but also external influences G and autonomous trends. *H* is the exploitation/harvest rate. The *maximum sustainable yield* (MSY) or use is defined as the maximum value of *H* for which the resource does not vanish. This occurs for $dR/dt = 0$, which is the case for F[R(t);G;t] = H(R). If the resource dynamics is assumed to follow a logistic growth path as given, then F = αR(1-R/K) in equation A12.1 and the maximum value of H is found for the maximum value of dR/dt. Calculus tells us that this happens for $d^2R/dt^2 = \alpha(1\text{–}2R/K) = 0$. This condition is fulfilled for a resource stock of R = ½K. Therefore, the MSY equals $H_{max} = {}^1/_4\alpha K$. For a harvest H, the value of dR/dt equals the difference between the two curves in Figure 12.4a. For values of R between R_A and R_B, the resource will increase (dR/dt>0). For R exceeding R_B, the resource will decrease towards the stable attractor B with R = R_B. For R less than R_A, the resource will decrease and become extinct (R = 0), as it moves away from the unstable attractor A with R = R_A. At or below the sustainable yield level, the resource can – if it were to behave according to this simple description – be exploited indefinitely. If the world were only that simple, this would reflect a clear measure of sustainable resource use.

If the resource is an environmental sink, which is used for its capacity to absorb and break down pollutants, the same formalism is used. H is the disposal rate, and the function F represents the 'cleansing' capacity of the sink. For instance, emitted carbon dioxide (CO_2) disappears from the atmosphere in about 100 years, on average, and certain chemicals such as polychlorinated biphenyls (PCBs) take millennia to be broken down if ever at all.

Appendix 12.2 Resource Use in the Simple Model

If the resource is exploited as an open-access resource by individuals, it is possible to derive the tragedy of the commons logic. If the harvest H is assumed to be a function of effort E and resource size R, equation A12.1 can be written as:

$$\frac{dR(t)}{dt} = \alpha R(1 - R/K) - \varepsilon E^a R^b \qquad \text{(A12.2)}$$

For simplicity, we assume that there are no decreasing returns for an increasing harvest rate (a = b = 1). The parameter ε is a measure of catch effectiveness (catch per unit effort). The system is in steady-state equilibrium for dR/dt = 0, which implies:

$$\alpha R(1 - R/K) = \varepsilon ER \qquad \text{(A12.3)}$$

Rewriting shows that this is the case either for R = 0 (no resource) or for E = (α/ε) (1–R/K). In other words, for each harvesting effort level E, there is a corresponding stock size R for which the system is in a steady-state. Graphically, this implies that the system is somewhere on the dotted isocline sloping downwards in Figure A12.1.

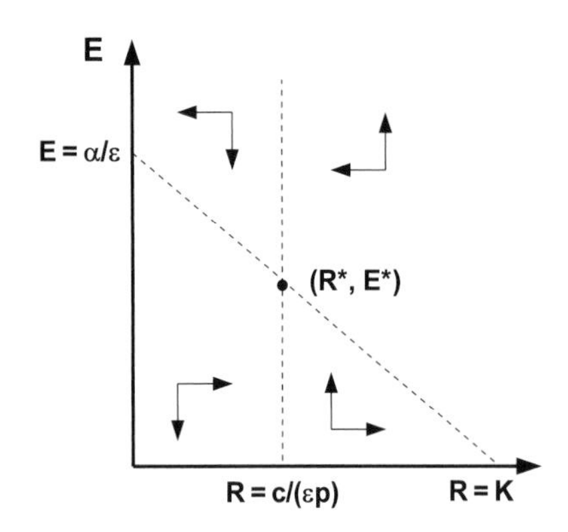

Figure A12.1. The points for which dE/dt = 0 and dR/dt = 0 (isoclines) in an open-access fishery according to eqn. A12.2. At the intersection, the system is in dynamic equilibrium and the last entrant will have zero profit.

What determines the economically efficient effort level E? Let the market price of fish, p, and the costs per unit effort, c, be constant. Then, the profit of a fishing firm equals the revenues $B = pH$ minus the costs $C = cE$. Applying the open-access principle of zero profit for the last entrant, the condition $B = pH = C = cE$ translates into $E = pH/c$. Because we assume proportionality, $H = \varepsilon ER$. Substitution gives $R = c/(\varepsilon p)$ as the steady-state situation where the last entrant has a zero profit. Graphically, it is the dotted vertical isocline in Figure A12.1. It proves that open-access exploitation tends towards a stationary state of effort level E* and a stock size R*. This situation is at the crossing point of the two isoclines. This is for $E = (\alpha/\varepsilon)(1\text{-}c/(\varepsilon pK))$. The higher the regeneration rate (α) and/or the price (p) and the lower the cost (c), the higher the steady-state effort level will be.

If the system is somewhere off the isoclines drawn in the graph, it will move towards the attractor. How it moves is indicated with the arrows and can be seen from the conditions:

- If R is on the right-hand side of the vertical isocline, $R > c/(\varepsilon p)$, which implies that revenues exceed costs, $B > C$, and new firms will enter such as $dE/dt > 0$;
- if E is on the right-hand side of the sloped isocline, $E > (\alpha/\varepsilon)(1\text{-}R/K)$, which implies that the harvest exceeds the natural growth rate, $H > \alpha R(1\text{-}R/K)$, and the resource will shrink as in $dR/dt < 0$.

The position of the attractor depends, in this simple analysis, on the ratio between natural growth rate α and catch effectiveness ε. It can be interpreted as the relative force of resource (fish, tree and so on) and predator (people). The lower it is, the more vulnerable the resource is for overexploitation. Or its corollary: *The more effective people become ($\varepsilon\uparrow$), the more the resource is at risk.* It also depends on the ratio $c/(\varepsilon p)$, which is for given ε equal to the cost-price ratio. The higher the cost ($c\uparrow$) and the lower the price ($p\downarrow$), the lower the risk of overexploitation. In this model, high prices protect the resource and cost-reducing innovation do the opposite.

Appendix 12.3 Modelling Different Harvesting Strategies

Let us consider in more detail an investigation of different harvesting strategies (Brede and de Vries 2009). Let there be a distributed resource consisting of j = 1..P fishing zones, each modelled similar to the one introduced previously (equation A12.2). Each agent chooses where to fish in the next season. The choice is determined by a probability table that gives the chance that agent i fishes in zone j in the next season. The table mirrors the agent's view of the resource base in terms of catch and profit performance. After each fishing season it is updated, using the information about his and possibly also other agents' last and previous performances. How an agent builds this probability table is determined by the strategy it pursues. Therefore, we define four strategies:

- the agent bases resource selection only on individual profit (Minority Game (MG));
- the agent's choice is motivated by the performance of a team it belongs to and shares its catch with (Team Game (TG));
- agents use the impact of their catch on the performance of the whole community (COllective INtelligence (COIN)); and
- the agent selects the resource to be exploited randomly (RANDom (RAND)) similar to a stochast strategy.

MG is a very selfish, individualist strategy. The COIN strategy is more 'cooperative' because agents base decisions not on their own profit but on the impact of their effort for the community outcome. If played by many agents in the group, TG is a very cooperative strategy.

Appendix 12.4 The Geonamica Software

Started by physicists and geographers at the Université Libre in Brussels, the Geonamica (formerly RamCo) software is one of the tools to incorporate local specificities with global developments. It has been used to investigate a variety of resource and environment issues, such as the impact of climate change on the Caribbean Islands (Engelen *et al.* 1993), land use planning in Indonesia (de Kok and De Wind 2002), future land use patterns in the Netherlands, management of the Wadden Sea in the Netherlands and floods risks (Engelen 2004; Nijs *et al.* 2004). It is developed and applied at the Research Institute for Knowledge Systems (RIKS; www.riks.nl/products/geonamica) and at the Vlaams Instituut for Technologie onderzoek (VITO; Natuursimulator). Its basic methodology is cellular automata (CA). The strength of this and similar models is the integration inherent to the methodology and the interactive user-friendly mode.

The framework distinguishes three layers. The lowest layer contains the data on the biophysical variables, such as elevation, water bodies, soil features and rainfall patterns. It represents local (potential) suitability for various activities, for instance, accessibility for housing and shopping, good soils for wheat or rice, attractive beaches for tourism and so on. The highest level is an aggregate simulation of the driving forces, notably population growth and sectoral macro-economic activity growth. It is used as a scenario-generator. The intermediate level is based on a CA model. It simulates the relative attractiveness of different forms of land use, such as agriculture, tourism or industry, against the background of suitability (lower level) and driving forces (upper level). Changes in the driving forces cause changes in land use, such as a conversion from agricultural land into housing areas, depending on relative attractiveness. For instance, a premium price for a seashore area reflects its suitability and attractiveness for tourism, which, however, declines with distance

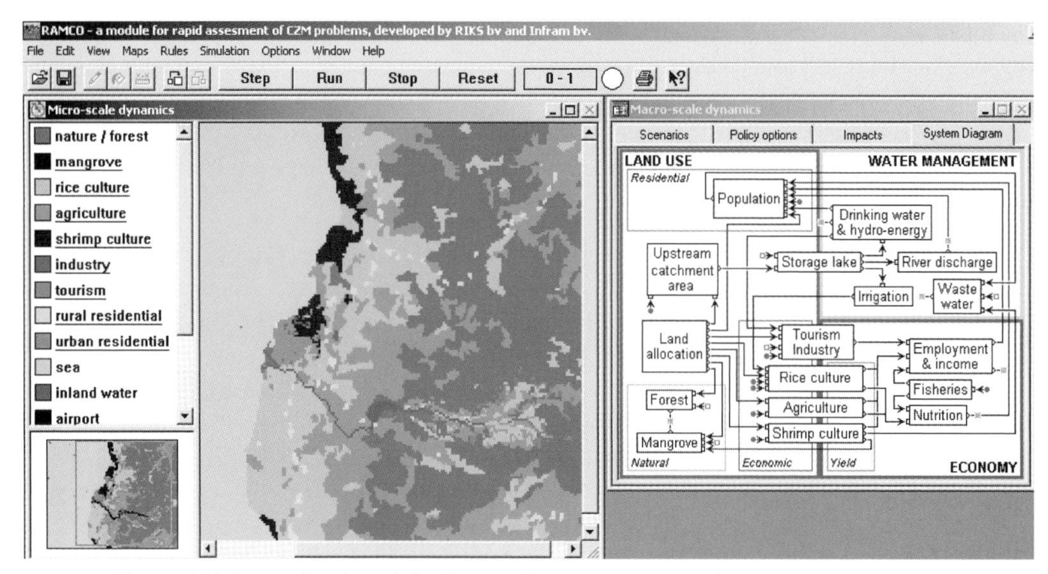

Figure A12.2. Application of the Geonamica software to explore sustainable development in western Sulawesi (Indonesia). The right side shows the high-level dynamic simulation model; the left side shows the underlying base maps. The CA mechanisms relate both.

from the sea. The attractiveness potentials can interactively be adjusted to reflect local expertise. Tools like these can identify the challenges and opportunities for sustainable development of local/regional resources. An illustrative application is the analysis of the impact of agricultural expansion on mangrove forests on the island of Sulawesi in Indonesia (Engelen *et al.* 1995; Figure A12.2). The lower layer contains a map with the actual use/cover and suitability for various activities of grid cells. The upper layer is a system dynamics model that simulates regional demographic and economic developments. The middle layer is the CA model, that contains the rules according to which land use/cover change in response to regional drivers, on the one hand, and biogeographical characteristics, on the other.

13 Non-Renewable Resources: The Industrial Economy

13.1 Introduction: The Industrial Regime

Until the 18th century agriculture dominated the human economy, with almost everywhere an important role for landed aristocracy, urban military and merchant elites and religious institutions. Around 1750, a new social-ecological regime began: the industrial regime. The beginnings were small, and hardly noticeable to most contemporaries. While human history over the past 10,000 years has been the history of the agrarianisation of the world, the history over the past 250 years has been the history of *industrialisation*. The metabolic profile of the agrarian socio-ecological regime and the industrial one are quite different (Schandl *et al.* 2009; §4.3). In the industrial regime, energy and material use per capita is three to five times higher than in agrarian societies. Population densities tend to be three to ten times higher, energy and material use densities even ten to thirty times higher than in agrarian societies. The fraction of biomass in energy supply is a factor three to then lower.[1]

The industrial regime operates on finite stocks that are produced in nature at rates close to zero on a human timescale, hence the name non-renewable resources. They are often referred to as minerals, from the Latin *minerale*, which means something mined, although later it broadened to 'substances neither animal nor vegetable'. Most importantly and already known and used in antiquity are sand, salt, glass, limestone and of course metals, and for a few centuries also fossil fuels.[2] Mining and processing minerals to make metal objects has a history of several millennia. Ancient Egypt was renowned for its rich gold, copper, silver and tin mines. There has probably been silver (Ag) mining in Greece since 3500 Before Present (BP): In the 5th century, between ten and thirty thousand miners were at work in the mines of Laureion. A reconstruction of the exploitation history of lead (Pb), from measurements of lead concentrations in Swedish lake sediments, put the first mining back to the period 4000–3500 BP. Concentrations rose towards a peak during the

[1] In the industrialised countries, the distinction between agriculture and industry has largely vanished. Agri- and horticulture, livestock, fisheries and forestry have become almost completely industrialised.

[2] Minerals are usually understood as a chemical element or compound, which is also a (mineral) ore in the sense that it contains a valuable constituent, often one of the elements defined as metals. Mineral is the broader class, but I do not make a strict distinction between mineral and metal.

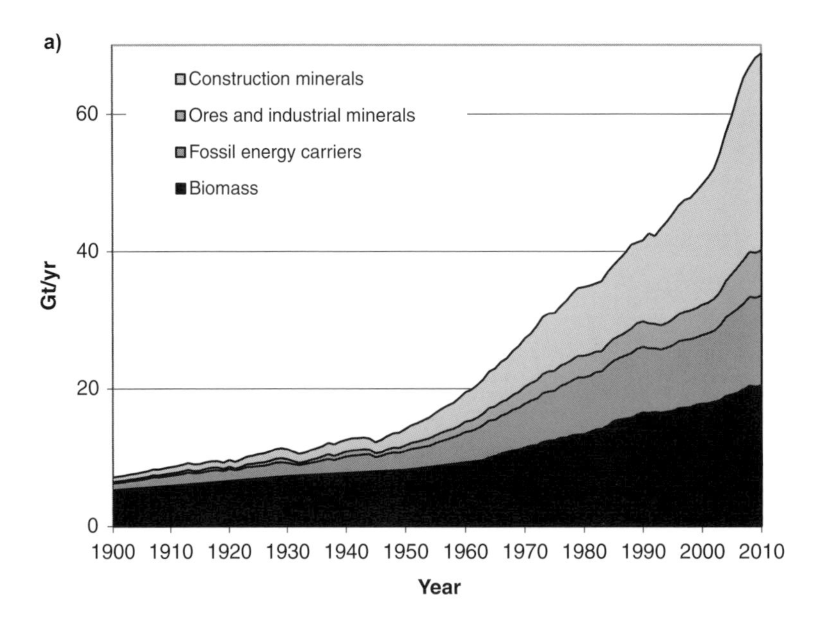

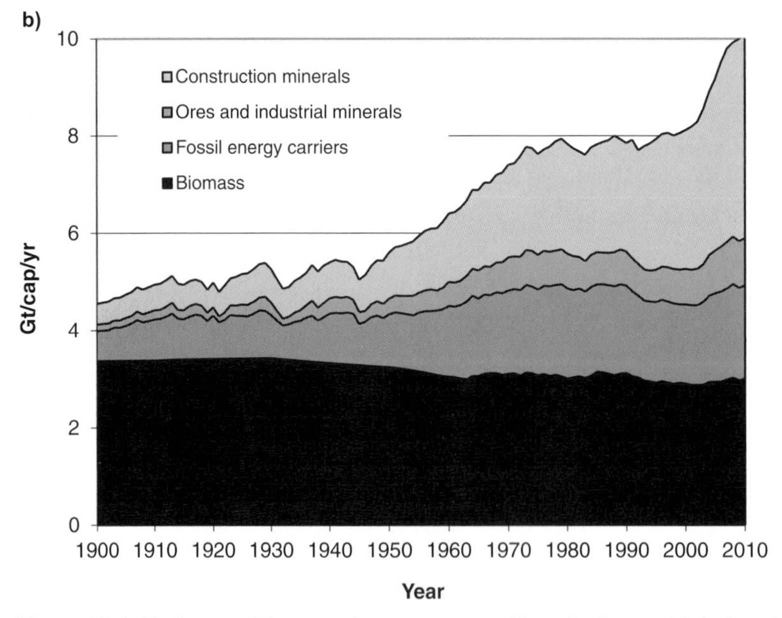

Figure 13.1. Estimate of the most important mass flows in the world during the industrial era, in absolute amounts (upper) and in per capita (lower) amounts (Krausmann *et al.* 2009).

Roman period (2150–1550 BP), declined thereafter and started to rise again with the industrial revolution. It mirrors the European history of lead use of about 300 tons per year (t/yr) in 2700 BP to an estimated 80,000 t/yr around 1950 CE. Mining and processing of minerals for iron (Fe), copper (Cu) and other metals has also increased exponentially. Since 1900, the mass flows associated with fossil fuel use, ores and industrial minerals and construction minerals has increased more than tenfold, whereas the biomass fraction in it has shrunk to one-third (Krausmann *et al.* 2009; Figure 13.1). The worldwide use of five key materials: cement, steel, paper, aluminum and thermoplastics, has increased four- to sixfold between 1960 and 2005 and the exponential growth trend continues (Allwood *et al.* 2011).

The industrialisation process not only changes the material basis but also the economic and social-cultural aspects of life. Unlike the rather static agrarian societies, the industrial economy has a strong endogenous growth tendency stemming from the accumulation of capital and the desire for high returns in combination with pure (science) and applied (technology) knowledge, as is discussed in Chapter 14. Average income rose exponentially, and colonial states and later multinational corporations became the most important institutions. The rational mind and secular values came to prevail in a world of economic rationalisation, a utilitarian ethic and a liberal ideology – which also bred radical communism and fascism (§6.4).

The industrial economy is an everyday experience for an estimated 30 to 40 percent of the present world population. Its centre of gravity is shifting from Europe, the United States and later the Former Soviet Union (FSU) and Japan to emerging economies such as China. Some of its typical characteristics are:

- a delinking between regional biogeography and economic activities, except for natural resource exploitation (minerals and fuels);
- a transition from local and diffuse land-based energy sources (wood-water-wind) to concentrated and globally traded fossil fuel–based energy carriers;
- association of economic growth with core industrial sectors and physical throughput: steel, cement, fertiliser and other materials, and intermediate capital goods;
- societal dynamics driven by resource and capital ownership and the forces of a technological-industrial complex run by urban elites of bureaucrats, managers and technocrats;
- a tendency towards concentration and homogeneity of economic activities, largely driven by economies of scale and scope and preferential attachment mechanisms;
- continuing competition between three dominant institutional clusters: tradition-based communities, nation-states, and private enterprise corporations.

The signs of a next, postindustrial regime are already visible, with a crucial role for information and communication technology (ICT). But large populations in the world are still in the first stages of industrialisation. Therefore, it is widely expected that, in conjunction with increasing population and economic activity, non-renewable resource use and the associated emission flows to the environment will continue to grow in the 21st century.

13.2 Non-Renewable Resource Chains: Extraction

13.2.1 Biogeochemical Element Cycles

The stocks and flows of materials and energy for human use should be seen in the larger context of System Earth. Figure 13.2a shows a conceptual model of the human population and economic subsystem within the finite biosphere. It highlights the role of high-quality solar energy inflow and the Earth and cosmic environment as a sink for low-quality energy outflows. The sun's role in creating chemically stored exergy (fossil fuels and metal-enriched mineral ores) and 'closing the loop' in the form of recycling materials are also shown. From a sustainability perspective, it is not so

Table 13.1. *Estimates of average metal abundance in crust and seawater, of total mobilisation flows and the anthropogenic fraction therein.*[a] *Also, indications of richest ore and specific mineralisability are given.*[b] *The elements for which the distribution is drawn in Figure 13.4 are shown in bold*

Metal	Avg conc in ppm: crust g/Mg (ppm)	Avg conc in ppm: Seawater g/Mg (ppm)	Total mobilisation rate (TMR) Tg/yr	Anthropogenic fraction in TMR % of total	Richest ore grade in crust wt %	q spec mineralisability
Aluminum (Al)	77,440	0.002	309	26	49.2	0.041
Iron (Fe)	30,890	0,002	848	90		0.087
Magnesium (Mg)	13,510	1,290	1,944	3		
Carbon (C)	3,240	28	118,450	9		
Titanium (Ti)	3,117	–	11	54	24.7	0.127
Sulphur (S)	953	905	766	22		
Phosphorous (P)	665	0,1	538	6		
Chlor (Cl)	640	19,354	6,264	2		
Manganese (Mn)	527	0.0002	40	22	43.7	0.201
Zirconium (Zr)	237	–	1.6	78		
Nitrogen (N)	83	150	6,067	7		
Cerium (Ce)	**66**		**0.2**	**56**		***0.15***
Vanadium (Va)	53	0.003	1.6	72		
Zinc (Zn)	52	0.8	20	47	4.3	0.215
Chromium (Cr)	35	0.0003	15	99		0.287
Niobium (Nb)	26	–	0.1	59		
Lithium (Li)	22	0.2	0.2	51		
Nickel (Ni)	19	0.001	2.3	69		0.153
Lead (Pb)	**17**	**0.00004**	**3.9**	**84**	**5**	**0.286**
Copper (Cu)	14	0.001	16	85	4	0.2
Gallium (Ga)	14	–	0.05	54		
Cobalt (Co)	12	0.00005	3.5	6		
Tin (Sn)	2.5	0.00001	0.3	78	1	0.277
Tungsten (W)	1.4	0.0001	0.04	95	2	0.287
Molybdene (Mo)	1.4	0.01	0.8	24		
Uranium (U)	2.5	0.003	0.05	90	0.18	0.2
Arsenic (As)	2	0.004	0.1	63		
Antimony (Sb)	0.3	0.0002	0.1	90		0.375
Bismuth (Bi)	0.1	–	0.01	97		
Silver (Ag)	0.055	0.00004	0.03	58	0.056	0.294
Mercury (Hg)	**0.06**	**0.00003**	**0.07**	**95**	**0.6**	**0.378**
Palladium (Pd)	0.004		0.001	99		
Gold (Au)	0.003	0.000004	0.003	100	0.002	0.298
Platinum (Pt)	0.0004	–	0.001	100	0.0006	0.198

[a] Klee and Graedel 2004.
[b] de Vries 1989.

much the absolute amount of anthropogenic flows but the amount relative to the natural flows that matter. How much do humans interfere and does the interference entail risks for ecosystems and human systems and, if so, how can they be avoided or mitigated?

The elements are chemically or otherwise mobilised in natural and human-induced processes. One can distinguish *sequestration* reservoirs in which nature stores material for long periods and *mobilisation* reservoirs in which material is transferred for much shorter periods (Figure 13.2b). Natural mobilisation flows are associated with crustal weathering, sea spray and primary production by vegetation,

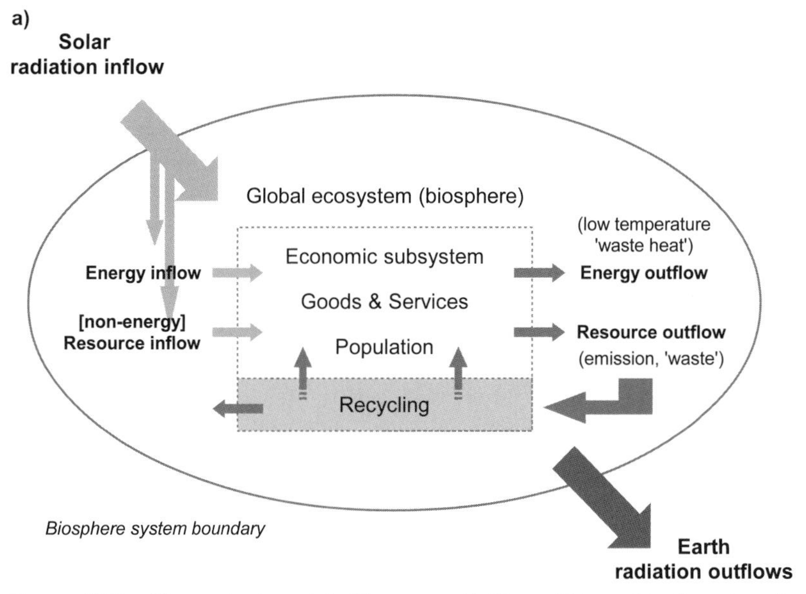

Figure 13.2a. Conceptual model of human activities within an Earth system framework.

whereas anthropogenic mobilisation flows are from mining, fossil fuel combustion and biomass burning. Klee and Graedel (2004) have analysed the stocks and flows of seventy-seven out of ninety-two elements with the aim 'to reflect the average rate at which the state of a material is transformed from passive (e.g., in rock or soil) to potentially interactive (e.g., in industrial products or in vegetation), with a focus on the pedosphere . . . and the near-surface ocean'.

Not surprisingly, the mobilisation rates of the rare 'precious metals' such as platinum (Pt) and gold (Au) and somewhat less rare elements like tin (Sn) and tungsten (W) are dominated by human-induced, not natural flows (Table 13.1). The anthropogenic mobilisation of elements of medium abundance, like vanadium (V), zinc (Zn), chromium (Cr), nickel (Ni), lead (Pb) and copper (Cu) is significant and

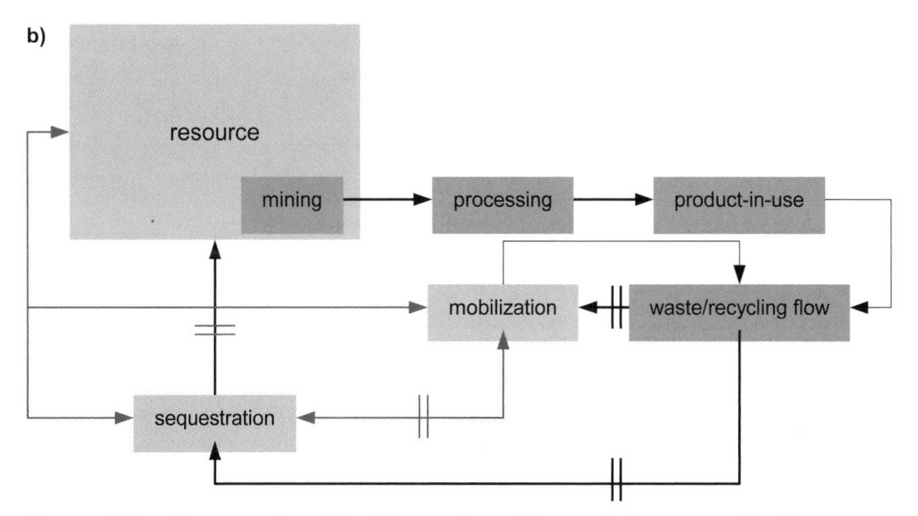

Figure 13.2b. Conceptual model of the stocks and flows of elements on Earth.

more than 65 percent of the total mobilisation flow. Anthropogenic interference with abundant elements like aluminium (Al), titanium (Ti), sulphur (S) and manganese (Mn) are also significant (22 to 54 percent). Some other abundant elements like magnesium (Mg), carbon (C), phosphorous (P) and nitrogen (N) are part of such large natural mobilisation flows that even their massive anthropogenic mobilisation, notably in agriculture and fossil fuel combustion, represents only a small fraction of the total mobilisation rate. Iron (Fe) is the exception here: Human use is an extreme intervention (90 percent) in the natural flows. One possible explanation of the large differences is that nature tends to rely for nutrients on water soluble elements – as in the case of Ca in $CaCO_3$ – whereas the mineral ores and metals used in human structures are predominantly water-insoluble. Data such as these give an idea of the extent of human interventions: The ratio of anthropogenic material flows and natural stocks and flows is one example of a high-level sustainability indicator.

13.2.2 Classification

Two resources are crucial in the industrial transition: non-renewable mineral and fossil fuel resources and environmental sinks.[3] *Non-renewable resources* are exhaustible or finite stocks of a substance, which are depleted upon exploitation. Their natural rate of formation happens at a geological timescale of thousands or millions of years. Examples are mineral and oil deposits. Unsustainable use is, in essence, exhaustion. *Environmental sinks* are natural systems such as soils, river flows and the atmosphere, which are used by humans for 'waste' disposal. They are finite in the sense that the disposed material is not immediately broken down or immobilised. In the meantime, the disposed material may affect the system's functions, often in complex and only partly understood ways. In these respects, environmental sinks are similar to renewable resources like fish and forest (Table 12.1). If the inflow of a particular substance exceeds the rate at which the substance is eliminated and the system regenerates, the sink is used unsustainably. Note that an environmental sink can at the same time be a renewable resource, as for instance in the case of water reservoirs. The two forms of use are usually in conflict.

Mining is at the 'front-end' or 'upstream' part of the resource chain. Mineral and fuel exploitation ('mining industry') make up less than 1 percent of gross world product (GWP) and the share in employment is much less. But the physical impact is disproportionately large. Mining activities account for an estimated 7 to 10 percent of global energy use, emit about 13 percent of global sulphur dioxide emissions, not to mention the numerous other emissions, and impact upon 5.3 million square kilometres (km^2) of forest (WRI 2003). Even more than in agro-food systems, activities are concentrated in a few large corporations. It is said that Glencore, the world's largest commodities trading company in 2010, controls 60 percent of zinc, 50 percent of copper, 30 percent of aluminum and 25 percent of coal through its mining

[3] What is considered a resource reflects the needs, skills and values in a society. For subsistence farmers, farmland and pastures are the relevant resources and bauxite is useless if one does not know about aluminium production and use. Uranium only became a resource once the discovery and control of fission technology had developed in the 20th century. The same holds for environmental sinks. Rivers were only recognised for their cleansing potential when water quality deteriorated.

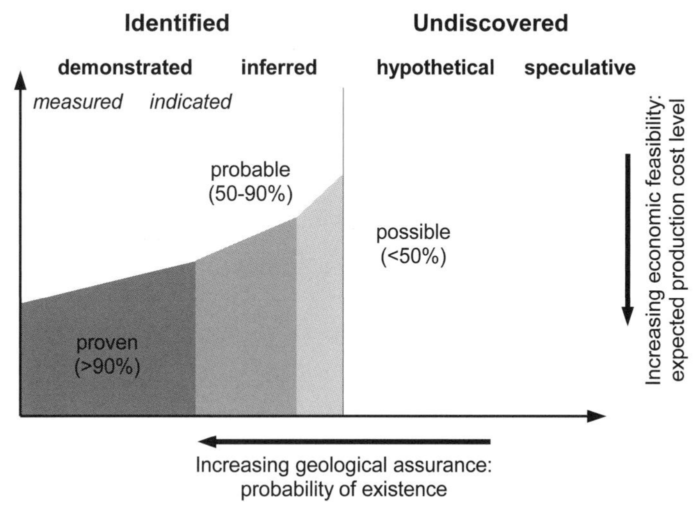

Figure 13.3. Resource classification. The x-axis represents the probability that a certain amount of resource is in place, the y-axis is an indication of the cost at which it is recoverable.

and trading interests.[4] Mining mineral ores and fossil fuels is a complex process of exploration and exploitation in a dynamic environment of markets, prices, technologies, regulations and politics and with large uncertainties and risks (Harris 1984; Yergin 1991). Complete certainty about how much resource and at which cost can be extracted is only known when it is depleted. Added to this are the risks of market disruptions and political conflicts. Investors, therefore, often want – and usually get – high returns on their capital investment.

Resource assessments focus on two *resource characteristics*: the probability of existence and the cost of extraction. Figure 13.3 shows a widely used classification scheme. On the one extreme are the identified reserves. These refer to deposits that are 'proven', meaning they do exist with >90 percent probability, and 'economic', meaning profitable extraction is possible. The other extreme are the deposits that are not yet discovered but which may be inferred from data and models of other resource provinces. The estimates of the extraction costs are, at best, educated guesses. If the exploration history of a region unfolds, the probability that the inferred resources actually exist increases and their status is upwardly appreciated into possible (>10 percent) and later on probable (>50 percent) reserves.

13.2.3 Availability, Exploration and Extraction: Two Models

In the context of sustainable development, the emphasis is usually on depletion and a key question is: How big is the resource or, in expert jargon, what is the *geological resource* base? First, I look at mineral ores. Solid evidence on resource ore quantity and quality is limited by its very nature to those regions that have been explored. However, it is possible to infer more speculative occurrences with hypotheses about

[4] The company, founded in 1974, has been and still is surrounded by stories of fraud, tax evasion and violation of human rights and environmental legislation. It is also an important player in world grain trade and suspected of speculation on the grain market.

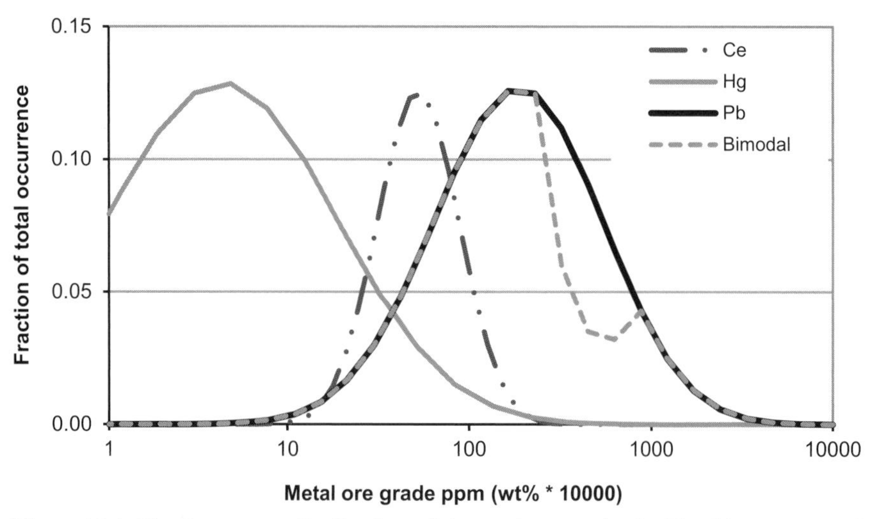

Figure 13.4. The frequency distribution of three elements in the Earth's crust as calculated with the CAG model. The curves depict the fraction of the estimated total amount at a given concentration interval. The dashed curve shows the possibility of a bimodal distribution.

how ores have formed in geological time. In 1954, the geologist Ahrens proposed a 'fundamental law of geochemistry': The concentration or grade of an element is lognormally distributed in a specific igneous rock. This – not uncontroversial – crustal abundance geostatistical (CAG) model makes it possible to estimate the geological resource in the form of a *grade distribution curve* (Appendix 13.1). This is important because the extraction cost, in money and energy units, is to a large extent determined by the ore grade. The outcome of such an estimation is the lognormal distribution shown in Figure 13.4 for the relatively abundant lead (Pb), the very rare mercury (Hg) and the rare earth element cerium (Ce). You remember the notion of exergy introduced in §7.3? The high-concentration lead deposits have a chemical potential with respect to the average crustal abundance that gives them a non-zero exergy content. In other words, natural processes, fed by nuclear energy on the Sun and Earth, did the separating work we benefit from (Appendix 13.1; Table 7.2).

The CAG model can put resource scarcity in a long-term perspective. For instance, a news item such as 'China's export restraints on rare earth elements has inflamed trade ties' is clearly to be interpreted in a short-term market and trade perspective and not (yet) considered a long-term scarcity concern. But there are at least two caveats. First, the analysis does not include elements in seawater. Although the concentrations are low, the amounts are huge (Table 13.1).[5] Second, the evidence from mining metal ores up to now does not exclude the possibility that, for chemical or physical reasons, the distribution differs from a binomial one. It can turn out to be a bimodal one, for instance – the dashed curve in Figure 13.4. This would lead to a very different estimate of the amounts of metal available at low concentrations.

[5] Some elements in seawater get concentrated by natural processes, such as manganese (Mn) nodules on the ocean floor. Sometimes their increase is substantial on the scale of decades, which make such occurrences nearly a renewable resource. They are a prime example of common pool resources (CPRs) (§5.4).

There are other ways to estimate the geological resource base and its life cycle. A widely known one is the *logistic growth life cycle model* (Appendix 13.2). It was first stated in 1956 by the American geologist King Hubbert and is based on the exploration and exploitation history of a region and has been applied mostly for oil and gas. Using past oil exploration and exploitation statistics for the United States (48) (onshore and offshore of the lower 48 states without Alaska), King Hubbert found that the discovery rate per $ spent on exploration declined and forecasted that output in the United States (48) would peak around 1970. Since then, United States (48) oil production has declined within 5 percent of this prediction. The discovery and subsequent exploitation of huge oil fields in Alaska retarded the decline for a few years. But since 1986, the decline has continued and the United States has become ever more dependent on imports – from Canada and Mexico, amongst others. Accumulated oil output in the United States can accurately be simulated with a logistic trajectory, with minor deviations during the period of the oil crises and high oil prices of the 1970s and 1980s (Figure 13.5a). There is a delay of only ten years between accumulated production with and without reserve, which means the reserves have always been equal to about ten years of production.

In retrospect, King Hubbert's prediction looks surprisingly correct, and his model seems a useful metamodel. However, the explanation of the peak is not necessarily a geological constraint due to depletion. It may also reflect the strategy of U.S. oil companies to look abroad for cheaper crude oil supplies and to import oil and oil products. For instance, in the Cold War era, the U.S. government restricted for military reasons oil imports to less than 10 percent of oil use, which led to a relatively deep depletion of U.S. resources and a cheap Middle East oil glut in Europe (Yergin 1991).

Table 13.2 contains data on fossil fuel reserves and production for a couple of countries, and there are clear signs of the occurrence of a life cycle profile in reserve and production. For instance, production since 2000 in four countries outside the Middle East (Mexico, Indonesia, United Kingdom and Norway) is probably past their peak. Their share in world oil production declined from 23 percent in 1995 to 15 percent in 2009, although this too may partly be a reflection of market circumstances and strategic reasons. For these countries, a decline in production and reserves may pose serious challenges for economic development and force them into a transition to other sources.

In principle, the logistic model is also valid for natural gas. Empirical data for reserves and production of natural gas in The Netherlands, however, deviate significantly from a logistic growth path (Figure 13.5b). The profile is more plateau-like and this is probably due to the fast upward appreciation of the gas reserve in the giant Groningen field in the 1960s. For coal and mineral resources, the logistic model has limited validity, because local exploitation depends to a large extent on other factors than depletion.

Many experts and nonexperts speculate these days about the time of maximum world oil production – the global 'oil peak'. One reason is that almost two-thirds of world conventional oil production is from so-called giant fields (>500 mln barrel of oil equivalent (bbl) or about 3 exajoule (EJ)), most of these having been discovered decades ago and already for years declining in output. King Hubbert's model is often applied to calculate the world peak oil date. However, there are reasons to be

Table 13.2. *Resource characteristics for some countries/regions in the world, 2009*[a]

2009	USA	Japan	EU-27	Korea, Rep	S. Arabia	Russian Fed
Population (mln)	307	128	499	49	25	142
Income (PPP U.S.$/cap, 2009)	**46,437**	**32,433**	**30,543**	**27,195**	**23,421**	**18,938**
Oil, proved reserve (EJ)	13.5		3.0		126.1	35.4
Oil reserve as % of world proved reserve	2.1		0.5		19.8	5.6
Oil, production rate (EJ/yr)	3.4	0	1.0	0	4.6	4.8
RPR (yr) – Oil	10.8		8.2		74.6	20.3
Natural gas						
Coal						
Import/use ratio – Oil						
Natural gas						
Coal						
Military expenditures (% of GDP)	4.3	0.9	1.8	2.8	8.2	3.5

2009	Mexico	Brazil	China	India	Vietnam	Nigeria
Population (mln)	107	194	1.332	1.155	87	155
Income (PPP U.S.$/cap, 2009)	**14,341**	**10,429**	**6,838**	**3,275**	**2,956**	**2,149**
Oil, proved reserve (EJ)	5.6	6.1	7.1	2.8	2.1	17.7
Oil reserve as % of world proved reserve	0.9	1.0	1.1	0.4	0.3	2.8
Oil, production rate (EJ/yr)	1.4	1.0	1.8	0.4	0.2	1.0
RPR (yr) – Oil	10.8	17.4	10.7	21.1	35.7	49.5
Natural gas						
Coal						
Military expenditures (% of GDP)	0.5	1.5	2.0	2.6	2.4	0.8

[a] *Source*: BP, en.wikipedia.org/wiki/List_of_countries_by_military_expenditures.

cautious. First, unlike the United States, several regions in the world are relatively unexplored and the model gives no clue about the size of the ultimately recoverable resources. For instance, under the ice layers, Greenland may have 10 to 30 percent of the as yet undiscovered oil and gas resources.[6] Second, in several regions there are large reserves of unconventional oil and gas deposits, the so-called oil and gas shales and tar sands. Canadian oil sands alone amount to the equivalent of at least 10 percent of total conventional oil reserves. Another reason to be cautious is the limited reliability of the statistical data on exploration and production. Major stakeholders have strategic reasons to manipulate the data: oil companies for reasons of stock market value and price speculation, and governments of Saudi Arabi, Russia and other countries in order to optimise their revenues. What can be said with near certainty, however, is that the world community faces in the next half century a transition away from oil and with gas as an important transient fuel.

[6] A dramatic story is unfolding in Greenland and other parts of the Arctic, where oil companies, governments and NGOs are preparing themselves for the future battle when large oil and gas deposits are discovered. The first international disputes about territorial claims have started. Will global warming facilitate the exploitation? Can oil spills be avoided? Are indigenous cultures threatened?

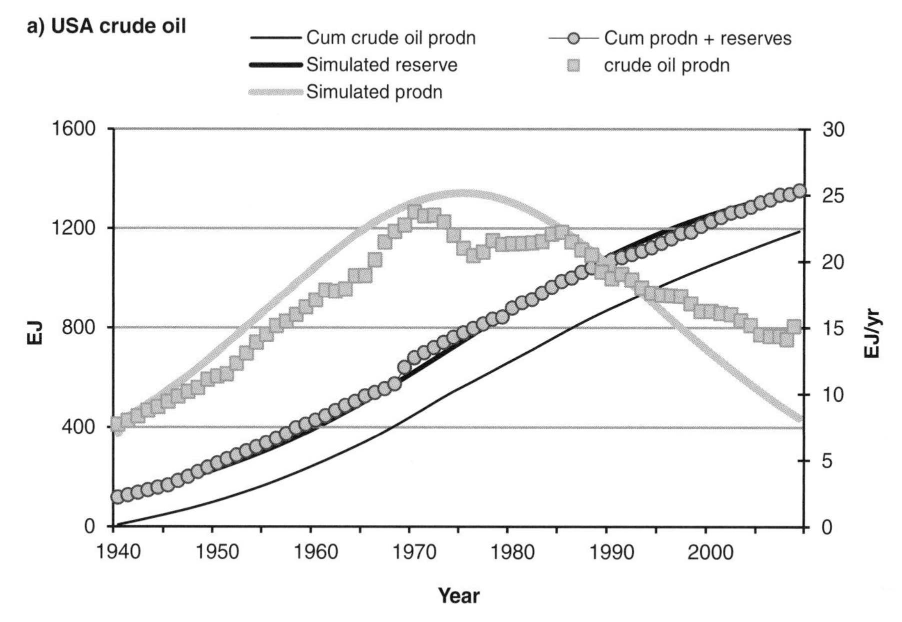

Figure 13.5a. Oil accumulated production, reserves and annual production rates for the United States (source of data: TEPD 2010, www.bp.com).

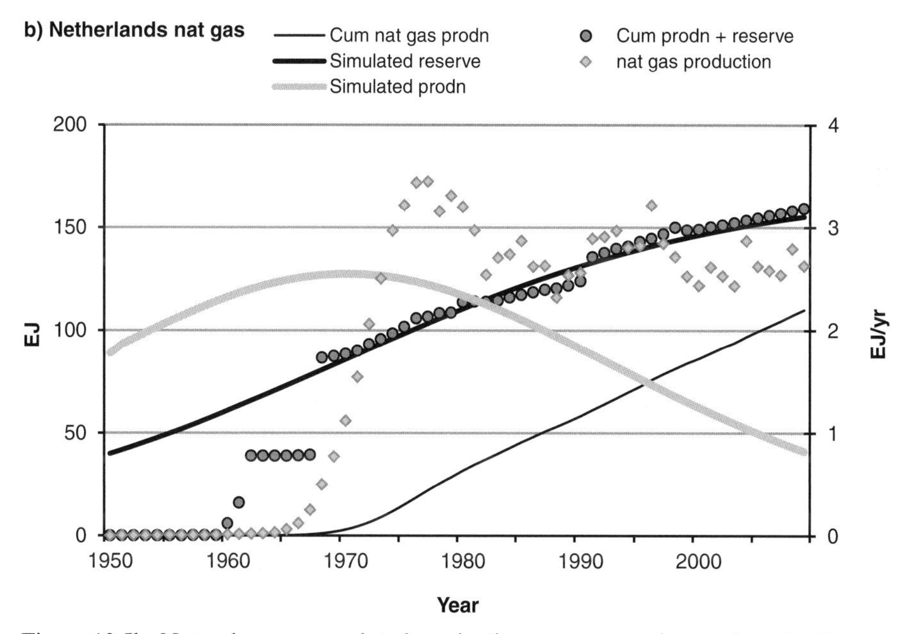

Figure 13.5b. Natural gas accumulated production, reserves and annual production rates for The Netherlands and some other countries (source of data: BP and Ministry of Economic Affairs).

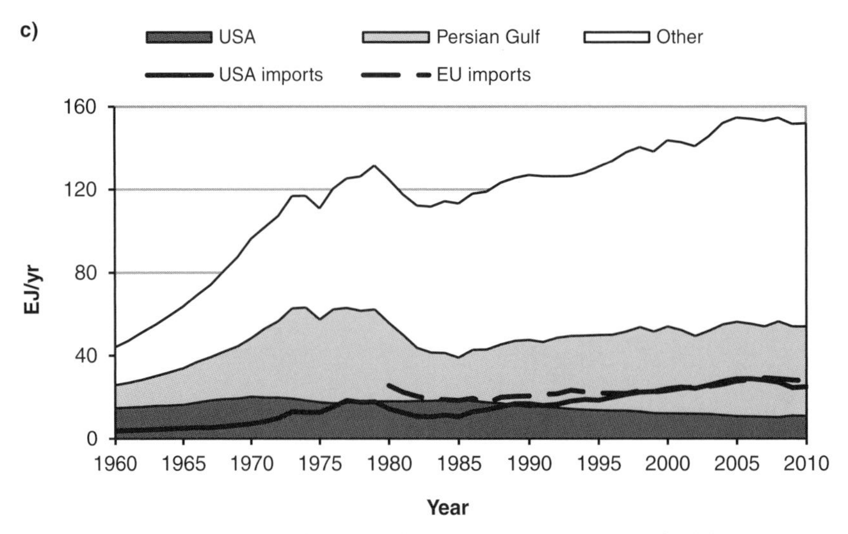

Figure 13.5c. World crude oil production 1960–2010 for the United States, the Persian Gulf and other countries. The lines show the crude oil imports for the United States and the EU. Data for 2010 are provisional (source of data: TEPD 2010, www.bp.com).

13.3 Elementary Resource Economics

13.3.1 Supply Cost Curves

Minerals and fuels exist as deposits: rock layers with varying ore grades, subsurface layers of coal or subsurface porous layers filled with trapped oil or gas. Exploration produces knowledge about the probability of a deposit to exist and, in combination with production tests, about the *cost* at which it can be extracted. Mining corporations assess the economic value of a deposit along at least two dimensions:

- *technology and production cost*: given the existing exploitation technology, at which cost can the resource be produced? Costs depend on depth and composition of an ore deposit, thickness and depth of a coal layer and depth and size distribution of oil and gas fields. The separate deposits or fields in a region are usually ranked in order of increasing extraction cost, which results in the so-called long-term supply cost curve (SCC)[7];
- *processing and marketing*: the resource, once extracted, requires processing, transport and upgrading. Cost and price of marketable products will, therefore, depend on product quality and standards, distance to markets and safety and environmental regulations.

The larger picture has to be constructed from the, mostly confidential, data on ore and coal deposits and oil and gas fields by mining corporations. There is a difference, though, with food: Ore and fuel exploitation is inherently more global because of the concentration of 'value' (exergy) by natural processes. It explains the early globalisation and concentration of non-renewable resource markets and business.

[7] The resource deposit, technology or substitute at which cost no longer increases upon extraction is called the 'backstop' resource. In economics, the increase of the extraction cost is associated with differential (or Ricardian) rent, as introduced in the 19th century by Ricardo for land resources.

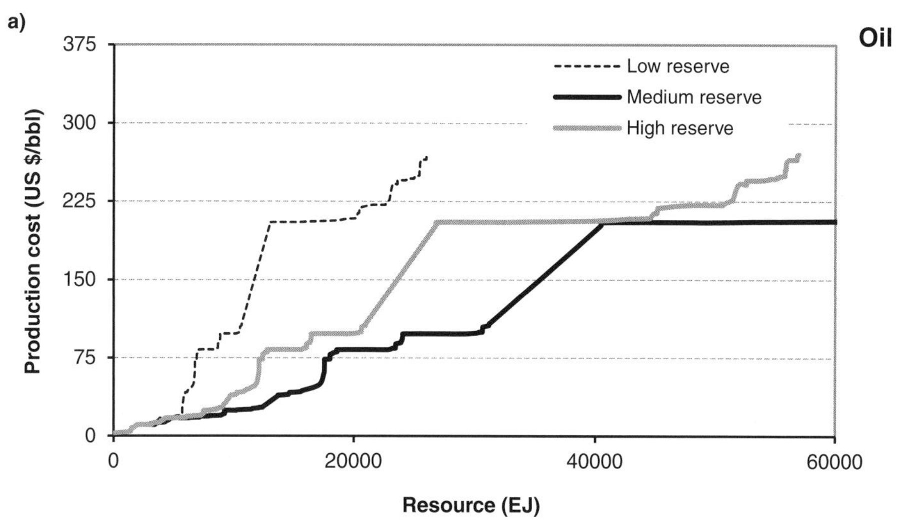

Figure 13.6a. Long-term cost supply curve estimates for world oil (1 US$/GJ ≈ 7 US$/bbl). World oil use in 2005 amounted to over 0,16 ZJ (van Vuuren 2007/USGS).

In practise, resource availability and depletion is often evaluated in terms of the *reserve production ratio* (RPR), which is the ratio between (proven) reserves and (actual) annual production. It is measured in years and used by countries and corporations. Table 13.2 gives some RPR estimates. However, the use of the RPR is misleading as a scarcity indicator because reserves are a function of exploration, which is often driven by price (expectations). If prices go up, exploration is intensified and new discoveries and re-appraisal of previously uneconomic deposits add to the reserves.[8] A better way to evaluate the resource situation of a country or of the world at large is the aforementioned SCC. In Figure 13.6, it is shown for conventional and unconventional world oil and gas resources as of 2005. For any given resource size estimate, the curve indicates the estimated marginal cost. The uncertainty is indicated by showing upper and lower estimates. The right-hand parts in Figure 13.6 are highly uncertain and speculative, which is reflected in the larger spread in the estimates. Besides geological uncertainties, there is the ignorance about as yet unknown technologies and environmental constraints.[9]

What do these curves tell us? In 2005, world crude oil extraction amounted to about 155 EJ or about 27 billion barrels (bbl) (Figure 13.5c). The SCC indicates that crude oil reserves at <50 $/bbl would last at the 2005 extraction rate for another seventy years.[10] It may also be 100 (lower SCC) or 35 (upper SCC) years (band in Figure 13.6a). The reserves at <100 $/bbl would last 70–190 years at the 2005 extraction rate. But, of course, at these costs several alternatives become very competitive. Organisations like the International Energy Agency (IEA) expect in the 2011 World

[8] Reappraisal is the process of re-evaluating the resource in existing and abandoned mines and oil/gas fields, with the new insights and techniques. For instance, novel techniques to recover more oil out of the carrier sediment or better seismic exploration techniques to more accurately assess the 3-D reservoir characteristics.

[9] The curves in Figure 13.4b can also be considered SCC inasfar as there is an inverse proportional relationship between cost and metal concentration.

[10] In other words, the RPR is 70 years at a cut-off extraction cost level of 50 $/bbl.

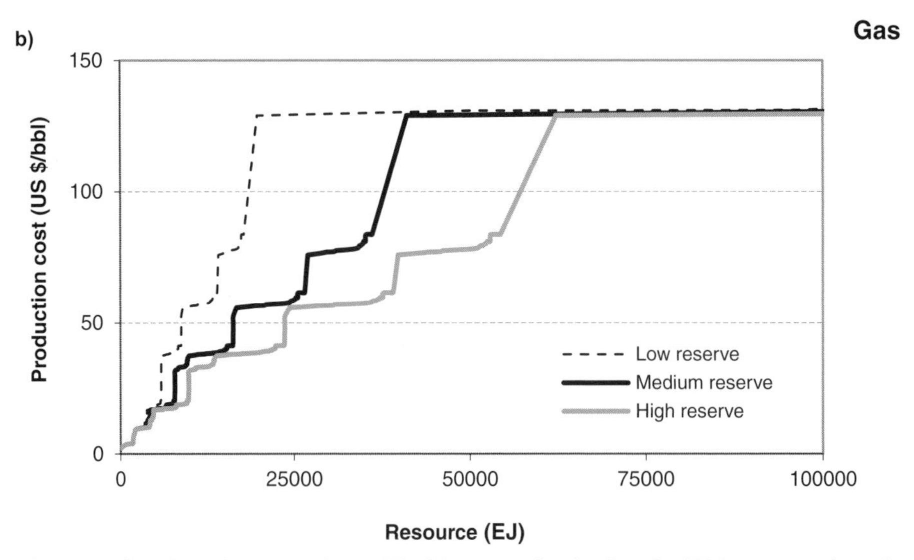

Figure 13.6b. Idem, for natural gas. World gas production/use in 2005 amounted to almost 0,11 ZJ in 2005 (van Vuuren 2007, www.usgs.gov).

Energy Outlook (WEO) that world oil production rises 15 percent between 2010 and 2030, with most expansion coming from unconventional deposits. Other agencies are more pessimistic and expect a world oil peak to come before 2020 (TDBT 2004; ZTB 2010). The uncertainties give room for controversy and speculation and different interests and worldviews influence the analyses and expectations.[11] Whatever may turn out to be correct, there will be rising tensions in the coming decades in the form of volatile and high oil prices and intense strategic games and political conflicts.

A similar situation exists for natural gas, but it is widely believed that the resources are still larger, that there are interesting opportunities to convert gas to liquids to curb oil shortages, and that there may be a backstop in the form of huge amounts of unconventional gas (Figure 13.6b). Most experts in business and government count on natural gas as a key transition fuel for the next decades because of its relative abundance, the potential it offers for further efficiency gains and its relatively low carbon content. There are downsides, too. The necessary infrastructure, in the form of long-distance pipelines and liquid natural gas (LNG) plants and ships, is capital-intensive, and the production of unconventional gas in the form of shale gas – and similarly of shale oil and of oil from tar sands – is not only more costly but is also expected to cause serious environmental risks from large use of water and chemicals.

The dynamics of resource exploration and exploitation is summarised in Figure 13.7. Demand drives investment in capital stocks for exploitation. This relative short-term mechanism is supplemented with three longer-term loops. The first one is *depletion*: The unit cost tends to go up with accumulated production, which has a negative influence on demand via the price elasticity. Simultaneously, companies invest part of the revenues and profits into cost-decreasing *R&D and innovations*. This tends

[11] For instance, a 2010 WikiLeaks rumour indicated that Saudi Arabian oil reserves are overestimated with an amount in the order of 2,000 EJ.

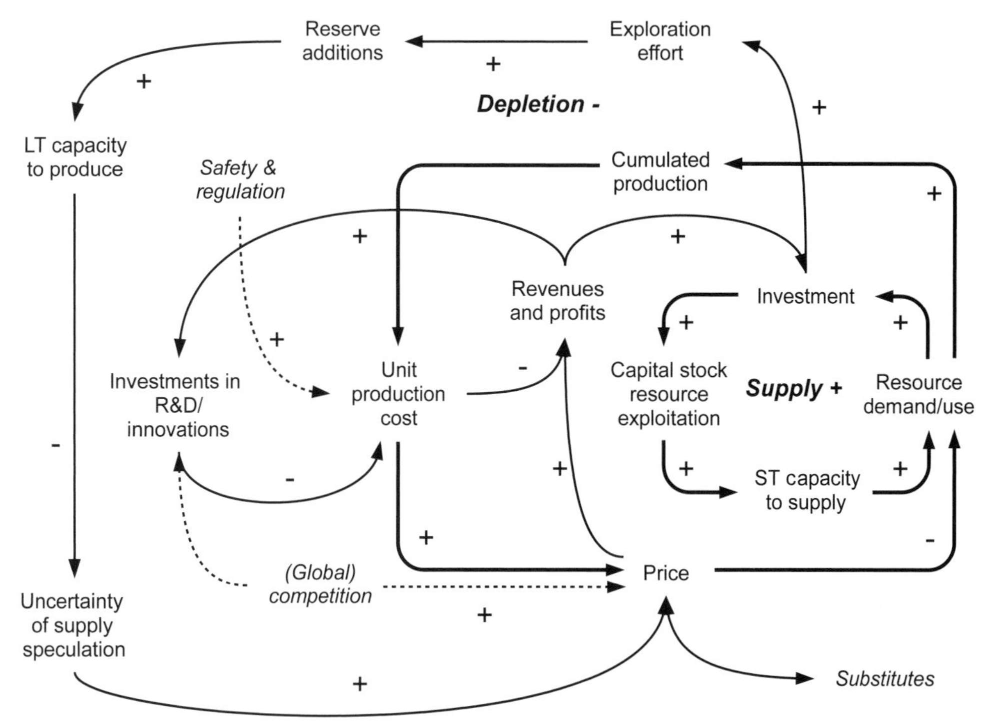

Figure 13.7. The most important dynamic mechanisms that determine long-term trends in resource supply, costs and prices. The items in italic are influences from other subsystems.

to stabilise or lower unit cost, if there is sufficient competition on (global) markets. Another part is invested in *exploration*, when the reserve base is considered too low or when newcomers enter the market. This permits a constant or increasing identified reserve. Fourth, the more stringent *safety and environment regulations* tend to increase unit costs, although these are difficult to separate from the other mechanisms. The resulting price changes will probably be upward and have an influence on substitutes such as biomass-based fuels or plastics. If the reserve base becomes less than desired by companies or governments and there is also limited competition, one can expect price fluctuations because of uncertainty and speculation. Geopolitical and technological developments make the story still more complex.

13.3.2 Innovation: The Learning-By-Doing Mechanism

The SCC not only reflects the observation that costs of minerals and fossil fuels increase because of depletion but also, in the last decades, because environmental and social costs are increasingly internalised in the cost, for instance, via *environmental and safety regulations*.[12] Therefore, one expects costs and prices of non-renewable resources to rise. Why then have the production cost of most metals and fossil fuels gone down in the course of the 20th century, despite an accelerating extraction rate and more stringent regulation (Allwood *et al.* 2011)? One answer is: *exploration*. Resources are not necessarily discovered in order of cost and newly discovered fields may turn out to be cheaper to exploit than those already under

[12] Indeed, it is argued that lax implementation in a competitive and greedy industry are the cause of disasters like the BP oil spill in the Mexican Gulf in 2010.

Box 13.1. *Carbon sources and sinks.* As discussed later on in this chapter, one of the environmental issues is the change in the radiation balance of the Earth as a consequence of rising concentrations of so-called greenhouse gases, notably carbon dioxide (CO_2). Because CO_2 is emitted upon burning fossil fuels, notably coal because it has a two times higher emission per unit of energy than natural gas, mankind faces a somewhat weird source-sink problem. On the one hand, people desperately long for cheap carbon in order to sustain or enter the industrial era with its cars, appliances and so on. On the other hand, every kilogram (kg) of coal, oil or gas discovered is almost certain to cause an emission in the order of 2 to 4 kg of CO_2, which will remain in the atmosphere for on average 100 years.

An affordable and profitable energy supply system requires more carbon and oil companies and governments frantically search for carbon all over the world in order to increase the existing proven reserves of about 3,400 gigaton carbon (GtC). Reducing the risk of large-scale damage from climate change, for instance, according to the 2°C target of the EU, implies that the cumulated carbon emissions in the 21st century should not exceed the range of 800 to 1,000 GtC. The reconciliation of this source-supply quandary may become the ultimate challenge to human aspirations.

exploitation. An equally important answer is: *technology*. Extraction costs tend to fall over time because of innovations ('technical progress'). Innovation-driven cost reductions are often spectacular, as in offshore oil and gas production – although the move towards offshore fields was itself the consequence of depletion of onshore resources.

The cost-decreasing effects of innovations, both incremental and breakthrough, have been generalised into the concept of the *learning or experience curve.* It is an empirical fact that key performance indicators of industrial processes – such as cost per unit of output – decline over time because of the incremental technical and organisational innovations and increases in scale.[13] In-depth analysis shows the complexity and contingency of the underlying processes (Wene 2000; Junginger *et al.* 2010). Learning-by-doing, as it is called, is incorporated in many models according to a simple power law:

$$\frac{y}{y_0} = \left(\frac{x}{x_0}\right)^{-p} \tag{13.2}$$

for a cost/performance variable y and an learning-by-experience variable x. The suffix 0 indicates the time at which learning is supposed to have started. Usually, the cost/performance variable y is specific investment or operational cost and the variable x is the cumulated output over a certain period. The learning rate p can also be expressed by the *progress ratio* ρ, which indicates the factor with which the cost

[13] Scale in industrial production can be either the size of plants, as with power plants, or the number of items being produced, as with solar photovoltaic (PV) panels. Both aspects contribute to cost reductions because of standardisation, shared overhead costs and so on.

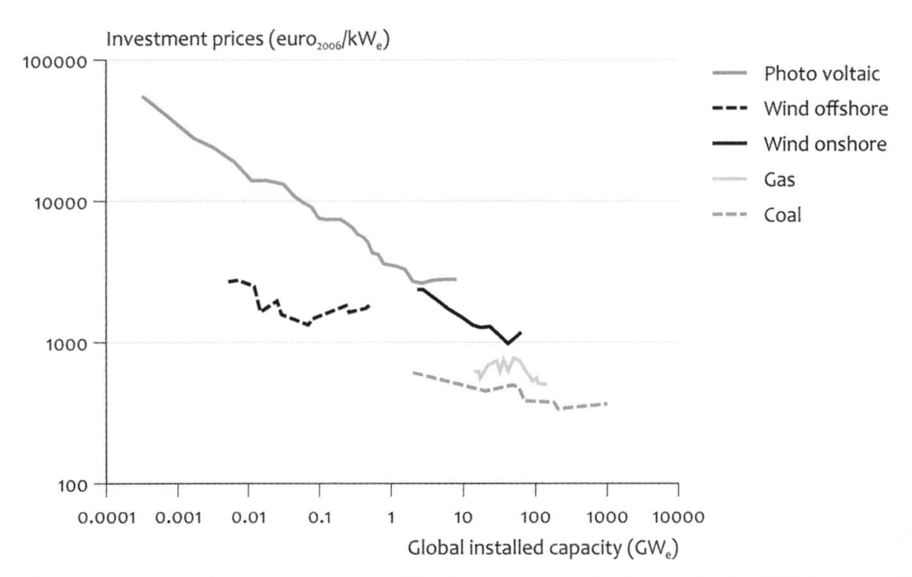

Figure 13.8. Learning-curve relationships for energy technologies in the EU (source: PBL, Junginger *et al.* 2010).

measure y decreases on a doubling of x.[14] Equation 13.2 has gotten the status of a law, although it rather is a metamodel.

Data for seven electric power generation technologies are drawn in Figure 13.8 in a log-log plot. It approaches a straight line over a large domain of cumulated installed capacity, as it should for a power law, but the slope differs for different technologies and changes over time. Progress ratios are for these technologies in the range of 0.65 to 0.95. The implications are significant. If the cost of solar-photovoltaic (PV) panels, for instance, decreases with 10 percent ($\rho = 0.9$) for the first doubling of output (in MWpeak), then you need one more doubling of the cumulated output for another 10 percent decline in costs. If the first doubling took 1 year and output remains constant, the next 10 percent cost reduction takes two years. But the next 10 percent cost reduction will take four years at constant output – and the next one eight years! This explains why learning-by-doing is only fast in time if production grows fast as well and why otherwise massive subsidy is needed to stimulate demand.

Thus, long-term cost trends are the effect of two mechanisms: depletion and learning (Figure 13.7). Because the costs of most metals and fossil fuels have declined during most of the 20th century, the learning effect appears to have dominated. Recent trends suggest that for several minerals/metals and fossil fuels the two effects increasingly offset each other and costs remain more or less constant for long periods. But reliable cost data are scarce and most available data are about *price*. Ideally, there is an equilibrium between demand and supply and the price is closely correlated with (marginal) cost. But the world market for metals and fuels fluctuates between a sellers' market when demand exceeds (perceived) supply and prices tend to rise,

[14] If the logarithm is taken on both sides, a plot on a log-log paper gives a straight line. It is also seen then that $\rho = 2^{-p}$ ($p = -\log(\rho)/\log(2) > 0$).

and a buyers' market when supply exceeds demand and prices fall. One reason is the inevitable system delays due to long construction times of new installations and mines (5 to 10 years) and fluctuations in demand in response to changing economic activity (§2.3). For instance, the rapidly increasing demand for steel and oil in China was a major cause of oil price volatility in the early 21st century. Such volatility spreads throughout the economy, because prices of natural gas and oil are coupled and fossil fuels are important inputs for most industrial products. Speculation makes it worse.

In the longer term, most experts anticipate higher prices, not only because of depletion but also in view of (geo)political tensions and environmental policies. Political upheavals such as the fall of the Iron Curtain in 1989 and the Gulf Wars and Middle East uprisings create uncertainty and influence business profits and economic policy. Important from a sustainability perspective are the trends in environmental indicators and policies. More stringent regulation to protect biodiversity, in combination with human rights protection, affects metal prices. Strong and consistent greenhouse gas emission taxes raise the fossil fuel prices directly, but also indirectly (§14.3).The effects of these factors are mitigated or exacerbated by trade and it is not surprising that mineral/metal and fossil fuel prices experience significant and hardly predictable fluctuations.

13.3.3 Optimal Depletion: The Resource Curse and Resource Security

Resource depletion should be judged in the broader context of the transition towards sustainable metal and energy use systems. Economists have examined the optimal resource extraction rate from a macroeconomic point of view (Dasgupta and Heal 1979; Ströbele 1984). In a capitalist setting of private ownership, an optimal extraction path H(t) to deplete a finite amount of resource R_{ult} is found for:

$$\max \int_0^T e^{-rt} \left[(p - c(CP)).P \right] dt \tag{13.3}$$

under the condition that $dCP/dt = -P$ and $CP \leq R_{ult}$, with CP cumulated production and P production. T is the planning horizon, r is the discount rate. In words, the optimal extraction path is when the discounted net profit $e^{-rt}(p-c)P$ is maximised over the project duration (§10.4, §12.2)· If costs are known as a function of depletion, the optimal path can be calculated.

In an ideal world, it can be shown that an economically optimal resource extraction path requires an annual percentage change in the resource price equal to the discount rate during the exploitation period.[15] In formula: $dp(t)/dt = r$. This outcome has become known as the *Hotelling rule*, after the economist Hotelling who formulated it in the 1930s. Therefore, the price should rise exponentially unless the extraction cost goes down or a cheaper alternative or a constant cost 'backstop' option shows up within the exploitation period. For low-risk projects and in

[15] The price p is the net-price which can be considered to represent rent or royalty and is given by $P = p + MC$ with P the gross market price and MC the marginal cost of production, such as the cost to produce the last unit of the annual flow (Perman *et al.* 2003).

Box 13.2. *Theory and observations.* In economic theory, the relation with empirical observations is a difficult one. The Hotelling rule has never been unambiguously confirmed from empirical data. One reason is, as often in economic and social sciences, that neither the discount rate nor the net profit can be measured – they are both aggregates to be estimated with proxies from incomplete data. 'The Hotelling rule is an economic theory... a theory is not necessarily correct... [a theory] may fail to "fit the facts" because it refers to an idealised model of reality that does not take into account some elements of real-world complexity. However, failing to fit the facts does not make the theory false; the theory only applies to the idealized world for which it was constructed... The history of attempts to test the Hotelling principle is an excellent example of the problems faced by economists... many of the variables used in our theories are unobservable or latent variables. Shadow prices are one class of such latent variables. The best we can do is to find proxy variables for them. But if the theory does not work, is that because the theory was poor or because or proxy was not good?' (Perman *et al.* (2003) 527–529). This situation is often dealt with in the form of metamodels and stylised facts, as is discussed in §8.6.

situations of abundant and cheap capital (small r), the price increase will be smaller than for high-risk projects and capital scarcity (large r). The Hotelling rule points at the existence of *resource scarcity rent*, which comes on top of the cost to extract and produce it. Such rents are difficult to measure and sometimes huge, and their appropriation is a continuous battle between national and international business elites and governments.

Optimal resource exploitation models are also applied for society at large, with the optimal path being the one that maximises intergenerational social welfare (§14.1). It remains mostly academic but raises a couple of normative issues such as the form of a societal welfare function and the value of the discount rate (Dasgupta 2008; Stern 2008). But the relation between economic performance, resources and technology is far more complex than can be dealt with in these models. Nevertheless, optimal resource algorithms are widely used, for instance, in estimating the consequences of climate change impacts and policies, but the methodology, assumptions and outcomes are controversial (IPCC 2007).

Real-world behaviour is often far from economically optimal and rational (§10.4). Therefore, theoretical analyses should be complemented with stories from the real world. The extraction of rich metal ores and fossil fuel deposits was and still is as much a matter of military adventure and political power struggles as it is of economically rational behaviour (Yergin 1991). The major mining companies are from countries like Australia and Canada, but many of the richest ores and, therefore, most profitable mining opportunities are currently in countries in the periphery of the industrial centres. The mining activities often generate large revenues for the host country, but they also disrupt local cultures and communities and cause environmental havoc. In countries with authoritarian or weak governance structures, the revenues and job opportunities tend to go for the larger part to the ruling elites – at least, this is what journalists found out about the mines in Papua, Niger and other places and what is broadcast about the bank accounts of the former dictator families of some Asian, African and Middle East countries. This and the bad working conditions of the miners and world market price fluctuations cause regularly social unrest. This too is part of unsustainable development.

One particular hypothesis has received much attention, namely that a state's reliance on either oil or mineral exports tend to make it less democratic.[16] This is called the *resource curse* (RC) hypothesis. An analysis of 113 countries in the period 1971–1997 suggests a link between the large oil wealth of a nation and its governance: '[there are] three causal mechanisms that link oil and authoritarianism: a rentier effect, through which governments use low tax rates and high spending to dampen pressures for democracy; a repression effect, by which governments build up their internal security forces to ward off democratic pressures; and a modernisation effect, in which the failure of the population to move into industrial and service sector jobs renders them less likely to push for democracy' (Ross 2001). Recent econometric analyses for a large sample of countries for the period 1970–1995, confirm that natural resource abundance, with mineral production share in GDP as a proxy, correlates negatively with economic growth. It suggests a resource curse. However, the effect reverses sign if other possible explanatory variables such as corruption, investment, openness, terms of trade and schooling are included. In a long-term perspective, there is evidence that abundant natural resources tend to crowd out other income-supporting activities, notably savings (Papyrakis and Gerlagh 2004, 2006).

Not surprisingly, a peculiar relationship has evolved between natural resources, in particular oil and gas, and *arms*. Huge amounts of money are flowing into the Middle East region from oil and gas export revenues. This influx, itself unstable and a source of dependency, is part of an arms race. In the 1990s, between 20 percent and 40 percent of annual oil export revenues in Saudi Arabia were spent on weapon imports, mostly from the European Union (EU) and the United States. In 2009, military expenditures of Middle East countries as % of GDP were between 3.4% for Syria and 8.2% for Saudi Arabia (Table 13.2). Buying and selling arms stems from the understandable need of protection, but it may be one of the great pseudo-satisfiers (§6.2).

Another issue that enters the equation is the concern about *energy security* in industrialised and industrialising countries (Kruyt *et al.* 2009). The dependency of a number of states on oil and gas imports has been growing for decades and keeps increasing (Table 13.2; Figure 13.5c). The interests at stake are huge, even apart from environmental concerns like climate change. For instance, the EU's dependence on energy imports is increasing again, because the indigenous resources have rapidly been exploited after the 1972 and 1979 oil crises. In 2005, the EU imported about 50 percent of its energy with 45 percent of oil from the Middle East and 40 percent of natural gas from Russia. It is expected to rise to 70 percent by 2025 (OECD/IEA 2010). The monetary value of energy imports is still only 6 percent of total imports, but the physical dependency can have serious economic and social risks in the face of global depletion. A similar situation exists for the United States,

[16] The high-income country version is the Dutch disease, according to which the easy revenues from a large and cheap to exploit resource cause governments to overspend, make citizens complacent and incur inflation. Of course, one may assume that the equivalent of windfall profits for the mining corporations have similar, though less publicised effects.

where oil imports amounted to twice the indigenous production. Japan is even more dependent on fossil fuel imports. China and India are probably soon to follow. The energy transition to noncarbon energy sources such as hydropower, nuclear power and renewable sources (wind and solar) reduces the tensions, but oil price volatility slows the transition down because investors avoid the risks in the absence of clear signals of high future oil prices. The same is true for the materials transition, where efficiency and recycling initiatives suffer from uncertain and volatile primary metal prices. In the longer term, the use of finite resources is unsustainable. So what are they used for? That is the topic of the next section – but first some stories.

13.4 Stories

13.4.1 Oil and Power[17]

Natural resources such as metal ores, coal and oil have been throughout history at the centre of power struggles and ideologies. 'The rapid rise of Russian production, the towering position of Standard Oil, the struggle for established and new markets at a time of increasing supplies – all were factors in what became known as the Oil Wars. . . . there was a continuing struggle involving four rivals – Standard, the Rothschilds, the Nobels, and the other Russian producers. At one moment they would be battling fiercely for markets, cutting prices, trying to undersell one another; at the next they would be courting each other, trying to make an arrangement to apportion the world's markets among themselves; at still the next, they would be exploring mergers and acquisitions. On many occasions they would be doing all three at the same time, in an atmosphere of great suspicion and mistrust . . . ' (Yergin 1991). This sounds familiar, but it describes the 1890s – not much has changed.

In March 2001, U.S. oil majors were the leading foreign investors in Kazakhstan's hydrocarbon riches. Six months before September 11, 2001, U.S. Ambassador to Kazakhstan Jones said the United States wanted 'Kazakhstan to develop its energy resources . . . and have better access to world (oil) markets. . . . a stable central Asia – and Kazakhstan in particular – would allow stable exports of Caspian oil to international markets . . . the US government backed the export pipeline from Kazakhstan to Russia's Black Sea port of Novorossirsk . . . Washington also supported two other planned pipelines, which will eventually link the Caspian region . . . on the Mediterranean. But . . . Washington would continue to oppose plans to export Caspian oil to neighbouring Iran, which it accuses of sponsoring international terrorism . . . the US Congress was likely to extend sanctions against Iran after they expire in August.'

It had long been known that Afghanistan had significant deposits of gemstones, copper and other minerals, but in the summer of 2010, U.S. officials said they had discovered and documented major, previously unknown deposits, including copper, iron, gold and industrial metals like lithium. A Pentagon team, working with geologists and other experts, had shared its data with the Afghan government. It was working with the Afghan Ministry of Mines to prepare information for potential investors. A few weeks later, Afghan officials said they believed that the American

[17] The quotes are from the site www.nytimes.com/2010/06/18/world/asia/18mines.html.

estimates of the value of the mineral deposits – nearly $1 trillion – were too conservative. They could be worth as much as $3 trillion. The Ministry of Mines also announced that it would take the first steps towards opening the country's reserves to international investors. Two hundred investors from around the world have been invited to offer suggestions for how to develop the iron ore deposits at the Hajigak area of Bamian Province, according to the principal mining specialist for Afghanistan for the World Bank.

13.4.2 The Promise of Gold: Tambogrande[18]

In the 1990s, a controversial mining project was proposed for Tambo Grande, a town of roughly 18,000 people in northern Peru. The government of Peru conceded three blocks of land totaling 87,000 hectares (ha) to the Canadian Manhattan Minerals Corporation. One 10,000-hectare concession, right in the town of Tambo Grande, is particularly attractive to Manhattan Minerals. It is a massive polymetallic deposit known as TG-1 that has gold sitting closest to the surface. Manhattan Minerals can mine the gold first and finance the underlying copper and zinc extraction with the revenues generated early in the project's life.

The Tambo Grande valley is a green valley on Peru's desert coast, which came into being in the 1950s after building a reservoir and irrigation system with World Bank money. It is now producing 40 percent of Peru's mangoes and limes. These and other crops create about U.S. $2 billion in revenue annually and permanently employ roughly 15,000 people, and more during the harvest. A majority of the 65,000 people living in the area make their living directly or indirectly from growing fruit.

If the project were to get the green light, it would deeply affect the life of the people in the province of Piura, changing it from a region dominated by agriculture to one dominated by mining. Roughly half the townspeople in Tambogrande will have to relocate to make room for the one kilometre long open-mine pit next to the Piura River. A conflict about the mine plans is growing since the Fujimori government granted the concession without consulting the local population. Fruit growers in the arid region say that toxic mine wastes from the project will contaminate the (recent and World Bank-funded) irrigation system. Environmentalists wish to protect the fragile dry tropical forest and local officials fear the influx of labourers. 'In the San Lorenzo valley the people are not rich but they aren't paupers. They have families and animals and roads and schools, and they can live', says Ulisses Garcia, 35, leader of Tropico Seco, a youth environment organisation. 'If Tambogrande is exploited it will rape the nucleus of this area. It is a demented project. If they carry it out we will have to leave the area. Once the mine happens, who will want to purchase produce from this area? It will ruin everything. It is crazy to do this mine from a social point of view'.

Opposition against the project increased over the years, despite promises by Manhattan Minerals to build new houses and water and sewer infrastructure. With the Peru mining ministry holding a 25 percent stake in the mine project, the people

[18] The sources of this story are www.oxfamamerica.org, www.foei.org and www.ichrdd.ca/site/. See also icarusfilms.com/new2007/tam.html about the film Tambogrande: 'A compelling account of how the global can become painfully local.'

of Tambogrande do not feel protected by their own national government. This combination of uncertainty and vulnerability has led to frequent demonstrations in the town, one of which turned violent in February of 2001. Since then, both the Mayor of Tambogrande and the Archbishop of Piura have spoken out against the mine. The mayor circulated a petition against the mine and got over 70 percent of the town to sign it. In 2008, the government decided to block the project, informing the mining company that it had failed to meet financial criteria necessary for digging the proposed open-pit gold mine. There are similar stories all over the world, many with a less fortunate outcome.

13.5 Resource Chains: Material Use and Efficiency

13.5.1 Assessment Methods

The exploration and extraction process is the 'upstream' part of the resource chains. It operates to satisfy the *demand* for and make possible the *use* of materials in the 'downstream' part of the resource chains[19] (Figure 13.2b). They make up the metabolism of industrial societies and are the topic of industrial ecology. In this part of the chain, the resource is transformed in myriad ways and becomes in-use stocks in appliances, vehicles, houses, offices and so on. Ultimately, it ends up in landfills and air, water and soils. Several methods exist in resource analysis, from primary resource to final product use and disposal. Usually, the flows are expressed in *physical terms*. For heterogeneous product groups (such as machinery), high-value products (such as pharmaceuticals) and aggregate sectoral output such as services (such as health and education, financial or consultancy), *monetary units* are more suitable. The level of analysis can be the unit process (such as pumps, mixers, furnaces, evaporation and cooling units) or a combination of processes (such as a factory). Because of limited data availability and other reasons, many analyses are done at economic (sub)sector level, with particular interest in technological change. Also, the entire economy can be the unit of investigation, for instance, in international comparisons for benchmarking with regard to efficiency and dependency issues.

There are many analyses of resource use chains for a variety of products and processes, including agricultural ones. Most analyses are stationary state descriptions for stocks (inventories) and flows (in- and outputs) in a given year. Increasingly, dynamic analyses are made. I refer the reader to the Suggested Reading and Useful Websites for details and results. The most widely used methods are:

- *Substance Flow Analysis (SFA)*: the flows of individual elements and compounds through society. A typical application is for toxic substances;
- *Material Flow Analysis (MFA)*: the flows of bulk materials through society, which are faced with depletion or cause environmental impacts during production, consumption and waste management; and

[19] Usually, the words resource use and resource demand are interchangeable. Quantitative historical analysis equates use and demand by definition. Resource use is the observable quantity in the statistical data bases. Resource demand is by implication equal to use for the past and a nonobservable expectation and projection for the future. The latter is considered equal to projected use for all practical purposes (such as investment decisions).

- *Life Cycle Assessment (LCA)*: the resource and environmental impacts of a product or a service, using MFAs and SFAs for the relevant compounds and a methodology for weighing inputs, outputs and impacts.

While SFA and MFA usually study material and energy flows at product/process level, the system boundaries in an LCA are drawn broader. An LCA will usually include an *impact analysis*, which is an evaluation of impacts in various categories such as resource scarcity, ecotoxicity and global warning (Baumann and Tilman 2004). Dedicated software packages for LCAs are available, such as the Chain Management by Life Cycle Assessment (CMLCA) package. In view of its importance, *energy analysis* has turned into a separate discipline with specific principles and methods (§7.4; Blok 2006).

System boundaries should be carefully specified in order to avoid confusion about and misinterpretation of the outcomes. Common system boundaries are *cradle-to-factory gate* and *cradle-to-grave*. The system *cradle-to-factory gate* covers all steps from the extraction of resources from the environment up to the final product as it is leaving the factory gate. The system *cradle-to-grave* goes one step further and includes the product use and its treatment in a waste management system. If the flow to the grave is minimised or even disappears, because almost all of it is reused within the same or other manufacturing processes, one speaks of a *cradle-to-cradle* approach. There are increasing numbers of firms that actively pursue cradle-to-cradle strategies. They hope to exploit first mover advantages and gain a competitive advantage when resources become scarce, as part of a long-term strategy. The emphasis shifts from delivering goods to the provision of services and from a linear chain to a circular metabolism, but the long-term macroeconomic effects are still difficult to assess.

A recent analysis of the global tin (Sn) cycle illustrates, as one example out of many, the MFA (Izard and Müller 2010). Tin is already known and used for more than two millennia, but it is quite rare. The known economic reserves are only enough for twenty-two years of global extraction at the present rate and are concentrated in only three countries. The authors have constructed a dynamic model of the global tin cycle, which is calibrated on the available data for the period 1927–2005. It simulates the usual stages of production, fabrication and manufacturing, use and waste management. It is difficult to trace where the tin ends up in the final stages because it is used in small quantities in many applications: cans and containers, parts for vehicles, plumbing in construction and solder and tinning in electronics, which are the most important ones. Tin has accumulated in different stocks throughout the world economy (Figure 13.9). In total, 17 million ton (Mt) of tin have been mined from an estimated 28 Mt of resource. Of this, 13 Mt or nearly 70 percent has ended up in landfills. Much tin is used in short-lived products, notably cans and electronics, which also end up in dissipated form in the landfills. This explains why only 2.2 Mt is in products-in-use (plus another 1.9 Mt in association with steel) and why the recycling rate is quite low. MFAs like this one show that it matters for key aspects of the life cycle for which products the refined metal is used. The specific features of resource chains make it necessary to investigate each metal metabolism separately.

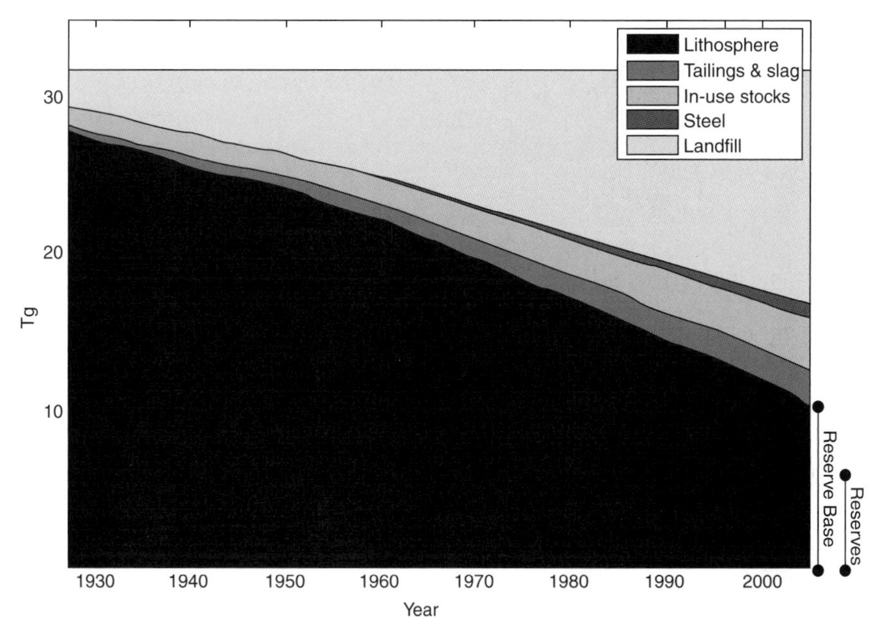

Figure 13.9. Cumulative global tin stocks 1927–2005 (1 Tg = 1 Mt) (Izard and Müller 2010).

13.5.2 Dematerialisation: The Intensity-of-Use Hypothesis

What is driving use of materials and fuels?[20] Who are the users and for what purposes and how efficiently do they use it? And, what is the role of material and fuel substitution and material reuse and recycling? Usually, income (GDP/cap) is considered the driving force behind the use of materials and fuels. Econometric analyses focus on income and price elasticities to better understand the determinants (§11.3; Common and Stagl 2005). The following introduces two metamodels, which are derived from country-based empirical analyses and are used to extrapolate future trends.

As early as the 1950s, it was observed for the United States that the inputs of materials and semi-finished goods becomes relatively less important vis-à-vis nonmaterial inputs such as labour, capital and services. The concept 'intensity of use' was introduced to draw attention to the phenomenon that over time (time-series) and across nations (cross-country) the use of materials and fuels per unit of GDP tends to rise and then fall as function of income (Malenbaum 1978). This bell-shaped form of the intensity-of-use curves has subsequently been called the *intensity-of-use (IU) hypothesis* and is associated with the notion of dematerialisation. It has two postulates (Bernardini and Galli 1993):

- the IU of a given material (in t/€) tends to follow the same pattern for all economies: at first increasing, then reaching a maximum and declining as a function of income[21];

[20] This paragraph consistently speaks about materials (from minerals) and (fossil) fuels. Because fossil fuel is still the dominant source of energy, resources can still be equated to material and fuel stocks and flows. If renewable energy gains in importance, this is no longer a valid assumption.

[21] The inverse of intensity of use is resource productivity (in €/ton). Some materials and most fuels are widely traded, and there can be substantial differences between production and consumption in a

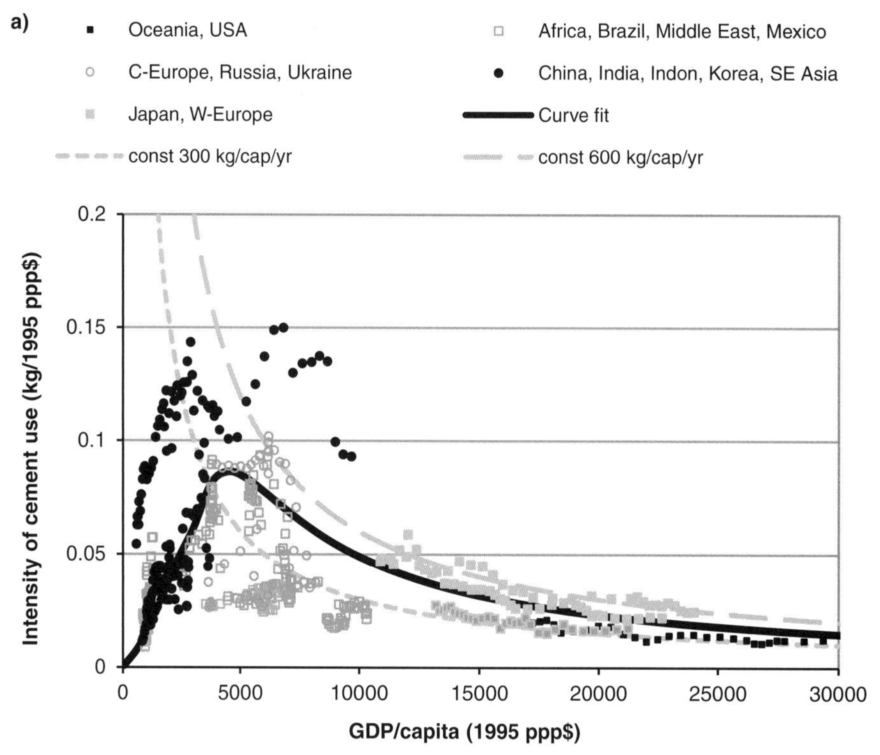

Figure 13.10a. The relationship between cement use intensity (t/$) and income (GDP/cap). The dots indicate the cement use intensity (t/1995 $) for some regions in the period 1971–2000. The solid line is the IU hypothesis curve; the dotted curves are two lines of equal cement consumption per capita (t/cap) (source of data: PBL).

- the maximum IU tends to be lower for countries, which had a later economic takeoff, such as Japan and South Korea and the presently emerging economies.

The empirical data for cement use in the period 1971–2000 in world regions indicate an overall trend in line with the IU hypothesis (Figure 13.10a). The data can be fitted to a function of income.[22] However, there are large differences between the regions. United States–style countries have higher IU values than European countries and Japan, and there are large differences amongst low-income countries; for instance, a factor five between Korea and Brazil, countries with comparable incomes. A similar proximate confirmation of the IU hypothesis is found for steel (Figure 13.10b) and other metals, but for paper and plastics there is no leveling off yet. The inverted U shape is also visible for total commercial fuel use (coal, oil or gas), but it breaks down when traditional fuels (biomass) are included (Grübler 1999).

country. The data in this paragraph are all based on consumption or use. Note that the IU hypothesis implies an income elasticity that declines with the increase in income.

[22] The curve used to describe the intensity-of-use and also, in the next paragraph, the emission intensity as function of income is used in the IMAGE energy model (Van Ruijven *et al.* 2009).

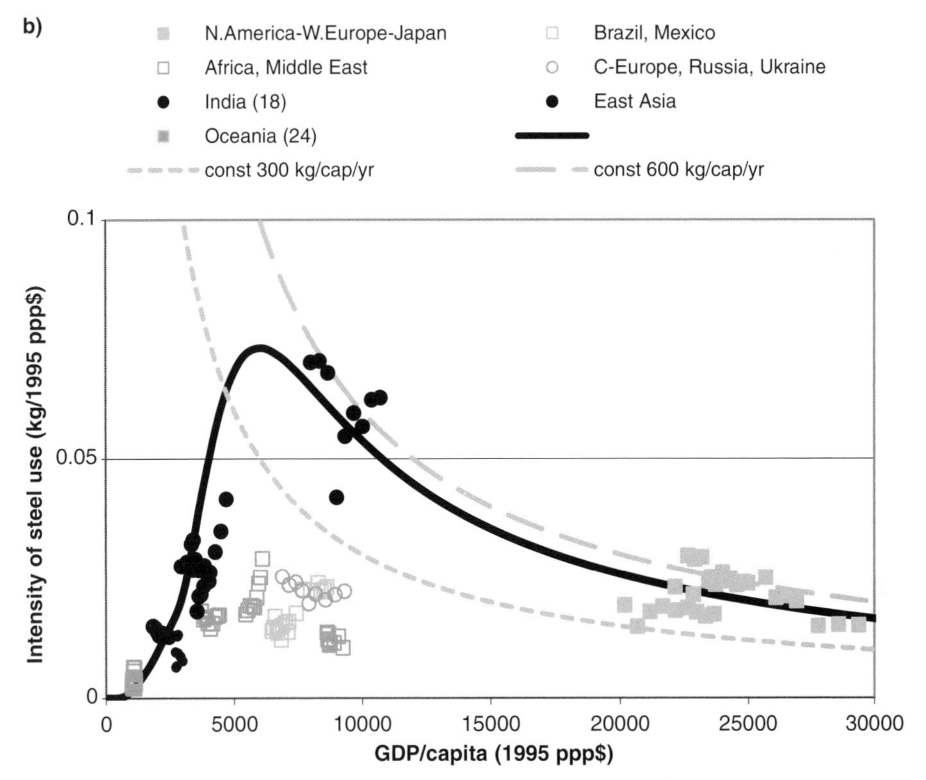

Figure 13.10b. Idem, for steel use intensity (source of data: PBL).

Several explanations have been suggested. The first one is the *structural change* in economic activity (§4.3). In the early stages of industrialisation, there is a rapid expansion of material- and energy-intensive infrastructure (roads, railways, dwellings and so on). Economic growth in affluent postindustrial societies, on the other hand, stems largely from technology-driven, less material-intensive service activities (education, medical care, hygiene and safety, design and so on). A second explanation is the role of *prices and technology*: Industrial sectors with high-material inputs have a large incentive, particularly in a situation of high or volatile prices, to develop and implement innovations that make resource use more efficient or provide attractive substitutes such as plastics for steel. Conversely, the lack of substitutes is also part of the explanation why the IU hypothesis does not (yet) hold for paper and plastics. A third, related factor is that rising prices make *recycling* more attractive for the producer while making the solid waste problem more tractable. Because innovating and recycling require a techno-economic and social-political infrastructure, the high-income countries are leading in dematerialisation. However, globalisation in its various manifestations allows the application of the latest technologies in low-income countries, at least in principle and with the necessary basic infrastructure – a phenomenon called 'leapfrogging'. This explains why the IU curves are lower for the latecomers in industrialisation.

Because so many forces are simultaneously at work, one cannot expect more than a correlation between intensity-of-use and income. For instance, the intensification

Box 13.3. *More than mass alone?* Environmental NGOs as well as corporations and bureaucracies often equate sustainability with the technical and behavioural options to reduce a person's use of materials and energy while simultaneously embracing a culture of consumption. Their sites and brochures are flooded with data on physical stocks and flows and how to reduce them. An example of such an obsession with facts is the site energyfacts.bp.com/ze/recycling.aspx, which, on the virtues of recycling, states that 'recycling aluminium cans saves 95 percent of the energy used to make the cans from virgin ore.... Recycling one glass bottle can save enough energy to light a 100-watt bulb for four hours... Recycling one tonne of aluminium saves the equivalent in energy of 2,350 gallons of gasoline. That's equivalent to the electricity used by the typical home in a 10-year period... Recycling one glass jar can save enough energy to run a television for 3 hours.... Recycling one aluminium can saves enough energy to run a TV for three hours – or the equivalent of a half a gallon of gasoline. We use over 80,000,000,000 aluminium soda cans every year. The impact of such data floods on human behaviour are uinclear – but you should in any case avoid 100-watt light bulbs.'

of trade flows makes it more difficult to find reliable statistics and less probable to find country-based correlations. Also, the differences in variables other than income, such as climate and culture, and the continuous change in material use because of novel applications – think of computers and mobile phones, which were absent before 1985 – cause country-specific deviations from the trend.

Often, the IU hypothesis is interpreted as a sign that economies will converge to roughly equal 'best practice' levels of material intensity and that materials and fuels pose no limits to economic growth. Such a generalisation is unwarranted. First, the material and energy use per unit of economic activity declines, but in a growing economy, the *absolute use* of materials and energy keeps increasing. Second, the high-tech innovations that are part of dematerialisation require often *new and rare materials*. A well-known example is the platinum used in catalytic converters in cars and another one is the use of rare earth elements for electronic applications and batteries. Finally, dematerialisation involves the use of *substitutes* that may have scarcity problems of their own. The obvious case here is the substitution of metals by oil- and gas-based plastics. Mineral and fuel scarcities change but do not end. Their depletion remains an issue in sustainable development.

It seems correct to conclude that 'while qualitatively correct, the dematerialisation model is quantitatively an oversimplification' (Bernardini and Galli 1993). But there is another unsustainable part of the industrial economy. The mining and processing of materials and fuels creates also large flows of 'waste', burdening environmental sinks. Upstream, huge amounts of solid waste are generated in mining and processing processes. One-fifth (20 percent) of global energy- and process-related CO_2 emissions arise from the production and processing of only five materials: steel, cement, plastic, paper and aluminium (Allwood *et al.* 2011). Downstream, the discarded materials and burnt fuels accumulate in landfills and air, water and soils. Let us examine another metamodel.

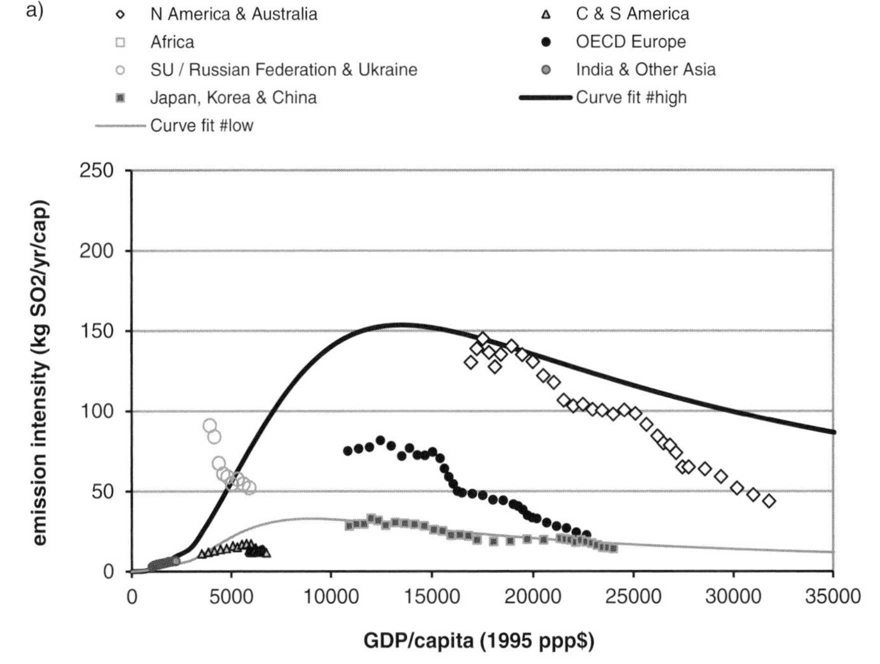

Figure 13.11a. The relationship between SO_2 emission intensity (t/cap/yr) and income (GDP/cap) for some regions in the period 1971–2008. The dots indicate the empirical SO_2 emission intensity (t/cap/1995$) and the solid line is a possible curve fit (source of data: PBL).

13.5.3 Richer and Cleaner?

In 1955, the economist Kuznets suggested that income inequality will initially increase with rising income per capita and then decline – another bell-shaped or inverted U curve. The hypothesis was that political mechanisms such as voting by the poor for redistributive taxes and financial mechanisms such as credit markets could explain it. But research at the World Bank did not find evidence for a Kuznets curve in a 1960–1990 seventy-three country database. What the data do indicate is that overall economic growth coincides with growth in the income of the poorest segments of society. This is the 'trickle-down percolator' hypothesis, which states that economic growth will automatically benefit the poorest segment of society.

In the early 1990s, a Kuznets-type relationship was hypothesised between the per capita emission of pollutants (in t/cap/yr) and income. It is called the Environmental Kuznets Curve (EKC) hypothesis. Early work for the World Bank Development Report in 1992 popularised this idea to the extent that an influential economist claimed that 'there is clear evidence that, although economic growth usually leads to environmental degradation in the early stages of the process, in the end the best – and probably the only – way to attain a decent environment in most countries is to become rich' (Beckerman, in Stern 2004). Again, such a generalisation is not justified.

What do the facts say? The empirical sulphur dioxide (SO_2) emission intensity as a function of income for the years 1971 to 2008 seems to confirm the EKC hypothesis (Figure 13.11a). The early industrialised and recent industrialising regions in the world exhibit an inverted U shape. Explanations are economies of scale in

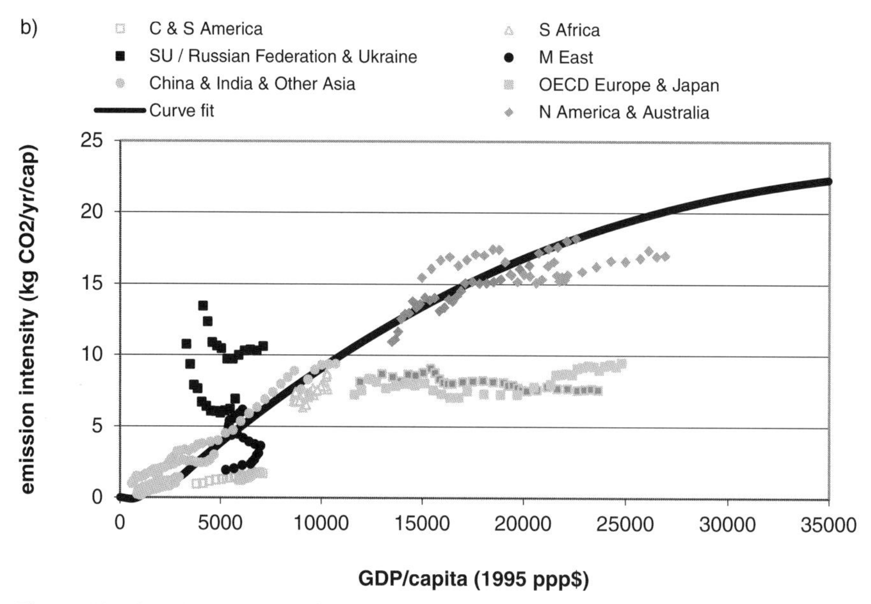

Figure 13.11b. Idem for CO_2 (source of data: PBL).

abatement techniques, more strict regulation as people become more affluent and changing sector structure and trade patterns. Per capita emission in newly industrialising countries are significantly lower than for the older ones, probably because of the availability of cheap abatement technologies and less coal use. For CO_2, the curve does slope downwards, but there is as yet no maximum (Figure 13.11b). One explanation is that CO_2 emissions are from essential and rather dispersed sources; another one is that international competition initiates a *race to the bottom* in emission standards that obstructs or at least postpones the decline phase. For both SO_2 and CO_2 emissions, there is a large spread in the data and, again, a significant difference between North America and Australia on the one hand and the more densely populated Europe and Japan on the other. Several researchers have investigated the EKC hypothesis in an econometrically rigorous way and some find the methods invalid and the evidence spurious (Wagner 2008). Overall, 'the majority of studies have found the EKC to be a fragile model suffering from severe econometric misspecification . . . it seems unlikely that the EKC is an adequate model of emissions or concentrations' (Stern 2004).

Evidently, the validity of the EKC hypothesis depends on the definition and type of pollutants and on the specification of causes. Using urban pollutant concentration data, for instance, leads to a different conclusion than using national SO_2 and CO_2 emission data. Once again: such a metamodel has some validity in specified places and periods. Both the IU and the EKC hypothesis may be an aggregate feature of advancing economies, but for a deeper understanding of mechanisms and policies one has to broaden the scope towards the *material and energy transition*, with an explicit consideration of use efficiency, substitution and reuse and recycling. For most new chemicals released on the market, the intensity-of-use and per capita emission probably increase with higher income. The brings us to the very end of resource chains: the sinks. But again first some stories.

13.6 Stories

13.6.1 Pearl River Estuary, China[23]

There are numerous environmental histories, tracing human interference with the natural environment throughout the chain from mining to accumulation over time and space. Measuring the concentrations of certain compounds in soils and sediments yields an archeological print of industrial activity. The sediment quality in coastal wetlands of the Pearl River Estuary in Guangdong Province in China gives such a print – and reason for concern. The wetlands are used for land reclamation, aquaculture and wildlife protection, but meanwhile they also serve as one of the main ultimate sinks for large amounts of heavy metals discharged from the rapidly developing Pearl River Delta. QuSheng Li *et al.* (2007) have investigated the concentrations of a series of heavy metals and their chemical speciation. The results show that the sediments are significantly contaminated by Cd, Zn and Ni with concentrations in the range of 3–5, 240–346 and 25–122 milligram per kilogram (mg/kg), respectively. The sediment quality is no longer meeting the requirements of the current wetland utilisation strategies and notably Cd and Zn pose a potential health risk. The sediments have effectively become a metal resource.

The pollution causes social tensions between farmers and city dwellers because the latter are rich enough to pay for wastewater treatment plants but do not allow the farmers downstream in the Delta to use the treated water. 'If we [also] continue to pollute the water at its source, all the efforts of cities downstream will be wasted', according to Liu Chen, director of the Protection Bureau of the Pearl River Water Resources Commission. Water pollution may be particularly damaging during floods, when it can contaminate soils and accumulated pollution can be released ('chemical timebomb').

Many suggestions have been made for a cleanup of the water but none has been implemented so far. In many areas in the world, such events unfold – it is one of the well-known syndromes (§4.3). It also appears to be part of a transition: The high-income regions are increasingly cleaning up and applying stricter emissions standards, whereas most low-income regions are still in the early, rather high-polluting stage. The latter do benefit, however, from the awareness and technological development in the high-income regions.

13.6.2 Water As a Commodity: Ban on Bottled Water in Australian Town[24]

The market has discovered scarcity as something to be desired. In 2007, the Stock Fund Utilities Department of the Dutch bank ING announced that 'prospects remain good. In the states of California and New York energy shortages threaten and it drives the price up'. Similar advertisements are seen for water. Drinking water used to be a service 'freely' offered by nature. With growing populations and hygiene

[23] This story is largely based on the site www.asianews.it/news-en/Pearl-River-pollution-a-serious-concern-3264.html.

[24] This story is based on www.norlandintl.com/blog/2009/04/bottled-water-market-share-volume_28.html and an article in Le Monde 22 juillet 2009. See also www.circleofblue.org/waternews/2010/world/peter-gleick-bottled-and-sold-whats-really-in-our-bottled-water/.

standards and increasing food demand, water demand is rapidly becoming a scarce commodity in many places – or at least, that is a widespread perception. A process of 'commodification' has started. Every scarcity offers an opportunity for value added: Profit and employment, and its exploitation, therefore, tends to be supported by business people and politicians.

In the first decade of the 21st century, an estimated 200 billion water bottles are *annually* consumed and bottled water has become worldwide a U.S. $60 billion industry. Total bottled water consumption amounted to some 35 billion litres, largely in high-income regions. In 2007, a U.S. citizen drank, on average, 125 litres of bottled water per year, up from 7 litres per year in 1976. Massive marketing efforts have led to this change. Selling one litre of water at a price many times the production cost (and many times the price of the equivalent from the public water system) is undoubtedly a marketing success. The bottled water market is expanding at the expense of soda drinks, as these are increasingly seen as a cause of obesity. 'Drinking water instead of three sugary drinks per week for a year will spare you seven pounds of fat', says one of the Nestlé ads.

The industry worries that increased activism on the alleged environmental impact of bottled water can affect sales negatively. The billions of mostly plastic bottles, also from soft drinks, are now causing waste problems on local to global scales. This is pushed aside by the International Bottled Water Association (IBWA): 'There's little if any measurable evidence that activists have had an impact upon bottled water sales. Bottled water is well-established and popular with consumers who rely on its convenience, healthfulness and refreshing taste... [but] Consumers must also be made aware of the bottled water industry's outstanding record of environmental stewardship, protection, and sustainability... Bottled water containers are 100 percent recyclable.' It is promised that the fully biodegradable bottle is next in line.

But not everyone is getting in line. In Bungadoo, a 2,000-inhabitant town 120 kilometres (km) south of Sydney, the community council decided to put a ban on plastic bottled water. The decision was taken in response to a project of a large company to pump water in Bungadoo and then bottle and sell it. Permission for the project has been refused, but the decision is reconsidered in the Land and Environment Court. 'We are a small community in favour of environmental sustainability', says a local shopkeeper. 'Bundy on tap' has become the slogan, now that the community council has decided to install tapwater systems. Australians pay more than 2 A$/litre for bottled water, whereas tapwater is almost free. The industry fights the idea with the arguments of unemployment and obesity.

13.6.3 Organotin Compounds As Antifouling Agents[25]

There is little doubt that the accumulation of marine fouling organisms on vessels and man-made structures at sea increases drag and vessel fuel consumption, with substantial consequences in terms of economics and emissions. Tributyltin (TBT) compounds and the less widely used triphenyltins are extremely effective and

[25] This story is based on Santillo *et al.* (2001), Tributyltin (TBT) antifoulants: a tale of ships, snails and imposex. In: EEA, *Late lessons from early warnings: the precautionary principle* (2001): 1896–2000.

relatively economical as antifouling biocides. Use of TBT in marine antifouling paints dates from the 1960s and accelerated greatly in the 1970s with the rapid take up of organotin-based paints by the shipping industry and small boat owners. Some decades later, the use of organotin antifoulants has now resulted in widespread and sometimes severe environmental effects. Unusual in the field of ecotoxicology, the evidence linking cause and effect is irrefutable. The phenomenon is by some considered as 'the best example of endocrine disruption in invertebrates that is causally linked'.

Two regional case histories were instrumental in identifying low-dose effects of TBT and initiating development of the first regional controls: the collapse of the shellfish industry in Arcachon Bay (off the Atlantic coast of France) and the reporting of widespread imposex in dogwhelks, a kind of snail, from southern UK coastal waters. Early predictions that TBT would degrade rapidly in surface waters were partly falsified. Half-life estimates in the order of days for eutrophic surface waters are contrasted with up to several years for residues in nutrient-poor waters and marine sediments, especially anaerobic sediments. The persistence of TBT, and its toxicity to sediment communities, raises the prospect of delayed recovery of damaged ecosystems and represents an unpleasant legacy for the authorities responsible for dredging operations. Organotin bioaccumulation was also underestimated, and the specific role of biofilms in enhancing bioaccumulation and toxicity has only recently been recognised. Accumulation in top predators was simply not envisaged.

13.7 The Sink Side: Environment and the Industrial Economy

13.7.1 Resource Chains: The Sink Side

Human activities have always interfered with 'undisturbed nature'. It shows up most conspicuously in the legacy of deforested lands and abandoned mining areas in Europe and the United States. Copper and other metal mining pits scar many landscapes in South America, Africa and Asia. Oil spills did and do happen, polluting the oceans and coastal waters and rivers. Less visible are the large emission flows of carbon, sulphur, nitrogen, phosphorus and other compounds into air, water and soil compartments. Previously unknown substances enter into the natural environment. The following looks in more detail further downstream the resource chain.

Table 13.3 lists the number of substances/compounds that are emitted into the biosphere as regular part of the industrial economy. At the global level, many emission rates follow the familiar trend of exponential growth, with doubling emission rates every two or three decades, but in the high-income regions some emission rates are going down since the 1990s (cf. EKC). Some substances have a long residence time, because they stay for long periods in the atmosphere and are only slowly absorbed in soils, lakes and sediments and, ultimately, in oceans and rock. Others may decompose within days, or rain out onto rivers and land and accumulate in organisms. Natural cleaning mechanisms help to mitigate harmful impacts by breaking down or immobilising the substances. It is one of the ecosystem services (§9.7). Are these substances pollutants? The word *pollution* is rooted in Latin-Greek words for filth, dirt and mud. It originally meant to contaminate or desecrate and was only in the 19th century for the first time used in regard to the natural environment.

Table 13.3. *Some important anthropogenic emission in industrial societies and an indication of their emission rate in the EU*[a]

Compound ('pollutant')	Emission rate EU27 (kton/yr) 2009	Main impacts
Acidifying compounds (SO_2, NO_x, NH_3...)	SO_2: 5015 NO_x: 9374 NH_3: 3770	Damage to animals and vegetation, and buildings
Eutrophying compounds (N, P...)	N: 156[b] P: 31[b]	Decrease in water transparency (increased turbidity); fishing harvest loss; algae growth; ecological impacts such as on phytoplankton
Aerosols, particulate matter (PM_x)	$PM_{2.5}$: 12093 PM_{10}: 1971	Adverse impacts on human health, depending on particle size
NMVOC	7761	Smog (from VOC and NO_x) causes tropospheric ozone (O_3) concentration increase with damage for humans, ecosystems, crops and materials
Heavy metals (Cd, Pb...) and persistent organic compounds (PCBs, TBT)	Cd: 0,1 Pb: 2,05 PCB: 0,003	Adverse impacts on human health and ecosystems (Pb, Cd, Cu, Hg, dioxins and so on)
Greenhouse gases (CO_2, CH_4, N_2O...)	CO: 24073	Enhanced radiative forcing, causing rise in temperature and sea level
Radioactive compounds		Adverse impacts on human health and ecosystems

[a] *Source*: www.eea.eu.
[b] Discharges of nutrients from urban waste water treatment plants for eight EU countries.

Pollution is sometimes a too value-laden term. All human activity interferes with the natural environment starting with breathing, which converts oxygen (O_2) into CO_2. But the industrial economy has intensified some of the waste flows already characteristic of agricultural economies, such as some metals. It has enormously increased the flows of fossil fuel related elements (C, S and N) and it has created a large series of new substances and compounds, many of which did not occur on Earth before. The monitoring and investigation of substances and their fate and effects is the essence of environmental science. How to evaluate this part of the industrial economy in the context of sustainable development?

To understand the consequences of the anthropogenic emissions into air, water and soils for the function and quality or health of ecosystems and human populations, one usually has to go down to the local scale. It requires model simulations of the spread of substances in space and time, the chemical reactions involved and the harmful interferences caused. In environmental science jargon, it is about emission, immission and exposure rates and about the probability of harmful impacts. Within a pressure-state-impact-response (PSIR) framework (Figure 8.5), one can then identify responses to adapt to and prevent or reduce the consequences. It makes sense to distinguish between relatively short-term and local forms of pollution and long-term global interferences. A crude classification of the physical outflows of substances in human activities can be based on four dimensions:

- impact concentration: what is the ratio between the spatially distributed substance concentration and the local natural concentration at which undesirable interference with natural and human life occurs;

- degradation rate: at which rate is the substance broken down or otherwise made inert in comparison with the relevant natural processes;
- 'strangeness': to what extent is the substance alien to natural stocks and flows and, therefore, *a priori* difficult to evaluate as there is no natural equivalent; and
- 'waste': at what net cost and effort can the substance be kept inside the industrial system by reconfiguration, reuse or recycling?

Measured along these criteria, one can distinguish the extremes. Radioactive plutonium (Pu) qualifies as an extreme, where it is known to cause cancer at very low doses, occurs in nature at extremely low concentrations and is toxic for millions of years. At the other extreme are discarded aluminium objects, as it is abundant in the Earth's crust and the human use rate is still rather small in comparison with natural processes. In between are thousands of substances. This book does not intend to give a complete overview, but instead focuses on some problems of importance in a sustainability context. I refer the interested reader to the Suggested Reading.

13.7.2 Enduring Environmental Problems

In the 1960s, air and water pollution became dramatically visible in North America and Europe, because in many places the natural removal mechanisms were no longer sufficient to balance/buffer pollutant emissions. Amongst the first substance known as industrial pollutants with long-range and long-term effects are the *acidifying compounds* (sulphur dioxide SO_2, nitrogen oxides NO_x and ammonia NH_3). The wind disperses these substances, that mainly result from burning fossil fuels, into the air and cause harm, damaging the health of people and trees. They cross borders and give rise to international transboundary long-term environmental problem. It became all too obvious that the atmosphere is a common pool resource (CPR) (§5.4).

Evidence in the 1980s suggested that British emissions were to blame for the acidification of Swedish lakes. Similar long-range effects were identified in the United States. This led to successful policies to reduce acidifying emissions. Model-based negotiations were instrumental in the design and implementation of these policies (Haas *et al.* 1993; Hettelingh *et al.* 2009). In the Netherlands and Europe, large reductions have been achieved for SO_2, but reduction of NO_x and NH_3 turns out to be more difficult. Scientists have established safe levels with respect to soils ('critical loads') and these will still be exceeded by 2020 in parts of northwestern Europe. Emerging economies such as China and India are rapidly facing the same problem, as their use of fossil fuels increases and the health and environmental impacts become evident.[26]

It has been said, in retrospect, that the risks and damage of acid rain in Europe are less severe than initially feared. But what would have happened without these policy-induced reductions is hard to say. A recent, careful analysis of 30-year acid rain history in the Netherlands concludes that 'in the past, no unnecessary measures

[26] The maps on the sites www.geiacenter.org and www.eea.europa.eu/data-and-maps are good illustrations of past changes in emissions in Europe and in Asia.

have been taken to counter acidification. The seriousness of the acidification problem has been visibly reduced, although it has definitely not been solved yet' (PBL 2010c). In retrospect, policies were an adequate application of the precautionary principle. Although some impacts, for instance, on forests may have been exaggerated, the underlying science has been proven right. Also, the evidence is there that the benefits of avoided damage are substantial. Acid rain did inflict damage to life in tens of thousands of lakes and streams in Norway and Sweden, many previously with salmonids in great numbers. Natural conditions come back only very slowly despite the billions of Swedish kronor spent on liming these lakes for over twenty years. At the same time, it appears that the costs of emission reductions were overestimated because the success of cost-reducing innovations was underestimated. It has become a famous case study about the interactions between pricing, regulation and innovation.

Finally, and important in a sustainability science context, the emissions of sulphur compounds and also, particularly, nitrogen compounds are shown to have other than acid rain effects and not only at the regional, but also at the global scale. For instance, exposure to nitrogen in the air reduces life expectancy, deposition of nitrogen reduces biodiversity and too high concentrations in drinking water increase the risk of cancer according to the European Nitrogen Assessment.[27] The emission of NO_x in the stratosphere by airplanes is affecting the Earth's radiation balance. In view of the expected impacts and the increase in fossil fuel combustion, stringent international acidification policy measures are rightly an integral part of sustainable development.

Another well-known pollution story is about *chlorofluorcompounds* (CFCs). In the 1930s, it was discovered that CFCs are useful in a host of applications: propellants and foams, refrigerators, air conditioning apparatus and fire extinguishers (Haas *et al.* 1993, Bryson 2003). In the 1970s, it was discovered that the concentration of ozone molecules in the upper atmosphere (stratosphere) above Antarctica had fallen dramatically (the 'ozone hole').[28] A major consequence is increased UV radiation on the Earth's surface, which has harmful effects for human health and crop growth – the two best known effects. By 1980, it was convincingly proven that CFCs were the culprit. By 1990, the industry in the industrialised countries no longer resisted plans to stop CFC production. The Montreal Protocol is considered one of the great successes in international environmental negotiations (Parson 1993). But it is also a warning: History shows the long delays in understanding the problem and reaching agreement and how long it takes before the consequences show up. In this case, preindustrial ozone concentrations are not restored before the year 2060, the incidence rate of skin cancer in places like Australia will continue to rise during the 21st century and adaptation is the only option for the next generation. Both the health risks and the fear of it are reducing the quality of life of humans on Earth.

Some effects, such as the genetic effects of increased UV radiation from ozone depletion, are so difficult to assess and so beyond imagination that it is hard to

[27] www.nine-esf.org/ENA.

[28] Initially, it was overlooked because the computer had truncated the observation as an impossible outlier.

agree on any kind of collective action. The result is an ongoing controversy that reflects the divergence in worldviews. This is most obviously the case with the largest environmental issue of our times: the risk of climate change due to emission of CO_2 and other greenhouse gases (GHG) into the atmosphere (IPCC 2007; PBL 2009). This so-called *(enhanced) greenhouse effect* – or its popular term 'global warming' – is dominating many sustainability debates and environmental policy discourses.

It is now beyond doubt that there is a steady increase in the atmospheric concentrations of gases which are known to 'trap heat', such as CO_2, methane (CH_4) and dinitrogen oxide (N_2O). Deforestation reduces the natural fixation rate of CO_2 in photosynthesis and has a similar effect. The resulting change in radiative forcing is expected to cause a rise in global average surface temperature. How much and at which rate is still quite uncertain because of divergent estimates of the climate sensitivity, amongst others. But there is broad agreement that stabilisation of the temperature rise to 2°C since the pre-industrial era – which is the stated EU target – can only be reached with a reasonable probability if world greenhouse gas emissions are drastically reduced (Meinshausen 2006). Climate change will have a variety of impacts on local temperature and precipitation patterns, but here too are still many uncertainties. While the trends in global average near-surface temperature are in agreement with scientific expectations, a possible change in frequency and intensity of extreme events in the last decades is more difficult to establish and still controversial. The severe floods and droughts in parts of Asia and Africa in recent times are an indication of the suffering that changes in rainfall patterns can cause for millions of people. In the long term, significant sea level rise is anticipated, with devastating consequences for the large coastal megacities.

People can respond basically in two ways: *adaptation* to actual or imagined impacts and *mitigation* in the form of emission reductions. Some adaptation will happen anyway, because some further rise in temperature is already inevitable. Both the impacts of climate change and the cost at which it can be mitigated, that is, at which greenhouse gas emissions can be reduced, differ amongst countries. This makes it a global social dilemma with many uncertainties and different perceptions and interests (§5.2; de Vries 2010). The payoff matrix has a complicated and dynamic structure that can give rise to a rich variety of strategies for the different actors (Table 10.2b). It spells a continuation of the already difficult negotiations. It also means a proliferation of interpretations and controversies.

In 1992, the United Nations Framework Convention on Climate Change (UNFCCC) was signed. In 1997, the Kyoto Protocol was adopted, and in 2005 it entered into force. Its major objective was to stabilise worldwide emission of greenhouse gases, as a first step to much more drastic cuts later on. The argument was and is that the costs made now prevent presumably larger and irreversible damage later. The value of the discount rate becomes a crucial point in this reasoning (§10.4; Dasgupta 2008). Another argument is that emission reduction has a series of cobenefits, such as reduced negative health impacts from air pollution, and helps realise other sustainable development goals (Metz *et al.* 2002). Since then, the climate science justifying such drastic and, in the public perception, costly measures suffered from a loss of legitimacy. This was fuelled by active campaigning by fossil fuel interests, but it could happen because scientists and NGOs underestimated the

Box 13.4. *Reuse: the road towards a circular economy.* Following an earlier systems-oriented definition of sustainability, material use chains have to become closed to the extent that finite sources are not depleted without generating substitutes and inflows in the environment do not exceed the natural decay rate. This is currently almost never the case. Metals and plastics in 'waste' flows are dispersed across the planet and accumulate to often hazardous concentrations in parts of the biosphere. For instance, polyethylene, the most widely used plastics, is produced at a rate of 80 Mt/yr and is showing up in many places, including in large soup-like accumulations in the oceans. Recycling rates are still small. The post-industrial economy has to be a circular economy, in which many manufactured goods are reused somewhere along the chain. From an exergy point of view, it is preferred to prevent dissipation and introduce reuse as much as possible upfront, near the mines and the factories, or as secondhand goods. This is not easily organised in an economy oriented towards profits and novelty, so most reuse is done at the end of the chain by recycling (glass, paper or plastics) and incineration.

In the Netherlands, households produced 9 million tons of 'waste' in 2010, double the amount produced in 1970. The total levels off, but the fraction of plastics has tripled in this period. About half of it is collected separately. A number of towns are now experimenting with plastics recycling because cleaning up litter is costly. Citizens are responding positively, as they did with glass and paper recycling. The plastic is compressed in the trucks that empty the containers and then transported to large installations for separation and reuse in construction materials and other applications. It is only accepted if it satisfies certain quality criteria. One of the innovations is to put sensors in the containers and only collect waste from full containers. A recently opened installation in Rotterdam applies high-tech separation technology to produce PET, PE and PP fractions that are sold to automotive and other manufacturers. Waste collection and treatment is still largely in the hands of the municipalities, but it is one area where public-private partnership works. As in metal and other recycling activities, raw material price fluctuations are one of the obstacles. A deposit-refund scheme, particularly for the huge quantities of bottles, is perhaps a better solution, because it is more upfront and captures a larger fraction of value-added. Glass bottles with deposit is often still the most environmentally friendly option – and aluminum cans the least. But the main maxim should be: reduce before reuse, expressing the challenge of using materials more efficiently while containing the rebound effect.

intricacies of postmodern science (§8.6; Metz 2009). In 2009, at the Conference of Parties (COP-15) in Copenhagen, the clash between the different views of and solutions for the problems and the certainties and the uncertainties was so vigorous that 'global warming' has again received a more modest position in the manifold trends and events that make up the sustainability challenge. With many countries not having met the targets for the first Kyoto commitment period (2008–2012), a new global treaty has to be negotiated unless the world community decides to wait for the results of local bottom-up initiatives and measures. For a more in-depth discussion, I refer the reader to the Suggest Readings and Useful Websites.

13.7.3 Persistent Chemicals

More difficult to assess but possibly equally serious in their long-term consequences are the numerous emissions, leaks and disposals of toxic persistent substances. They range from oil spills and mine waste to industrial emissions in water and air, from dissemination of food additives and exotic metals to garbage in the streets and plastic in the oceans. One of the best documents regarding pollution in the context of sustainable development is a study for the European Environmental Agency (EEA) that was published in 2001 called *Late Lessons from Early Warning: The Precautionary Principle 1896–2000* (EEA/Harremoes *et al.* 2001). It contains a series of investigations into the use, neglect and possible misuse of precaution in dealing with occupational, public and environmental hazards. The fourteen cases provide interesting and instructive insights about how one may try to deal with risk, scientific uncertainty and ignorance. Two dozen environmental scientists tried to answer three questions for these case studies:

- When was the first credible scientific early warning?
- When and what were the main actions and inactions on risk reduction taken by regulatory and other responsible bodies?
- What were the resulting costs and benefits of the actions or inactions?

A fourth question was: What can we learn from the answers? The study takes explicitly a long-term perspective.

The issues are complex and the impacts only visible in the course of decades, so it is difficult to get strong knowledge (§8.3). There are always unexpected and unintended consequences and conflicts between economic and social interests. The assessment of (avoided) costs and benefits is therefore difficult and controversial and a 100-year perspective is needed for the evaluation of the consequences of the persistent chemicals manufactured at ever larger rates. The theoretical framework for the study was the *precautionary principle*:

> The main element of the *precautionary principle* . . . [is] a general rule of public policy action to be used in situations of potentially serious or irreversible threats to health or the environment, where there is a need to act to reduce potential hazards *before* there is strong proof of harm, taking into account the likely costs and benefits of action and inaction.

Its application reflects differences in worldviews and, more specifically, in people's attitudes towards risks, technology and nature (§6.3). The principle can be viewed as a necessary consequence of postnormal science.

There is often a gap of decades between the first 'early warning' and decisive action. A vivid illustration is the observation by a UK factory inspector made in 1898 about the evil effects of *asbestos* and a UK government decision, one century later, to ban asbestos on the evidence that past asbestos exposure was expected to lead to 250,000–400,000 asbestos cancers in Western Europe in the period 2000–2035. Also in the other cases, the early warning was in retrospect justified. Ionising *radiation* was discovered by the end of the 19th century and the perception of its effects has slowly developed ever since. Protection standards are firmly established in legislation, but the long-term impacts in the form of cancers are still uncertain because they are only known from undesirable bombing and accidents. There are now over 400 nuclear power plants, many of them in need of dismantling and replacement. The handling of the associated stocks and flows of radioactive material will probably be one of the exacting tasks in the first half of the 21st century. Since 1897, *benzene* is known to be a powerful bone marrow poison. Increasingly used for tyre and explosives manufacturing, benzene had caused by 1939 widespread poisoning of workers with leukaemia as one of the consequences. After more convincing evidence of benzene as a cause of leukaemia, the standards became much more stringent with the Benzene Decision in 1980 in the United States in the context of the Occupational Safety and Health Act (OSHA).

Another case study is about polychlorinated biphenyls (PCBs). First synthesised in 1881, its damage to the skin was known by 1899. Mass production started in 1929, and by 1965, it had become a major public issue because of its danger to humans and animals. PCBs were never intentionally spread into the environment. They were used to replace more flammable, unstable or bulky products, in particular, ingredients of electrical apparatuses. The first crucial observation was the discovery in 1966 that the Baltic Sea fauna had remarkably high concentrations, leading to the hypothesis that these compounds were apparently persistent in living tissues with bioaccumulation as a result. Experiments in the 1970s showed clear correlations between pathological uterine changes in female seals and concentration of contaminants, particularly PCBs. A link with another pollutant, DDT, was suspected. Other evidence came from an incident with rice oil in Japan: PCBs were found to be the cause of the 'rice oil disease' (yusho). Their persistency led to accumulation in river sediments, with concentrations in the range of 12 to 24 mg/kg in Rotterdam Harbour. Tests in 1977 and 1988 indicated significant PCB accumulation in eels and other fish. The Dutch population was found in the 1980s to have the highest PCB contamination levels in the world.

In the 1970s and 1980s, new observations and experiments – sometimes unintended, as in another rice oil poisoning accident in Taiwan – led to a refinement of the scientific models. For instance, it was found that the number and position of the chlorine atoms played an important role, that the PCBs change during bioaccumulation and biodegradation, which makes them more dangerous, and that the dose and timing of exposure, in particular in pre- and postnatal children, are important factors. By now, most of the PCB chain is understood. The main producer, Monsanto, initially denied any health impacts, but mounting evidence led to a gradual phasing out of PCB production between 1972 (Sweden) and 1995 (world). But there will be a toxic legacy for decades to come: The 1.5 million tons produced between 1929 and 1988 (excluding USSR and China) have spread via the atmosphere to even the remotest places and are still spreading from the equipment in which they were used.

The other case studies in the report are about fisheries, halocarbons and several other compounds, antimicrobials and hormones and the mad cow disease. They show that the dangers did not only affect workers but also the public at large, as with a carcinogenic compound such as benzene that is a hazard for all those exposed to gasoline. Their study also identified cases of a (near) false alarm. In their recommendations, the authors of the study stress the importance of long-term monitoring, evaluating alternative options, engaging also 'lay' and local knowledge and ensuring regulatory independence, amongst others.

The stories reveal a pattern in the political process. When explicit risk analysis and evaluation became part of legislature and mandatory, corporations began to hire consultants to downplay or put in doubt the scientific observations about toxicity and impacts. The conflictual nature of this feature of the industrial economy became clear in corporate lobbying with antipublic health campaigns and policies to circumvent environmental laws. This is possible and understandable in view of the social dilemma character and the complexity and value-ladenness. How much are the individual and the collective willing to pay to reduce the risk that – as yet unknown – members of the population prematurely die of cancer? How much certainty is possible, to what extent can medical advances repair or mitigate the health damage, how far in space and time should the effects be considered? These questions will remain part of the quest for a sustainable future.

13.8 Perspectives on the Industrial Economy

The interpretations and expectations about non-renewable resources and environmental degradation differ widely amongst the various stakeholders. It can be illustrated with news items, that are a mixed bag of good and bad stories:

- Over 300 factories in Brazil produce ethanol from sugarcane on about 8 million ha of land, resulting in over 22 million tons in 2008. Ethanol is a cheap substitute for imported oil, does not compete with food and offers the prospect of further expansion. It makes up one-fourth of gasoline use in Brazil – is it the solution to fossil fuel depletion and climate change?
- 'Congolese are learning to sleep with an empty stomach', says a laid-off mine worker. The Congo has abundant mineral resources. Since 2003, the mining industry grew spectacularly because of the booming demand from Asia. But with the financial crisis in 2008, investors left the country and a quarter of a million people lost jobs and income. The mineral wealth is also a cause of continued war and corruption.
- High oil prices make it economical to squeeze crude oil out of shale oil deposits, and U.S. oil production is expected to increase in the coming decade. But the new techniques, such as hydraulic fracturing that fuel the shale gas boom, cause large greenhouse gas emissions and create other environmental risks. In the search for the 'black gold', the North Pole becomes Oil Pole and Kazakhstan oil becomes part of a global scramble.
- 'Garbage searchers in Rio hard hit by falling prices of recyclable materials.' Thousands of people search the megacities' municipal waste dumps for recyclable material, as if it were the stomach of a huge organism. 'Their lives are hard and vulnerable... they have no security nets to fall back on.' They started to collect frying oil in response to falling prices.

Controversial supply-side issues are: How much resource is geologically available? How much can be extracted and at what cost? How large are the possibilities to substitute scarce resources with more abundant ones? Less controversial, but still uncertain are the availability of oil and gas and of rare minerals and the distribution of the benefits from resource exploitation. Controversial too are the seriousness of enduring environmental problems such as greenhouse gas–induced climate change and the spread of persistent chemicals. Appreciation and interpretation of the need and cost of effective waste and reuse/recycling and of environmental and social impacts of mining also differ, sometimes widely.

Governments and corporations tend to see resource and environment issues as primarily global. They propose rational solutions, in line with the modernist worldview, that are based on some, sometimes hotly debated, mix of market incentives and government regulation. For many citizens, it is mostly the local aspect that counts: Am I affected by air pollution? Do I have to spent ever more money on gasoline for commuting? Is my health affected by the chemicals in the water? Can I insure my house against floods? The larger context is not or at best partly visible and full of uncertainties. Plus, who is to blame? In a highly connected world, a fair distribution of the costs to reduce risks and help the victims is no easy matter. Negotiations will be tough and it is easy to create a sense of unfairness and scapegoats. As in previous chapters, I present a table with statements and their evaluation from the four different worldviews introduced in §6.4 (Table 13.4). The reader is again invited to give his or her own evaluation.

13.9 Summary Points

The industrial regime brought profound changes to the world and a significant part of the world population is still in the midst of it. Cheap energy based on fossil fuels provided the engine, but it becomes less accessible now that ever more people aspire for the industrial form of modernity. Features of the industrial economy to remember in a sustainable development context are:

- the Earth's crust contains huge amounts of elements that are exploited for industrial purposes. There are reasons to assume that resources are large and that the sink side – interference with natural stocks and flows – is a more serious constraint than the (re)source side;
- basic elements of resource dynamics include the classification of resources according to size and probability, the supply cost curve as a representation of resource size as function of extraction cost, and the interplay between depletion and learning;
- fossil fuels are essential for today's advanced and emerging industrial economies. The transition to less accessible deposits in combination with growing demand will cause economic and geopolitical tensions. The role of combustion products in changing the Earth's climate is slowly becoming another stress factor;
- resource use trends depend on a complex mix of factors. There are clear signs of lower resource intensity (t/€) with rising incomes (IU hypothesis), but the causes are only partly understood. The hypothesis that a society can outgrow its environmental problems (EKC hypothesis) is at best partially valid;
- the industrial economy is causing large outflows of substances in the natural environment. Because some substances have long residence times, are not or only partly broken down and move across compartments and borders, they pose serious challenges for sustainable development. Examples are the enduring environmental problems of acidification, ozone depletion, the enhanced greenhouse effect and persistent chemicals.

Table 13.4. *Worldviews and non-renewable resource futures*

Non-Renewable Resources			
Statement 1: Before 2025, the world is faced with severe oil shortages and associated economic crises.			
It may happen, but not because oil resources are depleted but because of geopolitical tensions and lack of production investments.	Probable so we should focus on energy security from other sources (nuclear, coal).	Probable so let us prepare for it. Rising oil prices will stimulate energy efficiency and frugality.	Probable and possibly dramatic for the low-income regions that struggle with oil price–induced inflation.
Statement 2: Ensuring and safeguarding access to rare earth and other strategic mineral resources should be a strategic policy priority.			
Correct. Strategic materials are essential for modern industry, so governments should make it a top priority to safeguard their access and supply.	I don't know. Business and government should take timely action – beware of the Chinese.	I don't know. Society can no longer function without high-tech and this vulnerability is now becoming manifest. Simplify!	It is a UN responsibility to arrange multilateral agreements on scarce resources. Limiting state sovereignty is to be considered.
Statement 3: Recycling of metals such as aluminium and copper can be left to the market.			
In principle, yes. But citizens have to be informed and secondary metal prices stabilised. The government has a role here.	It is more difficult for new products like computers. It is a regional but difficult market.	We should first stop to consume and waste so much! Reuse, buy secondhand. And if recycling is needed, it is a community task.	This is an important strategy. International coordination is needed to ensure quality standards for labour conditions and product quality.
Statement 4: Attention should be given to cancer from air pollution, more than to cancer from smoking.			
Most important is the choice of the consumer: if he wants a cleaner car, car manufacturers will respond with lower specific emissions.	There are too many environmental regulations. The links between air pollution and lung cancer are still unclear.	Modern man poisons his interior and exterior in the name of progress and money. Regulation and market will not be enough – our attitude has to change.	Both forms of cancer should be prevented. But air pollution is more of a social dilemma, so stricter international standards for traffic are to be applied.
Statement 5: The Earth is on its way to the next Ice Age – global warming is at most a temporary slowdown.			
It is an illusion that we can control the future evolution of the world. Let's trust our ingenuity and adaptability!	It is good that some scientists put climate change in perspective, if only to calm down the alarmists. Most climate policies are costly and unnecessary.	An irresponsible attitude. Life is precious, and humans do not have the moral right to interfere in this way.	Irresponsible. There is quite some evidence that human-induced climate change may trigger catastrophic system change. This will have large consequences for our (grand)children.

> Geologists have estimated that if a single thimbleful of water were poured into a river, after only a few years the circulation on a global scale would be so complete that a similar thimbleful of water taken from anywhere on Earth would contain some molecules from the original thimbleful. Others have expressed much the same idea regarding atmospheric circulation by estimating that every breath of air we inhale contains molecules that were once breathed by Galileo, Aristotle, and the dinosaurs.
> – Chaisson 2001:193

> If we look around we can see evidence on all sides of the wonderful revolution which oil has wrought upon our very existence. It has brought in the motorcar for pleasure and commerce... it has contributed to the conquest of the air, and is even revolutionising movement across the seven sees... it has directly provided us with the fastest and most formidable capital warship afloat... It may be said in very truth that the community is becoming daily more and more dependent upon petroleum; it promises to survive as the one force which will drive the world.
> – Talbot, *All About Inventions and Discoveries*. Cassell and Company Ltd., London, 1916

SUGGESTED READING

Collection of papers in industrial ecology with theory and practice.
Ayres, R., and L. Ayres, eds. *A Handbook of Industrial Ecology*. Edward Elger: Cheltenham, UK, 2002.

An introduction to the concepts and methods of energy analysis.
Blok, K. *Energy Analysis*. Amsterdam: Techne Press, 2006.

Some chapters deal with industrialisation, energy and materials, and life cycle assessments.
Boersema, J., and L. Reijnders. *Principles of Environmental Sciences*. New York: Springer Science and Business Media B.V., 2009.

One of the first, largely theoretical texts on resource economics.
Dasgupta, P., and G. Heal. *Exhaustible Resources and Economic Theory*. Cambridge: Cambridge University Press, 1979.

Comprehensive treatment of learning curves for a series of energy supply and demand technologies.
Junginger, M., W. van Sark and A. Faaij, eds. *Technological Learning in the Energy Sector*. Cheltenham, UK: Edward Elgar, 2010.

An extensive treatment of natural resource economics in a largely neoclassical framework.
Perman, R., Yue Ma, J. McGilvray and M. Common, eds. *Natural Resource and Environmental Economics*. Harlow, UK: Pearson Education Ltd, 2003.

This paper is a good example of a strategic scarcity analysis of the world market for metals (copper).
Rosenau-Tornow, D., P. Buchholz, A. Riemann and M. Wagner. Assessing the long-term supply risks for mineral raw materials – a combined evaluation of past and future trends. *Resources Policy* 34 (2009): 161–175.

An entertaining account of the history of oil in all its aspects
Yergin, D. *The Prize – The Epic Quest for Oil, Money and Power*. London: Simon & Schuster, 1991.

USEFUL WEBSITES

Resources

- www.globalpolicy.org/security/docs/minindx.htm is a site on the exploitation of natural resources (diamonds, oil and gas, water, timber and minerals).
- webelements.com/ is an informative site about the elements in the periodical system.

- www.peakoil.com is the site on world (peak) oil production; there are many sites on oil and gas (peak) production, for instance, www. aleklett.wordpress.com.
- www.iea.org is the site of the International Energy Agency in Paris, publisher of the annual *World Energy Outlook* (WEO) and with extensive reporting on energy trends and outlooks. See also www.bp.com for the Statistical Review of World Energy.
- minerals.usgs.gov/minerals/ is the Minerals Information site of the U.S. Geological Survey (USGS), with well-organised data on on a commodity and country basis.
- cml.leiden.edu/software/software-cmlca.html is the site where the Chain Management by Life Cycle Assessment (CMLCA) package can be downloaded free of charge.
- www.oxcarre.ox.ac.uk is the site of the Oxford Centre for the Analysis of Resource Rich Economies, on the relationship between resources, economic welfare and governance.

Environment

- www.geiacenter.org/ is the site of the Global Emissions Inventory Initiative (GEIA) with large datasets and maps on emissions worldwide.
- www.unep.org/ is the site of UNEP with data and documents on global environmental issues, in particular the *UNEP Yearbook* on www.unep.org/yearbook/2011/.
- glossary.eea.europa.eu/ is the site of the European Environmental Agency (EEA) with brief explanation of terms in the area of environmental economics and policy and downloads of reports including the report *Late lessons from early warning* can be downloaded.
- www.milieuennatuurcompendium.nl/ is the site with environmental data operated by the Netherlands Environmental Assessment Agency (PBL) (in Dutch).
- themasites.pbl.nl/en/themasites/ is the site of the environmental models operated at PBL and gains.iiasa.ac.at/ is the site of the GAINS model on air pollutants.
- www.ipcc.ch/ the site of the Intergovernmental Panel on Climate Change (IPCC), with downloads of the five Special Reports and other IPCC publications.
- www.atmosfair.de/en/about-us/what-is-atmosfair/ is an example of a site that offers an emission calculator for individual air travel and offers to compensate emissions.

Appendix 13.1 The Crustal Abundance Geostatistical (CAG) Model

The *Crustal Abundance Geostatistical (CAG) model* is constructed from three assumptions (de Vries 1989):

- ore deposit formation processes (igneous activity, erosion-sedimentation, solution-precipitation) can be viewed as sequences of discrete separation events;
- with each separation event, a mass of M units of material with average concentration G_{av} is divided into two equal parts, one enriched, one depleted; and
- the separation events can be stylised in a series of steps, with each separation step characterised by a certain average efficiency q ($0 < q < 1$).

The average concentration of element i, $G_{av,i}$, is called the crustal abundance. The separation efficiency q_i of element i is called the specific mineralisability. The value of q tends to be higher for the oxidation- and corrosion-resistent 'noble metals' like gold (Au) and platinum (Pt). Using the estimates of $G_{av,i}$ and the data on known deposit size and concentration, it is possible to estimate q_i. Estimates of $G_{av,i}$ and q_i are shown in Table 13.1.

For a specific mineralisability q, for each separation step, one half is enriched q% and the other half is impoverished q%. This is graphically shown in Figure A13.1. In formula:

$$TM = M \cdot G_{av} = 0.5 \cdot G_{av}(1 + q_i) + 0.5 \cdot G_{av}(1 - q_i) \qquad \text{(A13.1)}$$

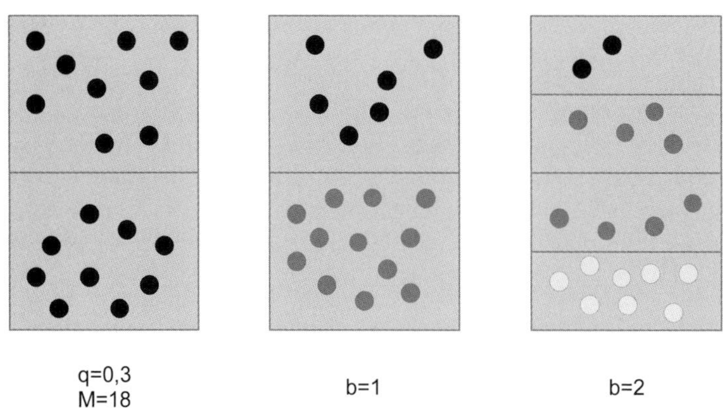

Figure A13.1. Natural enrichment of mineral ore: With each step, there is an enrichment in one part and an impoverishment in the other part.

After b steps, one gets:

$$TM = \sum_{k=0}^{b} \frac{1}{2^b} M \cdot C_k^b \cdot G_{av}(1+q_i)^{b-k} \cdot (1-q_i)^k \tag{A13.2}$$

with C_{kb} the binomial coefficient and $(1/2^b)M$ the unit deposit size. The amount of element i in concentration class k, M_k, equals:

$$M_k = \frac{1}{2^b} M \cdot C_k^b \left(\frac{1-q}{1+q}\right)^k \cdot G_{ref} \tag{A13.3}$$

with G_{ref} the grade of the richest occurring ore. If you plot the resource size as a function of the ore grade, you get the lognormal distribution shown in Figure 13.4.

For instance, lead (Pb) has an estimated average concentration of 17 parts per million (ppm) and an estimated q-value of 0.288 (Table 13.1). The high-concentration deposits are the easiest to find and exploit. For lead, they contain in the order of 5 percent by weight (wt%) or 50,000 ppm Pb. The CAG model generates the probability distribution shown in Figure 13.4, which with a reasonable accuracy represents those reserves that are known to have existed in the past and still exist nowadays. The other two elements for which the distribution is shown are mercury (Hg) and cerium (Ce). Mercury production is declining because of the discovery of toxic effects, but it used to be produced almost exclusively from an exceptional 0.6 wt% (6,000 ppm) deposit in Spain. Cerium is the most abundant of the rare earth or lanthanide elements, which are essential for the manufacturing of cell phones, computers, hybrid cars and military and other equipment. Its average crustal abundance is estimated in the order of 66 ppm. It is rather reactive and supposedly has a q-value of 0.15. The CAG-model suggests that the richest ores – it is usually mined as a by-product – are in the order of 0.1 wt% (1,000 ppm) and that rather large amounts are available at slightly lower concentrations.

The enrichment process represents a form of exergy because it implies a change in the chemical potential with respect to the earth environment. (§7.4). If you consider the Earth's environment as a sink with a certain composition (78.1 percent N_2, 20.95 percent O_2, 59.1 percent SiO_2, 15.8 percent Al_2O_3, 6.6 percent FeO and so on), then every system with these substances in a different concentration or composition represents an amount of exergy. Why? Because if it were brought reversibly into contact with the Earth's biosphere, work could be extracted – even if it were done irreversibly, although less than the maximum.

Appendix 13.2 The Logistic Growth Life Cycle Model

The logistic growth life cycle model is based on the following assumptions (King Hubbert 1971):

- oil fields in an oil province contain a finite amount of recoverable oil, R_{ult}, which is identified by exploratory drilling;
- the oil found per unit of effort will decrease – or the reverse: drilling effort per unit of oil found will increase – and the per unit cost to discover the oil will increase;
- oil once found will be extracted within a rather short period thereafter, as long as there are no competitive substitutes.

Initially, there is an exponential growth in the amounts of oil discovered (discovery rate) and, somewhat later, in the amounts of oil extracted (extraction rate). There is a buildup of oil reserves, which is the technically and economically recoverable oil that is discovered but not yet extracted. Within an oil province, the cost to discover and to produce oil rises upon depletion. The oil from this province can no longer compete with oil from other regions and the discovery and extraction rate start falling.

The model proposed by King Hubbert to describe this process is the logistic growth equation with the accumulated oil production CP as the state variable (equation 2.5):

$$dCP/dt = \alpha CP\left(1 - \frac{CP}{R_{ult}}\right) \text{ with } CP_t = \sum_{i=0}^{i=t} P_i(i) \qquad \text{(A13.5)}$$

P_i is the annual extraction rate in year i and R_{ult} is the ultimately recoverable oil resource. The accumulated production CP follows a logistic path over time. The oil reserves follow the same pattern but a couple of years earlier, and not as function of accumulated oil produced but of oil discovered. The oil production rate $dCP/dt = P$ is a bell-shaped form with a peak halfway (Figure 2.10).

14 Towards a Sustainable Economy?

14.1 Introduction

14.1.1 An Archetypical Model

The industrialisation process, and the transition to a next regime, which is in full swing but still not easily recognised and interpreted, consists of several changes in succession (§8.6). I consider the *postindustrial transition* process a change from low income and low resource use levels towards a new equilibrium at high- and stable income and resource use levels. In stylised form, it can be summarised in seven items:

1. a positive feedback process drives the growth of income (€/cap/yr), which leads to (exponential) growth in activity chains;
2. the chains consist of exploration and extraction of resources, followed by processing into manufactured goods, use and discarding;
3. fossil fuels are an essential high-quality energy resource, that enables in combination with capital and knowledge accumulation a steady increase in labour productivity;
4. the resource intensity of income (mass/€) first rises and then declines with income (IU hypothesis);
5. the externalities[1] along the chain: resource depletion, pollution impacts and waste accumulation, are initially not perceived and not priced;
6. with a delay, externalities are incorporated in the resource price in response to complaints of the more affluent citizens (EKC hypothesis);
7. the rising resource price stimulates a process of cost- and pollution-reducing innovations and the development of substitutes and alternatives.

The affluent countries are in the later stages of the transition; in the low-income countries, the first stages can be identified although the details depend on local circumstances and there may be jumps ('leapfrogging'). The possibilities to import

[1] The word externality is used in economics to denote those costs or benefits that are not transmitted through prices. With costs, it is a negative externality; with benefits, a positive one.

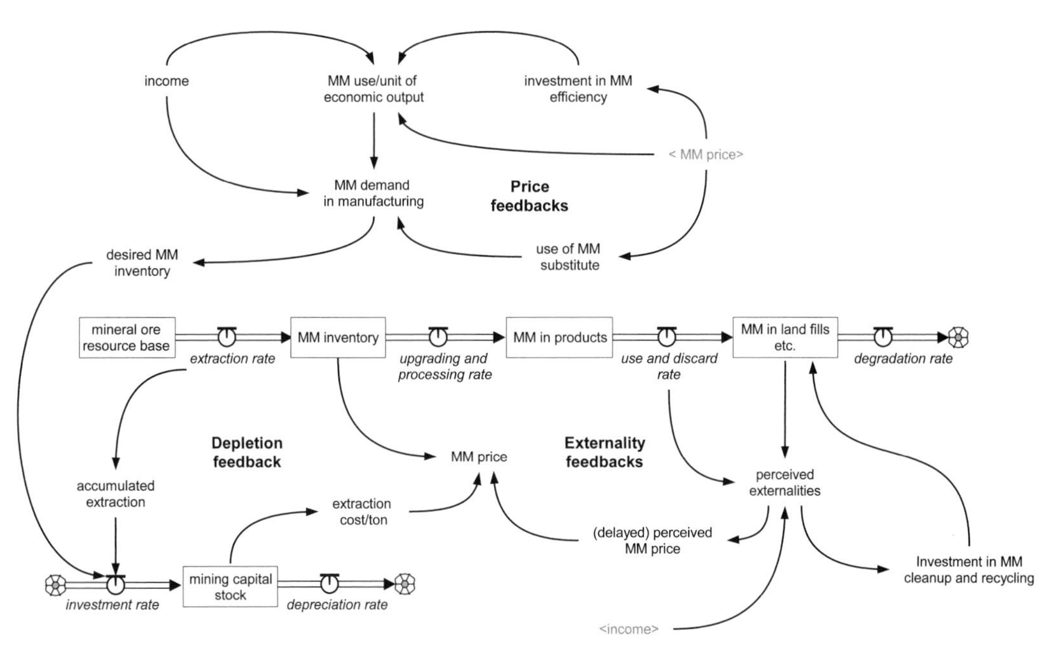

Figure 14.1. Causal loop diagram (CLD) of the resource chain process in an industrial economy.

resources from abroad, the vulnerability of the natural environment and the worldviews and behaviour of people are important determinants of how the transition will unfold. In this chapter, I focus on the techno-economic aspects.

An influence diagram of the archetypical transition process in an isolated economy is sketched in Figure 14.1. The core element is the *physical resource chain*, with the *source* and the *sink* sides as discussed previously. Upstream are the non-renewable resources that are extracted for industrial upgrading and processing and then an input for manufacturing and marketing. When the products are sold on the consumer market, it accumulates as resource-in-use. After the lifetime of the product, its constituents end up via various routes in one of the *sinks*: landfills, air, rivers, groundwater and soils. It accumulates and, for most compounds, is degraded and eliminated. The route from resource-in-use can be prolonged in a round of reuse or recycling and its extent and composition can be changed in waste treatment plants.

A possible storyline is as follows. The non-renewable resource is an aggregate of minerals/fuels and indicated with MF. Income growth causes an increase in MF use, initially at an accelerating pace. The resource is still cheap, but the rapidly rising extraction rate causes a cost increase: the *depletion feedback* loop. The MF market price goes up and this induces more efficient MF use in the *price feedback* loop. When income increases further, the MF intensity starts to decline (IU hypothesis). With rising income, citizens start to protest against the pollution from resource use, because they experience it as loss of quality of life. Government and industrial elites are forced to invest in pollution abatement and in resource use efficiency: the *externality feedback* loop (EKC hypothesis). Additional costs are made in the economy for waste treatment, material efficiency and recycling and for the associated physical

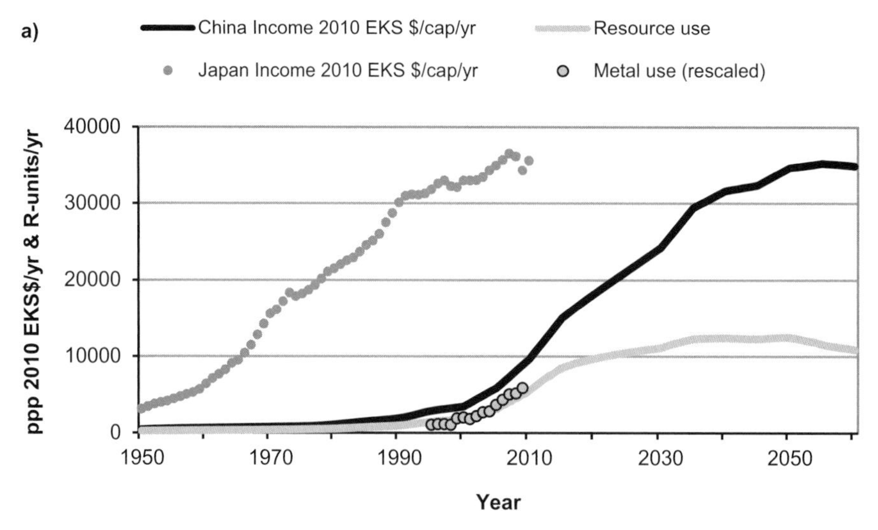

Figure 14.2a. Simulation of the stylised postindustrial transition, with the historical income growth in China between 1950 and 2010 and its extrapolation to 2060 to the level of Japan in 2010. The graph shows income and resource use; the dots indicate normalised use of the major metals (steel, aluminium, copper, lead and zinc) in China.

and institutional infrastructure (regulation, enforcement and so on). It shows up in higher cost and price levels and reinforces the price and externality feedbacks. The new, high-income equilibrium can be sustained as long as there are enough resources available at the higher cost level or if a substitute is available.

A stylised postindustrial transition can be simulated with a simple model based on the mechanisms sketched in Figure 14.1. For illustrative purposes, I use the historical income growth path of China between 1950 and 2010 as an exogenous input and assume that it continues until 2060 at the rate that the Japanese people experienced since 1965.[2] The transition period, between 2010 and 2050, is one of *steady increase in affluence* and one of *social and environmental stress*. A characteristic outcome is shown in Figure 14.2a-b. Resource prices go up and the expenditures on the resource as fraction of income increase significantly (from 5 percent to 21 percent in the simulation). A decoupling between income and resource use occurs after the year 2015, induced by depletion and externality pricing effects. Nevertheless, by 2060, the indigenous resource is depleted by two-thirds and, despite environmental regulations, there are serious negative environmental consequences. Mining and infrastructure cause pollutant emissions – apart from large-scale displacement of rural populations. Waste disposal affects the food chain and have negative health impacts. What the Chinese Deputy Minister of Environmental Protection in 2006 said comes true: 'China has entered a phase of frequent environmental accidents. With a fast growing economy, systematic risks of environmental accidents brought about by unscientific placement and industrial structure will replace individual pollution to become the No. 1 threat to our country's environmental safety'. Infrastructure falls short of

[2] In other words, in 2060, the average Chinese person will have the same income as the average Japanese person in 2010. Interactions between income and population growth and differences in the composition of production and consumption are not considered.

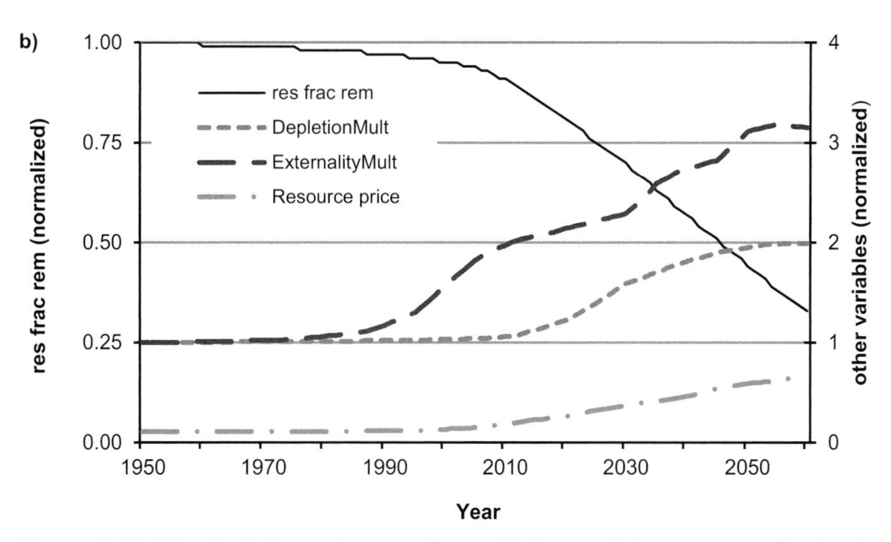

Figure 14.2b. The depletion and externality multipliers and the resource price for the simulation in Figure 14.2a.

expectations and income inequalities increase, which gives rise to social tensions. Also, there are long-term impacts, such as changes in climate and biodiversity, but these tend to be ignored or denied in the face of more direct challenges and threats. From a source perspective, this economic growth path is unsustainable because the resource is depleted before the year 2100 despite a stabilised extraction rate. From a sink perspective, it entails the risk of irreversible losses in quality of life. But such a growth path is possible, as countries such as Taiwan and Korea have shown (TCEPD 1989).

The stylised transition sketched previously is a widely believed, official outlook on the future. It presumes that countries can maintain economic growth rates and thus, implicitly, can overcome a number of social, resource and environmental constraints. The real-world transition can differ from the simulated one for at least two specific reasons. First, the IU and EKC hypotheses have limited validity. The IU hypothesis presupposes the existence of a substitute, such as plastic from oil for steel, that is itself part of another chain with similar dynamics. The EKC hypothesis may be invalid or too late for indirect and long-term impacts, because citizens do not experience the negative consequences themselves.[3] A second reason is that no country operates on its own. Historically, resource shortages have been countered by military conquest, import of resources and outmigration of people. Nowadays, trade is the preferred choice and mineral ore and fossil fuel are the largest bulk trade flows in the world. Resource-scarce countries will try to exchange manufactured goods and services for resources.[4] Availability on the world market will come under stress

[3] Negotiations about greenhouse gas emissions and loss of biodiversity and military adventures to safeguard resource imports and waste exports testify that such a response is deeply ingrained.

[4] Land-scarce countries did and do exactly this: importing food as soon as they are able to exchange it for manufactured goods. The UK has done it since the 19th century, and some Asian countries just started doing it. The phenomena of virtual water use and 'land grabbing' are in the same category of responses.

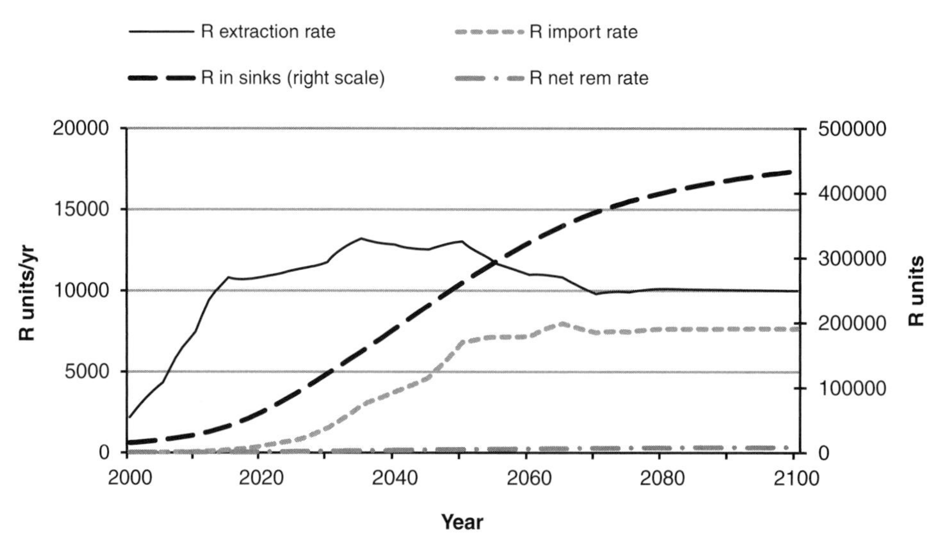

Figure 14.3. As in Figure 14.2a, but extending the simulation beyond 2060 with the assumption of zero income growth after 2060 and the possibility to import resource.

and cause political tensions and fluctuating prices. A counterforce is that resource extraction costs, at least temporarily, decrease because of innovations, but this will partly be undone by oligopolistic price setting and safety and environmental regulations (Figure 13.7). Real-world resource prices can also be expected to fluctuate significantly, which in combination with mismatches between demand and supply from system delays, feedbacks and noise effect the overall process. History abounds with examples.

Nevertheless, our simple archetypical model highlights a few points. First, imports are from a global perspective merely displacing the problem. This is seen if we lengthen the time horizon to the year 2100 and open the country for resource imports. Next, we assume that the country starts importing the resource at a price that is initially higher than the price of the indigenous resource. Then, we also suppose that after 2060 the economy does not grow anymore or, if it grows, at zero marginal resource intensity, so resource *use* remains constant thereafter. With the substitution elasticity assumed in this simulation, imports start around 2010 and make up more than 70 percent of the total flow by 2060. In this longer time resource-trade perspective, the country has successfully postponed depletion of indigenous resources: Even by 2150, still one-third of the initial resource remains (Figure 14.3). But this has been possible at the expense of resource use elsewhere – an example of shifting the burden (§2.4). On the sink side, the situation has aggravated because without a decline in resource use, the stock of pollutants keeps rising towards a dynamic equilibrium of high levels of accumulated pollution with local and global feedbacks. In this long-term perspective, economic growth is dependent on imported, finite resources and creates large and rising environmental impacts. It is, again though differently, unsustainable (Ho and Vermeer 2006). Truly long-term solutions are more intense resource efficiency and recycling and further development of substitutes. They are needed, and without delay, for a path that can be considered a *sustainability transition*. In some countries, signs of this are visible. In others, only the resistance against it has a voice. In many countries, short-term priorities still overwhelm all long-term concerns.

14.1.2 Substitutability, Technology and Optimality

An industrial economy can make the transition to a sustainable state, provided the average income stabilises and there are sufficient resources, substitutes and sinks. During the transition, price-based market forces, government-coordinated regulatory policies and behavioural changes are all three needed. Even then, there will be periods during which parts of the population suffer from negative environmental and social impacts. Episodes of severe pollution on many locations appears to be part of the overshoot dynamics, just like the stress of overpopulation during the demographic transition, the increase of welfare diseases during the health transition and loss of fertile land to degradation and urbanisation during the land transition. The experiences in the United States and Europe show that part of the damage is reversible and can be undone. Pollution-related diseases can be addressed with medical expenses and technology; rivers and soils can be cleaned up or will slowly recover if no longer under stress. But it takes time and effort. Some impacts are irreversible – notably the expected change in climate and the loss of biodiversity. The social impacts show up as a further increase in income and wealth inequality, decline of community life and tensions in the provision of public goods. These can be remedied by proper social security schemes and public planning, but here the experience shows that it is difficult in times of high expectations, rapid globalisation and large uncertainties.

Not surprisingly, the assumptions of resource substitutability and of technological change are the critical ones in macroeconomic analyses of growth and sustainability. Per capita consumption can continue to grow until the investments needed to compensate the declining resource quality become so large that the output available for consumption makes further *net* growth impossible. Beyond that point, per capita utility inevitably declines. If environmental constraints are also considered, the long-term feasibility of sustainable pathways is further narrowed down. Such source and sink constraints, notably in mining efficiency, land yields and pollutant elimination, are the causes of collapse in the *Limits to Growth* model (Meadows *et al.* 1971). There are three solutions to such a 'doomsday' scenario. First, the existence of a *backstop resource*, which is a substitute resource that is available in nearly infinite amounts. A second solution is *technology* that makes it possible to perform economic activities at vanishing resource and environmental intensity, as is the case in extreme forms of the IU and EKC hypotheses. A third and officially rarely discussed possibility is a change in lifestyle and values that breaks by a mixture of necessity and creativity, the link between the use of finite sources and sinks and the quality of life.[5] For many people, this is what the sustainability transition

[5] Such an extreme form of dematerialisation is known to economists as (resource-growth) decoupling and features in the debate also as delinking, degrowth, *décroissance* and green growth. It might happen in the form of the rapid expansion of cyberspace on the basis of ICT we are currently witnessing. But this is in no way certain: Non-material space is also exploited in favour of more consumption, in a mixture of commercial battles, directionless innovation and meaningless content and similar to what happened in physical space with the European conquest of the Americas.

is about. It is implied in Stuart Mill's famous statement: 'I sincerely hope, for the sake of posterity, that [the population] will be content to be stationary, long before necessity compels them to it'. It would coincide with a less materialist worldview (Figure 6.3).

Macroeconomists have explored the conditions for sustained economic growth with finite resources primarily in the form of an optimal control problem (Dasgupta and Heal 1979; Ströbele 1984). The idea is to maximise the discounted utility over the time horizon considered under two constraints: First, the macroeconomic production function discussed later, and second, the finiteness of one or more sources and sinks (§12.3 and §13.3). In formula:

$$\max \int_0^T e^{-rt} [U(C)]\, dt \tag{14.1}$$

with U = U(C) the social utility function and C consumption. T is the social planning horizon. It can be shown analytically that, under particular and simplifying assumptions, macroeconomic output can be sustained forever on the condition that the resource is completely substitutable (Perman *et al.* 2003).[6] In economic science, this assumption is known as *weak sustainability*. It implies that natural capital can in the long term be substituted completely by man-made capital and the incorporated ingenuity. In theory, complete source substitution cannot be excluded. It would be an economy with complete recycling and an energy supply based wholly on a resource that can be produced in near infinite amounts at constant cost. Solar power is probably the only option that qualifies as such a backstop resource. Complete sink substitution is more difficult to imagine. If sustainability is defined as the situation in which production- and consumption-related waste flows ending up in environmental sinks do not exceed the rate at which they are absorbed and broken down, then weak sustainability implies that ever larger waste flows must be treated and eliminated. This itself will require large amounts of exergy and materials. One only has to consider the energy balance of the Earth to realise that this will pose limits to growth at some point in time and place.

One final point about these rather academic modeling exercises. In macroeconomic models, supply and demand are assumed to equilibrate in the time period considered (usually one year). Therefore, responses to source depletion and sink overexploitation are timely and smooth. Moreover, the response activities are assumed to share in the innovation-driven productivity increase that is supposed for the economy at large. Therefore, the calculated costs of adaptation are in most macroeconomic model calculations small and declining and the simulated economic growth trajectories hardly differ from the ones without resource constraints.[7] For instance, most integrated economy-climate models suggest that policies to stabilise

[6] Many other formulations and models of the optimisation problem have been explored, for instance, incorporating technical change and resource and environmental constraints, feedbacks and costs.

[7] In these simulations, there is also hardly any or no consideration of the economic damage from changes in, for instance, climate and biodiversity. The overall costs of such policies are, therefore, unknown, because the costs of no policies in the form of non-avoided damages cannot be known. Some economists argue that this is irrelevant, because any future costs vanish with the discount rates of 3%/yr and more in such simulations.

atmospheric concentrations would postpone income growth with only a few years (Azar and Schneider 2002; IPCC 2001, 2007). For these reasons, most participants in the sustainability debate prefer *strong sustainability*, which implies limited substitutability between natural and man-made capital. Long-term concerns about climate change, biodiversity and other issues in relation to economic prosperity becomes then a matter of (very) small risks of (very) large and irreversible damage – a world far away from deterministic optimality (Meinshausen 2006).

In a broader context, one can reframe the situation in three questions. How much decline in quality life can a social-ecological system sustain before it collapses (*resilience*)? How productive and circular can the industrial system become (*efficiency*)? How can a decent quality of life be secured at low and stationary levels of physical throughput (*sufficiency*)? The answers to these questions are to be discovered 'along the road' and it is unclear whether and when to speak of a sustainability transition. Any assessment is complicated by the fact that, in the new postindustrial economy, new issues emerge. What is the role of the information or experience economy? How can we sustainably manage the commons? Can there be a human capitalism (Pine and Gilmore 1999; Barnes 2006; Tasaka 2009)? To find answers, I continue with a brief discussion on what economists and economic theory have to say about sustainable development.

14.2 Theories of Economic Growth

14.2.1 Classical Theories

The history of economic thought is associated with a number of mainly British and American scholars. The history of economic science is in itself important and revealing if one wants to understand present positions in the debate on sustainable development. Unlike in the natural sciences, the subject of analysis is so complex and dynamic that no economic theory consolidates into natural law. Economic theory more or less starts with the late 18th-century scholar Adam Smith and his book *Inquiry into the Causes and Nature of the Wealth of Nations* (1776). It explained how increasing trade and market size leads to specialisation, which causes labour productivity to increase. This in turn increases income per capita and, subsequently, market size. This positive feedback loop only starts to operate when market size exceeds a certain threshold, as it did in Smith's time in response to novel techniques, colonial conquests, falling transport costs and other factors. Smith also looked for another mechanism than church and state that could tie together society, and he found it in the famous *invisible hand* of the *market mechanism*. If the individual is given the opportunity to pursue his own self-interest, in an environment of likely motivated individuals, this leads to competition. This in turn produces the goods society wishes to have and at prices that people are willing to pay. Thanks to competition, selfish motives yield the unexpected outcome of social harmony.

A second line of thinking in classical economics is associated with the name of Thomas Malthus. In his *Essay on the Principle of Population* (1798), Malthus showed himself more sceptical about societal progress than Smith. He argued that unconstrained population growth will inevitably outstrip food production because land is a finite resource and food production has diminishing returns, which means that twice as many labourers on the land produce less than twice as much food. The consequence is declining food per person, followed by a rise in mortality and a stabilisation with overshoot (§9.5). In other words, natural processes tend to put limits to growth. Malthus concluded that population control was necessary. By breaking the link between population growth and output growth, he felt an escape was possible, but only if it was engineered.

The evidence in support of Malthus' theory is quite convincing for most of history and most countries – although not for Europe in the past 200 years because key assumptions turned out to be less and less valid. First, Malthus failed to recognise the importance of capital and technology, which allow farmers to escape the diminishing returns of land and to increase yields significantly. Second, better hygiene and sanitation, medical advances and birth control options in combination with rising income led to a reduction in birth as well as death rates in the European countries. Two hundred years of sustained growth of economic output in Western countries suggests that Malthus was wrong. However, 10,000 years of pre-industrial history as well as the large group of countries that still seem locked in a no-growth poverty trap show that rejection of 'Malthusianism' is premature. More inclusive models of the various stages of and transitions in economic growth will have to explore the dynamics in more detail (Galor and Weil 2000).

With Smith and Malthus, discussion focused on two early and extreme positions in the debate. One optimistic, suggesting that growth will accelerate forever and more or less automatically, and one pessimistic, predicting that growth cannot endure no matter what is done. The 'perfect world of Adam Smith', as Heilbroner (1955) calls it, did, of course, only exist in small pockets in time and space. Much of 19th- and 20th-century political economy was about the world's imperfections: the permanent threat of scarce resources for a growing population (Malthus and Ricardo), the exploitation of labourers (Marx), the absence of coordinated planning and redistribution (Saint-Simon and Stuart Mill), the elimination of competition by large oligopolistic enterprises and the psychopathology of the capitalist class society (Veblen) and the creative destruction of technological change (Schumpeter). Nevertheless, Smith's idea of the market as the coordinator has become, in more or less idealised forms, a pillar of the Postmodernist worldview (§6.5).

The early political and ethical works of classical economic theory are gradually narrowed to a body of highly abstract and rather esoteric concepts and hypotheses. One explanation for this is the desire, in the context of the Modernist worldview, to become an objective 'natural' science (§8.3). Modern economic (growth) theory has a narrow perspective and is largely associated with (Anglo-American) capitalism. Ethical issues such as a fair reward distribution between the 'production factors' capital(ists) and labour(ers) and, later, between the rich and the poor and between present and future generations were formalised and rationalised or disappeared altogether from mainstream economic literature. Many social scientists, including many political and welfare economists, are critical of this development and its consequences and of the validity of the underlying 'image of man' (§10.4). Nevertheless, economic growth theory has generated insights that are relevant for the sustainable development discourse and some of these are introduced in subsequent sections. Again, this is a vast topic, so I refer the reader to the Suggested Reading for further study. The focus is on models first: 'Every discipline that I am familiar with draws

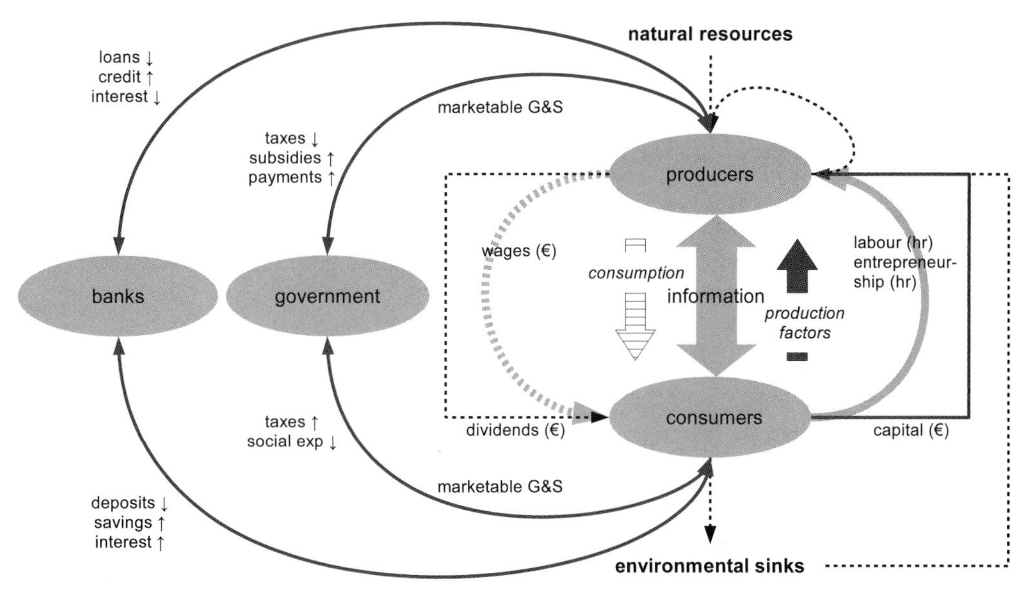

Figure 14.4. Flows in the economic system and its relations with the outside world. The black arrows represent money flows. The dashed arrows represent their physical and informational equivalents.

caricatures of the world in order to make sense of it. The modern economist does this by building models, which are deliberately stripped down representations of the phenomena out there' (Dasgupta 2007).

14.2.2 Economic Growth Theory

The starting point for our investigation is the scheme in Figure 14.4. It is an extended version of the standard flow diagram of economic textbooks. Transactions between producers and consumers are at the centre of the (market) economy. Governments and banks are two other important subsystems. In the core part of the industrial economy, labour and capital are the production factors and the rewards (wages and dividends) are used for consumer expenditures. Money is the numéraire in this system. The monetary flows are attached and add value to the underlying physical stocks and flows. Besides, there are large information flows to facilitate transactions between producers and consumers and between entrepreneurs, labourers, governments and bankers. Most economic activities, regulations and data are defined at the level of the 'sovereign' states, that are mutually linked through exchanges of goods and services, information and people. Not all human activities are registered as part of the formal economy: The monetary economy is continuously, and in manifold ways, interacting with people living in subsistence and informal economic systems. Both formal and informal economy are embedded in the larger life support system (Figures 8.6 and 13.1). In a sustainability context, it is important to have this more inclusive perspective on 'the economy'.

It was long known that the use of tools and machines make human labourers more productive. It was formalised by Solow (1957) in the *neoclassical model of economic growth*, in which capital accumulation plays a central role. Capital K and labour L are the production factors. Land and resources, including fossil fuels, were initially left out because of their small and declining role. The increase in the capital-labour ratio allows output per worker, that is, labour productivity, to increase.[8] Because it implies that the same output can be produced with less labourers, the process is referred to as capital-labour *substitution* because workers are replaced by machinery. The capital stock K is built up from investments I. It wears out and is taken out of production and depreciates at a presumedly fixed rate m.[9] Thus:

$$\frac{\Delta K}{\Delta t} \approx \frac{dK}{dt} = \dot{K} = I - mK \tag{14.2}$$

In simulation models, the equation is treated as a discrete equation, but in its derivations, it is usually expressed as a differential equation. The equation describes a pseudo-dynamic process, during which the system jumps from one equilibrium state to the next. If the population increases at rate n and the fraction of workers in the population remains constant, the growth of the labour force is given by $dL/dt = nL$. Using the macroeconomic bookkeeping identity that output Y equals the sum of consumption C and investment I and normalising variables with respect to L, equation 14.2 becomes:

$$\dot{k} = y - c - (m + n)k \tag{14.3}$$

with $y = Y/L$ output per worker and $k = K/L$ capital per worker. This equation states that the stock of capital per worker equals the per worker investments minus the per worker capital depreciation and the dilution from population growth.

You can specify the model if you postulate a *production function* $Y = Y(K,L)$ that represents all the possible combinations of production factors {K,L} that produce a certain output Y.[10] Following the aggregate trends in both agricultural and industrial economies, the production function is assumed to exhibit constant returns to scale and diminishing returns.[11] Using these properties and assuming that the production function is normalised ($Y_0 = 1$, $K_0 = 1$, $L_0 = 1$ at time $t = 0$), a suitable production function is of the form:

$$Y = K^{\alpha} L^{1-\alpha} \tag{14.4a}$$

[8] The production factors capital K and labour L can be expressed per unit output. It is called the factor intensity. The inverse is called the factor productivity and tells how much output is produced per unit of factor input. More specifically, the output per unit of labour is called labour productivity and indicated with a small letter: $y = Y/L$.

[9] The depreciation rate is often approximated with 1/L with L the economic lifetime of the capital stock (Chapter 2). I suppress the index t (time) in this and subsequent equations. The assumption is that the period Δt is long enough to let the system come into equilibrium – usually one year.

[10] Normally, only the input combinations that give the highest possible output are considered. In that sense, it represents a *production frontier* and the best available technology. Real-world performance will be lower. According to economists, this justifies the assumption of *perfect rationality* in human behaviour: The models do not claim perfectly rational behaviour of real-world persons but, rather, explore the conditions and outcomes if such behaviour were to happen.

[11] Constant returns to scale is the property that total output increases with a factor n if all inputs increase with a factor n: $Y(nK, nL) = nY(K,L)$. If it is less (more), there are decreasing (increasing) returns to scale. This assumption is shown to be invalid in the service and information economy (Appendix 10.1). A production function exhibits decreasing or diminishing returns if doubling of one of the inputs gives less than a doubling in output. It is effectively equal to a stabilising feedback. The opposite of a positive feedback is referred to as increasing returns.

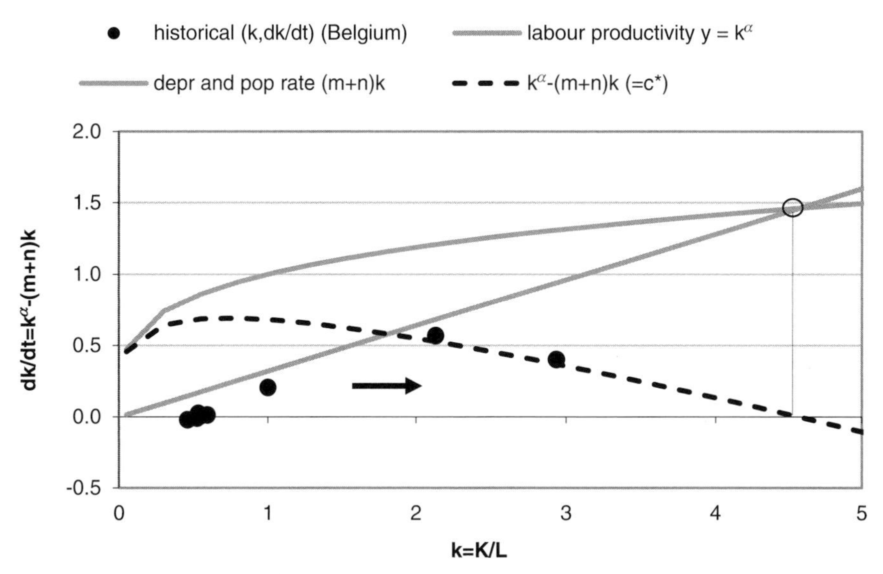

Figure 14.5. The neo-classical economic growth model. Labour productivity y increases in proportion to k^{α} (upper grey curve). Output partly goes to capital depreciation (mk) and to population growth (nk). The difference is available for per capita consumption c (dashed black curve). The steady-state levels k* and c* are at the intersection of the two curves (open dot). (alpha = 0.35; m = 25%/5 yr; n = 7%/5 yr). The black dots indicate estimated values for Belgian agriculture 1880–1980 (source of data: Swinnen *et al.* 2001).

or normalised with respect to L:

$$y = \frac{Y}{L} = \frac{K^{\alpha}L^{1-\alpha}}{L} = \frac{K^{\alpha}}{L^{\alpha}} = \left(\frac{K}{L}\right)^{\alpha} = k^{\alpha} \tag{14.4b}$$

This is the widely used formulation proposed in the 1920s by Cobb and Douglas (CD). It states that a rising capital-labour ratio k increases labour productivity y but with diminishing returns and with constant returns to scale. Inserting equation 14.4b into equation 14.3 gives:

$$\dot{k} = k^{\alpha} - c - (m+n)k \tag{14.5}$$

This is the basic equation of the neoclassical growth model (Barro and Sala-i-Martin 2004). For dk/dt = 0, there is a steady-state $\{c^*, k^*\}$ at which per worker consumption level c^* is given by:

$$c^* = k^{*\alpha} - (m+n)k^* \tag{14.6}$$

at the capital-labour ratio k*. Equation 14.5 is graphically shown in Figure 14.5 with the curves for k^{α} and $(m + n)k$ as a function of the capital-labour ratio k. For $c < c^*$ the economy will grow towards higher k-values.

Is there any empirical confirmation? By way of example, I inserted in Figure 14.5 the empirical data for agriculture in Belgium for the period 1880–1989 (Swinnen *et al.* 2001). Between 1880 and 1930, the capital-labour ratio k fluctuated between 0.46 and 0.54, and there was no substitution and no productivity increase. After World War II, the k-value started to increase rapidly and these data can be reproduced with a CD production function. It suggests a growth path towards a steady-state, but in reality, the k-value dropped significantly in the 1980s. Even in the relatively simple agricultural sector, other mechanisms than factor substitution were operating, such as innovation waves, international trade, government interference on behalf of consumers or farmers and the possibility of capital-land substitution and capital-labour complementarity in poor countries (Haley 1991).

From a sustainable development perspective, the neoclassical growth model suggests the best of possible worlds and a recipe for sustainable development. It predicts that per capita income levels can exceed subsistence levels permanently through capital accumulation, that is, rising k, in contrast to Malthus' conviction. It indicates that an economy has a natural tendency to reach a zero-growth steady-state and that per capita income differences between countries have a natural tendency to disappear.[12] Unfortunately, even apart from resource and environment constraints, the real world behaves differently. Capital accumulation and capital-labour substitution undoubtedly played an important role in economic growth. But time-series and cross-country analyses of economic growth in industrial economies during the 20th century show that they explain only half or less of the observed output growth and that no income convergence amongst the countries of the world takes place except in a few 'convergence clubs'. These shortcomings in the theory have been acknowledged since the 1980s (Barro and Sala-i-Martin 2004; Helpman 2004).

One reason of failure is the high aggregation level of description, interpretation and measurement of the production factors. *Labour* reflects a wide variety in skill and age levels. In advanced economies, fossil fuel is almost completely substituting for physical labour and most labour is involved in control and information. Economists have introduced the notion of human capital in order to incorporate the differences and changes in skill levels with proxies, such as number of years of schooling. It provides some additional explanation (Helpman 2004). *Capital* also represents a variety of items and undergoes quantitative and qualitative change. Table 14.1 shows data for the capital stocks in four countries as quantified in monetary units by the national statistical offices. Per person capital stocks have roughly linearly increased over the last decades in these countries.[13] The data indicate that in the order of two-thirds of the capital stock consists of dwellings, buildings and infrastructure and that the structural composition in monetary terms changes only slowly. But scale effects, factor returns and capital-labour substitution for dwellings, buildings and infrastructure are different from those in sectors such as information and communication technology (ICT). Indeed, this is one reason to use the simpler production function: Y = AK, introduced by Leontief. Defining the savings rate σ = I/Y and replacing $k^{\alpha} - c$ with σAk in equation 14.4, the growth equation becomes:

$$\frac{dy}{dt} = (\sigma A - m - n)y \qquad (14.7)$$

[12] The growth rate is lower for higher capital-labour ratio's, so poor countries with low capital per worker levels can grow fast and eventually converge with the rich countries at high k-levels (convergence).

[13] The increase in k, that is, in capital per worker, is probably lower in these economies because the labour force grew faster than the population. This has partly been compensated by less working hours (Chapter 4).

Table 14.1. *Structure of capital stocks in Switzerland (2006), Finland (2006) and Australia (1993–1994)*[a]

Switzerland 2006		%	Finland 2006	%	Australia 1993–1994	%
Construction	*Building*	55.8	Residential buildings	40.0	*Private*	
	Civil engineering	15.2	Non-residential buildings	26.7	Dwellings	32.5
As part of construction: Machinery and equipment		27.6	Other structures	12.6	Construction	17.2
Fabricated metal products and machinery		14.4	Other machinery and equipment	15	Equipment	15.5
Office machinery and computers		1.2			*Public*	
Electrical machinery and apparatus n.e.c.		2.3			Dwellings	1.9
T.V., radio and communication equipment and apparatus		2.1			Construction	23.6
Medical, precision and optical instruments, watches and clocks		3.8	Transport equipment	3.1	Equipment	4.8
Motor vehicles, trailers and semi-trailers		0.8				
Other transport equipment		2.9				
Cultivated assets		0.2	Major improvements to land and so on	1.1		
Computer and related services		1.3	Software, knowledge, entertainment, art and so on	1.5		

[a] *Sources*: www.rba.gov.au/statistics/op8_index.html, www.bfs.admin.ch/bfs/portal/en/index/themen/04/02/04/key/Stock_cap.html and www.stat.fi/til/pka/tau_en.html.

Assuming a closed economy in which the savings rate is constant and equal to the investment rate, this model stipulates a continuous exponential growth in output as well as in capital stock and consumption. Although this is the historical experience in certain periods and places, it lacks explanatory power. Already a century ago, economists turned to another determinant of growth: the knowledge incorporated in capital and labour or, in short, *technology*.

14.2.3 The Role of Technology, Learning and Behaviour

In the 1950s, technological change was already identified as a major source of long-term economic growth. But technology and innovation are elusive concepts that are hard to measure or observe directly. Investment in physical capital and equipment; in human capital and in research; and development (R&D) are all forms of technology and causes of productivity growth. There is not yet a validated theory of how each of them contributes and how they interact. Madsen *et al.* (2010) did an extensive econometric analysis of the determinants of economic growth in the UK in the period 1620–2006, including proxies for sectoral shares, coal availability, government expenditures, urbanisation and trade barriers. They conclude that 'the empirical estimates so far give support to the hypothesis that productivity growth

Box 14.1. *Is technology the ultimate force driving growth?* This book has discussed technology and transitions several times – but do we understand the underlying dynamics? In a rather narrow, techno-economic context, these phenomena are mostly substitution of one product or process by another in markets, as in the CD production function. Several model formulations have been proposed for the substitution process, one of the most widely known being the *logistic substitution model*. It derives from the predator-prey and logistic growth models (Fisher and Pry 1971).

If you change the state variable X from an absolute to a relative variable Y, with $Y = X/K$, you can rewrite it into its normalised form: $dY/dt = \alpha Y(1 - Y)$. Y can be interpreted as the occupancy of a particular niche. It represents the fraction *f* of the market occupied by a particular species or technology. It has been observed that *f* follows a S-shaped growth curve over time for many technological substitution processes (Grübler 1999). If the observed f-values are plotted as $\log[f/(1-f)]$ over time, it is found to approximate a (segmented) straight line for a variety of long-term historical techno-economic processes. The model can be applied to more than one variable. Examples are the rise and decline of various transport modes in France, the substitution of steel manufacturing processes and the gradual switch from biomass-based to modern fuels and then electricity for lighting in the United Kingdom (Figure 14.6a–c). Other data indicating such transition dynamics are the substitution of energy carriers and the replacement of steam engines by explosion engines.

in Britain, until the twentieth century, was predominantly a race between technological progress and population growth'. Income changes are well explained by the changes in population growth rate and the changes in research intensity, the latter being approximated by patent applications per worker. The rather high population growth before 1900 was a drag on the technology-driven income growth. On the other hand, the continuous research, patenting and implementation effort was a driver of economic growth. But trade openness, financial (de)regulation, inflation and unemployment rates, income distribution, political and civil rights, the size of government and public infrastructure investments all influenced productivity growth and continue to do so (Helpman 2004; Barro and Sala-i-Martin 2005).

The first and simplest way to include technology in the production function is with an exogenous factor, for instance:

$$Y = AK^{\alpha}L^{1-\alpha} \tag{14.8a}$$

The term A is called the *total factor productivity* (TFP) and can, not surprisingly, explain a large part of productivity growth.[14] 'Technology' becomes an all-explaining *deus ex machine*, a manna from heaven. One way to estimate A is through the idea

[14] For instance, in OECD countries, 16 to 47 percent of GDP growth in the period 1960–1995 is not explained by contributions from capital and labour (Barrro and Sala-i-Martin 2005).

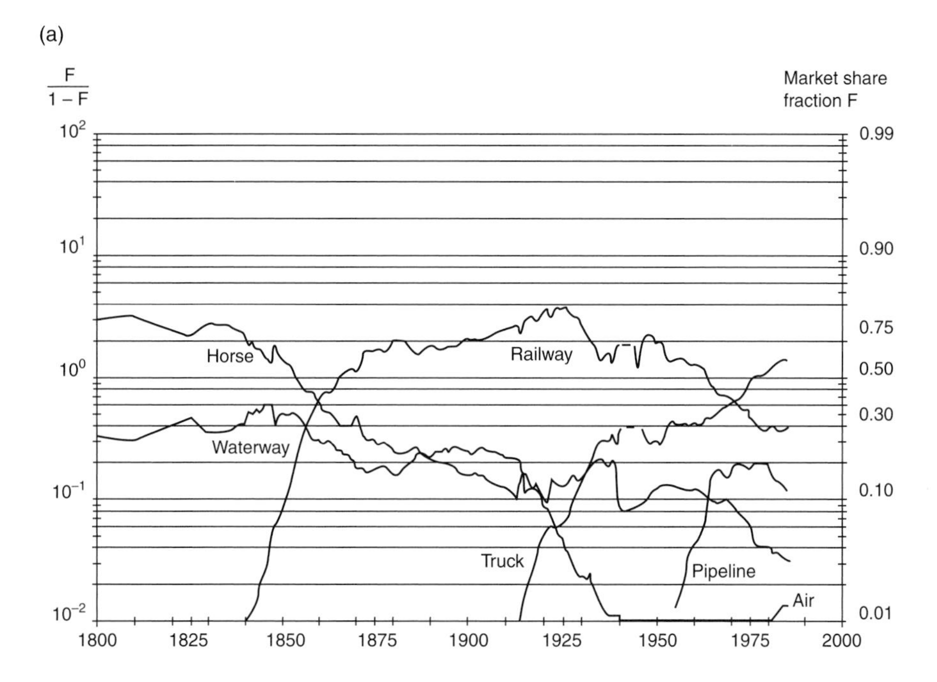

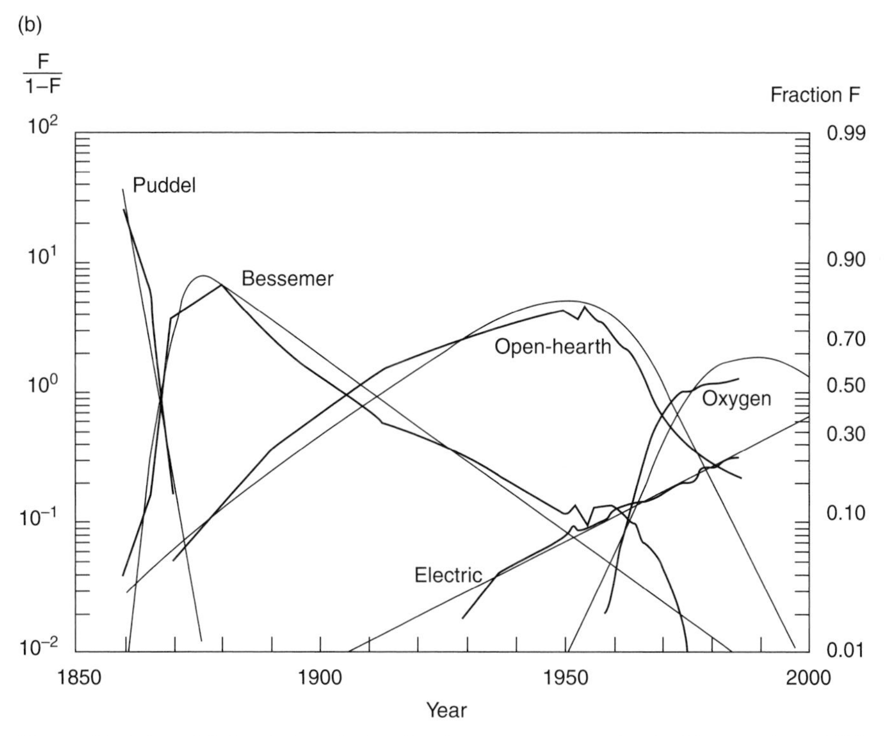

Figure 14.6a,b,c. (a) Logistic substitution in transport modes in France since 1800, (b) in global steel manufacturing process technologies since 1850 (c) and in lighting technology (*Source:* Grübler 1999, www.iiasa.ac.at).

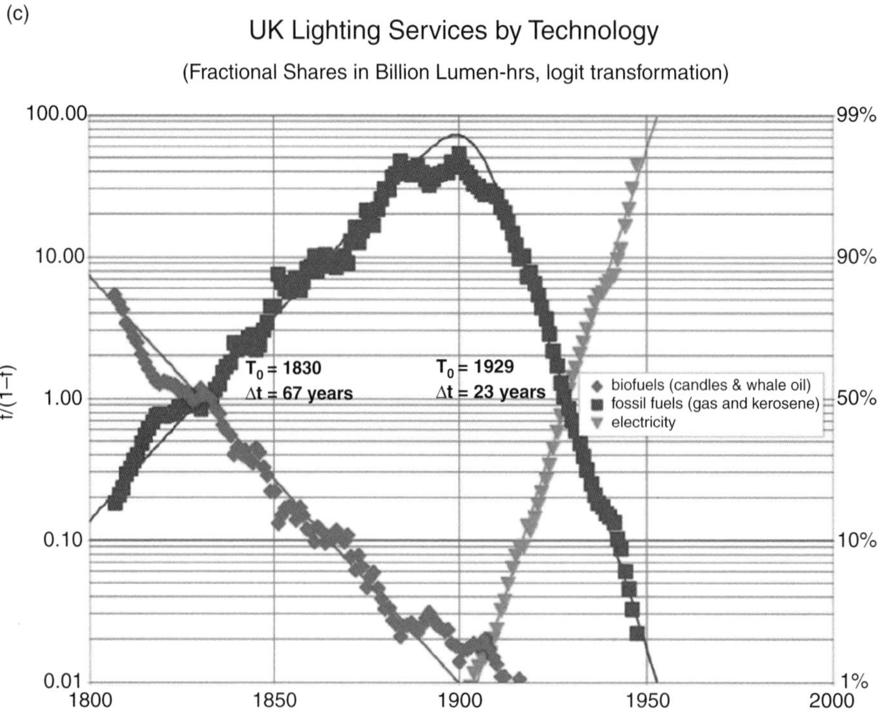

Figure 14.6 (*continued*)

of learning-by-doing, that expresses A explicitly as a function g of accumulated knowledge by workers (equation 13.2):

$$Y = g\left(\int_{\tau=0}^{T} Y_\tau d\tau\right) f(K, L) \qquad (14.8b)$$

Often, an S-shaped curve is assumed for g, with the argument that, within a certain regime, additional output no longer generates any productivity increase when the niche is filled (§13.3). Assuming a CD production function, the new formalism implies again a path towards a steady-state. Another way to endogenise technology is to emphasise the knowledge of workers and introduce human capital as a production factor: H = Lh. The variable h is a measure of the quality of the worker that can be estimated from education levels, amongst others (Barro and Sala-i-Martin 2004, Weber *et al.* 2005).

At least three more mechanisms link technology to output. The first one is the role of *factor price–induced innovation*. Production theory indicates that cost-minimisation implies a capital-labour ratio that is proportional to the inverse of the price ratio: $(K/L)_{opt} \sim (p_L/p_K)_{opt}$ (Samuelson 1947). Relative factor costs induce technical changes that allow substitution of the cheap for the expensive factor. It implies a *de facto* development and implementation of innovations away from the most expensive input, as in labour-saving mechanisation. Second, *R&D investments* generate innovations in a variety of ways, partly unplanned and unforeseen and with long-term consequences. Economists emphasise that private rewards for the innovator in the form of an entry barrier, for instance, through intellectual property rights (patents), are a necessary condition. But technological breakthroughs and lock-ins, organisational skills and other factors also play a role.[15] Third, and relatedly, *education* is a necessary ingredient of technology-induced productivity growth. Economists have introduced it in the rational agent model by making education a function of expected future earnings. In combination with demographic changes, it can provide a better understanding of the long-term population-economy nexus (§10.3; Galor and Weil 2000).

More in-depth insights into the mechanisms of economic growth come from innovation theory and business dynamics (§10.4). Since the 1980s, some economists have more radically incorporated the role of science and technology and of human behaviour in economic theory and models. Building on earlier work by Veblen, Schumpeter, Simon and others, Nelson and Winter laid the foundation of evolutionary economics in their book *An Evolutionary Theory of Economic Change* (1982). The *Homo economicus* is replaced by a pragmatic and adaptive individual or organisation, who is imperfectly informed, operates with bounded rationality and learns from previous experiences.[16] On the basis of reproduction, selection and mutation processes, agents choose or imitate strategies that are more rewarding and successul vis-à-vis their objective. The evolutionary mechanism of survival under selection pressure is thus central in these models (van den Bergh *et al.* 2007). It has the strength of the biosciences but creates a 'bottom-up' bias that tends to neglect or deny the 'top-down' coordination and regulation mechanisms (§10.5).

Evolutionary economists attempt to simulate two major groups of actors in an economy: producers and consumers.[17] A small group of agents (producers and consumers) 'drive' the system to more and novel products for consumption, exploiting the desire for profit, status and novelty. Technological change is incorporated in the investment decisions of the producers. If it is completely absent, the model economy is in a steady-state with zero growth. When it is turned on, the model economy starts to evolve in a permanent disequilibrium. Although the vocabulary of evolutionary economists is different, several model features (such as desired or anticipated versus actual values) resemble elements of the system dynamics models – and, not surprisingly, so do the conclusions. Two illustrative examples are summarised in Appendix 14.2.

System dynamic models of economic growth address some of the shortcomings in economic theory by introducing agents (mechanisms) that have local information about profitable opportunities for change but proceed, in the absence of global information about the (future) system, incrementally in a direction that improves the profit or another target or performance indicator (Sterman 2000). Change happens in incremental steps, in a gradient-following process. The engine of economic growth

[15] It may well be that – hardly predictable – long waves of general-purpose technologies (GPT), such as the steam engine, electricity and the computer, are more important for long-term TFP changes than the processes of incremental innovations

[16] In the process, the 19th-century metaphors of classical thermodynamics and mechanics are replaced by another set of – also 19th-century – metaphors from Darwinian biology (Döpfer 2005).

[17] What is called evolutionary economics has overlaps with innovation studies and behavioural and experimental economics. Methodologically, it uses the novel method of ABMs (§10.5).

is conceived of a series of connected feedbacks, or increasing returns in economist terminology. Important ones at the level of corporations are:

- *unit production cost:* there is continuous drive to lower unit cost through R&D (§11.6). Traditionally, the ways to reduce costs are through economies of scale, economies of scope and learning-by-doing. All three can work as positive feedback loops through which unit cost declines;
- *unit development cost:* in many modern knowledge-intensive industries, the upfront development costs, are a large fraction of total cost and the actual production cost are small or negligible (chips, software or music as examples). Once underway, there is an enormous drive to create large sales to recover the upfront cost;
- *product awareness:* firms will use advertising and sales efforts to promote their products. In combination with word of mouth and media attention, this may create a positive feedback towards ever larger sales and market share.

New product development, acquiring mono- or oligopolistic market power, mergers and acquisitions, promoting workforce quality and loyalty, and access to cheap capital by high profits and growth rates are other elements in the battle for growth of firms and corporations. A particularly important mechanism is the positive feedback from interaction synergies and network effects, with significant path dependencies (§10.5).

Still, there is no conclusive answer to the question whether economic growth tends towards a sustainable steady-state or not. Does it matter *what* is produced and consumed? Perhaps, the transition to a new service- and information-driven economic regime will relieve the tensions posed by finite resources, as examined in §14.1. Does *human behaviour* matter? Mainstream economic science sees individuals as rational agents, who immediately and intentionally adjust to whatever is economically optimal under new conditions (Appendix 14.1). Economic growth models can, therefore, not make meaningful statements about a change towards postindustrial values (§4.2). Do *nonlinear feedbacks* in the system influence growth patterns? The analytical growth models discussed so far are pseudo-static in the sense that the economic system is assumed to reach an internal price-volume equilibrium at any moment considered (§7.2). '[Equilibrium thinking has] so permeated economics that very few attempts have been made . . . to develop theories in which the existence of cycles is an integral feature of the economy' (Ormerod 1998). Real-world supply-demand mismatches are probably endogenous and can significantly influence the economic and financial system, but they are absent in economic growth models.

The economic system is far too complex to conceptualise in a single scheme, even if there were agreement on the mechanisms. Nevertheless, following the system dynamics approach, this book offers a causal loop diagram (CLD) of three subsystems in the economy that are discussed in this and Chapter 10: consumer behaviour, manufacturing, and capital and technology (Figure 14.7). The first one is the consumer side, in the upper part of the CLD. It is a reinforcing feedback representation of a complex process, that revolves around the individual needs and desires for positive experiences. In Figure 14.7, it is shown as the need or desire to drive a bicycle. In agreement with psychological research, the level of happiness/contentment is influenced by processes of comparison, competition, repetition and habituation (§6.2).

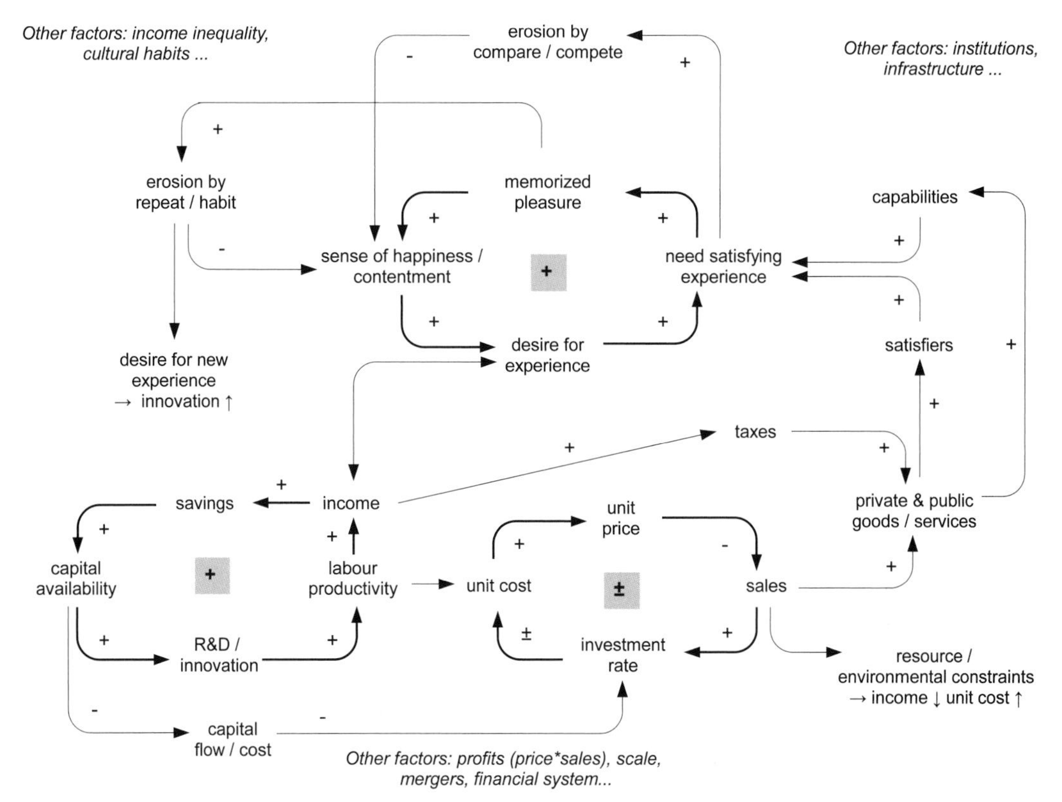

Figure 14.7. A simplified causal loop diagram representation of the economic system. It portraits three subsystems: the consumer side (above), the producer side (lower right) and the capital and technology side (lower left).

This is indicated with two negative feedbacks. The fulfillment of needs and desires is satisfied with help of private and public goods or satisfiers and capabilities. The capabilities are to a large extent supplied in the form of public goods (infrastructure, education or health), although there is no sharp distinction with privately supplied (pseudo-)satisfiers. Tax is one of the main mechanisms to finance public goods.

The goods and services that provide the satisfiers and capabilities have to be produced. This is represented with another feedback loop, in the lower right of the CLD. In essence, demand leads to investments from which costs and prices follow. The underlying mechanisms are complex market and institutional processes, but the widely shared assumption is that lower prices will lead to higher sales in the longer run in competitive markets. As we have seen, however, the dominant force is technology. It is represented in the feedback loop in the lower left of the CLD. Savings and profits provide capital for expansion and replacement investments, but also for R&D and innovation. In combination with capital-labour substitution and economies of scale/scope, it leads to higher labour productivity. This increases income and tends to decrease unit costs. In this way, technology drives the larger, positive feedback process of economic growth. It also satisfies the desire for new products that is a consequence of the repetition/habituation process amongst consumers and is stimulated by advertisements. Although the financial sector is an important or even crucial component of the economic system from a sustainable development perspective,

Box 14.2. *Are values driving growth?* Is economic prosperity following our values or the other way around? There is evidence of a link between the long Kondratiev waves and political and social values in Western nations during the last two centuries (Sterman 1986).[18] 'During periods of long-wave expansion, material wants are satisfied, and social concerns turn to civil liberties, income distribution, and social justice... As the expansion gives way to decline, conservatism grows, and political attention returns to material needs... During the downturn, the accumulation of wealth becomes the overriding concern, at the expense of civil rights, equity, and the environment... The variation of political values is primarily the result of entrainment by the economic cycle' (Sterman 1986). A deeper understanding of the interacting economic, technological and social forces is one of the great challenges for sustainability science, more so now that these interactions have spread globally and are faster than ever before.

I do not discuss it here because there is no coherent body of scientific knowledge about it yet. Instead, I continue with three other contributions of economic science to the sustainability discourse: physical inputs in the production function, the empirical input-output formalism and the dynamics of resource efficiency.

14.3 Source and Sink Constraints in the Economy

14.3.1 Structural Economics: The Input-Output Formalism

A large part of the cost of economic goods and services consists of rewards for the production factors labour L (wages or salaries) and capital K (dividends or rent), together making up the value added (VA). Labour – in essence, organisation and knowledge because most physical work has been substituted for by fuels and electricity – is usually the largest cost component. Capital, in the form of buildings, machinery and equipment, is an important second component.

Until the oil crises of the 1970s, energy was not considered in economic analyses and neither were materials. This is understandable because coal and later oil and gas became ever cheaper in the industrial economy and were considered an operational input, not a potential constraint. Since the oil crises in the 1970s and 1980s, attempts are made to introduce energy E and materials M explicitly into the production function, the so-called KLEM production function.[19] The results are, however, rather inconclusive because of the fact that energy can be both substitute and complement to labour and capital and that technological change is treated at too high an aggregation level.

[18] Including feedbacks, notably the inherent oscillatory tendencies of firms and reinforcing mechanisms among firms and labour and capital markets, is one of the ways in which long-term Kondratiev waves are explained (Sterman 1986).

[19] The costs of energy and materials as fraction of total factor costs are still small except for a few sectors such as mining and petrochemicals. With more pressure on land for food and biomass-based fuels, land may also return into the manufacturing production function.

From a natural science perspective, it has been argued since the 1930s that energy inputs, if properly measured, are a key explanatory factor of economic growth. More recently, the role of energy and, particularly, exergy in the economic production has been examined in detail. It appears that the inclusion of useful work ('exergy services') explicitly as an input in the production function yields an almost perfect correlation with GDP growth for the period 1900–1975 (United States) and 1960–1993 (United States, Japan and Germany) (Warr *et al.* 2002; Ayres and Warr 2005).[20] The apparent trend break around 1975 in the United States and other advanced economies is possibly explained by the oil price hike induced efforts to increase energy productivity and by the rise of ICT. The analyses make it clear that useful work is a *sine qua non* for the growth in economic output realised in the high-income regions in the 20th century and being underway in emerging economies in the 21st century.[21] It strongly suggests that GDP growth in the low-income regions of the world must concur with an increase in the use of exergy services, notably electricity.

National economic models distinguish economic sectors, each with their production function. Each sector delivers part of its output directly to consumers, including government and investments. This is called final demand. The remainder is delivered to other sectors and are called intermediate deliveries. Together with the primary inputs labour and capital, they make up the *input-output (I-O)* table (Appendix 14.3). From the I-O table, it is immediately seen how much input from an economic sector j is used per unit of final demand output from sector i, in €/€. This is the *direct intensity* of that input – and the input can be energy, steel or whatever specific entry the table contains. If you choose labour or capital as the input, you can calculate the direct labour or capital intensity. Most material- and energy-intensive sectors have a rather low labour intensity and a rather high capital intensity, at least in the high-income regions where wages are high. Not surprisingly, direct expenditures on energy are higher when you buy for €100 products from the chemical sector than when you spend it on childcare – twenty times more in the Netherlands in 2001. But those same €100 generate six times more employment when spent on childcare as compared to chemical products.

It is also possible to calculate the monetary flows associated indirectly with a final demand output purchase – the *indirect intensity* of that input. For instance, the energy used to deliver 1 kg of garbage bags has a direct energy intensity α_{ij} with i energy sector and j bag manufacturer, but there is also energy used for the production of the plastic used in the bag (α_{jk} with j plastic manufacturer and k bag manufacturer) (Appendix 14.3). The sum of direct and indirect intensity yields the total intensity. The differences in *total* intensities between sectors are smaller than the direct intensities. Take the example of childcare: Part of your €100 spent on childcare will soon be spent afterwards on gasoline by one of the employees travelling by car as part of her service sector job. This adds to the total energy intensity of those €100 of expenditures.

An illustrative outcome is given in Figure 14.8a,b. It is done on the basis of a time-series of annual consumer expenditures (final demand) and I-O tables and

[20] For commercial energy instead of exergy use the correlation with empirical data is much weaker.

[21] Two other conclusions are that electric power is of more than proportionate importance because of its high quality, and that there is a discrepancy between the elasticity of production and the cost

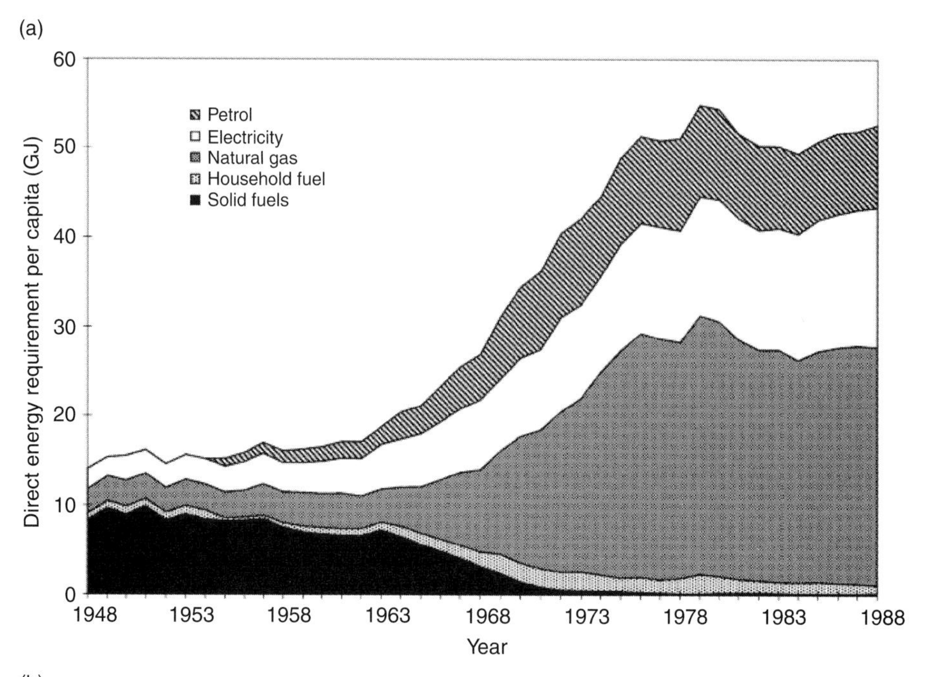

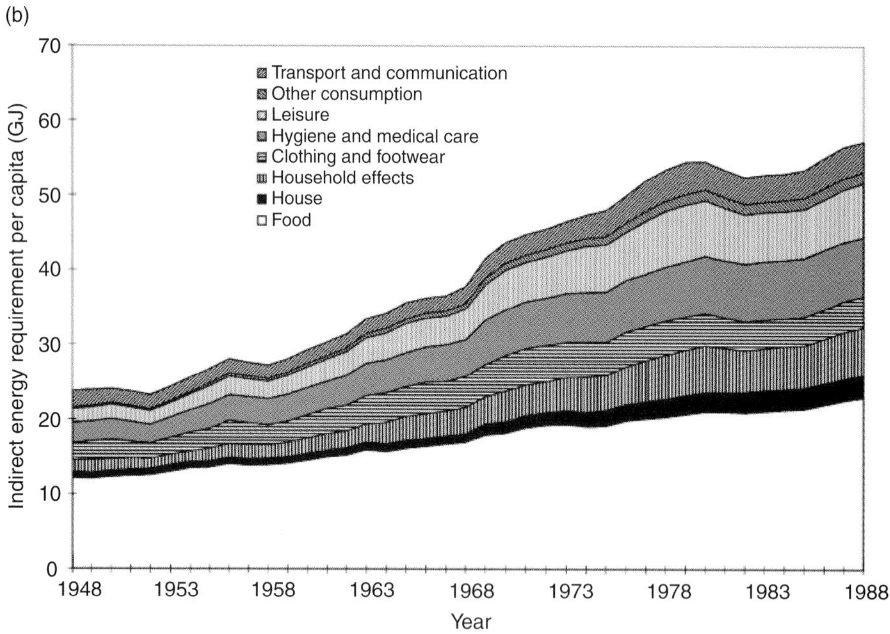

Figure 14.8a,b. The direct and indirect energy use per capita per year for Dutch households from 1948 to 1988. The indirect energy use is given for eight consumption categories (Vringer and Blok 2000).

some simplifying assumptions. The rise in income from about 7,000 to 18,000 NLG (Guilders) per person per year between 1948 and 1988 coincided with an increase of

shares of energy and labour (Kummel *et al.* 2002). The latter points at the necessity of higher energy prices and lower labour costs/prices.

3.4 percent per year (%/yr) in total energy use per year per person. Direct energy use started to rise rapidly after 1965, partly in response to the discovery of the large Groningen gasfield (Figure 14.8a). In 1990, less than half (44 percent) of total energy use was spent directly on energy in the form of fuel and electricity (Figure 14.8b). The remainder was indirect energy use, with a significant but declining fraction of it for food. I refer the reader to the Suggested Reading for more detailed methods and results.

There are some pitfalls in the use of I-O tables. One of them is how to deal with trade in open economies. An accurate treatment is to include the energy embodied in imports and in exports on the basis of transport and I-O tables for the trade partners. Although this is a rather tedious exercise, it can shed light on an element of unsustainability in globalisation, namely that stricter environmental regulation in high-income regions makes it, in combination with cheap labour in low-income regions, profitable for corporations to move their (material- and energy-intensive) sectors away from high-income countries towards countries with less strict regulation. Theoretically, this is simply using a comparative cost advantage. Practically, it can lead to a slowdown or even decline in environmental and other regulation.[22] Analyses based on I-O analyses with estimates of embodied energy in imports/exports indicate that at least part of the dematerialisation (in MJ/€) in the United States, Europe and Japan has occurred because of shifts in economic activity to Mexico, China and other low-income regions.[23]

I-O tables can be of great help in linking physical flows to monetary flows. The elements of the I-O matrix are converted to physical units with use of (average) sectoral prices and double-checked with statistical data on physical flows in so-called *satellite accounts*. These represent I-O tables in physical flows and are sometimes called physical input-output tables (PIOT) (Ayres and Ayres 2002). A systematic connection between monetary and physical I-O stocks and flows was first applied in energy analysis. Later, environmental accounts were used to assess the the use of materials and amount of emissions per unit of final demand (t/€) (Wilting *et al.* 2008; Peters *et al.* 2011). In this way, the direct physical input (energy or material) and output (emission) flows associated with consumer and government expenditures can be calculated. A further extension is to include more detail on the primary factor contributions, for instance, separating non-paid agricultural labour and distinguishing income classes. Such a framework is called a social accounting matrix (SAM). For low-income agrarian societies, this is a more adequate framework than the monetary I-O table (Morrison and Thorbecke 1990).

An I-O matrix gives a static picture or *snapshot* of the structure of an economy. One of the challenges is to connect economic growth models and I-O data in a transparent and effective way. One way to do this is to isolate investment flows as a separate final demand category. In this way, the empirical data on structure can be

[22] The country with the lowest standards in environmental regulation – and other regulation, such as labour conditions and nature protection – attracts industries in search of higher profits and force other industries to lobby for less strict regulation in their country. This is the *race-to-the-bottom* dynamic.

[23] This is one of the aspects that complicate the negotiations about carbon emission reductions in the context of international climate policy, because it invalidates a simple GJ/€ measure as indicator for emission targets.

Box 14.3. *Rebound effect.* One interesting feedback phenomenon is the so-called rebound effect (Polimeni *et al.* 2009). It is simply explained. If cost-effective measures to reduce energy and material use in industrial production help to lower production cost, they will also in the longer run reduce the product price, if competition is working. This in turn probably leads to a higher demand for the product and hence to production expansion and a higher overall material-energy use. As a consequence, a part of the benefits of the measure (material-energy saving) are annihilated by growth. It was famously stated by Stanley Jevons in his book *The Coal Question* (1865): 'It is wholly a confusion of ideas to suppose that the economical use of fuel is equivalent to a diminished consumption. The very contrary is the truth'.

A similar mechanism occurs in private households when money not spent on fuel because of better insulation or lower room temperature is spent in other ways, thereby leading to additional activity and hence additional use of energy. Some have called it the Torremolinos effect, suggesting that the saved money is used to fly with TransEasyRyan to a Spanish resort. Rebound effects – and other economic feedback effects such as carbon leakage – are studied extensively and their extent is still a matter of controversy.

combined with a standard economic growth model in order to simulate the system as it goes from one equilibrated situation at time t to the next at time $t+\Delta t$. In between, the technological coefficients are adjusted to represent trend- or expert-based expectations about (future) technologies (Sassi *et al.* 2010). This is probably one of the more solid ways to explore future demands on natural resources (Duchin and Lang 1994).

14.3.2 Resource Efficiency and Pollution Abatement: Economic Mechanisms

The dynamics behind the changes in material and energy intensity is an important topic in industrial economics/ecology and also in sustainability science (§13.5). Most goods and services are produced with a variety of inputs and consist also of many different 'products'. Engineers prefer to construct production functions on the basis of physical flows (engineering production functions), but this is only possible for a few basic processes (steel, cement and some others). Currently, complex manufacturing processes are investigated in dedicated technology assessments. Monetary analyses in production economics offer nevertheless some useful insights in the mechanisms of resource efficiency, or resource saving as it is popularly called, and provide some guidelines for a sustainable resource use policy.

Assume that a manufacturing plant produces P units/yr of a good with machinery valued as a capital stock K. It uses E energy units/yr. Energy use can be reduced by investments, and thus substitute capital for energy. At the level of a manufacturing plant, office or house, there are numerous examples of additional investments, for instance, more insulation, double glazing, larger heat exchangers and so on, that lead to a reduction in energy required to produce the same output.[24] It is described with a production function, for which we choose again the simple CD form:

$$P = P(K, E) = A_t P_0 K^{\alpha} E^{\beta} \text{ t/yr, m}^3\text{/yr}, \ldots \quad (14.9a)$$

A_t is the productivity-increasing technology factor. The total costs to produce P units is given by:

$$C = p_k K + p_E E \text{ €/yr} \quad (14.9b)$$

with p_i the price of factors i = K, E respectively. An equivalent formulation can be used for materials M. The formulation implies that an energy *flow* can be substituted with a capital *stock*. It is assumed that there are no positive or negative effects on output from increase in scale, such as constant returns to scale ($\alpha + \beta = 1$). Empirically, the production function is a set of points that represent combinations of inputs K and E in existing plants, long-term analyses or future plants.[25] They represent the *production frontier*.

What does this simplified model tell about energy efficiency? Suppose that a firm is producing on the production frontier. Any change in the p_E would induce a switch to a technology using less E and more K as long as the savings exceed the cost. In other words, as long as:

$$p_E \Delta E \geq p_k \Delta K \text{€/yr} \quad (14.9c)$$

with prices p_K and p_E the respective factor prices and under the assumption of cost rationality.[26] To understand this process, I express equation 14.9a-b in per unit of product P for a given output level P_0. Assuming $A_t = A_0$, this yields:

$$k = A_0^{-1/\alpha} e^{(\alpha-1)/\alpha} \quad (14.10a)$$

$$c = p_K k + p_E e \quad (14.10b)$$

with $c = C/P_0$, $k = K/P_0$ and $e = E/P_0$. Plotting the product cost c as a function of the energy input e shows the two components (Figure 14.9). The linear part are the energy cost and proportional to e (equation 14.10b). The downward sloping curve are the capital cost expressed as function of e (equation 14.10a). The black curve are the total costs at an energy price $p_E = 10$. The grey curve are the total costs at an energy price $p_E = 1$. The arrow indicates that an increase in energy price p_E should induce an additional capital investment of about 50 percent that reduce energy use with about 60 percent. This 'energy-saving' operation makes the unit cost about ten units less

[24] The empirical data have as yet not given unambiguous confirmation of capital-labour substitution at the aggregate level of a sector or an economy. As said before, this is because capital and energy are not only substitutes but also, and certainly in a longer time perspective, complements (mechanisation, automation or robotisation) (Frondel and Schmidt 2006).

[25] Note that the input of material or energy per unit of output is the material intensity (t/€) or energy intensity (MJ/€).

[26] It is assumed that the additional investment ΔK is depreciated over an economic lifetime 1/m (equation 14.1). This can be done in the form of an annuity, which includes depreciation and rent in the cost of capital p_k. The reduction in material and/or energy flows ΔME is on an annual basis.

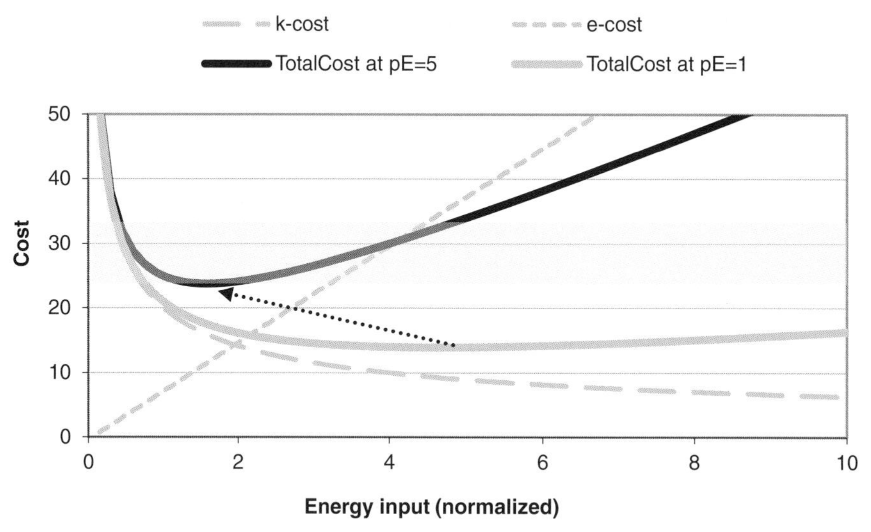

Figure 14.9. Substitution dynamics between capital and energy. The dashed curves indicate the per unit cost as function of the energy input ($A_0 = 1$, $\alpha = 0{,}5$, $p_K = 10$ and $p_E = 5$). The upper dashed curve are total costs. The linear curve are energy costs. If p_E goes up, the total cost and energy cost curves move upwards, and the minimum cost level will be at a lower energy input.

costly than doing nothing (the grey area). The plant is then operated at higher fixed capital cost but lower operational (energy) cost. It can be proven that the minimum cost in the total cost curve is for $E_{opt} = (p_K/p_E)K_{opt}$.[27] It represents the optimal investment in energy efficiency improvement and corresponds in engineering terms with, for example, a certain size of a heat exchanger surface, a certain thickness of insulation material or a certain number of steps in a distillation column. Or, in your house, with a certain thickness of insulation or triple-glazed window surface.

The example illustrates the importance of price responses in resource efficiency. It also offers two other insights. First, it is the *ratio* of energy and capital price, not the energy price *per se*, that induces efficiency gains. If the price of capital p_K increases too, for instance because the investor can get a higher return elsewhere, then the reduction in energy use will be lower. Second, the approach presumes a continuous set of techniques being available as energy-savings options. In reality, there will be a discrete set of available techniques on the production frontier which are always 'on the move' in the K-E-plane because of (expectations about) innovations, energy prices, environmental regulation, changing tax regimes, wage negotiations and so on. Indeed, these options are usually only developed when the energy price (is expected to) increase.

Manufacturers, as well as office managers and households, will often combine replacement investments with expansion investments. This makes it more difficult to find optimal estimates of ΔE and ΔK. In the practical evaluation of whether to

[27] Of course, energy also substitutes for labour and environment, for instance in automation and pollution abatement. This makes the analysis more complicated. The optimal factor allocation has historically been applied to capital and labour and was one of the controversies in economic history (Samuelson 1947).

switch from technology 1 → 2, often the simpler criterium of the *payback time* (PBT) is applied:

$$\frac{K_2 - K_1}{p_E(E_2 - E_1)_1} \leq PBT \tag{14.11}$$

This rule-of-thumb investment criterium says that a firm will only decide to switch to a more energy-efficient technique if the ratio of additional investment and annually saved expenditures is less than PBT years. The lower PBT, the stricter is the criterium. The higher p_E, the more efficiency-investments are made.

This type of analysis gives a feel for the energy and material savings potential and for the opportunities to reduce the physical flows in manufacturing. Also, it helps to distinguish different efficiency potentials. The *theoretical* potential – which is in physical terms but not in monetary terms determined by the laws of thermodynamics – cannot be realised. It is some vertical line to the left in Figure 14.9.[28] The *technical* potential at any given moment will rarely be realised, because it will run against the rising cost barrier and becomes too expensive. The *economic* potential is given by the optimal investment at the minimum overall cost. Note that the cost curve in Figure 14.9 is the equivalent of the supply cost curve (SCC) in energy supply (§13.3). Therefore, its empirical equivalent is called an *efficiency supply cost curve* (ESCC). The ESCC may fall over time when incremental innovations in the process of learning-by-doing bring down the production frontier and the economic potential approaches the technical potential.[29] Even more important are technological breakthroughs, such as fundamental innovations that lift the technical potential, thus creating a new branch in production space. In some situations, the total cost curve is almost flat, which suggests that energy efficiency is possible at no additional costs. Such possibilities are called *no-regret* or *win-win options*. But the common situation is that, after the 'low-hanging fruit' has been picked, more efficient technologies and organisational measures are accompanied by higher cost.

From this brief exposition, it is seen that the potential to reduce the resource intensity of a process is a *moving target*. It depends on the changes in factor prices and on the progress made in technology and organisation. Moreover, other product and process innovations and marketing dominate business decisions, because in most firms the costs of energy and materials are only one, often minor, element in a dynamic and competitive environment of capital and labour related decisions. Moreover, a large fraction of office managers and households still have only limited knowledge of energy/material costs and, if so, see them in relation to their income – and conclude, for instance, that a tough negotiation about wages brings in more money than investing in resource efficiency. Finally, there is the rebound effect. These barriers to sustainable resource use have to be overcome by better information and indicators and by gradual changes in perceptions and behaviour.

Upon widening the system boundary, the production space becomes more-dimensional and the production frontier is no longer a well-defined set. Other, complementary effects are identified that may stimulate but also hinder resource

[28] Strictly speaking, one should, therefore, use a production function of the form $Y \sim K^{\alpha}(E - E_0)^{1-\alpha}$.

[29] Innovation should be interpreted here in a broad sense: It refers to technical as well as organisational and legal skills, practices and measures.

Box 14.4. *The industrial economy growth paradigm.* Although most people are not familiar with economic growth models, the idea that economic growth is good and necessary is widespread. In April 2010, an editorial comment of a Dutch quality newspaper stated that 'a healthy economic growth is the best way to relieve the debt burden. Whatever other considerations, this should be a short-term priority'. One often hears mentioning the unemployment rate as an argument for economic growth. But in 2010 the debts accumulated from pro-growth financial policies and rescuing of the banking system became the argument for economic growth.

In February 2007, the same newspaper had as its headline that the Dutch Central Bureau of Statistics (CBS) announced 2006 as an excellent year for the economy 'with a growth of 2,9%... The growth in 2006 is however smaller than the 3,25% which the cabinet used as its starting point in its Miljoenennota... Germany announced this morning 2,7% over 2006 and France 2%. The Dutch economy may in the fourth quarter have been affected negatively by the very mild autumn. This caused a lower demand for natural gas by consumers and the energy sector therefore contributed less to the growth'. This article highlights at least three phenomena: the government (always) expects higher growth, a country judges its growth performance in competition with others, and less energy use is seen as bad for economic growth.

As a third example of the economic growth addiction: numerous are the newspaper items that state that *'economic growth is good for the environment'*. But growth-enhancing expenditures to combat pollution amongst the rich is very different from expanding economic output to satisfy basic needs of poor people. All three examples are symptoms of the fundamental dilemma: *with* growth the boundaries and feedbacks of a finite earth, *without* growth the risks of social instability – a genuine quandary for governments. Human social and economic systems are like 'somebody who is struggling not to fall forward and can only prevent this by walking forward' (Jantsch 1980).

efficiency investments. For instance, the actual realisation of a switch from technology 1 → 2 often turns out to incur additional 'costs' such as gathering information, surveying the work, maintenance arrangements and so on. Such costs are called *transaction costs* and they are often overlooked – making it less puzzling that efficiency investments are not made or lagging. Other obstacles to realising the economic potential are status-driven behaviour, incomplete information on prices and available techniques, and diverging values, for instance, between cultures. It can also work in the other direction, as with status, and if one includes the positive externalities such as lower emissions and other co-benefits, there may exist even negative cost options.

The preceding formalism can also be applied to pollution abatement. Investment decisions are made in order to replace part of operational costs for waste management and emission charges by fixed capital costs. Those operational costs now concern non-market priced environmental goods such as clean air or water, and society must agree on some price, via taxes or emission trading, or introduce and enforce regulation and standards in order to elicit the necessary investments (Pearce 1990; Perman *et al.* 2003).

14.4 Economic Growth and Sustainable Development

14.4.1 GDP and the Need for a Better Indicator

The usual measure for the standard of living is the *gross domestic product* (GDP) per person (Appendix 14.4). In combination with the expectations and aspirations of people, heightened in today's globalising world of communication and advertisements, there is an enormous momentum for ongoing economic growth of GDP. Such growth is hoped to satisfy the large still unmet basic needs of the population in low-income regions and provide jobs in order to distribute rising welfare and prevent social unrest. It is also driven by the desires of an emerging middle class for the luxuries of modern life. In the high-income countries, only a combination of steady innovations, advertisements and high levels of working hours can bring ongoing growth in GDP and income. Here, too, it is considered the recipe for jobs because employment is the core mechanism of wealth redistribution and meaningful social positioning. In both low- and high-income regions, growth is propelled by the desire for returns on capital, a continuous battle for corporate profits and market share, large advertising and sales efforts and the stimulants provided by the financial system. Governments are also keen on economic growth because it is a prominent way to satisfy the large needs for public goods (roads, schools, hospitals or army). Besides, politicians depend on budgets for their status as a person and as a nation – and occasionally for their personal wealth. At a psychological level, growth in income satisfies the needs for personal success, social distinction and novelty.

Therefore, income growth is expected to continue, driven by the spontaneous urge to activity which Keynes called *animal spirits*. From a sustainability perspective, one should ask whether continued economic growth is possible in view of the constraints of a finite planet. Will it not start to interfere with its very aim of a better quality of life? And, relatedly, is a growing income the best or only way to well-being? I leave it to the reader to construct his or her own answer, with the previous chapters as possible guidelines. But whatever the answer, the question needs to be addressed as to whether income measured as GDP/cap is an adequate indicator of well-being in a sustainable development framework. The answer is no, for various reasons (Hueting 1980; Daly and Cobb 1989; Daly 1996; van den Berg 2009).

Surveys suggest that income is only at low levels clearly correlated with well-being – and even there it depends on context (§6.1). The use of *average* income as the standard measure of well-being makes it worse, because it neglects the role of income differences and comparison. However, there is a more down-to-earth accounting objection: In the calculation of GDP, there is no distinction made between costs and benefits. This may cause a serious overestimation of GDP as a source of quality of life. Many market-related transactions contribute to GDP but are not benefitting anyone and are actually costs (negative externalities). Well-known examples are the expenditures associated with traffic accidents, pollution abatement and natural disasters. Advertisements are counted as benefit, although they largely reflect corporate battlecries. The experience of nature is to an ever larger extent associated with travel and accommodation expenditures. Financial transactions show up as contributions to GDP, although they are at least partly an indicator of excess greed and risk. And one can also wonder about the medical and psychiatric expenditures to combat the side effects of life in a (post)industrial society.

A second flaw is that there is no distinction between stocks and flows. Resource use is accounted for in GDP on the basis of the monetary value of the extraction flow, without accounting for the decrease of the resource stock. It is like counting only the money you withdraw from your account, without considering the debt you are building up meanwhile.[30] Similarly, the degradation of environmental sinks is not counted in GDP, whereas it incurs future costs in order to maintain quality of life that are counted in future GDP. Thus, many negative externalities are either counted as positive contributions or not counted at all or are counted showing up as 'contributions' to future GDP. This partly explains newspaper statements like 'Catastrophic floods stimulate economic growth' or 'Pollution boosts industrial investments'.

Two alternative indicators explicitly address these shortcomings (§5.5).[31] The first one is the Index of Sustainable Economic Welfare (ISEW), which tries to account for contributions to GDP that are actually costs (Daly and Cobb 1989). It has been constructed for many countries. Using a set of conventions and assumptions, well-being measured as ISEW stopped growing in most countries some decades ago. The second indicator is the genuine savings rate, which focusses primarily on the adequate inclusion of changes in natural and human capital. This one has also been constructed for many countries. The results suggest that GDP growth is significantly overestimated in countries that overexploit their oil and forest resources, whereas it is underestimated in those that have invested in education. I refer to the Useful Websites for more details. Interestingly, there are also reasons to suspect that GDP and income *under*estimate our well-being, or at least welfare and utility. This stems from the fact that GDP only counts activities that are associated with formally registered monetary transactions. In pre-industrial Europe and still in parts of non-industrialised countries, many goods and services are produced and consumed within the confines of family and village. In many places, labour is still paid *in natura* in exchange for protection and market transactions take place outside the monetary system (§3.3).[32] GDP underrepresents these economic activities. Some attempts have been made to calculate their monetary equivalent, resulting in more weight for the role of agriculture.

A second, overlapping category of activities that are not counted are household activities. Statistics and economic textbooks see households as places of consumption and leisure. Including non-monetised transactions in households by valuing working-hours at ongoing average wage rates would increase GDP with the non-market activities (cooking meals, cleaning clothes, rearing children, do-it-yourself repair and so on). For subsistence, for households in agrarian but even more so in industrial societies, the value of the production in households

[30] In this sense, the debt crisis in many countries if not different from the resource squandering in others.

[31] There are many additional indicators that are used to evaluate a country's economic situation. For instance, the Human Development index (HDI) and the Corruption Perceptions Index emphasise social-political aspects, whereas the Competitiveness Index and a host of financial indicators reflect supposedly the economic situation.

[32] In large parts of rural India, for instance, more than 90 percent of human activities is estimated to be in the informal economy and not recorded in monetary balances (Dasgupta and Singh 2005).

(gross household production or GHP) may be significant. In the United States, it has been estimated to be in the order of one-third of GDP. In Australia, the time spent on meals, laundry, childcare and shopping was in 1975–1976 equivalent to an estimated 60 percent of total GDP (Ironmonger and Sonius 1989).

The net result of these deficiencies in GDP are difficult to assess. On the one hand, ever more activities are incorporated in the formal economy through increasing employment of women, the desire for control and taxes by governments and most politicians, and the pressures and conveniences of ICT services offered to citizens by banks and governments.[33] On the other hand, the transition to a postindustrial service, information and experience economy generates a wealth of new activities – think of Internet websites and financial products – that are not or only indirectly included in the national statistics.

Yet another mismatch between income and well-being stems from the difficulty to measure the contributions of a non-ending wave of innovations to individual and collective well-being that are counted in GDP only with their monetary equivalent. I now get a thousandfold faster computer for 100 hours of paid work than thirty years ago for 100 hours of work. The opportunities and positive externalities of ICT have increased beyond imagination and exceed by far the growth of income per se. This is often not perceived because of habituation and real and imaginary inflation (§6.2). It seems inevitable that the usefulness and adequacy of GDP and income as an indicator of 'the good life' will further deteriorate.

The many facets and ambiguities of well-being and quality of life cannot be expected to be covered with a single indicator. Nevertheless, it is necessary and useful to look for indicators that can set targets for and use measurements of progress towards sustainable development (§5.4). The challenge is to find a proper balance between the universality and locality and between the material and the immaterial. Universality of an indicator is desirable in order to reflect strong knowledge of the world and to make comparisons possible, but locality is needed in order to respect the large diversity in people and their quality of life situations and experiences. The material aspects of quality of life have to be covered because aspects such as adequate food and access to health services are essential and measurable ingredients of the good life. But the immaterial aspects should not be forgotten, as access to education and the right to justice are equally important though harder to measure determinants of a good quality of life.

14.4.2 Beyond Models: Welcome in the Real World

Elementary models of economic growth, it seems, make up a consolidated body of theory, closely connected to prevailing attitudes and practices, and there is as yet no coherent alternative.[34] Yet, there are numerous shortcomings and criticisms of mainstream economic theory (§10.4). This is not surprising because the (world) economic system is extremely complex and strong knowledge about it is not to be expected. The inadequacy of economic theory is, however, partly because of an

[33] It has been estimated that the entrance of women to the formal labour market was responsible for up to one-third of GDP growth in the Netherlands in the 1990s.

[34] See, however, the list of Useful Websites for various new directions.

overemphasis on mathematical formalism: 'The human mind is built to think in terms of narratives, of sequences of events with an internal logic and dynamic that appear as a unified whole . . . It is generally considered unprofessional for economists to base their analyses on stories. On the contrary, we are supposed to stick to the quantitative facts and theory – a theory that is based on optimisation, especially optimisation of economic variables' (Akerlof and Shiller 2009). I, therefore, discuss in a more story-like fashion two issues that are particularly relevant in a sustainable development context and difficult, if not impossible, to incorporate in models: *public goods* and *governance*. For more sophisticated treatises, I refer the reader to the Suggested Reading on these topics.

Knowledge and infrastructure – including our cultural heritage – are in essence *public goods* and provide *public services*. It is evident that accumulation and diffusion of knowledge increases productivity. It is driven by educational institutions and R&D investments and has important positive externalities in an industrial economy. Infrastructure, for water, energy and transport and, more broadly, for education and health services and social security is equally essential for economic prosperity.[35] They generate positive externalities or spillovers: the effects benefit more people than just the ones who are directly involved. For reasons of planning, cost and fairness, they use to be provided by governments and are considered a 'natural monopoly'.[36]

It can be expected that public goods become more, not less, important in an era of increasing population, rising expectations and growing pressures on the natural life support system. Increasing densities of people and activities will naturally cause the need for more regulation and organisation (Figure 12.10). Health, education and transport services are constituents of the 'good life', in the sense of use and capability to use. Their inherently public goods character demands collective arrangements and planning, realisation and enforcement. Similarly, their rivalrous twin: common pool resources (CPR), are crucial for quality of life, as previous accounts of groundwater, fisheries and forests show. Identifying and governing the'global commons' such as the atmosphere and oceans is one of the greatest challenge for sustainability.

The neoliberal trend of the last decades towards deregulation and privatisation in the public domain is heading for the wrong direction. It has been legitimised with the argument that competition leads to more efficiency and lower costs, but the validity of this argument is one of the worldview-related controversies (§6.4). Innovations and infrastructural developments cannot and should not be directed by consumer markets only, because the needs of those who are less powerful and assertive in the market place – amongst them non-humans and future humans – are not met. Finding a balance between unregulated market-driven processes and state- and community-based planning is a precondition for most sustainable development initiatives. New social arrangements, entrepreneurial models and consumer lifestyles

[35] In an analysis of physical indicators (paved roads, telephones, electric power) for a set of countries over the period 1950–1992, there is 'clear evidence that in the vast majority of cases infrastructure does induce long run growth effects', but with a great deal of variation in the results across individual countries (Canning and Pedroni 2004).

[36] Knowledge and infrastructure have both a public good/service character, but there are also differences. Knowledge *per se* is not rivalrous, whereas most infrastructure has congestion effects that make their use often rivalrous (§5.4). See Table 5.1 with respect to various forms of management.

must be explored, representing a societal contract that bridges the extremes of full state control and anarchic individualism.[37]

One of the big questions is whether the necessary new forms of *governance* and *institutions* are possible within the current capitalist system and without a new 'grand story'. In the 1990s, the system of neoliberal capitalism with democracy as its political counterpart was by many seen as the 'end of history': Nothing better was to be expected (Fukuyama 1992). Economic theory and practice took market capitalism and its basic tenets as the starting point. History has proven its superiority – or so the myth goes. Property rights, competition, the right to free enterprise and free exchange of goods and services brought the rich countries today's prosperity. The reality is different, with its corporate oligopolies, alliances between business and government and militarily supported control of resources. In the course of the 20th century, the most destructive aspects of European and American capitalism were mitigated by government regulations and programs.

With the fall of communism, the forces that restrained capitalism were gone. In combination with the ICT/revolution, globalisation intensified competition, eroded the tax basis and the legitimacy of governments and gave room for free-riding and illegal and criminal practices (Castells 1996, Reich 2007). But the expansion of the global economy with large, new players (Brazil-Russia-India-China or BRIC) also brings a new world order with new rules. Whether and how democracy will fare in this situation is uncertain. Will the 'emerging economies' (BRIC) give the world new models of economic development and political governance? For the moment, it seems that, in the 21st-century Internet world, NGOs in all their diversity are probably the best if not only way to draw the divergent private interests in a permanent dialogue on how to safeguard the overarching public interests.

These events and trends influence the prospects for sustainable development. The following mentions five areas of concern and directions for solutions.

- *The struggle for resources*. Rising demand in combination with scarcity and environmental change lead to price volatility and speculation, which in turn trigger social unrest. It induces corruption and war in states with large resources but weak governance and institutions. World price and trade agreements for specific commodities are part of the solution. Transfer of technologies and introduction of standards and labels for more efficient resource use and recycling are another part.
- *The income and wealth gap*. A conventional view on economic inequality is that wealth 'trickles down' from the richer to the poorer strata of society with a rise in average income. 'A rising tide lifts all boats.' This view rationalises wealthy people's desire for more. But the measured effects of growth on the poorest segments of society are controversial. Often, a more appropriate aphorism is 'the winner takes it all'. The erosion of middle classes and taxpayer ethics is a matter of concern in this respect. There is probably an 'optimum' income inequality, when the incentive to take risks and create wealth is in balance with the fairness needed for citizen's compliance. Outside such a balance, social and

[37] Recent examples of such explorations are Matsutani (2006), Barnes (2006) and Jackson (2009).

economic stability is at risk. Restoration of government legitimacy and other forms of redistribution than through paid jobs and taxes are needed.[38]

- *Trade and globalisation*. More open economies are said to have higher economic growth rates. But trade effects are mixed up with trends in transport cost, capital mobility and credit and debt formation. In theory, trade can encourage or discourage the growth of income per capita. In practise, the already rich countries appear to benefit most.[39] There is a need for whole-chain analyses in order to understand and evaluate the effects of trade on local, regional and global sustainability. Globalisation is perhaps already in the overshoot domain and targeted forms of protectionism are needed.
- *Corporate responsibility*. In the absence of legitimate global governance, sustainable management of the 'global commons' cannot succeed. The short-term and quick and high returns still dominate most business. Design and implementation of more stringent rules for corporations, as laid out in, for instance, Corporate Social Responsability (CSR) and Earth Charter, are needed and some corporations already take the lead. The financial system in particular has to be transformed towards more responsible behaviour.

There are a few principles for a sustainable economy that have to be strengthened in the coming decades:

- internalise, encouraged by regulation and prices, an ethic that does not permit shifting the burden onto people far away in space and time and onto other species;
- as part of it, redirect consumerist lifestyles towards more immaterial lifestyles that emphasise moderation, intelligence, cooperation and sharing;
- follow basic principles of ecology: there is no such thing as 'waste' in nature, interdependence is the rule not the exception, (bio)diversity is the best insurance for adaptiveness and spread of risk – and there is only one energy source: the sun;
- redirect technical innovations towards a better balance between (rich) consumer wants and (poor) human needs and stimulate a 'green economy' of reuse, recycling and renewables;
- apply the subsidiarity principle: global governance where necessary and local governance where possible, in order to give room to the greatly needed dynamic and creative forces of civic communities and entrepreneurial capitalism;

Because the arguments are largely based on weak knowledge, there are uncertainties and controversies in the interpretation and valuation of what *is* and of what *ought to be*. Ethics and the image of man are inherent parts of the discourse. Worldviews inevitably play a role in the interpretation and appreciation of (models of) economic growth. As in previous chapters, I offer the reader at the end of this chapter four

[38] In this respect, initiatives around basic income (www.basicincome.org/bien/), local exchange and trade systems (LETS: www.gdrc.org/icm/lets-faq.html) and green business networks are interesting. Households are also an important agent in redistribution of income and wealth via transfers, grants and gifts. One estimate for the United States indicates these transfers are three times the government and private charity flows.

[39] There are, for instance, indications that protectionism was promoting economic growth before World War I and negative thereafter (Helpman 2004).

different perspectives on economic growth in reply to four questions (Table 14.2). More than in any of the previous chapters, the perspectives and questions about the economic system will be at the core of the search for a sustainable future for mankind.

In conclusion, mainstream economic theory and models give neither in a descriptive nor in a prescriptive way much insight about how a sustainable economic system looks like. The economic system must be transformed, preferably proactively for the better and not reactively for the worse. The germs of the transformation are around us. There is an enormous amount of and variety in ideas, practises and experiments for a transition to a sustainable economic system. I refer to the list of Suggested Reading and Useful Websites for further study.

14.5 Summary Points

Many of the sustainability issues regarding provision of food, water, minerals and fuels are in essence about human behaviour in the face of scarcity – the topic of economic science. Nearly all development has an economic aspect in the sense of trading off alternatives under boundary conditions – and sustainable development is no exception. Our brief exploration of macroeconomic growth models has given some insights about mechanisms and processes driving economic activities and growth in the long term. A simple archetypical model of the postindustrial transition shows possible pathways to a sustainable economy. But the complexities of the (world) economic system in relation to sustainability are such that a qualitative discussion in a worldview framework will always be needed. Here are some of the points to remember from this chapter:

- (neo)classical models of economic growth focus on the role of capital, labour and technology (capital-labour ratio's, substitution, income convergence, steady-state). The models are founded on oversimplified ideas about the physical and the social world and are poorly reproducing the empirical data on economic growth;
- technology and innovation dynamics, human behaviour and subsystems such as the financial system, government policies and the informal economy must be incorporated in the models to make them useful for sustainable development strategies and policies. Complex system science will advance such efforts;
- empirical input-output (I-O) tables and treatment of resource efficiency and pollution abatement as capital-resource substitution are useful approaches in assessing overall system effects and resource and environment aspects of economic activities and growth;
- we do need a more adequate indicator than GDP in order to measure well-being and quality of life and to formulate targets and evaluate progress towards sustainable development. Some promising efforts are underway;
- experimenting with novel arrangements and models of economic activity to redress perverse incentives for and addiction to material growth and to organise more fair (re)distributions of what the Earth can sustainably produce are amongst the great challenges for the 21st century. A proper balance between global governance and local creativity and community is a crucial ingredient.

Table 14.2. *Worldviews and directions for a sustainable economy*

A Sustainable Economy

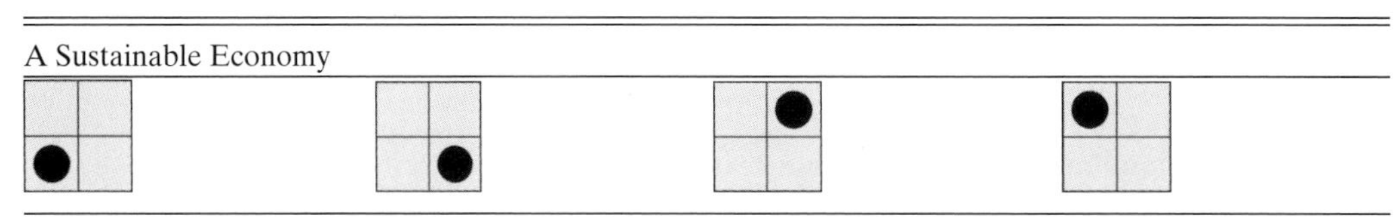

Statement 1: Eco-socialism, that is: the effort to centralise environmental decision making within a state apparatus, will prove to be the most unsustainable system of all.			
History has shown what comes from centralised state power. Business should join forces with govenments and civil society, in order to be an innovative leader in the world economy and do well in the 'Green Race'.	There is no need for planning: existence of unpriced goods is the course of the problem. Corporations and individuals seek out unpriced goods – that is neither surprising nor to be condemned.	Power corrupts and absolute power corrupts absolutely. A more sustainable world starts with each of us individually. The next step is to organise action-from-below at community level. Forget about the state.	Typical right-wing nonsense. As explained recently by Chinese officials, planning is necessary to grapple with the large-scale problems facing the world. Cooperation and coordination are a must.
Statement 2: The runaway spending at the top has been a virus that has spawned a luxury fever that, to one degree or another, has all of us in its grip (Luxury Fever, 1999).			
It may have been exaggerated, but you forget how important the purchases of the rich are for innovation and employment.	This is a moral sentiment that is based in envy. To appreciate luxury is an art, and everyone likes to taste its privileges – but that's impossible.	The extravagance of the rich shows the decline of Western society into materialism, egoism and hedonism. The punishment is inner emptiness and social falling apart.	Luxury spending of the rich sends off the wrong signal in a world plagued by financial and environmental crises. Governments need to regulate the extreme differences in wealth and income.
Statement 3: The existing system of market capitalism creates unemployment and inequity, that breed insecurity and instability. Sustainable development is out of reach without its transformation.			
It is true that poverty and inequity are critical challenges for global stability and sustainability. But the root problems are subsidies, lack of market incentives and incompetence of governments.	Typical left-wing nonsense. Thanks to market capitalism, most of us have a good life. Insecurity and instability come from the lazy, the losers and the thieves – amongst them the immigrants.	Correct. We need to find novel ways, for instance a *basic income*. It may be the only viable way of reconciling poverty relief and full employment. Also local money and exchange systems are part of a real solution.	History has shown the failures of unregulated market capitalism. The strengths of capitalism have to be combined with strict regulation of labour conditions, environmental practises, speculation and finance, trade a.o.
Statement 4: Innovations and open trade are the solution to development – and they are also the solution to sustainability problems.			
Correct. Economic models and data clearly show the importance of innovations and trade in promoting growth. Only then, the poor will be able and willing to support sustainability policies.	Correct. But in an increasingly crowded world, there may be not enough for everyone. Besides, one has to ensure intellectual property rights and a level playing field for trade.	This is the mantra of modern globalism and capitalism. Often, the opposite happens: local economies, cultures and communities are destroyed and replaced by uniformity and consumerism.	The evidence for an Environmental Kuzents curve is not convincing. It is dangerous to count on GDP growth as the way out.

The difficulty lies not with the new ideas, but in escaping the old ones.

– John Maynard Keynes

The important problem of steady state would not be production but distribution. You can no longer avoid the problem of relative distribution by resorting to growth.

– Herman Daly, personal communication

In industrial capitalism, the production of economic goods along with the system of allocating them has conditioned the type of satisfiers that predominate .... [it] leads to an alienated society engaged in a senseless productivity race. Life [is] placed at the service of artifacts, rather than artifacts at the service of life. The question of the quality of life is overshadowed by our obsession to increase productivity.

– Max-Neef, *Human Scale Development*, 1991

'... the modern economist has been brought up to consider 'labour' or work as little more than a necessary evil... Hence the ideal from the point of view of the employer is to have output without employees, and the ideal from the point of view of the employee is to have income without employment... The Buddhist point of view takes the function of work to be at least threefold: to give a man a chance to utilize and develop his faculties; to enable him to overcome his egocentredness by joining with other people in a common task; and to bring forth the goods and services needed for a becoming existence... the Buddhist sees the essence of civilisation not in a multiplication of wants but in the purification of human character... [which] is primarily formed by a man's work... While the materialist is mainly interested in goods, the Buddhist is mainly interested in liberation... [and] it is not wealth that stands in the way of liberation but the attachment to wealth; not the enjoyment of pleasurable things but the craving for them'.

– E.F. Schumacher, *Small is Beautiful*, 1973

Stretch a bow to the very full,
And you will wish you had stopped in time.

– Lao Tze, Tao Te Jing

SUGGESTED READING

This textbook gives a comprehensive overview of macroeconomic growth models with detailed mathematical analyses. The words resources and environment are not in the index.

Barro, R., and X. Sala-i-Martin. *Economic Growth*, 2nd ed. Cambridge: MIT Press, 2004.

A textbook with elementary notions in micro- and macroeconomics.

Common, M., and S. Stagl. *Ecological Economics – An Introduction*. Cambridge: Cambridge University Press, 2005.

A concise and humane introduction in the core concepts of an essential yet complex scientific discipline.

Dasgupta, P. *Economics – A Very Short Introduction*. New York: Oxford University Press, 2007.

An elementary textbook with both disciplinary and more general topics.

Folmer, H., H. Gabel, and H. Opschoor, eds. *Principles of Environmental and Resource Economics. New Horizons in Environmental Economics*. Cheltenham: Edward Elgar, 1995.

An in-depth investigation of the events, persons and forces of the industrial revolution.

Heilbroner, R., and W. Milberg. *The Making of Economic Society*, 12th ed. Boston: Pearson Prentice-Hall, 2005.

A critical exposition of the explanations of economic growth that are given in macroeconomic literature.

Helpman, E. *The Mystery of Economic Growth.* New York: McGraw-Hill, 2004.

A detailed historical description of the evolution of concepts and theories.

Pearce, D. (2002). An Intellectual History of Environmental Economics, *Annual Rev. Energy Environ.* 27 2002: 57–81.

A rather radical but in-depth critique of dominant financial-economic theory and practice.

Keen, S. *Debunking Economics.* London/New York: Zed Books, 2012.

An extensive treatment of natural resource economics in a broad neoclassical framework.

Perman, R., Yue Ma, J. McGilvray, and M. Common, eds. *Natural Resource and Environmental Economics.* Harlow, UK: Pearson Education Ltd., 2003.

Overview of economic theory regarding natural resource use, from a variety of angles.

van den Bergh, J., ed. *Handbook of Environmental and Resource Economics.* Cheltenham: Edward Elgar, 2002.

USEFUL WEBSITES (SEE ALSO CHAPTER 5)

Theory and Data

- www.beijer.kva.se/research/research.html is a site devoted to ecological economics research.
- www.unifr.ch/econophysics/ is a blog about the need and suggestions for alternative, physical science–oriented economic theories and models ('econophysics').
- www.neweconomics.org/ is the site of the New Economics Foundation, with non-standard news and critical views on poverty, environment, etc.
- ineteconomics.org/about of the Institute for New Economic Thinking, www.greeneconomycoalition.org/ and www.communityeconomies.org/Home are three of the many sites that engage citizens in the search for a more sustainable economy.
- www.basicincome.org/ is the Basic income Earth Network, exploring the idea of basic income as a solution to inequity and unemployment.
- www.itcilo.org is organised by the International Labour Organization (ILO) and has information on employment and corporations. See also www.ips-dc.org/reports/top_200_the_rise_of_corporate_global_power# on trends in global corporate power.
- www.gdrc.org/informal/ is the site of the Global Development Research Center (GDRC) with papers on the informal sector.
- www.wiod.org/index.htm is the site with the recent World Input-Output Database. Another widely used set of I-O data is available as part of the Global Trade Analysis Project (GTAP) at www.gtap.org.
- ec.europa.eu/environment/enveco/index.htm is the site on Environment and Economics of the European Commission with illustrative documents and links.

Indicators, Models and Games

- csls.ca/iwb.asp is the site of the Index of Economic Well-being (IEW).
- www.econmodel.com/classic is a site with a description and exercises of classic macro- and microeconomic models (neoclassical growth model or cobweb supply-demand mechanism)
- www.iiasa.ac.at/collections/IIASA_Research/idocs/Research/TNT/WEB/Software/LSM2/lsm2-index.html?sb=3 is about the logistic substitution model as developed and applied at IIASA.
- www.rug.nl/ees/onderzoek/ivem/publicaties/software/eap is on the interactive model EAP, which calculates on the basis of I-O analysis and the energy impact of your expenditures.
- www.btplc.com/Societyandenvironment/Businessgame/ gives the opportunity to experience society and environment dilemmas in a business environment (30-minute game).

Appendix 14.1 A Simple Behaviour Model of Saving

A counterpart of the production growth model is the standard model of savings behaviour of households (Barro and Sala-i-Martin 2004). The assumption is that a household decides the savings rate σ, that is: the fraction of income that is not consumed, on the basis of an intertemporal maximisation of the discounted utility. Assuming equilibrium on savings and labour markets, the savings S per household obey the equation:

$$\frac{dS}{dt} = rS + wL - C \tag{A14.1}$$

because the change in saved assets equals the inflow of interest at rate r and income w minus the outflow consumption c. The discounted utility can be shown to be at a maximum when:

$$\frac{dc}{dt} = \gamma(r - \rho) \cdot c \tag{A14.2}$$

with ρ the household's time preference and γ a measure of the marginal utility, that is ± how fast utility declines for rising consumption levels. This behavioural rule, which is associated with the British mathematician Ramsey, says that households will postpone consumption for future generations if they expect the interest rate to exceed the time preference ($r > \rho \rightarrow dc/dt > 0$). Otherwise, they will prefer to consume now and have a declining consumption level for future generations. This economic image of household behaviour expresses the importance of short-termism versus caring for the future. In short, the demands of capital markets and banks versus sustainability. In combination with a model of behaviour of firms, the savings rate becomes an endogenous part of the economic growth process. I refer to the Suggested Reading for more details.

Appendix 14.2 Evolutionary Models of Producers and Consumers

This appendix briefly introduces two evolutionary economics models for illustrative purposes (de Vries 2010). Dosi *et al.* (2008) have constructed an evolutionary, multi-agent model of an economy in line with the foundational work of Nelson and Winter (1982). There are F firms and L workers/consumers. The firms belong to either consumption-good firms F_1 or machine-tool firms F_2. The consumption-good firms (F_1) plan investment decisions, such as orders for the machine-tool firms, on expected demand, desired inventories and desired capacity utilisation. The key equations describing investment behaviour of firm j are:

- the desired production level Q in period t depends on expected demand and desired inventory;
- the desired capital stock K in period t is a function of the (desired) production level and the (desired) level of capacity utilisation;
- given the labour productivity of machine tool producer and vintage, the unit labour cost c is calculated as the ratio of wages and labour productivity.

The actual investment only takes place above a certain trigger. A crucial part is that new capital stock has a higher labour productivity and thus lower product cost.

A firm decides to scrap old capital according to a simple payback criterium and orders new capital from a subset of suppliers (F_2). Which machine-tools are bought depends on the price and productivity of this subset of suppliers. As a consequence of expectations and imperfect information, firms will perform differently in the consumption-good market. This is simulated on the basis of the replicator dynamics (Appendix 10.2). The authors identify a dozen of macro- and micro-empirical regularities about modern economies and distill from their model a number of variables/distributions for comparison with macro-statistics. The results do reflect most of the 'stylised facts' and generate fluctuations in investment, consumption, employment and other variables in agreement with empirical counterparts.

Another evolutionary economics model focuses on the direction and pace of innovations from the co-evolution between producers and consumers (Safarzynska and Van den Bergh 2010). On the producer side, it simulates the probability of technological lock-ins, path dependency and the role of quality improvements and marketing in the evolutionary 'survival of the fittest'. On the consumer side, it simulates network effects, such as the influence of other individuals (the social network) on the decisions of a person. Such effects can represent important feedback mechanisms, which are from an innovation. In the context of innovation policy for sustainable development, these feedback processes can be positive as in the bandwagon or herd effect and its opposite, the snob effect, or negative as in conformity behaviour. The existence of multiple equilibria in models like this one offers the prospect of understanding sudden large changes in economic systems (§9.6).

Appendix 14.3 Input-Output Tables

An input-output (I-O) table or matrix A contains intersectoral flows between economic sectors. The coefficients A_{ij} of the square matrix A indicate the monetary transactions between sectors i and j in the formal economy during a certain period, usually a calendar year. In network terminology, the economic sectors are the vertices and the transactions of buying from other sectors (input) and delivering to other sectors and to final demand (output) are the edges (§10.5). The level of detail depends on the available statistics and on the purpose of the analysis. If I-O tables have only a few sectors, much information remains concealed in the (large) diagonal elements of intrasectoral transactions. Lack of detail is because of lack of data and/or proprietary information. For some countries, there are matrices with hundreds of sectors while for other countries or regions the disaggregation is limited to a few sectors only.

The basic scheme is shown in Figure A14.1 and an example is given in Table A14.1. The columns indicate the payments for inputs to a sector, including the payments for the so-called primary production factors: wages for labour, dividends and interest for capital, and rent for land and other resources. The rows are filled with the deliveries in money, that is, ± sales, of a sector to other sectors and to the final demand. The latter comprises deliveries to *final demand*, that is, ± to consumers, the government and exports minus imports. The sum of the column elements equals the sum of the row elements. GDP is defined as the total payments to the factors of production and is equal to the value added (VA) in an economy, defined as the summed rows of primary inputs capital and labour. Inputs of resources

Table A14.1. *Example of an input-output (I-O) table: India in 1998. Coefficients larger than 0.01 are printed in bold, larger than 0.1 also in grey*[a]

	1	2	3	4	5	6	7	8	9	10	11	12
1. Food crops	**0.08**	0.00	0.00	**0.02**	**0.02**	0.00	0.00	0.00	**0.08**	0.00	0.00	0.01
2. Cash crops	0.00	**0.03**	0.00	**0.02**	**0.02**	0.00	0.00	0.00	**0.17**	**0.06**	0.00	0.01
3. Plantation crops	0.00	0.00	0.00	0.00	0.00	0.00	0.00	0.00	**0.04**	0.01	0.00	0.00
4. Other crops	0.00	0.00	0.00	0.01	**0.20**	0.00	0.00	0.00	**0.12**	0.00	0.01	0.01
5. Animal husbandry	**0.04**	**0.04**	**0.04**	**0.04**	0.00	0.00	0.00	0.00	**0.07**	0.01	0.00	0.00
6. Forestry and logging	0.00	0.00	0.00	0.00	0.00	0.00	0.00	0.00	0.01	**0.02**	0.00	0.00
7. Fishing	0.00	0.00	0.00	0.00	0.00	0.00	**0.01**	0.00	0.01	0.00	0.00	0.00
8. Primary minerals	0.00	0.00	0.00	0.00	0.00	0.00	0.00	**0.01**	0.01	0.01	**0.07**	0.00
9. Food processing	0.00	0.00	0.00	0.00	**0.02**	0.00	0.00	0.00	**0.09**	0.00	0.00	0.01
10. Other agroprocessing	0.00	0.00	0.00	0.00	0.00	0.01	**0.03**	0.01	0.03	**0.18**	**0.03**	**0.01**
11. Industrial manufact	**0.14**	**0.11**	**0.10**	**0.08**	**0.01**	**0.06**	**0.05**	**0.15**	**0.11**	**0.24**	**0.37**	**0.10**
12. Services	**0.04**	**0.03**	**0.03**	**0.02**	**0.04**	**0.02**	**0.02**	**0.03**	**0.15**	**0.14**	**0.13**	**0.11**
Primary inputs	***0.70***	***0.79***	***0.83***	***0.81***	***0.67***	***0.91***	***0.87***	***0.80***	***0.13***	***0.32***	***0.38***	***0.74***

[a] *Source:* de Vries *et al.* 2007.

and ecosystem services are not valued until the moment that people add value by digging, transporting or any other form of processing. The example of 1998 India (Table A14.1) reflects the great importance of agriculture with its many subdivisions and the distortion because of the high aggregation level of industrial manufacturing and services.

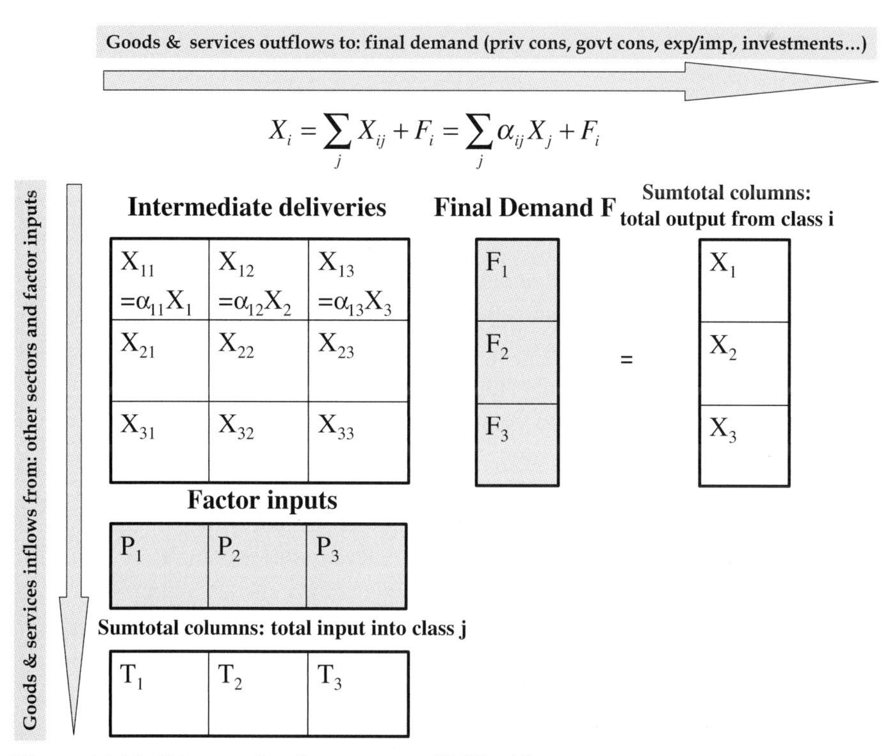

Figure A14.1. Scheme of an input-output (I-O) table.

In the 1940s, Leontief introduced the assumption that the inputs are proportional to the outputs. In other words, the production function is a simple linear relationship between output Y and the limiting production factor K or L (Y = min[AK,BL] with A and B constants). The basic identity in an I-O table is then:

$$X_i = \sum_{j=1}^{n} X_{ij} + F_i \qquad \text{(A14.3a)}$$

in an economy with n sectors and a final demand F_i for goods and services in sector i. It states that the output of any sector i equals the deliveries to all the other sectors and the deliveries to final demand. The assumption of a linear production function implies that the output X_i can be written as a linear function of the inputs X_{ij}:

$$X_{ij} = a_{ij} X_j \qquad \text{(A14.3b)}$$

In words, each matrix element is divided by the sumtotal of the column elements. The elements $\alpha_{ij} = X_{ij}/X_j$ are called the *technical coefficients*. For example, if sector j is the medical service sector and sector i is the electric power production sector, the ratio $X_{ij}/X_j = \alpha_{ij}$ indicates the fraction of every € spent on medical services that is used to purchase electricity. Inserting equation A14.3b into equation A14.3a and continue the substitution gives:

$$X_i = \sum_{j=1}^{j=n} a_{ij} X_j + F_i = \sum_{j=1}^{j=n} a_{ij} \left[\sum_{k=1}^{k=n} X_{jk} + F_i \right] + F_i = \cdots \qquad \text{(A14.3c)}$$

In matrix notation, this becomes $X = (I-A)^{-1} F$, with I the identity matrix and A the coefficient matrix. The matrix A is a concise representation of the economic structure of a country. For more in-depth discussion, I refer to the Suggested Reading (Ayres and Ayres 2002; Hoekstra and Van den Bergh 2002; Perman *et al.* 2003).

Appendix 14.4 Gross Domestic Product

Gross domestic product (GDP) and income (GDP/cap) are probably the most widely used indicator to compare a country's performance and its citizens' welfare. Because of its widespread availability, it is often used as a proxy driver. In fact, it does correlate in one way or another with almost everything. What is it?

The GDP is the value of all the goods and services produced in one year in a given economy as expressed in the monetary System of National Accounts (SNA). In other words, the value of the monetary transactions per year as measured in the official statistics. Divided by the number of people in a country, it is the (average) income in monetary units per year per person (GDP/cap). If income transfers with the rest of the world are included, it is gross national product (GNP). If the goods and services for investment are substracted, one gets the net domestic product (NDP).

GDP is defined from the production point of view as the sum of all value added (VA), which equals the payments for the primary production factors labour and capital. It equals gross domestic income (GDI), which is defined as the sum of all incomes (wages, salaries, entrepreneurial income and all rents). From the consumption point of view, the GDP is the sum of consumption expenditures (including government expenditures, gross capital formation and the transactions with abroad). If the system is considered to be in accounting equilibrium within the given time period, the domestic (or national) product being the sum of all expenditures equals the domestic (or national) income consisting of all production factor rewards.

15 Outlook on Futures

15.1 Introduction

When individual human beings began to experience the first flashes of consciousness, there must have been an emerging anxiety about death of the individual. Biology may have dictated the individual to put the survival of the species above that of the individual, but individual physical survival became the preoccupation that determined actions, emotions and ideas, albeit extended to the nearest members of family and tribe. Whenever threats to survival were absent, the individual and his kin could pursue other qualities of life such as improved shelter and clothing. It was here when development of the individual, and with it society, started.

Survival and reproduction of the individual and his kin has for ages been the main if not only sustainability concern for humans. Only a few individuals broadened their horizons and interests to larger areas and longer periods – the kings and priests in recorded history. But, as seen in Chapter 3, levels of quality of life above mere subsistence could rarely be sustained for more than a dozen of generations and for more than a few small elite populations. For the majority of people, individual suffering from illness, strenuous labour, oppressive overlords or natural disasters was never absent or far away. One response to these realities of life was military valour and conquest – the way of the warrior. Another one was transcending the individual self, in art, love, sacrifice, meditation and compassion – the way of artists, philosophers and priests.

The benefits of working together for a goal that transcended individual survival were clear from early times on (Wright 2000). Already thousands of years ago, Krishna told Arjuna in the Bhagavad Gita: 'The ignorant work for their own profit; the wise work for the welfare of the world' (Easwaran 1985). Concern for fellow men and for society is found in many ancient works of philosophy and religion (Armstrong 2006). They offer reflections on suffering and compassion, on good and evil and on endurance beyond individual birth, reproduction and death. Amongst those who held positions of power, the concerns usually extended to the tribe, the nation or the empire and ranged from years to decades. In philosophy and religion, the emphasis was on salvation in the hereafter or becoming (again) one with the cosmos. These were the early forms of inclusive thinking, and therewith of concern about sustaining quality of life for the collective and the long term.

In the 21st century, the world has become a crowded place, with huge and still growing numbers of human beings and of activities and connections. The finiteness of the planet is in sight. Inclusive thinking in space and time is an urgent necessity and possible pathways for the world and *la longue durée* have to be explored. This chapter probes deeper into what the data, models and insights in sustainability science can tell us about the future. I introduce scenarios as combinations of stories and models in order to be somewhat prepared for an uncertain and complex future. I also pose a few questions that remain, even though the answers have come closer. Can forecasts of key trends be made with any certainty? Is information a necessary step for transformation? What is the right mix of ideas, values and actions for a sustainability transition? And what does this imply for the sustainability science research and education agenda? I do not aim at an overarching synthesis or final answers because this book is itself an attempt at synthesis. I merely sketch a platform that can guide further thought and action in this new branch of science.

15.2 Outlooks

15.2.1 Sustainable Futures: Urban, Rural, Global

Cities, or more broadly, urbanised regions, are increasingly the centerpieces of the human world. How will a *sustainable city* look? Upon entering it, you become part of a centrally coordinated world of constrained individual freedom. This is a consequence of the high density of people and activities and the need for maximum resource efficiency. Physical resource flows are optimised for their use in a circular economy: urban gardens and 'living machines' recycle waste, water and most materials are reused, and the built environment is largely supplied from local, renewable energy sources. Remaining inputs from outside the system are mostly from the land: food, (bio)fuels, electric power. The cities accommodate the centres of knowledge and communication, of health and education, and of the latest accomplishments in resource-efficient building and mobility. They have turned into the largest 'complex adaptive system' on Earth, continuously reinventing and rejuvenating themselves. An apt metaphor of this manager-engineer world is the urban metro system: continuous adaptations with the latest technology in an infrastructure that lasts for centuries and is operated in seamless cooperation between city government and private enterprise. It is a world of intelligence and comfort, where the physical body has to follow the prestructured lanes, but the mind and the spirit can freely wander – at least in principle. People have accepted and internalised the trade-offs between physical and privacy constraints on the one hand and creativity, diversity and sustainability on the other.

The countryside, the *rural regions*, are different. Here, sustainable development means use of local resources to provide for own use and in exchange with the urban areas in the region or country. Farmers grow food and produce fuels and electricity for rural and urban areas on the basis of nutrient (re)cycling and local flow resources. Animal husbandry is limited and integrated in the natural cycles. Degraded lands in arid and semi-arid regions are restored, using the theory and the practice of ecosystem dynamics. Information and communication technology (ICT) plays a key role in making communication and transport appropriate for low population densities and for interaction with urban areas. It is here, also, that your aging parents live and where you go for a holiday or a rest. The experience of nature – its silence, beauty, diversity – provides the appreciated complement to urban life. Governance is mostly a village affair: a combination of local autonomy and bi-/multilateral agreements with urban areas in regional exchange networks.

In this future, urban and rural, city and village live in mutual and beneficial interaction, as was often but certainly not always the case in history. Some call it *rurbanism* and consider it feasible and sustainable. Certain forms of coordination and regulation are necessary, but quite different ways of life can be accommodated. Of course, there are always activities that exceed regional boundaries and sovereignty. A dense network of mining, processing and shipping raw materials and manufactured products has evolved and business travel and tourism connect people to an ever larger extent. They satisfy the need for basic inputs from and outlets for elsewhere and provide adventure, exchange, knowledge or, in short, development. But they have consequences that can make development unsustainable without more coordination and regulation. Think of acidification, the ozone hole, biodiversity and climate change, as well as of economic interdependence, food price volatility, income gaps and financial crises. In practise, it means the imposition of regional and global constraints on regional and local development with UN-led targets, standards and allocation mechanisms. Globally operating corporations and national governments, working together in supraregional and global institutions, have to take the lead and act. Is this a utopian illusion? Not really, because the outlines can be seen in some wealthy enclaves as well as in some poor places.

Instead of a rosy future, we can describe dystopian futures. Resources are squandered, its revenues are spent on armaments and used to provide luxury items. Corruption and crime are rampant. Institutions fail to deliver necessary health, educational, financial and legal services. Public transport and other infrastructure lags far behind. ICT is used for repression and manipulation. The features of this future can also be seen around the globe, in poor as well as in rich places. But there will never be a single story for the whole world. The future world will, like the world of today, consist of a variety of regions, some of which realise their own brand of sustainable development and others become more or less isolated islands of decline and misery. The bewildering richness and complexity of the world contains the seeds of all these possible futures – but how is their potential identified and how are those worldviews and forces that can bring about a more sustainable future identified?

One option is to do imaging and visioning as in the previous tale about sustainable cities and rural regions. We combine scientific models and data with stories and fiction.[1] And we ask ourselves which events, behaviours and structures can make a particular future happen and which ones prevent it from happening or make it even impossible. Is a story plausible and feasible? Are the models able to adequately represent the story elements? Are the data consistent and relevant? Besides plausibility

[1] We usually get our stories from the media and from personal friends and colleagues. The community level is often underrepresented and a site like *The Story Garden* (www.storygarden.ca/) tries to fill in the gap. It is one of the novel ways in which civic society can communicate and build up a reservoir of experiences and experiments.

and feasibility, there is desirability.[2] If you desire a sustainable future, what is it you desire? A good quality of life – but for whom, where and for how long (§1.2)? Are personal and societal developments in values and consciousness to be considered? And how is personal quality of life connected to systemic resilience and social and ecological fairness? Perhaps you have your own answers to these questions and, at the same time, accept the diversity in other people's answers. Of course, your answers are not determining history, or certainly less than those of the ones in power or those of the majority of the people. But scenarios can make each of us more effective in identifying and using the proper indicators and levers in the system that initiate change in the desired direction.

15.2.2 The Scenario Approach

The scenario method combines the qualitative aspect of stories (or narratives) with the quantitative of models. Practitioners have given various definitions, for instance:

> a *scenario* is a combination of qualitative story-telling and quantitative modelling, with the purpose to construct and explore with stakeholders different possible futures, each with their own logic.

Scenarios are supposedly a tool for (better) (strategic) decisionmaking. Common elements of the scenario method are (de Vries 2006):

- emphasis on the construction of alternative futures in order to prepare for divergent plausible futures;
- use of both qualitative ('story-telling', narrative) and quantitative ('modelling') approaches and challenging existing mental models;
- a training in finding key trends, recognising prevalent myths and imagining attitudes of key players.

It is important to know for whom scenarios are made and for which purpose. Credibility, legitimacy and creativity are core ingredients of process and product. The identification of the driving forces: 'what makes it going', and of predetermined elements, in particular slow changing variables, are at the core of a scenario construction process (Schwartz 1995). They provide the structure or *logic* of a scenario and give an idea of critical uncertainties. The scenario method, it is claimed, opens people to multiple perspectives on the world and offers a complement or even an alternative to the conventional languages of business, government and science in dealing with complex and ill-structured questions (Duke and Geurts 2004). Scenarios are a stepping stone to *strategy*, which is the art of deliberately recognising major trends, establishing one's own course of action and translating this into practical *plans*. Part of scenario construction is the process of *visioning*. 'Visioning means imagining, at first generally and then with increasing specificity, what you really want... not what you have learnt to be willing to settle for. Visioning means taking off all the constraints of assumed "feasibility"' (Meadows *et al.* 1991). In this sense, scenarios express the ethos of their times.

[2] The possible and the hopefully desirable overlap. To make things more complicated, they also interact.

Box 15.1. *The future and scenarios about the future.* The word future is from Latin *futurus* meaning that what is going to be, yet to be. The French equivalent avenir is clear: that which is to come, à venir. In German and Dutch, the words Zukunft and toekomst express the same: that which is to come. Much has been said about 'that which is to come or to be'. A few aphorisms:

> Prediction is very difficult, especially about the future.
>
> – Niels Bohr

> The trouble with our times is that the future is not what it used to be.
>
> – Paul Valery

> The function of science fiction is not always to predict the future, but sometimes to prevent it.
>
> – Frank Herbert

> Our view of the future affects the present as surely as do our impressions of the past.
>
> – Willis Harman

There is reason for skepticism about predictions and forecasts. In a large-scale experiment on future predictions, it was found that 'experts who did particularly badly were not comfortable with complexity and uncertainty . . . and also more confident than others that their predictions were accurate . . . Experts who did better than average . . . were comfortable seeing the world as complex and uncertain – so comfortable that they tended to doubt the ability of anyone to predict the future' (Gardner 2010, 26–27).

Governing elites have always been interested in anticipating future events and, often, priests were involved in order to let the gods participate in judging, legitimising and rationalising. It provided strategy as well as legitimation against competing groups with opposing views of what the present is and the future could or should be. Thomas More offered his Utopia as a visionary critique. Karl Marx offered a rationale for the demise of capitalism. Jules Verne expanded technical possibilities far beyond the known options and religious leaders have promised mixtures of catastrophe and salvation. Prevailing ideologies and their utopian challengers, variously considered rebels, visionaries or prophets, are in essence radicalised worldviews. Currently, there are hundreds of individuals and organisations who offer their view of the future, which ranges from alerts and warnings to technological paradise and fundamentalist doom.[3] Some are pragmatic, 'business-as-usual' trend extrapolations. Others are engaging, socially or technically radical utopias (Achterhuis 1998).

The scenario method has risen to prominence because old centres of authority were increasingly challenged in the 1960s and 1970s, and the control paradigm had met its limitations. New and powerful actors appeared on the scene: activist

[3] Warnings of overshoot in views or actions can often be identified from the titles: *Limits to Growth* (Meadows *et al.* 1971), *Social Limits to Growth* (Hirsch 1977), *Limits of Organization* (Arrow 1974), *Limits to competition* (Group of Lisbon 1995) and *Limits to Certainty* (Giarini and Stahel 1993) are examples.

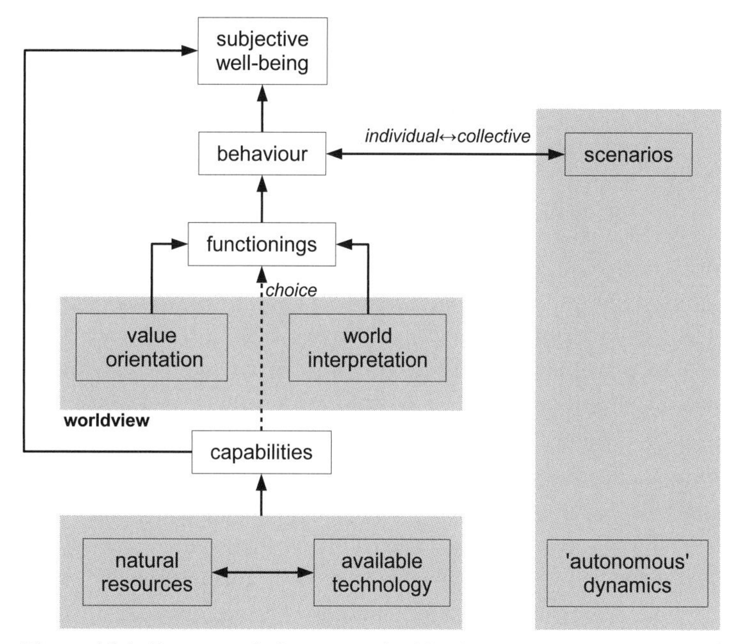

Figure 15.1. Framework for a sustainable development assessment (de Vries and Petersen 2009). See text for explanation.

scientists, innovative entrepreneurs and committed citizen groups. A new legitimacy with participation of stakeholders was needed. The scenario method, with its explicit consideration of uncertainties, multiple perspectives and stakeholders, addressed at least partly the new requirements of a postmodern era. Not everyone embraces it. Engineers may not subscribe to the qualitative elements and social scientists are suspicious of quantitative models. CEOs and politicians may also show signs of dislike, preferring command and control over participation and pluralism.

15.2.3 Sustainable Development in a Scenario Frame

Sustainable development faces macro-problems that are high in aggregate complexity and low on consensus in worldviews (Figures 8.5 and 8.6). Exploring future developments and constructing interesting scenarios for sustainable development must, therefore, consider both ends *and* means, both subjective *and* objective experiences and knowledge. Figure 15.1 sketches a framework for such a broad sustainability assessment. The lower part represents resources and technologies, both important determinants of societal change. It includes the 'autonomous' dynamics of processes such as resource depletion, pollution and degradation of the environment, and technological developments. The knowledge about it is rather strong, and its refutation erodes the very foundation of rational discourse. The upper part represents the individual and collective sense and view of well-being that has historically also been an important driver of change. In this domain, subjective experience matters more than objective fact. The issues are complex, the solutions are controversial – and knowledge about them is rather weak.

The upper and lower parts are connected through capabilities that arise from what resources and technologies can offer, and functionings that correspond to the behaviour that satisfies our aspirations for 'the good life'. The middle part constitutes the essence of a person's worldview. Explicit acknowledgement of *worldview pluralism* is the basis for legitimacy in a modern democracy and prevents us from falling in the trap of reducing the problem to a single solution for an – illusory – single constituency. Sustainability is lost when one worldview starts to dominate and society moves towards extremes (§6.5).

Scenarios are positioned in the upper right of Figure 15.1. They combine the social and cultural outlook with the resource and technology potential. Scenarios are, in the present context, constructed in order to examine possibilities and strategies for a (more) sustainability development. This can be done from different vantage points. One approach is to position oneself in the centre of the worldview space and to identify values and beliefs, hopes and aspirations (utopia) and fears and concerns (dystopia) in each of the different worldviews. It includes an evaluation of capabilities, resources and technologies in the form of financial, economic, social and media power and control. A scenario then becomes a policy program that is robust and resilient in the face of change, including sudden shifts in the political landscape and surprise events in the natural system (MNP 2004). Such an integral and non-ideological approach is what governments (should) do, in the public interest, in an attempt to identify and implement robust policies.

A scenario can also be constructed to explore possible futures with respect to the pros and cons for the stakeholders and their worldview. It serves the strategic objectives of stakeholders, such as providing warning signals and leverage points. Given the aim to strengthen one's worldview, important questions are: which values to promote, which ideas to communicate, which facts to mention and which risks and opportunities to spell out? If the particular worldview has support from powerful groups, it can effectively be marketed as a desirable view of the future, with other scenarios and associated worldviews as potential threats. In this way, scenarios become politicised and get ideological content. There are many examples of such partisan scenarios supported by government institutions, corporations or NGOs, for instance, in the areas of energy and food.

A third option is to start from policy goals and targets that have already been agreed upon, such as the Millennium Development Goals or the European Union's 2°C target for climate change, and investigate various model-supported storylines that might meet the goals and targets (PBL 2009). The worldview pluralism serves as a heuristic that can create new solutions or identify ignored trade-offs. It also permits an assessment of policy options in futures that are dominated by worldviews different from the existing one. This can, in combination with backcasting, enhance the likelihood of successful and robust policies.

15.3 Scenarios for a Sustainable World

15.3.1 Four Stories

Numerous scenarios have been made in the last decades. One of those is the Special Report on Emission Scenarios (SRES), that was written on request of the Intergovernmental Panel on Climate Change (IPCC) in order to provide possible greenhouse gas emission trajectories (Nakicenovic *et al.* 2000). Two dimensions are proposed, along which future developments might evolve: globalisation versus regionalisation

Box 15.2. *Decision making in politics.* A textbook on sustainable development should also address public decision making and politics. Politics can be described as the area in society where issues transcending mere individual needs and activities are debated and decided and where disagreement is legitimate. It has its own rules and rites. Miklos Persanyi, former Environment Minister of Hungary, shared with me and others some experiences about life in politics. A basic rule for politicians is:

> If there is no must to decide, it is a must not to decide.

When you are in the science-policy interface dealing with politicians, obey the following rules:

- draw/keep politicians' attention;
- speak their language: be stupidly simple/primitive, talk about money, talk about benefits for constituency/business;
- don't use words like future, climate . . . instead, talk about costs, floods;
- become a media star, marry his/her daughter, or go golfing.

It is tempting to reproach politicians for their often narrow-minded, myopic, self-interested behaviour. But they probably represent what a postmodern, media-hyped society deserves.

On a more serious note, the notion of *bounded rationality* was introduced by the sociologist Simon to indicate that people usually act on the basis of a few, rather simple rules. Cognitive sciences are bringing new insights into how human process information. Individual experiences are filtered by several subsequent layers (Morecroft and Sterman 1992):

- tradition, culture and the like;
- organisational and geographical structure;
- information, management and communication systems;
- operating goals, rewards and incentives; and finally
- people's cognitive limitations.

Illustrative examples of the first three layers in a sustainable development context are the role of tradition in the introduction of energy-efficient and solar-based cooking stoves in rural India; the role of the state in explaining why centralised nuclear power thrives in some cultures and is resisted in others; and the role of (perverse) incentives in an oil company that rewards its executives on the basis of the oil reserves at the end of the year. The rationality of decision making is perhaps even farther away from the real world than we imagined in our criticism of *Homo economicus* (§10.4).

and government versus market. For each of the four resulting quadrants, a *storyline* or narrative is developed that describes a plausible – but not necessarily probable – path of key variables such as population, economic activity and energy use. The keywords of the SRES storylines are summarised in Table 15.1, together with the scenario names (A1, B2, B1, A2). There is resemblance between the SRES scenario dimensions and those in the worldview framework presented in Figure 6.3.[4]

[4] Globalisation vs. regionalisation coincides with the horizontal axis of big world vs. small world. Government vs. market represents the difference between a less vs. more outspoken orientation on

Table 15.1. *Keywords for the four scenario quadrants and associated features, in the worldview framework of Figure 6.3. Besides the names B1, A1, A2 and B2, many other names have been proposed for the scenarios. One set is indicated in bold*

B1 *collective and immaterial*	**B2** *individual and immaterial*
(Absolute Idealism)	*(Subjective Idealism)*
Government/church authority offers universal truth	Truth is personal
Tech and organisation in the service of the soul	Low/appropriate tech and human scale
Hierarchist tendency in collectivist setting	Egalitarian in individualist setting
Welfare economics	Buddhist economics
Our Common Future	**Small is beautiful**
A1 *collective and material*	**A2** *individual and material*
(Objective Materialism/Modernism)	*(Subjective Materialism/Postmodernism)*
Science is universal truth	Truth is what can be sensed
Society high-tech and well-managed	Tech and organisation for material wellness
Hierarchist tendency in collectivist setting	Individualist tendency
State/corporate economics	Market economics
The End of History	**Clash of Civilisations**

The scenario *logic* is derived from hypotheses or metamodels about how the world works. For instance, it was in SRES hypothesised that more 'free market' forces of privatisation and deregulation lead to higher economic growth; that higher income causes a more rapid stabilisation of population growth; and that protectionism hinders trade and, therefore, economic growth. This led to consistent population and economic growth paths, that were introduced into the energy and land use models. The latter were then run with assumptions consistent with the storylines. The outcomes of the thus generated sets of scenarios were presented for world regions until 2100 in the form of energy use and greenhouse gas emission pathways.[5]

How does the world evolve according to these scenarios and stories? I give a brief caricature description. In the modernist *A1 world*, material wealth and high-tech business are essential for a good quality of life. Everyone with specialised skills and a competitive and risk-taking attitude can share in it. Market-driven economic efficiency, minimal government regulation and open trade are preconditions. The benefits accrue to a small part of the population, but wealth and novel goods and services will over time 'trickle down' to the poor. Post-war performance of the United States is the most visible proof of the correctness of this ideology.[6] The archetypical A1 citizen is convinced that society will become stagnant and backward if these values and qualities are lacking or obstructed. Indeed, he represents the successful transition from traditional to secular/rational values, that has been observed in many places during the 21st century (§4.3)

This worldview is not shared by people who value small-scale enterprises and cooperatives, social and cultural traditions, community, personal growth, spirituality

material welfare along the vertical axis. For a discussion of these scenarios and the proposed axes, names, interpretations and shortcomings, I refer to Nakicenovic *et al.* (2000), de Vries (2006), Riahi *et al.* (2007) and de Vries and Petersen (2009).

[5] The scenarios are caricatures in the sense that they presume a particular and rather extreme worldview to prevail over a long period of time.

[6] The emergence of China as an industrial nation with a command-and-control form of capitalism is interpreted by some as a proof that democracy may not be needed or even be harmful, whereas others see it as a temporary phase until postindustrial values will break through also in China.

and nature. They manifest their worldview by protecting the small and local world in the form of cooperation, solidarity and forms of enclosure. In this *B2 world*, citizens find their identity in cherishing what is local, be it their vernacular, the village church or a nearby forest. The focus is on local needs and livelihoods, to which technology and governance should be geared. Protectionism is seen as a survival option for their socio-economic as well as cultural way of life. They adhere to a civic society ideal, that often clashes with the logic of rationalist modernity and global capitalism.

The B2 citizens cherish autonomy, but they cannot exclude the larger world with its lures and necessities. Much to their discomfort, they are confronted with alarming stories about the large, evil world out there: dwindling fish stocks, water and air pollution, the threat of climate change, destruction of local jobs, immigrants and refugees and so on. Blaming the others is not helpful. Many of them, may, therefore, adhere to the *B1 vision*, of a world in which international cooperation and solidarity are needed in order to solve the large-scale and long-term social and environmental problems the world is facing. Persons adhering to the B1 vision support universal values and aspirations regarding global solidarity and peace, human rights and sustainable management of the global commons. Theirs is a world of global governance and value-inspired science. The Millennium Development Goals (MDGs) are an expression of this worldview.

But many people currently dismiss the B1 world as a failed ideal and a bureaucratic dystopia, pointing at the political quarrels, fraud and mismanagement in the UN organisation and in development aid. They prefer the 'realism' of the postmodern *A2 world*. They are joined by those who are disappointed by the promises and technocracy of the A1 world and by those who have anyway no interest in the societal and immaterial aspects of life. Their future brings a world of individualism and hedonism, in which citizens resort to a mix of consumerism, opportunism and fatalism. It offers the pleasures of the material and body culture and opportunities for local politicians and entrepreneurs, but injustice and lack of social coherence take their toll in the form of fear, disorder and crime. Politically, it manifests itself in protectionism, nationalism and clientelism. Bilateralism characterises trade, governments emphasise security and military strength. Prominent beliefs are that markets and money run the show and that governments waste tax money and cannot be trusted. In many places, there is a continuous crisis of legitimacy of political and financial elites, unless or until a military elite takes over.

As this brief description shows, the scenario method can be used to tell different stories about the future, with different values and beliefs dominating society. For the 21st century, it can be surmised that the trend towards the postmodern A2 world continues for some time although in parts of the world people rather experience A1 modernism. It may spell an era of intensified conflict between nations about scarce resources, environmental damages, novel (arms) technologies and cultural identities. When the need for collective action becomes sufficiently clear and urgent, there will be attempts at coalitions or even a world government to deal more effectively with the global problems that most seriously threaten the prospects for sustainable development (B1; Schrijver 2010). Meanwhile and along the edges, small-scale groups and initiatives that stress the spiritual, the community and the natural will grow in strength (B2; Hedlund-de Witt 2011). Can models be of any use in this exploration of world futures?

15.3.2 Growth within Limits . . .

In the last decades, the number of future scenarios have grown exponentially. For key variables such as population, economic activity, food supply, energy use and other system stock and flow variables, there are new forecasts every year. The scale is usually regional and the time horizon between ten and forty years ahead. In previous chapters, some possible future trends in demography, biodiversity, agriculture and energy are given However, the complexities and uncertainties of our future world are such that most scenarios are rather short-lived. They also increasingly reflect, with or without explicit storylines, a particular worldview or ideology. Therefore, this book refrains from giving more scenario outcomes. Most scenarios are made by professionals in institutions such as the UNFPA for population, the World Bank and IMF for economic activity, the FAO for food, the IEA and the WEC for energy and so on. Some institutions make integrated scenario analyses in a policy setting, for instance the World Development Reports by the World Bank and the Global Environmental Outlooks (GEO) by UNEP. Corporations and NGOs also make and sometimes publish scenarios, for instance, Shell's strategic planning scenarios or the green growth scenarios of the European Climate Forum. Institutions such as IIASA and PBL have developed Integrated Assessment Models (IAM) in order to incorporate the latest scientific insights in the analyses and scenarios. I refer to the Useful Websites for more information.

One thread of model-based scenario explorations is the *Limits to Growth* report (1972) and its subsequent appraisals and updates: *Beyond the Limits* (1991) and *The 30-Year Update* (Meadows *et al.* 2004; Meadows 2006). It is based on a system dynamics world model and can be considered the first IAM. The updated scenarios suggest, that the world is still following a path towards overshoot that results from the connected processes of resource depletion, soil degradation and pollution (Turner 2008). A transition to a sustainable society is still possible, but markets and technology alone will not make it happen. The underlying worldview manifests the anxiety about the A world and its technological and market optimism, and the necessary change towards a B world. The model outcomes are at a high level of aggregation (world), which gives them an inherent bias towards universalism and a tendency to overlook the real-world diversity and contingency and its surprises.

Inspired by the *Limits to Growth* report, a number of other long-term models have been constructed to investigate the world system past and future. One example is the TARGETS model, an IAM that simulates the most important subsystems of the global system and their interactions (Rotmans and de Vries 1997; de Vries 2001). The perspectives of Cultural Theory were used to incorporate the uncertainties and controversies (§6.3). By distinguishing between the dominant beliefs in society on the one hand and the, as yet unknown, world system on the other, utopias and dystopias were constructed. If the optimism of the modernist worldview is justified and its supporters are in power (the individualist utopia), the increasing pressure from exponentially growing population and gross domestic product (GDP) can be managed for the next hundred years because natural ecosystems turn out to be resilient, resources abundant and new technologies efficient. The future will then be high-tech and highly managed. If the optimism about the robustness and resilience of our life support system turns out to be unjustified, a more dystopian future can unfold in which tremendous suffering and damage – notably from climate change – is to be expected. The future then follows a Limits to Growth overshoot-and-collapse path. This is prevented if a more egalitarian perspective gets the upper hand, leading to deep reductions in energy and material use and corresponding reductions in environmental damage and destructive feedbacks.

A mainstream scenario is one in which the hierarchist perspective of official government institutions prevails. It is a world of 'muddling through': A series of incremental policy measures can avert major catastrophes, but large long-term environmental risks are not properly addressed. This is caused by waiting too long for more and more reliable knowledge and getting bogged down in bureaucratic ineffectiveness and political deadlocks. The only consolation in the face of repeated and spreading disasters is perhaps that the emergency actions are well organised. The simulation experiments carried out with the TARGETS model represent one approach to manage the large uncertainties in basic, quantifiable variables such as the size of population and economic activity, life expectancy, food intake and water and energy availability.

A recent scenario study for the world is the *Growing within Limits* report (PBL 2009). Two scenarios, Trend and Challenge, are investigated with a focus on greenhouse gas induced climate change and land use/biodiversity. The *Trend scenario* is the dominant 'business-as-usual' view of the future amongst government and corporate officials. It is a mixture of expectation and desire, largely following an A1 storyline. Its outline is sketched in Figure 15.2.[7] The graphs are indicative.

There is probably least uncertainty about population size. Most demographers expect a stabilising trend towards a population between 8 billion and 10 billion individuals for the period until 2050. Underneath are the demographic and epidemiological transition (Figure 10.6). The assumption is that there will be no massive catastrophes that could make the population decline. The band width of 2 billion people results from uncertainties in trends in income, food, education, availability of medical services and biodiversity. Food supply is also projected with rather large certainty, because presumedly it is well defined in relation to population and can be met in all but the worst places and periods. Biofuel plantations will not greatly affect the picture.

More uncertainty exists about the rate and nature of GDP and income growth. It is a dominant variable because it drives most other variables – at least in the models. The Trend scenario assumes more than a doubling of average income in the world between 2010 and 2050, with an uncertainty range that suggests it might be considerably less. Such an income growth path is an average across populations and the kind of economic activities and their global distribution is largely based on extrapolations and the assumption of income convergence. Consequently, the forward projections of resource use and environmental load have similar or higher uncertainty bands (Van Vuuren *et al.* 2008). For instance, the use of energy is expected to increase with a factor 3 ± 0.5. For more details, I refer to the Suggested Reading and Useful Websites.

[7] Such scenarios used to be indicated with business-as-usual, but the feeling behind the term has disappeared.

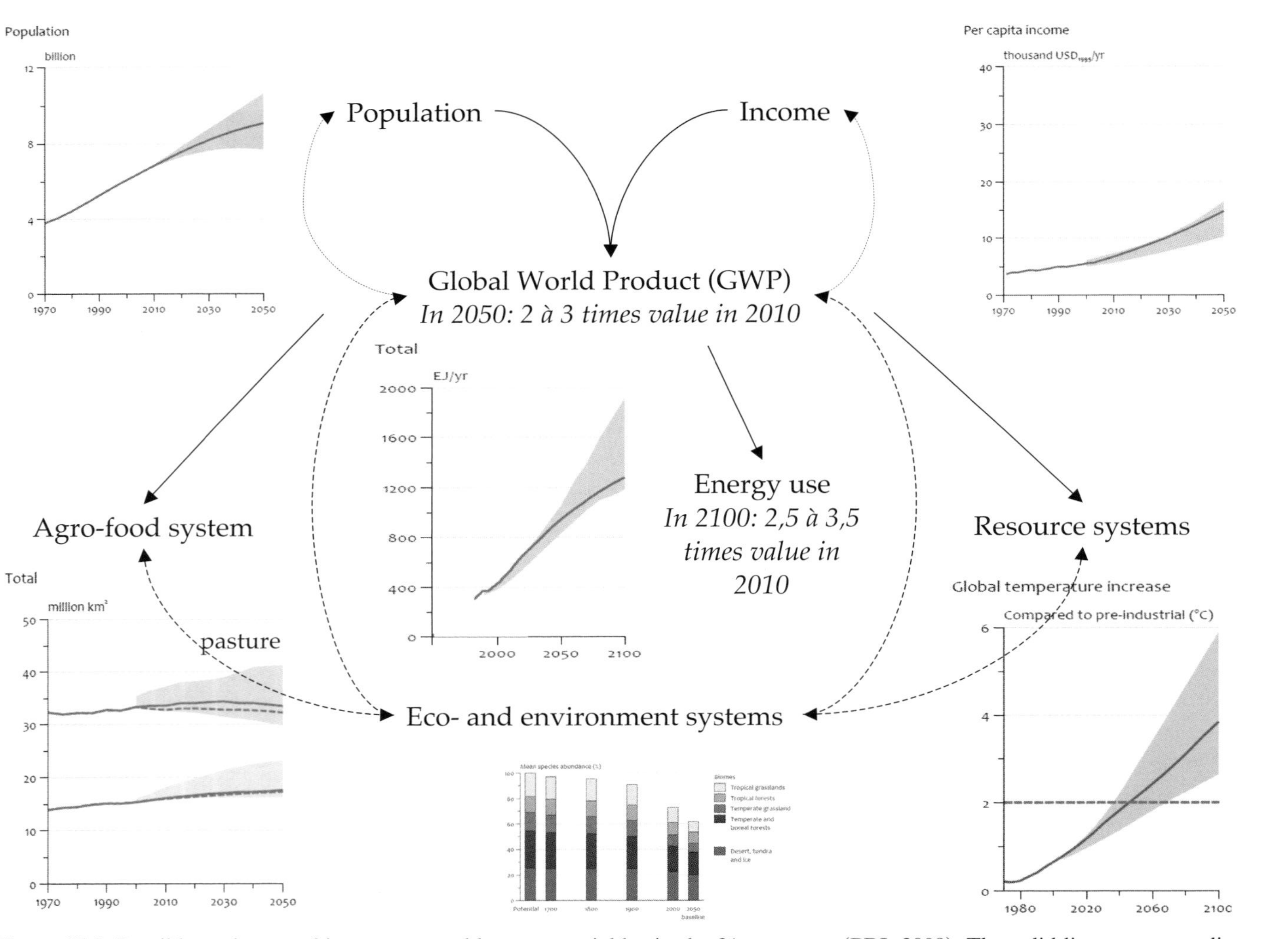

Figure 15.2. Possible pathways of important world system variables in the 21st century (PBL 2009). The solid lines are according to the Trend scenario; the coloured bands are the uncertainty margins from the literature. Solid arrows indicate fairly well-known, causal relationships. Dashed arrows indicate less-known feedback relationships.

Box 15.3. *Economic growth projections.* Most official organisations still use a rather straightforward neoclassical growth model to project future economic output. An example of a long-term model-based projection of world economic growth is the OECD report ECO/WKP (2009) by Duval and De la Maisonneuve. Its projection of economic output Y is 'based on a standard aggregate Cobb-Douglas production function with physical capital, human capital, and labour as production factors and labour-augmenting technological progress, and [assumes] that the production function is invariant both across countries and over time'. Output per capita Y/Pop can be decomposed as follows:

$$\frac{Y_t}{Pop_t} = \left(\frac{K_t}{Y_t}\right)^{\frac{\alpha}{1-\alpha}} A_t h_t \left(\frac{L_t}{Pop_t}\right)$$

where K_t/Y_t, A_t, h_t, and L_t/Pop_t denote the capital/output ratio, TFP, human capital per worker and the employment rate (defined here as the ratio of employmed and total population), respectively, and α is the capital share in aggregate output. From the graphs in Figure 15.3, it is seen that this official forward projection indicates, like most others, exponential growth and regional convergence in economic activity measured as GDP and income (GDP/cap) until 2050 in the world.

15.3.3 ... But Is It Sustainable?

Accommodating up to 9 billion people at twice the average income, as presumed in most official scenarios, puts large strains of the life support system. What are the implications for sustainability if it becomes reality? Combining the insights of the preceding chapters with the A1 worldview yields the following, plausible story of the Trend scenario. Development is guided by maximising productivity and efficiency through competition, innovation and abolition of trade barriers in the search for high returns on capital savings and decreasing consumer prices. Multinational corporations are the prime actors in an ongoing process of globalisation, whereas governments, including international organisations, focus on safeguarding a level playing field for the corporations and on securing access to resources and contain environmental side effects (Dicken 2009; Schrijver 2010). Much creativity, organisational skill and willpower is spent on overcoming resource scarcity and environmental degradation – the knowledge and money are available. The high income growth goes together with increasing energy use, but depletion and environmental impacts are met by resource efficiency, recycling, renewable energy and pollution abatement in response to rising world market prices. The postindustrial transition sketched in §14.1 succeeds. There are still uncertainties, especially around potential bifurcations. Will it be solar or nuclear or coal energy that wins the game? Will recycling go as deep as cradle-to-cradle? Are GMOs boosting a second green revolution? Whatever the answers, it will be a future of hard work, and in which we have to work hard to cope with the consequences of our hard work.

In the Trend scenario, bottom-up market processes in combination with corporate efficiency and organisation are expected to deal incrementally with upcoming problems and events unfold without a comprehensive and pro-active assessment to

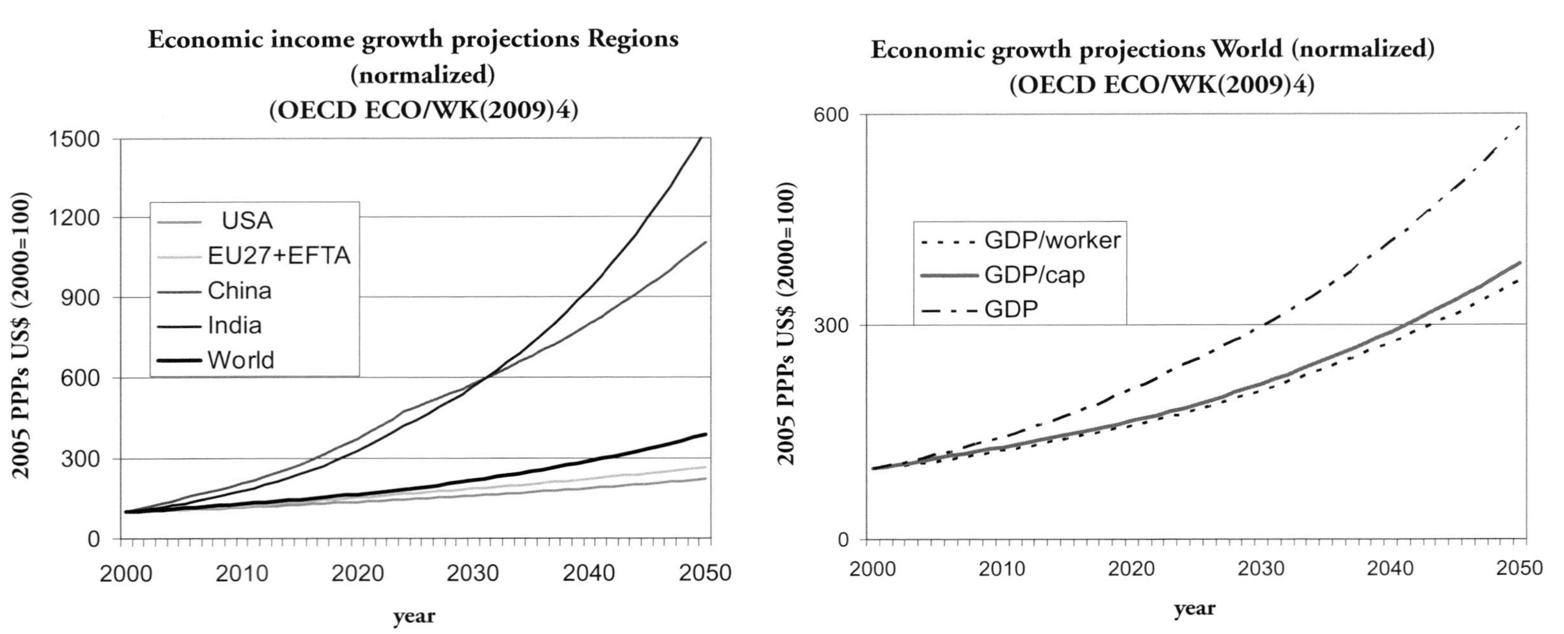

Figure 15.3. One projection of economic output until 2050 (OECD; /www.olis.oecd.org/olis/2009doc.nsf/LinkTo/NT00000AE2/$FILE/JT03260306 .PDF.

make the sustainability transition happen. There is a decline in long-term strategic planning and R&D investments in both corporate business and governments. Its flavour can be tasted in particular mixes of planning ideals and market mechanisms such as in the April 1990 issue of Scientific American: *Managing Planet Earth* and in the optimal control solutions to greenhouse gas emissions in *Managing the Global Commons* (Nordhaus 1994).[8] A precondition for its realisation is that the ruling elites must be sufficiently convinced of the seriousness of scarcity and degradation in order to act and sufficiently convincing to induce the investments and implement the policies that are needed.

The Trend future is exposed to two major risks, one external and one internal. The external one is that the life support system – nature – turns out to be more fragile than expected (Tables 6.2 and 6.3). The planetary boundaries are real indeed. A series of natural catastrophes disrupts large parts of the global system; fragmentation, egoism and protectionism hamper coordinated responses; and in many places, people start to rely increasingly on local resources and institutions. The worst crises are initialy relieved with large-scale aid, but people get weary of it and it often comes too little and too late. In the terminology of the narratives, the world moves towards some mixture of fragmented, regionalised A2/B2 worlds, because the forces behind a B1 world have failed in a predominantly A1 world.

The second risk is internal: The breakdown comes from the inside, because income inequality destabilises the necessary social contracts and because governments fail to organise the public goods needed for income growth and quality of life. In quite a few places, social revolts disrupt the growth engine, institutions fall apart and disorder spreads in a densely connected world. It is an erosion process that manifests itself in widespread distrust in government and retraction into a private consumer world. It is a move towards an A2 world, because the promise of rising prosperity for the masses is not fulfilled for the poor or no longer fulfilling for the rich, and partly driven by anticipated shortages and scarcities. Reputable analysts interpret the 2008–2011 financial crises as the onset of such a storyline. Both risks are exacerbated because of pernicious feedback loops. For those who can see, the signs are there. It is phrased in a penetrating way by Yeats in his post–World War I poem *The Second Coming*:

> Turning and turning in the widening gyre
> The falcon cannot hear the falconer;
> Things fall apart; the centre cannot hold;
> Mere anarchy is loosed upon the world,
> The blood-dimmed tide is loosed, and everywhere
> The ceremony of innocence is drowned;
> The best lack all conviction, while the worst
> Are full of passionate intensity.

It also shows up in the various names of recently constructed scenarios: Hyperindividualism, Fortress World, Tribal Society, Battlefield (de Vries 2006). Some argue

[8] Options to rescue the planet by large-scale geo-engineering reflect an even more tense marriage between the aim of large-scale engineering to 'save humanity' and a rejection of planning in favour of 'the market' and 'bottom-up' self-organisation.

that a 'great disruption' is inevitable and that it will bring the necessary awakening and a subsequent massive switch to more durable ways of life (Gilding 2010). Hopefully, mankind will accomplish sustainability, fairness and dignity for itself without the suffering and perishment of hundreds of millions of people.

The A1 world of the Trend scenario and its natural devolution into an A2 world is the wrong direction for sustainable development. It carries substantial environmental and social risks for the long term and confronts the world with some difficult trade-offs. The quantification of the environmental impacts suggest severe climate impacts long before the end of the century. Greenhouse gas emissions have to be halved at least, and before 2060, in order to reach the 2°C temperature target. In an alternative *Challenge scenario*, it is shown that this is technically possible with drastic improvements in energy efficiency and a deep penetration of non-carbon energy sources. Large-scale carbon sequestration and storage (CSS) is needed in the transition period. It implies massive and fast development and transfer of new technologies and adaptation of users and institutions at all scale levels. In storyline terms, it is only feasible in a B1-like narrative. The Challenge scenario is one of the hopeful prospects, but many have begun to doubt its feasibility.

A second and different narrative sees a reconnection to the immaterial and communal aspects of life as the essence. The emphasis is on personal growth and self-awareness, as a source of inner conviction and of behavioural change and in natural association with sustainable livelihood precepts. People experiment with sustainable communities in a variety of ways, with their own local resources and their own values and understandings. If such enclaves become mainstream, the B2 world with scenario names such as Local Stewardship, Voluntary Simplicity and Let Hundred Flowers Blossom gets a chance. Some have envisioned it as the transindustrial society (Harman 1976). Others deride it as utopian because it is ineffective in the face of global threats and the absence of global government. But, as Paul Tillich once remarked, 'Mensch sein heist Utopien haben'.

As discussed in Chapter 6, a genuine sustainable development should be built on the balanced manifestation of all four worldviews: the integral worldview. In practise, this means that we have to reorient existing values and beliefs in each of the four quadrants in such a way that they contribute in their characteristic way to the transition towards sustainability. Table 15.2 lists suggestions for reorientation. It is a tentative list, in an attempt to include all the local and global and material and immaterial initiatives that are taken every day all over the world. Some of these concern local habits and practises, others take by their very nature place in a more global arena of ideas and institutions. Yet others can only be successful in the small and idealistic worlds of committed individuals, from where some in retrospect turn out to be the seeds of large and strong trees. Some of them are most attractive to the young and adventurous, others will appeal to the elderly and prudent, and many will be rooted in the rationality and pragmatism of the sciences. Together, they provide the impetus for humanity to explore its wisdom, intelligence, empathy and sensuosity – the four elements fire, air, water and earth.

For each of us, as individuals and inhabitants of a finite planet, there is an invitation to contribute in various directions. Practise inclusive thinking and meditate upon the wisdom and compassion of enlightened ancestors. Confide in and

Table 15.2. *Elements of sustainable development strategies within the four worldviews*

B1 *collective and immaterial (Absolute Idealism)*	**B2** *individual and immaterial (Subjective Idealism)*
Engage religious leaders and ethics	Personal growth/consciousness
Address values of transcendence and stewardship	Experiment with autarchy, community
Principles and institutions for Global Commons	Stimulate arts, nature education, slowfood
Strengthen initiatives for fair trade and aid	Permaculture/organic farming for local markets
Nature: protect biodiversity hotspots	Experiment with forms of basic income
Design and implement subsidiarity principles	Stimulate reuse and secondhand markets
Revitalise science and society reflection	Distributed ICT-supported energy and transport
Experiment with public-private arrangements	Transparency and integrity in product information
Cyberspace as common/public space	Local autonomy in taxes, social security
Ecological restoration of degraded lands	Nature: green city design
Envision and expand sustainable infrastructure	Stimulate consumer labels and certificates
Participatory discourse on innovations	'Greening' banks and enterprises
Stimulate and direct long-term R&D	Strengthen small-medium scale enterprise/skills
Regulation of (global) financial and tax systems	Merge sustainability, cost and comfort
Engage business in transition technologies	From junk to health food
Corporate Social Responsibility	Address security and identity concerns
A1 *collective and material (Objective Materialism/Modernism)*	**A2** *individual and material (Subjective Materialism/Postmodernism)*

strengthen institutions that enforce global standards, regulations and laws to manage the commons. Contribute to the knowledge and techniques that help reduce resource inefficiency and fulfill the needs of the poor. Celebrate the body and its expressions, yet do not identify with it.

15.4 An Agenda for Sustainability Science

Sustainability science has a more narrow scope than sustainable development. Nevertheless, the list in Table 15.2 is also indicative of the topics to be researched and taught in sustainability science. Most of these are introduced, explicitly and implicitly, in the previous chapters. This book ends with a brief discussion of a research agenda, defining science as the aspiration to produce, individually and collectively, knowledge of the highest possible contextual validity (§8.3). It has its stronghold in the lower left of the quadrants in Table 15.1, but it has an open mind towards the other quadrants.

The natural and engineering sciences offer undoubtedly enormous prospects for contributions to a more sustainable future. It is their responsibility because they generate not only solutions but also problems. At present, their advances are largely driven by the search for profits and the desire for innovative consumer goods. Over time, this can lead to technologies that promote sustainability, as historical developments testify. However, the direction of science and technology must more intensely be steered in accordance with societal priorities. Here, sustainability science can give context and directions. Besides, there are sometimes real breakthroughs that may invalidate much of ongoing deliberations. What about genetic engineering – are we ethically and psychologically ready for it? What if man learns to manipulate nuclear fusion reactions – would democracy survive? What would be the consequence of rolling out geo-engineering, of deep control over individual mind and consciousness or of a technology to manipulate gravitational force? In these domains, sustainability science can offer critical reflection and initiate participatory technology assessments in the science and society tradition.

We live in physical space and the sciences of ecology and geography contribute in manifold ways to our understanding and organisation of natural and economic, social and cultural space. They are the material linchpin between our quality-of-life aspirations on the one hand and the resources to fulfill them on the other – in other words: between the subjective ends and the objective means. Our knowledge of the Earth is rapidly increasing in the three dimensions of space, time and structure. The tools of complex system science are disclosing new and fascinating panoramas that already play a large role in sustainable development–oriented design and implementation processes. It provides the necessary system perspective on the efficient, reliable and affordable supply and use of water, energy and transport that are at the core of sustainable cities. It can guide a more sustainable use of renewable resources, such as soils, fish, forests, and of nonrenewable resources, such as fossil fuels, phosphates and scarce minerals.

We live in mental and social space. Our interaction with the physical world and with other living beings is mediated through the senses, the emotions and the mind and (self-)awareness. We have created and learned habits, customs, languages, ideas, calculus that frame our actions as individual-in-collective. Efforts in the social sciences: economics, sociology, psychology and others, aim at the construction of an image of man that explains and, if possible, predicts in probabilistic terms the behaviour of the members of the species *Homo sapiens*. Insights from biology and demography are an essential background. These are areas where contributions to sustainability science are dearly needed. It represents another link between ends and means.

On the agenda is the refinement of the image of man in economic science – the *Homo economicus*. We should intensify the experiments and simulations in behavioural economics in order to understand the decision of individuals as consumers and producers. What determines the balance between egoism and altruism, between competition and cooperation in individuals? What is the position of economic behaviour in the broader setting of social rules, ethical considerations and inner consciousness? Other important research areas are the social construction of needs in relation to innovation dynamics, and the necessary and sufficient rules to manage common pool resources sustainably. Complex system science offers new vistas here, too. Novel tools and methods for involving stakeholders and expert knowledge, such as policy exercises and simulation games, are a helpful and timely development. If we find answers, we are better prepared for the inevitable changes that will come and the creativity that is needed. And for the possibly largest challenge of all: to sustain the balance between the freedom and dignity of the individual and the solidarity and brotherhood that is a necessity in a world of more than 7 billion human beings that long for the good life.

Sustainability science just started, it seems.

Since 1900, a great dream has faded. To many in the West, their civilisation appears to have gone wrong. Much that is unique about it has seemed to turnout to be weakness, or worse . . . There seems to be little left for the educated to believe in.

The intellectual hegemony of western science has also been buttressed by its sheer size and scale . . . [and] explains why science is now a major religion, perhaps *the* religion, of our civilisation. . . . Of course, there are suspicions of the new priesthood.

– Roberts, *The Triumph of the West*, 1985

Decadence is the subordination of the whole to the parts.

– Oscar Wilde

Whoever in this world overcomes his selfish cravings, his sorrows fall away from him, like drops of water from a lotus flower.

– Dhammapada

He who knows he has enough is rich.

– Tao Te Ching

In the lower knowledge doubt and scepticism have their temporary uses; in the higher they are stumbling blocks: for there the whole secret is not the balancing of truth and error, but a constantly progressing of revealed truth.

– Bhagavad Gita, Commentary in the Words of Sri Aurobindo

The candles are different, but the light is the same.

– Anonymous

SUGGESTED READING

A journalistic account of predictions of the future in history and present.

Gardner, D. *Future Babble: Why Expert Predictions Are Next to Worthless, and You Can Do Better*. New York: Dutton, 2010.

One of the books that presents the scenario method during its ascendency in the 1990s.

Heijden, K. van der. *Scenarios – The Art of Strategic Conversation*. New York: Wiley & Sons, 1996.

This report describes the SRES methodology and outcomes and can be downloaded from www.ipcc.ch.

Nakicenovic, N., and R. Swart (Eds.) *The IPCC Special Report on Emissions Scenarios (SRES)*. Cambridge: Cambridge University Press, 2000.

An evaluation and update of the Limits to Growth (1971) report on the basis of the trends between 1970 and 2000.

Meadows, D. L., J. Randers, and D. H. Meadows. *Limits to Growth – The 30-Year Update*. Vermont: Chelsea Green Publishing Company, 2004.

A model-based investigation of two scenarios in order to find out about the resource and environment constraints in a world with population and economic growth. It can be downloaded at www.pbl.nl/en/publications/2009/Growing-within-limits.-A-report-to-the-Global-Assembly-2009-of-the-Club-of-Rome.

PBL. Growing within Limits. Bilthoven: Netherlands Environmental Assessment Agency, 2009.

USEFUL WEBSITES

- www.gaia.org/gaia/ecovillage/ is a site with explanation and network of ecovillages.
- www.iiasa.ac.at/ and themasites.pbl.nl/en/themasites/image are the sites with descriptions of the models at IIASA and PBL used for scenario construction.
- econ.worldbank.org/ is the site where one can find the *World Development Report* operated by the World Bank.
- www.agri-outlook.org/ is the site of the *Agricultural Outlook* operated by the FAO and OECD.
- www.worldenergyoutlook.org/ is the site of the *World Energy Outlook* (WEO) operated by the IEA.
- www.unep.org/geo/ is the site of the *Global Environmental Outlook* (GEO) operated by the UNEP.
- www.shell.com/home/content/aboutshell/our_strategy/shell_global_scenarios/ is the site with scenarios made and used by Shell.
- www.futurescenarios.org is a site with futures scenarios about the oil peak.

Glossary

This dictionary contains a list of acronyms and abbreviations and a brief etymology and definition of key words and concepts introduced in this book. Etymology and definitions are based on international English dictionaries and websites such as www.etymonline.com/ and www.merriam-webster.com/dictionary. Wikipedia is also a source for further enquiries.

Acronyms

Acronym	*Description*	*Website*
ABM	Agent-Based Models	www.openabm.org
BP	Before Present; British Petroleum	www.bp.com
CA	Cellular Automata	cell-auto.com/
CAS	Complex Adaptive Systems	
CE	Christian Era	
CFC	ChlorFluorCompounds	
CNG	Compressed Natural Gas	
CSD	Commission on Sustainable Development (UN-based)	www.un.org
CSS	Complex System Science	
FAO	Food and Agriculture Organization (UN-based)	www.fao.org
GDP	Gross Domestic Product	
GEO	Global Environmental Outlook	www.unep.org
GIS	Geographical Information Systems	
GMO	Genetically Modified Organism	
GWP	Gross World Product	
IAM	Integrated Assessment Model	
IIASA	International Institute for Applied System Analysis	www.iiasa.ac.at
ICT	Information and Communication Technology	

Acronym	*Description*	*Website*
IEA	International Energy Agency (OECD)	www.iea.org
IFs	International Futures model	www.ifs.du.edu
IGBP	International Geosphere Biosphere Programme	www.igbp.ch
IHDP	International Human Dimensions Programme	www.ihdp.org
IMAGE	Integrated Model to Assess the Global Environment	www.mnp.nl/image themasites.pbl.nl/en/
IMF	International Monetary Fund	www.imf.org
IMR	Infant Mortality Rate	
IPCC	Intergovernmental Panel on Climate Change	www.ipcc.ch
IUCN	International Union for the Conservation of Nature	www.iucn.org
LCA	Life Cycle Analysis	
MA	Millennium Ecosystem Assessment	www.millenniumassessment.org
MAS	Multi-Agent Simulation models	
MDG	Millennium Development Goals	www.un.org/millenniumgoals/
MNC	MultiNational Corporation	
Netlogo 4.1	Netlogo agent-based simulation modelling software	ccl.northwestern.edu/netlogo
NPP	Net Primary Production	
OECD	Organization for Economic Cooperation and Development	www.oecd.org
PAGES	Past Global Changes network	www.pages.org
PCRASTER	PCraster environmental modelling language	pcraster.geo.uu.nl/
Phoenix	Phoenix Population modelling framework	themasites.pbl.nl/tridion/en/ themasites/phoenix/index.html
PBL	PlanBureau voor de Leefomgeving (Netherlands Environmental Assessment Agency, formerly MNP/RIVM)	www.pbl.nl
PPP	Purchasing Power Parity	
R&D	Research and Development	
SAM	Social Accounting Matrix	
SCC	Supply Cost Curve	
SD	Sustainable Development	
SDI(S)	Sustainable Development Indicator (System)	
SES	Social-Ecological Systems	www.resalliance.org

Acronym	*Description*	*Website*
SNA	System of National Accounts	
Stella®	Stella system dynamics modelling software	www.iseesystems.com
UN	United Nations	
UNCED	United Nations Commission on Environment and Development	('Brundtland Commission')
UNDP	United Nations Development Programme	www.undp.org
UNEP	United Nations Environment Programme	www.unep.org
UNFPA	United Nations Population Fund	www.unfpa.org/public/
Vensim®	Vensim system dynamics modelling software	www.vensim.com
WCDR	World Commission on Disaster Reduction	
WCED	World Commission on Environment and Development	
WEC	World Energy Council	www.worldenergy.org
WHO	World Health Organization (UN-based)	www.who.int/en/
WMO	World Meteorlogical Organization	www.wmo.ch
WRI	World Resources Institute	www.wri.org
WSSD	World Summit on Sustainable Development (Johannesburg 2002)	
WTO	World Trade Organization	www.wto.org
WWC	World Water Council	www.worldwatercouncil.org
WWF	World Wildlife Fund	www.wwf.org.uk

Adaptation

Synonym/associate: adaptability, adaptive capacity, coping capacity, resilience, vulnerability

Etymology: from Latin *adaptare,* from *ad* near, and *aptare* to fit, *aptus* meaning apt, fit.

In an evolutionary context, the modification of an organism or its parts that makes it more fit for existence under the conditions of its environment. In a short-term and local situation, adaptation means making or undergoing an adjustment to environmental conditions.

Agrarianisation

Synonym/associate: agricultural transition

Etymology: from Latin root *agrarius,* from *ager* field.

The gradual intensification of the relationship between groups of humans, their environment and each other. During this process, humans applied increasing amounts of labour and inputs to the land and raised agricultural output per unit area and per unit time by applying more inputs (labour, fertiliser) and new techniques (multi-/intercropping, irrigation).

Algorithm

Synonym/associate: arithmetic rule

Etymology: from medieval Latin *algorismus,* derived from Arabic *al-khuw Arizmi,* an Arabian mathematician (825 CE).

A procedure for solving a mathematical problem (for instance finding the greatest common divisor) in a finite number of steps that frequently involves repetition of an operation. The broader meaning is a step-by-step procedure for solving a problem or accomplishing some end, especially by a computer.

Analogy

Synonym/associate: resemblance, similarity

Etymology: from Greek *analogia* proportion.

If two things are similar in certain respects, the word analogy is used to infer that they may also be equal in other respects.

Analysis

Etymology: from Greek, from αναλυψειν to break up, to loosen.

The separation of a whole into its component parts and/or the identification or separation of ingredients of a substance – as, for instance, in chemical analysis. In a more abstract sense, it is proof of a mathematical proposition by assuming the result and deducing a valid statement by a series of reversible steps. Less formally, it is an examination of a complex, its elements, and their relations.

Anthropocene

Etymology: From Greek αντρηοποσ man, human being and καινοσ new, fresh, recent.

This word was coined by Störmer and Crutzen in their paper *The Anthropocene* (Global Change Newsletter 41(2000)17–18). It was meant to indicate that with the increase of human population and activities, the Earth has entered a new era.

Anthroposphere → see Sphere

Archetype

Synonym/associate: prototype, perfect example, generic structure

Etymology: from Greek *αρχηειν* to rule and *τψπος* type.

The original pattern or model of which all things of the same type are representations or copies. In the psychology of Jung, it means an inherited idea or mode of thought that is derived from the experience of the race and is present in the unconscious of the individual. In systems thinking, it refers to patterns of structure recurring again and again.

Attractor

Synonym/associate: prototype

Etymology: from the Latin verb *attrahere* to pull or draw to.

A point or state to which a system is drawn or attracted. In mathematics, it is the state in which one or more of the state variables do not change. In simple systems, the attractor is a point in phase space. If the attractor is a closed trajectory in phase space, it is called a *limit cycle*. If the attractor has a complex self-similar structure, one speaks of a *strange attractor*.

Bifurcation

Synonym/associate: branching

Etymology: from Latin *bifurcatus,* of the verb *bifurcare,* from *bi* two, and *furca* fork.

The process which causes a system to divide into two branches or parts or during which a system divides itself into two branches or parts.

Binomial distribution

Etymology: from Latin *bi* and *nomen* having two names.

A particular probability function, namely the one that develops from a series of independent yes/no experiments each of which has a probability of success (yes) p. The probability of getting k successes after n trials equals:

$$P(k; n, p) = \binom{n}{k} p^k (1-p)^{n-k}$$

with the first term the binomial coefficient.

Biodiversity

Synonym/associate: biological diversity, genetic diversity, species diversity, ecosystem diversity

Etymology: from Greek *βιος* life and from Latin *divertere* to turn in opposite directions.

The variety of ecosystems and species of plants and animals that can be found in nature. Biodiversity matters at the level of genes, species and ecosystems. Genetic diversity is the sum total of genetic information, contained in the genes of individual plants, animals and microorganisms. Species diversity refers to the variety of living organisms. Ecosystem diversity is about the variety of habitats, biotic communities, and ecological processes in the biosphere, as well as the diversity in habitat and ecological processes within ecosystems. Preservation of biodiversity usually refers to maintaining individuals or groups of species; conservation to maintaining the ability of species to evolve.

Biogeography

Etymology: International Scientific Vocabulary.

The study and interpretation of geographical distribution of organisms (animals and plants), both living and extinct.

Biome

Etymology (etymonline.com): from Greek βιος life.

A vegetation type that contains ecosystem communities with similar growth forms and interactions in animal, microbial and soil components and in above-ground plants and trees. There is usually a relationship with climate and other environmental factors. Biomes are abstractions: They have no sharp boundaries and mask lower-scale heterogeneity. The concept is less useful for aquatic and mountainous systems

Biosphere → see Sphere

Carrying capacity

Synonym/associate: equilibrium state, steady-state, attractor

The level of the population size, which the resources of the environment can just maintain ('carry') without a tendency to either increase or decrease. At this population density, birth rate equals death rate as a consequence of intraspecific competition. Mathematically, it is the equilibrium level or attractor in the corresponding system description.

Causality

Synonym/associate: reason, motive

Etymology: from Latin *causa* a reason, interest, judicial process.

The relation between a cause and its effect or between regularly correlated events or phenomena. A cause is a reason for an action or condition or a person or thing that brings about an effect or a result. The notion of causality has also been defined more formally. Let us assume that a situation is described at two given space-time points $\{X_1,t_1\}$ (called the initial state) and $\{X_2,t_2\}$ (called the final state). If there exists a dynamical equation of change that can be used to derive $\{X_2,t_2\}$ from $\{X_1,t_1\}$, one can speak of *causation* or causal linkage.

Cellular Automata (CA)

Synonym/associate: robot, machine

Etymology: from Latin *cella* small room and from Greek αυτοματοσ self-moving.
The name of a mathematical technique to describe change of a system element as a consequence of changes in neighbouring elements.

Coevolution → see Evolution

Collapse

Synonym/associate: catastrophe, overshoot-and-collapse, disintegration

Etymology: from Latin *collapsus,* of the verb *collabi: com* with, and *labi* to fall, slide.

To fall or shrink together abruptly and completely, to break down completely or to suddenly lose force, significance, effectiveness or worth.

Common Pool Resources (CPR)

Synonym/associate: 'the commons'

Natural or humanly created systems (resource, facility) that generate a finite flow of benefits, are available to more than one person and are subject to degradation as a result of overuse. They can be owned by a community or a government or be privately owned. When no property rights are established and its use is not regulated, the CPR is under an open-access regime. Examples are a grazing area, a fishing ground, a waterstream, the atmosphere, a natural park or a road.

Competition

Synonym/associate: rivalry

Etymology: from the Latin *competitio* and the Latin verb *competere* strive together, from *com* with, and *petere* to strive, seek, fall upon, rush at, attack.

The act or process in which two or more parties acting independently attempt to win or gain, for instance, two or more organisms competing for some environmental resource in short supply. More generally, it denotes a contest between rivals. In the World Economic Forum's *Global Competitiveness Report,* it is defined as 'the fitness of a country's economic institutions and structures to produce growth, in view of the overall structure of the global economy'.

Complexity

Synonym/associate: complex systems, complexity theory, complexity science

Etymology: from Latin *complexus* totality, and the Latin verb *complecti* to encircle or embrace.

The quality or state of being complex. As a noun, a complex is a whole being made up of interrelated parts. As an adjective, it refers to obviously related units of which the degree and nature of the relationship is imperfectly known. Nonlinearity and collective behaviour are characteristic features of a complex system. In the context of socio-natural systems, it means differentiation in social, political and/or economic structure combined with organisation that integrates diverse structural parts into a whole.

Complex Adaptive Systems (CAS)

Synonym/associate: agent-based modelling (ABM), rule-based modelling, multi-agent simulation (MAS)

Systems with inherent uncertainty in their dynamics that tend to have multiple stable states and that exhibit self-organisation. More generally, it refers to systems that show complex adaptive behaviour in response to environmental change – ecosystems but mostly systems with human actors.

Conspicuous consumption

Synonym/associate: striking, extravagant

Etymology: Latin *conspicuus,* from *conspicere* to get sight of, from *com* with, and *specere* to look.

Behaviour of extravagant and ostentatious displays of resources that function as a competitive strategy to demonstrate wealth and social status. The term was coined by the economist Thorstein Veblen. Conspicuous means obvious to the eye or mind, attracting attention or marked by a noticeable violation of good taste.

Contingency

Synonym/associate: accidental, possible, unpredictable

Etymology: from Latin *contingent-, contingens,* present participle of *contingere* to have contact with, befall, from *com* with, and *tangere* to touch.

The quality or state of being contingent means being likely but not certain to happen, of being not logically necessary, of happening by chance or unforeseen causes or of being subject to chance or unseen effects. Contingency in socio-natural system evolution is the dependence of conditions on the operation of previous processes.

Correlation

Etymology: from Latin *correlation: com* with, and *relatio* a bringing back or restoring.

A relation existing between phenomena or things or between mathematical or statistical variables which tend to vary, be associated, or occur together in a way not expected on the basis of chance alone.

Cosmology

Etymology: from Greek κοσμος meaning (good) order or orderly arrangement, and λογος discourse.

Life on earth, in this world, as opposed to the afterlife. Later its use was extended to the known world, and still later to the physical world in the broadest sense: the universe.

Dialectic

Etymology: from Greek διαλεκτικε (τεχηνε) (art of) philosophical discussion or discourse.

Originally, reasoning by dialogue as a method of intellectual investigation. In philosophy, it is a process of resolving or merging contradictions in character and in which a concept or development changes into its opposite while also being preserved and fulfilled. It got its meaning from the works of Kant and Hegel.

Discounting

Synonym/associate: disregard, discount rate, time preference

Etymology: from Latin *discomputare,* from *dis* not or lack of, and *computare* to count.

The verb to discount means, in everyday market speak, to make a deduction from usually for cash or prompt payment or to sell or offer for sale at a reduced price. It also means: to leave out of account, to disregard or minimise the importance, to make allowance for bias or exaggeration or to view with doubt. It is in this sense that it is used in taking into account future events in present (financial) calculations.

Dissipative structure

Synonym/associate: dissipation, dispersion, diffusion, dissolution

Etymology: from Latin *dissipatus,* from the verb *dissipare*, from *dis* not or lack of, and *supare* to throw.

The verb to dissipate, in a transitive sense, means: to break up and drive off (as a crowd), to cause to spread thin or scatter and gradually vanish, or in physics: to lose (as heat or electricity) irrecoverably. It may have a negative connotation: to spend or use up wastefully or foolishly. In an intransitive sense, it means: to break up and scatter or vanish.

Ecology

Etymology (m-w.com): from Greek οικοσ house and λογοσ reason or idea.

A branch of science concerned with the interrelationship of organisms and their environments as well as what this science describes: the totality or pattern of relations between organisms and their environment. In the last three decades, several derived scientific branches have emerged: human ecology, industrial ecology, social ecology.

Economy

Synonym/associate: economic science, economics

Etymology: from Latin *oeconomia,* from Greek *οικονομος* household manager, from *οικος* house and *νεμειν* to manage.

An archaic meaning of *economy* is the management of household or private affairs and especially expenses. From this meaning, it has gone into two directions:

- equating it with good (household) management: thrifty and efficient use of material resources, frugality in expenditures, or even an instance or a means of saving on resources as in 'economies of scale'. It may also concern the efficient and concise use of nonmaterial resources such as effort, language, or motion.
- equating it with the whole of production and consumption: the arrangement or mode of operation of something.

As with ecology, there are several scientific branches derived from 'the science of the economy': ecological economics, evolutionary economics and others.

Ecosystem

Etymology: from Greek *οικοσ* house.

A complex of a community of organisms and its environment functioning as an ecological unit. In the UN Convention on Biological Diversity, it is defined as 'a dynamic complex of plant, animal and micro-organism communities and non-living environment interacting as a functional unit'. Another definition is: any unit that includes all of the organisms in a given area interacting with the physical environment so that a flow of energy leads to . . . exchange of materials between living and non-living parts within the ecosystem.

Ecotone

Etymology: from Greek *εχ* out and *τονος* tension.

A narrow and rather sharp defined transition zone between two ecological communities; they are usually species-rich and may have come into existence either from natural or anthropogenic processes.

Emergence

Synonym/associate: rise

Etymology: from Latin *emergere,* rise out or up, bring forth, bring to light, from *ex* out, and *mergere* to plunge.

The verb to emerge refers to biological processes of various superficial outgrowths of plant tissue or of the penetration of the soil surface by a newly germinated plant. In complexity science, it has been used in relation to *emergent properties*: system properties at a certain level of observation and abstraction, that cannot be observed at a lower level of individual elements and are the product of the interactions between the elements (and "more than the sum of the parts"). As with complexity and self-organisation, the observer is involved in its definition.

Emic and etic

Etymology: from Greek ετικον pertaining to.

The ending emic means relating to or involving analysis of cultural phenomena from the perspective of one who *does* participate in the culture being studied, whereas the ending etic means relating to or involving analysis of cultural phenomena from the perspective of one who *does not* participates in the culture being studied.

Emulation

Synonym/associate: imitation

Etymology: from Latin *aemulatus,* of the verb *aemulari* to rival.

To imitate, with the connotation of ambition or endeavour to equal or excel others. This aspect is emphasised in the expression of competitive emulation – a process in which individuals, families, companies or states compete.

Entropy

Synonym/associate: (dis)order, chaos

Etymology: from Greek *εν* in, and the verb *τρεπειν* to turn.

In a natural science context, a well-defined quantitative measure of the microscopic disorder of a system. Entropy change is caused by heat transfer, mass flow and irreversible processes degrading energy from higher to lower order. In a broader and looser sense, entropy is the degree of disorder or uncertainty in a system and refers to a process of degradation or running down or a trend to disorder.

Epistemology

Etymology: from Greek *επιστημη* knowledge, from *επιστανaι* to understand, know, from *επι* with, and *ηιστανaιι* to cause to stand.

The study or a theory of the nature and grounds of knowledge especially with reference to its limits and validity. Philosophers always warn to distinguish between what can be known (epistemology) and what is (ontology).

Equilibrium (state)

Synonym/associate: attractor, steady-state, carrying capacity

Etymology: from Latin *aequilibrium,* from *aequi* equal and *libra* weight, balance.

In a mechanical sense, a state of balance between opposing forces or actions that is either static (as in a body acted on by forces whose resultant is zero) or dynamic (as in a reversible chemical reaction when the rates of reaction in both directions are equal). It is often used outside the mechanical context, as in systems theory: a state of adjustment between opposing or divergent influences or elements, or in psychology: a state of intellectual or emotional balance.

Eutrophication

Etymology: from Greek ευ good and τροφη nourishment.

The process of enrichment of water by nutrients, often stimulating plant growth and as a result depletion of oxygen in the water.

Evolution

Synonym/associate: co-evolution, development, unfolding, growth

Etymology: from Latin *evolutio* unrolling, from the verb *evolvere* to unroll.

A process of continuous change from a lower, simpler, or worse to a higher, more complex, or better state. In biology, it is specifically about the historical development of a biological group, as a race or species. The underlying evolution theory asserts that the various types of animals and plants have their origin in other preexisting types and that the distinguishable differences are because of modifications in successive generations. In a cosmological sense, it is a process in which the whole universe is a progression of interrelated phenomena. In a social science context, it means a process of gradual and relatively peaceful social, political, and economic advance – in contrast to a revolution with its suddenness and discontinuity.

Exergy → see Energy

Fractal

Synonym/associate: self-similar, self-symmetric

Etymology: from Latin *fractus* fragmented or broken.

A geometrical or physical structure having an irregular or fragmented shape at all scales of measurement between a greatest and smallest scale. They exhibit self-similarity. The mathematician Mandelbrot (1977) has proposed the name fractal geometry; it has been applied to such diverse fields as the stock market, chemical industry, meteorology and computer graphics.

Gradient

Etymology: from Latin *gradientem*, derived from *gradi* to walk.

A (steep) slope of a road or railroad. In mathematics, it refers to the curvature in a surface in phase or state space, such as the first derivative $\partial F/\partial X_i$ of a state variable X_i for $F = F(X_1 \ldots X_n)$.

Growth

Synonym/association: development

Etymology: from Old English *growan* and akin to Old High German *gruowan* to grow. It supplanted in medieval times the Old English *weaxan* (cf. German wachsen).

To grow, or the process of growth, means to spring up and develop to maturity, to be able to grow in some place or situation and to assume some relation through or as if through a process of natural growth. It also describes the phenomenon of increase in size by assimilation of material into the living organism or by accretion of material in a nonbiological process (such as crystallisation).

Heterogeneity

Synonym/associate: diverse, different

Etymology: from Greek *ετερογενες*, from *ετερ* other and *γενος* kind.

An area, a population and so on are said to be heterogeneous if they consist of dissimilar or diverse ingredients or constituents. The opposite is homogeneity: consisting of the same ingredients or constituents.

Heuristic

Synonym/associate:

Etymology: from Greek *ευρισκειν* to find, discover.

A device or procedure that involves or serves as an aid to learning, discovery, or problem solving by experimental and especially trial-and-error methods. In a computer context, it refers to exploratory problem-solving techniques. In a more general sense, it is a set of principles used in making decisions when not all possibilities can be fully explored and consisting of guided trial and error.

Hierarchy

Synonym/associate: order, ranking

Etymology: from Greek ιερος priest and αρχη, from *αρχηειν* to begin, rule, and *αρχη* beginning, rule, *αρχηος* ruler.

The first meaning of hierarchy in the English dictionary is a division of angels: the celestial hierarchy. These religious roots also show up in the second meaning: a ruling body of clergy organised into orders or ranks each subordinate to the one above it and, more general, church government by a hierarchy. Later on, it has become a wider term: a body of persons in authority, the classification of a group of people according to ability or to economic, social or professional standing, or a graded or ranked series as in a hierarchy of values.

Homogeneity → see Heterogeneity

Identity

Etymology: from Latin *identitas* sameness, and the word *idem* the same.

The property of remain the same, having essential or generic character in different instances. Identity can thus denote that which constitutes the objective reality of a thing or a person. In the latter, psychological sense, it is the distinguishing character or personality of an individual (individuality) and can serve as a way to establish relations between individuals.

Indicators

Etymology: from Latin *indicare* to point out, show, from *in* in, and *dicare* to proclaim and *dicere* to speak, to say.

An indicator is meant to provide information about part(s) of a system which is useful within the context of the receiver. The usefulness depends on the objective of the user, on the presumed relation between indicator and system, and on the (scientifically) acknowledged relation between indicator and system. The degree to which control over the system is possible in the form of choice (amongst alternatives) matters too.

Inference

Synonym/associate: conclusion, deduction, consequence

Etymology: from Latin infere, from Greek ειν in and φερειν to carry, bring.

The act of making a statement or judgment that is considered to be valid or true on the basis of another statement or judgment that one believes to be valid or true. It also denotes the step from statistical data to a generalisation.

In silico

Etymology: from the Latin word for the element silicium or silicon.

Something (done) in or on a computer or in computer simulations/with computer software. Coined recently after in vivo and in vitro, it stems from the notion that silicium is an important element in chips (and the core element of sand).

Institution

Synonym/associate: (self-)organisation, establishment

Etymology: from Latin from *instatuere* to set up, from *in* in and *statuere* to establish, cause to stand.

The word *institution* refers to a significant practice, relationship or organisation in a society or culture, as in the institution of marriage. In a political/organisational context, it means an established organisation or corporation, especially of a public character, as for instance a college or university.

Institutions can be viewed as a set of rules actually used by a set of individuals to organise repetitive activities that produce outcomes affecting those individuals and potentially affecting others. They structure human interaction, consisting of formal (such as law) and informal (such as behavioural norms) constraints and enforcement characteristics. Institutions are social constructs with their own normative, cognitive and regulative (goals, values, paradigms or instruments) dimensions.

Irreversibility → see Reversibility

Isomorphism

Synonym/associate: similarity

Etymology (m-w.com): from Greek ισος same and μορφος form.

In mathematics, a one-to-one correspondence between two sets of objects.

Limit cycle → see Attractor

Marginal Cost

Etymology: from Latin *margo* edge, boundary space.

In economics, it is the cost of one additional unit. Mathematically, it is the first derivative of the total production cost with respect to the coresponding input.

Metabolism

Etymology: from Greek *μεταβολη* change, from *μεταβαλλειν* to change, from μετα after, beyond, and βαλλειν to throw.

Metabolism is the set of chemical changes in living cells by which energy is provided for vital processes and activities and new material is assimilated. In biology, it is the thousands of biochemical reactions that sustain the processes of life and in particular of cells and form metabolic networks. By analogy, industrial metabolism refers in human/industrial ecology to the transformation of material fluxes in the natural environment into usable 'goods' and their subsequent excretion, discarding and/or recycling. In the social sciences, Marx and Engels introduced metabolism for the material exchange between man and nature on a fundamental anthropological level and as a critique of the alienating capitalist mode of production. It is a translation of the German word *Stoffwechsel*, introduced in the 19th century for the description of material exchanges within the (human) body.

Metaphor

Synonym/associate: comparison, analogue, symbol

Etymology: from Latin *metaphora,* from Greek *μεταφερειν* to transfer, from *μετα* after, beyond, and *φερειν* to bear.

A figure of speech in which a word or phrase literally denoting one kind of object or idea is used in place of another to suggest a likeness or analogy between them. A thing is given a name that belongs to something else, but not all of the characteristic properties are carried over together with the name. The purpose of using a metaphor is to elucidate a new and unfamiliar concept by taking support from a familiar one. Metaphors are violations of literalness – and it is from the form of violation that they draw their emotive and cognitive effects. A prerequisite for correct use is that the users have the same perception and experience of the named 'thing'.

Model

Synonym/association: archetype, copy, image, replica, analogue, metaphor

Etymology: from Latin *modulus* small measure, from *modus* way in which something is done.

Any physical or symbolic representation of certain properties and interactions of a system. A model is, in a way, a primitive theory. A theory is, in a way, a generalised model. The domain of a theory or model is the state space in which it is valid and usable. A domain is always limited.

Nature

Etymology: from Latin *natura* course of things, derived from *natus* born.

The original from Latin word *natura* it denotes the course of things, the universe. The medieval connotation of the word Nature is: essential qualities, innate disposition, and also the creative power in the material world. Used for humans, as in human nature, it refers to the inherent, dominating power or impulse of a person.

Noösphere → see Sphere

Observation

Etymology: from Latin *observare* to attend to, to keep, to follow.

A series of organised sensory experiences. Events are recorded within a certain time-space domain. Recording can be qualitative, using verbal descriptions or ordinal ranking, or quantitative, using measurement units. From a cognitive science perspective, observation is the interface between perception and learning.

Ontology

Etymology: from Greek *οντος* being and *λογος* word, reason.

A part of metaphysics concerned with the nature and relations of being or existence, or a particular theory about it.

Organisation

Synonym/associate: establishment, society

Etymology: from Latin *organum* instrument, organ.

An administrative and functional structure, as a business or a political party. It refers in a more formal sense to a set of entities that constitute a dynamical system and are organising or being organised with a collective purpose. Self-organisation refers to a set of dynamical systems, whereby structures appear at the global level of a system from interactions amongst its lower-level components. It coincides with a move towards lower probability and higher order, and increasing heterogeneity and differentiation. It involves structures with function or purpose and it requires a non-trivial fitness function. If a system involves only one or a few elements and/or follows rules that can be easily understood, one does not speak of self-organisation.

Panarchy

Etymology: from Greek παν everything, all and αρχηειν to rule, command.

A nested set of adaptive cycles at different scales, that exhibits cross-scale interactions. The concept is used by Holling and colleagues to capture the hypothesis that ecosystems evolve through adaptive cycles that are nested one within the other across space and time scales. Although many ecosystem observations are in conformity with the panarchy hypothesis, its validity and usefulness in ecological and other (economic, social) domains is still controversial.

Paradigm

Synonym/associate: example, pattern, archetype

Etymology: from Greek *παραδειγμα,* from παρα alongside, near, and δεικνψναι to show.

Originally, the word paradigm was simply synonymous with pattern or model. Later, it was used for an outstandingly clear or typical example or archetype. With the work of Kuhn, it acquired the meaning of a philosophical and theoretical framework of a scientific school or discipline within which theories, laws, and generalisations and the experiments performed in support of them are formulated. In a broad sense, the (dominant) paradigm is the basic way of perceiving, thinking, valuing and doing, associated with a particular vision of reality.

Pastoralism

Etymology: from Latin *pastorem* shepherd, also spiritual guide (shepherd of souls). Related to pasture, from Latin *pastura* feeding, grazing.

The practise of herding as the primary economic activity of a society.

Pedigree

Etymology: probably from a forked sign as from a bird's footprint, in old French *pied de gru* foot of a crane.

Ancestral line, descent, genealogical chart or table.

Perennial (wisdom)

Etymology: from Latin *per* through *annum* year.

Lasting through the years, enduring, permanent.

Persistence

Synonym/association: continuity, resilience

Etymology: from the Latin verb *persistere,* existing for a long or longer than usual time or continuously.

A good or substance is persistent if it is retained beyond the usual period, continues without change in function or structure, is effective in the open for an appreciable time usually through slow volatilising, degrades only slowly by the environment or remains infective for a relatively long time in a vector after an initial period of incubation.

Phase plane

A plane (or space) with all the dimensions needed to describe dynamical physical behaviour. For a single state variable X, it contains the dimensions X and dX/dt. For a system with two or more state variables, it is the n-dimensional phase space $\{X_1..X_n\}$.

Phenomenology

Synonym/associate: appearance

Etymology: from the Greek verb *πηαινεστηαι* to appear, and related to *πηαινειν* to show.

An observable fact or event, or an object or aspect known through the senses rather than by thought or intuition. As a philosophical school, it refers to the study of the development of human consciousness and self-awareness as a preface to or a part of philosophy.

Photosynthesis

Etymology: from Greek *φοτον* light and σψντηεσις composition, something put together.

The synthesis process of chemical compounds with the aid of radiant energy and especially light; in particular, the formation of carbohydrates from carbon dioxide and a source of hydrogen (as water) in the chlorophyll-containing tissues of plants exposed to light.

Power laws

A mathematical description applying when the size S of a certain phenomenon (event, property) and its frequency of occurrence N(S) are a straight line if plotted in a log-log graph, that is, $N(S) = aS^{-\alpha}$. They are common in nature. Examples are the energy release of earthquakes E and their frequency F(E). Power laws are scale invariant: if $S' = \beta S$, then $N(S') = aS'^{-\alpha} = a\beta^{-\alpha}S^{-\alpha} = a'S^{-\alpha}$.

Redundancy

Synonym/associate: superfluity, abundance, diffuseness

Etymology: from Latin *redundare* come back, contribute, literally: overflow, from *re* again, and *undare* rise in waves.

The property of abundance in the sense of being superfluous, but nevertheless useful in case of emergency or failure as a backup.

Resilience

Synonym/associate: buffer, stability, robustness

Etymology: from the Latin verb *resilire* to jump back, recoil, from *re* back, and *salire* to leap.

The capacity of a system to absorb disturbance, undergo change and still retain essentially the same function, structure, identity, and feedbacks. A measure of the resilience of a system is the full range of perturbations over which the system can maintain itself. Resilience is often viewed and used as a general term meaning either stability or robustness and linked to sensitivity and vulnerability.

Reversibility

Etymology: from Latin *revertere* to turn back.

The property of being capable of being reversed (in time). Irreversible is: not capable of being reversed (in time).

Robustness

Synonym/associate: sturdy, resilient

Etymology: Latin *robustus* oaken, strong, from *robur* oak, strength.

The property of having or exhibiting strength or vigorous health, being strongly formed or constructed or being capable of performing without failure under a wide range of conditions. Robustness is associated with having or showing firmness, such as having a solid structure that resists stress, not subject to change or revision, not easily moved or disturbed. If a system has a well-defined identity, robustness refers to the property to either absorb external perturbations from sheer size/mass or redundancy or to use controls on the basis of anticipation and power.

Scenario

Synonym/associate: screenplay, scenario method/approach, myth, strategy, plan, visioning

Etymology: from Latin *scaenarium* place for erecting stages, from *scaena* stage.

An outline or synopsis of a play, more spcifically, a plot outline used by actors of the commedia dell'arte. In a broader sense, it is a sequence of events especially when imagined. As a methodology, the scenario method has become widespread over the last few decades in futures research and strategic planning.

Science → see Chapters 1, 8 and 10

Sedentary

Etymology: from the Latin verb *sedēre* to sit.

Non-migratory; permanently attached.

Self-organisation → see Organisation

Self-similarity

Synonym/associate: self-resemblance, self-symmetry, scale invariance, fractal

Etymology: from Latin *similis* like.

A self-similar pattern or object is one whose component parts resemble the whole and that remains invariant under changes of scale, that is, they are scale-symmetric.

Sphere

Etymology: from Greek *σφαιρα* globe, ball. Biosphere from Greek *βιος* life. Pedosphere from Greek πέδον soil. Ecosphere from Greek οικοσ house, habitation. Anthroposphere from Greek αντροποσ man, human being. Sociosphere from Latin *socius* companion. Technosphere from Greek τεχηνη art, skill or craft. *Noösphere* from Greek νοος mind.

A ball or body of globular form. Later, it broadened to the range of something or a space where certain processes exist or dominate. In sustainable development context one distinguishes:

- The *biosphere*, introduced by in 1883 by the geologist Suess and later reconceived and refined by the chemist Vernadsky in the 1920s. Building upon insights from geology and agrochemistry, Vernadsky coined it to indicate the *'la région unique de l'écorce terrestre occupée par la vie'* (Vernadsky, La Biosphère (1929):19), and considered its investigation a necessarily transdisciplinary enterprise. The word *ecosphere* is used in similar ways.
- The *pedosphere* is the outermost layer of the Earth with its soils. It is where soil formation processes take place.

- The *anthroposphere* is the part of the Earth where humans have a significant impact on energy and material fluxes. The *sociosphere* is sometimes used to indicate a similar domain.
- The *technosphere* is used to distinguish the part which consists of human-made capital stocks and matter, energy and information flows.
- The *noösphere* was a word crafted by the palaeologist Teilhard de Chardin to indicate the collective of human minds.

Spirituality

Synonym/associate: incorporeal, transcendent (in English also: church, clergy)

Etymology: from Latin *spiritus* of breathing, of the spirit.

Relating to, consisting of, or affecting the spirit, relating to sacred matters, concerned with religious values or relating to supernatural beings or phenomena. In English, it also means ecclesiastical rather than lay or temporal. In the official definition of the British Department of Education, *spirituality* is the valuing of the non-material aspects of life, and intimations of an enduring reality. The French *spiritualité* denotes that which is from the *esprit* and is disconnected from any material manifestation.

Stability

Synonym/associate: resilience, firmness

Etymology: from Latin *stabilis,* from *stare* to stand.

The quality, state, or degree of being stable, such as having the strength to stand or endure. The adjective stable means firmly established; unvarying, permanent and enduring; not changing or fluctuating. It is also used in a more active, psychological sense: steady in purpose, firm in resolution, not subject to insecurity or emotional illness, as in a stable personality.

Strange attractor → see Attractor

Structure

Synonym/associate: construction, arrangement

Etymology: from Latin *structura* a fitting together, adjustment, building, from the verb *struere* to build up, heap.

Something that is constructed, arranged or organised in a definite pattern.

Sustainability → see Chapter 1

Syndrome

Synonym/associate: pattern

Etymology: from Greek *σψνδρομε* concurrence of symptoms, concourse, from *σψν* with and *δρομος* running, course.

In medical terminology, a syndrome is a complex clinical picture. Outside of medicine it refers to a combination of phenomena seen in association. The concept of syndrome has been introduced by the WBGU (1994) to understand Global Change as a co-evolution of dynamic partial patterns of unmistakable character or symptoms. Symptoms, derived from Greek and meaning accident or misfortune, is a departure from normal function or feeling and an indication of the presence of disease or abnormality.

Taxonomy

Synonym/associate: classification

Etymology: from Greek ταξις arrangement and νεμειν to manage.

The study of the general principles of scientific classification; more specifically, the orderly classification of plants and animals according to their presumed natural relationships.

Technosphere → see Sphere

Template

Synonym/associate: mold, example

Etymology: probably from French *templet,* diminutive of *temple* part of a loom, probably from Latin *templum* plank.

Literally, a short piece or block placed horizontally in a wall under a beam to distribute its weight or pressure (as over a door). It also is a gauge, pattern, or mold used as a guide to the form of a piece being made. In a figurative way, it refers to something that establishes a pattern or serves as an example, for instance, a molecule (such as DNA) that serves as a pattern for the generation of another macromolecule (such as messenger RNA).

Theory

Synonym/associate: hypothesis, speculation, conjecture

Etymology: from Greek *θεορειν* the analysis of a set of facts in their relation to one another.

The general or abstract principles of a body of fact or a plausible or scientifically acceptable general principle or body of principles offered to explain phenomena. Less strictly, it is a belief, policy, or procedure proposed or followed as the basis of action. It may also be an ideal or hypothetical set of facts, principles, or circumstances, as used in the phrase *in theory*.

Topology

Etymology: from Greek τοπος place and γραφειν to write.

A branch of mathematics concerned with those properties of geometric configurations (as point sets), which are unaltered by elastic deformations (as a stretching or a twisting).

Transaction cost

Etymology: from Latin *transactionem* agreement, accomplishment, from *trans* through, and *agere* to drive.

Generally speaking, it refers to costs that are made for an exchange in the market or for particpation in market exchanges. Examples are the costs to acquire information and to police and enforce contracts.

Utilitarianism

Etymology: from Latin *utilitas* usefulness, serviceableness, profit.

A doctrine that states that the useful is the good and that conduct should ethically be judged for its utility, that is, the usefulness of its consequences. In practice, it implies actions aiming at the largest possible balance of pleasure over pain or, in society, at the greatest happiness of the greatest number. Utilitarianism is associated with the British philosopher Bentham.

Utopia

Etymology: from Greek ου not and τοπος place.

The word was created by Thomas More for the title of his book *Utopia* (1516) about a perfect, imaginary island. It is used to indicate a perfect but infinitely remote and impossible to reach land ('nowhere') or situation.

Vicious/virtuous circle

Etymology: from Latin *vitiosus* faulty, defective, corrupt; from Latin *virtus*) moral strength, manliness, excellence, from *vir* man.

A vicious circle or cycle is a series of events that reinforces itself through a positive feedback loop, with a negative result. The response to a problem causes a new problem that aggravates the situation. A virtuous circle or cycle is the same but with a favorable outcome. Vicious circles can be converted into virtuous ones and vice versa.

Vulnerability

Synonym/associate: sensitivity, lack of resilience

Etymology: from Latin *vulnus* a wound, and *vulnerare* to wound.

A strict definition of vulnerability is able to be hurt or wounded. In a broader sense, it can be viewed as the potential for negative outcomes or consequences.

Three distinct clusters of definitions for vulnerability have been identified: risk of exposure to hazards, a capability for social response, and an attribute of places (such as vulnerability of coastlines to sea level rise). Other descriptions are the interface between exposure to the physical threats to human well-being and the capacity of people and communities to cope with those threats, and the degree to which a system or community is susceptible to, or able to cope with, adverse effects of such as climate change. The common element is the potential of a system to be damaged or deteriorate.

References

Aalbers, T., ed. Waardenoriëntaties, wereldbeelden en maatschappelijke vraagstukken: Verantwoording van het opinieonderzoek voor de Duurzaamheidsverkenning Kwaliteit en Toekomst. Report 550031002. Bilthoven, the Netherlands: Netherlands Environmental Assessment Agency (PBL), 2006. Available at www.pbl.nl/search/node/550031002.

Acevedo, M., F. Urban, and H. Shugart. Models of forest dynamics based on roles of tree species, *Ecological Modelling* 87 (1996): 267–284.

Achterhuis, H. *Utopia's Heritage (De erfenis van de utopie)*. Amsterdam: Ambo, 1998.

Adams, J. *Risk*. London: UCL Press Ltd., 1995.

Akerlof, G., and R. Shiller. *Animal Spirits – How Human Psychology Drives the Economy, and Why It Matters for Global Capitalism*. Princeton, NJ: Princeton University Press, 2009.

Alcamo, J., P. Döll, T. Henrichs, F. Kaspar, B. Lehner, T. Rösch, and S. Siebert. Global estimation of water withdrawals and availability under current and "business as usual" conditions, *Hydrological Science*, 48(2003):339–348.

Alkemade, R., M. van Oorschot, L. Miles, C. Nellemann, M. Bakkenes, and B. ten Brink. GLOBIO3: A framework to investigate options for reducing global terrestrial biodiversity loss, *Ecosystems* 12 (2009): 374–390.

Allen, P., and J. McGlade. Modelling complex human systems: A fisheries example, *European Journal of Operations Research* 30 (1987): 147–167.

Allwood, J., M. Ashby, T. Gutowski, and E. Worrell. Material efficiency: A white paper, *Resources, Conservation and Recycling* 55 (2011): 362–381.

Ananthu, T. *Gandhi's Hind Swaraj*. New Delhi: Gandhi Peace Foundation, n.d.

Anderson, C., K. Frenken, and A. Hellervik. A complex network approach to urban growth, *Environment and Planning A* 38 (2006): 1941–1964.

Armstrong, K. *The Great Transformation: The World in the Time of Buddha, Socrates, Confucius and Jeremiah*. New York: Random House, 2006.

Armstrong, K. *Twelve Steps to a Compassionate Life*. New York: Knopf Doubleday Publishing, 2010.

Arquitt, S., X. Honggang, and R. Johnstone. A system dynamics analysis of boom and bust in the shrimp aquaculture industry, *System Dynamics Review* 21 (2005): 305–324.

Arrow, K. *The Limits of Organization*. New York: W.W. Norton & Company, 1974.

Arthur, W. *Increasing Returns and Path Dependence in the Economy*. Ann Arbor: The University of Michigan Press, 1994.

Aslin, H., and D. Bennett. Wildlife and world views: Australian attitudes to wildlife, *Human Dimensions of Wildlife* 15(2000): 15–35.

Atkisson, A. *The ISIS Agreement – How Sustainability Can Improve Organizational Performance and Transform the World*. London: Earthscan, 2008.

Aurobindo. *Le Cycle Humain*. Pondichéry, India: Sri Aurobindo Ashram, 1972/1998.

Axelrod, R. *The Evolution of Cooperation*. New York: Basic Books, 1984.
Axelrod, R. *The Complexity of Cooperation: Agent-Based Models of Competition and Collaboration*. Princeton, NJ: Princeton University Press, 1997.
Ayres, R., and L. Ayres, eds. *A Handbook of Industrial Ecology*. Cheltenham, UK: Edward Elgar, 2002.
Ayres, R., and B. Warr. Accounting for growth: the role of physical work, *Structural Change and Economic Dynamics* 16 (2005):181–209.
Azar, C., and S. Schneider. Are the economic costs of stabilizing the atmosphere prohibitive? *Ecological Economics* 42 (2002): 73.
Bai, Z., D. Dent, L. Olsson, and M. Schaepman. Proxy global assessment of land degradation, *Soil Use and Management* 24 (2008): 223–234.
Bakan, J. *The Corporation: The Pathological Pursuit of Profit and Power*. New York: Free Press, 2004.
Bakkes, J., et al. Background Report to the OECD Environmental Outlook to 2030. Overviews, Details, and Methodology of Model-Based Analysis. Bilthoven, the Netherlands: Netherlands Environmental Assessment Agency (PBL), 2008.
Barash, D. *The Survival Game: How Game Theory Explains the Biology of Cooperation and Competition*. New York: Times Books, 2004.
Barnes, P. *Capitalism 3.0 – A Guide to Reclaiming the Commons*. San Francisco: Barrett-Koehler Publishers, 2006.
Barney, G., ed. *The Global 2000 Report to the President: Entering the 21st Century*. London: Penguin Books, 1980/1982.
Barro, R., and X. Sala-i-Martin. *Economic Growth*. (2nd ed.). Cambridge, MA: MIT Press, 2004.
Batty, M. *Cities and Complexity – Understanding Cities with Cellular Automata, Agent-Based Models, and Fractals*. Cambridge, MA: MIT Press, 2005.
Basavanna, in: *Speaking of Siva* (Translated with an introduction by A.K. Ramanujan). London: Penguin Books, 1973.
Batty, M. Rank clocks, *Nature* 444 (2006): 592–596.
Baumann, H., and A.-M. Tillman. *The Hitch Hiker's Guide to LCA. An Orientation in Life Cycle Assessment Methodology and Application*. Lund, Sweden: Studentlitteratur AB, 2004.
Baumol, W. Macroeconomics of unbalanced growth: The anatomy of urban crisis, *American Economic Review* 57 (1967): 415–426
Beder, S. *The Nature of Sustainable Development*. Newham, Australia: Scribe Publication, 1993.
Beinhocker, E. *The Origin of Wealth – Evolution, Complexity and the Radical Remaking of Economics*. London: Random House, 2005.
Benítez-López, A., R. Alkemade, and P. Verweij. The impacts of roads and other infrastructure on mammal and bird populations: A meta-analysis, *Biological Conservation* 143 (2010): 1307–1316.
Berkes, F., and C. Folke, eds. *Linking Social and Ecological Systems – Management Practices and Social Mechanisms for Building Resilience*. Cambridge: Cambridge University Press, 1998.
Berkes, F., J. Colding, and C. Folke. *Navigating Social-Ecological Systems – Building Resilience for Complexity and Change*. Cambridge: Cambridge University Press, 2003.
Bernardini, O., and R. Galli. Dematerialization: Long-term trends in the intensity of use of materials and energy, *Futures* 25 (1993): 431–448.
Beumer, C., and P. Martens. Noah's ark or world wild web? Cultural perspectives in global scenario studies and their function for biodiversity conservation in a changing world, *Sustainability* 2 (2010): 3211–3238.
Blanton, R., S. Kowalewski, G. Feiman, and L. Finsten. *Ancient Mesoamerica*. Cambridge: Cambridge University Press, 1993.
Blok, K. *Energy Analysis*. Amsterdam: Techne Press, 2006.
Blum, U., and L. Dudley. Religion and economic growth: Was Weber right? *Journal of Evolutionary Economics* 11 (2001): 207–230.

Boersema, J. Hoe Groen is het Goede Leven (How Green is the Good Life). Inaugural Address. Amsterdam: Free University, 2002.

Boersema, J., and L. Reijnders. *Principles of Environmental Sciences*. The Netherlands: Springer Science +Business Media B.V., 2009.

Boorstin, D. *The Discoverers – A History of Man's Search to Know His World and Himself*. New York: Vintage Books, 1985.

Boserup, E. *Population and Technological Change*. Chicago: The University of Chicago Press, 1966.

Bossel, H., and H. Krieger. Simulation model of natural tropical forest dynamics, *Ecological Modelling* 59 (1991): 37–71.

Bossel, H. *Modeling and Simulation*. Wellesley, MA: AK Peters and Wiesbaden: Vieweg, 1994.

Bossel, H. TREEDYN3 forest simulation model, *Ecological Modelling* 90 (1996): 187–227.

Bossel, H. *Earth at a Crossroads – Paths to a Sustainable Future*. Cambridge: Cambridge University Press, 1998.

Boucheaud, J.-P., and M. Mézard. Wealth condensation in a simple model of economy, *Physica A* 282 (2000): 536–545.

Boulding, K. *Ecodynamics – A New Theory of Societal Evolution*. London: Sage Publications, 1978.

Bourdieu, P. *Science de la science et réflexivité*. Paris: Raisons d'agir Editions, 2001.

Bourguignon, F., and C. Morrison. Inequality among world citizens: 1820–1992, *The American Economic Review* 92 (2002): 727–744.

Bousquet, F., and A. Voinov. Editorial Thematic Issue – Modelling with Stakeholders. *Environmental Modelling & Software* 25 (2010): 1267.

Bouwman, L., T. Kram, and K. Klein Goldewijk, eds. Integrated modelling of global environmental change – An overview of IMAGE 2.4. MNP Report 500110002/2006. Bilthoven, 2006. Available at www.rivm.nl/bibliotheek/rapporten/500110002.pdf.

Bouwman, F., A. Beusen, and G. Billen. Human alteration of the global nitrogen and phosphorus soil balances for the period 1970–2050, *Global Biogeochemical Cycles* 23 (2009): GB0A04.

Brander, J., and M. Taylor. The simple economics of Easter Island: A Ricardo-Malthus model of renewable resource use, *The American Economic Review*, 88 (1998): 119–138.

Braudel, F. *Civilization & Capitalism 15th–18th Century*. New York: Harper & Row, 1979.

Brede, M., and B. de Vries. Networks that optimize a trade-off between efficiency and dynamical resilience, *Physics Letters A* 373 (2009a): 3910–3914.

Brede, M., and B. de Vries. Harvesting heterogeneous renewable resources: Uncoordinated, selfish, team-, and community-oriented strategies, *Ecological Modelling* 25 (2009b): 117–128.

Brock W., and S. Durlauf. Discrete choice with social interactions, *Review of Economic Studies* 68 (2001): 235–260.

Brown, L. *Outgrowing the Earth – The Food Security Challenge in an Age of Falling Water Tables and Rising Temperatures*. New York: W.W. Norton & Company, 2004.

Bryson, B. *A Short History of Nearly Everything*. New York: Broadway Books, 2003.

Buchanan, M. *Nexus/Small Worlds and the Groundbreaking Science of Networks*. New York: W.W. Norton & Company, 2002.

Buckingham/USGS. Aluminium stocks in use in automobiles in the United States. United States Geological Survey (USGS), 2010. Available at pubs.usgs.gov/fs/2005/3145/fs2005_3145.pdf.

Buringh, P. The land resource for agriculture, *Philosophical Transactions of the Royal Society of London* B310 (1985): 151–159.

Capra, F. *The Hidden Connections – Integrating the Biological, Cognitive, and Social Dimensions of Life into a Science of Sustainability*. New York: Doubleday, 2002.

Carpenter, S., and W. Brock. Rising variance: A leading indicator of ecological transition, *Ecology Letters* 9 (2006): 311–318.

Case, T. *An Illustrated Guide to Theoretical Ecology*. Oxford: Oxford University Press, 2000.

Castells, M. *The Rise of the Network Society*. Oxford: Blackwell Publishers, 1996.
Castells, M. *The Power of Identity*. Oxford: Blackwell Publishers, 1997.
Cavana, R., and A. Ford. Environmental and resource systems, *System Dynamics Review* (Special Issue) 20 (2004): 2.
Cavane, R., P. Davies, R. Robson, and K. Wilson. Drivers of quality in health services: Different worldviews of clinicians and policy managers revealed, *System Dynamics Review* 15 (1999): 315–340.
Chaisson, E. *Cosmic Evolution – The Rise of Complexity in Nature*. Cambridge, MA: Harvard University Press, 2001.
Chakrabarti, D. K. *The Archaeology of Ancient Indian Cities*. Delhi: Oxford University Press, 1997.
Chalmers, R. *What Is This Thing Called Science?* (3rd ed.). Berkshire, UK: Open University Press, 1999.
Chisholm, M. The increasing separation of production and consumption. In *The Earth as Transformed by Human Action – Global and Regional Changes in the Biosphere over the Past 300 Years*, pp. 87–102, edited by B. Turner, W. Clark, R. Kates, J. Richards, J. Mathews, and W. Meyer. Cambridge: Cambridge University Press, 1990.
Cirera, X., and E. Masset. Income distribution trends and future food demand, *Philosophical Transactions of the Royal Society B* 365 (2010): 2821–2834.
Claassen, R. *Het eeuwig tekort*. Amsterdam: Ambo, 2004.
Clark, C. *Mathematical Bioeconomics – The Optimal Management of Renewable Resources*. Wiley Interscience, New York, 1990.
Clark, N., F. Perez-Trejo, and P. Allen. *Evolutionary Dynamics and Sustainable Development: A Systems Approach*. Cheltenham, UK: Edward Elgar, 1995.
Clark, W., and N. Dickson. Sustainability science: The emerging research program, *Proceedings of the National Academy of Sciences of the United States of America (PNAS)* 100 (2003): 8059–8061.
Common, M., and S. Stagl. *Ecological Economics – An Introduction*. Cambridge: Cambridge University Press, 2005.
Costanza, R., B. Low, E. Ostrom, and J. Wilson, eds. *Institutions, Ecosystems, and Sustainability. Ecological Economics Series*. London: CRC Press, 2001.
Costanza, R., L. Graumlich, and W. Steffen, eds. *Sustainability or Collapse? An Integrated History and Future of People on Earth (IHOPE). 96th Dahlem Workshop*. Boston: MIT Press, 2007.
Costanza, R., L. Graumlich, W. Steffen, C. Crumley, J. Dearing, K. Hibbard, et al. Sustainability or collapse: what can we learn from integrating the history of humans and the rest of nature? *Ambio* 36, (2007): 522–527.
Crosby, A. *Ecological Imperialism – The Biological Expansion of Europe, 900–1900*. Cambridge: Cambridge University Press, 1993.
Daly, H. *Steady-State Economics: The Economics of Biophysical Equilibrium and Moral Growth*. San Francisco: W. H. Freeman, 1977.
Daly, H., and J. Cobb. *For the Common Good – Redirecting the Economy toward Community, the Environment, and a Sustainable Future*. Boston: Beacon Press, 1989.
Daly, H. *Beyond Growth*. Boston: Beacon Press, 1996.
Daré, W., R. Ducrot, A. Botta, and M. Etienne. Repères méthodologiques pour la mise en oeuvre d'une demarche de modélisation d'accompagnement. ComMod www.cardere.fr (2009).
Dasgupta, P. S., and G. W. Heal. *Exhaustible Resources and Economic Theory*. Cambridge: Cambridge University Press, 1979.
Dasgupta, P. *Economics – A Very Short Introduction*. Oxford: Oxford University Press, 2007.
Dasgupta, P. Discounting climate change, *Journal of Risk Uncertainty* 37 (2008): 141–169.
de Geus, A. Modelling to predict or to learn? In Modelling for learning (special issue), pp. 1–5, edited by J. Morecroft and J. Sterman. *European Journal of Operational Research* 59 (1992): 1.
de Geus, M. *The End of Over-Consumption: Towards a Lifestyle of Moderation and Self-Restraint*. Utrecht, The Netherlands: International Books, 2003.

de Kok, J., and H. Wind. Designing rapid assessment models of water systems based on internal consistency, *Journal of Water Resources Planning and Management*, 128 (2002): 240–247.

de Roos, A., and L. Persson. Unstructured population models: Do population-level assumptions yield general theory? In *Dynamic Food Webs – Multispecies Assemblages, Ecosystem Development and Environmental Change*, pp. 31–62, edited by P. de Ruiter, V. Wolters, and J. Moore. *Theoretical Ecology Series*. Amsterdam: Elsevier, 2005.

de Ruiter, P., V. Wolters, and J. Moore, eds. *Dynamic Food Webs – Multispecies Assemblages, Ecosystem Development and Environmental Change*. Theoretical Ecology Series. Amsterdam: Elsevier, 2005.

de Vries, B., K. Uitham, and G. J. Zijlstra. Kernenergie in Nederland, *Acta Politica* 2 (1977): 161–206.

de Vries, B. Perspectieven op duurzame ontwikkeling, *Lucht en Omgeving* 6 (1989): 168–172.

de Vries, B. Effects of resource assessments on optimal depletion estimates, *Resources Policy* 15 (1989): 253–268.

de Vries, B., and R. Benders. Electric power planning in a gaming context, *Simulation and Games* 20 (1989): 227–244.

de Vries, B. SusClime – a simulation game on population and development in a resource- and climate-constrained two-country world, *Simulation & Games* 29 (1998): 216–238.

de Vries, B. Perceptions and risks in the search for a sustainable world – a model-based approach, *International Journal of Sustainable Development* 4 (2001): 434–453.

de Vries, B., and J. Goudsblom, eds. *Mappae Mundi – Humans and Their Habitats in a Long-Term Socio-ecological Perspective. Myths, Maps, and Models*. Amsterdam: Amsterdam University Press, 2002.

de Vries, B., M. Thompson, and K. Wirtz. Understanding: Fragments of a unifying perspective. In *Mappae Mundi – Humans and Their Habitats in a Long-Term Socio-ecological Perspective. Myths, Maps, and Models*, pp. 257–300, edited by B. de Vries and J. Goudsblom. Amsterdam: Amsterdam University Press, 2002.

de Vries, B. Scenarios: Guidance for an uncertain and complex world? In *Sustainability or Collapse? An Integrated History and Future of People on Earth (IHOPE) – 96th Dahlem Workshop*, pp. 378–398, edited by R. Costanza, L. Graumlich, and W. Steffen. Boston: MIT Press, 2007.

de Vries, B., A. Revi, G. Bhat, H. Hilderink, and P. Lucas. India 2050: Scenarios for an uncertain future. MNP Report. 2007. 550033002/2007. Available at www.pbl.nl/en/publications/2007/India2050_scenariosforanuncertainfuture

de Vries, B. Environmental modelling for a sustainable world. In *Principles of Environmental Sciences*, pp. 345–373, edited by J. Boersema and L. Reijnders. Netherlands: Springer Science +Business Media B.V., 2009.

de Vries, B., and A. Petersen. Conceptualizing sustainable development: An assessment methodology connecting values, knowledge, worldviews and scenarios, *Ecological Economics* 68 (2009): 1006–1019.

de Vries, B. Interacting with complex systems: Models and games for a sustainable economy. PBL Background Studies Report number 550033003. 2010. Available at www.globalsystemdynamics.eu; www.pbl.nl.

de Waal, F. *Primates and Philosophers – How Morality Evolved*. Princeton, NJ: Princeton University Press, 2007.

Dearing, J. Climate-human-environment interactions: Resolving our past, *Climate of the Past* 2 (2006): 187–203.

Demeny, P., and G. McNicoll. *Population and Development Review* 37 Special Volume on Demographic Transition (2011).

Dennett, D. *Kinds of Minds – The Origins of Consciousness*. London: Orion Books, 1997/2001.

den Elzen, M., N. Höhne, and S. Moltmann. The Triptych approach revisited: A staged sectoral approach for climate mitigation, *Energy Policy* 36 (2008): 1107–1124.

Devara, Dasimayya, in: *Speaking of Siva* (Translated with an introduction by A.K. Ramanujan). London: Penguin Books. London.

Diamond, J. *Collapse – How Societies Choose to Fail or Succeed*. New York: Penguin Books, 2006.
Dicken, P. *Global Shift – Mapping the Changing Contours of the World Economy* (5th ed.). London: Sage Publications, 2009.
Dietz, S., and E. Neumayer. Genuine savings: A critical analysis of its policy-guiding value, *Int. J Environment and Sustainable Development* 3 (2004): 276–292.
Dietz, T., A. Fitzgerald, and R. Shwom. Environmental values, *Annual Review of Environment and Resources* 30 (2005): 335–372.
Dirzo, R., and P. Raven. Global state of biodiversity and loss, *Annual Review of Environment and Resources* 28 (2003): 137–167.
Disperati, P., J. van de Steeg, P. van Breugel, A. Notenbaert, J. Owuor, and M. Herreror. *Environmental Security and Pastoralism*. International Livestock Research institute (ILRI) Report, Nairobi, 2009.
Döpfer, K., ed. *The Evolutionary Foundations of Economics*. Cambridge: Cambridge University Press, 2005.
Dosi, G., G. Fagiolo, and A. Roventini. The microfoundations of business cycles: An evolutionary, multi-agent model, *Journal of Evolutionary Economics* 18 (2008): 413–432.
Doucet, P., and P. Sloep. *Mathematical Modeling in the Life Sciences*. New York: Ellis Horwood, 1992.
Douglas, M., D. Gasper, S. Ney, and M. Thompson. Human needs and wants. In *Human Choice and Climate Change. Vol. 1: The Societal Framework*, pp. 195–264, edited by S. Rayner and E. Malone. Columbus, OH: Battelle Press, 1998.
Duchin, F., and G.-M. Lang. *The Future of the Environment*. Oxford: Oxford University Press, 1994.
Dudley, R. A basis for understanding fishery management dynamics, *System Dynamics Review* 24 (2008): 1–29.
Duke, D., and J. Geurts. *Policy Games for Strategic Management: Pathways into the Unknown*. Amsterdam: Dutch University Press, 2004.
Dumont, L. *Homo Hierarchicus: The Caste System and Its Implications*. Chicago: University of Chicago Press, 1984.
Dunham, J. An agent-based spatially explicit epidemiological model in MASON, *Journal of Artificial Societies and Social Simulation (JASSS)* 9 (2005). Available at jasss.soc.surrey.ac.uk/9/1/3.html.
Dunne, J., R. Williams, and N. Martinez. Food-web structure and network theory: The role of connectance and size, *Proceedings of the National Academy of Sciences of the United States of America (PNAS)* 99 (2002): 12917–12922.
Earle, C. The myth of the southern soil miner: Macrohistory, agricultural innovation, and environmental change. In *The Ends of the Earth – Perspectives on Modern Environmental History. Studies in Environment and History*, edited by D. Worster. Cambridge: Cambridge University Press, 1988.
Easterlin, R. Modernization and fertility. In *Determinants of Fertility in Developing Countries* (Vol. 2), edited by R. Bulatao and R. Lee. London: Academic Press, 1983.
Easton, T. *Taking Sides: Clashing Views on Environmental Issues*. Contemporary Learning Series (11th ed.). New York: McGraw-Hill, 2006.
Easwaran, E. *The Bhagavad Gita*. London: Arkana Penguin Books, 1985.
Edelstein-Keshet, L. *Mathematical Models in Biology*. New York: Random House, 1988.
European Environment Agency (EEA). The European Environment – Synthesis. Copenhagen: European Environment Agency, 2010. Available at eea.europa.eu.
EEA. *Late Lessons from Early Warnings: The Precautionary Principle 1896–2000*. Copenhagen, European Environmental Agency, 2001.
Elgin, D. *Awakening Earth: Exploring the Co-evolution of Human Culture and Consciousness*. New York: Morrow, 1993.
Elias, N. *The Civilizing Process. Sociogenetic and Psychogenetic Investigations*. Oxford: Blackwell, 1967/2000.

Ellis, E., and N. Ramankutty. Putting people in the map: Anthropogenic biomes of the world, *Frontiers in Ecology and the Environment* 6 (2008): 439–447.

Ellis, E., K. Klein Goldewijk, S. Siebert, D. Lightman, and N. Ramankutty. Anthropogenic transformation of the biomes, 1700 to 2000, *Global Ecology and Biogeography* 19 (2010): 589–606.

Elvin, M. Three thousand years of unsustainable growth: China's environment from archaic times to the present, *East Asian History* 6 (1993): 7–46.

Engelen, G., R. White, and I. Uljee. Exploratory modeling of socio-economic impacts of climate change. In *Climate Change in the Intra-Americas Sea*, edited by G. Maul. London: Edward Arnold, 1993.

Engelen, G., R. White, I. Uljee, and S. Wargnies. Numerical modeling of small island small socio-economics to achieve sustainable development. In *Small Islands: Marine Science and Sustainable Development, Coastal and Estuarine Studies* Volume 51, pp. 437–463, edited by G. Maul. Washington, DC: American Geophysical Union, 1996.

Engelen, G., R. White, I. Uljee, and P. Drazan. Using cellular automata for integrated modelling of socio-environmental systems, *Environmental Monitoring and Assessment* 34 (1995): 203–214.

Engelen, G. Models in policy formulation and assessment: The WADBos decision-support system. In *Environmental Modelling: Finding Simplicity in Complexity*, pp. 257–271, edited by J Wainwright and M. Mulligan. Environmental Modelling – Finding Simplicity in Complexity. London: John Wiley& Sons, Ltd., 2004.

Epstein, J., and R. Axtell, eds. *Growing Artificial Societies*. Washington, DC/Cambridge, MA: Brookings Institution Press/MIT Press, 1995.

Eriksson, A., and K. Lindgren. Cooperation in an unpredictable environment. In *Artificial Life VIII*, edited by R. Standish, M. Abbass, and H. Bedau. Boston: MIT Press, 2002.

Etienne, M., C. Le Page, and M. Cohen. A step-by-step approach to building land management scenarios based on multiple viewpoints on multi-agent system simulations, *Journal of Artificial Societies and Social Simulation (JASSS)* 6 (2003). Available at jasss.soc.surrey.ac.uk/6/2/2.html.

Etienne, M., M. Cohen, and C. Le Page. A step-by-step approach to build-up land management scenarios based on multiple viewpoints on multi-agent system simulations. *Journal of Artificial Societies and Social Simulation (JASSS)* 7 (2003):

Falkenmark, M. and Lindh, G. How can we cope with the water resources situation by the year 2015? *Ambio* 3 (1976): 114–122.

Falkenmark, M., and J. Rockström. The new blue and green water paradigm, *Journal of Water Resources Planning and Management*. (2006): 129–132.

Fay, T., and J. Greeff. Lion, wildebeest and zebra: A predator-prey model, *Ecological Modelling* 196 (2006): 237–244.

Feinstein, C., and M. Thomas. *Making History Count – A Primer in Quantitative Methods for Historians*. Cambridge: Cambridge University Press, 2002.

Ferber, J. *Multi-Agent Systems – An Introduction to Distributed Artificial Intelligence*. London: Addison-Wesley, 2000.

Ferber, J. Multi-agent concepts and methodologies. In *Agent-Based Modelling and Simulation in the Social and Human Sciences*, pp. 7–34, edited by D. Phan and F. Amblard. Oxford: Bardwell Press, 2007.

Fernández, J., and M. Selma. The dynamics of water scarcity on irrigated landscapes: Mazarrón and Aguilas in south-eastern Spain, *System Dynamics Review* 20 (2004): 117–137.

Feynman, J., and A. Ruzmaikin. Climate stability and the development of agricultural societies, *Climatic Change* 84 (2007): 295–311.

Finlayson, A., and B. McCay. Crossing the threshold of ecosystem resilience: The commercial extinction of cod. In *Linking Social and Ecological Systems – Management Practices and Social Mechanisms for Building Resilience*, pp. 311–337, edited by F. Berkes and C. Folke. Cambridge: Cambridge University Press, 1998.

Fischer, G., M. Shah, H. van Velthuizen, and F. Nachtergaele. Global Agro-ecological Assessment for Agriculture in the 21st Century. IIASA/FAO Report, 2001. Available at www.iiasa.ac.at/Admin/PUB/Documents/XO-01-001.pdf.

Fischer-Kowalski, M., and H. Haberl, eds. *Socioecological Transitions and Global Change: Trajectories of Social Metabolism and Land Use*. Cheltenham, UK: Edward Elgar, 2007.

Fisher, J., and R. Pry. A simple substitution model of technological change, *Technological Forecasting and Social Change* 3 (1971): 75–88.

Flannery, K. V. The cultural evolution of civilizations, *Annual Review of Ecology and Systematics* 3 (1972): 399–426.

Florax, R., and H. de Groot. Meta-analyse als hulpmiddel bij beleidsinstrumentatie, *ESB* 90 (2005): D8–D10.

Folmer, H., H. Landis Gabel, and H. Opschoor. *Principles of Environmental and Resource Economics – A Guide for Students and Decision-Makers*. Cheltenham, UK: Edward Elgar, 1995.

Ford, A. *Modeling the Environment*. Washington, DC: Island Press, 2009.

Ford, J., and R. Myers. A global assessment of salmon aquaculture impacts on wild salmonids, *PloS Biology* (2008). Available at www.lenfestocean.org/publications/plbi-06–02-07_Ford1REV.pdf.

Forrester, J. *World Dynamics* (2nd ed.). Cambridge, MA: Wright-Allen Press, 1973.

Frenken, K. *Innovation, Evolution and Complexity Theory*. Cheltenham, UK: Edward Elgar, 2006.

Frondel, M., and C. Schmidt. The empirical assessment of technology differences: Comparing the comparable, *The Review of Economics and Statistics* 88 (2006): 186–192.

Fukuyama, F. *The End of History and the Last Man*. New York: Avon Books, Inc., 1992.

Funtowicz, S., and J. Ravetz. *Uncertainty and Quality in Science for Policy*. Dordrecht, the Netherlands: Kluwer Academic Publishers, 1990.

Gales, B., A. Kander, P. Malanima, and M. Rubio. North versus South: Energy transition and energy intensity in Europe over 200 years, *European Review of Economic History* 11 (2007): 219–253.

Galor, O., and D. Weil. Population, technology, and growth: From Malthusian stagnation to the demographic transition and beyond, *The American Economic Review*, 90 (2000): 806–828.

Garcia, S., and R. Grainger. Gloom and doom? The future of marine capture fisheries, *Philosophical Transactions of the Royal Society B* 360 (2005): 21–46.

Garcia, S., and A. Charles. Fishery systems and linkages: From clockworks to soft watches, *ICES Journal of Marine Science* 64 (2007): 580–587.

Garcia, S., and A. Rosenberg. Food security and marine capture fisheries: Characteristics, trends, drivers and future perspectives, *Philosophical Transactions of the Royal Society B* 365 (2010): 2869–2880.

Gardner, D. *Future Babble: Why Expert Predictions Are Next to Worthless, and You Can Do Better*. New York: Dutton, 2010.

Geels, F. Technological transitions as evolutionary reconfiguration processes: A multi-level perspective and a case-study, *Research Policy* 31 (2002): 1257–1274.

Geels, F. Co-evolution of technology and society: The transition in water supply and personal hygiene in the Netherlands (1850–1930) – A case study in multi-level perspective, *Technology in Society* 27 (2005): 363–397.

Geels, F., and J. Schot. Typology of sociotechnical transition pathways, *Research Policy* 36 (2007): 399–417.

Geist, H., and E. Lambin. What drives tropical deforestation? IHDP-IGBP LUCC Report Series No. 4. Available at www.pik-potsdam.de/members/cramer/teaching/0607/Geist_2001_LUCC_Report.pdf.

Geist, H., ed. *Our Earth's Changing Land – An Encyclopedia of Land-Use and Land-Cover Change*. Westport, CT/London: Greenwood Press, 2006.

Giampietro, M. *Multi-scale Integrated Analysis of Agroecosystems. Advances in Agroecology*. Boca Raton, FL: CRC Press, 2004.

Gilbert, N., and J. Doran, eds. *Simulating Societies – The Computer Simulation of Social Phenomena*. London: UCL Press, 1994.

Gilbert, N., and K. Troitzsch. *Simulation for the Social Scientist*. Buckingham, UK: Open University Press, 1999.

Gilding, P. *The Great Disruption: Why the Climate Crisis Will Bring On the End of Shopping and the Birth of a New World*. New York: Bloomsbury Press, 2011.

Gilg, O., I. Hanski, and B. Sittler. Cyclic dynamics in a simple vertebrate predator-prey community, *Science* 302 (2003): 866–868.

Gintis, H. Beyond *Homo economicus*: Evidence from experimental economics, *Ecological Economics Special Issue* 35 (2000): 311–322.

Gintis, H. Behavioral game theory and contemporary economic theory, *Analyse & Kritik* 27 (2005): 48–72.

Giordano, M. Global groundwater? Issues and solutions, *Annual Review of Environment and Resources* 34 (2009): 153–178.

Gleick, P., and associates. The World's Water: The Biennial Report on Freshwater. Washington, DC: Resources. Island Press, 2011.

Gong, M., and G. Wall. On exergy and sustainable development – Part 2: Indicators and methods, *Exergy, an International Journal* 4 (2001): 217–233.

Gordon, H. The economic theory of a common-property resource: The fishery, *The Journal of Political Economy*, 62 (1954): 124–142.

Gordon, L., G. Peterson, and E. Bennett. Agricultural modifications of hydrological flows create ecological surprises, *Trends in Ecology and Evolution* 4 (2008): 211–219.

Goudie, A. *The Human Impact – Man's Role in Environmental Change*. Cambridge, MA: MIT Press, 1981.

Goudsblom, J. *Fire and Civilization*. London: Penguin, 1992.

Goudsblom, J., E. Jones, and S. Mennell. *Course of Human History: Economic Growth, Social Process and Civilization*. Armonk, NY: M. E. Sharpe, 1996.

Goudsblom, J. The past 250 years: Industrialization and globalization. In *Mappae Mundi – Humans and Their Habitats in a Long-Term Socio-ecological Perspective. Myths, Maps, and Models*, pp. 353–378, edited by B. de Vries and J. Goudsblom. Amsterdam: Amsterdam University Press, 2002.

Gourou, P. *Les pays tropicaux*. Paris: Presses Universitaires de France, 1947.

Grazi, F., J. Van den Bergh, and P. Rietveld. Spatial welfare economics versus ecological footprint: Modeling agglomeration, externalities and trade, *Environmental Resource Economics* 38 (2007): 138–153.

Greenblat, C. *Designing Games and Simulations: An Illustrated Handbook*. London: Sage Ltd., 1988.

Grigg, D. *The Agricultural Systems of the World*. Cambridge: Cambridge University Press, 1974.

Grigg, D. The industrial revolution and land transformation. In *Land Transformation in Agriculture*, edited by M. Wolman and F. Fournier. New York: John Wiley, 1987.

Grimm, V., and H. Wissel. Babel, or the ecological stability discussions: An inventory and analysis of terminology and a guide for avoiding confusion, *Oecologia* 109 (1997): 323–334.

Groenewold, H. Evolutie van kennis, waarden en macht. Technische Universiteit Twente, 1981.

Grove, A., and O. Rackham. *The Nature of Mediterranean Europe – An Ecological History*. New Haven and London: Yale University Press, 2001.

Grübler, A. *Technology and Global Change*. Cambridge: Cambridge University Press, 1999.

Grübler, A., and N. Nakicenovic. *Long Wave, Technology Diffusion, and Substitution*. Laxenburg, Austria: IIASA, 1991.

Gunderson, L., and C. Holling. *Panarchy – Understanding Transformations in Human and Natural Systems*. Washington, DC: Island Press, 2002.

Haas, P., R. Keohane, and M. Levy, eds. *Institutions for the Earth*. Cambridge, MA: MIT Press, 1993.

Haberl, H. Human appropriation of net primary production as an environmental indicator: Implications for sustainable development, *Ambio* 26 (1997): 143–146.

Haddeland, I. *Anthropogenic Impacts on the Continental Water Cycle*. Ph.D. Thesis. Oslo, Norway: University of Oslo, 2006.
Haley, S. Capital accumulation and the growth of aggregate agricultural production, *Agricultural Economics* 6 (1991): 129–157.
Hall, S., D. Held, and K. Thompson. *Modernity: An Introduction to Modern Societies*. Bodmin, Cornwall: MPG Books, 1996.
Handy, C. *The Hungry Spirit – Beyond Capitalism – A Quest for Purpose in the Modern World*. London: Arrow Books, 1998.
Hanjra, M., and M. Qureshi. Global water crisis and future food security in an era of climate change, *Food Policy* 35 (2010): 365–377.
Hardin, G. The tragedy of the commons, *Science* 162 (1968): 1243–1248.
Harman, W. *An Incomplete Guide to the Future*. Stanford, CA: Stanford Alumni Association, 1976.
Harris, D. *Mineral Resource Appraisal*. Oxford: Clarendon Press, 1984.
Harte, J. Toward a synthesis of the Newtonian and Darwinian worldviews, *Physics Today* 10 (2002): 29–34.
Hassan, F. Environmental perception and human responses in history and prehistory. In *The Way the Wind Blows: Climate, History, and Human Action*, pp. 121–140, edited by R. McIntosh, J. Tainter, and S. McIntosh. New York: Columbia University Press, 2000.
Hedlund-de Witt, A. The rising culture and worldview of contemporary spirituality: A sociological study of potentials and pitfalls for sustainable development, *Ecological Economics* 70 (2011): 1057–1065.
Heijden, K. van der. *Scenarios – The Art of Strategic Conversation*. New York: Wiley & Sons, 1996.
Heilbroner, R. *The Worldly Philosophers*. New York: Simon & Schuster, 1955.
Heilbroner, R., and W. Milberg. *The Making of Economic Society* (12th ed.). New York: Pearson Prentice-Hall, 2005.
Hein, P., C. van Kopen, R. de Groot, and E. van Ierland. *Spatial Scales, Stakeholders and the Valuation of Ecosystem Services*. Wageningen, the Netherlands: Wageningen University, 2005. Available at dx.doi.org/10.1016/j.ecolecon.2005.04.005.
Hein, L. *Optimising the Management of Complex Dynamic Ecosystems*. Ph.D. Thesis. Wageningen, the Netherlands: Wageningen University, 2005.
Hein, L., and E. van Ierland. Efficient and sustainable management of complex forest ecosystems, *Ecological Modeling* 190 (2006): 351–336.
Helbing, D., and W. Yu. The outbreak of cooperation among success-driven individuals under noisy conditions, *Proceedings of the National Academy of Sciences of the United States of America (PNAS)* 106 (2009): 3680–3685.
Helpman, E. *The Mystery of Economic Growth*. Boston: Harvard University Press, 2004.
Hettelingh, J.-P., B. de Vries, and L. Hordijk. Integrated assessment. In *Principles of Environmental Sciences*, pp. 385–420, edited by J. Boersema and L. Reijnders. The Netherlands: Springer Science +Business Media B.V., 2009.
Hilderink, H. *World Population in Transition – An Integrated Regional Modelling Framework. Faculteit der Ruimtelijke Wetenschappen*. Ph. D. Thesis. Groningen, the Netherlands: Rijksuniversiteit Groningen, 2000.
Hilderink, H., P. Lucas, and M. Kok, eds. Beyond 2015: Long-term Development and the Millennium Development Goals. Report 55002500. Bilthoven, the Netherlands: Netherlands Environmental Assessment Agency (PBL), 2009.
Hirsch, F. *Social Limits to Growth*. London: Routledge & Kegan Paul, 1977.
Hirsch, R. Mitigation of maximum world oil production: Shortage scenarios, *Energy Policy* 36 (2008): 881–889.
Hisschemöller, M., J. Eberg, A. Engels, and K. von Moltke. Environmental institutions and learning: Perspectives from the policy sciences. In *Principles of Environmental Sciences*, pp. 281–303, edited by J. Boersema and L. Reijnders. The Netherlands: Springer Science +Business Media B.V., 2009.

Hitlin, S., and J. Piliavin. Values: reviving a dormant concept, *Annual Review of Sociology* 30 (2004): 359–393.

Ho, P., and E. Vermeer. *China's Limits to Growth – Greening State and Society*. Oxford: Blackwell Publishing, 2006.

Hodgson, G. Decomposition and growth: Biological metaphors in economics from the 1880s to the 1980s. In *The Evolutionary Foundations of Economics*, pp. 105–148, edited by K. Döpfer. Cambridge: Cambridge University Press, 2005.

Hoekstra, R., and J. van den Bergh. Structural decomposition analysis of physical flows in the economy, *Environmental and Resource Economics* 23 (2002): 357–378.

Hofstede, G. Cultural dimensions in management and planning, *Asia Pacific Journal of Management* 1 (1984): 81–99.

Holland, J. *Hidden Order – How Adaptation Builds Complexity*. Reading, MA: Helix Books/Addison-Wesley, 1996.

Holling, C. Resilience and stability of ecological systems, *Annual Review of Ecological Systems* 4 (1973): 1–23.

Holling, C. The resilience of terrestrial ecosystems: Local surprise and global change. In *The Resilience of Terrestrial Ecosystems: Local Surprise and Global Change*, edited by W. Clark and R. Munn. Cambridge: Cambridge University Press/IIASA, 1986.

Hollis, M. *The Philosophy of Science – An Introduction*. Cambridge: Cambridge University Press, 2007.

Homer-Dixon, T. *The Upside of Down: Catastrophe, Creativity, and the Renewal of Civilization*. Washington, DC: Island Press, 2006.

Hueting, R. *New Scarcity and Economic Growth*. Amsterdam: North-Holland Publishing Company, 1980.

Hulme, M. *Why We Disagree About Climate Change*. Cambridge: Cambridge University Press, 2009.

IEA (2004). See www.iea.org

IEA (2010). See www.iea.org

IFPRI (2008). See www.ifpri.org

Inglehart, R., and C. Welzel. *Modernization, Cultural Change and Democracy: The Human Development Sequence*. New York: Cambridge University Press, 2005.

INFO Project. 2003. See Project Publications at info.k4health.org/about.shtml.

IPCC. *Climate Change 2001 (TAR)*. Cambridge: Cambridge University Press, 2001.

IPCC. *Climate Change 2007 (AR4)*. Cambridge: Cambridge University Press, 2007.

Ironmonger, D., and E. Sonius. Household productive activities. In *Households Work*, edited by D. Ironmonger. Sydney: Allen & Unwin, 1989.

IUCN-WWF (1991). Caring for the world: a strategy for sustainability. IUCN-UNEP-WWF Report. See www.iucn.org

Izard, C., and D. Müller. Tracking the devil's metal: Historical global and contemporary U.S. tin cycles, *Resources, Conservation and Recycling* 54 (2010): 1436–1441.

Jackson, T. *Prosperity without Growth – Economics for a Finite Planet*. London: Earthscan, 2009.

Jager, W., M. Janssen, B. de Vries, J. de Greef, and C. Vlek. Behaviour in commons dilemmas: Homo Economicus and Homo Psychologicus in an ecological-economic model, *Ecological Economics* 35 (2000): 357–379.

Jager, W., and H.-J. Mosler. Simulating human behavior for understanding and managing environmental resource use, *Journal of Social Issues* 63 (2007): 97–116.

Jaggard, K., A. Qi, and E. Ober. Possible changes to arable crop yields, *Philosophical Transactions of the Royal Society B* 365 (2010): 2835–2851.

James, W. *The Varieties of Religious Experience – A Study in Human Nature (The Gifford Lectures Edinburgh 1901–02)*. Glasgow: Collins Fountain Books, 1902/1977.

Janssen, M., ed. *Complexity and Ecosystem Management – The Theory and Practice of Multiagent Systems*. Cheltenham, UK: Edward Elgar, 2002.

Janssen, M., T. Kohler, and M. Scheffer. Sunk-cost effects and vulnerability to collapse in ancient societies, *Current Anthropology* 44 (2003): 722–728.

Janssen, P., P. Heuberger, and A. Tiktak. *Metamodelleren bij het MNP-RIVM*. MNP-Report, Bilthoven, 2005.
Jantsch, E. *The Self-organizing Universe*. Oxford: Pergamon Press, 1980.
Jarvis, P. *Ecological Principles and Environmental Issues*. Essex, UK: Prentice Hall/Pearson Education, 2000.
Jeong, H., B. Tombor, R. Albert, Z. Oltvai, and A. Barabási. The large-scale organization of metabolic networks, *Nature* 407 (2000): 651–654.
Jones, E. *The European Miracle* (2nd ed.). Cambridge: Cambridge University Press, 1981/2003.
Jung, C.G. *The Nature of the Psyche*. Princeton, NJ: Princeton University Press, 1960.
Junginger, M., W. van Sark, and A. Faaij, eds. *Technological Learning in the Energy Sector*. Cheltenham, UK: Edward Elgar, 2010.
Kahneman, D., E. Diener, and N. Schwarz, eds. *Well-Being – The Foundations of Hedonic Psychology*. New York: Russell Sage Foundation, 1999.
Kates, R., W. Clark, R. Corell, J. Hall, C. Jaeger, I. Lowe, et al. Sustainability science, *Science* 292 (2001): 641–642. Available at dx.doi.org/10.1126/science.1059386.
Kauffman, S. *At Home in the Universe – The Search for Laws of Complexity*. London: Viking, 1995.
Kay, J. *Culture and Prosperity: The Truth about Markets – Why Some Nations Are Rich but Most Remain Poor*. New York: HarperBusiness, 2004.
Kearney, J. Food consumption trends and drivers, *Philosophical Transactions of the Royal Society B* 365 (2010): 2793–2807.
Kefi, S., M. Rietkerk, C. L. Alados, Y. Pueyo, V. P. Papanastasis, A. ElAich, and P. C. de Ruiter. Spatial vegetation patterns and imminent desertification in Mediterranean arid ecosystems, *Nature* 449 (2007): 213–218.
King Hubbert, M. The energy resources of the Earth, *Scientific American* 225 (1971): 60–70.
Kirman, A. Ants, rationality, and recruitment, *The Quarterly Journal of Economics* 108, vol. 1 (1993): 137–156.
Klee, R., and T. Graedel. Elemental cycles: A status report on human or natural dominance, *Annual Review of Environment and Resources* 29 (2004):69–107.
Knappett, C., T. Evans, and R. Rivers. Modelling maritime interaction in the Aegean Bronze Age, *Antiquity* 82 (2008): 1009–1024.
Kok, M., and J. Jäger, eds. Vulnerability of People and the Environment – Challenges and Opportunities. PBL/UNEP Background Studies, 2007. Available at www.pbl.nl/en.
Kok, M., M. Lüdeke, T. Sterzel, P. Lucas, C. Walter, P. Janssen, and I. de Soysa. Quantitative Analysis of Patterns of Vulnerability to Global Environmental Change. PBL/PIK/NTNU Background Studies, 2001. Available at www.pbl.nl/en.
Krausmann, F., S. Gingrich, N. Eisenmenger, K.-H. Erb, H. Haberl, and M. Fischer-Kowalski. Growth in global materials use, GDP and population during the 20th century, *Ecological Economics* 68 (2009): 2696–2705.
Kruyt, B., D, van Vuuren, B. de Vries, and H. Groenenberg. Indicators for energy security, *Energy Policy* 37, vol. 6 (2009): 2166–2181.
Kuik, O., and H. Verbruggen, eds. In *Search of Indicators of Sustainable Development*. Dordrecht, the Netherlands: Kluwer Academic Publishers, 1991.
Kümmel, R., J. Henn, and D. Lindenberger. Capital, labor, energy and creativity: Modeling innovation diffusion, *Structural Change and Economic Dynamics* 13 (2002): 415–433.
Lambin, E., H. Geist, and E. Lepers. Dynamics of land-use and land-cover change in tropical regions, *Annual Review of Environment and Resources* 28 (2003): 205–241.
Lambin, E., and H. Geist, eds. *Land-Use and Land-Cover Change – Local Processes and Global Impacts. The IGBP Series*. Berlin: Springer, 2006.
Lansing, J. *Perfect Order: Recognizing Complexity in Bali*. Princeton, NJ: Princeton University Press, 2006.
Lao-Tzu. *Tao Te Ching* (Introduced by Burton Watson). Indianapolis/Cambridge: Hackett Publishing Company, 1993.

Layard, R. *Happiness: Lessons from a New Science*. London: Allen Lane/The Penguin Press, 2005.

Leemans, R., and A. Kleidon. In *Global Desertification: Do Humans Cause Deserts? 88th Dahlem Workshop*, edited by J. Reynolds and D. Stafford Smith. Berlin: Dahlem University Press, 2002.

Leiserowitz, A., R. Kates, and T. Parris. Sustainability values, attitudes, and behaviors: A review of multinational and global trends, *Annual Review of Environment and Resources* 31 (2006): 413–444.

Licker, R., M. Johnston, J. Foley, C. Barford, C. Kucharik, C. Monfreda, and N. Ramankutty. Mind the gap: How do climate and agricultural management explain the "yield gap" of croplands around the world? *Global Ecology and Biogeography* 19 (2010): 769–782.

Lim, M., R. Metzler, and Y. Bar-Yam. Global pattern formation and ethnic/cultural violence, *Science* 317 (2007):1540–1544.

Limburg, K., R. O'Neill, R. Costanza, and S. Farber. Complex systems and valuation. Special Issue: The Dynamics and Value of Ecosystem Services: Integrating Economic and Ecological Perspectives, *Ecological Economics* 41 (2002): 409–420.

Lindgren, K. Evolutionary phenomena in simple dynamics. In: *Artificial Life II, SFI Studies in the Sciences of Complexity*, Vol. X, edited by C. Langton, C. Taylor, J. Farmer, and J. Rasmussen. Redwood City, CA: Addison-Wesley, 1991.

Little, L., and A. McDonald. Simulations of agents in social networks harvesting a resource, *Ecological Modelling* 204 (2007): 379–386.

Lovejoy, T., and L. Hannah, eds. *Climate Change and Biodiversity*. New Haven/London: Yale University, 2005.

Lowenthal, D. Awareness of human impacts: Changing attitudes and emphases. In *The Earth as Transformed by Human Action – Global and Regional Changes in the Biosphere over the Past 300 Years*, pp. 121–136, edited by B. Turner, W. Clark, R. Kates, J. Richards, J. Mathews, and W. Meyer. Cambridge: Cambridge University Press, 1990.

Lüdeke, M., G. Petschel-Held, and H. Schellnhuber. Syndromes of global change: The first panoramic view, *GAIA* 13 (2004).

Lumley, S., and P. Armstrong. Some of the nineteenth century origins of the sustainability concept, *Environment, Development and Sustainability* 6 (2004): 367–378.

Lutz, W., W. Sanderson, and S. Scherbov. The end of world population growth, *Nature* 412 (2001): 543–545.

Lutz, W., B. O'Neill, and S. Scherbov. Europe's population at a turning point, *Science* 299 (2003): 1991–1992.

MacKay, D. *Sustainable Energy – Without the Hot Air*. UIT Cambridge, 2009. Available at www.withouthotair.com.

Maddison, A. *Monitoring the World Economy*. Paris: OECD Development Centre Studies, 1995.

Maddison, A. *The World Economy. Vol. 1: A Millennial Perspective; Vol. 2: Historical Statistics*. Paris: OECD Development Centre Studies, 2006.

Madsen, J., J. Ang, and R. Banerjee. Four centuries of British economic growth: The roles of technology and population, *Journal of Economic Growth* 15 (2010): 263–290.

Malenbaum, W. *World Demand for Raw Materials in 1985 and 2000*. New York: McGraw-Hill Inc., 1978.

Mäler, K.-G., A. Xepapadeas, and A. de Zeeuw. The economics of shallow lakes, *Environmental and Resource Economics* 26 (2003): 603–624.

Manson, S. Simplifying complexity: A review of complexity theory, *Geoforum* 32 (2001): 405–414.

Martens, P. Health transitions in a globalising world: Towards more disease or sustained health? *Futures* 34 (2002): 635–648.

Maslow, A. *Motivation and Personality*. New York: Harper, 1954.

Matsutani, A. *Shrinking-Population Economics – Lessons from Japan*. Tokyo: International House of Japan, 2006.

Max-Neef, M. *Human Scale Development – Conception, Application and Further Reflections.* London: The Apex Press, 1991.
May, R. Will a large, complex system be stable? *Nature* 238 (1972): 413–414.
May, R. Simple mathematical models with very complicated dynamics, *Nature* 261 (1976): 459–467.
May, R. Harvesting natural populations in a randomly fluctuating environment, *Science* 197 (1977): 463–465.
Maynard Smith, J. *Evolution and the Theory of Games.* Cambridge: Cambridge University Press, 1982.
Mazoyer, M., and L. Roudart. *Histoire des agricultures du monde.* Paris: Éditions du Seuil, 1997.
McGovern, T. The economics of extinction in Norse Greenland. In *Climate and History – Studies in Past Climates and Their Potential Impact on Man*, edited by T. Wigley, M. Ingram, and G. Farmer. Cambridge: Cambridge University Press, 1981.
McNeill, J. Social, economic, and political forces in environmental change: Decadal scale (1900–2000). In *Sustainability or Collapse? An Integrated History and Future of People on Earth (IHOPE) – 96th Dahlem Workshop*, edited by R. Costanza, L. Graumlich, and W. Steffen. Boston: MIT Press, 2007.
MEA (Millennium Ecosystem Assessment). *Ecosystems and Human Well-Being – Our Human Planet. Synthesis and Summary for Decision Makers.* Washington, DC: Island Press, 2005.
Meadows, D. H., D. L. Meadows, J. Randers, and W. Behrens III. *The Limits to Growth.* New York: Potomac Associates, 1971.
Meadows, D. H., D. L. Meadows, and J. Randers. *Beyond the Limits – Confronting Global Collapse, Envisioning a Sustainable Future.* Post Mills, VT: Chelsea Green Ltd., 1991.
Meadows, D. *Fish Banks Ltd.* Durham: University of New Hampshire, 1996.
Meadows, D. H. Indicators and Information Systems for Sustainable Development. Report to the Balaton Group. Sustainability Institute, 1998.
Meadows, D. L., J. Randers, and D. H. Meadows. *Limits to Growth – The 30-Year Update.* Post Mills, VT: Chelsea Green Publishing Company, 2004.
Meadows, D. L. Evaluating past forecasts: Reflections on one critique of *The Limits to Growth.* In *Sustainability or Collapse? An Integrated History and Future of People on Earth (IHOPE) – 96th Dahlem Workshop*, edited by R. Costanza, L. Graumlich, and W. Steffen. Boston: MIT Press, 2007.
Meadows, D. H. *Thinking in Systems.* Post Mills, VT: Chelsea Green Publishing, 2008.
Meinshausen, M. What does a 2°C target mean for greenhouse gas concentrations? In *Avoiding Dangerous Climate Change*, edited by J. Schellnhuber, W. Cramer, N. Nakicenovic, T. Wigley, and G. Yohe. Cambridge: Cambridge University Press, 2006.
Mertz, O., R. Wadley, U. Nielsen, T. Bruun, C. Colfer, A. de Neergaard, et al. A fresh look at shifting cultivation: Fallow length an uncertain indicator of productivity, *Agricultural Systems* 96 (2008): 75–84.
Metz, B., M. Berk, M. den Elzen, B. de Vries, and D. van Vuuren. Towards an equitable global climate change regime: Compatibility with Article 2 of the Climate Change Convention and the link with sustainable development, *Climate Policy* 2 (2002): 211–230.
Metz, B. *Controlling Climate Change.* Cambridge: Cambridge University Press, 2009.
Meyer, P., and J. Ausubel. Carrying capacity: A model with logistically varying limits, *Technological Forecasting and Social Change* 61 (1999): 209–214.
Migheli, M. Supporting the free and competitive market in China and India: Differences and evolution over time, *Economic Systems* 34 (2010): 73–90.
Miles, E., A. Underdal, S. Andresen, J. Wettestad, J. Skjaerseth, and E. Carlin. *Environmental Regime Effectiveness.* Cambridge, MA: MIT Press, 2002.
Mitchell, M. *Complexity – A Guided Tour.* Oxford: Oxford University Press, 2009.

MNP. *Duurzaamheidsverkenning (Sustainability Outlook).* Bilthoven, the Netherlands: Netherlands Environmental Assessment Agency (PBL), 2004.

Morecroft, J., and J. Sterman. Modelling for learning (special issue), *European Journal of Operations Research* 59 (1992): 1.

Morrison, C., and E. Thorbecke. The concept of agricultural surplus, *World Development* 18 (1990): 1081–1095.

Moxnes, E. Not only the tragedy of the commons: Misperceptions of bioeconomics, *Management Science* 44 (1998): 1234–1248.

Moyo, D. *Dead Aid – Why Aid Is Not Working and How There Is Another Way for Africa.* New York: Farrar, Straus and Giroux, 2009.

Mulder, K. *Sustainable Development for Engineers.* Sheffield, UK: Greenleaf Publishing, 2006.

Nakicenovic, N., Alcamo, J., Davis, G., de Vries, B., et al. *Special Report on Emissions Scenarios (SRES) for the Intergovernmental Panel on Climate Change (IPCC).* Cambridge: Cambridge University Press, 2000.

Nelson, R., and S. Winter. *An Evolutionary Theory of Economic Change.* Cambridge, MA: Harvard University Press, 1982.

Neumann, K., P. Verburg, E. Stehfest, and C. Müller. The yield gap of global grain production: A spatial analysis, *Agricultural Systems* 103 (2010): 316–326.

Newman, M. The structure and function of complex networks, *SIAM Review* 45 (2003): 167–256.

Niamir-Fuller, M., and M. Turner, eds. *Managing Mobility in African Rangelands: The Legitimization of Transhumance.* London: Intermediate Technology Publications, 1999.

Nicolis, G., and I. Prigogine. *Exploring Complexity – An Introduction.* New York: Freeman & Company, 1989.

Nieuwlaar, E., and W. de Ruiter. *Toegepaste Thermodynamica en Energieconversies.* Cursuscode GEO2–2212. Utrecht, the Netherlands: Utrecht University, 2010.

Nijs, T. de, R. de Niet, and L. Crommentuijn. Constructing land-use maps of the Netherlands in 2030, *Journal of Environmental Management* 72 (2004): 35–42.

Nilsson, S. Experiences of policy reforms of the forest sector in transition and other countries, *Forest Policy and Economics* 7 (2005): 831–884.

Norberg, J., and G. Cummings, eds. *Complexity Theory for a Sustainable Future.* New York: Columbia University Press, 2008.

Nordhaus, W. Geography and macroeconomics: New data and new findings, *Proceedings of the National Academy of Sciences of the United States of America (PNAS)* 105 (2005): 3510–3517.

Nowak, M., and K. Sigmund. Evolutionary dynamics of biological games, *Science* 303 (2004): 793–799.

Nowak, M. *Evolutionary Dynamics – Exploring the Equations of Life.* Cambridge, MA: Harvard University Press, 2006.

Nussbaum, M., and A. Sen, eds. *The Quality of Life.* Oxford: Clarendon Press Oxford, 1993.

Nussbaum, M. *Women and Human Development: The Capabilities Approach.* Cambridge: Cambridge University Press, 2000.

Odling-Smee, J., K. Laland, and M. Feldman. Niche construction – The neglected process in evolution. Princeton, NJ: Princeton University Press, 2003.

Odum, H. *Environment, Power and Society.* New York: Wiley, 1976.

OECD, 2008. Available at www.oecd.org.

OECD Factbook, 2007. Available at www.oecd.org

OECD/IEA. World Energy Outlook 2010. IEA Report. Paris: OECD. See www.worldenergyoutlook.org/.

Okin, G. Toward a unified view of biophysical land degradation processes in arid and semi-arid lands. In *Global Desertification: Do Humans Cause Deserts? 88th Dahlem Workshop*, pp. 95–109, edited by J. Reynolds and D. Stafford Smith. Berlin: Dahlem University Press, 2002.

Oldeman, L., R. Hakkeling, and G. Sombroek. *World Map of the Status of Human-Induced Soil Degradation* (2nd ed.). Wageningen, the Netherlands: ISRIC, 1990.

Opschoor, H. The ecological footprint: Measuring rod or metaphor? *Ecological Economics* 32 (2000): 363–365.

Ormerod, P. *The Butterfly Economy: A New General Theory of Social and Economic Behavior*. London: Basic Books, 1998.

Ostrom, E. *Governing the Commons – The Evolution of Institutions for Collective Action*. Cambridge: Cambridge University Press, 1990.

Ostrom, E. Collective action and the evolution of social norms, *Journal of Economic Perspectives* 14 (2000): 137–158.

Ostrom, E., T. Dietz, N. Dolsak, P. Stern, S. Stonich, and E. Weber, eds. *The Drama of the Commons. Committee on the Human Dimensions of Global Change*. Washington, DC: National Academies Press, 2002.

Ostrom, E., M. Janssen, and J. Anderies. Going beyond panaceas – Special feature, *Proceedings of the National Academy of Sciences of the United States of America (PNAS)* 104 (2007): 15176–15178.

Ostrom, E. A general framework for analyzing sustainability of social-ecological systems, *Science* 325 (2009): 419–422.

Papyrakis, E., and R. Gerlagh. The resource curse hypothesis and its transmission channels, *Journal of Comparative Economics* 32 (2004): 181–193.

Papyrakis, E., and R. Gerlagh. Resource windfalls, investment, and long-term income, *Resources Policy* 31 (2006): 117–128.

Parris, T., and R. Kates. Characterizing a sustainability transition: Goals, targets, trends, and driving forces, *Proceedings of the National Academy of Sciences of the United States of America (PNAS)* 100 (2003): 8068–8073.

Parris, T., and R. Kates. Characterizing and measuring sustainable development, *Annual Review of Energy and the Environment* 28 (2003): 559–586.

Parson, E. Protecting the ozone layer. In *Institutions for the Earth*, pp. 27–73, edited by P. Haas, R. Keohane, and M. Levy. Cambridge, MA: MIT Press, 1993.

Pauly, D., and M.-L. Palomares. Fishing down marine food web: It is far more pervasive than we thought, *Bulletin of Marine Science* 72 (2005): 197–211.

PBL. *Growing within Limits*. Bilthoven, the Netherlands: Netherlands Environmental Assessment Agency, 2009.

PBL. *Assessing an IPCC Assessment: An Analysis of Statements on Projected Regional Impacts in the 2007 Report*. Bilthoven, the Netherlands: Netherlands Environmental Assessment Agency, 2010a.

PBL. *Rethinking Global Biodiversity Strategies*. Bilthoven, the Netherlands: Netherlands Environmental Assessment Agency, 2010b.

PBL. *Zure regen – Een analyse van dertig jaar verzuringsproblematiek in Nederland*. PBL Beleidsstudies (Buijsman et al.). Bilthoven, the Netherlands: Netherlands Environmental Assessment Agency, 2010c.

Pearce, D. *Economics of Natural Resources and the Environment*. Baltimore: John Hopkins University Press, 1990.

Pearce, D. An intellectual history of environmental economics, *Annual Review of Energy and the Environment* 27 (2002): 57–81.

Pedersen, J., and T. Benjaminsen. One leg or two? Food security and pastoralism in the Northern Sahel, *Human Ecology* 36 (2008): 43–57.

Penn, D. The evolutionary roots of our environmental problems: Toward a Darwinian ecology, *The Quarterly Review of Biology* 78 (2003): 3.

Perez, P., and D. Batten, eds. *Complex Science for a Complex World – Exploring Human Ecosystems with Agents*. Canberra, Australia: ANU Press, 2003.

Perman, R., Yue Ma, J. McGilvray, and M. Common, eds. *Natural Resource and Environmental Economics*. Harlow, UK: Pearson Education Ltd., 2003.

Peters, G., R. Andrew, and J. Lennox. Constructing an environmentally-extended multi-regional input-output table using the GTAP database, *Economic Systems Research* 23 (2011): 131–152.

Petschel-Held, G., A. Block, M. Cassel-Gintz, J. Kropp, M. Lüdeke, O. Moldenhauer, et al. Syndromes of global change: A qualitative modelling approach to assist global environmental management, *Environmental Modelling and Assessment* 4 (1999): 295–314.

Phan, D., and F. Amblard, eds. *Agent-Based Modelling and Simulation in the Social and Human Sciences*. Oxford: Bardwell Press, 2007.

Pielke, R. Jr. *The Honest Broker*. Cambridge: Cambridge University Press, 2007.

Pimm, S. The structure of foodwebs, *Theoretical Population Biology* 16 (1979): 144–158.

Pine, J., and J. Gilmore. *The Experience Economy*. Boston: Harvard Business School Press, 1999.

Polimeni, J., K. Mayumi, M. Giampietro, and B. Alcott. *The Jevons Paradox and the Myth of Resource Efficiency Improvements*. London: Earthscan, 2009.

Pollock, S. *Ancient Mesopotamia*. Cambridge: Cambridge University Press, 2001.

Ponting, C. *A Green History of the World*. London: Penguin Books, 1991.

Proctor, R., and L. Schiebinger. *Agnotology: The Making and Unmaking of Ignorance*. Palo Alto, CA: Stanford University Press, 2008.

Pumain, D., ed. *Hierarchy in Natural and Social Systems*. Berlin: Springer, 2006.

Railsback, S., and V. Grimm. *Agent-Based and Individual-Based Modeling: A Practical Introduction*. Princeton, NJ: Princeton University Press, 2011.

Ramankutty, N., and J. Foley. Estimating historical changes in global land cover: Croplands from 1700 to 1992, *Global Biogeochemical Cycles* 13 (1999): 997–1028.

Ramankutty, N., A. Evan, C. Monfreda, and J. Foley. Farming the planet: I. Geographic distribution of global agricultural lands in the year 2000, *Global Biogeochemical Cycles* 22 (2008): doi:10.1029/2007GB002952.

Raquez, P., and E. Lambin. Conditions for a sustainable land use: Case study evidence, *Journal of Land Use Science* 1 (2006): 109–125.

Ratnieks, F., and N. Carreck. Clarity on honey bee collapse? *Science* 327 (2010): 152–153.

Ravetz, J. *The No-Nonsense Guide to Science*. Oxford: New Internationalist, 2006.

Rayner, S., and E. Malone, eds. *Human Choice and Climate Change* (4 vols.). Columbus, OH: Battelle Press, 1998.

Rayo, L., and G. Becker. Evolutionary efficiency and happiness, *Journal of Political Economy* 115 (2007): 2.

Read, D. W. Kinship based demographic simulation of societal processes, *Journal of Artificial Societies and Social Simulation* 1 (1998): 1.

Reader, J. *Man on Earth*. New York: Perennial Library, 1988.

Redman, C. L. *Human Impact on Ancient Environments*. Tucson: University of Arizona, 1999.

Reich, W. *Supercapitalism*. New York: Knopf Publishers, 2007.

Renfrew, C., and J. Cherry, eds. *Peer Polity Interaction and Socio-political Change*. Cambridge: Cambridge University Press, 1985.

Revi, A., N. Dronin, and B. de Vries. Population and environment in Asia since 1600 AD. In *Mappae Mundi – Humans and Their Habitats in a Long-Term Socio-ecological Perspective. Myths, Maps, and Models*, pp. 301–352, edited by B. de Vries and J. Goudsblom. Amsterdam: Amsterdam University Press, 2002.

Reynolds, J., and D. Stafford Smith, eds. *Global Desertification: Do Humans Cause Deserts? 88th Dahlem Workshop*. Berlin: Dahlem University Press, 2002.

Riahi, K., A. Grübler, and N. Nakicenovic. Scenarios of long-term socio-economic and environmental development under climate stabilization, *Technological Forecasting & Social Change* 74 (2007): 887–935.

Richardson, G. P. *Feedback Thought in Social Science and Systems Theory*. Philadelphia: University of Pennsylvania Press, 1991.

Rietkerk, M., S. Dekker, P. de Ruiter, and J. van de Koppel. Self-organized patchiness and catastrophic shifts in ecosystems, *Science* 305 (2004): 1926–1929.

RIVM. *Zorgen voor Morgen. Nationale milieuverkenning 1985–2010*. Bilthoven, the Netherlands: RIVM, 1988.
Robert, J. *Water Is a Commons*. Mexico City: Habitat International Coalition, 1994.
Roberts, N., D. Andersen, R. Deal, M. Garet, and W. Shaffer. *Introduction to Computer Simulation – A System Dynamics Approach*. New York: Addison-Wesley Publishing Company, 1983.
Roberts, P. *The End of Food*. London: Bloomsbury, 2008.
Robeyns, I., and R. van der Veen. Sustainable Quality of Life – Conceptual Analysis for a Policy-Relevant Empirical Specification. Report 550031005. Bilthoven, the Netherlands: Environmental Assessment Agency (PBL)/University of Amsterdam, 2007.
Robinson, J. *Ordinary Differential Equations*. Cambridge: Cambridge University Press, 2004.
Rockström, J., W. Steffen, K. Noone, A. Persson, F. Chapin, E. Lambin, et al. A safe operating space for humanity, *Nature* 461 (2009): 472–475.
Rodriguez-Iturbe, I., and A. Rinaldo. *Fractal River Basins*. Cambridge: Cambridge University Press, 1997.
Rokeach, M. *The Nature of Human Values*. New York: The Free Press, 1973.
Rosen, R. *Anticipatory Systems*. New York: Pergamon Press, 1985.
Rosenau-Tornow, D., P. Buchholz, A. Riemann, and M. Wagner. Assessing the long-term supply risks for mineral raw materials – a combined evaluation of past and future trends, *Resources Policy* 34 (2009): 161–175.
Ross, M. Does oil hinder democracy? *World Politics* 53 (2001): 325–361.
Rotmans, J. and B. de Vries. *Perspectives on Global Change – the TARGETS Approach*. Cambridge: Cambridge University Press, 1997.
Ruth, M., and B. Hannon. *Modeling Dynamic Economic Systems*. New York: Springer, 1997.
Sabates, R., B. Gould, and H. Villarreal. Household composition and food expenditures: A cross-country comparison, *Food Policy* 26 (2001): 571–586.
Safarzynska, K., and J. van den Bergh. Demand-supply coevolution with multiple increasing returns: Policy analysis for unlocking and system transitions, *Technological Forecasting and Social Change* 77 (2010): 297–317.
Sahlins, M. *Stone Age Economics*. London: Tavistock Publications, 1972.
Samuelson, P. *Foundations of Economic Analysis* (8th ed.). New York: Atheneum, 1947/1976.
Sassi O., R. Crassous, J.-C. Hourcade, V. Gitz, H. Waisman, and C. Guivarch. Imaclim-R: A modelling framework to simulate sustainable development pathways, *International Journal of Global Environmental Issues* 10 (2010): 5–24.
Satake, A., and T. Rudel. Modeling the forest transition: Forest scarcity and ecosystem service hypothesis, *Ecological Applications* 17 (2007): 2024–2036.
Schäfer, A., J. Heywood, H. Jacoby, and I. Waitz. *Transportation in a Climate-Constrained World*. Cambridge, MA: MIT Press, 2010.
SCEP. *Man's Impact on the Global Environment*. Cambridge, MA: MIT Press, 1970.
Schandl, H., and N. Schulz. Industrial ecology: The UK. In *A Handbook of Industrial Ecology*, pp. 323–333, edited by R. Ayres and L. Ayres. Cheltenham, UK: Edward Elgar, 2002.
Schandl, H., and F. Krausmann. The great transformation: A socio-metabolic reading of the industrialization of the United Kingdom. In *Socioecological Transitions and Global Change: Trajectories of Social Metabolism and Land Use*, edited by M. Fischer-Kowalski and H. Haberl. Cheltenham, UK: Edward Elgar, 2007.
Schandl, H., M. Fischer-Kowalski, C. Grunhubel, and F. Krausmann. Socio-metabolic transitions in developing Asia, *Technological Forecasting and Social Change* 76 (2009): 267–281.
Scheffer, M., S. Carpenter, J. Foley, C. Folke, and B. Walker. Catastrophic shifts in ecosystems, *Nature* 413 (2001): 591–596.
Scheffer, M. *Critical Transitions in Nature and Society*. Princeton, NJ: Princeton University Press, 2009.
Scheffer, M., J. Bascompte, W. Brock, V. Brovkin, S. Carpenter, V. Dakos, et al. Early-warning signals for critical transitions, *Nature* 461 (2009): 53–59.

Scheffer, V. The rise and fall of a reindeer herd, *The Scientific Monthly* 73 (1951): 356–362.
Schneider, S., and P. Boston, eds. *Scientists on Gaia*. Boston: MIT Press, 1991.
Schnellnhuber, G., A. Block, M. Cassel-Gintz, J. Kropp, G. Lammel, W. Lass, et al. Syndromes of global change, *GAIA* 6 (1997): 19–34.
Schnellnhuber, G., P. Crutzen, W. Clark, and J. Hunt. Earth system analysis for sustainability, *Environment* 47(2005): 10–27.
Schrijver, N. *Development without Destruction – The UN and Global Resource Management*. Bloomington: Indiana University Press, 2010.
Schumacher, E. *Small Is Beautiful*. New York: Harper & Row, 1973.
Schumacher, E. *A Guide for the Perplexed*. London: Harper Perennial, 1977.
Schwartz, M., and M. Thompson. *Divided We Stand – Redefining Politics, Technology and Social Choice*. Hemel Hempstead, UK: Harvester Wheatsheaf, 1989.
Schwartz, P. *The Art of the Long View: Planning for the Future in an Uncertain World*. New York: Currency Doubleday, 1995.
Schwartz, S. Beyond individualism/collectivism: New dimensions of values. In *Individualism and Collectivism: Theory Application and Methods*, edited by U. Kim, H. Triandis, C. Kagitcibasi, S. Choi, and G. Yoon. Newbury Park, CA: Sage, 1994.
Schweitzer, F., G. Fagiolo, D. Sornette, F. Vega-Redondo, A. Vespignani, and D. R. White. Economic networks: The new challenges, *Science* 325 (2009): 422–425.
Scott, J. *Seeing like a State – How Certain Schemes to Improve the Human Condition Have Failed*. New Haven, CT: Yale University Press, 1998.
Senge, P. *The Fifth Discipline – The Art & Practice of the Learning Organization*. New York: Doubleday Currency, 1990.
Seppelt, R., and O. Richter. It was an artefact not the result: A note on system dynamics model development tools, *Environmental Modeling & Software* 20 (2005): 1543–1548.
Shah, I. Learning how to learn – Psychology and spirituality in the Sufi way. London: Penguin Books, London, 1978/1985.
Shiva, Vandana. The practice of earth democracy, *Development Dialogue, What Next?* 52 (August 2009). Available at www.whatnext.org
Sieferle, R. *Ruckblick auf die Natur – eine Geschichte des Menschen und seiner Umwelt*. Munich: Luchterhand, 1997.
Sietz, D., M. Lüdeke, and C. Walther. Categorisation of typical vulnerability patterns in global drylands, *Global Environmental Change* 21 (2011): 431–440.
Simon, H. *The Sciences of the Artificial*. Cambridge, MA: MIT Press, 1969.
Small, C., and J. Cohen. Continental physiography, climate and the global distribution of human population, *Current Anthropology* 45 (2004): 269–277.
Smil, V. *Energy in World History*. Boulder, CO: Westview Press, 1994.
Sohngen, B., R. Mendelsohn, and R. Sedjo. Forest management, conservation, and global timber markets, *American Journal of Agricultural Economics* 81 (1999): 1–13.
Solé, R., and B. Goodwin. *Signs of Life – How Complexity Pervades Biology*. New York: Basic Books, 2000.
Soramäki, K., M. Bech, J. Arnold, R. Glass, and W. Beyeler. The topology of interbank payment flows, *Statistical Mechanics and Its Applications – Physica A* 379 (2007): 317–333.
Sorokin, P. *Social and Cultural Dynamics*. London: Transaction Publishers, 1957.
Steffen, W., and P. Tyson. *Global Change and the Earth System: A Planet under Pressure*. New York: Springer, 2004.
Steg, L., and C. Vlek. Social science and environmental behaviour. In *Principles of Environmental Sciences*, pp. 97–142, edited by J. Boersema and L. Reijnders. The Netherlands: Springer Science +Business Media B.V., 2009.
Sterman, J. The economic long wave: theory and evidence, *System Dynamics Review* 2 (1986): 87–125.
Sterman, J. *Business Dynamics – Systems Thinking and Modeling for a Complex World*. Boston: Irwin McGraw-Hill, 2000.

Sterman, J., and L. Sweeney. Cloudy skies: Assessing public understanding of global warming, *System Dynamics Review* 18 (2002): 207–240.
Stern, D., M. Common, and E. Barbier. Economic growth and environmental degradation: The environmental Kuznets curve and sustainable development, *World Development* 24 (1996): 1151–1160.
Stern, D. The rise and fall of the environmental Kuznets curve, *World Development* 32 (2004): 1419–1439.
Stern, N. The economics of climate change, *American Economic Review* 98 (2008): 1–37.
Stikker, A. *Water – The Blood of the Earth. Exploring Sustainable Water Management for the New Millennium*. New York: www.cosimobooks.com, 2007.
Ströbele, W. *Wirtschaftswachstum bei begrenzten Energieressourcen*. Berlin: Duncker & Humbolt, 1984.
Strogatz, S. *Nonlinear Dynamics and Chaos – with Applications to Physics, Biology, Chemistry, and Engineering*. Reading, MA: Addison-Wesley, 1994.
Sustainability Institute. Commodity Systems Challenges. A Sustainability Institute Report, April 2003. Available at www.sustainer.org.
Swinnen, J., A. Banerjee, and H. de Gorter. Economic development, institutional change, and the political economy of agricultural protection – an econometric study of Belgium since the 19th century, *Agricultural Economics* 26 (2001): 25–43.
Tainter, J. *The Collapse of Complex Societies*. Cambridge: Cambridge University Press, 1988.
Tainter, J. Problem solving: Complexity, history, sustainability, *Population and Environment* 22 (2000): 3–41.
TCEPD (Taiwan Council for Economic Planning and Development). *Taiwan 2000 – Balancing Economic Growth and Environmental Protection*. Taipei: Taiwan National University, 1989.
Tannahill, R. *Food in History*. New York: Stein and Day, 1973.
Tasaka, H. *Invisible Capitalism: Beyond Monetary Economy and the Birth of New Paradigm Economies*. New York: Jorge Pinto Books, 2009.
Taylor, M. S. Buffalo hunt: International trade and the virtual extinction of the North American bison, *American Economic Review* 101 (2011): 3162–3195.
TDBT (The Danish Board of Technology). Oil-based Technology and Economy Prospects for the Future. 2004. Available at www.tekno.dk/pdf/projekter/p04_Oil-based_Technology_and_Economy.pdf.
Teilhard de Chardin, P. *L'avenir de l'homme*. Paris: Editions du Seuil, 1959.
Teives Henriques, S., and A. Kander. The modest environmental relief resulting from the transition to a service economy, *Ecological Economics* 70 (2010): 271–282.
TEPD (2010). Transportation Energy Data Book 2010. Available at cta.ornl.gov/data/download29.shtml.
Tesfatsion, L. Introduction to the special issue on agent-based computational economics, *Journal of Economic Dynamics & Control* 25 (2001): 281–293.
Thirgood, J. *Man and the Mediterranean Forest – A History of Resource Depletion*. New York: Academic Press, 1981.
Thompson, M., R. Ellis, and A. Wildawsky. *Cultural Theory*. Boulder, CO: Westview Press, 1990.
Thompson, M. The dynamics of cultural theory and their implications for the enterprise culture. In *Understanding the Enterprise Culture*, edited by S. Heap and A. Ross. Edinburgh: Edinburgh University Press, 1992.
Thompson, M. Security and solidarity: An anti-reductionist framework for thinking about the relationship between us and the rest of nature, *The Geographical Journal* 163 (1997): 141–149.
Thornton, P. Livestock production: recent trends, future prospects, *Philosophical Transactions of the Royal Society B* 365 (2010): 2853–2867.
Thurstan, R., S. Brockington, and C. Roberts. The effects of 118 years of industrial fishing on UK bottom trawl fisheries, *Nature Communications* 1 (2010): doi:10.1038/ncomms1013.

Toth, F. L. Policy exercises, *Simulation & Games* 19 (1988): 235–276.
Totman, C. *The Green Archipelago: Forestry in Pre-industrial Japan*. Berkeley: University of California Press, 1989.
Toynbee, A. *Mankind and Mother Earth: A Narrative History of the World*. Oxford: Oxford University Press, 1976.
Turchin, P. *Historical Dynamics: Why States Rise and Fall*. Princeton, NJ: Princeton University Press, 2003.
Turchin, P. *Complex Population Dynamics*. Princeton, NJ: Princeton University Press, 2008.
Turner, B., W. Clark, R. Kates, J. Richards, J. Mathews, and W. Meyer, eds. *The Earth as Transformed by Human Action – Global and Regional Changes in the Biosphere over the Past 300 Years*. Cambridge: Cambridge University Press, 1990.
Turner, G. A comparison of *The Limits to Growth* with 30 years of reality, *Global Environmental Change* 18 (2008): 397–411.
Tylecote, A. *The Long Wave in the World Economy – The Present Crisis in Historical Perspective*. London: Routledge, 1992.
Ulanowicz, R. *Growth and Development: Ecosystems Phenomenology*. San Jose, CA: Excel Press, 1986.
UN (United Nations). *Indicators of Sustainable Development – Framework and Methodologies*. New York: UN, 1996.
UNAIDS. 2004. Available at www.unaids.org/en.
UNEP. *Global Environmental Outlook* 2. 2000. Available at www.unep.org/geo/GEO2000/english/Index.htm.
UNEP. *Global Environmental Outlook: Environment for Development (GEO–4)*. Nairobi: UN, 2007.
van Beek, L., and M. Bierkens. The Global Hydrological Model PCR-GLOBWB: Conceptualization, Parameterization and Verification. Department of Physical Geography, Utrecht University, Utrecht, the Netherlands, 2008. Available at vanbeek.geo.uu.nl/suppinfo/vanbeekbierkens2009.pdf.
van Beek, L., Y. Wada, and M. Bierkens. Global monthly water stress: 1. Water balance and water availability, *Water Resources Research* 47 (2011): W07517, doi:10.1029/2010WR009791.
van den Bergh, J., A. Ferrer-i-Carbonell, and G. Munda. Alternative models of individual behaviour and implications for environmental policy, *Ecological Economics* 32 (2000): 43–61.
van den Bergh, J., ed. *Handbook of Environmental and Resource Economics*. Cheltenham, UK: Edward Elgar, 2002.
van den Bergh, J., A. Faber, A. Idenburg, and F. Oosterhuis. *Evolutionary Economics and Environmental Policy: Survival of the Greenest*. Cheltenham, UK: Edward Elgar, 2007.
van den Bergh, J. The GDP paradox, *Journal of Economic Psychology* 30 (2009): 117–135.
van der Leeuw, S., ed. The Archaeomedes Project: Understanding the Natural and Anthropogenic Causes of Land Degradation and Desertification in the Mediterranean Basin. Brussels, Science Research Development European Commission Report EUR 18181 EN. Brussels, 1998.
van der Leeuw, S., and B. de Vries. The Roman Empire. In *Mappae Mundi – Humans and Their Habitats in a Long-Term Socio-ecological Perspective. Myths, Maps, and Models*, pp. 209–256, edited by B. de Vries and J. Goudsblom. Amsterdam: Amsterdam University Press, 2002.
van Egmond, N., and B. de Vries. Sustainability: the search for the integral worldview, *Futures* 43 (2011): 853–867.
van Praag, B., and A. Ferrer-i-Carbonell. *Happiness Quantified*. Oxford: Oxford University Press, 2004.
van Praag, B., and P. Frijters. The measurement of welfare and well-being: The Leyden approach. In *Well-Being – The Foundations of Hedonic Psychology*, edited by D. Kahneman, E. Diener, and N. Schwarz. New York: Russell Sage Foundation, 1999.

van Vuuren, D., B. Strengers, and B. de Vries. Long-term perspectives on world metal use – a model-based approach, *Resources Policy* 25 (2000): 239–255.
van Vuuren, D., B. de Vries, A. Beusen, and P. Heuberger. Conditional probabilistic estimates of 21st century greenhouse gas emissions based on the storylines of the IPCC-SRES scenarios, *Global Environmental Change* 18 (2008): 635–654.
van Zon, R. Duurzame ontwikkeling in historisch perspectief. Vakreviews Duurzame Ontwikkeling. RU Groningen/KUN, 2002.
Veenhoven R. Is happiness relative? *Social Indicators Research* 24 (1991): 1–34.
Vennix, J. Group model building: Tackling messy problems, *System Dynamics Review* 15 (1999): 379–401.
Verburg, P., A. Veldkamp, and L. Fresco. Simulation of changes in the spatial pattern of land use in China, *Applied Geography* 19 (1999): 211–233.
Verburg, P., W. de Groot, and A. Veldkamp. Methodology for multi-scale land-use change modelling: Concepts and challenges. In *Global Environmental Change and Land Use*, edited by A. Dolman, A. Verhagen, and C. Rovers. Dordrecht, the Netherlands: Kluwer Academic Publishers, 2003.
Verburg, P., K. Overmars, M. Huigen, W. de Groot, and A. Veldkamp. Analysis of the effects of land use change on protected areas in the Philippines, *Applied Geography* 26 (2006): 153–173.
Verburg, P., D. van Berkel, A. van Doorn, M. van Eupen, and H. van den Heiligenberg. Trajectories of land use change in Europe: A model-based exploration of rural futures, *Landscape Ecology* 25 (2010): 217–232.
Vervoort, J., K. Kok, R. van Lammeren, and T. Veldkamp. Stepping into futures: Exploring the potential of interactive media for participatory scenarios on social-ecological systems, *Futures* 42 (2010): 604–616.
Verweij, M., and M. Thompson, eds. *Clumsy Solutions for a Complex World: Governance, Politics, and Plural Perceptions*. New York: Palgrave Macmillan, 2006.
Vringer, K., and K. Blok. Long-term trends in direct and indirect household energy intensities: A factor in dematerialisation? *Energy Policy* 28 (2000): 713–727.
Vringer, K., T. Aalbers, and K. Blok. Household energy requirement and value pattern, *Energy Policy* 35 (2007): 553–566.
VROM/ABF. 2009. Statistical Data on Housing in the Netherlands. See www.cbs.nl.
Wackernagel, M., N. Schulz, D. Deumling, A. Callejas Linares, M. Jenkins, V. Kapos, et al. Tracking the ecological overshoot of the human economy, *Proceedings of the National Academy of Sciences of the United States of America (PNAS)* 99 (2002): 9266–9271.
Wada, Y., L. Van Beek, C. van Kempen, J. Reckman, S. Vasak, and M. Bierkens. Global depletion of groundwater resources, *Geophysical Research Letters* 37 (2010): L20402.
Wada, Y., L. van Beek, D. Viviroli, H. Dürr, R. Weingartner, and M. Bierkens. Global monthly water stress: 2. Water demand and severity of water stress, *Water Resources Research* 47 (2011): W07518, doi:10.1029/2010WR009792.
Wainwright, J., and M. Mulligan. *Environmental Modelling – Finding Simplicity in Complexity*. London: John Wiley& Sons, Ltd., 2004.
Walker, B., J. Anderies, A. Kinzig, and P. Ryan, eds. *Exploring Resilience in Social-Ecological Systems – Comparative Studies and Theory Development*. Canberra, Australia: CSIRO Publishing, 2006.
Wall, G., and M. Gong. On exergy and sustainable development – Part I: Conditions and concepts, *Exergy International Journal* 1 (2001): 128–145.
Walters, C. *Adaptive Management*. New York: Macmillan, 1989.
Warde, P. *Energy Consumption in England & Wales 1560–2000*. Consiglio Nazionale delle Ricerche, 2007.
Warr, A., H. Schandl, and R. Ayres. Long-term trends in resource exergy consumption and useful work supplies in the UK, 1900 to 2000, *Ecological Economics* 68 (2002): 126–140.

Wassenaar, T., P. Gerber, M. Rosales, M. Ibrahim, P. Verburg, and H. Steinfeld. Predicting land use changes in the neotropics: The geography of pasture expansion into forest, *Global Environmental Change* 17 (2007): 86–104.

Watts, D. *Small Worlds – The Dynamics of Networks between Order and Randomness*. Princeton Studies in Complexity. Princeton, NJ: Princeton University Press, 1999.

WCED (World Commission on Environment and Development). *Our Common Future*. Oxford: Oxford University Press, 1987.

Weber, M., B. Volker, and K. Hasselmann. A multi-actor dynamic integrated assessment model (MADIAM) of induced technological change and sustainable economic growth, *Ecological Economics* 54 (2005): 306–327.

Weisbuch, G., and G. Boudjema. Dynamical aspects in the adoption of agri-environmental measures, *Advances in Complex Systems* 2 (1999): 11–36.

Weisdorf, J. From foraging to farming: Explaining the Neolithic revolution, *Journal of Economic Surveys* 19 (2005): 561–586.

Welcomme, J., I. Cowx, D. Coates, C. Béné, S. Funge-Smith, A. Halls, and K. Lorenzen. Inland capture fisheries, *Philosophical Transactions of the Royal Society B* 365 (2010): 2881–2896.

Wene, K.-O./IEA. *Experience Curves for Energy Technology Policy*. Paris: IEA, 2000.

Westhoek, H., M. van den Berg, and J. Bakkes. Scenario development to explore the future of Europe's rural areas, *Agriculture, Ecosystems and Environment* 114 (2006): 7–20.

White R., and G. Engelen. Cellular automata as the basis of integrated dynamic regional modelling, *Environment and Planning B* 24 (1997): 235–246.

WHO. 2002. Available at www.who.org.

Wilber, K. *The Marriage of Sense and Soul: Integrating Science and Religion*. Dublin: Gateway, 2001.

Wilkinson, R. *Poverty and Progress: An Ecological Model of Economic Development*. London: Methuen & Co., 1973.

Williams, M. A new look at global forest histories, *Annual Review of Environment and Resources* 33 (2008): 345–367.

Wilson, E. *Sociobiology – The New Synthesis*. Cambridge, MA: Harvard University Press, 1975.

Wilson, E. *Consilience: The Unity of Knowledge*. New York: Knopf, 1998.

Wilting, H., A. Faber, and A. Idenburg. Investigating new technologies in a scenario context: Description and application of an input-output method, *Journal of Cleaner Production* 16S1 (2008): S102–S112.

Wood, M. *Legacy – A Search for the Origins of Civilization*. London: BBC Worldwide Ltd., 1999.

World Bank. *Expanding the Measure of Wealth – Indicators of Environmentally Sustainable Development*. New York: World Bank, 1997.

WRI. World Resources Report, Chapter 2: Taking stock of ecosystems. 2000–2001.

WRI (2003). World Resource Report.

Wright, R. *Nonzero –The Logic of Human Destiny*. New York: Pantheon Books, 2000.

Yates, R. War, food shortages, and relief measures in early China. In *Hunger in History – Food Shortage, Poverty, and Deprivation*, pp. 147–177, edited by L. Newman. London: Basil Blackwell, 1990.

Yeats, W. *The Second Coming: Collected Poems of W.B. Yeats*. New York: Scribner Paperback Poetry, 1939/1996.

Yergin, D. *The Prize – The Epic Quest for Oil, Money and Power*. London: Simon & Schuster, 1991.

ZTB (Zentrum fur Transformation der Bundeswehr). Peak Oil – Sicherheitspolitische Implikationen knapper Ressources. 2010. Available at www.zentrum-transformation.bundeswehr.de.

Zuo Dakang. The Huang-Huai-Hai Plain. In *The Earth as Transformed by Human Action – Global and Regional Changes in the Biosphere over the Past 300 Years*, edited by B. Turner, W. Clark, R. Kates, J. Richards, J. Mathews, and W. Meyer. Cambridge: Cambridge University Press, 1990.

Index